Exploring Integrated Science

Exploring Integrated Science

Belal E. Baaquie
Department of Physics, National University of Singapore
Frederick H. Willeboordse
Department of Physics, National University of Singapore

CRC Press is an imprint of the
Taylor & Francis Group, an **informa** business

CRC Press
Taylor & Francis Group
6000 Broken Sound Parkway NW, Suite 300
Boca Raton, FL 33487-2742

Printed in the United States of America on acid-free paper
10 9 8 7 6 5 4 3 2 1

International Standard Book Number: 978-1-4200-8793-2 (Hardback)

Visit the Taylor & Francis Web site at
http://www.taylorandfrancis.com

and the CRC Press Web site at
http://www.crcpress.com

Contents

Foreword

Everyone has witnessed the beauty and wonders of nature — be it the grandeur of mountain peaks, the subtlety of the butterfly's wing or the splendor of the night sky. Human beings understand nature instinctively — we know how to jump over an obstacle or how to avoid direct sunlight and it is this intuitive appreciation of nature which is the starting point for the adventure of science. It is also the starting point for the majority of the topics in this book where the authors begin with the direct and immediate awareness that we have of nature and proceed to show how the simplest questions that arise from our daily experience can lead us, step by step, through a chain of reasoning, to the wonders of science and to some of the most remarkable and elegant scientific principles.

The wonders of science truly never stop. From the very smallest to the very largest, new discoveries, be they a new nano-motor, a new giant black hole or a novel genetic drug, are made at an astounding rate. At times it may seem as though the onslaught of change leaves even the staunchest supporters of innovation in its wake. Yet at the same time, key scientific ideas originating in the sixteenth and seventeenth centuries continue to have validity. Indeed, modern science is a cooking pot of permanence, evolution and revolution, a mixture of evergreens, adaptations and disruptive ideas.

The authors do a wonderful job reflecting the status of modern science in this masterful textbook. They present evergreens like Newton's laws together with groundbreaking new ideas from evolution so as to present integrated scientific ideas that seamlessly blend biology, mathematics, chemistry and physics for the understanding of, for example, olfaction, and revolutionary paradigm shifting notions like complexity and information theories. Do these topics belong together?

Absolutely! All of science is intricately connected through labyrinths of unseen methodological tunnels, personal interactions and even prejudices. Extraordinary ideas in one part of science can lead to major discoveries in other parts even when the topics appear to have nothing in common. The explanations are unified by the process of scientific reasoning, which is the common thread in the analysis of all phenomena addressed in this remarkable book. *Exploring Integrated Science* shines in that it does not shy away from the difficulties this presents and manages to bring to the audience the most complex and intriguing ideas in a manner that is accessible, entertaining and accurate at the same time.

Indeed, the range of topics and their presentation is what makes this book unique. Each chapter poses questions, many of which originate in daily life, that are transformed step by step — using common sense and analogies — into well-defined concepts that lead to the answers, be it a mathematical algorithm, a biological macro-molecule or an exotic physical object. At every step, the authors go to great lengths to transition intuitively from one idea to the next. The chapters are organized along what the authors call "story lines" that are meant to help in understanding why things are the way they are and why scientists have chosen the paths they have chosen. This brings a rare perspective to the subject matter and increases the readability enormously.

In reading *Exploring Integrated Science* I have often found myself wondering "What's next?", "How is this going to be solved?" At times, the book is as captivating as a mystery novel and yet it never forfeits scientific accuracy following the adage: tell the truth, nothing but the truth but not necessarily *all* of the truth (as that would involve a level of mathematics and rigor that is far beyond the scope of a textbook).

The amazement and admiration we feel on witnessing nature fuel the drive of the scientist to understand and comprehend the wonders of science. Dr. Baaquie and Dr. Willeboordse are driven by this wonder and show the enthusiasm characteristic of explorers and adventurers — sharing their remarkable insights in an exciting and entertaining manner. The breadth of the book extends beyond topics and paradigms to the audience itself since it will have something for almost everyone, from high school student to university professor. It is neither too difficult nor too easy — the Goldilocks of general scientific textbooks. Sit back and enjoy!

Shih Choon Fong
President, King Abdullah University of Science and Technology
Past President, The National University of Singapore

Preface

Science surrounds and permeates our lives like never before. By and large it has improved our standards of living more than we could have imagined even a hundred years ago. Mobile phones help us keep in touch with loved ones and business partners alike, the Internet has created a global information repository and ever more diseases can be treated or cured. As so much has been achieved, it would seem that little remains to be done yet nothing could be further from the truth. It is only now, in the early part of the twenty-first century, that erstwhile arcane ivory-tower theories like quantum mechanics and relativity enter our daily lives in ways that were never imagined, planned nor intended. Indeed, not only have such theories entered the mainstream, they have also become much more intelligible now that their direct effects and consequences can be experienced in everyday life. New realms of knowledge are opening up almost daily in the fields such as nano- and bioscience, and what is fascinating is that not only do these new fields move science into new frontiers, they also fundamentally change the meaning of science itself.

The days of physics, mathematics, chemistry and biology being the almost exclusive domain of eccentric individuals with highly specialized "far from reality" topics are gone. Science is here, today and everywhere while questions from once diverse disciplines meet each other in unexpected ways. Science has not only become an everyday affair, it has also become an integrated whole. If we want to understand the world we live in, interact with it and enjoy its many fruits, we need to understand science in this form: integrated multidisciplinary, interdisciplinary and transdisciplinary knowledge.

Science is neither remote nor boring; it can show us how things work and arouse the curiosities of our minds. Science not only brings practical benefits but above all is a reflection of the human mind, an exciting intellectual adventure that is both undertaken by us and also makes us who we are. Exploring the terrain of integrated science is the main adventure undertaken in this book.

In this book, we ask questions that naturally arise in everyday life and answer them step by step on an adventure track of ideas that introduces new concepts as and when we encounter them on our path. At all points we would like the reader to be able to say "yes, this makes sense!" of the concepts and answers provided in the book. As a consequence, rather than trying to present traditional topics like energy or genetics concisely, we present the materials as suitably chosen adventure tracks without much concern of the usual discipline boundaries; some key concepts are revisited in new and different contexts. This is in stark contrast to the more traditional approach, where at a given level the required information regarding that topic is all bundled together at the expense of having an understanding during the learning process of why those things actually are the way they are. In our view, there is no point in having a general textbook that only makes sense after its contents have been thoroughly digested.

Acknowledgments

We would like to thank our colleagues for numerous comments and extensive help in the preparation of this book. Without pretense of completeness and with apologies for the many omitted, we express our gratitude to: Thomas Osipowicz, Sow Chorng Haur, Kang Hway Chuan, Dagomir Kaszlikowski, Chammika N.B. Udalagama, Rafi Rashid, Arzish Baaquie, Wang XueSen, Haw Jing Yan, Leong Hong Fai and Wee Phua Kuan Keith. We would like to especially thank our dedicated and tireless teaching assistant Setiawan who spent many hours looking for mistakes in the manuscript.

Special thanks to Gregory Yablonsky and Rob Philips for their support and for many useful discussions and to Andreas Keil for a critical reading of the manuscript.

We would like to thank James Wee for his help with some of the initial figures. Some symbols are courtesy of the Integration and Application Network (ian.umces.edu/symbols/), University of Maryland Center for Environmental Science.

One of us (BEB) would like to thank Yamin Chowdhury and Zahur Ahmad for many fruitful and insightful discussions and K.Z. Islam for encouragement; the inspiration of M.A. Baaquie is gratefully acknowledged. The precious support and understanding of his wife Najma and children Arzish and Tazkiah has been a constant source of strength and encouragement.

One of us (FHW) would like to thank his wonderful wife and children, Aegean Leung, Alpha, Beta and Gamma, for inspiration and support throughout.

We would like to express our thanks to Bal Menon for his invaluable support, especially at a crucial juncture in the writing and publication of this book; his conviction that our book is a worthwhile undertaking is greatly appreciated.

We are deeply indebted to Shih Choon Fong for his unstinting support over the many long years — and through the many twists and turns — that the writing of this book has gone through. His support and guidance have been instrumental in the completion of this book and it is with great pleasure that we express our heartfelt thanks to him.

We acknowledge the unwavering and valuable support of the Department of Physics, the Faculty of Science and the National University of Singapore.

And last but not least, we would like to thank Sunil Nair and Leong Li-Ming of Taylor & Francis for their effort in making publication of this book possible.

Road Map

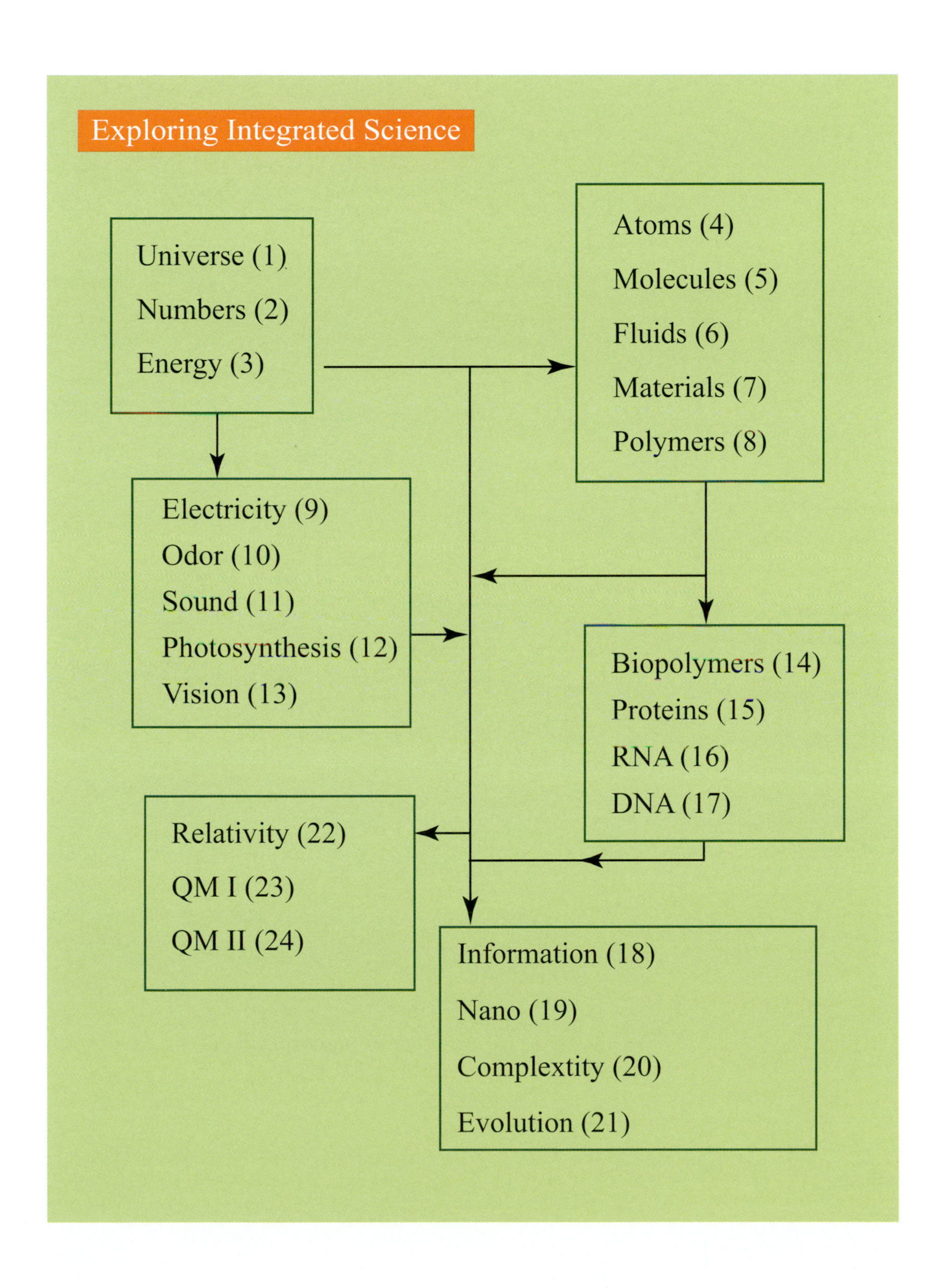

Exploring Integrated Science
Universe (1)
Numbers (2)
Energy (3)
Atoms (4)
Molecules (5)
Fluids (6)
Materials (7)
Polymers (8)
Electricity (9)
Odor (10)
Sound (11)
Photosynthesis (12)
Vision (13)
Biopolymers (14)
Proteins (15)
RNA (16)
DNA (17)
Relativity (22)
QM I (23)
QM II (24)
Information (18)
Nano (19)
Complextity (20)
Evolution (21)

1: Our Universe

Q - Where Are We?

Chapter Map

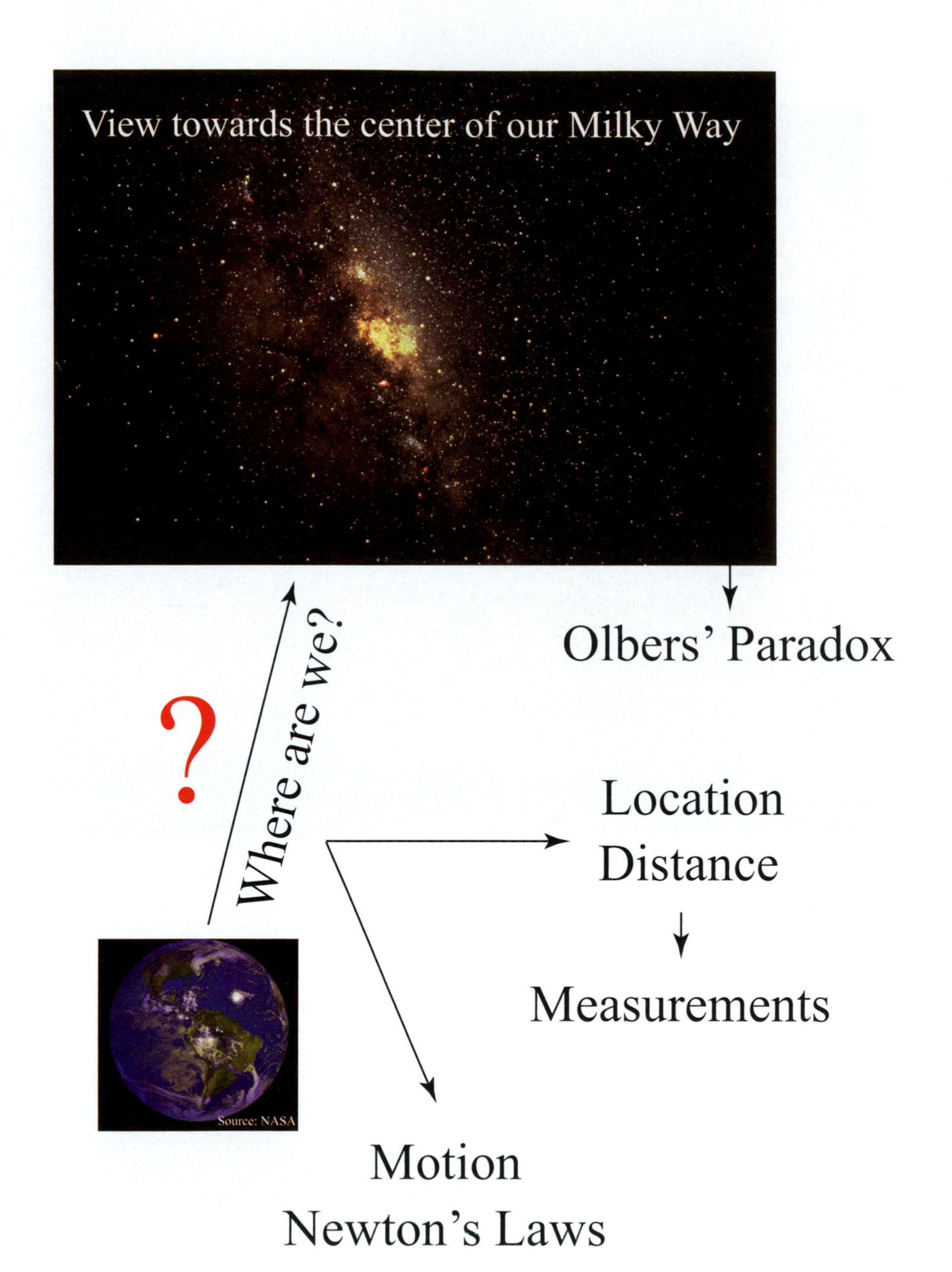

View towards the center of our Milky Way
Olbers' Paradox
Where are we?
?
Location
Distance
Measurements
Source: NASA
Motion
Newton's Laws

1.1 The question

"Where are we?" is a question that has likely been asked since the earliest days of mankind. During the day, we can observe that we are on land, that the land extends beyond the directly visible range, that we are not living in water and that we are not part of the sun. Although this tells us something, it is not all that much. The mystery of where we are becomes much greater during a clear night. What we then see is that there are many objects in the sky. Some objects are very bright, and some not so bright, some move very fast while others proceed slowly, some appear pretty close and others farther away.

Figure 1.1: Montage of Jupiter and four of its moons. Source: NASA.

Ancient civilizations already had a deep understanding of one important aspect of the movements in the sky: their surprising regularity. First, there is the pattern of night and day (a rhythm much multicellular life is dependent on), then there is the gradual change in the shape of the moon as well as the yearly pattern in the direction from which the sun rises. It is no surprise then that early mathematics was heavily or should we say heavenly influenced by astronomical observations.

Indeed, one of the greatest milestones demarcating the advent of modern science was the discovery of the laws which govern the motion of the planets by Newton in the seventeenth century. Even without the laws, however, the mere recognition of the patterns had an enormous impact on the development of our society. Besides stimulating early mathematics, it allowed for an accurate measurement of the time of the year and that in turn for an enormous improvement in agriculture. Better agriculture then made it possible for humanity to relieve larger parts of the population from food production or acquisition thus paving the way for urbanization. Urbanization finally was the basis for the emergence of highly specialized skills such as those of an astronomer.

But let us now return to the question of where we are. If we want to know that, we need to know where the other celestial objects are and that means we need to first know what is where relative to our position and second how far away the observed objects are. In the following two sections, we therefore first have a look at what can be found in the sky and then at how the distance to celestial objects can be measured.

1.2 Structure of the universe

Fig. 1.2: The galaxy cluster Abell 1689. Source NASA.

Of course, if we want to know the structure of the universe, we need to observe it. Until about 400 years ago, this had to be done with the naked eye. Even so, a lot of information can be gleaned by just looking at it. First, it is clear that there is a difference between the moon, the planets and the stars. The stars, besides their daily circular motion, appear basically fixed in position. Relative to the stars, the planets clearly move over the course of a year while the moon not only moves on a daily basis but also changes shape on a monthly basis. Second, the stars seem to be grouped together into what occasionally appears to be a pattern. Some of these patterns were given names such as "Orion" or "Ursa Minor" in the Greco-Roman tradition or "heart" and "ox" in the Chinese tradition. Named celestial patterns are generally referred to as constellations and all ancient high cultures had their own names and patterns, some of which overlap and some of which don't. Oftentimes, magic powers were ascribed to the various constellations but from the viewpoint of modern science, the patterns in constellations are purely coincidental. Nevertheless, even an untrained observer can immediately see that the number of stars is enormous and hence having some reference points is very handy. Consequently, even present-day cosmology commonly refers to the constellations.

As good as it is, the naked eye has its limitations. To obtain a better idea of the planets and the stars, it would clearly be of great benefit if there were a way to see them up closer. This became possible with the invention of the first practical telescope by the German-Dutch lens maker Johann Lippershey in 1608 (the likely purpose for this telescope was the spotting of ships). After hearing of Lippershey's design, Galileo Galilei built the first telescope to be used for astronomical observations in 1609. Ever since, better and better optical telescopes have been built, leading up to the Hubble Space Telescope being launched into space. This has led to a wealth of information about the structure of the universe and we now know that it is roughly organized in the following way:

- Stars are organized into galaxies
- Galaxies are organized into groups or clusters
- Groups or clusters are organized into superclusters

Galaxies contain from about 10^7 to about 10^{12} stars that orbit a common gravitational center (generally believed to be a giant black hole).

Our sun is a star in a galaxy called the **Milky Way** since the other stars in our galaxy leave a white hazy trail across the night sky as can be seen in the panoramic image shown in Figure 1.3. Up to about 50 galaxies may form into groups such as the group of over 50 galaxies which our own Milky Way is a part of. Galaxies may also be organized into larger clusters containing from about 50 to 1000 galaxies such as the Abell 1689 cluster shown in Figure 1.2, while superclusters contain several to dozens of galaxy groups or clusters often arranged as a long thin strand. Clusters and superclusters do not appear to have a common gravitational center like galaxies. The largest known structures in the universe are the so-called great walls that consist of huge filaments of galaxies.

Figure 1.3: Panoramic view of our Milky Way from Death Valley's Racetrack Playa. Source: US National Park Service.

For example the Sloan Great Wall is over 1 billion light-years long while the CfA2 Great Wall is about 500 million light-years long, 300 million light-years wide and 15 million light-years thick. Space between the superclusters and great walls is almost completely empty forming enormous voids with only very few galaxies.

Although most galaxies are members of a larger structure, even the largest structures are still relatively small when compared to the size of the universe. At very large scales, it is believed that the **cosmological principle** is applicable which states that the universe is both homogeneous (i.e. its average density is the same everywhere) and isotropic (i.e. it is the same in all directions). This principle implies that the universe looks the same in any galaxy. It is of course also assumed that the fundamental laws of physics are everywhere the same (but this is true at any scale not just large scales).

Thus far we have considered the patterns in the motions of the planets and the stars and the structure of the universe as it can be observed with the help of telescopes. At first sight, as was the case only a few hundred years ago, it appears as if it is the celestial objects that are moving and that earth is standing still. Purely from a mechanical point of view (think of the centripetal force that would be needed to keep the sun in its orbit if it would be circling the earth), this is already impossible to justify. Also, it is much easier to describe the motions of the sun, planets and stars if the earth is rotating and orbiting the sun. Still, it would be nice if besides astronomical evidence, effects of the earth's motion could be measured directly. An experiment that greatly contributed to clinching the debate about who is moving (especially amongst the general public) is due to the French physicist **Léon Foucault**. In his most famous demonstration, he created a huge pendulum by suspending a 28 kg bob from the ceiling of the Pantheon in Paris with a 67 m long wire as shown in Figure 1.4.

Fig. 1.4: The Foucault Pendulum in the Paris Pantheon. Source: Wikipedia.

Since a pendulum will keep swinging in the same direction unless a force acts on it (see also Newton's laws), a rotating earth means a relative change of the path the pendulum traces with respect to the floor.

Now that we have an idea of the structure of the universe, if we wish to know "where we are", we need to be able to determine the distances to and between the celestial objects. Let us therefore investigate how distances can be measured next.

1.3 Measurements of distance in the universe

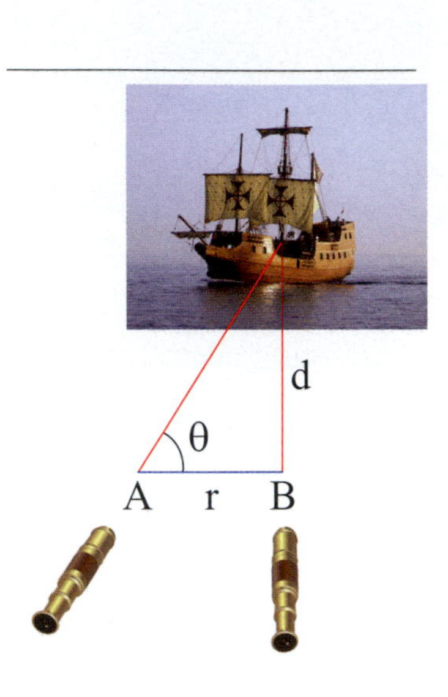

Fig. 1.5: The distance of a ship can be determined by triangulation.

When measuring small distances, one can use a ruler or a tape measure. For larger distances one may perhaps use an odometer if a suitable path exists between the start and end points. But, how about cases when there is no path, for example a ship near the horizon? Furthermore, the distances between earth and celestial objects can vary greatly with the moon being rather close and quasars very far away (quasars are extremely bright galactic cores of distant young galaxies). Consequently one may expect that there is not really one single method that works equally well in all the cases and we therefore wonder how the various distances can be measured.

The first method, which also works for ships is based on the geometric principle of **triangulation** where one constructs one or several triangles and uses known or directly measurable quantities and then solves for the unknown distance.

Triangulation

Let us consider the situation in Figure 1.5 where we would like to know how far away the ship is. If we make it so that there is a right angle between r and d, then given the distance r between the points A and B and the angle θ, we can easily find the distance d to the ship from $\tan\theta = d/r$.

While this method works fine, it is somewhat impractical since it requires a right angle between r and d. Fortunately, there is a somewhat more practical approach that uses a phenomenon called **parallax**. Parallax is the apparent shift in the position of an object due to a change in the viewpoint of the observer.

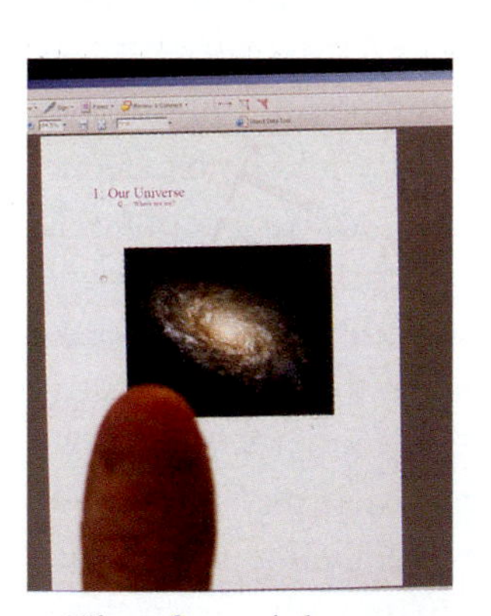

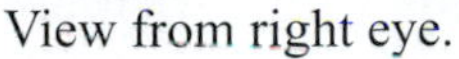

View from right eye.

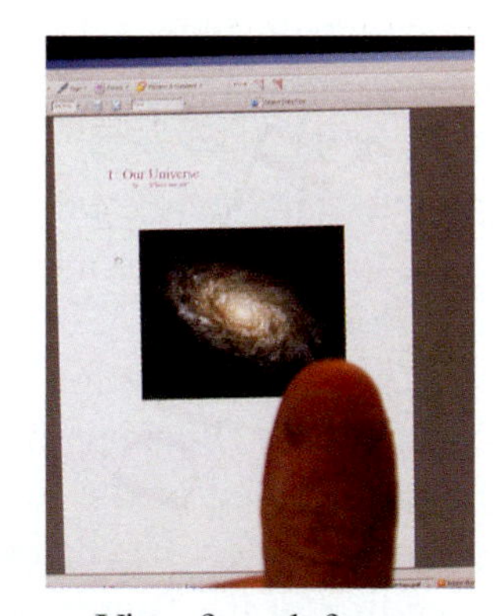

View from left eye.

Figure 1.6: Parallax of the thumb when closing first the right and then the left eye.

Parallax can be readily observed with one's own thumb as illustrated in Figure 1.6. The experiment works as follows: Hold your thumb about 20 to 30 cm in front of your nose. Next observe the position of the thumb relative to the background with one eye closed, and then observe its position with the other eye closed. Even though both the thumb as well as the background remain still, the thumb appears to move due to the change in viewpoint between the left and right eye. As can be imagined, the larger the baseline, the bigger the effect.

Parallax is the apparent motion of an object due to a change in viewpoint.

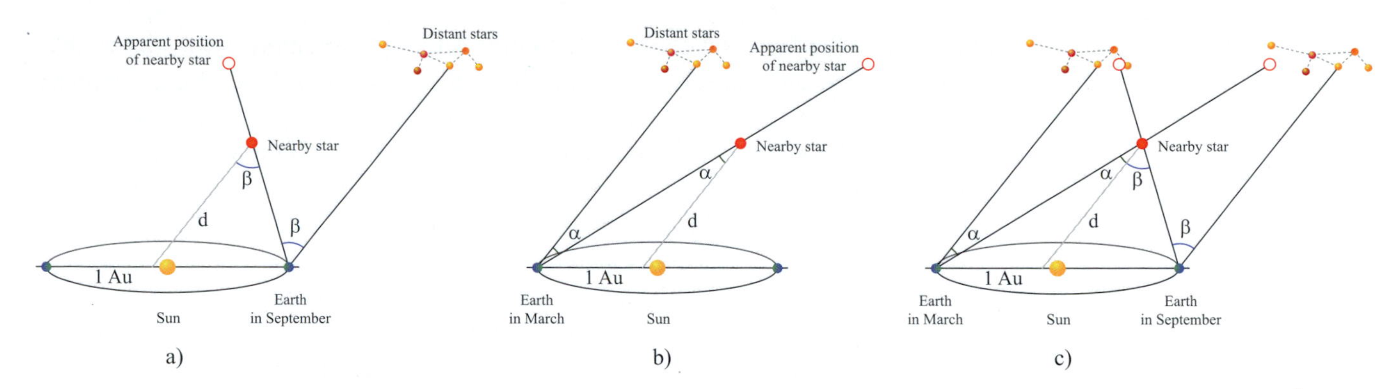

Figure 1.7: Using triangulation and parallax, the distance d to a nearby star can be measured. a) The star whose distance is to be measured appears to the left of some far away stars. b) Six months later, the star appears to the right of the far away stars. c) Merging the situations in a) and b) shows the geometry. Note that not having a right angle between the baseline and the line to the distant star induces only a small error that can easily be corrected for. Note furthermore that the line for the distance d does not (have to) pass through the sun as it is parallel to the line to the distant stars. The figure is not to scale, in reality d $\gg$ 1 Au.

Parallax can be used quite effectively to measure the distances to celestial objects as long as they are not too far away but it is important to distinguish between two basic scenarios.

If an object like a planet or a moon moves relatively fast with respect to the earth, the observations from different viewpoints need to be made at the same time or in a relatively brief interval. If the object such as a nearby star does not move very fast, then the observations do not need to be at the same time, and this last fact makes it possible to use the orbit of the earth around the sun as the baseline. The radius of this baseline, in other words the (average) distance from the earth to the sun, then makes for a convenient unit to measure distances in space. It is called the **astronomical unit (Au)** and roughly equal to 150 million kilometers.

The **parallax angle** p is defined as half the sum of the measured angles from the two viewpoints (i.e. $p = (\alpha + \beta)/2$). As illustrated in Figure 1.7, the distance to a nearby star can then be found as

$$\begin{aligned} \tan(p) &= \frac{1\text{Au}}{d} \simeq p \quad \text{(in radians)} \\ \Rightarrow d &= \frac{1\text{Au}}{p} \end{aligned} \tag{1.1}$$

where the fact that $\tan p \simeq p$ for small p in radians is used (contrary to the example shown in Figure 1.6, the parallax of nearby stars is very small so that the approximation shown in Figure 1.8 is valid).

Returning to Figure 1.6, if we estimate the distance between our eyes to be 0.1 m and the parallax angle to be 10°, then the distance to the thumb can be calculated as $d = 0.1\,\text{m}/(2\tan 10°) \approx 0.3$ m which is about right. (Note the factor 2: Equation 1.1 is based on the radius while the distance between the eyes is a diameter.)

While the definition of the parsec relates well with the observations astronomers make and is thus pretty handy, its disadvantage is that it does not directly

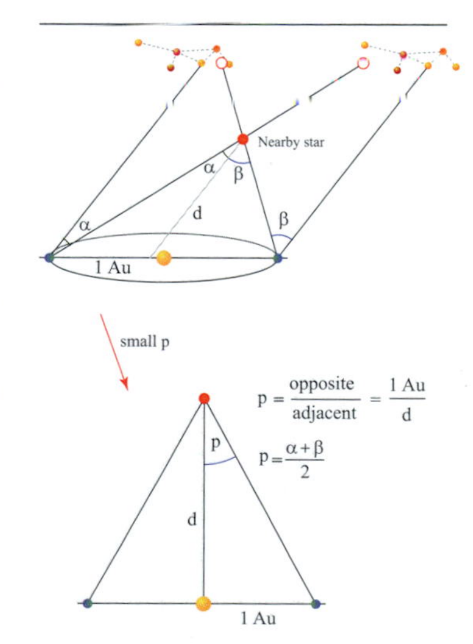

Fig. 1.8: The distance d to a star can be found as 1 Au/p for small parallax angles p.

A light-year is a distance of approximately 9.46×10^{15} *m.*

relate to any fundamental constant in nature. Although it is more or less inevitable to define some units arbitrarily, as much as possible, it is preferable to express quantities in terms of nature's constants. For large distances, a natural measure is the distance light travels in vacuum in a year and called a **light-year**. Taking the velocity of light to be 299,792,458 m/s, a light-year equals about 9.46053×10^{15} m, which is about 9.5 trillion kilometers. Since a parsec is about 3.09×10^{16} m, it readily follows that one parsec equals about 3.26 light-years.

The closest star Alpha Centauri is at a distance of $1.34\,\text{pc} \simeq 4.37$ light-years and this distance can nowadays quite easily be determined with the help of triangulation. However, even though it is the closest star, the parallax angle is already very small (being less than 1/3600th of a degree) and stars farther away, let alone distant galaxies, imply impossibly small parallax angles.

Definition 1.0: Parsec

Often distances in the universe are measured with a unit called the **parsec** (pc). It is defined as the distance for which the angle p equals exactly one arcsecond (an arcsecond is defined as 1/3600th of a degree). From Equation 1.1 we then find that

$$\Rightarrow \quad \frac{1}{3600}\frac{2\pi}{360} = \frac{1\text{Au}}{1\text{pc}}$$

$$\Rightarrow \quad 1\text{pc} = 3.09 \times 10^{16}\text{m} \tag{1.2}$$

Triangulation is the only direct method to establish distances in the universe. For objects that are too far away to measure the parallax angle, an indirect approach becomes necessary. The first such method we will discuss here is based on a fact well known from daily life: if we know the brightness of a light (for example a ship's lantern), then the apparent brightness gives us an idea of its distance — this is why we assume that a ship whose lantern is very dim is far way. Although this may be nice, how could this be applied to a star? How do we know how bright a star really is? Let us now look at some special stars called **Cepheids** whose brightness can be used to determine distances beyond what is possible with triangulation.

Cepheids

From daily life we know that a dimmer light often indicates a greater distance. The reason why we make this inference is that we know how bright certain types of light sources should be. For example, all the street lanterns in a town may use the same light bulbs operated identically. Then by standing right under a street lantern, we can find out how bright the light is and consequently if we see a dimmer lantern we conclude that it is farther away (unless it's broken of course).

Looking at the night sky, we are dealing with similar phenomena. There are many different stars with many different brightnesses. But the problem is, the stars are clearly not all the same so how do we know which one is farther away and which one is closer? What we need is some kind of reference and of course we also need some measure for its brightness.

The brightness of a star is expressed in terms of its luminosity, and we need to distinguish between two types: the **absolute luminosity** that gives the amount of energy per unit time radiated from the surface of the star and the **apparent luminosity** that gives the amount of energy per unit time per surface area which reaches our detectors. The apparent luminosity varies inversely with the square of the star's distance as shown in Figure 1.9. Knowing the distance we can then calculate the absolute luminosity, or vice versa, if we know the absolute luminosity, we can calculate the distance.

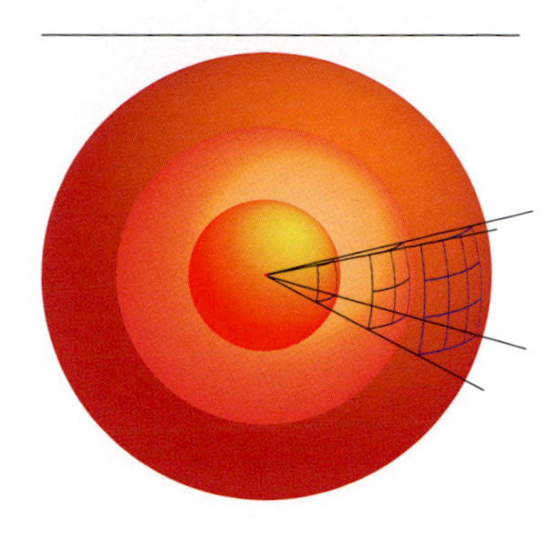

Fig. 1.9: The amount of radiation per unit area decreases as $1/r^2$.

The question then will be: How can we possibly know the absolute luminosity? This is where the Cepheids come in. Cepheids are stars with a combination of two special properties. First, their apparent luminosity varies regularly due to a periodic expansion and contraction, and second, the period of the expansion and contraction is a function of their absolute luminosity as illustrated in Figure 1.10. Although there are two types of Cepheids (Type I and Type II), the principle is the same in both cases. With the help of nearby Cepheids whose distance can be determined by means of triangulation, the function relating the absolute luminosity of a Cepheid to its periodicity can be determined.

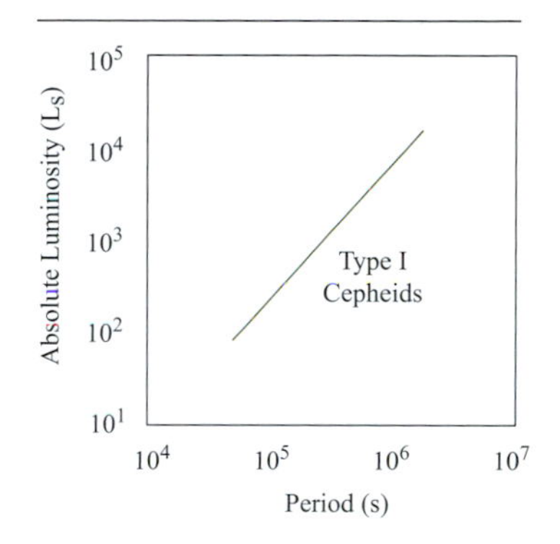

Fig. 1.10: The Cepheid Variables. Standard luminosity L_S is equal to 3.86×10^{26} W.

The mass of Cepheids is between 5 and 20 times that of our sun (hence they are yellow giants) and their luminosity is around 10^3 to 10^4 times that of our sun (the luminosity of the sun often referred to as standard luminosity L_S is 3.86×10^{26} W). As the periodicity can be measured very precisely, we thus have a tool to rather accurately determine distance: first we measure the periodicity in the apparent luminosity, from this we infer the absolute luminosity and we then use the inverse square law to calculate the distance.

Our galaxy, the Milky Way has been studied exhaustively using the Cepheid star method. The Milky Way is a disk of about 200 billion stars with two major spiral arms. The radius of the disk of stars in the Milky Way is about 100,000 light-years across and the disk is about 1,000 light-years thick.

All stars in the galaxy rotate around the galactic center but not with the same period. Stars at the center have a shorter period than those farther out. The sun is located in the outer part of the galaxy at the tip of one of the gigantic spiral arms about 28,000 light-years from the galaxy's center. The speed of the solar system due to the galactic rotation is about 254 km/s, and based on a distance of 28,000 light-years, the sun orbits around the center of the Milky Way once every 220 million years. This period of time is called a cosmic year. Consequently, the sun has orbited the galaxy core only 20 times during its lifetime. Figure 1.11 shows an artist's impression of the structure of our Milky Way from above. The location of our sun is indicated by the yellow arrow.

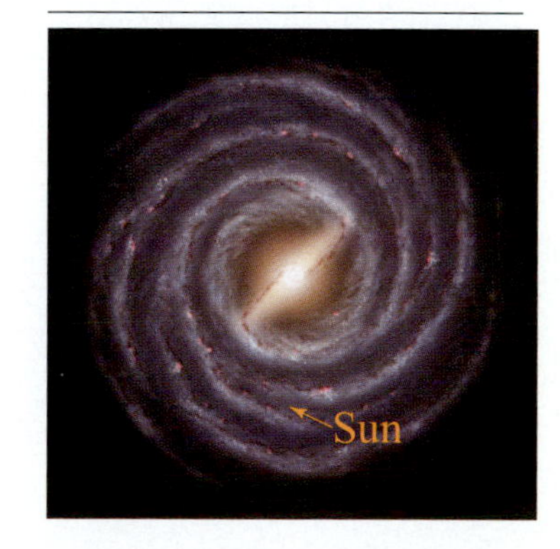

Fig. 1.11: Reconstruction of the top view of our Milky Way. Source: Spitzer Space Telescope.

Cepheid stars are bright enough to measure distances up to about 1 Mpc but that is not enough to reach much beyond the Milky Way as the closest galaxy to our own, M31 in the Andromeda nebula, is already 778 kpc $\cong$ 778 kpc $\times$ 3.26 light-year/pc = 2.54 million light-years away. Hence a different method is necessary for measuring greater distances. One such method, invented by the American astronomer Edwin Hubble (1889–1953), is based on the assumption that similar types of objects are the same throughout the universe and uses the luminosity of certain stars or galaxies.

Example 1.1: Distance by the Cepheid method

The period of a Type I Cepheid is measured to be 10^6 s with apparent luminosity $L_A = 2 \times 10^{-12}$ W/m^2. What is the distance of the star?

Answer: From Figure 1.10 we find the absolute luminosity L as

$$L = 8{,}000\, L_S = 8{,}000 \times 3.86 \times 10^{26}\,\text{W} = 3.088 \times 10^{30}\,\text{W}. \qquad (1.3)$$

The apparent luminosity is

$$L_A = \frac{L}{4\pi d^2} \qquad (d = \text{distance of star})$$

$$\Rightarrow d = \sqrt{\frac{L}{4\pi L_A}} = 3.5 \times 10^{20}\,\text{m} = 11.3\,\text{kpc}. \qquad (1.4)$$

Luminosity

What Hubble did was to measure the apparent luminosity of supergiants, a class of stars that are the brightest in each galaxy, and are visible over much greater distances than Cepheids. Hubble made the assumption that the supergiants in all galaxies have roughly the same absolute luminosity, and hence by measuring the apparent luminosity of distant supergiants and comparing it with the luminosity of nearer supergiants whose distance was known, Hubble could infer their distance. This technique works up to distances of 10 Mpc.

Hubble then extended his idea by assuming that not only the brightest stars in most galaxies have a similar absolute luminosity but that in turn, the brightest galaxies in galaxy groups or galaxy clusters also have similar absolute luminosities. Thus he was able to estimate distances up to 500 Mpc!

On a side note, it should not come as a surprise that there are different kinds of galaxies. Based on the spatial distribution of the stars, Hubble described four main types of galaxies shown in Figure 1.12 as ordinary spiral, barred spiral, elliptical and irregular.

Doppler effect

Probably, the best known example of the Doppler effect is the phenomenon that the pitch of sound of an approaching car is higher than that of a receding car. The Doppler effect holds not only for sound waves but for wave phenomena in general and that includes light — which is also a wave. Consequently, the color of an approaching light source appears to be shifted in the blue direction (shorter wave-

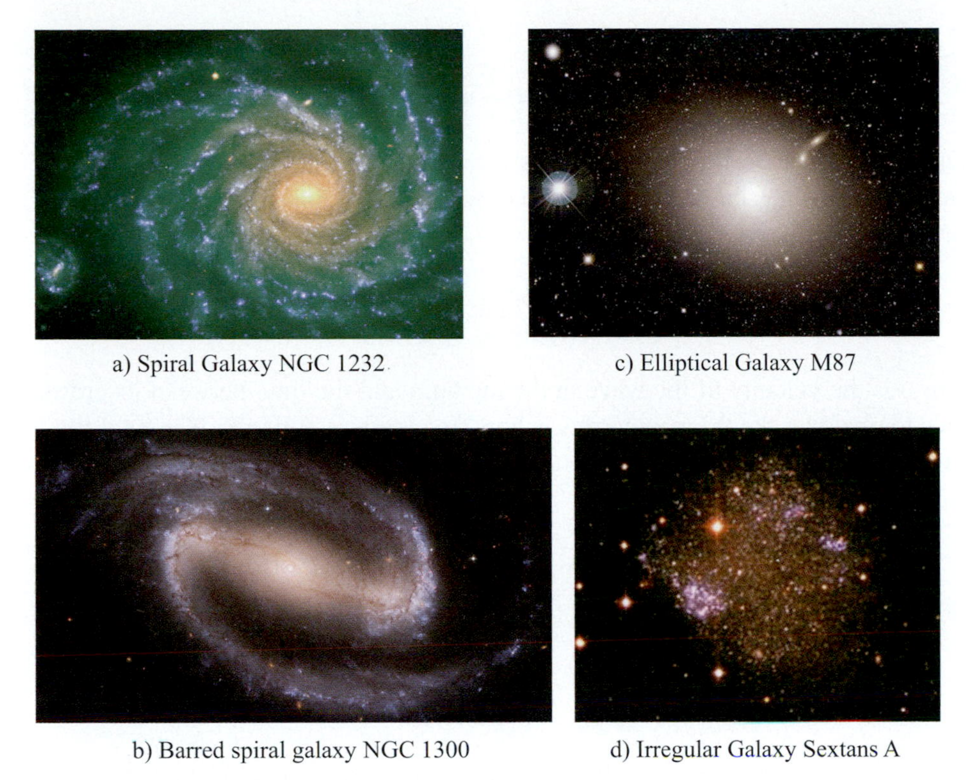
a) Spiral Galaxy NGC 1232
c) Elliptical Galaxy M87
b) Barred spiral galaxy NGC 1300
d) Irregular Galaxy Sextans A

Figure 1.12: Hubble's four main types of galaxies. Source: NASA.

Hubble's four main types of galaxies:

ordinary spiral

barred spiral

elliptical

irregular

length) and that of a receding light source shifted to the red (longer wavelength) — see also Figure 1.13. As we will need this effect for the next method to measure distances, we briefly have a look at it now.

Although being able to measure distances up to 500 Mpc is pretty impressive, compared to the size of the universe 500 Mpc is still not that far. Hence an even better method for the furthest celestial objects is needed. This method is based on the Doppler effect which we will therefore elaborate on next.

Up to a very high degree of accuracy, the speed of a light wave only depends on the medium in which it is propagating and not on the speed of the source which is emitting the light wave (this is perfectly so in vacuum). In general, it is reasonable that the frequency of the wave emitted is constant (e.g. to the driver, the pitch of the motor of the approaching car does not change or to an observer near a distant star, the color of that star remains the same). A wave is an oscillatory phenomenon with a given time interval between the crests as measured in its frequency. Now when a source is moving away, the distance between the crests increases because each successive crest is emitted a bit farther away and hence requires more traveling time to arrive at the observer. A greater distance between the crests means a lower frequency (vice versa, the frequency of a wave emitted from an approaching source will increase). Indeed, the extra time from crest to crest will be equal to the distance moved by the source during one period of the wave divided by the speed of the wave.

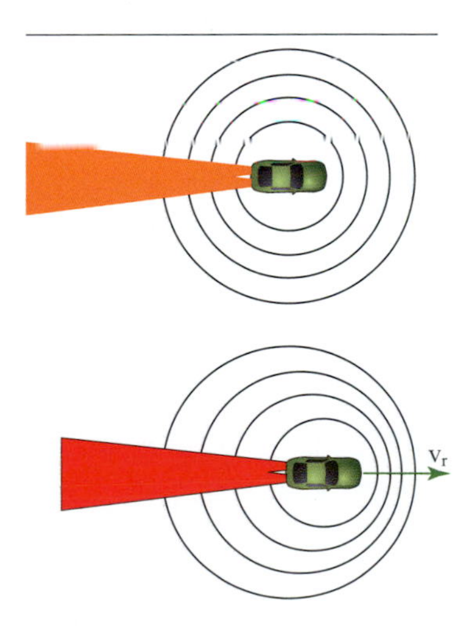

Fig. 1.13: Illustration of the Doppler effect. The tail light of a (very) fast moving space car will appear redder than when the space car stands still.

For a wave with frequency of oscillation given by f, the period of one oscillation is $1/f$. Hence, the distance Δx moved by the source in one period is given by

$$\Delta x = \frac{1}{f} v_r \tag{1.5}$$

where v_r is the velocity of the source as measured by the observer (a positive v_r indicates a receding source relative to the observer). This means that the extra time to arrive at the observer is given by

$$\Delta t = \frac{\Delta x}{v} = \frac{v_r}{f v} \tag{1.6}$$

where v is the velocity of the wave in the medium and the time between its crests equals to one over the frequency (this is how frequency is defined). The total time between crests at the observer is then this extra time plus the original time between crests, or

$$\frac{1}{f'} = \frac{v_r}{f v} + \frac{1}{f} = \frac{v_r + v}{f v} \tag{1.7}$$

which can readily be rewritten in the form

$$f' = f\left(\frac{v}{v + v_r}\right). \tag{1.8}$$

If the source were approaching, then the sign in front of v_r would be negative. In the case of a light wave, Equation 1.8 becomes

$$f' = f\left(\frac{c}{c + v_r}\right) = f\left(\frac{c + v_r - v_r}{c + v_r}\right) \approx f(1 - \frac{v_r}{c}) \tag{1.9}$$

where in the last step c is the speed of light and the fact that $v_r/(c + v_r) \approx v_r/c$ for $v_r \ll c$ is used.

Here, we are interested in determining the velocity v_r as this will tell us how fast the source of the wave is receding from the observer. Since we do not want to use the approximation at this point, we rewrite the left most equality in Equation 1.9 to obtain

$$v_r = \frac{(f - f')c}{f'} = \left(\frac{1}{\lambda} - \frac{1}{\lambda'}\right)\lambda' c \tag{1.10}$$

where the wavelength λ is given by $\lambda = c/f$.

Thus we see the relationship between recession speed and the observed wavelength, i.e. color. However, from a conceptional point of view, it is often easier to consider the *shift* in the color, i.e. the **redshift**. Generally, the redshift is denoted by z and for $v_r \ll c$ is given by

$$z = \frac{\lambda' - \lambda}{\lambda} = \frac{f - f'}{f'}. \tag{1.11}$$

Combining Equations 1.10 and 1.11, we can then express the recession velocity as

$$v_r = zc \quad \Rightarrow \quad \textbf{recession speed = redshift times speed of light.} \tag{1.12}$$

Now we know how to calculate the recession speed of a star if we know its original color. But how can we possibly know its original color? Perhaps considering the composite color of a star is a bit too much since it contains light of many different wavelengths as can easily be verified with the help of a prism. Still, even if we look at a single wavelength, how do we know that it is shifted? Indeed that is very hard to know but if one looks at the light of our sun through a prism, something rather peculiar can be noticed: The spectrum is full of black lines! These lines, called Fraunhofer lines after the German physicist Joseph von Fraunhofer (1787–1826), are shown in Figure 1.14. We now know that these lines are due to absorption of light by the elements in the sun's outer atmosphere and their location in the spectrum is exactly determined by those elements (for more details on this phenomenon, see Chapter 23). Indeed, each element gives rise to a certain very specific pattern of lines which makes it identifiable just like a fingerprint uniquely matches a person. Since the elements are identical in the entire universe they lead to black lines at the same point in the spectrum everywhere. By comparing the spectrum of the sun (or more precisely the spectrum of some common element like hydrogen) with that of a remote celestial object, we can then see how far a black line has shifted, thus showing us exactly how big the redshift is.

Fig. 1.14: The spectrum of the sun is full of black lines.

In the top part of Figure 1.15, some of the lines visible in the spectrum of the sun are depicted, together with the elements that cause these lines and their wavelengths. The rightmost black line is due to hydrogen and corresponds to a wavelength of 656 nm. The bottom part of the figure shows where the lines are located for an object that is receding with a velocity of 0.05 c. For example, the first hydrogen line can now be found at 689 nm.

While we have thus far shown how the redshift can be used to quite accurately

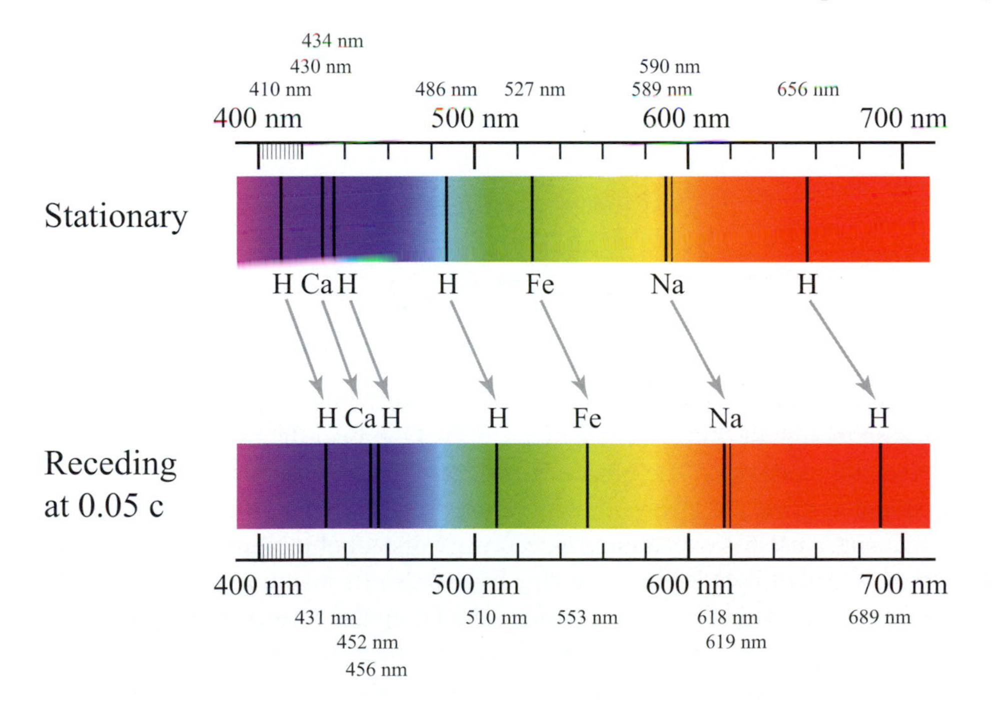

Figure 1.15: The shift in the spectrum of an object receding at 0.05 c.

determine the recession velocity of a celestial object as long as the shift is not too small, one can wonder what this has to do with distance. After all, on earth, whether a car goes fast or slow, is approaching or moving away has basically nothing to do with how far away it is from, for example, the great pyramids of Giza. This brings us to an astonishing finding by Hubble, nowadays referred to as Hubble's law.

Hubble's law

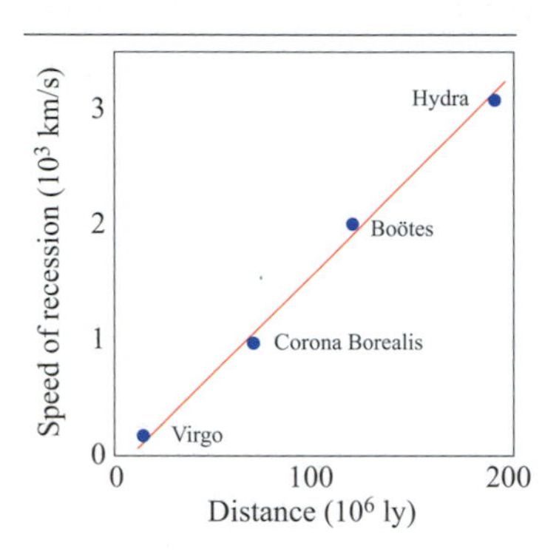

Fig. 1.16: Hubble's law.

When Hubble measured the redshift of many galaxies as well as their distance d, he discovered the rather amazing fact that for all sufficiently distant galaxies there is a linear relationship between recession velocity and redshift as illustrated in Figure 1.16! This relationship is called Hubble's law and given by

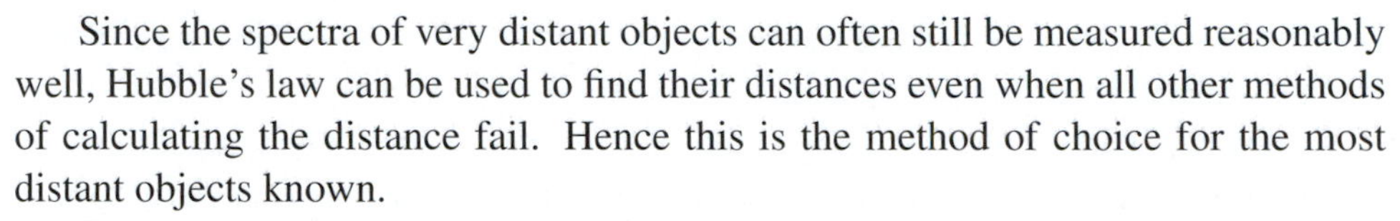

$$v = H_0 d \quad : \quad \textbf{Hubble's law} \tag{1.13}$$

Since the spectra of very distant objects can often still be measured reasonably well, Hubble's law can be used to find their distances even when all other methods of calculating the distance fail. Hence this is the method of choice for the most distant objects known.

The relation given in Equation 1.13 for very large values of d implies velocities that are very large as well, indeed they are approaching the velocity of light c. For such large values of the velocity of recession v, Equation 1.11 is no longer accurate and a modified expression derived from Einstein's theory of relativity needs to be employed. It is given by

$$\begin{aligned} z \equiv \frac{\delta\lambda}{\lambda} &= \sqrt{\frac{1+\beta}{1-\beta}} - 1 \\ \Rightarrow \beta &= \frac{(1+z)^2 - 1}{(1+z)^2 + 1} = \frac{v}{c} \\ \Rightarrow v &= \frac{(1+z)^2 - 1}{(1+z)^2 + 1} c \end{aligned} \tag{1.14}$$

where $\beta = v/c$.

Now it is also clear why above we only considered the shift due to receding objects even though the physical principles of the Doppler effect work just as well for approaching galaxies. The objects whose distance is measured with the help of the Doppler effect are always moving away from us. The ultimate reason for this is that space itself is expanding. Indeed, very distant objects recede from us at close to the speed of light leading to a very large redshifts that require the more advanced formula given in Equation 1.14 for calculating the correct recession speed. Hubble's law is not used for nearby objects not so much because the redshift may be too small to accurately measure (measurement apparatuses may be improved) but because it doesn't hold. For nearby stars, the effects of the expansion of space are not large enough to overcome their relative velocities.

At present, the value of Hubble's constant is estimated to be

$$H_0 \simeq \frac{7.2 \times 10^4 \mathrm{m/s}}{\mathrm{Mpc}} \tag{1.15}$$

In other words, a galaxy at a distance of 1 Mpc is receding from us at the velocity of 7.2×10^4 m/s. Note H_0 has the unit of $(\text{time})^{-1}$ (i.e. per second).

Example 1.2: Distance calculation employing Hubble's law

For a galaxy with z = 1, calculate its distance from us (this is a large z so we need the relativistic expression). **Solution:** With the help of Equation 1.14 we find that

$$\beta = \frac{(1+z)^2 - 1}{(1+z)^2 + 1} \tag{1.16}$$

For $z = 1, \beta = 3/5, v = 3c/5$. Hence

$$\begin{aligned} d &= \frac{v}{H_0} = \frac{3c}{5H_0} \\ &= \frac{(0.6) \times (3 \times 10^8 \mathrm{ms}^{-1})}{(7 \times 10^4)(\mathrm{ms}^{-1}\mathrm{Mpc}^{-1})} \end{aligned} \tag{1.17}$$

or, $d \approx 2600$ Mpc.

In summary, Hubble's law holds true for all galaxies in the universe that are not too close to each other, and thus states that all galaxies not close to each other are receding from all others due to the expansion of space. Galaxies in a local cluster do not need to recede from each other due to gravitational binding. However, all clusters are receding from all other clusters! It is interesting to note that the age of the universe is currently believed to be around 13.7 billion years, the oldest stars in the Milky Way are about 13.2 billion years old and the earth is about 4.56 billion years old.

Table 1.1 gives the distances to various astronomical objects and the methods used for measuring these distances.

Where are we now?

"Where are we?" is the question we like to address in this chapter. In order to answer that question, we need to discuss the structure of the universe and find that we cannot do that without having effective means to calculate the distances to celestial objects. Now that we finally have a full array of tools to calculate distances, we can come to a partial answer: We live in a huge universe full of galaxies made up of even larger numbers of stars. The universe appears to be well described by the cosmological principle stating that all parts of the universe are equivalent and consequently we can surmise that we live in an ordinary part of space with nothing special about it. Our star, the sun, is part of a spiral galaxy

Method	Distance (pc)	Object
Hubble's law	10^9–10^{10}	Quasars
Apparent luminosity - Galaxies	10^8	Virgo Cluster
Apparent luminosity - Supergiants	10^6–10^7	Andromeda Galaxy
Cepheid Variables - Type II	10^5	Andromeda Galaxy
Cepheid Variables - Type I	10^1–10^4	Center of Milky Way
Parallax	10^0	Alpha Centauri

Table 1.1: Methods used to measure distances in the universe.

called the Milky Way that is part of a cluster of over 50 galaxies called the **local group** which in turn is part of the Virgo supercluster. The Andromeda galaxy is the nearest spiral galaxy at a distance of about 2.54 million light-years and the largest member of the local group.

Within the Milky Way, the sun is located near the inner rim of a spiral arm called the Orion arm at a distance of about 26,000 light-years from the center of the galaxy. The nearest star system at a distance of about 4.37 light-years is **Alpha Centauri**.

If we look in our neighborhood, we see that we live on a planet that is part of a solar system consisting of a single star (our sun), eight planets (Mercury, Venus, Earth, Mars, Jupiter, Saturn, Uranus, Neptune), three dwarf planets (Ceres, Pluto, Eris), over 160 moons, and a number of other celestial objects such as asteroids, meteoroids and comets.

1.4 Motion of the planets

Although we have used the fact in our discussion on parallax, we have not paid much attention to one rather important aspect of where we are: we never stay at the same point! But our motion is not just random. The earth orbits the sun and the sun together with all the objects in the solar system orbits the center of the Milky Way. Therefore, if we want to understand where we are, we need to understand this motion. Hence, let us now investigate the motion of the planets.

Motion is usually associated with velocity, that is, the rate at which the position changes, while the rate of change in velocity is given by either acceleration or deceleration (see also section 2.6). But how can one describe velocity and acceleration more accurately? The average velocity is easy to determine by taking the distance traveled and dividing by the time taken to travel this distance. But when accelerating or decelerating, the velocity changes at every instant in time! Then what is the right way to calculate it? To do this, a whole new field of mathematics is necessary! And Newton created it — he invented calculus (note that calculus was independently also developed by Leibniz around the same time).

With the help of calculus, velocity is defined as the first derivative (rate of change) of the position with respect to time, and acceleration as the first derivative of the velocity with respect to time which equals the second derivative of the position with respect to time. In other words,

$$v(t) = \frac{dx(t)}{dt}, \qquad a(t) = \frac{dv(t)}{dt} = \frac{d^2x(t)}{dt^2}. \tag{1.18}$$

As is clear by dropping rocks from a tower, they accelerate while gaining velocity. An increasing velocity means that there is an acceleration and one can then wonder how big it is. The simplest assumption would be that the acceleration is constant (careful experimentation shows that this is true to a high degree of accuracy near the surface of the earth). So what would its value be? To measure it, we would need an expression relating acceleration, the distance traveled (height of the tower) and the time it takes to travel this distance (time until the rock hits the ground). Fortunately, such an expression is easy to obtain from the definition of acceleration in Equation 1.18 by integrating the acceleration twice with respect to time. We then find that

$$x(t) = \frac{1}{2}at^2 + v_0t + x_0 \tag{1.19}$$

where v_0 is the initial velocity and x_0 the initial position. When dropping a rock from a tower, the initial velocity and position can be taken to be zero. Denoting the gravitational acceleration as g we then have

$$x(t) = \frac{1}{2}gt^2. \tag{1.20}$$

Note that $g \approx 9.8$ m/s^2, which we can obtain by measuring the height of the tower (giving us $x(t)$) and the time it takes for the rock to reach the ground (giving us t and hence t^2).

What is crucial now is that once we have g, we can predict the trajectories of some moving objects. This was (and is) of great importance in military applications where until Newton, the answers to two key questions were based on trial and error rather than a rigorous scientific analysis. The questions are: if we shoot an arrow with a certain speed and angle, how far will it fly (see Figure 1.17), and if we want an arrow to land at a certain spot (e.g. the enemy lines), at what angle does it need to be shot (see Figure 1.18)?

Fig. 1.17: Finding the final postion x_f.

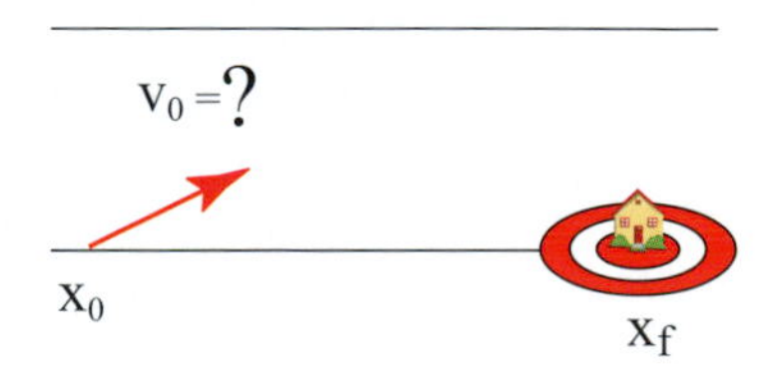

Fig. 1.18: Finding the initial velocity v_0.

Let us examine Figure 1.17 (Figure 1.18 is part of the exercises). Say we shoot off an arrow at a 45° angle with a velocity of 28.2 m/s, how far will it fly? Here we take our position $x_0 = 0$. The first thing we do is calculate the time the arrow spends in the air by solving $-1/2gt^2 + v_yt = 0$ where v_y is the vertical component of the velocity given by $v/\sqrt{2} = 20$ m/s (note g is in the opposite direction of v_y, hence the minus sign). This gives a flight time of $t = 4.1$s which we can then use together with the vertical velocity $v_x = 20$ m/s to calculate the impact point as $x_f = tv_x = 4.1\,\text{s} \times 20\,\text{m/s} = 82\,\text{m}$.

Now that we had a brief look at velocity and acceleration, let us return to the planets. When considering motion, we know from daily life that mass plays an important role in that it is easy to stop a light moving object such as a gently thrown tennis ball by hand but that it is very hard to stop by hand a moving truck

Fig. 1.19: The motion of the planets is well predicted by Newton's laws. Image source: NASA.

approaching with the same speed. Hence, a better measure of motion is not just velocity, but something which also takes into account the mass of the object as well.

This measure is called **momentum** and we expect that

$$\text{Momentum} \propto \text{mass} \times \text{velocity}$$

where $\propto$ means "proportional to". Since we have not yet defined the units for mass, we can absorb the proportionality constant into the definition of mass. For a particle with mass m and velocity v, the momentum p is consequently defined to be the following:

$$\begin{aligned} \text{Momentum} &= \text{mass} \times \text{velocity} \\ \Rightarrow p &= mv. \end{aligned} \tag{1.21}$$

Now we have a definition of motion but since the planets orbit the sun, their direction of motion is changing constantly. How do we deal with that? From daily life, we know that changing the path of an object in motion takes some effort, some "force". Indeed, we all have some intuitive idea of force. We know that for instance, to move a heavy object, we have to exert a lot of force. But then we need to ask, what happens to an object in motion upon which no force at all is exerted, exactly how much force do we need to change the direction in the motion of an object and what happens to the force if it is applied to an object? Finally, of course, we need to wonder what is it that exerts this force on the planets. It was Newton who first answered these questions in a scientific manner and consequently the resulting physical laws are called Newton's laws of motion.

Let us start with the first point: what happens if no force at all is applied to an object? Although when daily life is the starting point it is quite a stretch of imagination, the very simplest thing that could happen would be "nothing"! It seems that nature likes simplicity and indeed nothing happens. This fact is formalized in Newton's first law, the law of inertia.

Newton's first law: The law of inertia.

A body remains at rest or in constant motion along a straight line unless an external force acts upon it.

Although we have used the term "force", in fact it has not been defined yet but let us return to that after the next law.

Now the second point: how much force do we need to change the motion of an object? Following the reasoning above, it is clear that rather than velocity by itself, we need to consider the momentum of an object. In classical physics, the mass of an object is a constant so the only thing that can change is the velocity. Consequently force should be proportional the change in velocity, with the proportionality constant given by mass. The change in the velocity v is defined as the acceleration $a = dv/dt$. This leads to Newton's celebrated second law, the law of resultant force.

Newton's second law: The law of resultant force.

$$F = m\frac{dv}{dt} = ma = \frac{dp}{dt} \tag{1.22}$$

The rate of change of the momentum of an object is proportional to the force acting on it.

Note that formulation of the second law is in essence reversed from its conceptual justification (this is mathematically equivalent). The ubiquitous formula $F = ma$ defines the force in terms of the change in the momentum of an object but the description of the second law states what happens to an object when a force is exerted on it. But, importantly, we now have a rigorous definition for the term "force". So what would the unit for force be? Mass is measured in kg, acceleration is measured in ms^{-2}; hence the unit of force, called a newton, is given by $\mathrm{kg\,m\,s}^{-2}$.

Noteworthy 1.1: Newton's second law is a vector equation

In general, Newton's second law is a vector equation, namely

$$\mathbf{F} = m\,\mathbf{a} \tag{1.23}$$

Next, we need to investigate what happens to a force when it is applied to an object. Let us consider the object that is used to exert the force. For example, suppose we use a tennis ball to exert a certain force on a basketball by throwing the tennis ball at the basketball. The second law tells us what happens to the basketball but not what happens to the tennis ball. Conceptually, however, we know that from a different viewpoint, one could just as well argue that the basketball was moving toward the tennis ball so we could say that it is the basketball that causes a change in the motion of the tennis ball. Considering the symmetry of the situation, it appears that the simplest and most reasonable assumption should be that the force exerted on the basketball by the tennis ball is equal and opposite to the force exerted by the basketball on the tennis ball. Note that this does not mean that one of the two balls will be at rest after the interaction, only that the forces are equal and opposite. We can now formulate Newton's third law: the law of reciprocal actions.

Newton's third law: The law of reciprocal actions.

Forces always occur in pairs, for each force there is an equal and opposite force.

More precisely, let the force exerted by the first object on the second object be denoted by $\mathbf{F}_1$; then Newton's third law states that the force exerted by the second object on the first object, denoted by $\mathbf{F}_2$, must be equal and opposite. In other words

$$\mathbf{F}_1 + \mathbf{F}_2 = 0. \tag{1.24}$$

But from Newton's second law, for the first body having momentum $\mathbf{p}_1$ and the second body having momentum $\mathbf{p}_2$, Equations 1.22 and 1.24 yield

$$\frac{d}{dt}(\mathbf{p}_1 + \mathbf{p}_2) = 0$$

$$\Rightarrow \mathbf{p}_1 + \mathbf{p}_2 = \text{constant} \quad : \quad \text{Momentum conservation} \tag{1.25}$$

Newton's third law rephrased: Momentum conservation.

Newton's third law is equivalent to the statement that the total momentum of an isolated system is conserved (is a constant).

Some reflections on Newton's three laws

The first law took an enormous leap of imagination, since we know from experience that nothing can continue to be in motion forever. The reason being that frictional forces slow down any moving body. Newton, however, could imagine the idealized situation when no force acts on a body and understood the underlying principle of motion. The famous experiment of Galileo where he dropped two different masses — and showed that they fall at the same rate — also needed a leap of imagination, since again friction due to air had hindered a clear understanding of the workings of the force of gravity.

The second law is the backbone of the predictive power of Newton's laws. At first, one may (correctly) object that this law has no predictive power, since the moment one sees a particle accelerating, one can simply multiply it by the mass of the particle and determine the force that is causing this acceleration. And if one did this, then Newton's second law would indeed have no predictive power. The genius of Newton lies in understanding that nature has a quantity called force, and once we can determine these forces *independently* from Newton's second law, we can then use this force as input to predict the motion of particles experiencing this force. And sure enough, there are multifarious forces to be found in nature, from gravitational and electrical to subnuclear forces such as strong and weak interactions — which can be then be plugged into Newton's second law to predict the future.

The third law is often stated in popular literature as action equals reaction. There is a crucial concept buried in the third law, and that is the concept of an isolated system. We will use this concept time and again, and so it is best to say a few words on it. In reality, no system in the universe is perfectly isolated, since, if it were, it would not be a part of our universe. What we mean by an isolated system, be it in mechanics or thermodynamics, is that we can isolate and shield the system from the forces in the environment to a sufficient degree of accuracy.

A number of subtle assumptions have been made in stating Newton's three laws of motion. For example, what is mass? In which frame of reference are position, velocity and force being measured? If I am stationary and you are moving very fast, and we both observe the same particle, clearly we will observe very different forces and velocities. Newton's response was that all observers who are moving with constant velocity — called inertial observers — will observe the same physics. The question then arises is whether all inertial observers will measure the same flow of time. Although Newton's answer was yes, we now know that this is only approximately true. The ideas of time, position, velocity, mass and acceleration were all radically changed by Albert Einstein in 1905, and form the basis of the special and general theories of relativity.

Newton's universal law of gravitation

So now we have seen how the definition of acceleration can be used to calculate an arrow's flight path, and how the motion of objects is determined by Newton's laws. Back to the planets, what is the force that changes their motion? Well, since rocks are attracted to the earth by gravitation, it appears reasonable that the moon is attracted to the earth by gravitation as well. But then, wouldn't the same be true for the planets and the sun? We also know from daily life that bigger rocks are heavier than smaller rocks thus exerting a greater force on a scale. From Newton's second law, we already know that $F = ma$ which in the case of gravitation at the earth's surface is given to a good approximation by $F = mg$: a formula which apparently only takes the mass of the falling rock into account.

Fig. 1.20: The attraction between the earth and the moon is well described by Newton's universal law of gravitation. Image source: NASA.

A more general law of gravitation should really contain the masses of both bodies involved. In $F = mg$, the mass of the earth can be thought of being subsumed in the constant g which could therefore be proportional to m_e if nature is again as simple as possible (m_e the mass of the earth). That would give $F \propto m m_e$ but clearly this formula has a problem. It does not account for the distance between the two masses. It appears reasonable that the gravitational attraction decreases with distance but then by how much?

To answer that, let us start with assuming that the gravitational attraction of a spherical homogeneous body is a constant value on a surface at a distance r from the center of that body. Now if, just like in the case of light in Figure 1.9, gravitation is some sort of physical quantity (even though we may not exactly know what it is), then it makes perfect sense that it decreases proportional to the increase of surface area, or in other words as $1/r^2$.

Thus we can speculate a law of gravitation to look like $F = Gm_1m_2/r^2$, where G is a universal gravitational constant and is required to make the equation dimensionally consistent. Indeed this is exactly the form of Newton's universal law of gravitation:

Newton's universal law of gravitation.

$$F = G\frac{m_1 m_2}{r^2} \tag{1.26}$$

The gravitational attraction between two bodies is proportional to the product of their masses and inversely proportional to the square of the distance between them.

where the universal gravitational constant $G = 6.67 \times 10^{-11}\mathrm{m^3kg^{-1}s^{-2}}$, m_1 and m_2 are taken as point masses and r is the distance between the masses. It can be shown that acceleration due to the earth's gravity near its surface, namely g is given by

$$g = G\frac{m_e}{r_e^2} \simeq 9.822\, m\, s^{-2} \tag{1.27}$$

where m_e, r_e is the earth's mass and radius respectively.

Newton's universal law of gravitation employs what is commonly referred to as "action at a distance" since it is instantaneous, with no reference being made to what might be between the attracting bodies. Newton did not like this very much, but the equation works incredibly well, so he left it at that. It was not until Einstein that a more physically satisfactory theory of gravitation was developed.

Now we have some idea of position and motion and this gives us a pretty good idea where we are and will be in the foreseeable future. But thus far we have completely ignored one aspect in the observation of the night sky. Clearly we are in a dark universe. Why would that be so? There are so many stars and all these stars are all very bright. Then why is the night sky dark? Shouldn't it be bright? This is the famed **Olbers' paradox** which we will discuss next.

1.5 Why is the night sky dark?

Fig. 1.21: The sky is dark at night. Why? Image source: NASA.

Suppose the universe is of infinite extent, in a stationary (equilibrium) state, and that the stars are uniformly distributed with density n.

Consider an observer located on earth and stars in a shell with thickness dr at a distance r from the earth as illustrated in Figure 1.22. Under the assumption of a constant density, for increasing r, the number of stars in the shell will increase proportionally to the surface area of the shell and hence as $4\pi r^2$. At the same time, for increasing r, the luminosity decreases proportionally to the surface area of the shell, again by $4\pi r^2$. In other words, when increasing the radius of the shell, on the one hand the number of stars increases by a factor $4\pi r^2$ but the luminosity per star decreases by a factor $4\pi r^2$. These two factors exactly cancel each other out and the amount of light received from a shell is constant and independent of r. More precisely, let the intensity of light received on earth due to stars located in a shell at a distance from r to $r + dr$ be denoted by dO, and the absolute luminosity of a single star by L. We then have

$$dO = \frac{(Ln)4\pi r^2\, dr}{4\pi r^2} = Ln\, dr \;:\; \text{independent of distance r!} \tag{1.28}$$

where the numerator represents the light emitted in the shell (luminosity times density of stars times the volume of the shell) and the denominator how much of the emitted light reaches earth. The brightness O on the earth due to stars out to some large distance $R_{\max}$ is then given by

$$O = \int_0^{R_{\max}} Ln\,dr = LnR_{\max}. \tag{1.29}$$

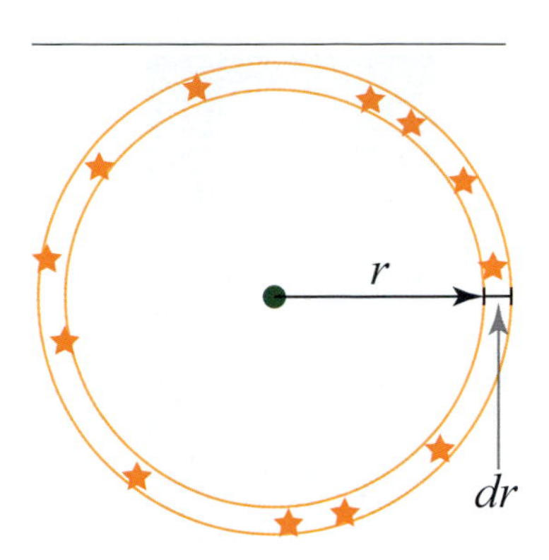

Fig. 1.22: Stars on a shell at distance r.

We see that all the stars along line of sight out to infinite distances ($R_{\max} \to \infty$) will contribute equally to O and hence the brightness O on earth should be infinity.

Clearly, this is not the case. The night sky is not bright but at the same time, the calculation above appears rather reasonable. This is Olbers' paradox. What could be the solution? Some of the suggested explanations are the following:

1. The universe has a finite size and hence only a finite number of stars contribute starlight to the night sky. A size of about 10 billion light-years would be enough to explain the dark night sky.

2. Stars have a finite life.

3. The universe has a finite age.

It turns out that all three of these factors contribute to the darkness of the sky. However, the finite lifetime of the universe is likely a significant contributor to this phenomenon. Note all three explanations require a dynamic and changing universe negating the assumption of Olbers of a stationary universe that always looks the same.

A simple fact like the sky being dark immediately leads one to a universe that is dynamic and changing!

1.6 Where are they?

Thus far we have considered our location in the universe. One of the things we found is that the universe is not only very large but also has a huge number of stars. A fundamental assumption made in studying the universe is that the laws of physics are identical everywhere. One consequence is that if the sun has planets, then unless the sun is very special which appears to be rather unlikely, many stars should have planets. If there are many stars with planets, some of these planets should be hospitable to life and on some of the planets with life an advanced civilization would be expected to emerge.

Since we have a reasonably good idea of how many stars there are in our galaxy, one can then make some assumptions on the probabilities that a star has a habitable planet, that the planet has life, that life is intelligent, etc. and formalize this in an equation. One of the most well-known of such equations is the **Drake equation**. It was first published in 1960 and estimates the number of civilizations N in the Milky Way with which communication might be possible as

$$N = R^* \times f_p \times n_e \times f_l \times f_i \times f_c \times L, \tag{1.30}$$

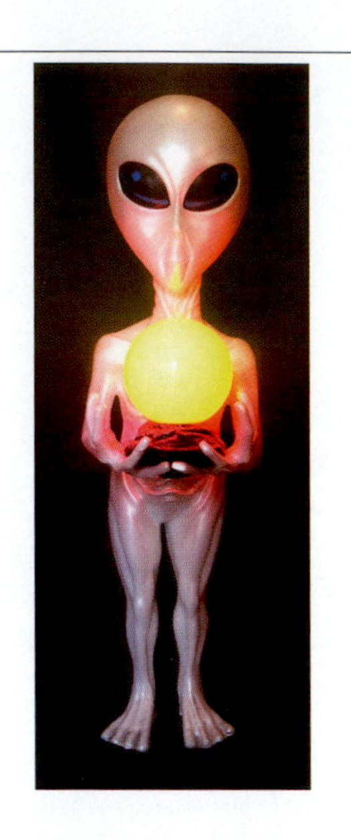

Fig. 1.23: Where are the aliens?

where Drake used $R^* = 10$/year for the number of stars formed per year, $f_p = 0.5$ for the probability that a star has planets, $n_e = 2$ for the number of habitable planets that a star with planets has, $f_l = 1$ for the probability that a habitable planet will develop life, $f_i = 0.01$ for the probability that life will be intelligent, $f_c = 0.01$ for the probability that the intelligent life will actually be able to communicate and $L = 10{,}000$ for the number of years that a civilization transmits detectable signals. Thus Drake obtained a value of $N = 10$ civilizations whose signals we might be able to detect.

Of course, the exact form of the Drake equation can be debated and even more so what values make for reasonable assumptions. However, the key issue is that any set of assumptions based on the universe and certainly the Milky Way being sort of the same everywhere, will lead to at least some and possibly a large number of advanced civilizations. The astronomer Carl Sagan once estimated that more than a million technical civilizations could exist in our galaxy.

The problem is that we have not seen any signs of intelligent life besides our own. The apparent contradiction between our observations (namely finding nothing) and what appear to be reasonable assumptions is often called the **Fermi paradox** named after the Italian Physicist Enrico Fermi who discussed this in 1950 but didn't follow up on the matter in great detail.

Instead of speculating about the likelihoods of planets having certain conditions, we can turn the discussion around and wonder how long would it take mankind to colonize the Milky Way, given currently available technology. Of course, besides technology, there are many other factors such as psychological, political, economical or religious matters. But we will completely ignore these and simply consider what would as such be possible with the technology we have.

Let us make some assumptions that are technologically speaking at the pessimistic side: We spend 200 years building two spaceships large enough to send colonists into space, the spaceship will accelerate to 0.1 c and on average travel for 100 years to find a suitable location. Then the colony takes 700 years to establish itself before sending out two new spaceships to continue the colonization process (one of the two spaceships always tries to maximize the distance to earth). This means that in 300 years we'll have two colonies at 10 light-years from earth, in 1,300 years we'll have four colonies, two of which will be about 20 light-years from earth etc.

Effectively, this means that we can explore our galaxy at a speed of about 0.01 c per year. Taking the diameter of our Milky Way to be about 100,000 light-years, then it should take at most about 10 million years to colonize most parts (and that is without technological advance). Now 10 million years may sound like a long time but from a cosmological view point, it's really very short. On earth, from the emergence of life to the ability to colonize space took about 4 billion years. Ten million years more or less (or even 100 million or perhaps even a billion years) is nothing but a tiny fluctuation in chance. Put in a different way, any space-faring civilization should be able to colonize the galaxy nearly instantly once it emerges.

The closest galaxy is about 2.54 million light-years away. A journey to the closest galaxy, at an effective speed of 0.1 c (no stopping on the way), will take about 25.4 million years. It is unlikely that any life form could undertake a journey of such a long duration through empty space. Hence, it is most likely that (intelligent) life cannot spread from one galaxy to another.

If there are many civilizations, then it seems unlikely that we're the first space-faring one. It also seems very unlikely that the capability to explore space emerged at roughly the same time. Then why aren't there any aliens here? One possible answer is that the emergence of life may not be such a common phenomenon; or that there are limitations to space travel that we have not yet become aware of. In any case, this is an intriguing question.

1.7 The answer

The answer as to where we are turns out to be as fascinating as it is bland. **We are on a planet in an ordinary solar system, in an ordinary part of a galaxy, in a ordinary galaxy, in a large expanding universe that keeps the night sky dark**. Fascinating it is because this implies that there's so much to be discovered that humanity will likely never run out of amazing phenomena to find and observe in our marvelous universe.

1.8 Exercises

1. What is a distance of 5 kilo parsec in light-years?

2. Would it be a good idea to measure the distance to the closest galaxy outside our own by triangulation?

3. If we have a Type I Cepheid with the period of luminosity being approximately 96 hours and an apparent luminosity of 4.63×10^{-9} W/m^2, what is this star's distance from us in light-years?

4. A girl (whose mass is 70 kg) is running at a constant speed of 10 km/h down a slope with an angle of 6°. What is her acceleration? What is her momentum? (Use standard units!)

5. Give the value of the Hubble constant in terms of seconds only (i.e. no other units).

6. If we know that a star's spectrum is blue shifted, what do we know about its distance?

7. Consider Figure 1.18 in the textbook. If I shoot a bullet into the air at a 30° angle and want it to fly 100 m, at what speed does it need to leave the barrel of my pistol?

8. How do we know that the units for the universal gravitational constant are $\text{m}^3\text{kg}^{-1}\text{s}^{-2}$?

9. Do you think that it is realistic to accelerate a large spaceship to 0.1 c?

10. Roughly, how long would it take us to transverse the thickness of the Milky Way's disk at a speed of 0.1 c?

11. Derive the volume of the shell in textbook Equation 1.28.

2: Numbers in Our World

Q - How Do We Reach Infinity and Beyond?

Chapter Map

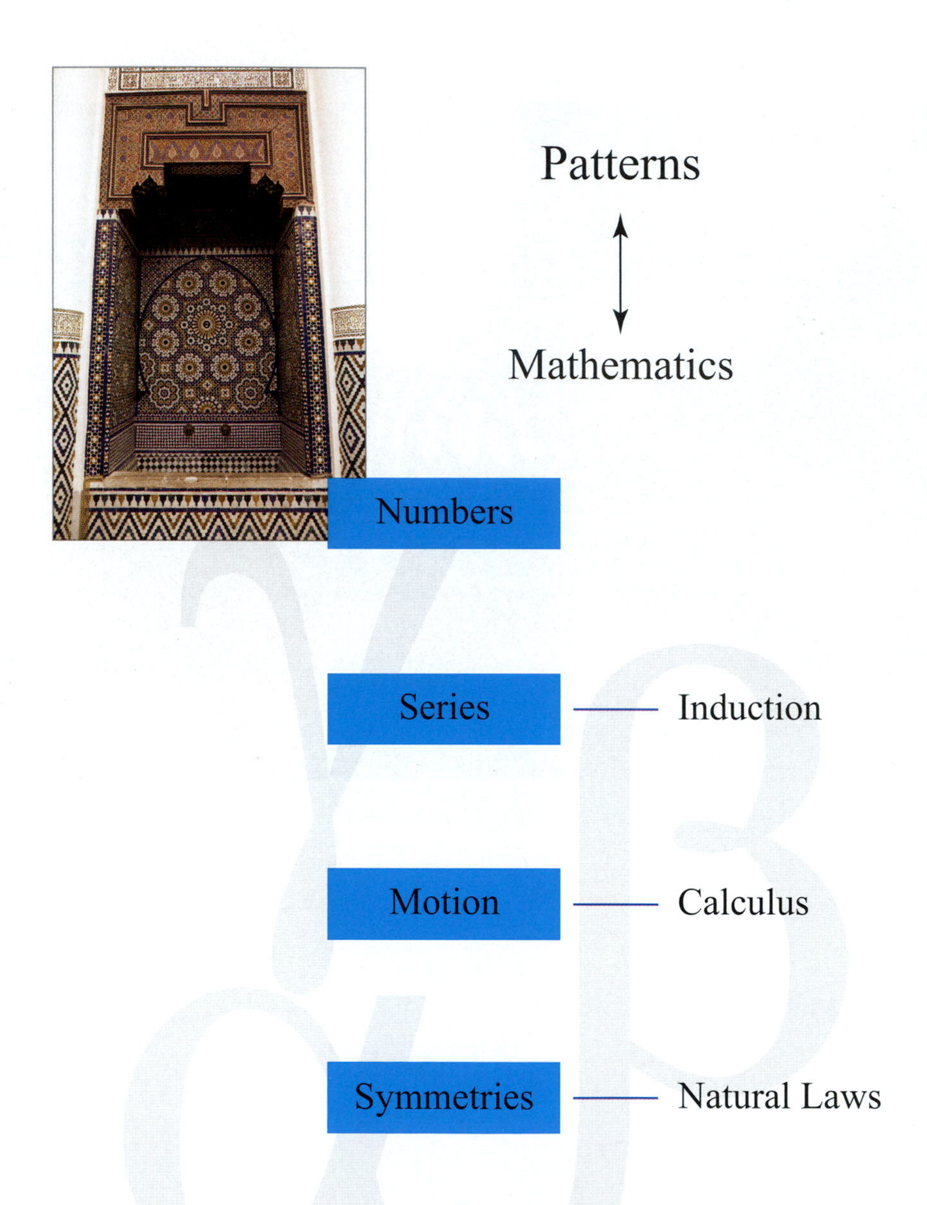

Patterns
Mathematics
Numbers
Series
Induction
Motion
Calculus
Symmetries
Natural Laws

2.1 The question

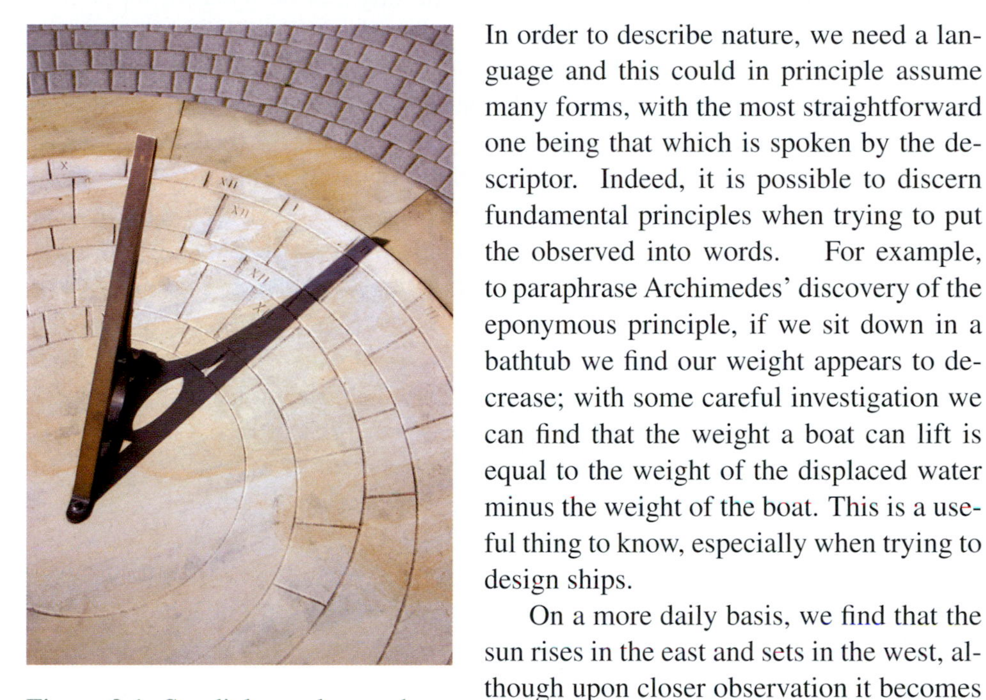

Figure 2.1: Sundials can be used to capture the pattern of the sun's movement.

In order to describe nature, we need a language and this could in principle assume many forms, with the most straightforward one being that which is spoken by the descriptor. Indeed, it is possible to discern fundamental principles when trying to put the observed into words. For example, to paraphrase Archimedes' discovery of the eponymous principle, if we sit down in a bathtub we find our weight appears to decrease; with some careful investigation we can find that the weight a boat can lift is equal to the weight of the displaced water minus the weight of the boat. This is a useful thing to know, especially when trying to design ships.

Fig. 2.2: Patterns are ubiquitous in nature. Some are easier to analyze than others.

On a more daily basis, we find that the sun rises in the east and sets in the west, although upon closer observation it becomes clear that it doesn't always do so in exactly the same fashion but in yearly cycles roughly in sync with the seasons. This is good to know when trying to run a farm.

What can we see from these two examples? First, that it would be useful to have some kind of special notation to condense long cumbersome sentences into precise statements and second, that the observation of patterns is essential to gain an understanding of many phenomena. Hence, we need a language with symbols and methods for analyzing patterns. This language and its associated methods is mathematics, and indeed the perhaps best way to define mathematics is that it is the science of patterns.

The ability to recognize patterns is an enormous power. Usually, if asked where the sun will rise tomorrow we'll hardly blink an eye and respond "In the east!" True of course, but what is actually going on here? If we give an answer like that what we are really doing is making a **prediction** based on a recognized pattern. If we had not recognized the pattern, we would not have been able to make the prediction. And this is in essence the source of the accomplishment of modern science: prediction through the recognition of patterns.

But what do we need for making some of the most basic predictions (like how the planets move, how far a golf ball will fly and so on)? To find what we need, we have to consider how the patterns progress, even ones like the rising and setting of the sun, and we see that there often is no end. Taking account of all such patterns soon leads us to the notion of infinity. But clearly infinity is a strange beast. After all, we can't just write out the all the patterns (that would take infinitely long!). We therefore need to ask: **How do we reach infinity and beyond?**

Mathematics is the science of patterns.

2.2 Numbers and symbols

Fig. 2.3: Numbers are an abstraction.

Before we continue our investigation on how infinity can be reached, let us first delve somewhat deeper into the symbols that are used in the description of patterns.

Among the very first symbols used were the numbers. One could say that these describe patterns such as "oneness" or "twoness". Why is this a pattern? It is in the sense that the notion of a number of items applies not just to specific instances like "three apples" but to any collection of three items, it is the pattern of there being three things. A number is an abstract idea. The concepts of oneness and twoness naturally lead to the next pattern, namely that of a sequence of numbers where each successive number is one larger than its predecessor. In other words: counting. Counting may seem like a very basic activity but it is not, as anyone who has seen children grow up can attest. Perhaps, the mental leap necessary to go from the direct perception of items like "one apple" to the abstract concepts of numbers and counting is as large as going from counting to string theory.

If one looks at numbers carefully, there turns out to be a great "number" of patterns and indeed a whole field of mathematics not surprisingly called number theory is devoted to it. When we write down a concrete number like 134, we do so by using a combination of the ten symbols 0, 1, 2, 3, 4, 5, 6, 7, 8, 9. One can then define rules on how such concrete numbers can be combined in like, for example addition and subtraction. If, then, we would like to go further in the level of abstraction, we could ask questions along the line of: What kind of number do we get when we calculate the circumference of a circle from its radius? For this we'll need several things.

First, we need to have certain units to measure the radius. Most units are arbitrary though clearly some units make more sense than others. The standard unit for length or distance in science is the meter. Second, we need to be able to have fractions of the units since it would be rather inconvenient to allow only circles with certain radii. Third and conceptually the most far reaching, we want to move up the abstraction ladder one more step in that we don't really want to be specific and enumerate the relationship between radius and circumference along the lines of: a circle with a radius of one unit has a circumference of 6.28 units, a circle with a radius of two units has a circumference of 12.56 units, etc. We want to be able to make statements for all circles and that means that we should have a symbol to represent *all* possible radii and a convention on how this symbol should be related to the circumference. For example, we can say that the symbol "r" represents *any* radius of a circle and we can create a symbol "=" to indicate that what appears on the left side of this symbol equals what is on the right of this symbol. If we then write known quantities on the right hand side, we can calculate the left hand side for any specific value of the radius. Thus we obtain the famous equation $c = 2\pi r$, where c is the circumference and r the radius of the circle.

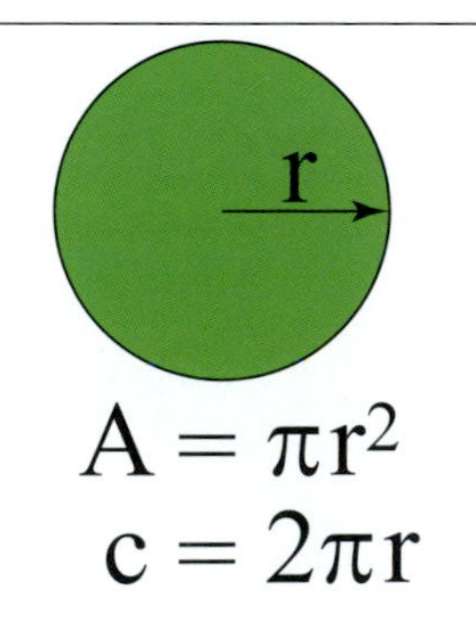

Fig. 2.4: Equations are a further abstraction.

Although we all learn this early in our school lives, it is important to reflect on the enormous level of abstraction required for these steps. It should also serve as somewhat of a relief that these steps comprise the essence of the level of mathematics required for this book.

We just mentioned that having only whole units is rather inconvenient and that therefore we need to use fractional units. For a long time, it was thought that all fractional numbers can be expressed as the ratio of two whole numbers like $0.3 = 3/10$. Numbers that can be expressed as the fractions of whole numbers are referred to as rational numbers and even a short consideration shows that there must be infinitely many of them even between 0 and 1. How can we see this? Probably the simplest way is to just take the sequence of fractions 1/2, 1/3, 1/4, etc. Since we can count to infinity, there are infinitely many of these fractions, one for every whole number. Then of course one can construct the rational numbers 2/3, 2/4 , 2/5, etc. some of which like 2/4 being equal to a previously obtained number. Nevertheless, there is again an infinite number of them. Then there's 3/4 ... and so on and so on. The general description of a rational number is that it is given by the fraction p/q where p and q are integers. (Note that the whole numbers generally only are the positive integers 0, 1, 2, 3 etc., the use of negative numbers became only widespread in the west around the seventeenth century! Then again, what exactly is "minus one apple"?)

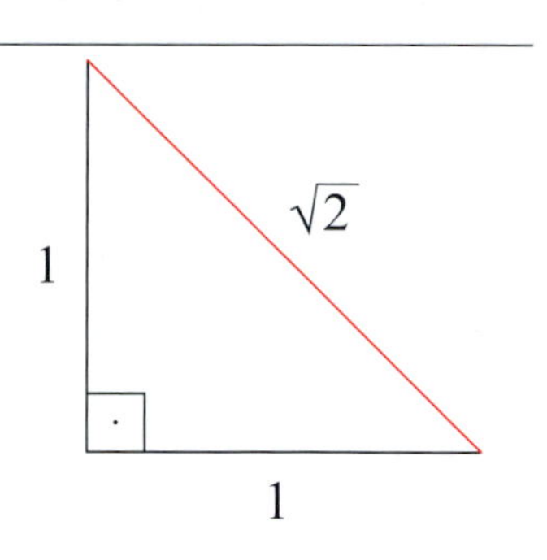

Fig. 2.5: The length of this hypotenuse is irrational.

There are two notable things. First, even though our first series of fractions (1/2, 1/3,..) yielded infinitely many rational numbers between 0 and 1, there's still room to place infinitely many other numbers in this interval. So having infinitely many numbers alone is not enough to fill up even a small interval like 0 to 1. Second, it seems that one can express any number between 0 and 1 with the help of a fraction p/q. For example 0.384 could simply be written as 384/1000. However, a classic proof credited to a mathematician in the Pythagorean school of ancient Greece by the name of Hippasus showed that there are other numbers besides rational numbers. The proof looks at the length of the hypotenuse in a right isosceles triangle with a side of length 1 which is given by the Pythagorean theorem as $\sqrt{2}$ (see Figure 2.5). The proof proceeds as follows.

Suppose there are numbers p and q such that $\sqrt{2} = p/q$ and that all common factors have been canceled out. Squaring this identity we obtain

$$2 = \frac{p^2}{q^2}, \tag{2.1}$$

or in other words

$$p^2 = 2q^2. \tag{2.2}$$

From this we see that p^2 is an even number. Now since in general the square of an even number is even and the square of an odd number is odd, we can conclude that p is an even number as well. As p is even we can write it in the form $p = 2r$. If we substitute this into Equation 2.2 we obtain after dividing out a factor 2

$$2r^2 = q^2 \tag{2.3}$$

and hence conclude that q^2 and consequently q are even. So we see that our assumption of $\sqrt{2} = p/q$ clearly implies that both p and q are even, and so have the common factor 2. This now contradicts our starting point as we had canceled out all the common factors from p/q. The only conclusion that can be drawn is that the initial assumption, namely that $\sqrt{2} = p/q$, with p and q being whole numbers, is wrong. Thus we see that there are numbers that cannot be expressed as the fractions of integers.

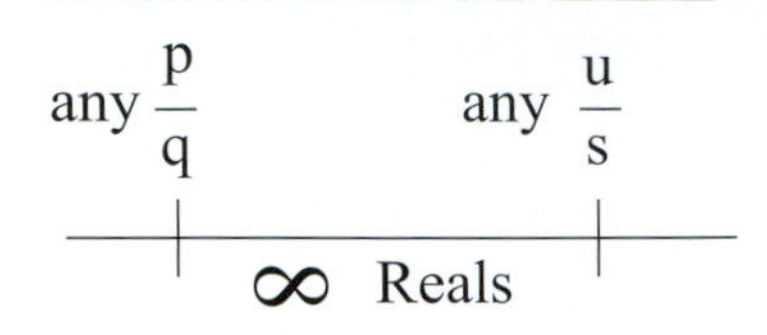

Fig. 2.6: There are infinitely many reals between any pair of rational numbers.

Numbers that are not rational are perhaps somewhat surprisingly called irrational even though there's really nothing "irrational" about them. Irrational numbers like $\sqrt{2}$ and also π fill up the number line (i.e. all the gaps between the rational numbers without leaving any gaps themselves) and it turns out that there are infinitely many of those as well. Together, the rational and irrational numbers form the so-called "real" numbers.

Hence we see that numbers can be represented by symbols like 1, 2, 3 that originate from a direct one-to-one correspondence with counting objects but also by symbols like p and q that can represent any of a certain type of number. Either way we soon encounter the notion of infinity when considering how far one can count or how many rational numbers fit into an interval. Since we clearly never can actually count very far let alone to infinity, we should wonder how we can deal with infinities. Also, we should wonder whether this has any relevance at all. Wouldn't it be enough to be able to count to, say, one trillion? Next, let us look at how to deal with infinities. Then in Section 2.6, it will be shown that the quantitative study of nature requires being able to handle infinites.

2.3 Induction

In the section on numbers above, we have seen the essential role that the recognition of patterns plays. What would be really good is if the structure of a pattern itself could be used to prove certain properties of the pattern. In order to see whether this is possible, let us consider the observation that the sum of the first n odd numbers is n^2, that is,

$$\begin{aligned} 1 &= 1 = 1^1 \\ 1+3 &= 4 = 2^2 \\ 1+3+5 &= 9 = 3^3 \\ 1+3+5+7 &= 16 = 4^2 \\ 1+3+5+7+9 &= 25 = 5^2 \end{aligned}$$

In other words, the pattern that we find is

$$P(n) \equiv 1 + 3 + \ldots + (2n-1) = \sum_{k=1}^{n}(2k-1) = n^2 \qquad (2.4)$$

where the symbol Σ indicates that we have to sum up what follows by substituting successive values of its subscript (in this case the argument is $2k - 1$ and the subscript k; the notation $k = 1$ means that the starting point is 1). For convenience, we have defined $P(n)$ to denote the statement "the sum of the odd numbers up to $2n-1$ equals n^2". For example $P(3)$ stands for the statement $1+3+5 = 3^2$. What we now have to do is to show that $P(n)$ is true for all whole numbers n. Again, it is important to stress that simply verifying $P(n)$ is not sufficient. Although doing so up to a large number like one million would be pretty convincing, it nevertheless would not be a proof. Let us now show how this pattern can be proved for *all* whole numbers, however big, all the way to infinity.

First, one has to show that $P(1)$ is true. Then, if we can show that *in general* a true statement $P(n)$ *implies* the true statement $P(n+1)$, we can safely conclude that the statement $P(n)$ is true for all n. Why is this so? Well, if we show that $P(1)$ is true and show that a true $P(n)$ implies a true $P(n+1)$, then we can conclude that $P(1+1) = P(2)$ is true. But if we know that $P(2)$ is true we can likewise conclude that $P(2+1) = P(3)$ is true, and so on and so on. This method of proof is called mathematical induction and one of the most fundamental proof strategies in mathematics. To summarize:

The principle of mathematical induction

$$\text{1) If} \quad P(1) \quad \text{is true; } and \tag{2.5}$$

$$\text{2)} \quad P(n) \quad \text{is true} \quad \text{implies} \quad P(n+1) \quad \text{is true} \tag{2.6}$$

$$\text{Then} \quad P(n) \quad \text{is true for all } n. \tag{2.7}$$

Now let us apply mathematical induction to the problem at hand in Equation 2.4. For $P(1)$ we have $1 = 1^2$ which is obviously true. Now let us look at $P(n)$. If $P(n)$ is true we have

$$1 + 3 + 5 + \ldots (2n-1) = n^2. \tag{2.8}$$

Now one may object that we can't be sure that Equation 2.8 is actually true. Indeed, we can't be sure of that and it very well could be wrong. However, all that we want to do at this point is to show what $P(n)$ implies *if* it is true. So we make the assumption that Equation 2.8 is true and then find out what that means for $P(n+1)$ to complete this step. To complete the proof we need step 1 (Equation 2.5) in the method of mathematical induction. To continue, let us add the next odd number to both sides of Equation 2.8 and use Equation 2.4 to obtain

$$P(n+1) = P(n) + (2n+1) = n^2 + (2n+1). \tag{2.9}$$

The right hand side of this equation equals to $n^2 + (2n+1) = (n+1)^2$ and therefore Equation 2.9 actually is

$$P(n+1) = (n+1)^2 \tag{2.10}$$

which is just the statement for $P(n+1)$. Hence we see that if $P(n)$ holds for a certain n, then we can conclude with certainty that $P(n+1)$ also holds. All that is needed now is a valid first step, i.e. a proof for the validity of $P(1)$ which can often be done easily by inspection. With the help of the principle of mathematical induction we therefore see that the sum of the first n odd numbers is indeed equal to n^2 for all n.

Since induction is such a useful method, let us have a look at another example. Suppose we have a chocolate bar with a grid of N rectangular squares as shown in Figure 2.7. How many breaks will it take to have N individual squares left? By considering a small chocolate bar with two ($N = 2$) or four ($N = 4$) squares we see that we need one and three breaks respectively. So the pattern we can discern is that it takes $m = N - 1$ breaks where the symbol m represents the number of breaks. Let us call this pattern $P(N)$, i.e. $P(N) \equiv m = N - 1$. How can we prove the general truth of $P(N)$ with the help of induction?

Figure 2.7: Typical chocolate bar.

A single square cannot be broken into two so we have zero breaks and indeed $P(1) \equiv 0 = 1 - 1$ and hence $P(1)$ is true. As a first try, assuming that $P(N)$ is true and hence having $m = N - 1$ we could add a 1 to both sides giving $m + 1 = N - 1 + 1$. This is of course the same as $m + 1 = (N + 1) - 1$ and exactly $P(N + 1)$. But is this reasoning correct? Not really! Although there is no calculation error in our adding the 1, what we should have done but haven't done is to relate that to breaking the bar. Indeed, adding a 1 to both sides has nothing to do with breaking (note in contrast that in the example above, Equation 2.4, what we added to both sides was in fact the next number in the series which is why it worked there). So, how can we have a better proof?

Fig. 2.8: Enlarged chocolate bar after a first possible break.

The thing we have to do is to break the enlarged $N + 1$ piece chocolate bar into two not necessarily equal halves — see Figure 2.8. If one half has N_1 pieces and the other half has N_2 pieces (so that $N_1 + N_2 = N + 1$), the question becomes how many breaks do we need for each side. Let us call the number of breaks for the side with N_1 pieces m_1 and the number of breaks for the side with N_2 pieces m_2. According to the principle of induction 2.6, we now make the assumption that the observed pattern is true (i.e. that in general $m = N - 1$ for any N), and then verify that this is true when replacing N by $N + 1$. If the assumption is true we have for $m_1 = N_1 - 1$ for the half with N_1 pieces and $m_2 = N_2 - 1$ for the half with N_2 pieces. Using the fact that $N_1 + N_2 = N + 1$ and that the total number of breaks is $m = m_1 + m_2 + 1$, i.e. the contributions from the two halves plus the initial break that created the two halves, we have

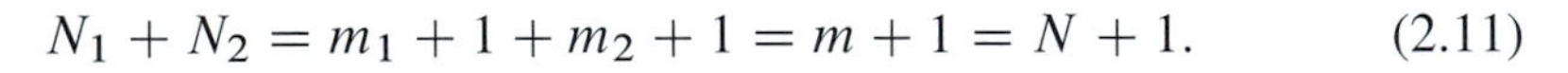

$$N_1 + N_2 = m_1 + 1 + m_2 + 1 = m + 1 = N + 1. \qquad (2.11)$$

Hence we have

$$m = (N + 1) - 1 \qquad (2.12)$$

and thus shown that the truth of $P(N + 1)$ follows from the truth of $P(N)$. Since we have already shown that $P(1)$ is true we can now conclude with the help of the principle of induction that $P(N)$ is true for all N however big.

We have seen how we can use certain patterns to make statements about arbitrarily large numbers that are in principle possible for those patterns. And if a number can be arbitrarily large, it can be infinitely large. If we have proved the pattern with the help of induction, we can be sure that it will hold even in that case. We are therefore in a position to safely move away from one-to-one correspondences to real objects or from explicit verification of as many instances of a pattern as possible and can have confidence in the power of abstract mathematical reasoning.

Counting does not appear to have an end (indeed it doesn't) so it may be somewhat trivial to ask what will happen if we add the numbers to obtain the sum

Leonhard Euler

Figure 2.9: Portrait of Leonhard Euler by Johann Georg Brucker.

Leonhard Euler was one of the eighteenth century's most influential scientists. He was born in the Swiss town of Basel in 1707 and after receiving his doctorate in 1727 spent most of his life in Berlin and St. Petersburg where he passed away in 1783, leaving behind what is perhaps the largest body of original scientific papers ever produced by a scientist (over 800 papers and books).

He introduced much of the modern mathematical notation and for example was the first to write a function f applied to a variable x as $f(x)$, $\sin(x)$ for the sine, introduced the number e as the base of the natural logarithm, the symbol i for the imaginary unit in complex numbers, the symbol $\sum$ for summations, and so on.

Euler was a frequent user of and major contributor to the development of power series. He discovered for example that e can be expressed as

$$e = \sum_{n=0}^{\infty} \frac{1}{n!}. \tag{2.13}$$

One of Euler's most well-known discoveries is the relationship between complex exponential numbers and trigonometric functions which immediately implies the stunning identity named after him:

$$e^{ix} = \cos(x) + i\sin(x) \quad \textbf{Euler's formula} \tag{2.14}$$

$$e^{i\pi} + 1 = 0 \quad \textbf{Euler's identity} \tag{2.15}$$

In Equation 2.15 five of mathematics' fundamental entities are related in a simple and elegant fashion. With the help of series (a subject Euler was a master in), Euler's formula can readily be derived by considering:

$$\begin{aligned} e^x &= 1 + x + \frac{x^2}{2!} + \frac{x^3}{3!} + \dots \\ \sin(x) &= x - \frac{x^3}{3!} + \frac{x^5}{5!} - \frac{x^7}{7!} + \dots \\ \cos(x) &= 1 - \frac{x^2}{2!} + \frac{x^4}{4!} - \frac{x^6}{6!} + \dots . \end{aligned}$$

If one replaces x by iz in Equation 2.15, Euler's formula immediately follows.

Euler made major contributions in diverse fields such as calculus, number theory, geometry, algebra in mathematics, and optics and classical mechanics in physics. Indeed, there are so many things named after Euler in different fields that it can be quite puzzling when those fields meet and the same term consisting of his name means something completely different. Perhaps, having been so prolific (including 13 children) it is of no surprise that he also left a bit of confusion.

$s = 1 + 2 + 3 + \ldots$. Since the largest term in this sum is infinity, it is rather reasonable to conclude that sum s is infinitely large. Well, how about the sum of all the integers, $s = \ldots\ldots -3 + -2 + -1 + 0 + 1 + 2 + 3\ldots$? A quick calculation of these numbers gives us $s = 0$. But is this correct? Since the pattern extends to minus and plus infinity respectively, we could group some numbers together as in $s = 0 + 1 + (2 + -1) + (3 + -2)\ldots$ or in other words $s = 0 + 1 + 1 + 1\ldots$. A sum of infinitely many 1s should be infinity shouldn't it? Well that's odd! First we conclude that the sum of all integers is 0 and then that it's infinity. Which result would be right? In order to get a better grip on this matter, let us delve a bit deeper into series.

2.4 Series

In mathematics, a series is defined as the sum of a sequence of numbers. Usually, but not necessarily, successive numbers are related to each other by some well-defined mathematical operation like addition.

Let us consider the sum

$$s = 10 + 1 + \frac{1}{10} + \frac{1}{100} + \frac{1}{1000} + \cdots \tag{2.16}$$

or written in the proper mathematical notation

$$s = \sum_{k=0}^{\infty} 10^{1-k} \tag{2.17}$$

where the symbol ∞ denotes infinity. Clearly, it is impossible to calculate this sum by adding together all the terms. In fact even adding up the first few terms is already quite tedious. But if we recognize the pattern we see that we simply have a succession of 1s, namely $11.1111\ldots$. That's good, a result without a calculation. However, in order to prepare us for the case when the numbers in the sum are not so nice as to exactly be powers of 10, how can we "calculate" the result without actually summing up the terms? One way, again by observing a pattern, is the following: Each successive term is obtained by dividing the previous term by 10. Therefore, if we divide both sides in Equation 2.16 by 10, we obtain exactly the same series with the exception of the first term

$$\frac{1}{10}s = 1 + \frac{1}{10} + \frac{1}{100} + \frac{1}{1000} + \cdots. \tag{2.18}$$

If we subtract this second sum (Equation 2.18) from the first sum (Equation 2.16) we find

$$s - \frac{1}{10}s = 10 \tag{2.19}$$

which is easily solved as $s = 100/9 = 11.11\ldots$, the same result we obtained before. Now of course the question has to be whether it is actually justifiable to multiply and subtract entire series, but fortunately a rigorous analysis shows that it is fine in this case.

Encouraged by this result and inspired by our previous discovery of the power of successive abstraction, it is natural to wonder next whether we can rewrite the series 2.16 in a generic way using symbols like r or a to represent any number that can be summed in the same fashion. Perhaps the easiest way to approach this problem is to sort of "reverse" engineer the formula. Let us say that $1/10$ is a specific example of a number r and substitute r everywhere we have a factor $1/10$ in Equation 2.16 to obtain

$$s = r^{-1} + r^0 + r^1 + r^2 + \cdots . \tag{2.20}$$

Perhaps this is not so nice. Who would want to start a series with the power -1? One way to solve this is to separate out a factor 10 on the right hand side. That is to say we rewrite Equation 2.20 as

$$s = 10r^0 + 10r^1 + 10r^2 + 10r^3 + \cdots . \tag{2.21}$$

That looks better already but of course replacing $1/10$ by the symbol r but not the factor 10 is not particularly elegant, and therefore let us replace the factor 10 by the symbol a. We then have for our sum

$$s = a + ar + ar^2 + ar^3 + ar^4 + \cdots = \sum_{k=0}^{\infty} ar^k . \tag{2.22}$$

This series is called a **geometric series**. Would it be possible to have a general solution for this series? Yes it is possible! Well, at least if $|r| < 1$. Employing the same trick as above, we take Equation 2.22, multiply it by r and subtract the result to obtain

$$s - rs = a \tag{2.23}$$

which is easily solved as

$$s = \frac{a}{1-r}. \qquad \textbf{solution of the geometric series} \tag{2.24}$$

So for example if we want to know what the sum is of the series

$$s = 1 + \frac{1}{3} + \frac{1}{9} + \frac{1}{27} + \frac{1}{81} + \cdots + \frac{1}{3^n} \cdots , \tag{2.25}$$

we identify $a = 1$ and $r = 1/3$, insert these values into Equation 2.24 and almost instantaneously find $s = 1.5$.

It is interesting to note that in this series subsequent terms become ever smaller and hence one may suspect that series with this property are always finite. Perhaps the best known series with ever shrinking terms is the so-called **harmonic series** given by

$$s = 1 + \frac{1}{2} + \frac{1}{3} + \frac{1}{4} + \frac{1}{5} + \frac{1}{6} + \cdots \frac{1}{n} + \cdots \qquad \text{Harmonic series} \tag{2.26}$$

Srinivasa Ramanujan Iyengar

Figure 2.10: Srinivasa Ramanujan Iyengar.

Ramanujan was one of the greatest mathematicians and was born in Chennai, India on December 22, 1887. He had no formal training in mathematics. Ramanujan taught himself mathematics and showed signs of his genius from the age of 10. In 1912–1913 Ramanujan sent some of his results to three mathematicians in the University of Cambridge, but it was only G. H. Hardy, a leading English mathematician of the twentieth century, who recognized his brilliance.

Ramanujan's results were so novel and unconventional that Hardy concluded that these "must be true, because, if they were not true, no one would have the imagination to invent them." Hardy arranged for Ramanujan to join Cambridge University in 1914 and, in 1918, he was the second Indian to become a Fellow of the Royal Society.

The results that Ramanujan discovered were highly original. Unlike those of conventional mathematicians, Ramanujan's results were usually given without any proof. It has taken many decades to prove that many of his results are correct and a few being incorrect; many other results still remain a mystery. When asked how he discovered his equations, Ramanujan's reply was that they were given to him in his dreams. A Hindu goddess, named Namakkal, would appear and present him with mathematical formulae, which he would verify after waking. Ramanujan related the following: "*While asleep I had an unusual experience. There was a red screen formed by flowing blood as it were. I was observing it. Suddenly a hand began to write on the screen. I became all attention. That hand wrote a number of results in elliptic integrals. They stuck to my mind. As soon as I woke up, I committed them to writing...*"

To get a flavor of Ramanujan's identities, consider the following formula for π given in Ramanujan (1913–1914)

$$\frac{4}{\pi} = \sum_{n=0}^{\infty} \frac{(-1)^n(1123 + 21460n)(2n-1)!!(4n-1)!!}{882^{2n+1}32^n(n!)^3}$$

where $n! = 1 \cdot 2 \cdot \ldots \cdot n$ and $(2n-1)!! = 1 \cdot 3 \cdot 5 \ldots \cdot (2n-1)$.

Ramanujan discovered over 50 such identities for π. Formulas like these are important for finding the value of π to great accuracy and provide a check to the workings of supercomputers. Ramanujan's formula has been used to compute one billion digits of π.

Ramanujan made outstanding contributions to mathematical analysis, number theory, infinite series and continued fractions and left behind 3900 identities. It was Hardy's view that Ramanujan could be compared "only with [Leonhard] Euler or [Carl Gustav Jacob] Jacobi". Ramanujan's work has taken a while to enter the mainstream of scientific thought and now appears in such diverse fields as crystallography and string theory.

Ramanujan fell ill soon after reaching England and died in Chennai on April 26, 1920, at the early age of 32. Hardy came to see him during one of his illnesses and remarked that he had taken a taxi that had the uninteresting number of 1729. "No" replied Ramanujan,"it is a very interesting number; it is the smallest number expressible as the sum of two cubes in two different ways". Note $1729 = 1^3 + 12^3 = 9^3 + 10^3$.

with the name of the series due to the fact that the first few terms represent ratios that also occur in a commonly used musical scale. Now if we add the first one thousand terms we get approximately 7.49 and if we add the first one million terms approximately 14.36. Clearly, numbers that aren't too big and at first sight seem to confirm the suspicion that series with terms that get ever smaller are finite. Surprisingly, the fourteenth century philosopher/scientist Nicole Oresme (who was also the bishop of the French area Lisieux) proved the series to have an infinite value. The proof is as elegant as it is simple and again based on the observation of a pattern. First, note that the third term is larger than the fourth term (as is the case for every term to the left of another term). Therefore, since the fourth term is $1/4$, the sum of the third and the fourth term is larger than $2 \times 1/4 = 1/2$ and we have

$$s > 1 + \frac{1}{2} + \frac{1}{2} + \frac{1}{5} + \frac{1}{6} + \frac{1}{7} + \frac{1}{8} + \frac{1}{9} + \cdots . \tag{2.27}$$

But now if we look at the three terms to the left of $1/8$ we know that each one of these is larger than $1/8$ and hence that the sum of the terms from $1/5$ to $1/8$ is larger than $4 \times 1/8 = 1/2$ giving

$$s > 1 + \frac{1}{2} + \frac{1}{2} + \frac{1}{2} + \frac{1}{9} + \cdots . \tag{2.28}$$

As the series is infinitely long, we can continue this process of combining terms to end up with the infinite series $1 + 1/2 + 1/2 + 1/2 + \cdots$ which adds up to infinity. Since the harmonic series is actually larger than this, we conclude that the harmonic series is infinite.

Over the years, many infinite series with special values or properties have been discovered. For example, the Swiss mathematician Leonhard Euler discovered that

$$\frac{\pi^2}{6} = \frac{1}{1^2} + \frac{1}{2^2} + \frac{1}{3^2} + \frac{1}{4^2} + \frac{1}{5^2} + \cdots . \tag{2.29}$$

2.5 Convergence

Infinite series have properties that are not deducible from a finite series like the one we used in Equation 2.4. Let us now return to the strange result mentioned at the end of Section 2.3. There we found that (in the given examples) the outcome of an infinite series can be 0 or infinity depending on how the terms are grouped. The series discussed in Section 2.4 on the other hand do not seem to have that problem. Then what is the difference?

The key concept turns out to be that of **convergence**. A series whose sum is less than infinity is said to be convergent and there are two types of convergence: **conditional convergence** and **absolute convergence**.

In the case of absolute convergence we have

$$\sum_{n=0}^{\infty} a_n < \infty \qquad \text{while also} \qquad \sum_{n=0}^{\infty} |a_n| < \infty \tag{2.30}$$

while in the case of a conditionally converging series we have

$$\sum_{n=0}^{\infty} a_n < \infty \qquad \text{while} \qquad \sum_{n=0}^{\infty} |a_n| = \infty. \tag{2.31}$$

If we inspect the series at the end of Section 2.3, it is readily found that they are conditionally and not absolutely converging. Indeed, it is shown in the so-called Riemann series theorem that in any conditionally convergent series the terms can be rearranged to give infinity as a result (the theorem also shows that the terms can be arranged to converge to any given value). On the other hand, absolutely converging series only sum up to one value, regardless of how the terms are arranged.

We are now at the point that we have some grip on infinity. Not only did we find that we can deal with it by the recognition and analysis of patterns, with the help of those same patterns, that is to say with the help of series, we also have a means to reach it. Although interesting in themselves, the question we should ask is whether these insights are of any practical use.

To answer that, let us begin the next section by considering the classical **paradox of Zeno**.

2.6 Motion

The ancient Greek philosopher Zeno of Elea devised a set of paradoxes in order to illustrate and support his teacher Parmenides' notion of all-in-oneness according to which among other things all motion is an illusion. Since much of modern science is very much about motion, Zeno's paradoxes are of extraordinary relevance for our understanding of the world and an excellent illustration of how the development of mathematics has resolved this issue.

Perhaps the most famous of Zeno's paradoxes is that of Achilles and the Turtle as reported by Aristotle:

> *"In a race, the quickest runner can never overtake the slowest, since the pursuer must first reach the point whence the pursued started, so that the slower must always hold a lead." (Aristotle Physics VI:9, 239b15)*

The argument is as follows: Achilles is to race a tortoise for one round in a stadium. As is well known, Achilles is a fast runner and tortoises are, well, not so fast. Generously, Achilles gives the tortoise a headstart of 100 pous (a Greek foot measuring approximately 30.8 cm) not realizing that he will never be able to catch up with it and hence lose the race. Why is this so? After the race starts, it will take Achilles only a short time to reach the point where the tortoise started but in the mean time, the tortoise has progressed too so it is still ahead. Next it will take Achilles even less time to reach the point where the tortoise was when he reached its starting point but nevertheless again the tortoise made some progress and is still ahead. Thus, Achilles gets ever closer to the tortoise but will never catch up since every time he reaches the tortoise's previous position the tortoise has moved a bit.

Fig. 2.11: Can Achilles overtake this speeding tortoise?

Of course, it didn't elude Zeno of Elea that this argument doesn't hold in the world we perceive as daily life. We all know that it will take Achilles only a small amount of time to overtake the tortoise. What Zeno wanted to illustrate is that the real world is different from the world we perceive and he did so by presenting an apparently absurd conclusion. This type of reasoning is called *reductio ad absurdum* and a powerful method for mathematics too.

So what is the modern solution to this paradox? Infinity! Or more precisely, the infinite series. What the Greeks didn't realize is that infinite series can have finite results. Let us say that Achilles runs 10 times as fast as the tortoise and that it takes him 10 seconds to reach the tortoise's starting point. Then it will take him 1 second to reach the point where the tortoise was when he passed the tortoise's starting point, one tenth of a second to reach the next point etc. The total time for Achilles to catch up with the tortoise is then given by the series

$$t = 10\,\mathrm{s} + 1\,\mathrm{s} + \frac{1}{10}\,\mathrm{s} + \frac{1}{100}\,\mathrm{s} + \frac{1}{1000}\,\mathrm{s} + \cdots . \tag{2.32}$$

Indeed, this is exactly the same geometric series that we've encountered in the previous section and we know the result: 100/9 seconds. Achilles will therefore catch up with the tortoise in approximately 11.1 seconds. Thus we see that as such the argumentation that ever smaller distances need to be covered is correct but that the conclusion from this infinity is *not* that the gap can never be covered but, *instead*, that infinity needs to be handled properly.

We have seen in Zeno's paradox that infinity and motion are intrinsically related. In order to further investigate this relationship and its implications for science, we need to have some way of describing motion. Before, we didn't really describe Achilles' motion but only the time it takes him to get from point a to b. If we want to describe his motion we need to express his position as a function of time. For example, if he has run 10 pous after 1 second, 20 pous after 2 seconds and so on, we can capture this pattern in the formula

$$x(t) = 10\,\frac{\mathrm{pous}}{\mathrm{s}}\,t, \tag{2.33}$$

where $x(t)$ is the position and t is the time elapsed. For the tortoise, the pattern of the distance covered would then look like $x(t) = 1\,\mathrm{pous/s}\,t$ and we therefore see that in the formulae the difference in the speed or velocity lies in the factor in front of the t.

In general, **velocity** is defined as the distance covered divided by the time it takes to cover this distance. Then for Achilles we have a velocity of 10 pous/s and for the tortoise 1 pous/s. From this, we see that velocity really is the *rate of change of position* and since we like to make general rather than specific statements, we replace the specific number 10 in Equation 2.33 with the general symbol v so that we have

$$x(t) = vt + x_0 \tag{2.34}$$

where we added x_0 to indicate the starting point (in the case of Achilles 0 and in the case of the tortoise 100 pous). Now, what if Achilles would have slowly accelerated instead? For example having run 2 pous after 1 second, 8 pous after 2 seconds, 18 pous after 3 seconds? Then we could describe the pattern as

$$x(t) = 2\,\frac{\text{pous}}{\text{s}^2}t^2. \tag{2.35}$$

What would be the meaning of the 2 in front of the t^2? Well if the 2 were a 0, Achilles would stand still and if it were a 5, he'd be accelerating faster, so the 2 is related to the "pace" of the acceleration (though it isn't exactly the acceleration itself as we shall see shortly — there's a factor 2). As before, we rather replace the 2 with a generic symbol, for example b. Of course, as such, one does not need to start the description of the motion from when someone is standing still, and as with the tortoise, one does not need to start at the origin. So, if we'd like to have a very generic description of the pattern in the position we can write this as

$$x(t) = bt^2 + v_0 t + x_0. \tag{2.36}$$

Before continuing, let us briefly pause and see what happened in Equation 2.36. We have expressed the position x in terms of the time t and some generic numbers representing acceleration b, the initial velocity v_0 and the starting position x_0. So for every specific time t, we can calculate the value of x. Or in other words, x is a function of t, and not surprisingly an equation of the type Equation 2.36 is called a **function**.

While Equation 2.36 can tell us the position at a time t, it doesn't tell us the velocity at this time (note that the number v_0 in the formula is the velocity when the description of the pattern began at time $t = 0$ and may have changed since). But the question of what the velocity would be is an interesting and important one and leads one to the calculus developed by Newton and Leibniz.

So, how can one get the velocity? The easiest point to start with is to take $v_0 = x_0 = 0$ and return to the description of velocity as the distance covered divided by the time taken to cover this distance. What we then can do is to take some time t and calculate the position x_t at this time (here the subscript denotes x at the time t), take some later time $t + h$ and calculate the position x_{t+h}, calculate the difference in time (time taken to cover the distance) and difference in position (distance covered) and divide them to obtain the velocity as

$$v(t) = \frac{x_{t+h} - x_t}{(t+h) - t} = \frac{b(t+h)^2 - bt^2}{h} = \frac{2bth + bh^2}{h}. \tag{2.37}$$

If we cancel h from the last fraction we obtain for the velocity (for $v_0 = 0$)

$$v(t) = 2bt + bh. \tag{2.38}$$

As an approximation, Equation 2.38 is pretty good but it is important to realize that it gives neither the velocity exactly at time t nor at time $t + h$ but the *average* velocity over the time interval from t to $t + h$. Although in hindsight it may seem obvious, the enormous mental leap that Newton and Leibniz made was to realize that h could be made smaller and smaller going toward 0. In Equation 2.38, we can just take $h = 0$ in the rightmost term as the limiting case but it is essential to note that this could not have been done from the outset as then we would have to divide by 0 in Equation 2.37 since $(t + 0) - t$ is obviously 0 — this illustrates the importance of taking the limit of h approaching 0 correctly. Thus we obtain for the velocity exactly at time t

$$v(t) = 2bt. \tag{2.39}$$

If we would have had a starting velocity v_0, it is intuitively clear that we can just add that. An initial positon x_0, however, in this scenario should not have any effects on the velocity so we don't need to add it. Of course, going through the calculations above with these terms included will give the same result. Also, it is convenient to replace the factor $2b$ with the single symbol a, where a is the acceleration. Thus we obtain for position and velocity the following two functions

$$x(t) = \frac{1}{2}at^2 + v_0t + x_0 \tag{2.40}$$

$$v(t) = at + v_0. \tag{2.41}$$

If we look closely at the top and bottom equations, we observe that the bottom equation can be obtained from the top equation by the following rule: terms without a t disappear and in terms with a t the exponent of t becomes a multiplicative factor while the exponent itself is reduced by 1 (i.e. t^2 becomes $2t$ and t become $1t^0 = 1$). Indeed this rule turns out to be generally true even for larger exponents.

As this method is based on taking differences (in x and t), it is called *differential calculus*. Going from Equation 2.40 to Equation 2.41 is called differentiation and the notation for indicating which variable one differentiates is to precede the function by d/dname-of-variable (here the variable with respect to which the function is differentiated is t, and the function being differentiated is the one which determines x). Hence in this notation we have

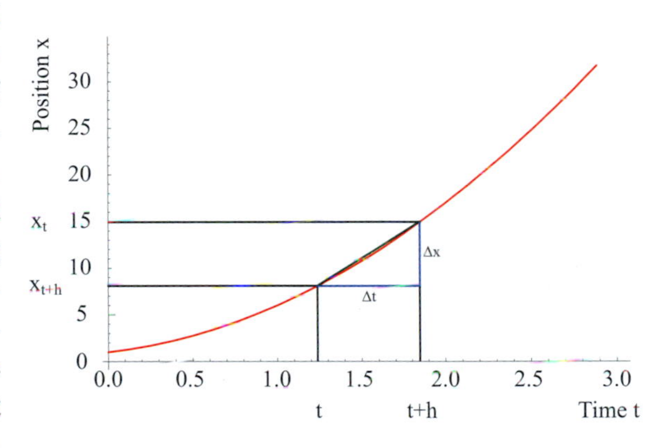

Figure 2.12: Derivative of a function.

$$v(t) = \frac{dx}{dt} = at + v_0. \tag{2.42}$$

Applying our differentiation rule of reducing exponents again we also see that

$$a(t) = \frac{dv}{dt} = \frac{d^2x}{dt^2}. \tag{2.43}$$

Above we used symbols for describing how the position of x changes as a function of the time t. Another good way to describe how one variable behaves as the function of another is a graph. For example $x = 3t^2 + 2t + 1$ is plotted in

Figure 2.12. Now if we consider the ratio of $\Delta x = x_{t+h} - x_t$ and $\Delta t = (t+h) - t$, we see that this represents the slope of the green line in the graph. If we imagine h to be ever smaller we find that the green line more and more looks like it's touching the red line and indeed in the limit for infinitely small h, this is what it does. Then the line is called the **gradient**. Therefore, obtaining the gradient is done exactly in the same way as we did above for the velocity if the axes are the time t and position x respectively, which leaves us with the conclusion that the velocity is given by the slope in a graph that plots the position as a function of time.

2.7 Symmetry

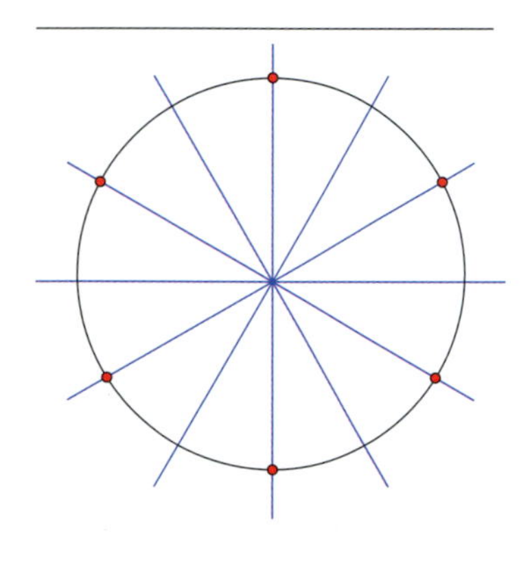

Fig. 2.13: This marked circle has a six-fold symmetry.

Thus far we have seen how infinity can be handled in infinite series by recognizing patterns and how the necessity of considering infinitely small intervals when trying to determine the velocity from a function for the position leads to the development of the differential calculus, one of the foremost mathematical tools in modern science.

Another key notion in science is that of the universality of natural laws. This notion is based on the recognition that the pattern of some things repeats, over and over again. For example, the velocity of light in vacuum is the same in all directions everywhere in the universe. Now, the universe is big, in fact it is so big that parts of it can never be observed from earth (see also Chapter 1). How, then, can we be confident that the speed of light is the same there, at a point that in a sense is beyond infinity? To investigate this matter, we need to study transformations, which is done in the mathematical field called symmetry (the reason for this name will become apparent shortly).

A **symmetry** of an object is a transformation which does not change the appearance of that object, or in more mathematical terms, is a transformation which leaves the equation describing the object invariant. That is to say a symmetry transformation does not change the position, shape, orientation or other specified qualities (like colors) of the object. Common transformations are for example rotations, reflections, translations, shrinkings and stretchings as depicted in Figure 2.14.

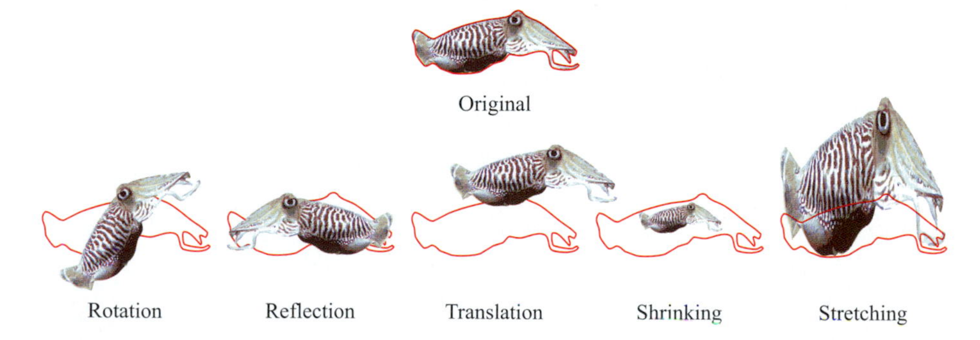

Figure 2.14: Some common transformations. In this case not symmetry transformations.

Clearly the squid in Figure 2.14 does change its appearance in all the transformations depicted. What kind of object should one look at then? As could be inferred from the name of the field (symmetry), symmetrical objects are the kind of objects we are interested in. One of the simplest symmetrical objects is the circle which is invariant under rotations about its center and reflections about any diameter. Note that for these symmetries of the circle, all the points on the circle are considered equivalent such that the object as a whole does not change when rotated or reflected. If, however, some points were marked as in Figure 2.13 where a sixfold symmetry remains, then some or all of the symmetries are lost.

The collection of *all* the transformations which leave an object invariant is called the **symmetry group** of that object. The symmetry group of the circle is the collection of all possible combinations of rotations about the center and reflections in any diameter. Since any of the transformations in the symmetry group leave the object invariant, several transformations can be carried out in sequence without affecting the invariance of the object. Thus we see that irrespective of the symmetry group concerned, we already have found a far-reaching pattern: transformations within the symmetry group can be combined to yield another transformation of that group. This is somewhat reminiscent of addition and hence one can wonder what kind of properties the symmetry transformations on the circle have. It turns out that associativity holds, i.e. that for arbitrary transformations S, T, W of the circle's symmetry group we have similar to the addition and multiplication of integers

$$(S \circ T) \circ W = S \circ (T \circ W), \tag{2.44}$$

where the symbol $\circ$ means "after", i.e. $S \circ T$ means apply the transformation S after the transformation T. Furthermore, it is clear that an identity operation exists (a rotation over an angle 0) which plays a similar role as the 0 in addition and the 1 in multiplication. Calling the identity operation I we have

$$T \circ I = I \circ T = T. \tag{2.45}$$

We also have an inverse (every rotation can be undone by an equal rotation in the opposite direction and every reflection can be undone by applying it a second time). In general, if T is an arbitrary transformation of the circle's symmetry group, then an inverse transformation S that is also a member of the circle's symmetry group exists such that

$$T \circ S = S \circ T = I. \tag{2.46}$$

Thus we see that circle's symmetry group consisting of any rotation about the circle's center and any reflection in a diameter has the three arithmetic properties of *associativity*, *identity* and *inverse*. Also we see that the outcome of any operation is again an element of the group. This is called *closure*. Although the details may be different, the same reasoning can also be applied to other symmetrical objects. Indeed, the properties of closure, associativity, identity, and inverses are so important that, in general, the collection for which these properties hold in mathematics is called a **group**.

In other words, if we have a collection of entities for which the following three conditions hold, then that collection of entities forms a group:

Definition of a group:

G0: (closure) for all x, y in G, $x * y$ is also in G.

G1: (associativity) for all x, y, z in G, $(x * y) * z = x * (y * z)$.

G2: (identity) there is an element e in G such that $x * e = e * x = x$, for all x in G.

G3: (inverse) for each element x in G there is an element y in G such that $x * y = y * x = e$, where e is the identity from condition G2.

With the help of this definition we see that if the $*$ operation is addition and the entities are integers, then the collection of integers forms a group. But if the operation $*$ is multiplication instead, the collection of integers is not a group (e.g. no inverse for 5 within the group in violation of condition G3 since 1/5 is not an integer). However, if we choose the entities to be the collection of rational numbers excluding 0, then we do have a group if the operation $*$ is multiplication.

Groups can be found in many areas of science and were originally developed by Evariste Galaois (who regrettably was killed in a duel at age 21) when he tried to find ways to solve certain polynomial equations (i.e. equations of the type $ax^2 + bx + c = 0$, or more generally, $a_1x^n + a_2x^{n-1} + \ldots + a_nx + c = 0$). In chemistry, they can be used to analyze crystal and molecular structures. The theory of groups, usually simply called group theory, is of great importance to modern physics due to the symmetries that can be found in the laws of physics.

Perhaps, the most easily observable symmetry in daily life is that of translational invariance. The results of a general experiment do not change if the lab is moved from one university to the other. In other words, there is no preferred stationary coordinate system. Coordinates are always relative and it is a matter of convenience which point is chosen to be 0.

Another invariance that can be observed in daily life was first described by Galileo Galilei in 1632, and hence usually called **Galilean invariance**. It is the fact that the laws of nature remain the same for constantly moving bodies. That is to say if one performs a set of experiments in a train moving with a constant velocity, one will obtain exactly the same results as if those experiments would be conducted in a lab fixed to the surface of the earth (of course, we have to be a bit careful here, the train has to be an ideal train that doesn't vibrate, doesn't move too fast or have some velocity fluctuations and we ignore such things as the rotation of the earth). As a consequence, a group of scientists working entirely inside an ideal box can never determine whether the box is moving or not. In other words, there is no preferred "box". Motion is relative and it is a matter of convenience which box is chosen to be "standing still".

Indeed the notion of an isolated box is so important that it has a special name in physics: **Inertial frame**. It has "inertial" in the name in reference to Newton's first and second laws of motion (see Section 1.4 for more details). Inertial motion of an object means that this motion can solely be changed by forces acting upon it. The term "inertia" originates from the latin words *in* and *ars* meaning unskilled or artless. Although it was Keppler who first used the word in a physical sense to describe an object at rest, it was Newton who defined its modern sense when he wrote in the Principia: "A body, from the inert state of matter, is not without difficulty put out of its state of rest or motion. Upon which account, this vis insita may, by a most significant name, be called vis inertia, or force of inactivity... ." That is to say, **inertia** is the property of an object to resist *changes* in its motion (including the property of being at rest, that is, of "no motion" in the absence of force). The second part of the term "inertial frame" refers to a set of coordinates relative to which observations are made and can hence be thought of as a "frame of reference."

This brings us to one of the most fundamental principles in physics: **All inertial frames are equivalent**. This principle is thought to be true for the entire universe even beyond the realm that could ever be observed. Indeed one of Einstein's great contributions to physics was the realization that the speed of light is the same in all inertial frames and hence that the speed of light in vacuum is a natural constant.

2.8 The answer

In this chapter we have seen how mathematics is about the recognition of patterns and how a systematic analysis of the methods dealing with these patterns leads one to statements whose validity reach far beyond the directly observable or testable. Series can be extended to infinity and motion can be described with the help of differential calculus while the notions of the inertial frame helps us extend the reach of our understanding beyond the observable universe. We can therefore answer our initial question of how do we reach infinity and beyond as **with the help of patterns and its science — mathematics**.

2.9 Exercises

1. With the help of mathematical induction, prove that $\sum_{k=1}^{n} k = \frac{n(n+1)}{2}$ for any n.

2. With the help of mathematical induction, prove that $(ab)^n = a^n b^n$ for any n.

3. With the help of mathematical induction, prove that $\frac{1}{3} + \frac{1}{15} + \frac{1}{35} + \cdots + \frac{1}{(2n-1)(2n+1)} = \frac{n}{2n+1}$ for any n.

4. How could a circle be marked such that both the rotational and reflectional symmetries are lost?

5. What is the symmetry group of a square?

6. Write the series 2 + 5 + 10 + 17 + 26 + ... with the help of the summation notation (i.e. use a Σ).
7. How far can a cannon ball maximally roll on a flat surface if its initial speed is 7 m/s and its deceleration due to friction etc. is 1 m/s?
8. How many real numbers are there in the interval [0.01,0.02]?
9. Do all the even numbers form a group? Explain your answer.
10. What is the derivative of t^3?

3: Energy

Q - Why Can Sunlight Power the World?

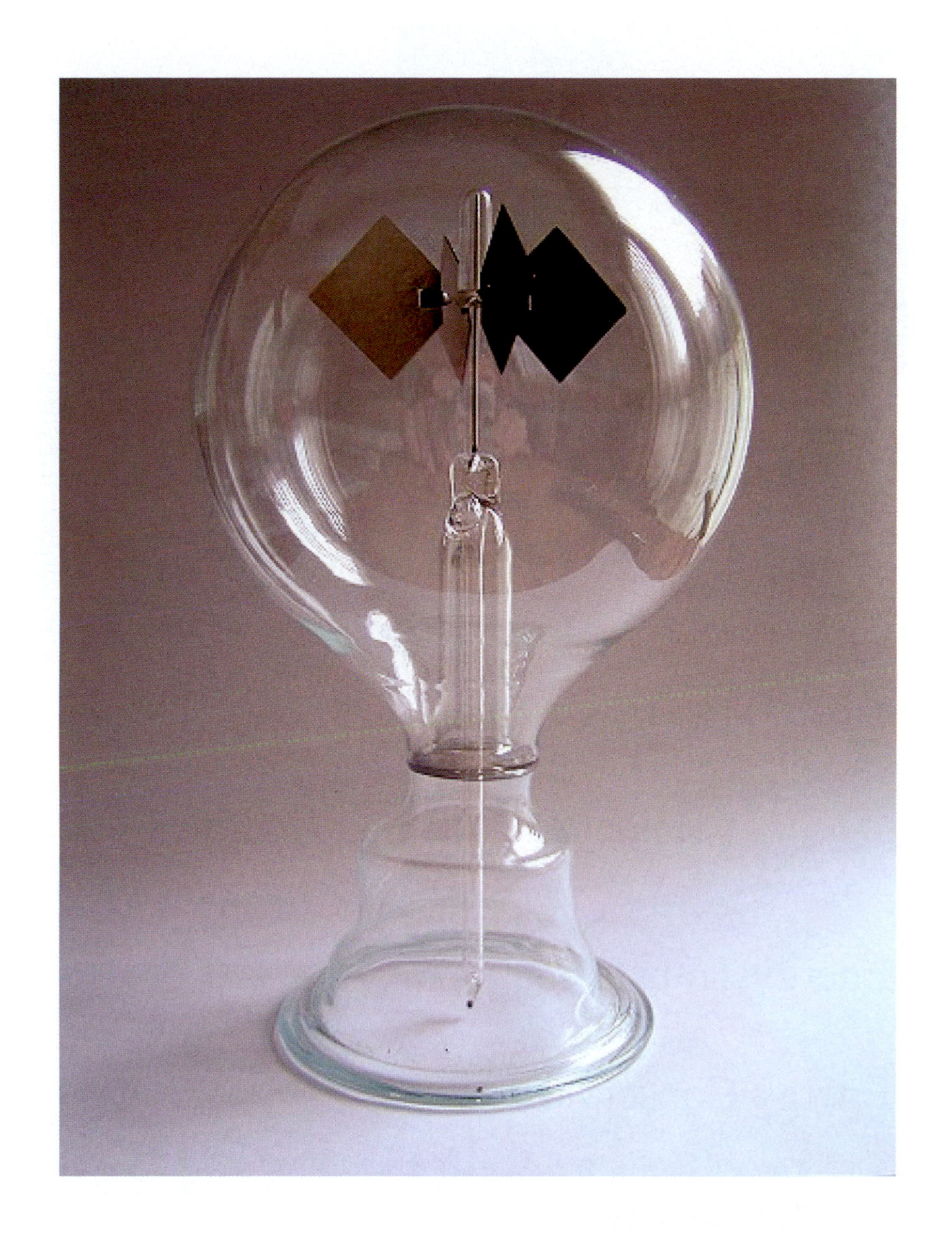

Chapter Map

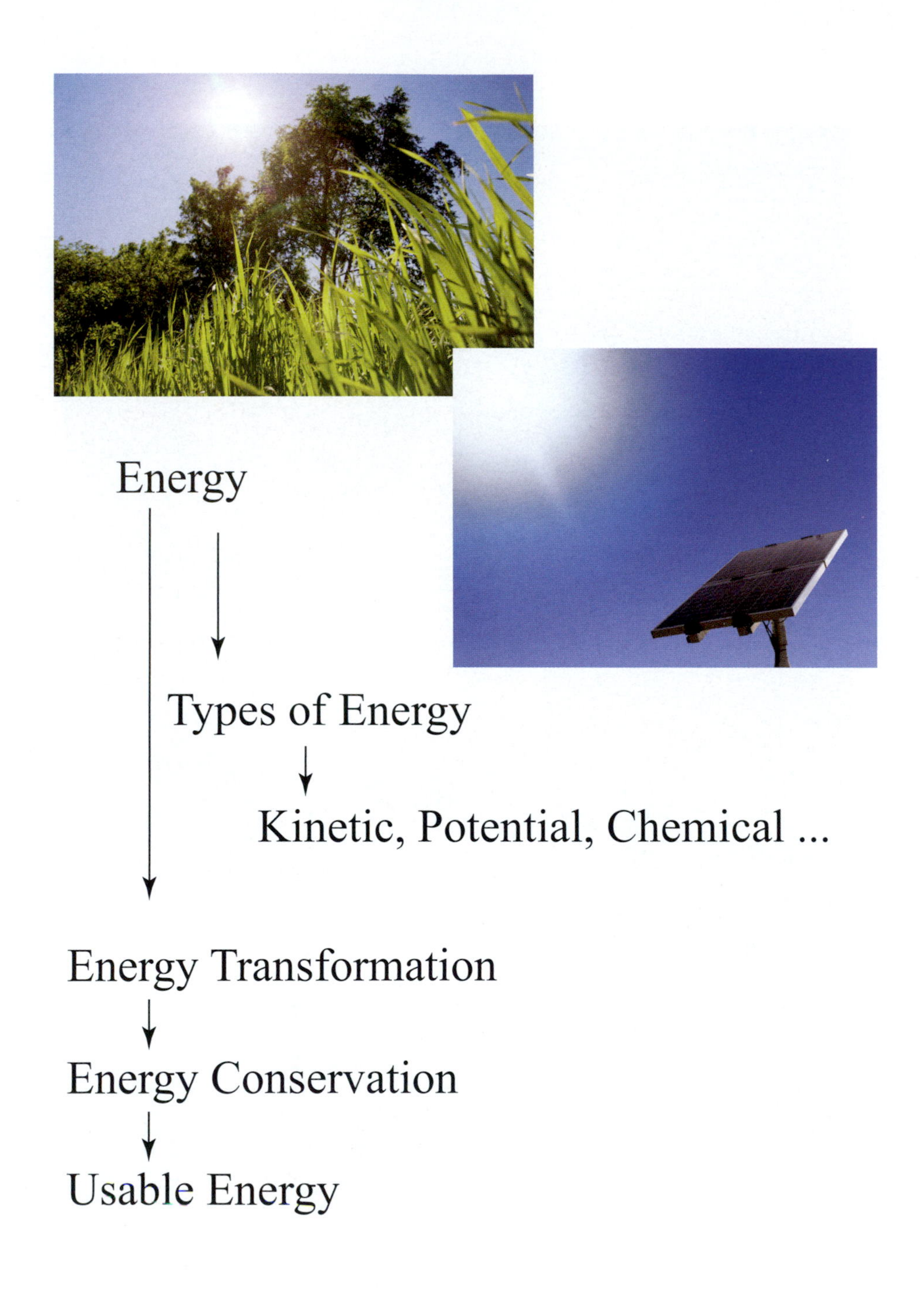
Energy
Types of Energy
Kinetic, Potential, Chemical ...
Energy Transformation
Energy Conservation
Usable Energy

3.1 The question

Figure 3.1: Lizard taking in energy from the sun.

Lizards commonly bask in the sun to increase their body temperatures thus using energy from the sun to warm up. Indeed, there are numerous direct examples of the energy contained in sunlight — from hot cars, to green houses to sunburns. We also know that sunlight provides energy to plants, for if we put them in a dark room for a long time they will die. Hence it is immediately clear that the sun provides energy for many processes and phenomena and a more careful analysis shows that the sun's energy plays a crucial role in nearly all dynamical processes near the earth's surface. However, warmth and life not only are different from each other, they are also entirely different from sunlight! Therefore we need to ask: **Why is it that sunlight can power the world?**

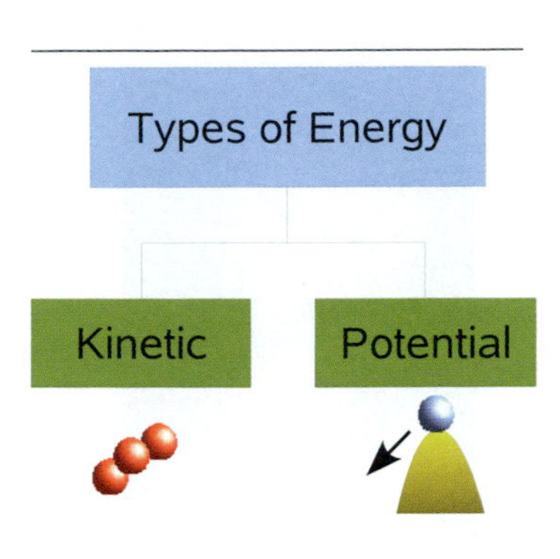

Fig. 3.2: The two key types of energy: kinetic and potential.

In order to understand this question, we first need to understand what we mean by energy. Originally, the word energy comes from the Greek word *energeia* which is translated as *activity* and indeed life is certainly some sort of activity. In daily life too, we clearly link activity with energy. Have you ever said "Today I feel really energetic!" to indicate your belief that you can achieve lot, or in other words that you can do or move a lot? In the scientific context it therefore comes natural to ask "How much energy does something have?" or "How much energy does it take to do this?"

These two questions directly lead one to consider two of the most basic concepts in science: motion and position. That motion is a kind of activity seems clear but how can position be relevant? The thing to realize is that position can lead to motion. Think of a ball on top of a mountain: give it a little nudge and it will roll down the slope gaining speed. Where does the energy come from for all this speed? It isn't the nudge, it's the position!

Energy due to motion in space is called kinetic energy, generally denoted by the letter T, and energy due to position or the internal configuration of the material body is called potential energy, generally denoted by the letter U. Classically, the sum E of the potential and kinetic energies is called the total energy — i.e. we have

$$E = U + T. \quad \textbf{Total energy = potential energy + kinetic energy} \tag{3.1}$$

An important type of potential energy is the energy contained in chemical bonds. When atoms form a molecule (for more details, see Chapter 5), in some cases (e.g. the formation of sugars from carbon, oxygen and hydrogen) the configuration of the molecule is such that a breakup of the molecule can provide some energy just like water flowing down a hill can drive a water mill.

Fig. 3.3: By flowing downward, potential energy of water is converted into kinetic energy. The waterwheel captures some of this.

What is the key idea of having a water mill, or perhaps more generally to have energy? It is to get some "work" done like the grinding of grains into flour. Therefore, next we will look into the scientific definition of work.

3.2 Work

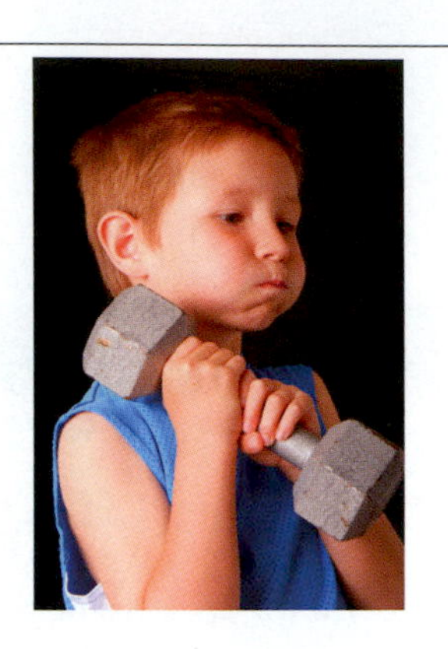

Fig. 3.4: Holding up a heavy weight costs a lot of energy even if the weight keeps the same position.

A key concept in science is force and, as illustrated with the waterwheel in Figure 3.3, kinetic energy can be used to exert a force. The general notion that links energy to force is called **work**.

How does one increase or decrease the energy of an object? Intuitively, one would think that by exerting a force on a body, one should be able to change its energy. For example, by exerting a force we can increase the velocity of a ball, thus increasing its kinetic energy. So clearly, there is an intimate connection between force and energy.

Work is a measure of a force acting over a distance. Think of pushing a car; the heavier the car, the more force is required to move it. In other words, the more force required, the more work. Also, the longer the distance the car needs to be pushed, the larger the amount of work that needs to be done. So it seems reasonable to define Work, denoted by W as

$$\begin{aligned} W &= \textbf{force} \times \textbf{distance moved along direction of force.} \\ &= F \times d. \end{aligned} \tag{3.2}$$

An illustration of how work is used to convert between potential and kinetic energy is given in Figure 3.5.

If the force is not constant then the expression for work is obtained by integrating along the path where the force acts

$$W = \int_{x_i}^{x_f} F(x)dx. \tag{3.3}$$

Intuitively, however, there is nevertheless a bit of a problem with this definition of work. We all know that if we hold up a heavy weight, even when the weight is

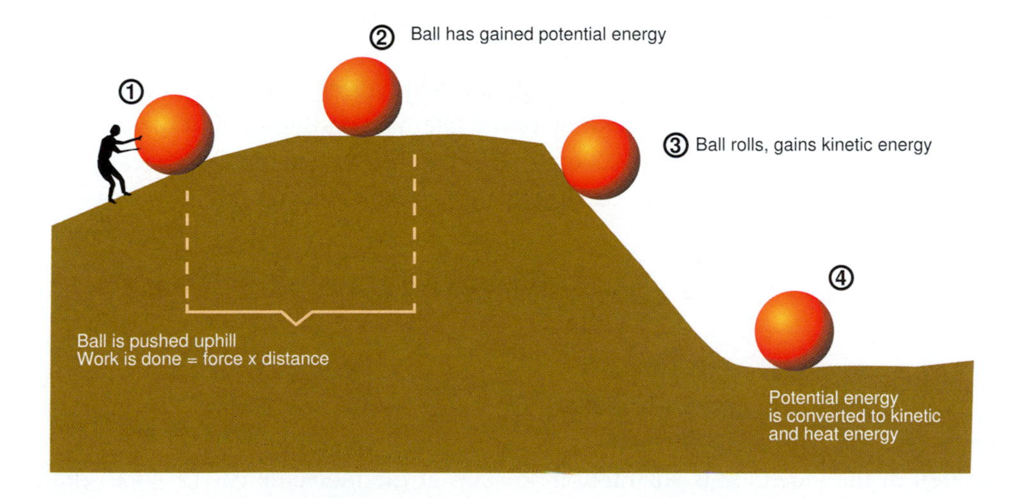

①
② Ball has gained potential energy
③ Ball rolls, gains kinetic energy
④
Ball is pushed uphill
Work is done = force x distance
Potential energy
is converted to kinetic
and heat energy

Figure 3.5: Work is needed to push a ball uphill. This increases its potential energy. When the ball rolls downhill, potential energy is transformed into kinetic energy.

not moving, and consequently no work is being done, we will soon break out into sweat from the exertion that we are undergoing. So what's going on? How can there be no work done, even though we have to exert ourselves? The answer to this counterintuitive result lies in the physiology of the human body.

The way our bodies have evolved is such that to keep a weight in position, our muscles are in a nonrelaxed state and need a constant supply of energy to remain that way. Hence we are doing internal biological work, and not work on the weight, when we hold it in air. In contrast, the muscles in a clam are such that once it is open, only very little energy is needed to keep it that way.

The fact that no work needs to be done to keep a weight in place can easily be verified by placing the weight on a flat table and leaving it there. As is clear from daily life, it will just stay put, for the same reason high-rise buildings can stand without any work being done. All the high floors are stationary, and hence do not require any expenditure of work (energy) to hold them up. Of course overloading may cause the floor to break, and this then becomes a problem of materials science rather than one of mechanics.

One cannot store work, since once the body ceases to move, no more work is done. However, unlike work, we can store energy. When work is done on, or by, an object, the result is to increase, or to decrease, its energy. In other words, the deep connection between work and energy is that force results in work, which in turn increases or decreases the energy of a body.

Thus far we have discussed potential and kinetic energy and the fact they are connected to work and hence force but we do not have any concrete expression for them yet. The theorem connecting the two concepts is generally called the **work-energy theorem** and given by

$$\Delta W = \Delta E : \qquad \textbf{work-energy theorem}, \tag{3.4}$$

where the ΔW and ΔE indicate "work done" and "change in energy" respectively.

Now that we have an expression that connects work and energy, we will need one for energy. Therefore, let us now look at kinetic energy and how it can be derived from work. Then we will quickly look at power as the rate at which work is done before delving into potential energy in Section 3.5.

3.3 Kinetic energy

We all know that a truck moving at high speed is something to be avoided; a head-on collision is violent enough to demolish a wall. The damage that a truck can do is due to the large kinetic energy it has. This kinetic energy comes not only from the fact that it is heavy, but also from the speed with which it is moving (a speeding truck does more harm than a slow one). No matter which direction the truck moves in, for a given velocity it has the same amount of kinetic energy. But what would be a good expression for the kinetic energy? mv? That would be strange since this is the momentum. In order to derive the correct formula for kinetic energy, let us have a look at what happens when work is done.

Fig. 3.6: A moving truck has a great deal of kinetic energy.

Consider a moving particle as in Figure 3.7 with mass m that at initial time t_i has a position of x_i and speed of v_i. Let a constant force F act on it from time

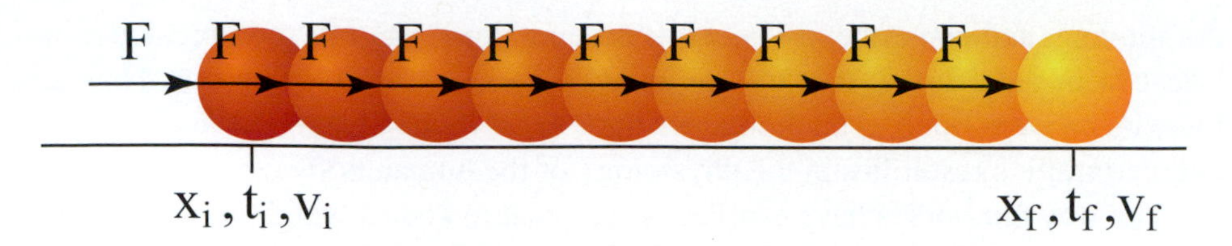

Figure 3.7: A force constantly acting on particle.

t_i to t_f, during which time it travels, along the direction of the force, to the final position of x_f. At the end time t_f, it has increased its velocity to v_f.

Using Newton's second law Equation 3.3 can be written as

$$W = \int_{x_i}^{x_f} F(x)dx = m\int_{x_i}^{x_f} \frac{dv}{dt}dx = m\int_{v_i}^{v_f} v dv \tag{3.5}$$

where we used

$$a = \frac{dv}{dt} = \frac{dv}{dx}\frac{dx}{dt} = \frac{dv}{dx}v$$

in the last step. The integral over v can now easily be carried out and we obtain

$$W = \frac{1}{2}mv_f^2 - \frac{1}{2}mv_i^2. \tag{3.6}$$

If there are no other forces at "work", all of the work should end up in kinetic energy; hence it is reasonable to define the kinetic energy as

$$T = \frac{1}{2}mv^2 \qquad \textbf{Kinetic Energy}. \tag{3.7}$$

The unit of energy can be deduced from the above equation as kg m^2 s^{-2}, and is called a joule, abbreviated as J.

For a related result, let us briefly consider the case of a constant force F acting on a particle. The force causes the particle to move with constant acceleration a, which from Newton's second law given in Equation 1.22 is given by $a = F/m$. The work done by the force in moving the particle from the initial position x_i to its final position x_f is given by Equation 3.3 as follows

$$W = \int_{x_i}^{x_f} F(x)dx = ma\int_{x_i}^{x_f} dx = ma(x_i - x_f) \tag{3.8}$$

Hence, from Equations 3.6 and 3.8 we find

$$v_f^2 - v_i^2 = -2a(x_f - x_i),$$

a relationship that is usually obtained by integrating Newton's second law.

As an illustration, let us consider a gun and a bullet. A natural question to ask is how much of the kinetic energy released by the explosion of the gunpowder when firing a shot ends up in the bullet. If we have a bullet of mass m_B being fired with some velocity v_B from a gun of mass m_G with a recoil velocity given by v_G, then, applying momentum conservation (Newton's third law, Equation 1.1) yields

$$m_B v_B = -m_G v_G. \tag{3.9}$$

This doesn't tell us much, since we are not able to translate this information into how much energy must be provided by the gunpowder.

We do know, however, that the total kinetic energy expended in shooting the bullet is given by

$$T = \text{Kinetic energy of bullet} + \text{Kinetic energy of gun}$$
$$= \frac{1}{2}m_B v_B^2 + \frac{1}{2}m_G v_G^2.$$

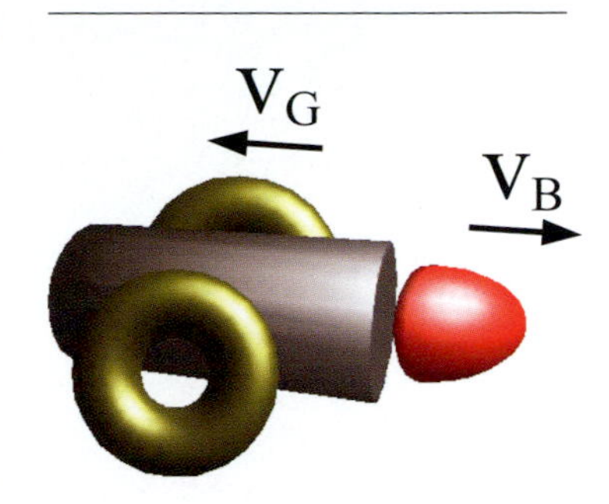

Fig. 3.8: Cannon and bullet both move.

Using Equation 3.9 we have the result that to design a gun which specifies the velocity of the bullet to be v_B, we need to spend an amount of energy given by

$$T = \frac{1}{2}m_B v_B^2 + \frac{1}{2}m_G\left(-\frac{m_B v_B}{m_G}\right)^2$$
$$= \frac{1}{2}m_B v_B^2 + \frac{1}{2}\frac{m_B^2}{m_G}v_B^2.$$

For most guns, the mass of the gun is much greater than that of the bullet, that is, $m_G >> m_B$, and hence we can see that most of the kinetic energy T ends up in the bullet.

3.4 Power

Consider a big pile of stones that we need to move up a hill. Clearly if we have enough energy we can move them and the amount of work can be calculated if we know the mass of the stones and the distance that they need to be moved. If we are patient enough, we can move them one by one but otherwise, having a bulldozer may be handier. In daily life we would say that the bulldozer has a lot more power than a pair of hands and it seems rather sensible to have this notion. So what would be the scientific way to define power?

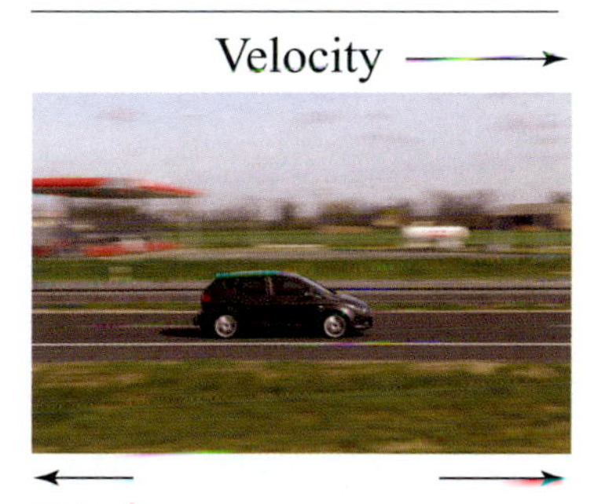

Fig. 3.9: Power is required to overcome friction. It is the idea of **power** rather than force that is required to explain the motion of a car.

Power is defined as the *rate* at which work is done. If we consider an object moving under constant acceleration (on a flat frictionless surface), its total energy is equal to its kinetic energy, and work done on the particle is equal to the change in its kinetic energy. Hence, in this case power is equal to the rate of change of kinetic energy of the particle.

Power, denoted by P, is then defined as

$$\textbf{Power} = \textbf{Rate of doing work}$$
$$\Rightarrow P = \frac{\Delta W}{\Delta t}. \tag{3.10}$$

From Equation 3.2 we have that the work done in time Δt is given by

$$\Delta W = \Delta(Fd). \tag{3.11}$$

In the case of a constant force like the gravitational force near the surface of the earth and noting that $\Delta d = \Delta x$, power then becomes

$$P = \frac{\Delta W}{\Delta t} = F\frac{\Delta x}{\Delta t}. \tag{3.12}$$

Since the time interval over which the kinetic energy has changed is Δt, we have $\frac{\Delta x}{\Delta t} = v$, and hence

$$P = Fv. \quad (3.13)$$

This is an important equation, since it tells us that the rate of change of kinetic energy of a particle, that is the power being expended on the particle, is equal to the force times the velocity of the particle. The unit of power is the watt (W) which is defined to a joule per second, hence having the base units kg m^2 s^{-3}.

Example 3.1: Power and Energy

Car manufacturers need to know how to build cars which can accelerate, coast at a constant speed and, of course, decelerate as well. Newton's laws are useful in telling us how much force is required if we want to accelerate, and also how much force the brakes must exert to decelerate. But Newton is silent on the great difference between the force that accelerates the car, and the brakes that decelerate it (one needs fuel, the other doesn't). However, with the help of the notion of power we can understand the difference. When the car accelerates, F and v are in the same direction, and hence the Power $= Fv$ required is positive and has to be supplied to the car. To provide this power is the reason why the car has to burn fuel. When we brake, F and v are in opposite directions, with the brake causing acceleration in the direction opposite to the velocity v. Hence the required Power $= -Fv$ is negative. In this case the car provides the power to the brake, causing the brake to heat up, and hence no fuel needs to be burned (of course it would be much better to use regenerative braking).

Suppose a passenger car is moving at 25 ms^{-1} (90 km/h) on a level surface and requires a force of 1,000 N to keep it moving. The power required is $P = Fv = 1{,}000\,\text{N} \times 25\,\text{m/s} = 25{,}000\,\text{W} = 33\frac{1}{3}$ horsepower. Since most car engines have much more than this amount of power, the car can easily cruise at constant speed. When the car is not accelerating, the force of resistance is exactly equal to the power P supplied by the engine, that is $P = vF_{\text{friction}}$. Consequently, we see that the velocity of the car $v = P/F_{\text{friction}}$.

Noteworthy 3.1: Units

The metric unit for force is the newton (N), the unit for energy and work is the joule (J) and the unit for power is the watt (W). For historical reasons, another unit of energy is often used, especially in the food industry. This is the **kilocalorie**: the amount of energy (heat) required to raise the temperature of 1 kg of water by 1 °C from 14.5°C to 15.5°C. The kilocalorie is sometimes called the Calorie (with a capital C). The relationship between joule and kilocalorie is

$$1\,\text{kcal} = 4.184\,\text{kJ}. \quad (3.14)$$

3.5 Potential energy

Fig. 3.10: Potential energy in the chemical bonds of the corn's starch can power many biological processes.

Kinetic energy is easy to visualize. Potential energy U is a more complicated concept, since potential energy means energy which is in a latent ("hidden") form, and capable of being "released". An example of potential energy is the energy stored in the physical shape and configuration of a body, and is different from kinetic energy in that it is present in the body regardless of motion or movement.

There are many different forms of potential energy but let us first have a look at the so-called gravitational potential energy.

We all know that a stone dropped from a great height picks up a lot of speed, and hence kinetic energy, before hitting the ground. However, it is also clear that it started with zero kinetic energy and consequently its position at a great height somehow must have endowed it with "potential" energy which then became "actual" (kinetic) energy on the way to hitting the ground.

So what would be a good way to express this potential energy? The hint is already given by the connection of potentiality with height. The higher the elevation, the greater must be its potential energy. At the same time, it also seems to be apparent that, the heavier the body, the greater its potential energy as can easily be tested by dropping, for example, a tennis ball and a cannonball from a building. Both these intuitive expectations reflect everyday experience when we see objects fall — if they fall from a greater height, they have higher impact, and similarly the impact is greater for a heavier object. If we consider a body with mass m at a height h, the most straightforward expression for (gravitational) potential energy that matches this daily experience looks like

$$U \propto mh.$$

If we call the proportionality constant g, we can then guess the expression

$$\Rightarrow U = mgh. \qquad \textbf{Gravitational Potential Energy} \qquad (3.15)$$

Clearly g is linked to gravity, since if there were no gravitational force, the particle would not fall toward the earth to start with. Rather amazingly, careful testing shows that this expression is, in fact, approximately correct. As discussed in Equation 1.27, g is the acceleration on the earth's surface due to earth's gravity and amounts to approximately 9.81 m s^{-2}.

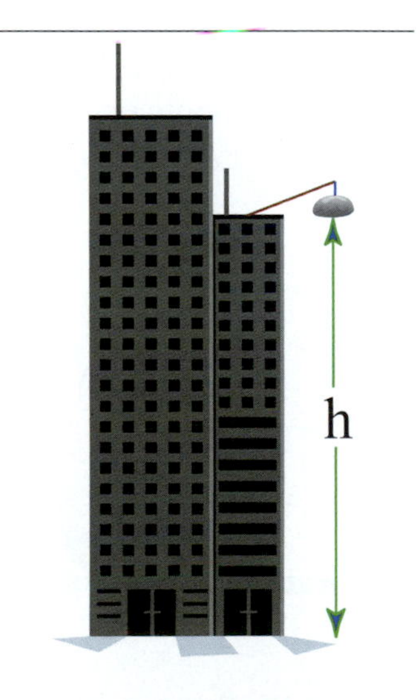

Fig. 3.11: Potential energy increases with height.

Consider a particle falling under the force of gravity. If at a height of h, it has velocity v, then its total energy is given by

$$E = T + U \qquad (3.16)$$

$$\Rightarrow E = \frac{1}{2}mv^2 + mgh. \qquad (3.17)$$

As the particle falls, its velocity increases while its height decreases. Intuitively, if nothing else is going on, it is clear that the change in E should be 0 (naturally such a statement requires extensive experimental verification). Hence we have the testable expression

$$\Delta E = \Delta\left(\frac{1}{2}mv^2 + mgh\right) = 0$$

implying that

$$\Delta v^2 = -2g\Delta h.$$

The gravitational potential given in Equation 3.15 is an approximation, valid only for bodies that are close to the earth's surface. At greater distances from the earth's surface, the correct expression for U is obtained from Newton's universal law of gravitation.

3.6 Elastic potential energy

Fig. 3.12: Elastic potential energy keeps the watch moving.

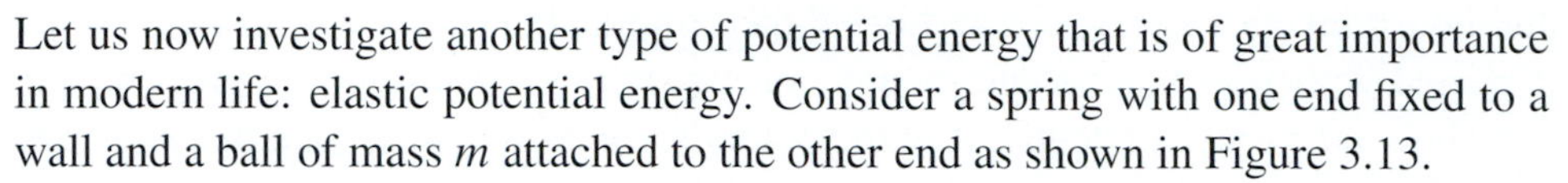

Let us now investigate another type of potential energy that is of great importance in modern life: elastic potential energy. Consider a spring with one end fixed to a wall and a ball of mass m attached to the other end as shown in Figure 3.13.

The system is in equilibrium when the ball is at position x_E. What do we expect for the potential energy of the spring? Whether we compress or stretch the spring, in both cases it gains energy.

Hence, at least in a first approximation, the potential energy should be equal if the new position x of the ball is compressed or stretched by the same distance from its resting position x_E. This would suggest that the potential is U-shaped or U-like shaped. It turns out that the U-shape doesn't work so well but the simplest U-like shape does, a parabola as shown in Figure 3.14 and given by

$$U = \frac{1}{2}k(x - x_E)^2 \tag{3.18}$$

where k is the **spring constant** that reflects the stiffness of the spring (the units of k are kg s^{-2}).

The total energy of the ball at position x moving with velocity v is then given by

$$\begin{aligned} E &= T + U \\ &= \frac{1}{2}mv^2 + \frac{1}{2}k(x - x_E)^2. \end{aligned} \tag{3.19}$$

Let the ball be at some position $x \neq x_E$ in rest as shown in Figure 3.15. We can then ask the following question: What is the force required to displace it by a small distance d? Let the ball have zero velocity. To change the energy of the ball by displacing it, an external force F_{external} has to do work on it.

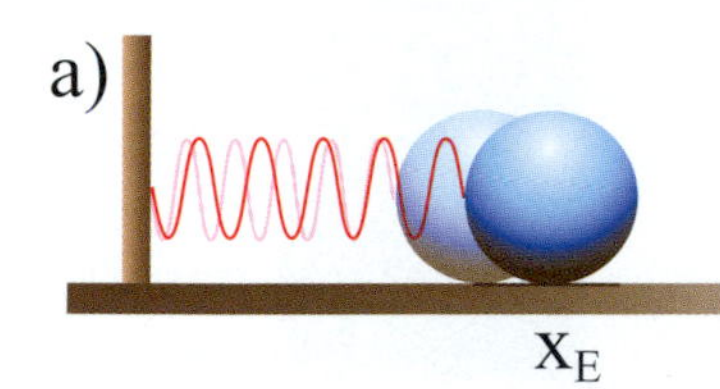

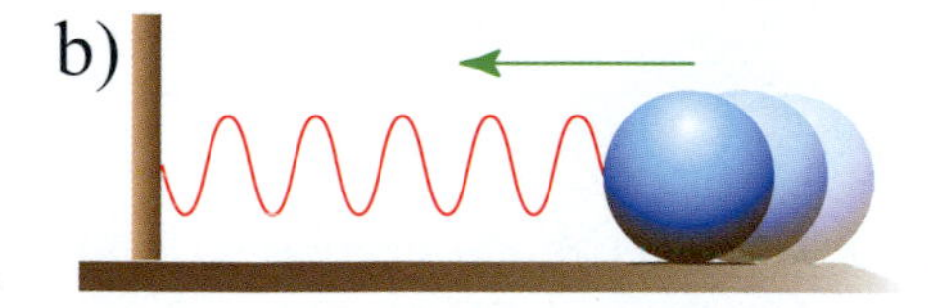

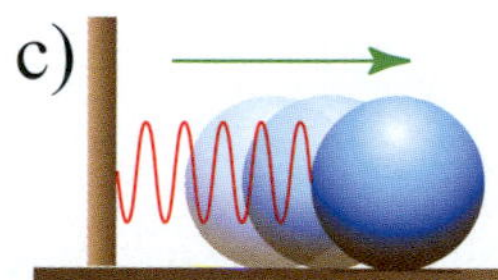

Figure 3.13: Force exerted on a ball by a spring: a) when the spring is relaxed, b) when the spring is stretched, c) when the spring is compressed.

If we approximate F_{external} to be constant (often a realistic assumption), we have

$$\begin{aligned} F_{\text{external}} \times d &= \Delta U = \frac{1}{2}k(x+d-x_E)^2 - \frac{1}{2}k(x-x_E)^2 \\ &= kd(x-x_E) + \frac{1}{2}kd^2 \qquad (3.20) \\ \Rightarrow F_{\text{external}} &= k(x-x_E) + \frac{1}{2}kd \\ &\simeq k(x-x_E) \qquad (3.21) \end{aligned}$$

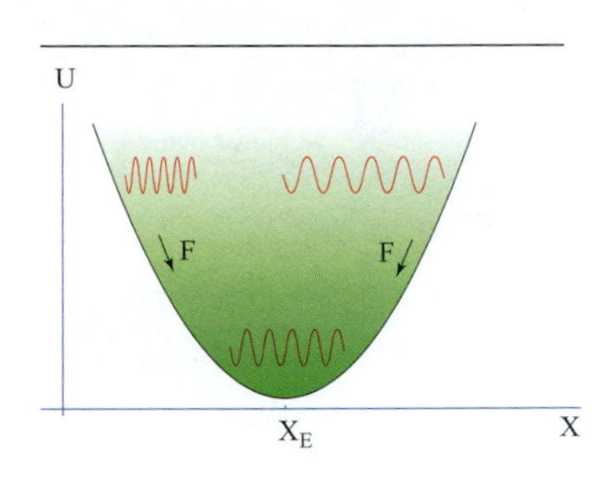

Fig. 3.14: Potential energy versus position.

where in obtaining the last equation we have taken d to be so small that it can completely be ignored. For equilibrium, the restoring force due to the spring has to exactly cancel the external force. Hence, if the ball is moved to a position $x \neq x_E$, we have a restoring force acting on the ball that is given by $F = -F_{\text{external}}$ and obtain

$$F = -k(x-x_E) \quad \textbf{Hooke's law} \qquad (3.22)$$

This is the famous **Hooke's law**.

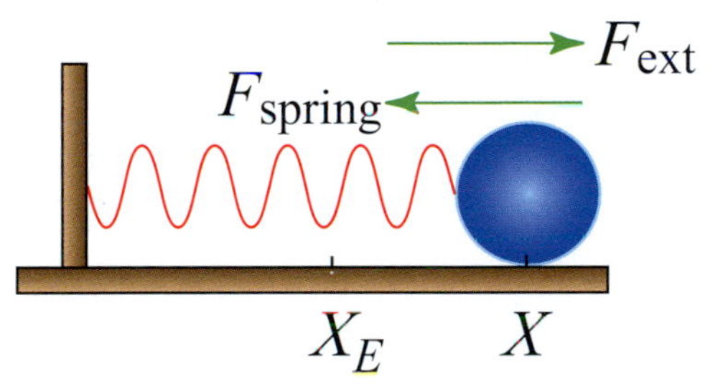

Figure 3.15: The forces acting on the ball.

In other words, the force required to stretch the spring is proportional to the amount of stretching, namely $(x - x_E)$. Note that the spring is always a restoring entity since it acts against any change of position of the ball. For $x > x_E$ the force is negative, that is, it acts to pull x back to x_E, whereas if $x < x_E$ then the force acts to push the ball back to x_E.

Now that we know what the forces on the ball are we would of course like to find out what its motion is, i.e. we consider x as dependent on t or in other words we consider $x(t)$. By rewriting Equation 3.19 we have for a particle moving with energy E the following expression for the velocity

$$v(t) = \frac{dx(t)}{dt} = \sqrt{\frac{2E - k(x(t)-x_E)^2}{m}}. \qquad (3.23)$$

However, this expression is not that handy since we need $x(t)$. It would be better if we could have an equation that describes the motion of the ball solely in terms of constants. This can be done in the following way.

Let us start by finding an equation for the position by combining Hooke's law given in Equation 3.22 with Newton's second law $F = ma$ as

$$\begin{aligned} ma(t) &= -k(x(t)-x_E) \\ \Rightarrow \frac{d^2x(t)}{dt^2} &= -\frac{k}{m}(x(t)-x_E). \qquad (3.24) \end{aligned}$$

Fig. 3.16: Elastic potential energy is released when the arrow is shot.

If, for the moment, we forget about the constants (noting that x_E is a constant too), we see that we have to solve an equation of the type

$$\frac{d^2x(t)}{dt^2} = -x(t). \tag{3.25}$$

Hence we need to find a function $x(t)$ whose second time derivative is itself multiplied by a minus sign. The cosine and sine functions fulfill this requirement. Hence we can try the following general cosine function

$$x(t) = x_E + A\cos(\omega t + \phi) \tag{3.26}$$

where A, ω and ϕ need to be determined. By differentiating Equation 3.26 with respect to time and satisfying Equation 3.19 we find that

$$\omega = 2\pi f = \sqrt{\frac{k}{m}} \tag{3.27}$$

$$A = \sqrt{\frac{2E}{k}} \tag{3.28}$$

$$\Rightarrow E = \frac{1}{2}m\omega^2 A^2. \tag{3.29}$$

Recall x_E is the position of equilibrium of the particle. Suppose the particle starts its oscillations at time $t = 0$ from the position $x(0)$. Then it follows from Equation 3.26 that ϕ is fixed to

$$\phi = \cos^{-1}\left(\frac{x(0) - x_E}{A}\right). \tag{3.30}$$

We can also differentiate Equation 3.26 with respect to time to obtain a somewhat more convenient expression for the velocity than Equation 3.23

$$v(t) = \frac{dx}{dt} = -\omega A\sin(\omega t + \phi). \tag{3.31}$$

Note from Equations 3.26 and 3.31 that the motion is *periodic*, with the period of one oscillation given by $P = 2\pi/\omega$. The system obeying Hooke's law is said to undergo **simple harmonic motion** and appears in almost all branches of science.

The average, over one period P, of the kinetic energy $\langle T\rangle$ is equal to the potential energy $\langle U\rangle$. To see this note

$$\langle T\rangle = \frac{1}{P}\int_0^P \frac{1}{2}mv^2(t)\,dt = \frac{1}{4}m\omega^2 A^2 \tag{3.32}$$

$$\langle U\rangle = \frac{1}{P}\int_0^P \frac{1}{2}k(x(t) - x_E)^2\,dt = \frac{1}{4}m\omega^2 A^2 = \langle T\rangle \tag{3.33}$$

$$\Rightarrow E = 2\langle T\rangle = 2\langle U\rangle. \tag{3.34}$$

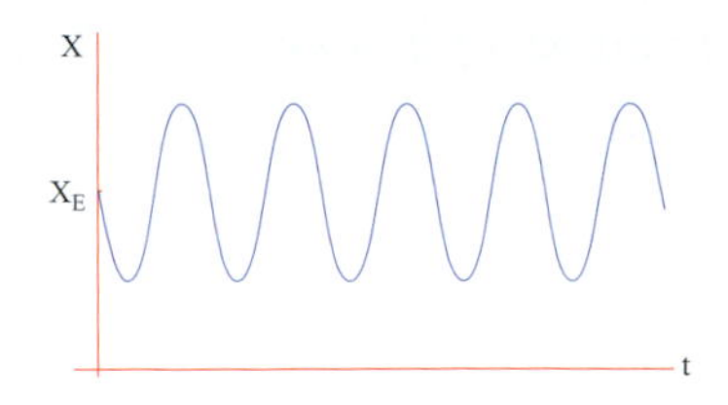

Figure 3.17: Position of the ball versus time.

The position of the ball versus time when released from a starting point $x(0) \neq x_E$ is shown in Figure 3.17 where the ball is undergoing oscillations about the equilibrium position x_E with period of oscillation given by $P = 1/f = 2\pi\sqrt{m/k}$. The maximum departure from x_E is called the amplitude of the oscillation, and is given by A.

The phase ϕ tells us where the oscillations begin at $t = 0$. The amplitude of the oscillation A is fixed by the ratio of the total energy and spring constant, while the period of oscillation of the particle, namely P, is given by the ratio of the spring constant k and the mass of the particle. In other words, the stiffer the spring is, the larger is the value of k and consequently the faster the particle oscillates.

The particle will oscillate forever if not for frictional forces, which we have ignored in our discussion. Due to the oscillatory motion that the elastic potential energy term produces, it is also called the harmonic oscillator potential.

3.7 Conservation of energy

Our examination of gravitational potential energy and the resulting kinetic energy of a falling object has guided us toward one of the key notions of moderns science: the conservation of energy. To illustrate this idea, let us start with the example used by Feynman in his famous Lectures on Physics.

Figure 3.18: Conversions of biological, potential and kinetic energies.

Consider the case of a 10 year old who has 28 indestructible playing blocks, the weight of each block being 0.1 kg. Every day, the boy's mother counts the blocks, and it always adds up to 28. One day she finds only 27 blocks. As she knows that blocks do not just vanish into thin air, she looks around and finds one lying under the bed.

A few days later she finds only 26 blocks. On searching the room she cannot find the two missing blocks but she does find a special toy box that can only contain blocks and no other toys; when she tries to open the toy box, the boy screams "Don't open my toy box". Being a smart mother, one day when she has all the 28 blocks so that she knows that the box must be empty, she weighs it and determines its weight to be 0.5 kg. She also weighs one of the blocks to find it to be 0.1 kg. Henceforth, when she cannot find all the 28 blocks, she weighs the toy box to figure out how many blocks it contains.

The mother now has the following formula for the number of blocks

$$\text{Total number of blocks} = \text{number of blocks seen} + \frac{(\text{weight of toy box}) - 0.5 \text{ kg}}{0.1 \text{ kg}}. \tag{3.35}$$

After a few days, she finds that the blocks no longer add up to 28 and that the toy box is too light to contain all the missing blocks. What now? Inspecting the house, she finds that the bathtub is full of water covered with a thick layer of white soap bubbles. She suspects that there are some more blocks submerged in the bathtub. How could she figure out whether this is so without removing the bubbles? Given that the original height of the bathwater was 0.5 m, and that each block raises the height of water by 0.01 m, she now has a new formula for computing the number of blocks, namely

$$\text{Total number of blocks} = \text{number of blocks seen} + \frac{(\text{weight of toy box}) - 0.5 \text{ kg}}{0.1 \text{ kg}} + \frac{(\text{height of water}) - 0.5 \text{ m}}{0.01 \text{ m}}. \tag{3.36}$$

We see that the calculation for the number of blocks is becoming more and more abstract, and has less and less to do with counting the blocks themselves. Measuring the weight of a toy box or the height of bathwater has no direct bearing on the blocks themselves.

The application of this example is straightforward to energy. Energy itself is represented by the indestructible "blocks". Unfortunately in nature we do not know what is energy *per se*, and hence we do not know what is analogous, in nature, to the blocks. The analogy of the heavier toy box is one form of energy and the higher water level is analogous to another form of energy. In other words, there are many, many forms of energy. Just as the total number of blocks was constant, the total amount of energy is constant (i.e. energy is absolutely conserved).

Energy can neither be created nor destroyed! All that a physical process can do is to transform energy from one form into another, analogous to, for example, the weight of the toy box (one form of energy) being reduced at the expense of the bathwater rising in height (another form of energy), so that the total number of blocks remains the same.

Fig. 3.19: We learn from a very early age that the number of blocks is conserved.

Hence, for energy we have the following expression:

$$\begin{aligned}\text{Total energy of an isolated system} &= \text{Form A of Energy} \\ &+ \text{Form B of Energy} \\ &+ \text{Form C of Energy} + \text{etc.} \\ &= \text{Constant.}\end{aligned} \tag{3.37}$$

Newton's second law and energy conservation

Newton's second law of motion states that the acceleration that a body undergoes is caused by a force F acting on it. For the position of a particle at time t denoted by $x(t)$ we have

$$m\frac{d^2x(t)}{dt^2} = F(t). \tag{3.38}$$

The nineteenth century saw a shift away from the Newtonian concept of force to that of energy. In particular, it was realized that in the context of a conservative force such as gravity we have

$$F = -\frac{dU}{dx}. \tag{3.39}$$

(A conservative force is a force where the change in the potential energy only depends on the start and end points and not on the path between them. A typical example of a conservative force is the gravitational force.) A special case of this equation, namely when $U = \frac{1}{2}k(x - x_0)^2$ leads to Hooke's law $F = -k(x - x_0)$ where the restoring force is linear .

Multiplying both sides of Newton's second law (Equation 3.38) by the velocity $v = dx/dt$ and using the chain rule of differentiation yields

$$\begin{aligned} m\frac{dv}{dt}v &= -\frac{dU}{dx}\frac{dx}{dt} \\ \Rightarrow \frac{d}{dt}\left[\frac{1}{2}mv^2\right] &= -\frac{d}{dt}[U(x)] \\ \Rightarrow \frac{d}{dt}\left[\frac{1}{2}mv^2 + U(x)\right] &= 0 \\ \Rightarrow \frac{1}{2}mv^2 + U &= E = \text{constant: Energy Conservation.} \end{aligned}$$

This derivation shows that Newton's equation of motion, which forms the cornerstone of Newtonian dynamics and, in particular, the ability to predict the future motion of an object, is a consequence of energy conservation. Newton's laws are a way of realizing energy conservation at each instant of a particle's motion. In Newton's framework, E is a constant of motion. Once all the constants of motion are specified, energy conservation is completely equivalent to Newton's equation of motion. It is presently thought that all fundamental forces in nature are conservative so that at the most fundamental level, Newton's law has nothing to add to energy conservation.

Does Newton's law have any utility above and beyond energy conservation? It turns out that the answer is yes for a rather unusual reason. In dissipative systems, such as the flow of viscous fluids, the system loses energy to friction and turbulence. If one wants to describe the dynamics of only the dissipative fluid without taking into account the total system, then one needs the concept of a nonconservative force that is not reducible to a potential. For such systems, the approximate description is best achieved by using Newton's law of motion and, in particular, it gives rise to the Navier-Stokes equation of fluid mechanics as discussed in Chapter 6.

A powerful idea in physics is that of **invariants**, that is, things that do not change as a system evolves in time (see also Chapter 2). The existence of invariants and their use allows us to place constraints on the possible dynamics of a system even though we might be ignorant of the details.

Denoting total energy by E we have the fundamental (classical) relation

$$\begin{aligned} \text{Total energy of system} &= \text{Kinetic Energy} + \text{Potential Energy} \\ \Rightarrow E &= T + U. \end{aligned} \tag{3.40}$$

Suppose the system has energy E_1 at time t_1 and energy E_2 at a later time t_2, then, the change in energy is $\Delta E = E_2 - E_1$. Conservation of energy implies that

$$\Delta E = 0 \tag{3.41}$$

$$\Rightarrow \Delta T + \Delta U = 0. \tag{3.42}$$

Note an important fact that since all we know is that $\Delta E = 0$, the absolute value of E has not been fixed. Hence, energy is only defined up to a constant, since E and E + constant would both be equally conserved.

3.8 Free energy

Fig. 3.20: Coal provides us with "useful" energy.

Energy is everywhere! Einstein's famous formula $E = mc^2$ tells us that even the mass of a body is a form of congealed energy (see Chapter 22). All material things are different forms of energy. If energy is indeed everywhere, why is there always a fear that our society is "running out energy", or that there is a shortage of fuel? Why are we asked to reduce, reuse and recycle? We all intuitively know that energy is precious and that possessing energy is of high value. So we need to wonder: Is all energy equal? Or is there a certain energy that is more desirable than another?

One of the main developments of science in the nineteenth century was the realization by Sadi Carnot, Rudolf Clausius and others that useful energy — energy that can do mechanical work, energy that can be "controlled" and directed — is a very special kind of energy. This special form of energy is called "free energy" to differentiate it from energy in general.

Let us consider the forms of energy that we find useful. A large piece of rock on a hill can do useful work; by tying a load to the rock via a pulley and then making the rock roll downhill, the rock can lift a load up the hill. Burning coal provides useful energy for generating electricity, heating homes and so on. In fact, efficiently obtaining, storing and utilizing useful energy is the one of the major preoccupations of engineering and technology.

So what is useful energy and how is it different from "useless" energy? The example of using a rock to raise a load is different from burning coal; in the case of the rock, the load raised up the hill in turn can be made to do useful work by itself, if it is made to come down tied to yet another load that it pulls up. In contrast, once coal is burned, the ashes that remain cannot be burned again. The difference between energy carried by a rock on a mountain top and that carried by coal is expressed by the concept of free energy.

To quantitatively understand the concept of useful energy, consider a system which can be a tiny thing such as a living cell or a large system like the air in a room. All things are at some temperature, denoted by T (note this symbol should not be confused with kinetic energy — with which it, indeed, has a fundamental link as is discussed in Equation 10.16). All objects are made out of atoms and molecules; temperature is a macroscopic manifestation of the microscopic and random motion of atoms and molecules. Since atoms and molecules are moving, they have kinetic energy; since their motion is random, they are equally likely to go in one direction as in the opposite direction. The microscopic "thermal" energy carried by atoms and molecules, at a given temperature T, cannot be extracted by any device which is at the same temperature T; any attempt to make movable parts of the device extract energy would result in the parts having net zero movement and hence no useful work would be extractable.

Fig. 3.21: The chemical energy stored in the bonds of the sugar molecules is essential for many living organisms.

If a system has total energy E, the amount of energy that can be converted to useful work is called **Helmholtz free energy** and denoted by F. One needs to subtract the microscopic energy, as reflected in temperature T, to obtain F from E. The precise formula is given by

$$F = E - TS. \tag{3.43}$$

The higher the temperature T, the greater is the energy locked into microscopic motion. The quantity S is called **entropy** and is one of the central concepts of science. Free energy, more precisely, is defined for a system at a fixed volume V and temperature T, namely $F = F(T, V)$. Helmholtz free energy is the amount of energy in a system that can be converted into useful work — at constant temperature and volume.

Entropy S, multiplied by T, encodes the precise amount of energy locked into microscopic motion. Similar to free energy, S is also a function of volume V and temperature T and one writes $S = S(T, V)$.

The transformation of energy has a direction, with some transformations being reversible and others being irreversible. The example of a rock moving in Earth's gravity is an example of a reversible energy transformation (friction ignoring) in contrast to the burning of coal, which is irreversible. Careful examination shows that a process transforming energy is reversible if the entropy S is a *constant*; otherwise, the process is irreversible. The fact that entropy must never decrease for all processes that are allowed in nature goes under the name of the **second law of thermodynamics**, and which states the following:

For all isolated processes in nature S either remains constant or increases; in other words, no natural process for an isolated system can reduce the entropy S, Hence

$$\Delta S \geq 0 \quad \textbf{: Necessary for all allowed (isolated) processes} \tag{3.44}$$
$$\textbf{:The Second Law of Thermodynamics.}$$

For *all* physical processes, we must always have

$$\Delta F \leq 0 \textbf{ : Necessary for all allowed (isolated) processes.} \tag{3.45}$$

Fig. 3.22: Left alone, a cottage decays thus increasing its entropy.

It can be shown that $\Delta F \leq 0$ is simply another statement of the second law of thermodynamics. If we consider the system together with its environment to be the universe, then the principle that $\Delta F \leq 0$ simply expresses the second law of thermodynamics that the total entropy of the entire universe must always increase.

Consider a system at constant temperature T and let a process causes a small change in the free energy of the system; hence

$$\Delta\left(\frac{F}{T}\right) = \Delta\left(\frac{E}{T}\right) - \Delta S \quad (3.46)$$

$$\Rightarrow \frac{\Delta F}{T} = \frac{\Delta E}{T} - \Delta S. \quad (3.47)$$

The second term ΔS is the change in the entropy of the system.

Equation 3.45 is one of the most fundamental equations in physics, chemistry and biology. Only those chemical reaction which lower net free energy are possible. In biology, Equation 3.45 forms the basis of how life organizes and sustains itself.

Consider for example a living organism. It is certainly not a random and homogeneous collection of molecules. Indeed from its conception to maturity, it develops into a highly complex and ordered structure. Of course, we recognize that it is not an isolated system to which the conclusion of the second law of thermodynamics applies. In fact a living organism feeds, breathes, and dissipates heat and other waste products: It takes in low-entropy energy (nutrients) and dissipates disordered energy (heat) in such a way that its own entropy decreases while that of the whole "universe" (consisting of the organism plus the environment) increases in accordance with the second law of thermodynamics, which states that $\Delta S > 0$. Note that the increase (or constancy) of entropy implied by the second law applies only to closed or isolated systems (here a closed system is a system that can exchange heat but not matter with its environment while an isolated system can exchange neither matter nor heat).

The cell has to carry out chemical reactions that constantly lower its entropy to offset the effects of thermal random motion that are always at present and that tend to increase the entropy inside the cell. The only way that a cell can lower its entropy is by increasing the entropy of the environment, so that net entropy of the universe always increases.

There is an *important difference* between living organisms and inorganic systems. A living organism is in fact not really in equilibrium because it continually requires an intake of nutrients, continually dissipates energy and often undergoes changes in size or shape. We can apply the second law (meant only for the end points of an equilibrium system) by taking a short time interval during which we may assume the organism to be in equilibrium. However, we know that in the long run, if the organism was to try to maintain this equilibrium continuously (with no intake and/or dissipation) it would soon degenerate (die) and this would result in its highly organized state decaying into disorder. Thus an out of equilibrium (or nonequilibrium) situation is actually required for living organisms to maintain their order and complexity!

In order to stay alive, a cell needs to be able to lower its entropy. Hence if $\Delta S_{\text{cell}} < 0$, the increase in free energy of the cell must be compensated by the cell losing energy to the environment (the sign convention is that energy lost by the system has a negative sign), resulting in the environment gaining entropy $\Delta E_{\text{cell}}/T$. The net effect of this whole process is the lowering of free energy for the cell. Using Equation 3.47, we have

$$\begin{aligned}\frac{\Delta F_{\text{cell}}}{T} &= \frac{\Delta E_{\text{cell}}}{T} - \Delta S_{\text{cell}} \\ &= -\frac{|\Delta E_{\text{cell}}|}{T} + |\Delta S_{\text{cell}}| \\ &< 0 : \text{Physically allowed.}\end{aligned} \tag{3.48}$$

The energy that is spent in mechanical work by a biological entity is supplied by the food taken. But even more importantly, a biological entity is usually at a higher temperature than the environment, and constantly loses heat energy to it. The heat energy that the cell loses to the environment stems from the conversion of food to useful work. We eat food, which has low-entropy, not only to regain lost energy, but also to lower our entropy. By losing heat energy we in effect lower our entropy. If the cell fails to obtain food, its entropy increases leading to the destruction of its highly ordered low-entropy structures and eventually to cell (or organism) death. Hence we have the paradox that we need energy to lose energy!

Entropy has an interesting application in evolution. Some people have argued that the emergence of life contradicts the second law of thermodynamics since life is a highly organized state, with very low entropy, that emerges from a high entropy environment. This argument is incorrect, since a living entity is not an isolated system; if one takes into account the total entropy of the earth and sun, then it can easily be shown that the total entropy of the total system including the living entity, together with the earth and sun, always increases.

3.9 The answer

Now we are able to answer the question why sunlight can power the world.

Energy is a physical quantity that has a numerical value which is assigned to every object. One can only measure specific forms of energy, with a set of experiments that is unique to that particular form. For example, heat energy can be measured using a thermometer and solar energy using semiconductors.

Let us elaborate a bit further. Energy is an idea that runs through the entire gamut of science, and defines what is meant by physical reality. Everything that exists has energy and everything that has energy exists. Energy is also central to the functioning of human civilization; to reshape nature in accordance with human

Fig. 3.23: Richard Feynman: "It is important to realize that in physics today, we have no knowledge of what energy *is*."

needs involves the expenditure of energy. Hence, engineering and technology are intimately concerned with the properties of energy that make it amenable for human manipulation and utilization. And of course, without the transformation of one form of energy into another, life itself would not be possible.

We know from experience that standing out in the open on a bright day can make us feel warm: The brightness is due to sunlight which is also responsible for the heat we experience. This heat from the sun is also responsible for our weather, and most of the other geographical changes that take place on the surface of the earth.

We also know from experience that sunlight is essential for the growth of plants. Just as water and minerals are required plant nutrients, so is light (photosynthesis). The plants in turn are essential nutrients for a large number of animals, the herbivores. The herbivores themselves are the food intake of the carnivores.

Moreover, sometimes when plants or animals die, over time, their remains become useful fossil fuels such as coal and petroleum that are used by mankind to drive various machines and industries. Indeed, a main concern of the industrial development in the nineteenth century was the efficient conversion of heat into useful work.

What can we surmise from the well-known facts mentioned above? First, that almost all activity on the surface of the earth is ultimately driven by sunlight. One can then ask whether in the various changes that take place, from the absorption of light by plants, their growth, their consumption by animals, death, fossilization, use as fuel sources, motion of machines, is there some property that remains unchanged ? It is far from obvious that such a property exists, but careful observations and investigations by many researchers in diverse fields and over many centuries revealed that one can think of energy as an abstract quantity that *manifests in many forms and that energy undergoes changes of form in various processes* but which remains unchanged as a quantity.

Every form of energy has its specific features, and although some may at first seem unrelated, as one progresses deeper into nature's laws, the various forms of energy turn out to be interconnected in all sorts of strange and unexpected manners. The amounts of energies involved in various phenomena span a truly enormous range as illustrated in Figure 3.24.

3.10 Exercises

1. A basketball is released from a height of 3 m onto a hard surface.
   (a) What height will the ball rebound to and why?
   (b) Describe all the energy conversions that take place from the moment the ball is released to the time it finally comes to rest.

2. What is the kinetic energy of a sprinter of mass 70 kg running at 10 m/s?

3. What is the kinetic energy of a ball rolling on a boulevard at sea level that weighs 1.96 N and travels at 50 m/s? (Note: as always, SI units are used here.)

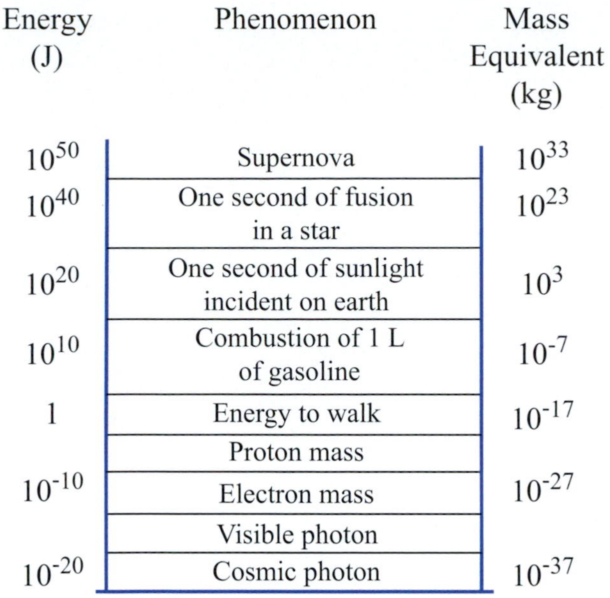

Energy (J)	Phenomenon	Mass Equivalent (kg)
10^{50}	Supernova	10^{33}
10^{40}	One second of fusion in a star	10^{23}
10^{20}	One second of sunlight incident on earth	10^{3}
10^{10}	Combustion of 1 L of gasoline	10^{-7}
1	Energy to walk	10^{-17}
	Proton mass	
10^{-10}	Electron mass	10^{-27}
	Visible photon	
10^{-20}	Cosmic photon	10^{-37}

Figure 3.24: Energy scales in nature.

4. It is sometimes stated that "wind energy is another form of solar energy."
   (a) Explain the above statement.
   (b) Give practical examples of harnessing wind energy.

5. (a) If the asteroid that hit the Yucatan Peninsula was traveling at 10 km/s when it released an energy of 10^{23} J, estimate its mass.
   (b) One kilogram of the explosive TNT releases about 4.184×10^6 J of energy. Estimate the TNT equivalent of the energy released above.
   (c) The atom bombs that destroyed the cities of Hiroshima and Nagasaki during World War II were about the equivalent of 15 kilotons of TNT each. Estimate the energy released by the meteor in terms of the equivalent number of atom bombs of this size.

6. Show that the gravitational potential energy given in Equation 3.15 can be derived from Equation 1.26. Determine the value of g in terms of the gravitational constant G. (Hint: Consider the whole mass of the earth to be as if it were located at the center of the earth.)

7. Derive Equation 3.9 from Newton's third law.

8. Derive the fact that the total momentum is conserved from Newton's third law.

9. Why are the units of k in Equation 3.18 kg s^{-2}?

10. How much work needs to be done to lift a cannonball with a mass of 20 kg up to a height of 8 m?

4: Atoms

Q - Why Are the Elements So Different from Each Other?

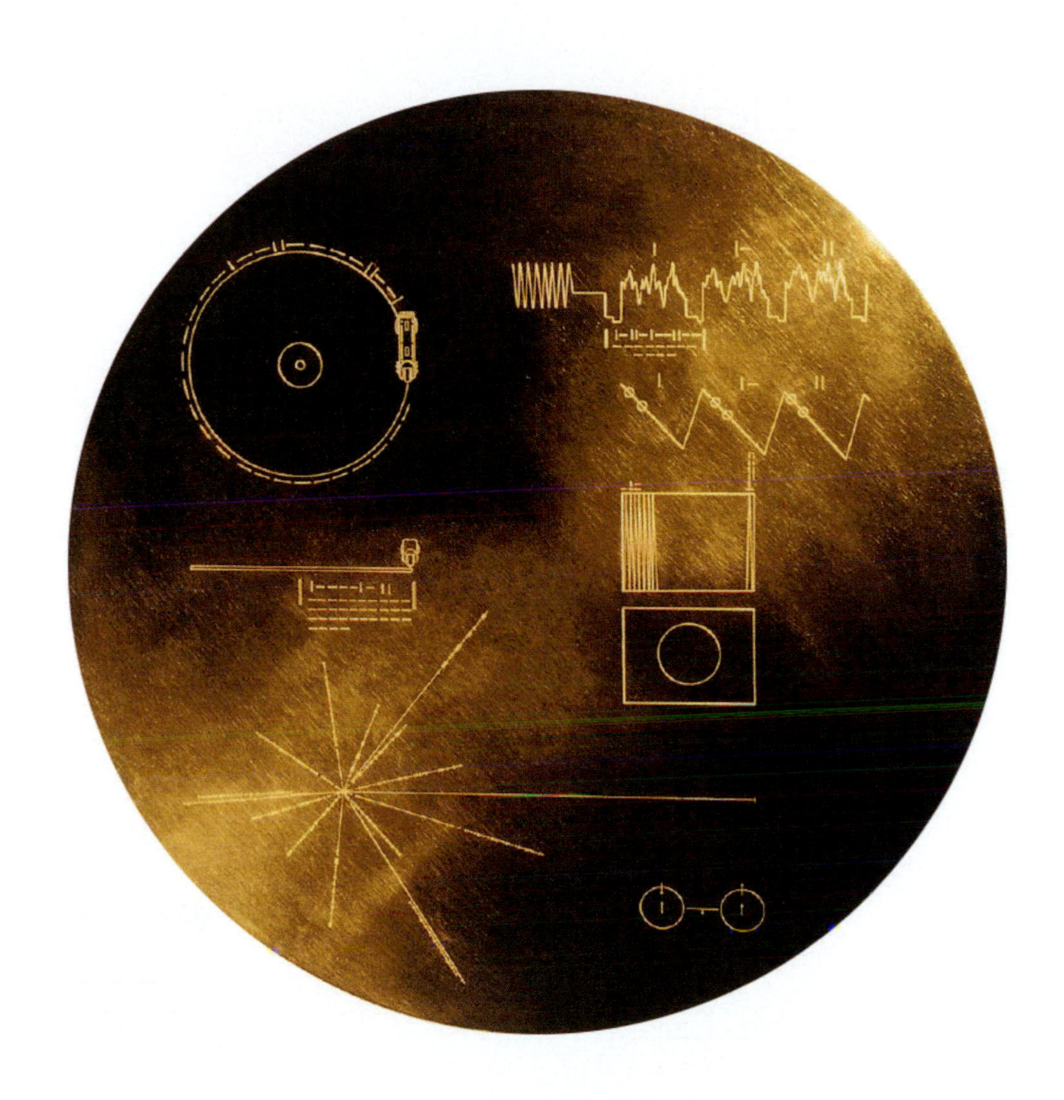

Chapter Map

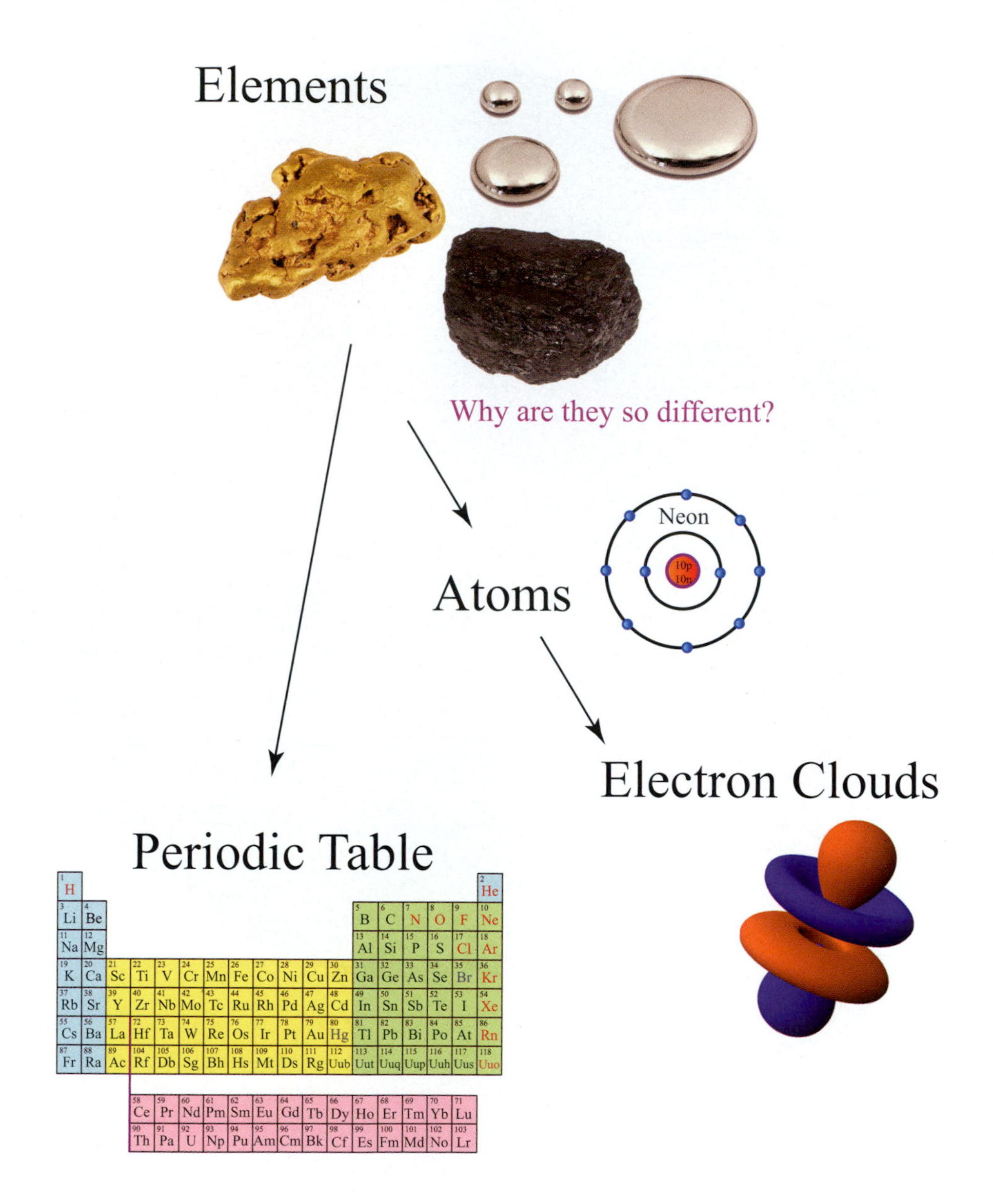

Elements
Why are they so different?
Neon
10p
10n
Atoms
Electron Clouds
Periodic Table

4.1 The question

Figure 4.1: Ultimately, what would this vase be made of?

We encounter materials every day. In fact, we are made out of materials ourselves and it is clear from direct observation that there are many different materials. Why would that be so? In order to obtain an answer, we'll first need to ask what materials are made of, a question that has intrigued mankind for thousands of years.

Indeed, ancient Indian and Greek philosophers already pondered the very same thing. For example, ultimately, what would the beautiful vase in Figure 4.1 be made of? Of course, on the macroscopic level, we know that it's clay and paint and so on. But then what are clay and paint made of? Even if there is still an obvious answer to these questions, one can keep on asking "What is this made of?" Doing so, it was already conjectured in the ancient world that materials are made up of some form of basic units.

The earliest references to the concept of atoms date back to ancient India in the sixth century BC. The Nyaya and Vaisheshika schools developed elaborate theories of how atoms combined into more complex objects (first in pairs, then trios of pairs). The references to atoms in Europe emerged a century later from Leucippus whose student, **Democritus**, systematically presented his views. Indeed, the word "atom" comes from the Greek word "uncuttable" coined by Democritus around 420 BC. He argued that if one started dividing some piece of matter repeatedly, then finally a piece would be obtained which could not be divided further. He called this smallest hypothetical piece an **atom**. In the eleventh century, the Asharite school of Arabic philosophers developed ideas about the atomic that represented a synthesis of both Greek and Indian atomism. They introduced new ideas, such as the possibility of there being particles smaller than an atom. Atoms were taken to be the only perpetual material entities and the world was taken to be contingent and lasting for only an instant.

In contrast to Democritus, other Greek philosophers such as Anaxagorus felt that matter consists of a layered structure, and that one could never reach the final constituents of matter. On the other hand, Buddhist philosophers — most notably Nagarjuna around 200 AD — reasoned that all material entities are composite. They therefore concluded that matter must be infinitely divisible, and cannot have any ultimate constituents.

The ancients did not have any experimental basis for their ideas of the atom, so it was no more than speculation. For the ancients, the atom was only an idea without any material entity that could be identified as an atom; the idea of the atom did not have specific properties that could be empirically tested. For this reason, the premodern idea of the atom cannot be taken to be a scientific hypothesis. Similarly, the Buddhist idea of the infinite divisibility of matter had no empirically testable consequence and, hence, in this case too we are not dealing with a scientific hypothesis.

Figure 4.2: The Alchemist. Painting by William Fettes Douglas.

Although not with the rigor of the modern scientific method, more or less systematic analysis of materials had been carried out by alchemists for many centuries. Already in ancient Egypt, it was discovered how to make cement and pitch (the latter used in ship building for caulking seams). In their desire to transform easily available materials into gold, medieval alchemists prepared the path for the advent of modern chemistry and little by little it became clear that all known materials are made up of a limited number of elements.

In order to explain what elements were made out of, in the latter half of the eighteenth century, chemists returned to the idea of the atom. In studying how to combine different materials, chemists found that there were two distinct forms of combination. The first case was of a mixture such that basically any amount of the two materials could be combined without leaving residues, for example as is the case when salt is combined with water. The second case was of a chemical reaction, where fixed proportions of two materials had to be combined in order to ensure that all the ingredients are used up. For example, when dissolving copper in sulphuric acid, an exact proportion of both materials was needed, and an excessive supply of, say, copper results in part of the copper being left undissolved. Chemists then went on to conclude, by the early 1800s, that there must be elementary constituents that were being combined in chemical reactions, and finally came up with the idea of irreducible elements, made up of atoms, and went on to infer that compounds are made from combinations of these atoms. Eventually, the desire to arrange the various different elements in an informative table led to the development of the periodic table.

Even when considering elements one finds that they are very different and this brings us to the chapter question: **Why are the elements so different from each other?**

Physicists in the nineteenth century took up the idea of the atom to provide a microscopic explanation of heat and thermodynamics. The kinetic theory of gases is based on the idea of a gas being composed of large collection of atoms, and quantities such as pressure and temperature are understood to result from the properties of the atoms that comprise the gas. Although many insights were gained, this didn't tell us all that much about why the elements are so different in their properties.

By the end of the nineteenth century, the following questions were left unanswered: what is the size of an atom? how much does it weigh? and most importantly, what are the laws that describe atomic phenomena? The first great success of quantum mechanics was to provide a complete theory to explain the existence, as well as the properties, of atoms.

It is clear that atoms are small (otherwise we could see them with the naked eye) and consequently macroscopic amounts of matter must contain a large number of atoms. Let us therefore first briefly digress to obtain an estimate of roughly how many atoms some visible quantity of, for example, gold contains. After that we will look at the evidence for atoms, the basic properties of the elements and how they can be arranged in a periodic table in order to obtain an idea of how the elements are different. Knowing first how they are different will then help us find out why they are different so that we can answer the chapter question.

4.2 Avogadro's number and atomic masses

Any object that can be directly perceived by our five senses consists of a large collection of atoms or molecules, called a macroscopic collection. **Avogadro's number**, experimentally determined to be to 6.022×10^{23}, is a typical measure of how many atoms are contained in a macroscopic object, for example in a tablespoon of salt. Avogadro's number is defined as the number of atoms in 12 g of carbon. Of course, one may ask why was 12 g of carbon chosen? The reason is that the nucleus of carbon has six protons and six neutrons and therefore has an atomic mass of approximately 12 u. Then, if one has as many grams of a substance as the numerical value of its atomic mass indicates, one will have approximately N_A atoms or molecules; the actual number needs to take the masses of the electrons into account as well as the binding energy of the protons and neutrons inside a nucleus. For example, 1 g of hydrogen, 56 g of iron, 16 g of oxygen, and 18 g of water (2 g from hydrogen and 16 g from oxygen) all have Avogadro's number of molecules.

For convenience, the quantity of a substance containing exactly N_A atoms or molecules is called a **mole** (mol). In other words, the mole is defined as an amount of substance with mass in grams numerically equal to its atomic (molecular) mass. Thus simply by definition, 1 mol of any substance contains the same fixed number, Avogadro's number, N_A, of atoms (molecules).

To get an idea of how large N_A is, note that 1 mol of tennis balls has a volume equal to the volume of the moon!

Here is one way of obtaining a crude estimate of Avogadro's number: consider 1 mol of water. Since the molecular mass of H_2O is $2 + 16 = 18$, this means that 1 mol of water has a mass of 18 g. The density of water is 1 g/cm^3, so the volume occupied by 1 mol of water is 1.8×10^{-5} m^3. Suppose each molecule of water has a size of the order 10^{-10} m in radius, which is the approximate size of a hydrogen atom and discussed in Chapter 23. Assuming a tight packing of N_A such water molecules, implies

$$1.8 \times 10^{-5} \sim N_A \frac{4}{3}\pi \left(10^{-10}\right)^3 \tag{4.1}$$

giving the very rough estimate $N_A \sim 10^{25}$. This number is a bit on the large size because we assumed a tight packing of the molecules which of course cannot be true in a liquid, but nevertheless the huge size of the number is the main indication we wanted to see. A more careful calculation shows that to obtain the correct value for Avogadro's number, each water molecule occupies a cube having a volume of about 7×10^{-20} m^3.

Thus far we have discussed how scientists came to the conclusion that elements consisting of atoms are a sensible hypothesis. However, we have not seen any direct evidence yet. Let us therefore now briefly look at the experiments that established the atomic theory.

4.3 Evidence for atoms

By the nineteenth century chemists had observed that some substances could be broken down into simpler chemicals while others maintained their uniqueness. The seemingly indivisible substances were called **elements**. It was also noticed that the elements always combine in the same integer ratios to form compounds, for example eight parts by weight of oxygen combines with one part by weight of hydrogen to form water. These empirical facts led **John Dalton** in 1808 to propose the modern theory of atoms: Atoms were considered to be indivisible "billiard ball" type building blocks of the elements. Atoms then combined to form molecules of compound substances. However, the arguments of the chemists, though persuasive, were not convincing to everyone as they were deemed too indirect.

Further indirect evidence for atoms was provided by the observations of the botanist **John Brown** in 1827. While watching pollen grains suspended in water through his microscope, he noticed that they engaged in complicated erratic motion, apparently wandering about at random. He eliminated the possibility of some living organism in the pollen as the cause of the motion by repeating his observations with varied inorganic dust particles. It was thus confirmed that something *in the water* was responsible for the motion of the suspended particles. This type of motion is nowadays called **Brownian motion** (see also Chapter 13).

Many believed that Brownian motion was due to molecules of water hitting the suspended particles, but it was only in 1905 that **Albert Einstein** gave a mathematical description of the phenomenon. Einstein's quantitative predictions of Brownian motion were verified in careful experiments conducted by **Jean Perrin** in 1909. As a result of those experiments Perrin obtained a value for Avogadro's constant (N_A), the number of atoms in 1 mol of a substance. The number agreed with values obtained previously using different methods and were a very strong confirmation of the atomic nature of matter. Perrin was awarded the Nobel Prize for his work in 1926.

While there was a significant amount of (indirect) evidence for atoms as tiny little billiard balls, this picture does not explain their properties. How would these billiard balls be different from each other to make all the materials? How could the various properties of the 76 elements known in 1897 be ascribed to smooth round billiard balls of various weights? Surely, weight alone could not explain the differences. Perhaps atoms aren't really like billiard balls at all! In 1897, while studying cathode rays, John Thomson discovered the electron. He then theorized atoms to look somewhat like depicted in Figure 4.3 where the electron swims in a continuous sea of positive charge.

That was a good start. In 1911, however, Ernest Rutherford and his students Hans Geiger and Ernest Marsden performed an experiment that would completely change our perception of the structure of atoms (this experiment is often referred to as the Rutherford or gold-foil experiment but sometimes also as the Geiger-Marsden experiment). First, they took the radioactive compound $RaBr_2$ to obtain positively charged **α-particles** (nowadays we know that an α-particle is the same as a helium nucleus but that was not known at the time). Then they shot a beam of these α-particles at a thin gold foil as shown in Figure 4.4. If an atom would have been accurately described by the Thomson model, then one would expect most of the α-particles to go straight through the foil and some to be slightly deflected. The reason for this expectation is based on elementary mechanics: It was known that α-particles are relatively heavy and that electrons are very light and consequently, just like throwing a very heavy ball at a very light ball will push the light ball aside and at worst deflect the heavy ball a bit, the electrons should affect the α-particles only a little bit if the Thomson model is correct.

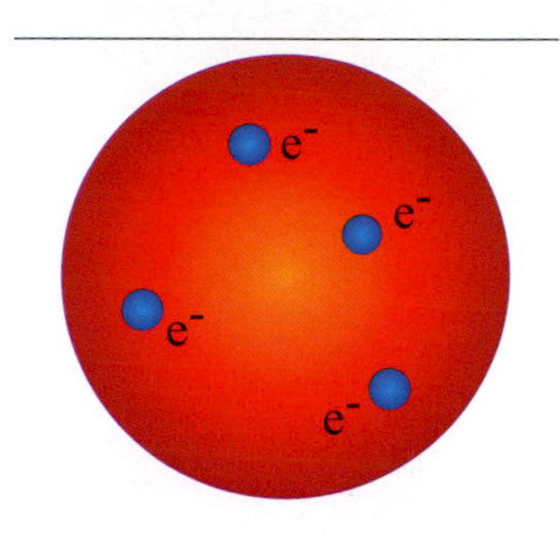

Fig. 4.3: Thomson's vision of an atom (Thomson model, also called plum pudding model).

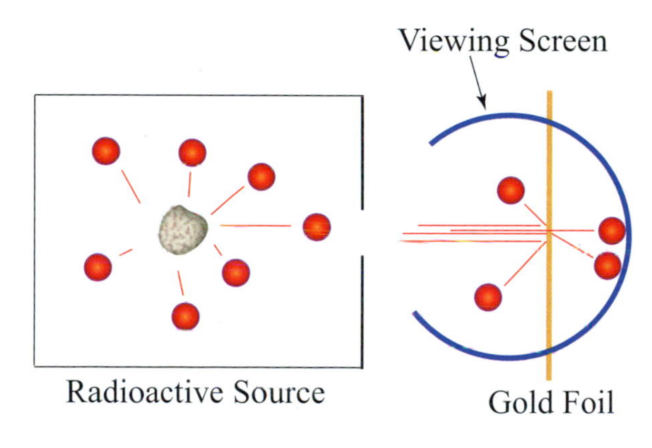

Figure 4.4: Rutherford's experimental setup.

The outcome of the experiment was nothing but shocking to Rutherford and his students. They made the following startling observation: Although, as expected, most of the particles went straight through the foil, those that did not go through were scattered over a very large range of angles. Indeed, some of the α-particles were basically scattered back. In order to explain his experimental results, Rutherford proposed that an atom consists of a very small positively charged nucleus with negatively charged electrons circling around it leading to the "planetary" **Rutherford model** of the atom (see also Figure 4.6 below).

After Rutherford, the existence of atoms was established beyond any reasonable doubt. However, even though the planetary model was very convincing with regards to explaining experimental observations, it was in conflict with the classical laws of electromagnetism which predict that an accelerating charge must radiate energy. Since the electron changes its direction all the time in its orbit around the nucleus, it is accelerating and must therefore radiate all the time. An estimate shows that the electron would radiate off all its kinetic energy in a very short time and spiral into the nucleus, causing the atom to collapse. Since this does not happen, something was clearly wrong with the Rutherford model.

The Rutherford model was replaced with an improved model during the early stages of the development of quantum mechanics by the Danish physicist Niels Bohr (see also Chapter 23) but a fully consistent picture only emerged much later.

In many cases, modeling an atom as a billiard ball is quite effective. Although it does not explain why different elements have different properties, it can account for some of their behaviors. Let us therefore now look at the billiard ball model of atoms.

4.4 Atoms as billiard balls

As we know from the game of billiards, when a billiard ball moves on a billiard table there is no interaction between the balls until they collide, when there is an apparently instantaneous interaction that leads to the balls bouncing off each other. Would this be similar for atoms? A key difference between atoms and billiard balls is that billiard balls have no electrical charge distribution while atoms, though electrically neutral at a distance large compared to their size, consist of a positively charged nucleus surrounded by negatively charged electrons.

The electrical structure of the atom has direct impact on how it interacts with other atoms. When two atoms come very close, there must be a strong electrostatic repulsion between the negatively charged electrons. One would then expect that when the two atoms are not too close, the repulsion to be still there but not as strong as when they are very close, just like in the case of macroscopic charges. Fascinatingly enough, this is not the case at all! When two atoms are at intermediate distances from each other, their electrons will synchronize to some degree with the consequence that the electrons start to "feel" the other atom's positive nucleus. Since positive and negative charges attract, the atoms will then be (slightly) attracted to each other.

The interaction between two atoms can be described quite well by the **Lennard-Jones (LJ) potential** depicted in Figure 4.5 and given by:

$$U_{LJ}(r) = 4\epsilon\left[\left(\frac{\sigma}{r}\right)^{12} - 2\left(\frac{\sigma}{r}\right)^{6}\right] \tag{4.2}$$

where the two atoms' positions are r_1 and r_2 so that they are at a distance $r = |r_1 - r_2|$ apart; ϵ is a measure related to the depth of the potential well and σ a measure related to the hard sphere radius of an atom; both parameters ϵ and σ are determined by experiment. There is a minimum value in the interatomic potential at a distance of r_0 from the atom.

The resulting force is given by the derivative of the potential with respect to r as (see also Equation 3.39 on p. 63)

$$F_{\mathrm{LJ}}(r) = -\frac{dU_{\mathrm{LJ}}(r)}{dr} = 48\epsilon\left[\frac{\sigma^{12}}{r^{13}} - \frac{\sigma^{6}}{r^{7}}\right]. \tag{4.3}$$

The two atoms are in equilibrium when the force they exert upon each other is zero, or what is the same thing, when the potential is at a minimum. Hence the equilibrium separation r_0 between the atoms can be obtained from Equation 4.3 as

$$F_{\mathrm{LJ}}(r_0) = 0 \Rightarrow r_0 = \sigma.$$

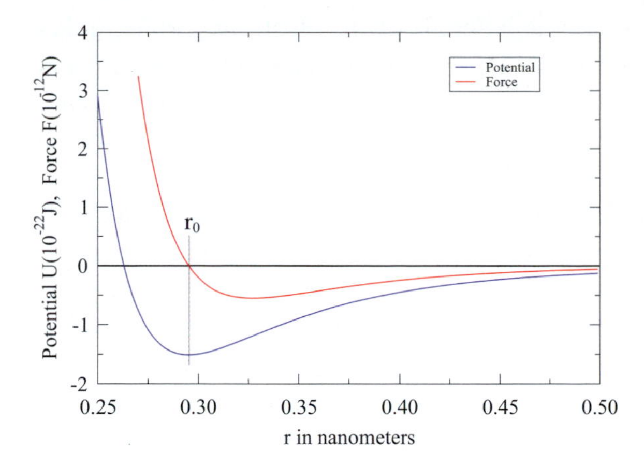

Figure 4.5: The Lennard-Jones potential and its associated force for typical values of $\epsilon = 1.51 \times 10^{-22}$ J and $\sigma = 0.29$ nm.

A negative value of the potential means that the atoms are in a bound state; a negative value of the force reflects that the force is attractive since it will tend to bring the atoms together; a positive value of the force similarly signifies a repulsive force. From Figure 4.5 one can see that the force is attractive as long as the atoms are separated by a distance larger than r_0; if one tries to push the atoms to have a separation less than r_0, the repulsive force rapidly increases. It follows from Equation 4.3 that equilibrium separation r_0 equals the parameter σ.

From Figure 4.5 it can be seen that pushing two atoms closer together than this equilibrium separation rapidly starts to require a lot of energy and σ therefore effectively acts as the radius of a billiard ball-like atom. Therefore, σ is often used as a measure of the classical size of an atom and referred to as the **Lennard-Jones radius**.

Experiments show that for typical atoms the LJ-radius is around 1 to 5Å (Å≙ Angstrom = 10^{-10}m). For example, the argon atom has a LJ radius of 3.5Å. While it is useful for obtaining an idea of the size of an atom, and sufficient as an approximation in a great number of situations (especially when considering gases), there are also many circumstances when the classical radius σ cannot be used.

The size of an atom is around 10^{-10} m.

For example, at very low temperatures and high densities the atoms are much closer to each other than the radius σ, and hence treating atoms as hard spheres is not valid — one needs to use quantum mechanics. At the other extreme of very high temperatures, the inner structure of the atoms, composed as it is out of a nucleus and electrons, is excited and hence treating the atom as a billiard balls again breaks down. Outside these two domains, however, atoms behave pretty much as hard spheres.

In conclusion, as long as air is at temperatures and densities that are not very high or very low and the atoms of the air are not squeezed together closer than the distance of the LJ radius, we can treat the atoms as classical billiard balls.

4.5 Atoms as solar systems; nucleus and electrons

From Rutherford's experimental results, we have found that an atom is an electrically neutral object composed of a heavy, positively charged nucleus, surrounded by a number of negatively charged electrons. As stationary electrons would be attracted by the positively charged nucleus and then fall into it, Rutherford theorized that the electrons would somehow be circling around the nucleus so that the centrifugal force would counteract the electrical attraction as shown in Figure 4.6.

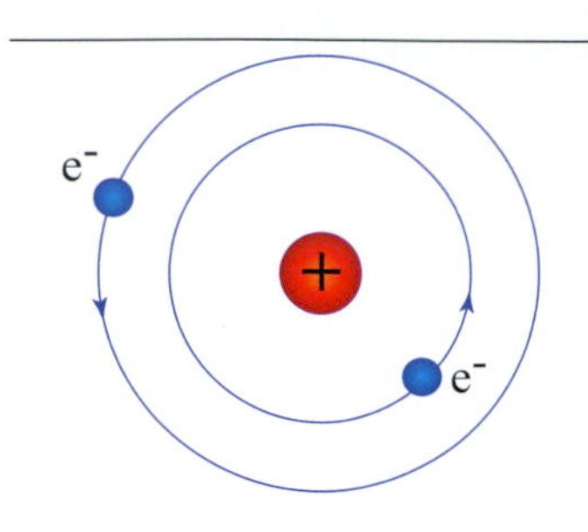

Fig. 4.6: Rutherford's planetary model of the atom.

Although Rutherford did not work out any details of the electrons' orbits, in his 1911 paper, he did refer to a model by the Japanese physicist Hantaro Nagaoka in which there is "a central attracting mass surrounded by rings of rotating electrons". Hence, in the Rutherford model, one imagines the nucleus being like the sun and the electrons like planets in various orbits.

We have found the (classical) LJ-size of the atom. How big is the nucleus? A rough idea for the size of the nucleus can be obtained by the following consideration based on Rutherford's original paper: If an α-particle heads straight for the nucleus, it will come to a halt when the potential energy from the electric repulsion (see also Equation 9.4) is exactly equal to the kinetic energy of the α-particle. That point must be quite close to actual radius if the nucleus is somewhat billiard ball like.

Let the charge and mass of the α-particle be q_α and m_α respectively and the initial velocity of the α-particle be v_α; let the charge of the nucleus be q_Z. The smallest distance of the α-particle to the nucleus is given by the distance at which the potential energy of the Coulomb potential between the α-particle and the nucleus is approximately equal to the kinetic energy of the incoming α-particle. To approach the nucleus any closer the α-particle would need greater energy. Hence we have,

$$\frac{1}{2}m_\alpha v_\alpha^2 = k_e \frac{q_\alpha q_Z}{r}. \tag{4.4}$$

Estimating the velocity and mass of the α-particle as 2×10^7 m/s and 6.7×10^{-27} kg respectively, and using the fact that the charge of the α-particle is minus two electron charges while the total charge of the gold nucleus is equal to minus 79 electron charges, we obtain for r the value 2.7×10^{-14} m (the charge of the electron is -1.6×10^{-19} C). What we have obtained from the example of the α-particle is only an upper bound on the size of the nucleus since an α-particle with higher velocity could approach the nucleus even closer. Careful experiments show that the size of the nucleus is around 10^{-15} m, but the simple calculation using α-particles does give a reasonably good idea of the nucleus' size.

Although the size of the electron is still unknown, it is ascertained by experiments to be smaller than 10^{-18} m. One can see that an atom, having a size of approximately 10^{-10} m, is mostly empty space as its radius is about 100,000 times larger than that of the nucleus. Hence, when studying the atom, it is often a very good approximation to consider both the nucleus and the electron to be point-like, and for example, the Coulomb potential between the nucleus and the electrons is written as if they were point-like particles.

The size of a nucleus is around 10^{-15} m.

Just how empty the atom is can be visualized like so: Suppose a nucleus is the sun and the electron is the earth. The diameter of the sun is roughly 10^9 m. If the distance of the earth to the sun is 100,000 times the diameter of the sun, then the diameter of our earth's orbit would be 10^{14} m. In fact, however, the diameter is only around 3×10^{11} m, with Pluto's being about 10^{13} m. In other words, if our earth would be as far from the sun as the (classical) electron is from the nucleus, it would be far beyond Pluto!

Most phenomena that we encounter in chemical, biological, thermal and macroscopically observed processes result solely from the behavior of an atom's electrons. The reason is the following: to affect the state an electron belonging to an atom may take only a few electron volts (eV) of energy (1 eV $\approx 1.602 \times 10^{-19}$ J). Energies in this order of magnitude can be supplied by such simple acts as shining light on the material, making it react chemically with other materials or by hitting it with a hammer. To alter the state of the nucleus on the other hand takes at least a million electron volts; such high energies are not available by heating or hitting or chemical reactions. In fact millions of electron volts of energy per nucleus are generally only available inside nuclear reactors and nuclear bombs.

It should also be noted that almost all of the mass of an atom is in its nucleus. Practically speaking, the mass of the electrons is therefore often ignored unless exact measurements or calculations are needed.

Up to this moment, we have found that atoms consist of a very small nucleus surrounded by electrons. Different kinds of atoms, e.g. hydrogen and gold, have different masses and this mass is mostly in the nucleus. Then, how is this mass distributed in the nucleus? Are the nuclei of different kinds of atoms billiard balls of different sizes, or perhaps billiard balls of different densities? Would they be built up from something?

4.6 Nuclei

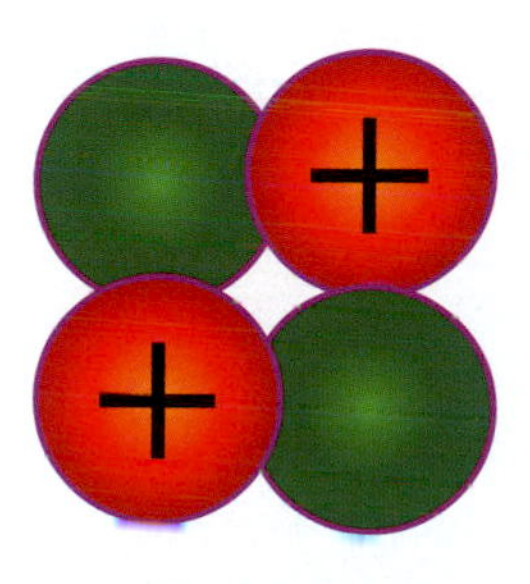

Figure 4.7: Nuclei consist of protons and neutrons.

In 1918, it was Rutherford again who carried out a breakthrough experiment. This time he shot α-particles at nitrogen gas and found that his detector displayed results characteristic of hydrogen (which by that time was known to be the lightest element). As he could not find any other source for the hydrogen, he correctly concluded that its source was the nitrogen gas and that the nitrogen nucleus therefore contains the hydrogen nucleus. From this stunning discovery he then concluded correctly that the hydrogen nucleus is a positively charged elementary particle now called a **proton**. In the process of his experiment, Rutherford changed nitrogen into oxygen (which has one proton less) and was thus in a sense the first "true" alchemist. Well, if part of the nitrogen nucleus is the hydrogen nucleus, what would the rest be? The most straightforward answer would be "more hydrogen" nuclei. Indeed, that is almost true, except for there being only about half as many protons as one would expect from their weight. As it turned out, the other half are **neutrons**, elementary particles with properties very similar to protons but without an electrical charge.

As atoms are electrically neutral in their normal state (if this were not so a gold bar would be charged and we know that we don't get zapped if we touch it), it is therefore clear that the number of protons and electrons in an atom exactly match. As a first go, one could therefore say that the identity of an element is determined by the number of electrons or equally by the number of protons. An obvious idea then would be to somehow arrange the elements in a table by the number of protons

Ernest Rutherford

Born in New Zealand in 1871, Ernest Rutherford was a British scientist who played a pivotal role in the development of modern physics. After his undergraduate studies in New Zealand he went for postgraduate studies to the University of Cambridge in England where he made his first important contribution by distinguishing and naming the three best-known forms of radiation: α, β and γ rays.

Figure 4.8: Ernest Rutherford.

At age 27, he was appointed as chair of physics at McGill University in Canada where he showed that radioactivity was the result of decaying atoms — work that earned him the 1908 Nobel Prize in chemistry. In the meantime he had returned to England in 1907 to take up the position of chair of physics at the University of Manchester where he carried out seminal experiments which discovered the structure of atoms.

He showed this by having his students Hans Geiger and Ernest Marsden bombard a thin gold foil with α-particles. The only way the experimental results could reasonably be described was by introducing a mostly empty atom with a heavy nucleus at the center. Rutherford then proposed a planetary model (often called the Rutherford model) for the structure of the atom.

In the Rutherford model, a heavy positively charged nucleus sits at the center of the atom and a number of negatively charged electrons circle this nucleus bound to it by electrostatic forces. As electromagnetism was well understood when Rutherford did his experiments and introduced his atomic model, it was clear immediately that the model had a fatal flaw: it would collapse since a circling electric charge radiates and hence loses energy. Still the model did capture essential experimental results and hence it became a key question as to why atoms are stable. Answering this question was one of the greatest successes of quantum mechanics.

In science, a successful approach often begs for repetition. Having found a massive nucleus at the center of the atom leaves open the question of whether the nucleus again has a structure or not. When Rutherford bombarded nitrogen gas with α-particles in 1918, he found that it did since his experiment resulted in the production of hydrogen. As hydrogen was known to Rutherford to be the lightest element, he correctly concluded that the nitrogen nucleus contains the hydrogen nucleus (i.e. the proton). With the production of hydrogen from nitrogen, nitrogen is transformed into oxygen and Rutherford thus became the world's first true alchemist.

Another remarkable aspect of Rutherford's career is the large number of Nobel laureates among his students.

Radioactivity

Figure 4.9: Henri Becquerel discovered radioactivity in 1896.

Protons, being positive, electrically repel each other but they are held together by the strong force that only affects certain subatomic particles such as protons and neutrons. The strong force of protons alone, however, is not quite strong enough to overcome the electric repulsion but with the help of some neutrons it is, as long as the nucleus is not too big. For atoms to have stable nuclei, the ratio between protons and neutrons needs to have a certain value or be in a (narrow) range. In the case of the lighter elements, the ratio is 1:1 while in the case of the heavier elements the ratio approaches roughly 5:8.

If an atom has a ratio between protons and neutrons such that the nucleus is not stable, it will tend to a stable state by the emission of particles. This process is called **radioactivity**. There are three main types of emitted particles, α-particles which are the same as helium nuclei, β-particles which are electrons, and γ-particles which are photons. The emission of a particle from an unstable nucleus is a random process and it is impossible to predict exactly when an emission occurs. However, for a given time interval, this probability for decay is a constant proportion of the not-yet-decayed atoms. In other words, the number of not-yet-decayed atoms N is described by

$$\frac{dN}{dt} = -\lambda N \Rightarrow N(t) = N_0 \exp(-\lambda t) \tag{4.5}$$

with λ an isotope dependent constant.

In practice, it is useful to measure how long it takes for half of the atoms to decay, a time called the half-life. From Equation 4.5 it then follows that the half-life time is $t_{\frac{1}{2}} = \frac{\ln 2}{\lambda}$.

Radioactivity can change the number of protons and hence transform one element into another. But if this is the case, then why are there (naturally occurring) radioactive elements in the first place? Shouldn't they all long have been transformed into stable elements? There are two reasons for this: first, a radioactive element like uranium-238 (i.e. the uranium isotope with 146 neutrons) has a half-life of 4.5 billion years, a very long time, and second, some radioactive isotopes like carbon-14 are constantly produced (carbon-14 is produced in the upper atmosphere from nitrogen with the help of cosmic rays).

Figure 4.10: Marie Curie discovered the elements polonium and radium.

An interesting application of carbon-14 is carbon dating. Carbon-14 has a half-life time of about 5730 years and has a relatively well-known concentration in the atmosphere but is very rare in the earth's crust. Since plants use up atmospheric carbon during photosynthesis but obtain very little carbon otherwise, the concentration of carbon-14 in plants is roughly equal to that of the atmosphere. Consequently animals that eat those plants also have about the same concentration of carbon-14. Once a plant or animal dies, no new carbon-14 is added so that the amount present in its remains gives an indication of how old the remains are.

or electrons. Since materials can be charged, however, it is known that electrons can sometimes escape from or hop onto an atom so that its total charge changes. Electrons are therefore not quite suitable for this table and hence protons are a somewhat better choice.

The table in which the elements are listed according to their number of protons, grouped such that their chemical properties are similar, is called the **periodic table** of elements. Now once protons were known, the idea of a table of elements looks pretty straightforward but in fact it was first proposed long before the proton's discovery. This reflects the enormously important and deep relationship between the properties of elements and their number of protons. It is, however, important to realize that the mass difference as such does not explain the different chemical properties. The mass difference between elements does suggest the order in which the elements are arranged in the table but it does not explain why this table is "periodic".

It is interesting to note that the notion of the element is founded on a conservation law. Under normal circumstances, i.e., those encountered in general chemistry, elements can neither be transformed into each other, created or destroyed. The conservation of elementary constituents is one of the most important principles of chemistry. Since it is such an important part of modern science and since there may be more clues as to why there are different kinds of materials, let us now have a closer look at the periodic table.

4.7 The periodic table

The first modern type of periodic table was drawn by the Russian chemist **Dimitri Mendeleyev** who arranged the elements according to their masses and some of their chemical properties. One of Mendeleyev's important insights was that the most logical way for the elements to be arranged meant that there were gaps in his periodic table. He then drew the correct conclusion that the gaps must correspond to unknown elements thus making one of science history's greatest predictions: The existence of hitherto unknown elements. Figure 4.11 shows his original publication of 1869. In 1871, Mendeleyev published an improved version of the table which looks very similar to the current one.

The nucleus of an atom, in general, consists of Z number of protons with net positive charge of $+Z$ and an approximately equal number of electrically neutral neutrons. The number of neutrons is larger than the number of protons for many nuclei, and **isotopes** denote atoms of the same element whose nuclei have different numbers of neutrons. To indicate which isotope one is dealing with, the name of an element is often followed by a number indicating the total number of protons and neutrons: e.g carbon-14 has six protons and eight neutrons. When the chemical symbol is used, the number of protons and neutrons is indicated with a superscript prefix: e.g. ^{14}C. The nucleus is surrounded by Z number of electrons with a total electrical charge of $-Z$, resulting in electrically neutral atoms. The type of atom (i.e. which element it is) is determined by the number of protons in the nucleus.

The number of stable elements found in nature is about 90, and it is estimated that the maximum number of elements that can exist is around 126. If one studies the physical and chemical properties of the various elements, a pattern is seen to emerge, wherein the physical and chemical properties of the elements seem to approximately *repeat* themselves. The word "periodic" expresses this repetitive (periodic) structure present in atoms.

Ueber die Beziehungen der Eigenschaften zu den Atomgewichten der Elemente. Von D. Mendelejeff. — Ordnet man Elemente nach zunehmenden Atomgewichten in verticale Reihen so, dass die Horizontalreihen analoge Elemente enthalten, wieder nach zunehmendem Atomgewicht geordnet, so erhält man folgende Zusammenstellung, aus der sich einige allgemeinere Folgerungen ableiten lassen.

			Ti = 50	Zr = 90	? = 180
			V = 51	Nb = 94	Ta = 182
			Cr = 52	Mo = 96	W = 186
			Mn = 55	Rh = 104,4	Pt = 197,4
			Fe = 56	Ru = 104,4	Ir = 198
			Ni = Co = 59	Pd = 106,6	Os = 199
H = 1			Cu = 63,4	Ag = 108	Hg = 200
	Be = 9,4	Mg = 24	Zn = 65,2	Cd = 112	
	B = 11	Al = 27,4	? = 68	Ur = 116	Au = 197?
	C = 12	Si = 28	? = 70	Sn = 118	
	N = 14	P = 31	As = 75	Sb = 122	Bi = 210?
	O = 16	S = 32	Se = 79,4	Te = 128?	
	F = 19	Cl = 35,5	Br = 80	J = 127	
Li = 7	Na = 23	K = 39	Rb = 85,4	Cs = 133	Tl = 204
		Ca = 40	Sr = 87,6	Ba = 137	Pb = 207
		? = 45	Ce = 92		
		?Er = 56	La = 94		
		?Yt = 60	Di = 95		
		?In = 75,6	Th = 118?		

1. Die nach der Grösse des Atomgewichts geordneten Elemente zeigen eine stufenweise Abänderung in den Eigenschaften.

2. Chemisch-analoge Elemente haben entweder übereinstimmende Atomgewichte (Pt, Ir, Os), oder letztere nehmen gleichviel zu (K. Rb, Cs).

3. Das Anordnen nach den Atomgewichten entspricht der *Werthigkeit* der Elemente und bis zu einem gewissen Grade der Verschiedenheit im chemischen Verhalten, z. B. Li, Be, B, C, N, O, F.

4. Die in der Natur verbreitetsten Elemente haben *kleine* Atomgewichte

Figure 4.11: Periodic table as published by Mendeleyev in Zeitschrift für Chemie, 1869.

The periodic table of elements is given in Figure 4.12. Chemically similar elements are arranged in columns called groups that are numbered from 1 to 18. The periodicity of the table is in the horizontal direction. The first problem that needs to be addressed is why atoms appear to have a periodic structure. If one goes down vertically along any column, the elements have similar chemical properties. What would be the fundamental reason for that?

Well, we already suspect that the nucleus does not play a role in general chemistry (see Section 4.5) so it is a sensible idea to consider the other atomic constituent, namely the electrons.

A first crude approximation to the multi-electron system of an atom is to ignore the interactions among the electrons. We assume that the multi-electron system is such that the electrons interact only with the nucleus, and not with each other. In a sense this is similar to the situation of the hydrogen atom.

For the hydrogen atom, consisting of one proton and one electron, the Danish physicist Niels Bohr found an expression for the energies the electrons can attain that describes the experimental findings rather well. It is given by

$$E_n = -\frac{1}{n^2} E_{\text{Rydberg}} \tag{4.6}$$

with

$$E_{\text{Rydberg}} = \frac{m k_e^2 e^4}{2\hbar^2} = 13.6\,\text{eV}. \tag{4.7}$$

Why is there a minus sign on the right hand side of Equation 4.6? First, we need to consider the kinetic and electrostatic energies assigned to a nonmoving free electron that is not under the influence of any electric charges, infinitely far away from a proton. It would seem reasonable to set its energy to zero. If an electron is not quite infinitely far from a proton, due to the attraction between positive and negative charges, energy is lowered when an electron approaches a proton. Putting it the other way around, if we have an electron orbiting a proton and try to remove this electron from the vicinity of a proton, energy is required. Since adding the removal energy implies a positive sign, and adding this energy leads to an electron with zero kinetic and electrostatic energy, the electron's energy must have been negative when it was still near the proton. This is very much in line with the general notion in science that systems tend to their lowest available energy state. What is important to remember is the reference point.

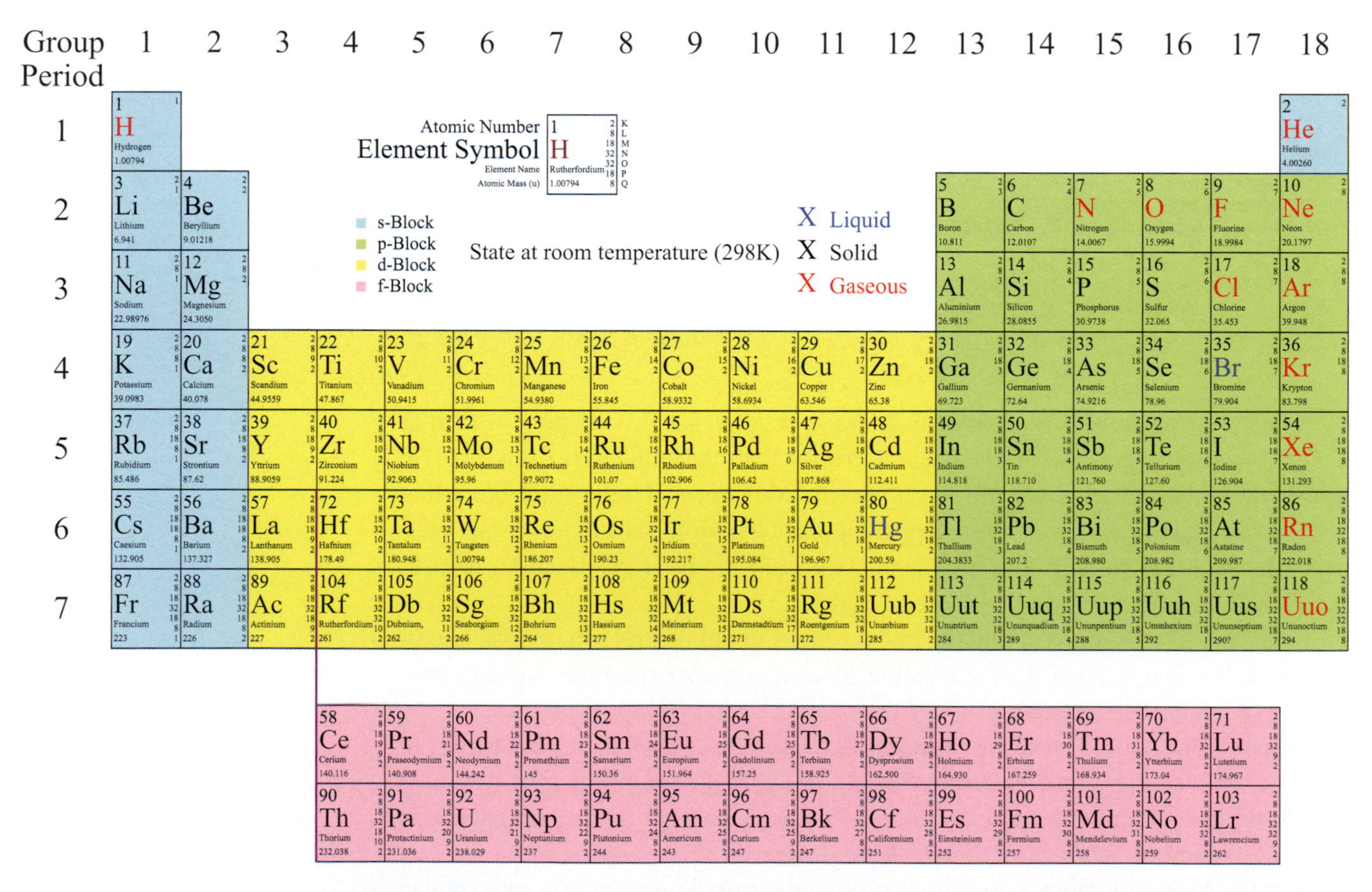

Figure 4.12: The periodic table of elements. The colors indicate the type of electron in the highest energy state.

In the case of the electron the reference point is the free electron which is why the electron near the proton has a negative energy. In the case of gravitational potential energy, the reference point is the surface of the earth and hence a ball on a mountain top has a higher potential energy than a ball in the valley. In general, a particle in a bound state, as is the case of the electron in an atom, is assigned negative energy.

In the approximation that we are considering, the energies of the electrons in the multi-electron system are given by Bohr's expression for energy, with the only difference being that the charge of the proton in the hydrogen atom is replaced by Z, the charge of the nucleus in question. Hence, from Equations 4.6 and 4.7, we have

$$E_n = -\frac{E_Z}{n^2} = -13.6\frac{Z^2}{n^2}\text{eV} \; ; \; n = 1, 2, ...\infty \tag{4.8}$$

$$E_Z = Z^2 E_{\text{Rydberg}} = \frac{mk_e^2 Z^2 e^4}{2\hbar^2} = 13.6 Z^2 \,\text{eV}. \tag{4.9}$$

In nature, systems always try to be in the lowest energy state possible. Consequently, one might expect that all the electrons are in the lowest orbit, namely the $n = 1$ orbit. This, however, turns out not to be the case. Due to a special quantum mechanical property called the Pauli exclusion principle, no two electrons can be in exactly the same state. Here, a state refers to the values of the properties that are necessary to describe an electron bound to an atom. One such property is the energy but there are others outlined below. As a consequence, it turns out that only two electrons can occupy the lowest energy state with the other electrons occupying the states with higher energy. Note that the lowest energy state still has two possible values for the so-called spin — hence having two electrons in the lowest energy state does not violate the Pauli exclusion principle.

The main energy levels are labeled by the number n and referred to as **shells** in analogy to Rutherford's planetary model of the atom, but it is important to realize that this is only a mnemonic — in quantum mechanics, electrons are characterized by a set of numbers called *quantum numbers*. Although these numbers have names that are sometimes based on the classical notion of the atom, this does not reflect the actual quantum mechanical picture. Each shell can accommodate $2n^2$ electrons. These $2n^2$ electrons are then further characterized by three more (quantum) numbers. First, there is a number called spin which has two values, $+1/2$ or $-1/2$ (often referred to as up and down). There is also a so-called azimuthal quantum number ℓ, which is the quantum mechanical analogue of angular momentum. ℓ can take the values $\ell = 0, 1, \ldots, n-1$. Finally, there is the so-called magnetic quantum number m (also denoted as m_ℓ) that can take integer values from $-\ell, -\ell+1, \ldots, 0, \ldots, \ell-1, \ell$, with a total of $2\ell+1$ possible values.

Hence one can see that the total number of energy states available for a given n is

$$\begin{aligned}\text{Number of quantum states for } n &= 2\sum_{\ell=0}^{n-1}(2\ell+1)\\ &= 2\sum_{k=1}^{n}(2k-1) = 2n^2,\end{aligned}$$

where we have used Equation 2.4 derived earlier to do the summation.

An atom is in its lowest energy state when all the electrons in the atom are distributed in the lowest energy states available to them as per the Pauli exclusion principle. Since we know that successive elements have more and more protons, from the fact that in their usual state atoms are electrically neutral, it is clear that successive elements have more and more electrons. How would those electrons fill up the consecutive shells? This is what we investigate next.

4.8 Shell filling

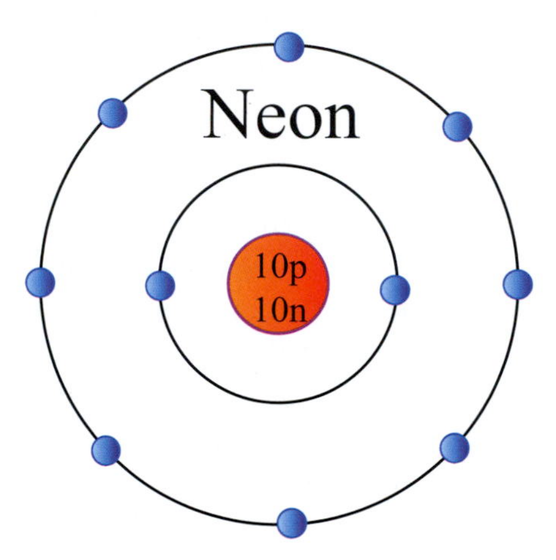

Figure 4.13: Arrangement of 10 electrons in two shells for neon.

If we want to know how the electrons fill up the shells, it is a good idea to look at the periodic table. After all, in the periodic table, elements are grouped both with regards to their number of protons (and consequently electrons) as well as their chemical properties. The chemical properties are almost exclusively determined by the outermost electrons and so the periodic table may be a good indicator of how the shells are filled.

Of course, initially we should suspect that the electrons do what appears to be the easiest: fill up the shells one by one. To check whether this is so, let us look at the noble gases (He, Ne, Ar, Kr, Xe and Rn). These can be found in the right-most column of the periodic table.

For helium, we have two protons and two electrons. So its $n = 1$ shell is completely filled up (recall that a shell can have $2n^2$ electrons in total). The next noble gas is neon depicted in Figure 4.13, with its 10 electrons distributed in two energy shells given by $n = 1$ (giving $2n^2 = 2$ electrons) and $n = 2$ (giving $2n^2 = 8$ electrons). This time the $n = 2$ shell is completely filled up. Experimentally, it is easy to establish that the noble gases do not generally react (this is why they are called noble gases in the first place), and hence we can suspect that inert atoms have full shells.

Then we would expect that the next noble gas (argon) has a full $n = 3$ shell with 18 electrons. But that is not the case! Argon has 18 protons and hence 18 electrons: two in the first shell, eight in the second shell and therefore only a further eight in the third shell. So the third shell is far from full, yet argon does not react and is a noble gas. The next element is potassium, it has one more electron than argon. Where would this electron go? From potassium's chemical similarity with lithium and sodium which both have one electron in the outermost shell, we are led to believe that the most logical place would be the next shell (i.e. the $n = 4$ shell) rather than the electron being the ninth electron in the the $n = 3$ shell. Indeed this turns out to be the case. The shells do not fill up exactly one after another. Table 4.1 shows the actual number of electrons in the outermost shell versus the maximum number possible for the noble gases.

To obtain an accurate prediction of the filling order of the shell, a detailed quantum mechanical calculation is necessary. However, without any detailed computation, we can qualitatively see where the deviations from the simple picture based on the hydrogen atom start to take place.

n^{th} Energy Shell	1	2	3	4	5	...
Corresponding noble gas	He	Ne	Ar	Kr	Xe	...
Maximum number of electrons $= 2n^2$	2	8	18	32	50	...
Observed number of outer electrons	2	8	8	18	18	...

Table 4.1: Expected and observed number of outer electrons in noble gases.

In the approximation that we considered above, we completely neglected the interactions among the electrons. The electrons repel each other due to the Coulomb potential. This leads to a change in the states that are obtained from the hydrogen atom. Hence, instead of completely filling up one shell before going on to the next, it may so happen that certain states for some values of n have a higher energy than the first few states belonging to the next $(n + 1)$ shell.

This is in fact precisely what happens by the time one reaches the $n = 3$ shell. Neon ($Z = 10$) is the noble gas that is the last element of the $n = 2$ shell, and, as expected, is inert and nonreactive. After neon, we begin to populate the $n = 3$ shell with electrons; we expect $2 \times 3^2 = 18$ atoms with electrons in the third shell. One electron in the third shell gives sodium and two electrons in the third shell yields magnesium, and which is then followed by six more elements (aluminum through argon). This accounts for eight atoms in the $n = 3$ shell. Following argon, we "should" be able to put in 10 more elements according to the hydrogen-like counting that we started with. However, at this point, due to the interactions of the electrons, and the fact that the electrons in the filled shells start to "shield" — thus effectively reducing — the charge of the nucleus felt by the outer electrons, the counting changes significantly. The electrons in both potassium ($Z = 19$) and calcium ($Z = 20$) choose to occupy the $n = 4$ shell as it has a lower energy than the next $n = 3$ states.

Since the shells fill up in a nontrivial manner, it is useful to have a way to denote which of the possible states are occupied by an electron. For historical reasons, the azimuthal quantum number ℓ is often referred to as angular momentum or subshell and its various possible values have alphabetic labels originating in spectroscopic observations as shown in Table 4.2.

Angular Momentum	Label
$\ell = 0$	s
$\ell = 1$	p
$\ell = 2$	d
$\ell = 3$	f
.....	...

Table 4.2: Alphabetic labels for angular momentum quantum numbers.

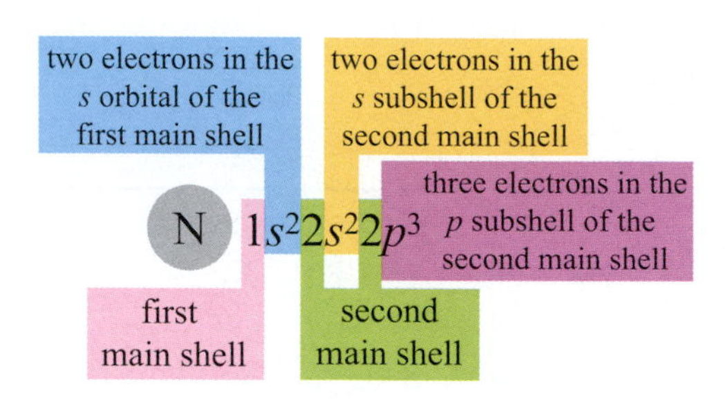

Figure 4.14: Illustration of the labeling of the electron orbitals for nitrogen.

We can then code for which states are occupied by electrons in an atom as follows. First write down the shell number and follow this by the azimuthal quantum number raised to the power of the number of electrons having this azimuthal quantum number. For example lithium is $1s^2 2s^1$, which means that in the $n = 1$ energy level there are two electrons in the s angular momentum state and in the $n = 2$ shell there is one electron in the s state. Oxygen is given by $1s^2 2s^2 2p^4$. In Figure 4.14 the labeling for nitrogen is shown graphically.

Although the filling of the shells is not straightforward, it is nevertheless subject to a very regular pattern as shown in Figure 4.15.

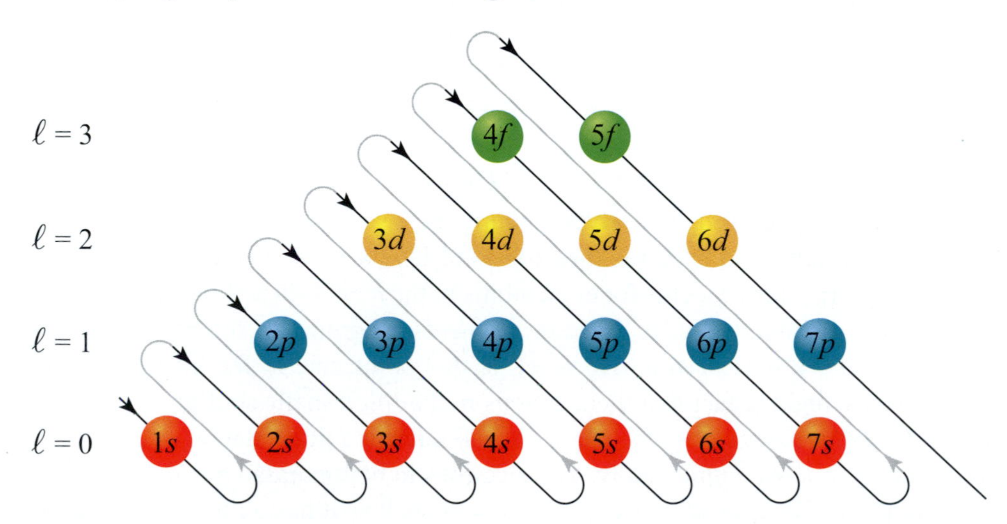

Figure 4.15: The order in which the shells of the first 108 atoms in the periodic table are filled with electrons.

Table 4.3 gives the electronic structure for the first 20 elements of the periodic table. As mentioned before, the chemically most relevant electrons determining almost entirely the chemistry of the elements are the outermost electrons. The outermost electrons have a special name due to their importance: **valence electrons**. In the table, valence electrons are indicated in boldface if they are chemically active and in italic if not (noble gases).

One can imagine how intricate the actual structure of the atoms becomes as one goes to larger and larger Z atoms with more and more electrons.

The chemical and physical properties of atoms are primarily determined by the number of electrons in their outermost shell, since these are the electrons that are free to interact with forces external to the atom. Hence, we expect atoms with the same number of electrons in the outer shell to have similar physical and chemical properties. This is the explanation of why the elements in the vertical column of the periodic table have similar properties. Let us now have a closer look at some of the notable groups in the periodic table.

Atoms with different atomic number Z, but with only one electron in the outer shell, comprise the first column of the periodic table, and are called the **alkali metals** (Li, Na, K, Rb, Cs, Fr) with hydrogen taking a special place. All the alkali metals have similar chemical properties: they are highly reactive, and are easily ionized (i.e. they can easily let go of their single outermost electron). Figure 4.16 shows the electronic configuration of **sodium (Na)**, with a single electron in the outermost shell. All the alkali metals have one valence electron.

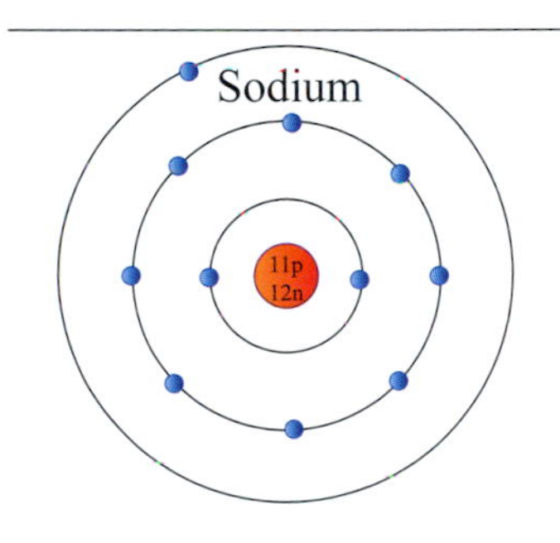

Fig. 4.16: Sodium is an alkali metal.

In the same fashion, if two atoms are, say, one electron short of completely filling up their outermost shell, they should be chemically similar. The **halogens** (F, Cl, Br, I and At), comprising the seventh column of the periodic table, are one electron short in filling up their outermost shell, and all have similar chemical properties.

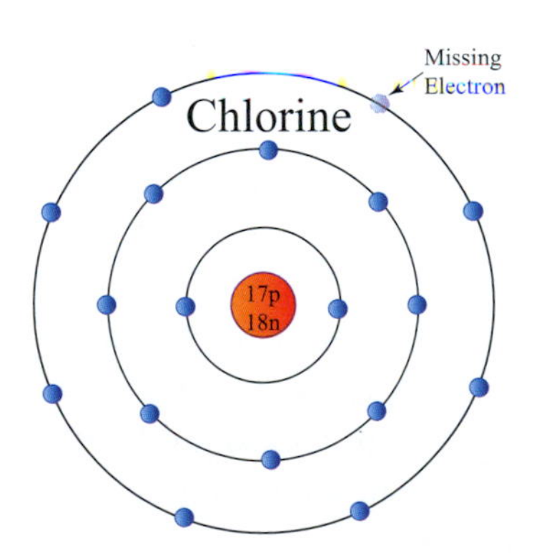

Figure 4.17: Chlorine is a halogen.

An example of a halogen is **chlorine (Cl)** and its electronic configuration is shown in Figure 4.17.

Thus far we have seen that elements are made up of atoms which in turn are made up of a nuclei containing protons and neutrons surrounded by electrons. The electrons are arranged in shells, characterized by a set of quantum numbers n, ℓ, m and s, and subjected to the Pauli exclusion principle which states that no two electrons in an atom can have the same set of quantum numbers.

We have also seen that chemistry is by and large determined only by the valence electrons (i.e. the outermost electrons). Would this be enough to understand why the elements have such differing properties?

On the one hand, one could argue that since all the elements either have a different number of valence electrons or a different number of shells, we now know the reason for the differences. For example lithium and sodium both have one valence electron but the valence electron of lithium is in the $n = 2$ shell while the

Name of Atom	Atomic Number	Electronic Configuration
Hydrogen	1	$\mathbf{1s^1}$
Helium	2	$1s^2$
Lithium	3	$1s^2\mathbf{2s^1}$
Beryllium	4	$1s^2 2s^2$
Boron	5	$1s^2\mathbf{2s^2 2p^1}$
Carbon	6	$1s^2\mathbf{2s^2 2p^2}$
Nitrogen	7	$1s^2\mathbf{2s^2 2p^3}$
Oxygen	8	$1s^2\mathbf{2s^2 2p^4}$
Fluorine	9	$1s^2\mathbf{2s^2 2p^5}$
Neon	10	$1s^2 2s^2 2p^6$
Sodium	11	$1s^2 2s^2 2p^6\mathbf{3s^1}$
Magnesium	12	$1s^2 2s^2 2p^6\mathbf{3s^2}$
Aluminum	13	$1s^2 2s^2 2p^6\mathbf{3s^2 3p^1}$
Silicon	14	$1s^2 2s^2 2p^6\mathbf{3s^2 3p^2}$
Phosphorus	15	$1s^2 2s^2 2p^6\mathbf{3s^2 3p^3}$
Sulfur	16	$1s^2 2s^2 2p^6\mathbf{3s^2 3p^4}$
Chlorine	17	$1s^2 2s^2 2p^6\mathbf{3s^2 3p^5}$
Argon	18	$1s^2 2s^2 2p^6 3s^2 3p^6$
Potassium	19	$1s^2 2s^2 2p^6 3s^2 3p^6\mathbf{4s^1}$
Calcium	20	$1s^2 2s^2 2p^6 3s^2 3p^6\mathbf{4s^2}$

Table 4.3: The electronic configuration for a few atoms. The chemically active valence electrons are indicated in boldface.

valence electron of sodium is in the $n = 3$ shell. Or we can consider nitrogen and oxygen; in this case, both nitrogen and oxygen have all their valence electrons in the $n = 2$ shell but nitrogen has five valence electrons and oxygen has six valence electrons.

On the other hand, if we imagine the electrons to be arranged in concentric rings as we have done for simplicity in Figures 4.16 and 4.17 above, then really, elements like chlorine and bromine, though both halogens, should be even more similar than they are. After all, they both have seven electrons in the outermost shell and are nearly of the same physical size, the only difference being that bromine has an extra shell inside the valence electrons but that is chemically speaking hardly noticeable. Yet chlorine is a gas and bromine a liquid at room temperature while chlorine is essential for life and bromine hardly plays a role.

Though useful as a mnemonic, in reality, electrons are not arranged in neat concentric rings. As mentioned above, a planetary system with the nucleus as the sun and the electrons as the planets is electrostatically unstable. Due to the attraction between protons and electrons and the fact that accelerating charges radiate an hence lose energy, a planetary atom would collapse. Let us now look at a more complicated but better way to picture the electrons around a nucleus.

4.9 Electron "cloud"

The correct way to describe electrons is with the help of quantum mechanics. While the detailed calculations can be rather complicated, what is important here is that in quantum mechanics, electrons do not actually travel on nice racetrack-like paths. Rather, the position of an electron in an atom is uncertain, with the electron being in a quantum mechanical state in which it has the likelihood of being anywhere inside the atom. The electron forms an electron "cloud" around the nucleus; we use the term "cloud" as a shorthand for representing the probability of finding an electron at different positions inside the atom. If this sounds strange, that is because it is strange! Each electron has its own cloud and it is impossible to determine the exact position of the electron in its cloud — one can only predict the probability of finding it at some definite position. For an atom with many electrons, there is a single cloud for all the electrons and that yields the probability of finding all the electrons at various positions inside the atom.

Now the key thing about these electron clouds is that they are in general not just spherelike but have very specific shapes that can exactly be calculated depending mostly on the numbers ℓ and m. The number n in a way, being the main indicator of their energy, represents their size. If one wants to obtain an idea of how the electron clouds look, one can plot a so-called equiprobability surface. That is to say, one can take a certain probability of finding an electron and then plot all the points around the atom that have the same probability of finding the electron. Some of these shapes are illustrated in Figure 4.18.

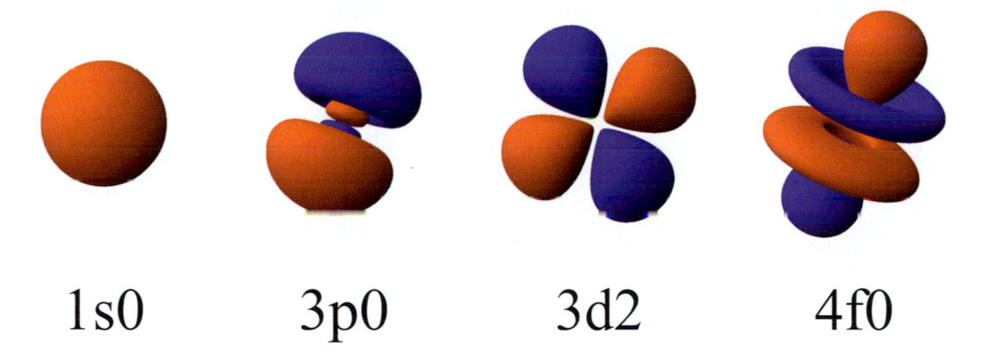

Figure 4.18: Electron clouds for several types of electrons. The labels indicate the quantum numbers n, l, m.

The electron clouds for $\ell = 0$ are always spherical regardless of the shell n they belong to. However, the probability of finding an electron on these spheres does depend on the distance from the nucleus. In the case of $n = 1$, the probability simply decreases outwards. In the case of $n \geq 2$, the probability varies periodically in a wavelike fashion going outwards from the center.

For $\ell \neq 0$, the electron clouds can have rather intricate shapes as illustrated in Figure 4.18.

In an atom, the electron clouds of all the different electrons coexist, and consequently the addition or removal of an electron can have a rather drastic impact on the overall shape.

This now allows us to explain why adding an electron to chlorine is quite different from adding an electron to bromine. In the latter case, there are an additional 10 electron clouds all having specific shapes.

Thus we see that it is not only the number of electrons, protons and neutrons that are different between elements, in fact the very shape of the atoms is different due to the various combinations of the electron clouds. As such we now have the main ingredients necessary for an explanation, but before ending the chapter let us briefly have a look at five elements that play a key role for life on earth.

4.10 Five special elements

If we look at the elementary composition of life on earth we find that in many organisms upwards of 99% of their mass is made up of only five elements: carbon, hydrogen, oxygen, phosphorus and nitrogen (a nice mnemonic is CHOPiN — the name of a famous classical composer). Bony animals further may also have around 1% to 2% of calcium. Some other elements such as sodium, potassium, chlorine or iron fulfill important biological functions but are only present in (mass-wise) small quantities.

Figure 4.19 illustrates the quantity in terms of mass of the most common elements in humans. By mass, our bodies consist of roughly 65% oxygen, 18% carbon, 10% hydrogen, 3% nitrogen, 1.5% calcium, 1.2% phosphorus, 0.2% potassium, 0.2% sulfur, 0.1% sodium, and smaller quantities of magnesium, iron, cobalt, copper, zinc, iodine, selenium and fluorine plus traces of some other elements.

The reason for oxygen being such a big part of us stems from the fact that our bodies are mostly water. Although water has twice as many hydrogen atoms as oxygen atoms, one oxygen atom weighs roughly 16 times more than one hydrogen atom and consequently by mass we contain much more oxygen than hydrogen.

At this point we should ask whether there is a special reason for the abundance of certain elements in living organisms. First, one may surmise that the abundance in organisms simply reflects the abundance of the elements in the environment. Well, there is certainly a lot of water and it is likely that life evolved in or near a pool of water, so the abundance of oxygen and hydrogen are perhaps of little surprise. Nitrogen is plentiful in the atmosphere but only found in very low concentrations in water. Furthermore, seawater does have quite a bit of sodium and chlorine in salt (about 3% by mass) as well as reasonable amounts of sulfur (about 0.1% by mass), calcium (about 0.05% by mass) and potassium (about 0.05% by mass). However, notably absent in the environment are carbon (needed in almost all molecules essential to life) and phosphorus (among others essential for the genetic materials DNA and RNA).

Therefore, abundance is not the sole key determining factor as to why an element is common in life. Of course, life may have first emerged in an environment where those elements were common but this is currently unknown. Even so, life quickly spread out and hence it is reasonable to surmise that carbon and phosphorus have some properties that are particularly desirable.

One special property of phosphorus is that it forms phosphates that can be strung together to store or retrieve energy. A special property of carbon is that

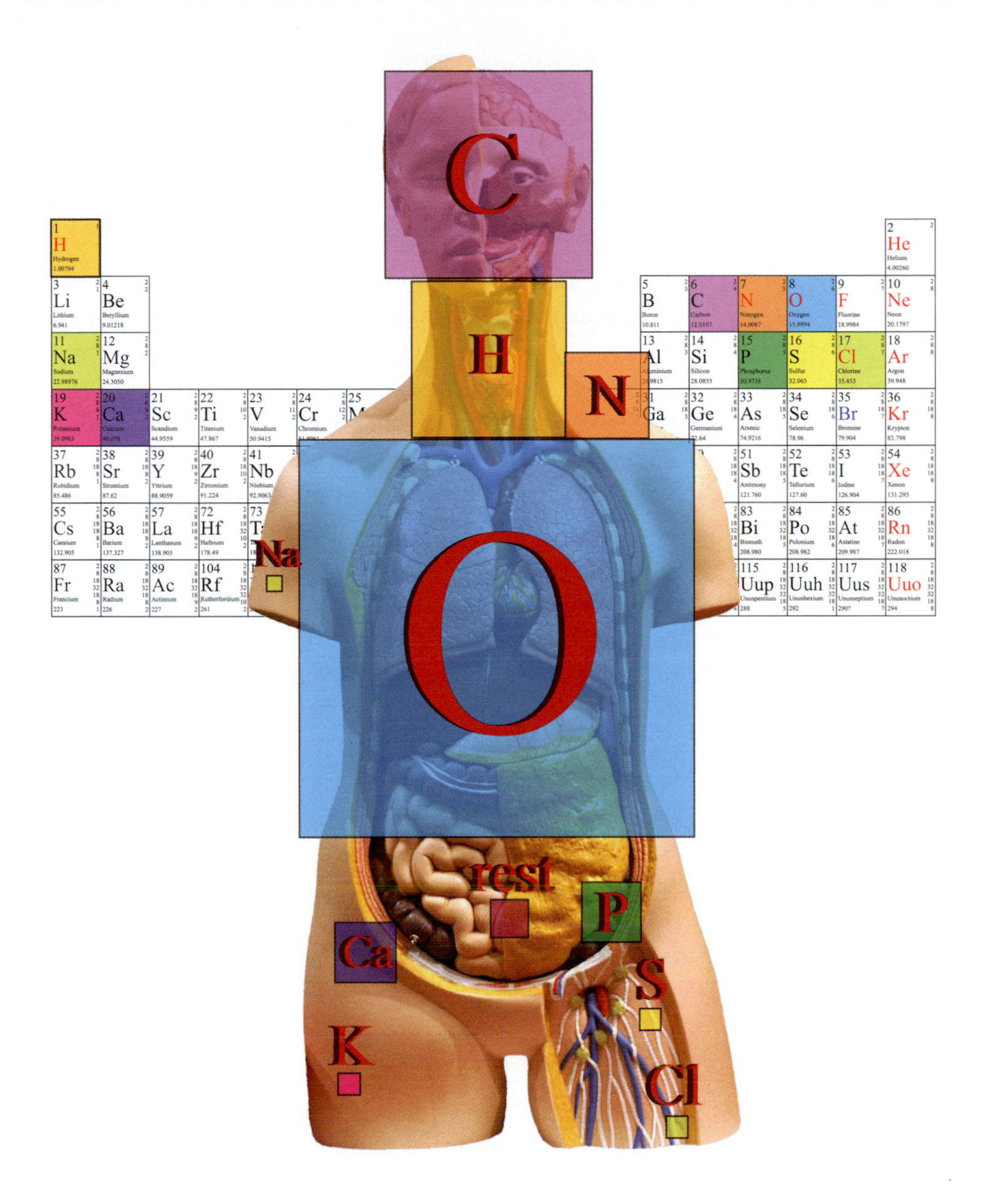

Figure 4.19: Humans are made mostly of only a few elements. Indeed, by mass, 93% is made up of the elements oxygen, carbon and nitrogen.

it has four valence electrons and can bond in many different arrangements. For example, carbon atoms can form long chains while maintaining the ability to bond to oxygen and hydrogen atoms giving enormous molecular diversity (for details, see Chapter 8).

Looking at the periodic table, we see that silicon is below carbon and arsenic below phosphorus. They therefore have somewhat similar properties and indeed one can speculate that on other worlds (if life is not exclusive to earth) silicon-based life-forms may exit. Indeed, without knowledge of a sufficient number of extraterrestrial life-forms, it is probably impossible to make a definite statement on how important certain atomic properties are for life in general. Nevertheless, we can see why on earth, the elements common in living matter do seem to be advantageous.

4.11 The answer

We can now answer our question why the elements are so different from each other. We have found that atoms are made of protons, neutrons and electrons, and that atoms are mostly empty space. The chemical properties are mostly determined by the outer shells of the electrons in the atom; the reason being that in all chemical reactions the electrons in inner shells are more or less inert and it is only electrons in the (not completely filled) outer shell that are chemically active. All materials are made out of a limited number elements whose atoms only differ in the quantity of protons, neutrons and electrons. Thus we have an astounding hierarchy: There is a sheer infinitely large number of materials, all of which are made of a relatively small number of elements that in turn are made of only three "elementary particles", namely the electron, proton and neutron. The elements themselves are different from each other because **each element is associated with a specific type of atom that has a number of electrons matching the number of protons in the nucleus. The electrons form probability clouds that are different for each element**.

The idea of the atom as being the irreducible constituent of matter is central to our current understanding of nature. According to Feynman, if there was one sentence that he could communicate to a future civilization that has lost all scientific knowledge, it would be the atomic hypothesis, namely that "*all things are made of atoms — little particles that move around in perpetual motion, attracting each other when they are a little distance apart, but repelling upon being squeezed into one another*".

4.12 Exercises

1. If the half-life of a radioactive substance is 12 years, how much of the substance will have decayed after 100 years?
2. Why is it reasonable to approximate atoms as billiard balls when discussing a gas at room temperature and one atmosphere pressure?
3. How many neutrons does the $^{16}O^{2+}$ ion have?
4. How could one turn alpha-rays into helium?
5. What is the difference between an H^+ ion and a proton?
6. Are elements that in the bulk state are in a gaseous form (at standard pressure and temperature) lighter than those in the solid state? If so, discuss what that means. If not, discuss why they are gases then.
7. Compare the magnitude of the gravitational force between the electron and proton in a hydrogen atom with its electrostatic force.

8. The atom is quite "hollow" so matter is mostly empty space. Then what is it that gives matter its "solid" feel?

9. If we strip lithium of its valence electron (this can quite easily be done), it has two electrons left whose configuration is exactly the same as that of helium. Would you expect the two-electron lithium to behave like a heavy version of helium?

10. Make a drawing of the electron configuration of silver "planetary model" style.

11. If bombarding nitrogen with α-particles leads to the production of hydrogen, then why is the other element produced oxygen and not carbon? (After all, removing or splitting off a proton from nitrogen should give a lighter element [carbon] and not a heavier element.)

5: Combining Atoms

Q - How Do Atoms Bond?

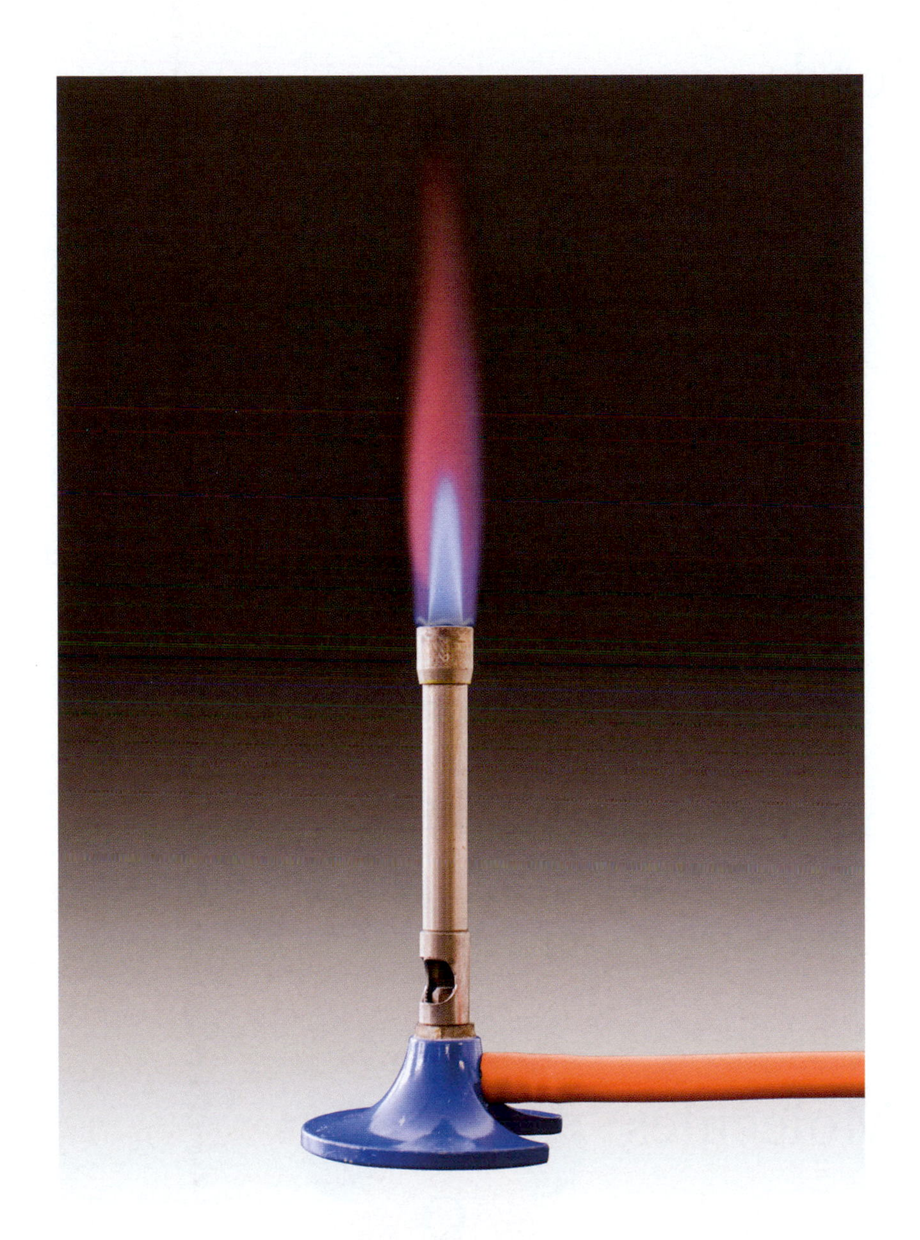

Chapter Map

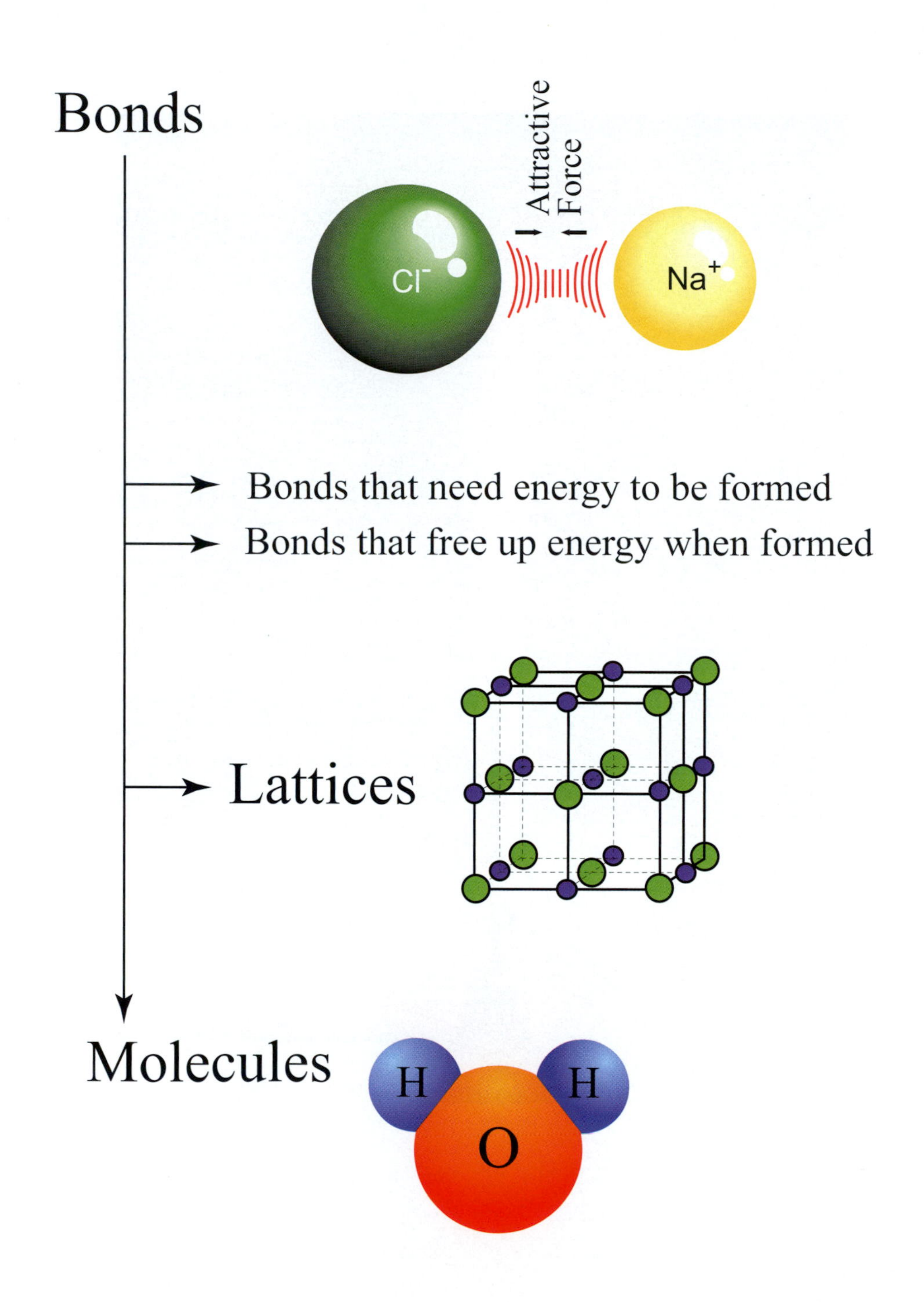

Bonds
Attractive Force
Cl⁻
Na⁺
Bonds that need energy to be formed
Bonds that free up energy when formed
Lattices
Molecules
H
H
O

5.1 The question

The world around us has a bewildering variety of materials. From the atomic point of view, discussed in our study of atoms in Chapter 4, all materials are made out of an enormous number of atoms; a typical object that we can sense has about 10^{23} atoms. We have to unravel how do trillions upon trillions upon trillions of atoms come together to form materials that have simple macroscopic properties such as toughness, plasticity, malleability and so on.

Clearly, atoms can only be held together if there are forces of attraction between them. The binding of atoms takes place at two different levels. Atoms can combine at the atomic scale to form molecules but atoms can also combine in vast numbers to form bulk matter that we see around us. The same atomic forces that bond atoms into molecules are at work in forming bulk material.

In this chapter we focus on how atoms combine to form molecules and briefly discuss how the bonding of atoms gives rise to a diverse range of solids and liquids. To understand the properties of liquids we focus on studying water since it is a prime exemplar of liquids and also has many remarkable properties. The macroscopic properties of bulk matter, mostly solids, are discussed in Chapter 7 on materials.

5.2 Atomic bonds

Atoms bond to form microscopic structures such as molecules on the one hand and bulk objects like solids and liquids on the other. The concept of a bond revolves around the behavior of electrons that are inside an atom. Bonds can explain why and how atoms combine to form molecules as well as why and how atoms combine to form solids and liquids. The role of bonds is not very relevant for dilute gases and the reason for this is discussed in a later section.

Atomic binding is the name given to the combination of atoms. One may ask: why do atoms combine in the first place? What is it that drives them to form molecules? The answer to this question is related to a simpler question: why is there any attraction between any two objects in the first place? Newton's answer is that objects are attracted to each other if there is an attractive force between them. In response to this attractive force, objects lower their total energy by moving toward each other and stay bound due to the force of attraction. A ball rolls down the hill because the force of gravitation attracts it toward the center of the earth.

In Chapter 3 we discussed how the concept of force is replaced by a more fundamental concept, namely that of energy. Energy is the sum of potential and kinetic energy; an object always tends to move in a direction that lowers its energy. An object comes to rest when it is in a state in which it has minimum potential energy. This same principle is at work in the bonding of atoms. A collection of atoms will tend toward a configuration that minimizes its energy. This minimum may not be an absolute minimum, but is a minimum locally, in its immediate neighborhood.

We consider a few specific cases of atomic bonds. Atoms bond into molecules due to the following forces of attraction:

1. Covalent bonds arising from the force of attraction caused by quantum mechanical effects due to the sharing of electrons.

2. Ionic bonds due to electrical Coulomb attraction between opposite electrical charges.

3. Metallic bonds in metallic solids.

4. Polar bonds are due to slight spatial charge distributions. The polar bond, of which the hydrogen bond is a special case, plays an essential role for understanding key features of the molecules necessary for life such as the DNA molecule.

5. Van der Waals bonds due to a weak electrical attraction caused by a dipole moment of the atoms.

6. Forces of attraction arising from a mixture of the above bonds.

Large collections of atoms form into solids due to forces of attraction. Based on the underlying structure and forces responsible for their formation, many solids can be classified as either crystalline or amorphous. As the name indicates, in crystalline solids the atoms composing them occur in a fundamental unit cell that is replicated to form the cyrstal lattice, whereas the atoms composing an amorphous solid have no underlying regular structure.

Amorphous and crystalline materials have some general properties in common. For example both have a well-defined shape, and are generally either insulators or conductors of electricity and heat. Crystalline solids consist of individual unit cells, each with a well-defined shape; they cleave to give well-defined faces and melt at definite temperatures. Examples of crystalline solids are sodium chloride, sucrose, and metals. Amorphous solids on the other hand do not consist of individual unit cells, they break to give curved or irregular faces, and soften and then melt over a temperature range. Examples of amorphous solids are asphalt, paraffin, window glass, obsidian, and glassy forms of glycerol.

5.3 Hydrogen molecule: covalent bond

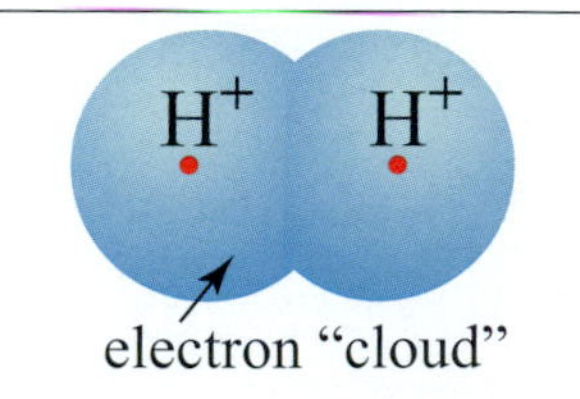

Fig. 5.1: Covalent bond.

The simplest possible molecule is two protons bonded into an (ionized) hydrogen molecule H_2^+ due to the sharing of one electron as illustrated in Figure 5.1. One can think of this molecule being formed by the following process: A hydrogen (H) atom is neutral; let it approach a proton that has a single positive charge, denoted by H^+. As it approaches the proton, the H-atom is polarized by the presence of the proton, with the orbit of the electron being deformed and stretched toward the proton. This deformation creates a tiny net attraction between the H-atom and the proton. On approaching even closer, the two protons feel an even stronger attractive force by sharing the electron, and in doing so considerably lowering the energy of the system.

If one tries to push the protons even closer together the electrostatic repulsion between the two positively charged protons takes over. Hence, the potential of the H_2^+ system is shown in Figure 5.2, with a *minimum* where the electron is equally shared by the two protons. The ionized hydrogen molecule is in stable equilibrium at the minimum of the potential and the two protons are said to be bonded by sharing an electron. This is an example of the *covalent bond*, one of the main types of chemical bond that exists between atoms.

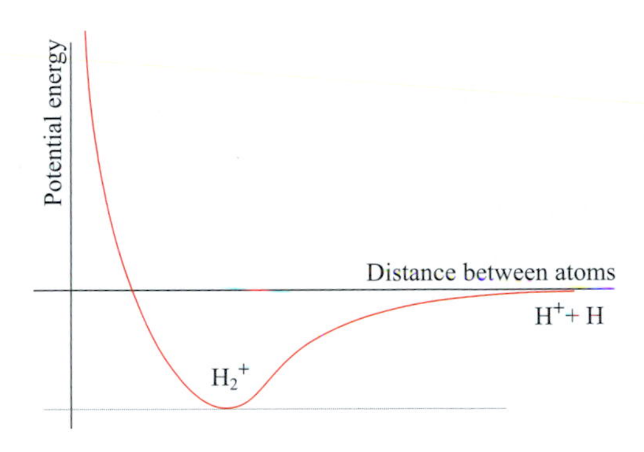

Figure 5.2: Potential energy for the $H_2{}^+$ ion.

Why should a covalent bond due to the sharing of an electron lower the energy of the H_2^+ molecule? Covalent bonds arise purely due to quantum mechanical effects, and a heuristic explanation is the following: Consider two children passing a ball to each other. They have to keep close to each other since otherwise the ball will not reach them. In effect due to passing a ball to each other, a net attractive force arises between them. In a similar manner, for example, the two hydrogen nuclei in H_2^+ have an attractive force due to the sharing of a single electron. The two hydrogen nuclei, although both positively charged, have a net attractive force between them as result of sharing an electron, and form a bound state of the hydrogen molecule H_2^+.

The bond due to the sharing of one or more electrons is called a covalent bond.

5.4 Salt: ionic bond

Ionic bonds arise from the force of electrical attraction between atoms that carry a net electrical charge. Since atoms are usually electrically neutral, the atoms need to lose or gain some electrons to have a net amount of electrical charge; this process of gaining or losing electrical charge is called **ionization**. Atoms and molecules need to be ionized in order to form ionic bonds. There are various types of ionic bonds that we will discuss.

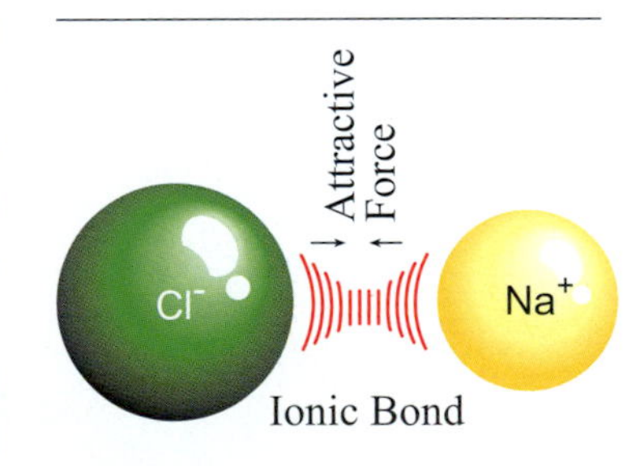

Fig. 5.3: Ionic bond.

Salt is composed of one atom of sodium (Na) and one atom of chlorine (Cl) and is written as NaCl. One can ionize Na by removing an (outer) electron of Na by giving the electron 5.14 eV energy, and consequently the sodium atom becomes positively charged, written as Na^+; the chlorine atom captures a free electron and liberates 3.62 eV of energy, and becomes negatively charged, written as Cl^-. There is now a net electrical attraction between the ionized sodium and chlorine atom, leading to a salt molecule that is a bound state of sodium and chlorine, designated by NaCl, and is illustrated in Figure 5.3. A bound state of two atoms, with some binding energy, is the same concept that explains why the earth does not fly away from the sun. A bound state means that the lowest energy state for the system is when the two atoms have combined to have net negative energy and one has to supply an energy equal to the binding energy for the atoms to be separated. It is for this reason a bound state is said to have negative energy.

A net energy of $5.14 - 3.62 = 1.52$ eV needs to be supplied to neutral Na and Cl atoms to form the ions which constitute NaCl. One may ask: if energy needs to be supplied to form ions, then why is salt so common rather than elementary sodium and chlorine?

The reason is the following: The binding energy is due to the Coulomb attraction, which for the distance between the sodium and chlorine ions in salt (NaCl) of 0.236 nm works out to be

$$\frac{-k_e e^2}{r} = \frac{-1.44 \text{ eVnm}}{0.236 \text{ nm}} = -6.1 \text{ eV}, \tag{5.1}$$

where $k_e = 1.44 \times 10^{-9}$ Vm/e^2 is the electrostatic constant.

Consequently, if a neutral sodium atom gets close enough to a neutral chlorine atom, it becomes energetically favorable for an electron to jump from the sodium to the chlorine over a potential that has the value of 1.52 eV and then falling into a potential well with a minimum of -6.1eV binding energy; the net amount of energy lowered as a result of the electron migration is -6.1 eV + 1.52 eV = -4.58 eV. Interestingly enough, an experimental measurement of the binding energy of the salt molecule yields only -4.26 eV leaving an energy of 0.32 eV unexplained. This extra repulsion (positive) energy of 0.32 eV turns out to be due to a quantum mechanical effect (generally called Pauli repulsion).

All the facts discussed about the molecule NaCl can be encoded in the potential energy between the Na and Cl atoms. When the atoms are far apart, they are neutral and do not attract and the potential is a constant. When they are closer to each other, due to the polarization of their charges — similar to the case of H_2^+ — there is a net attraction and the potential slopes downwards. The potential reaches a minimum when they are neutral since electrical repulsion sets in. However, if one can ionize Na into Na^+ by stripping it of an electron and donating it to Cl, thus making it Cl^-, then there is again attraction between the atoms.

The energy needed to ionize Na forms a hump, called a potential barrier; if the barrier can be crossed by externally supplying the atoms with energy, then the salt NaCl can form. As calculated above, the potential barrier is 1.52 eV and the system then falls into a potential well with a minimum of -6.1 eV binding energy; any distance closer faces Coulomb repulsion due to the positive charge of the nuclei. The potential energy has two minima, as shown in Figure 5.4, the deeper minimum being where the molecule NaCl forms.

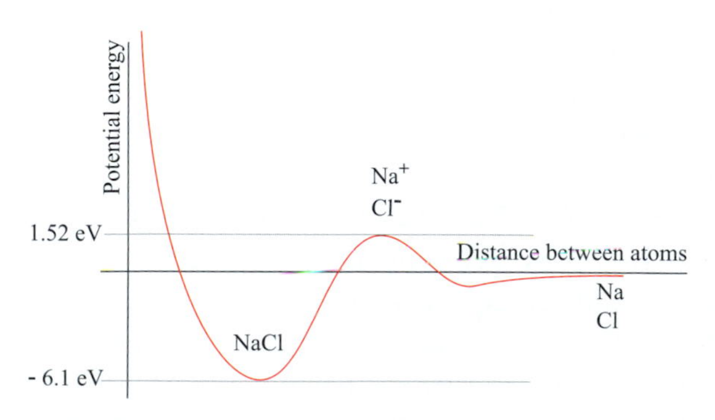

Figure 5.4: Potential energy for sodium chloride (table salt).

5.5 Iron: metallic bond

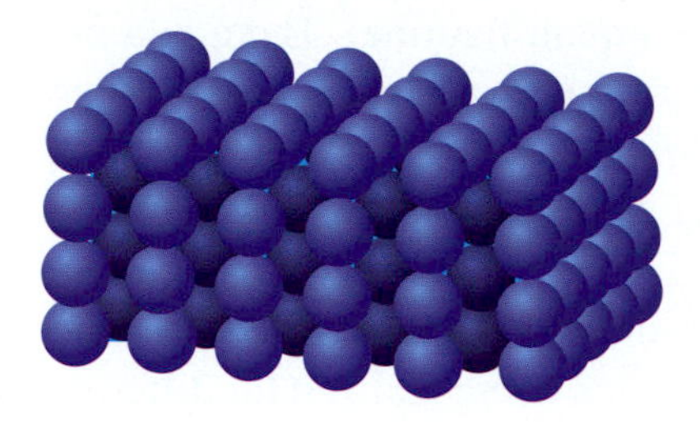

Figure 5.5: Iron atoms in a ferrite are arranged in a body centered cubic lattice.

Iron (Fe) forms a crystalline solid with the fundamental unit being a body centered cubic (bcc) lattice, which consists of eight iron atoms sitting at the vertices of the cube, and a ninth one sitting at the center of the cube; iron with this lattice configuration is called a **ferrite**. This is shown in Figure 5.5.

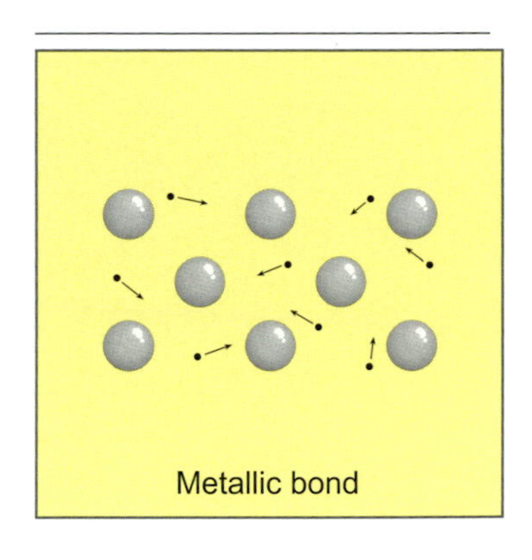

Fig. 5.6: Metallic bond.

Some elements, like the transition metals (i.e. elements in groups 3–12 with incomplete d subshells such as iron and cobalt), have electron configurations in which electrons from their inner shells can also act as valence electrons; these elements can have several different oxidation states. For example, iron can have two or three valence electrons. Hence, each Fe atom in the lattice contributes two to three electrons to the solid, and in doing so becomes an ion with a net positive charge. The electrons freely propagate throughout the lattice as shown in Figure 5.6 and form a sort of sea of electrons, called the conduction band electrons.

The Fe atoms form a lattice due to what is called a metallic bond: a metallic bond is a combination of the ionic and covalent bond, since there is a net attractive force on the Fe atoms both due to electrical attraction with the electrons (ionic) and also due to the sharing of electrons with other Fe atoms (covalent). The metallic bond drives the Fe atoms to bunch up and draw closer to other Fe atoms as this lowers the energy of the electron sea. Note that because the metallic bond is due to the sea of electrons, it is quite insensitive to the relative positions of the Fe atoms, and is also nondirectional.

5.6 Polar covalent bond

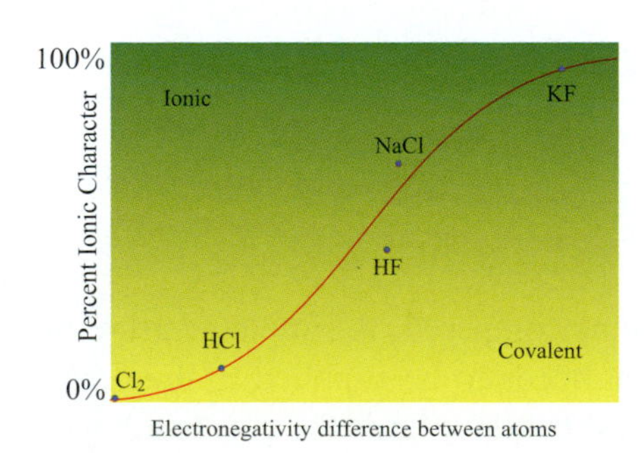

Figure 5.7: Percentage of ionic character for some molecules.

A nonpolar covalent bond has a uniform distribution of electron charge between the bonded atoms. The simplest nonpolar covalent bonds exist in "homonuclear diatomic" molecules like H_2 and Cl_2. Both atoms in the molecule share the electrons equally. There is no permanent nonzero net electric charge built up in the molecule. A **polar covalent bond** has a non-symmetric electron distribution and lies between mostly ionic and mostly covalent bonds as illustrated in Figures 5.7 and 5.8. The electrons forming the covalent bond are not shared equally by the atoms forming the molecule; the electrons responsible for the covalent bond are more strongly localized around one of the atoms. Such polar bonds occur when one of the elements attracts the shared electrons more strongly than the other element.

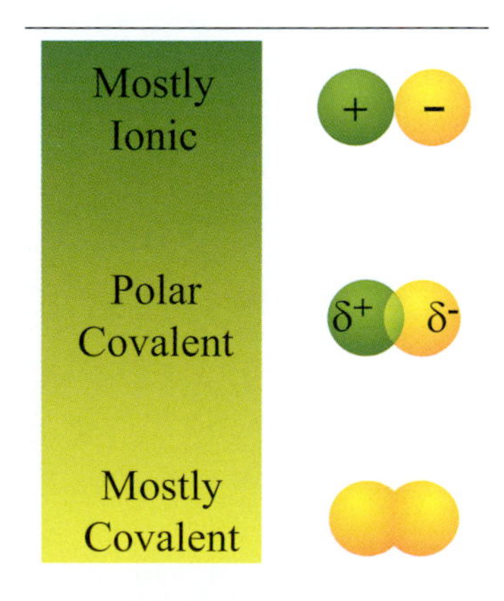

Fig. 5.8: Polar covalent bonds range from mostly covalent to mostly ionic.

In hydrogen fluoride, for instance, the electrons creating the covalent bond are more strongly attracted by the fluorine nucleus than by the hydrogen atom's nucleus, leading to the electrons being localized around fluorine. There is a net charge distribution of negative charge near the fluorine atom and net positive charge near the hydrogen nucleus creating a *permanent dipole moment*. The atoms in the molecule are bonded by both the covalent sharing of electrons as well as the force of Coulomb attraction between opposite charges. The various molecules that are polar bonded have the binding energy that vary from 100% covalent to 100% ionic bond, with all percentages between these limiting cases being realized for different molecules.

5.7 Van der Waals bond

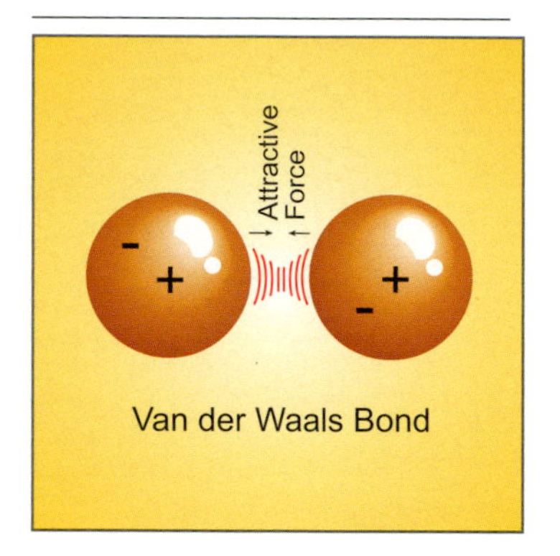

Fig. 5.9: Van der Waals Bond.

The van der Waals bond is due to the attractive forces between molecules (or between parts of the same molecule) and does not arise either from the sharing of electrons or from electrostatic attraction as is the case for the ionic bond. The origin of the van der Waals force is dynamical and can be thought of in the following manner: When two atoms approach each other, their mutual potential energy is lowered if the negatively charged electrons can stay as far as possible from each other. One way of achieving this is for the electrons to spin synchronously away from each other. This synchronous spinning induces a dipole moment in both the atoms resulting in an attractive force, and a heuristic representation is shown in Figure 5.9.

If the atoms are brought even closer, they repel each other due to the Pauli exclusion principle. The attractive van der Waals force and the repulsive "hard billiard ball" nature of atoms is well approximated by the 6-12 Lennard-Jones potential given below (see also Section 4.4):

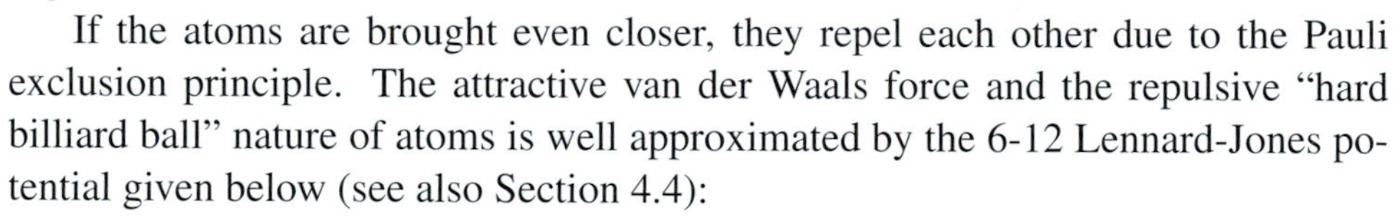

$$U(r) = 4U_0\left[\left(\frac{r_0}{r}\right)^{12} - 2\left(\frac{r_0}{r}\right)^{6}\right].$$

The minimum of the potential is at r_0 and the two atoms sit at this separation, bonded by the van der Waals force.

5.8 Strengths of bonds

The relative strength of the various bonds can be compared. A simple measure of bond strength is by the amount of energy that needs to be supplied to *break* the bonds. Table 5.1 below shows that the strongest bond by far is the covalent bond, followed by the ionic bond and so on.

5.9 Crystalline solids

Atoms bond to form molecules; these molecules are usually formed out of a relatively small number of atoms. Polymer- and bio-molecules, to be discussed in Chapter 8, are often made up of large numbers of atoms, and some of them even of

	Approximate binding energy per bond					
	Covalent	Ionic	Metallic	Polar	van der Waals	Hydrogen
Strength (in eV)	1.5–5	2–4.5	1–5	1.5–2	0.01–0.5	0.05–0.5

Table 5.1: The relative strength of the bonds; energy required to break the bonds.

billions of atoms. In contrast, trillions upon trillions upon trillions of atoms — in the order of Avogadro's number 10^{23} — combine to form solids, sticking to each other through the same bonds that combine atoms to form molecules. Solids come in three main varieties, namely crystalline, amorphous and polymeric. We now briefly discuss crystalline solids.

One of the simplest arrangements to form a large object is to repeat a fundamental unit many times, similar to a jungle gym. The structure of a crystalline solid is similar — a simple unit, called the fundamental cell, is repeated innumerable times. The fundamental cell for a solid has a very small size, with a typical distance between the corners of a fundamental cell being 10^{-9} m or 1 nm (nanometer).

Figure 5.10 illustrates three of the fundamental cells, namely the simple cubic lattice, the body centered cubic lattice and the face centered cubic lattice. The positions of atoms in crystalline solids are located at the vertices and other symmetrical positions in the lattice that is formed by the fundamental cells. The typical size of an atom is about 0.1 nm so the atoms are well separated in a solid. There are about 1 million atoms in 1 mm length of a solid.

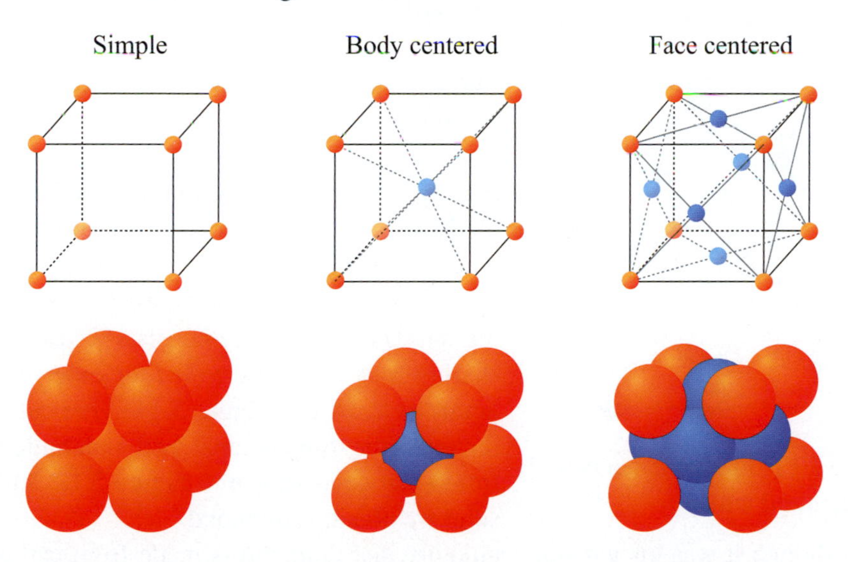

Figure 5.10: Fundamental cells: Simple, body centered and face centered cubic lattices. The colors for the atoms are used for indicating their positions.

Many of the familiar substances around us, such as salt or steel, are crystalline solids with a simple fundamental cell as shown in Figure 5.10. There are 14 possible fundamental cells in three dimensions and a lattice made up of a repetition of one of these fundamental cells is called a **Bravais lattice**.

Atoms combine to form crystalline solids as this leads to the minimization of energy for the vast collection of atoms in a macroscopic amount of substance. It is a mystery why nature chooses an exact lattice with, for example, the fundamental cell being repeated a countless number of times to form crystalline solids, and theorists are still trying to find an explanation. Atoms in a solid are bound to each other by the same bonds that form molecules, namely covalent, ionic and so on. We discuss the following crystalline solids to illustrate the diverse manner in which bonds operate in nature.

- Diamond, formed by the covalent bonding of carbon atoms.
- Salt, formed by the ionic bond of sodium and chlorine.
- Water and ice, formed by the hydrogen bonding of water molecules.
- Graphite, formed by the covalent and van der Waals bonding of carbon atoms.

Real crystalline solids never form a perfect lattice (crystal); nature randomly introduces all sorts of defects and imperfections into the crystal, and a typical crystalline solid is full of imperfections. However, underlying all the defects and impurities is the regular lattice and many of the properties of these solids are a direct result of the underlying fundamental lattice.

5.10 Diamond: covalent bond

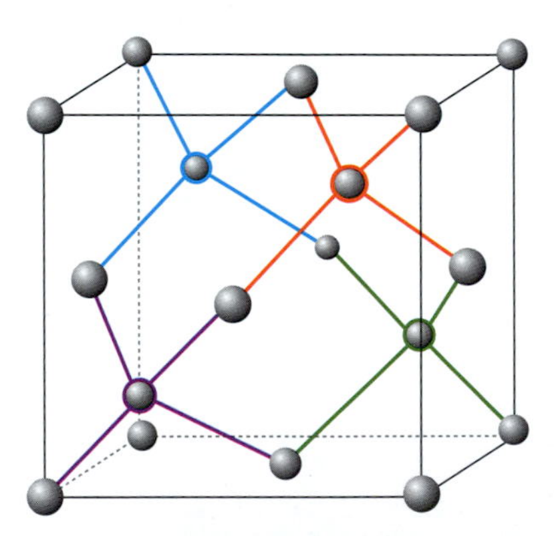

Figure 5.11: Fundamental cell of the diamond lattice.

The example of diamond is instructive. Each carbon atom, denoted by C, can share four electrons and if it shares these bonds with four hydrogen atoms, we obtain the molecule of methane CH_4. However consider another case where each C atom shares a single electron each of four other carbon atoms C. The resulting tetrahedral structure consists of each atom sharing four electrons with four other carbon atoms as shown in Figure 5.11. One can now replicate the fundamental cell indefinitely to form a lattice as shown in Figure 5.12, and the solid we obtain is diamond.

Although it was known since antiquity that diamond is made from carbon, a modern proof was only established in 1772 when French chemist Antoine Lavoisier burned diamond (ca. 150 mg) in a sealed container. Diamond does not burn readily, so Lavoisier used the high heat provided by a large lens placed in sunlight. By a simple analysis of the gases produced by burning, Lavoisier determined that the gas we now know as CO_2 was produced. Furthermore, the same amount of this gas was produced when 150 mg of amorphous carbon in the form of charcoal was burned. Thus, Lavoisier concluded that charcoal and diamond are the same.

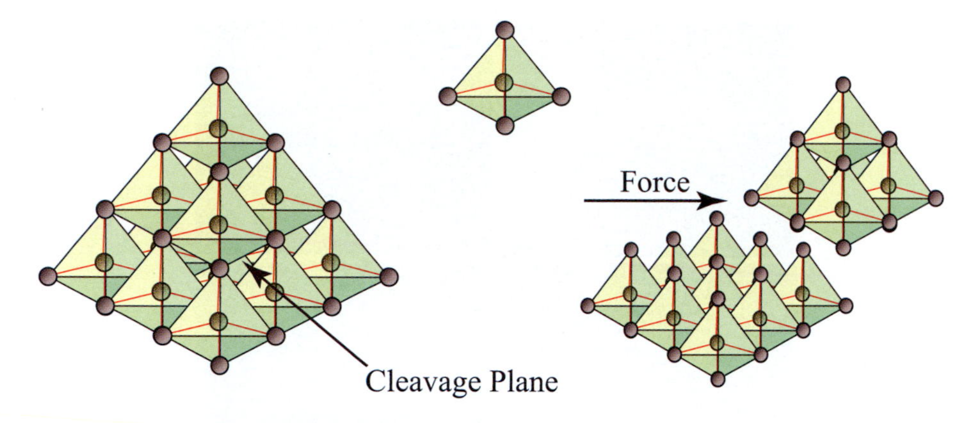

Figure 5.12: The structure of the diamond lattice.

All the exceptional properties of diamond — its extreme hardness (diamonds are the hardest natural substance known), resistance to tarnishing, exquisite clarity, clean faceting and lack of color — all arise due to the covalent bond that forms it.

Carbon has other forms of bonding and that give rise to other solid forms of carbon such as graphite, charcoal, buckyballs, and so on.

5.11 Salt crystal

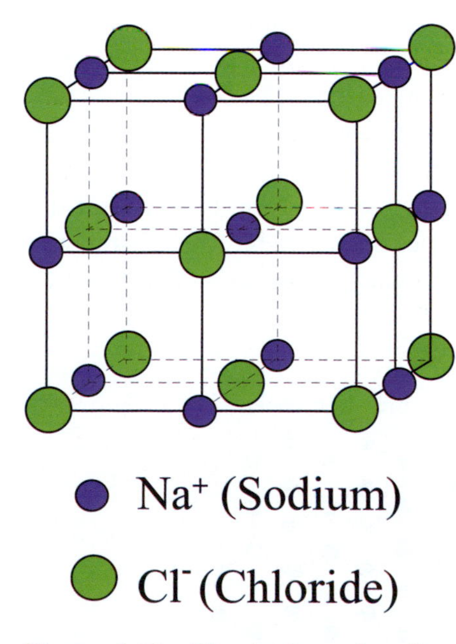

Figure 5.13: The lattice of sodium chloride.

Ionic attraction is responsible for the bonding of trillions of ions of sodium and chloride into the solid form that is salt. The new feature of salt is that the Na and Cl atoms all arrange themselves into a cubic lattice. This arrangement of atoms is called a face centered cubic (fcc) lattice, with Na atoms sitting at the mid-point of the sides and the Cl atoms sitting at the vertices and center of the faces of the cube. The fundamental fcc arrangement is repeated countless times to form the bulk form of salt, as shown in Figure 5.13 with a scanning electron microscope image given in Figure 5.14.

The fcc lattice of Na^+ and Cl^- forms due to electrical attraction and repulsion. The Na^+ ions attract the Cl^- and at the same time repulse other Na^+ atoms. The fcc lattice is the equilibrium arrangement in which all the atoms are occupying a state that minimizes the potential energy of the entire collection of Na^+ and Cl^- atoms. This minimum energy is about -8.2 eV and hence somewhat lower than the binding energy in a single NaCl molecule of -6.1 eV. Of course, the fact that the binding energy in the lattice is a bit smaller

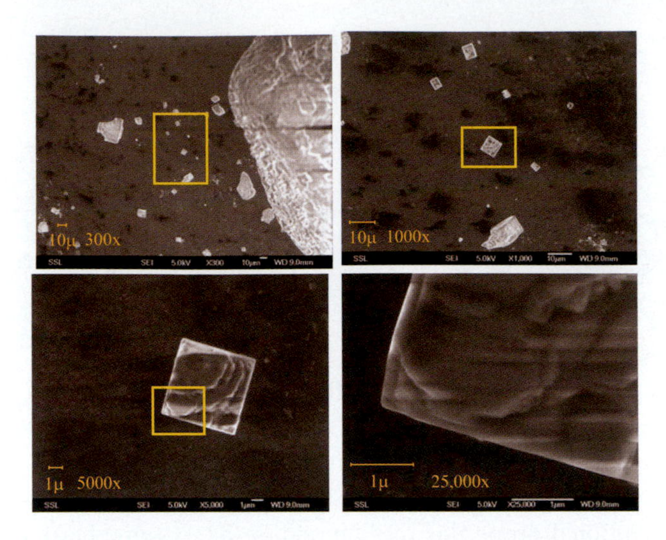

Figure 5.14: Scanning electron microscope image of crystalline salt.

is not a surprise since otherwise there would be no reason for NaCl molecules to form the crystals that one can easily observe.

5.12 Graphite: covalent and van der Waals bonds

Graphite has an underlying planar substructure in the following sense: Recall that in the formation of diamond, all the four bonds of carbon are with other carbon atoms; in graphite, each carbon atom covalently bonds with three carbon atoms in one plane, forming a planar hexagonal ring that looks like a chicken wire — there is no fourth covalent bond due to quantum mechanical reasons. The bonding structure of graphite is discussed in Section 19.7 and it is shown that each carbon atom in graphite donates one electron to the sea of electrons in the plane; the electron is delocalized and free to move as a quantum mechanical wave in the plane of the carbon atoms. There are such freely moving electrons in each plane of graphite, and result in van der Waals bonding between the planes, as shown in Figure 5.15.

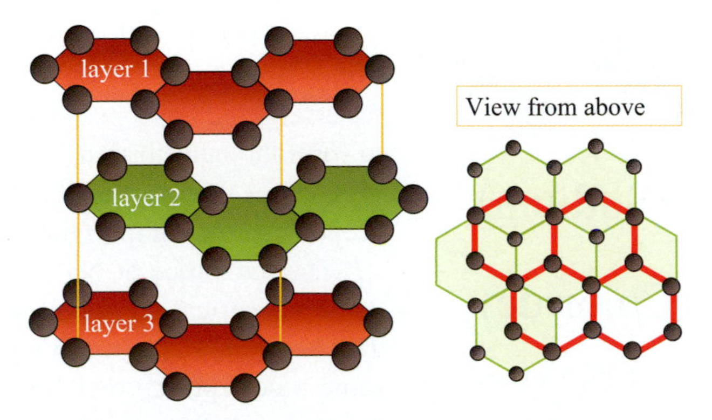

Figure 5.15: Layered lattice structure of graphite.

Hence the bonding of the carbon atoms in a plane is far greater than those with atoms in the adjacent planes. A scanning electron microscope image of graphite is shown in Figure 5.16.

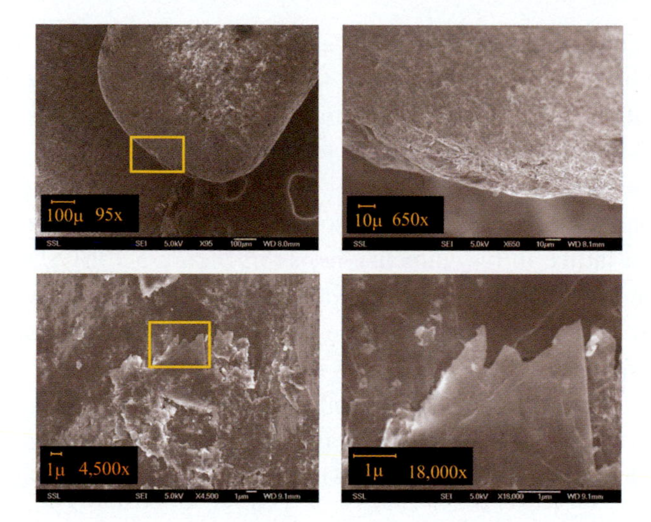

Figure 5.16: Scanning Electron Microscope image of graphite.

We can see why slip appears from the microscopic structure of graphite. A shear force is applied to a piece of graphite, for example, by drawing the tip of a graphite pencil across a piece of paper, as shown in Figure 5.17. Due to the strong intra-planar and weak inter-planar bonding of the carbon atoms of graphite, the shearing force more easily breaks the inter-planar bonds causing a whole plane of graphite to loosen itself from the lattice and slip off the lattice onto the piece of paper. In contrast, one cannot even scratch diamond let alone use it to mark paper; the reason lies in the three-dimensional arrangement of its covalent bonds: one has to break all the bonds in the whole crystal to cause any slippage — and this requires a large force, hence the hardness of diamond.

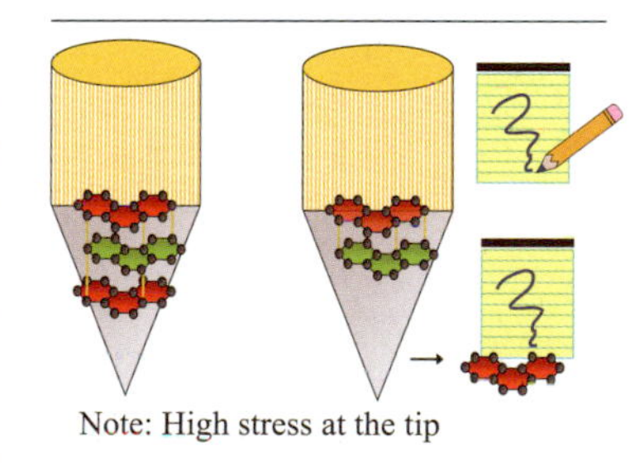

Fig. 5.17: Graphite — Pencil.

5.13 Amorphous solids: glass

Figure 5.18 shows the irregular structure of an amorphous solid in contrast to a crystalline solid.

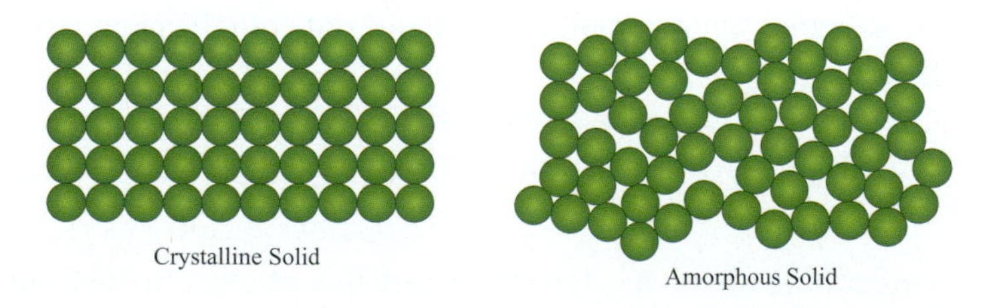

Figure 5.18: A regular crystalline solid and an irregular amorphous solid.

Silicon dioxide glass

We examine in some detail the property of window glass which is the most commonly occurring form of glass. Window glass is mostly made of silicon dioxide (SiO_2), the main constituent of common sand, with some additives that make it easier to manufacture.

A silicon dioxide molecule (also called silica) results from the covalent bond between the silicon and oxygen atoms. Crystalline silica forms a regular three-dimensional lattice, with each silicon atom forming a covalent bond with four oxygen atoms, yielding the fundamental cell to be a tetrahedron.

To simplify our analysis, we replace the bonding of silicon with four oxygen atoms with bonding with two oxygen atoms, and this allows us to represent the simplified silica lattice in two dimensions, as in Figure 5.19a. Oxygen acts as a bridge between the silicon atoms by covalently bonding with two silicon atoms. The real lattice is a three-dimensional version of the simplified two-dimensional lattice. When silica is melted by heating it above the temperature of 1713°C, the silicon atoms still stay covalently bonded with the oxygen atoms, but the liquid state has an irregular structure in which the orderly lattice of crystalline silica is replaced by a disordered interlinked network of irregular units of silica, as shown in Figure 5.19b.

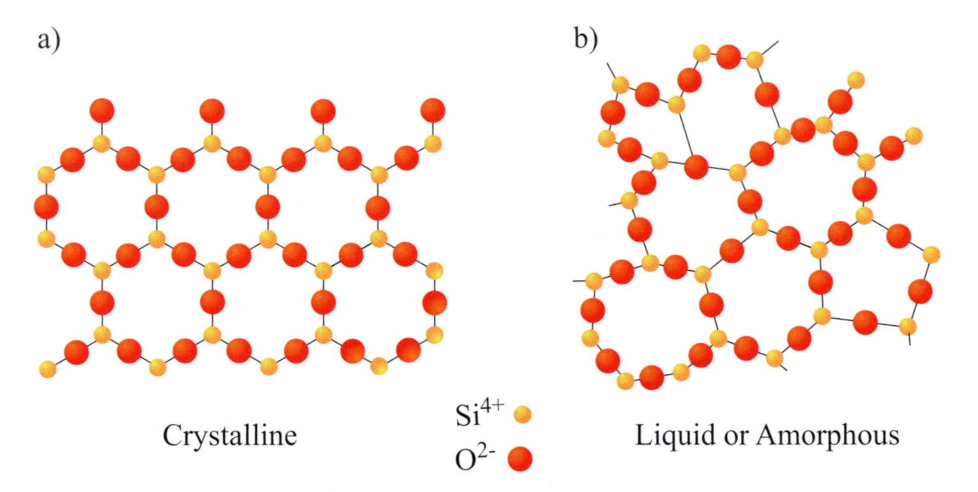

Figure 5.19: a) Simplified two-dimensional representation of the lattice of SiO_2. b) Liquid state of SiO_2 composed of irregular (disordered) rings.

When silica is cooled back to room temperature, due to the high viscosity of liquid silica, the irregular arrangement of the liquid state of silica persists, and takes the form of glass.

Soda-silica glass

The melting temperature of silica is far above even that of most metals, and hence it is difficult to mold and shape silica glass. It is therefore of immense technological benefit to find a way to lower the melting temperature of silica. It turns out that this can be achieved by adding various chemicals. The main chemicals that are added to silica are soda Na_2O, soda ash Na_2CO_3 and some others. The molecule soda is formed by the ionic bonding of oxygen with sodium.

When 25% soda is added to silica, the mixture melts at a temperature of only 793°C. The reason for the low melting temperature of the soda-silica mixture is due to the change in the atomic structure of glass as shown in Figure 5.20.

The sodium atoms in soda donate two electrons to the oxygen atom, leaving sodium positively charged and oxygen two negative charges namely O^{2-}. The negatively charged oxygen forms an ionic bond with the two sodium atoms and a covalent bond with one atom of silicon, and hence breaks the ring structure of glassy silica, as shown in Figure 5.20. A scanning electron microscope image of glass is given in Figure 5.21.

The atomic arrangement of silica resulting from the addition of soda is full of nonbridging oxygen atoms, with the ring network of silica being disrupted by a "dangling" oxygen atom. The breakage of the network greatly reduces the binding energy of the atoms, weakens the glass and lowers the melting temperature of the soda-silica mixture.

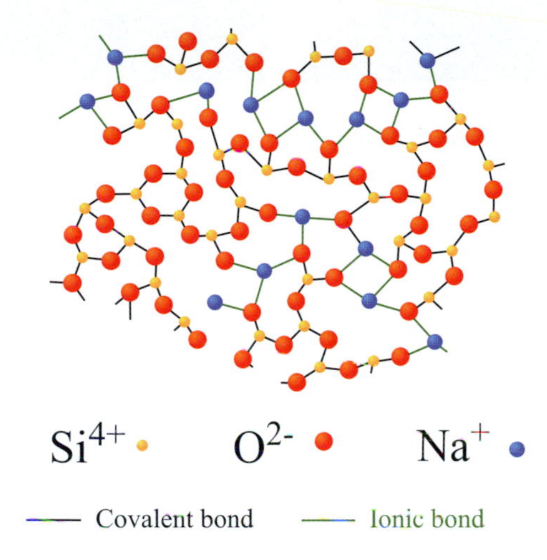

Figure 5.20: Broken rings of glassy SiO_2 due to addition of soda (or soda ash).

The brittleness of glass is due to the irregular network of glassy silica, and adding soda to silica makes the glass even more brittle.

A great variety of glass can be made from silica by adding various other chemicals to it.

5.14 Fuels: quasi-stable bonds

In general, when molecules are formed, the atoms that constitute them are in the state of minimum potential energy. There are, however, many molecules in nature that have energy stored in them and release this energy when properly activated. Petroleum is a prime example of such a case: On burning oil, heat is released; this means that the petroleum molecule is in a state such that by combining it with oxygen — which is what burning does at the atomic level — energy is released. For biological systems, glucose carries energy that living systems release by breaking the bonds in glucose. So how do we understand these molecules? What is the energy diagram for these?

Molecules in a quasi-stable state have their atoms held together by one of the bonds discussed above. In the energy diagram, they are at a local minimum of the potential with a potential barrier keeping them in their quasi-stable state. On

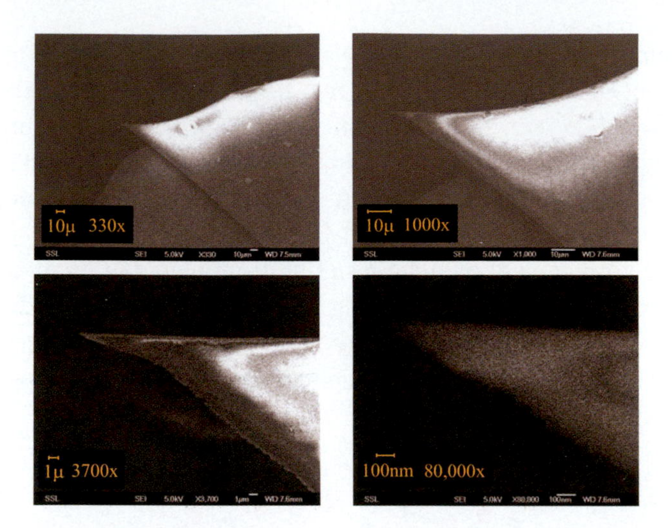

Figure 5.21: Scanning electron microscope image of glass.

external energy being supplied, the atoms comprising the molecules can cross the potential barrier and break apart, going to a new minimum of the potential as shown in Figure 5.22. For petroleum, igniting it with a matchstick supplies the molecule the energy required to cross the potential barrier. Once ignited, a chain reaction starts, with heat generated by the combustion of molecules providing energy for the combustion of other molecules and so on until all the petroleum is consumed.

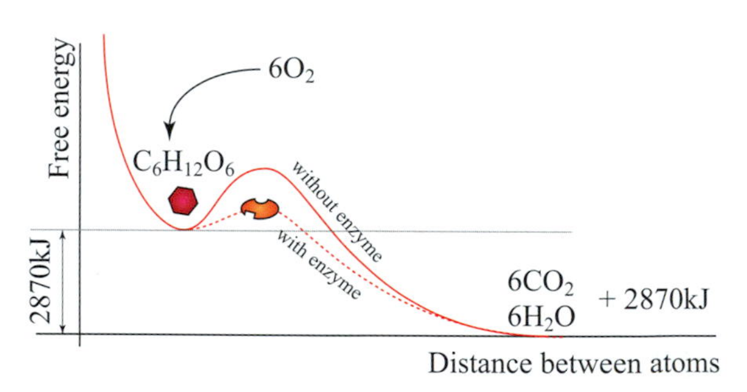

Figure 5.22: Potential energy for glucose.

Figure 5.4 for NaCl and Figure 5.22 for glucose look similar, but there is a very significant difference between the two. When the NaCl molecule is formed, it is in a deep minimum and hence net energy from outside has to be supplied to break it up and separate Na from Cl. In contrast, the local minimum that glucose is sitting at has net energy *higher* than the state in which the molecule is broken and the constituent atoms are separated. Hence, unlike NaCl, glucose $C_6H_{12}O_6$ will *release* energy on being broken up by oxidation; the amount is indicated in Figure 5.22. *Enzymes*, which are a type of protein, play a central role in making the energy stored in glucose available to biological organisms.

The potential barrier is too high for the thermal motion to supply enough energy for glucose to oxidize and cross the potential barrier. Enzymes lower the potential barrier by catalyzing chemical reactions in glucose (at certain steps with the help of other energy-carrying molecules). The oxidation reaction of glucose is given by

$$C_6H_{12}O_6 + 6O_2 \rightarrow 6CO_2 + 6H_2O + 2870 \text{ kJ}.$$

Hence, the oxidation of glucose releases 2870 kJ amount of energy.

One may question as to how did glucose arrive at its chemical structure so that it has a net amount of energy stored in it. It is not clear whether primordial glucose was formed before life emerged on earth. However, once life emerges, glucose is made inside plants by photosynthesis in the following process:

$$\underset{\text{Carbon dioxide}}{6CO_2} + \underset{\text{Water}}{6H_2O} + \text{Light} \rightarrow \underset{\text{Glucose}}{C_6H_{12}O_6} + \underset{\text{Oxygen}}{6O_2} .$$

It should come as no surprise that the energy stored in glucose ultimately comes from sunlight. Cells use glucose to store energy. Glucose is broken down using enzymes such that the stored energy is released and the end product is water and carbon dioxide.

5.15 Bonds in fluids

A fluid is a common term for both liquids and gases, since they share many important properties. The concept of a bond has a very different role in explaining the behavior of liquids and gases and in this section we briefly explain why.

Note: Fluids include both gases and liquids. In an ideal gas the atoms or molecules don't bond at all.

Atoms in a solid have an average position and zero average velocity; so it is meaningful to assign a location to an atom, be it at regular lattice points as in a crystalline solid or at irregular points as in an amorphous solid. One of the fundamental differences between a solid and a fluid is that, unlike a solid, atoms in a fluid have an average velocity that is nonzero; in fact, a measure of the average velocity is the temperature of the fluid. At absolute zero temperature fluids need to be described by quantum mechanics and continue to be unlike solids.

In dilute gases, atoms are well separated and interact only weakly. Atoms do not form any stable configurations and to a first approximation can be considered to be free particles. If one increases the density or lowers the temperature of a gas, it starts to behave more and more like a liquid. One can smoothly transform all gases into a liquid by going around its "critical point", or one can simply cool a gas until it undergoes a phase transition and becomes a liquid. In both cases, the description of a gas, which is a collection of weakly interacting particles, is akin to liquids and very different from solids.

The atoms in a liquid have two levels of structure; at the constituent level, the unit of a liquid is a molecule, made out of atoms that have various forms of bonds that we have discussed. In most liquids, the bonds have a half-life of 10^{-11} to 10^{-12} seconds. In contrast to other liquids, recent studies have shown that water molecules move more than 100 times faster than most liquids, and lose all memory of the hydrogen bonds in about 5×10^{-14} seconds. The reason that the bonds have a half-life of 10^{-11} to 10^{-12} seconds is because this is time taken by the nuclei of the atoms to change their positions; electrons vibrate in the atom at a much faster rate, changing their position in about 10^{-15} seconds.

Due to the changing position of the constituents of a liquid and a gas and the almost continuous forming and breaking of bonds, the force of bonding does not lead to any permanent structures: the intermolecular bonding forces are too weak in liquids to give rise to a stable ordered structure with long range order, as is the case for solids. For this reason, the concept of a bond has to be viewed in a dynamic context and the techniques developed for explaining the bonding of solids and molecules cannot be used for describing or explaining the behavior of liquids and gases.

5.16 Liquids: microscopic structure

To understand the properties of water, which are still not well understood, it is illustrative to compare it with the liquid state of argon, which is a simpler fluid compared to water. The liquids that are the simplest are the ones made from the atoms of the noble gases such as argon. The argon atom has a nucleus containing 18 protons (atomic number 18) and its 18 electrons fill the shells as $1s^2 2s^2 2p^6 3s^2 3p^6$; the argon atom's electron charge is perfectly symmetrical and resists forming any bonds with other argon atoms, existing at most temperatures as a gas.

However, due to random changes in the virtual position of its electrons, argon spontaneously generates a minute and short-lived charge asymmetry — yielding a net negative and a net positive charge distribution. The asymmetry gives rise to an evanescent van der Waals force, which has enough attractive power, at the positively freezing temperature of $-186°C$, to cause argon to condense into a liquid; argon is bonded into a tertramer by the van der Waals force. In contrast to water, on freezing, the volume of liquid argon contracts by 12%. Because the van der Waals force is so weak, liquid argon is relatively easy to model mathematically and by the 1970s most of the properties of liquid argon were understood from a microscopic point of view.

Water is an entirely different matter. The hydrogen bond means that the electrical imbalances are not some shimmering and evanescent phenomena as in the case of argon, but rather the electrical dipole is hardwired into the very molecular structure of the water. Hence, water molecules are subject to strong electrical forces, with each water molecule attracting other water molecules. Due to hydrogen bonding, each water molecule can bind up to four other water molecules. The water molecules form a vast network of bonded molecules. The network is constantly dissolved and formed on time scales of 5×10^{-14} seconds. A typical arrangement of water molecules in liquid water is shown in Figure 5.23.

In terms of the number of hydrogen bonds linked to a particular molecule and the strength of the bond, it is thought that liquid water is not very different from ice. In fact, liquid water has almost about as much hydrogen bonding as does ice. However, it is thought that the liquid quality of water actually depends more on the half-life of the hydrogen bonds rather than the actual number of hydrogen bonds. In liquid water, the water molecules change their bonding to other nearby molecules almost 100 times faster than other liquids.

5.17 H_2O: the water molecule

Water is composed of the water molecule, namely H_2O. The oxygen atom has atomic number 8 (eight protons in the nucleus); from our earlier discussion on the periodic table, it is known that the eight electrons arrange themselves in the

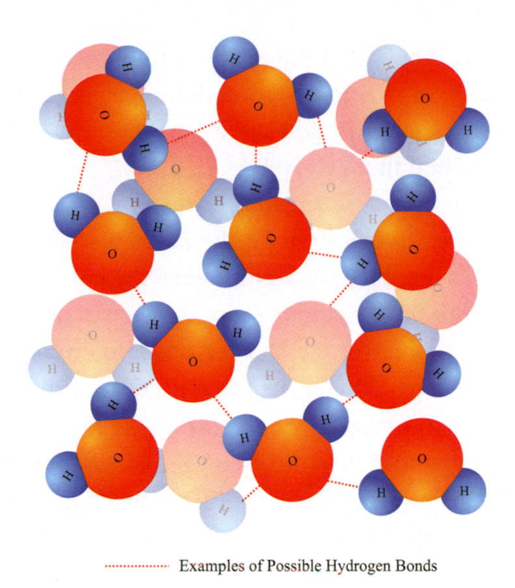

Figure 5.23: Hydrogen bonds constantly form and break up in liquid water.

following configuration: $1s^2\mathbf{2s^22p^4}$; what this notation means is that two electrons are in the lowest energy state of $1s$ orbitals, and another two electrons are in the next available energy states, namely the $2s$ orbitals. That leaves four electrons in the $2p$ orbital; now from our general formula, we know the second energy shell $n = 2$ closes once it has $2n^2 = 8$ electrons. Hence oxygen has a vacancy of two electrons in its second shell. All the electrons in the $n = 2$ energy level can interact with other atoms or molecules and hence are valence electrons.

The water molecule H_2O has a bent structure, with the angle H-O-H being $104.5°$. Due to its bent shape H_2O has a dipole moment permanently embedded into its molecular structure. The simplest explanation of the structure of H_2O is that each hydrogen atom forms a covalent bond with oxygen that involves the oxygen atoms' $2p$ electrons. There are four electrons in the 2p orbital of oxygen; it shares two of these electrons with the hydrogen atom, each of which has one electron in the 1s orbital. Crudely speaking, since each electron in the $2p$ state has orbital angular momentum of $\hbar$, it is "spinning" about an axis. The Pauli exclusion principle requires that the two electrons in the 2p orbital be in distinct states; in quantum mechanics, two states that are orthogonal are completely distinct (independent) of each other; hence, from this crude picture, we would expect that the two electrons should spin about directions that are perpendicular.

More precisely, choose the coordinate axis so that the first electron is spinning about the x-axis, the second electron about the y-axis and the remaining two electrons are in an orthogonal combination spinning about the z-axis. One of the hydrogen atoms forms a covalent bond with the $2p$ electron spinning about, say, the x-axis and the other hydrogen atom bonds with the electron spinning about the y-axis. Forming a covalent bond means that the two electrons, which are being shared by the oxygen and the hydrogen atoms, combine to form an s state (a zero orbital angular momentum state) that **lowers** the combined energy of the two atoms. The binding energy of the covalent bond between the hydrogen nucleus and the oxygen nucleus is -5.2 eV.

Our analysis seems to imply that the angle H-O-H should be 90°; hence, to get the correct answer one needs to improve on the approximation. The next refinement is to note that since the hydrogen nuclei are protons, they repel each other, increasing the angle between them to $104.5° \simeq 105°$.

The two electrons in the $1s$ state usually do not participate in the chemical properties of oxygen. The six electrons in the second shell all have the same energy, of which two are in the $2s$ state having zero orbital angular momentum and the four $2p$ electrons each have angular momentum equal to $\hbar$. Now it turns out that it is energetically favorable for the oxygen atom to combine the four electrons in the $n = 2$ energy shell in a phenomenon called hybridization, so that the resulting four states, which no longer have a definite angular momentum, have lower energies. The fully hybridized states of the oxygen atom form covalent bonds with the hydrogen atoms, and predict an incorrect H-O-H angle of 109°. The actual angle of 105° in the water molecule is produced by the partially hybridized states of the oxygen atom.

The four $2p$ electrons form a tetrahedral structure for the oxygen, of which two of the electrons are covalently bonded to the two hydrogen atoms. The covalently bonded O-H has a separation of 0.1 nm.

Polar molecule

Due to the bent shape of the H_2O molecule the hydrogen nuclei, carrying positive charge, are on one side of the oxygen atom with the compensating negative charges of the electrons concentrated on the opposite side of the oxygen atom. This asymmetry in charge distribution leads to a large and permanent dipole moment of the H_2O molecule in the symmetry plane pointing toward the more positive hydrogen atoms, as shown in Figure 5.24.

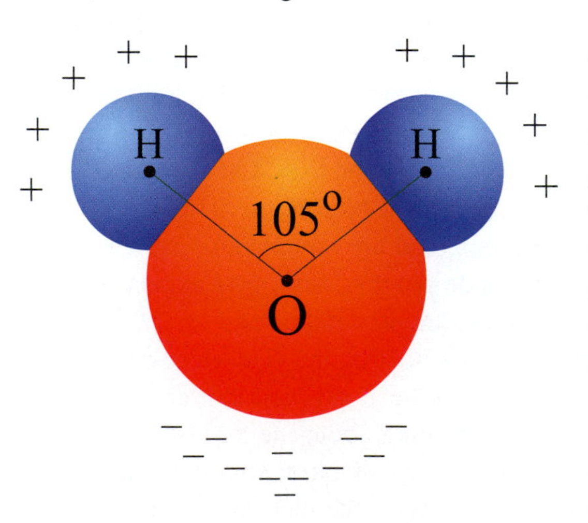

Figure 5.24: Water is polar.

The measured magnitude of this dipole moment equals $p = 6.2 \times 10^{-30}$ C m. We can find what is the classical dipole that corresponds to the dipole moment of the water molecule. If one considers H_2O to be a collection of positive charges with charge equal to $q = +10e$ separated by distance d from a negative charge of equal and opposite amount $-10e$, then setting $p = dq$ to the observed value yields $d \simeq 3.9 \times 10^{-12}$ m = 0.003 nm, which is much less than the "Bohr" radius of 0.05 nm, the typical size of an atom. The actual distance between the atoms in the water molecule is much larger than the distance required for a classical dipole moment, showing that quantum mechanical effects are at the root of the permanent dipole moment of the water molecule.

Due to its permanent electric dipole moment, H_2O is called a dipolar, or sometimes just a polar, molecule. The dipolar nature of the water molecule leads to the hydrogen bonding of the water molecules, shown in Figure 5.25. The dipole moment of H_2O is the reason that microwave ovens work. The microwave radiation is a time-varying electric (and magnetic) field inside the oven that couples to the dipole moment of water molecule H_2O in food and causes it to oscillate. The oscillations impart energy to the (cold) food, which gets heated by absorbing the energy.

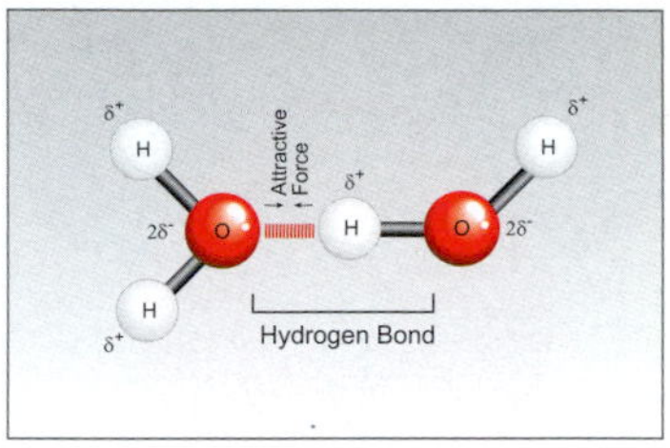

Fig. 5.25: Hydrogen bond.

Why is the color of water blue?

The blue color of water can be readily observed by looking at water in a transparent container. Color can originate in many different ways. In particular most of the colors that we see around us in daily life are the result of the way photons (quanta of light) and the electrons of an atom or molecule interact, with the atom's electrons absorbing and emitting photons, by a) resonant scattering, b) nonresonant free electron Rayleigh scattering, c) Thompson scattering, d) diffraction, e) interference and f) refraction. In all these cases photons interact primarily or exclusively with electrons.

The color of water does not come from reflecting the blue color of the sky and neither does it come from the interaction of photons with the electrons of the H_2O molecule. Instead, the intrinsic blue-green color of water is the only known case, from all the materials studied, that arises from the molecular vibrations of the water molecule! Water absorbs photons in the infrared and ultraviolet ranges and emits the energy absorbed by vibrational oscillations of its covalent bond in the blue end of the spectrum. The deep blue color arises from the symmetric bond stretching oscillations with a wavelength of 660 nm and the asymmetric bond stretching that yields vibrations of wavelength 605 nm.

The observed color of water has contributions from reflected sunlight as well as the intrinsic absorption of light by water molecules. Light reflected by suspended matter is required for the blue light emitted by water to return to the surface. Such scattering can also shift the emitted photon's energy toward the green color as often seen when water is polluted by suspended particles. The color of reflected light seen emerging from water strongly depends on the angle at which this light is observed.

5.18 The hydrogen bond and liquid water

The concept of atomic bond is discussed for the case of liquid water. The positions of water molecules in the liquid state are in a state of flux, changing at a time scale of about 5×10^{-14} seconds. Due to these constant changes, bonds between molecules are made and broken at a rapid rate.

The hydrogen bond plays a central role in the structure of both liquid water as well as ice. Due to its great importance in chemistry and biology, the structure of water is discussed.

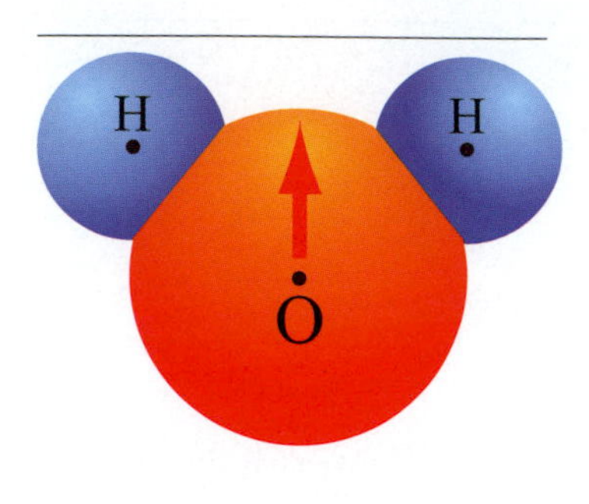

Fig. 5.26: Net permanent dipole moment of the water molecule.

What are the intra- and intermolecular forces between water molecules?

The water molecule itself, namely H_2O, is formed by covalent bonding of two hydrogen atoms with a single oxygen atom — with binding energy of -5.2 eV. The angle between the two hydrogen atoms in a water molecule is about 105°, as shown in Figure 5.24, and this asymmetry of the water molecule leads to a *permanent* dipole moment as illustrated in Figure 5.26. The dipole points toward the slightly positive hydrogen atoms. The measured magnitude of this dipole moment is 6.2×10^{-30} C m.

The dipole moment gives rise to the hydrogen bond between water molecules. A hydrogen bond is formed due to the permanent dipole moment of the water molecule. The hydrogen has a partial positive charge due to the permanent dipole moment, and for the same reason the oxygen has a partial negative charge. Hence, there is an attractive force between the hydrogen (in one water molecule) and the neighboring oxygen (in another water molecule).

The net positive charge carried by the hydrogen nucleus is attracted to the net negative charge carried by the oxygen atom, and is illustrated in Figures 5.25 and 5.27. The final configuration that results is that the hydrogen atom covalently bonded to the oxygen molecule is attracted to the oxygen of a *nearby* water molecule due to the (electrostatic) Coulomb force, as shown in Figure 5.27.

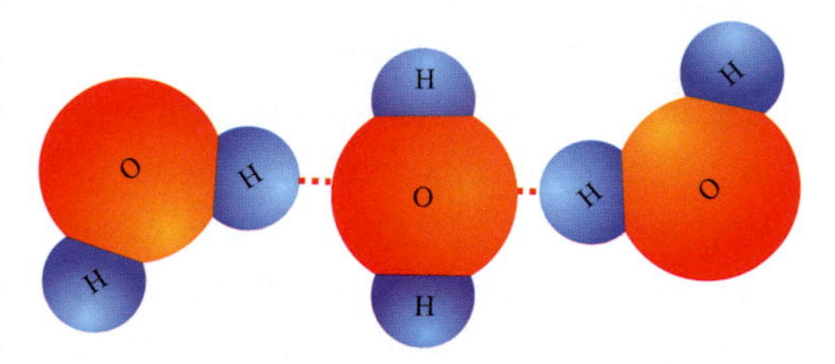

Figure 5.27: Hydrogen bonds in water molecules.

Each hydrogen bond has a binding energy of -0.2 eV, much smaller than the covalent bonding of the H_2O. The structure of water is a result of the hydrogen bond, and is illustrated in Figure 5.27. Denoting the covalent bond by $-$ and the hydrogen bond by $\cdots$, a hydrogen atom is bonded to two oxygen atoms by a covalent and hydrogen bond, which is denoted by $O^{-}H\cdots O$; recall the distance between $O^{-}H$ is 0.1 nm and that between $H\cdots O$ is 0.176 nm, and hence the two oxygen atoms are separated by 0.276 nm.

In conclusion, *each oxygen atom* is attached to *four* hydrogen nuclei, two via covalent bonding (that forms the H_2O molecule) and two via hydrogen bonding leading to liquid and ice structures of water. The hydrogen bond lasts for only 10^{-16} seconds before it is broken by the movement of the water molecules and reformed again. It is this constant making and breaking of the hydrogen bond that is the reason for the many remarkable properties of liquid water.

The hydrogen bond is especially important in chemistry and biology. Within macromolecules such as proteins and DNA (deoxyribonucleic acid), the hydrogen bond exists between two parts of the *same* molecule; in particular, the DNA is composed of two strands of biopolymers, with the two strands being held together by the hydrogen bond between the nucleic acid monomers that constitute the strands. The hydrogen bond is an important force determining the overall shape and structure of such macromolecules.

Although stronger than most other intermolecular forces, the hydrogen bond is much weaker than both the ionic bond and the covalent bond. Note that the hydrogen bonds between polar molecules are much stronger than the very weak bonds formed due to the van der Waals force, which have a binding energy of only -0.01 eV per bond. The dynamic nature of the hydrogen bond and the large value of its binding energy is the reason why liquid water has so far defied all attempts to understand it from the first principles of quantum mechanics.

In both water and ice, each H_2O molecule is surrounded by four other H_2O molecules, bound to each other by the covalent and hydrogen bonds. In ice, the crystalline lattice is dominated by a regular array of hydrogen bonds which cause the water molecules to be farther apart than they are in liquid water. This accounts for water's decrease in density upon freezing. In other words, the presence of hydrogen bonds enables ice to float, because this spacing causes ice to be less dense than liquid water.

Noteworthy 5.1: Microwave oven

Most food placed inside a microwave oven usually has plenty of water molecules. Food can be heated with a microwave because the water molecule has a dipole moment.

Microwave radiation inside the oven creates a rotating electric (and magnetic) field. The dipole moment of the water molecule couples to the rotating electric field and rotates synchronously with the field. As shown in Figure 5.28 for the H_2O molecule to rotate (with the microwave electric field), its hydrogen bond with other water molecules needs to be broken. The energy of microwave radiation is sufficiently high to provide the 0.2 eV energy required to break the hydrogen bond. The breaking of the hydrogen bond leads to the absorption of microwave energy thus heating up the food.

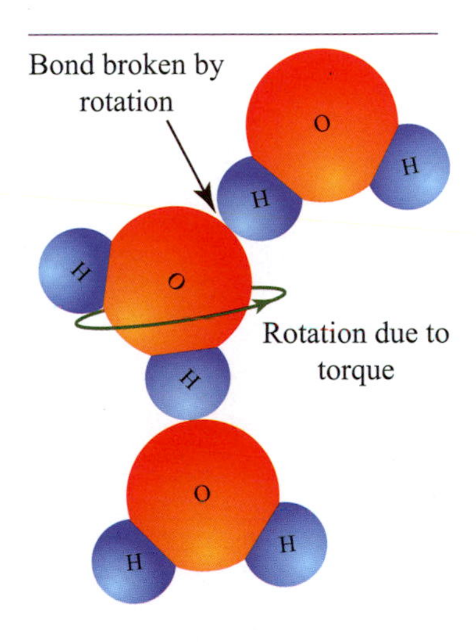

Fig. 5.28: Effect of microwave energy.

5.19 Properties of water

Water is a rather unique substance having many unusual properties. We list a few properties of water to illustrate its remarkable nature.

Ionic dissociation and the pH of water

An acidic solution is one that has a surplus of positive hydrogen ions H^+ and a basic solution has a surplus of negative hydroxide ions OH^-, as shown in Figure 5.29. One may ask the question: are there ions present in water? Is water perfectly neutral, or is it acidic or basic? It turns out that both H^+ and OH^- ions are present in water in equal concentrations.

Figure 5.30 shows how water molecules dissociate into positively and negatively charged ions through the reaction

$$H_2O \leftrightarrow H^+ + OH^-.$$

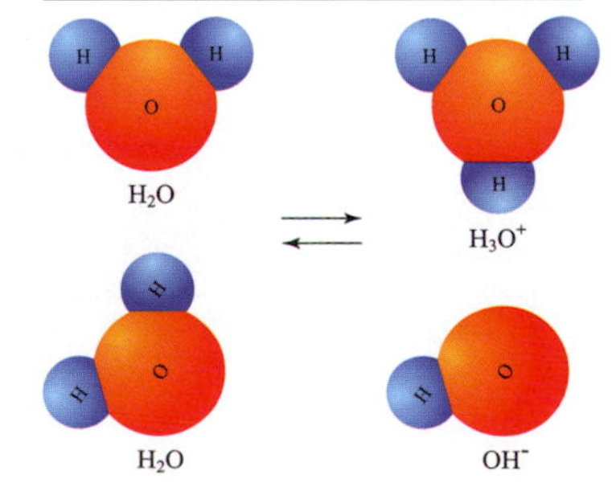

Fig. 5.29: Ionized forms of water.

Another spontaneous process that creates ions in neutral water is the capture of one of the covalently bonded hydrogen nuclei by a water molecule by the reaction

$$2H_2O \leftrightarrow H_3O^+ + OH^-$$

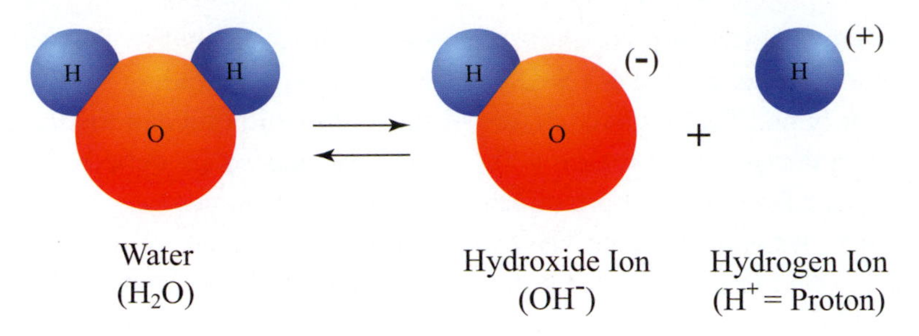

Figure 5.30: Water constantly goes back and forth between ionized and non-ionized forms.

and is shown in Figure 5.29. Water that is free from all forms of contaminations, i.e. pure water, contains hydrogen ions with a concentration of 10^{-7} mol/L, with hydroxide ions with the same concentration; the ions are created by spontaneous ionization reactions. The pH of a liquid measures the density of hydrogen ions, with pH $= 7$ for pure water. The more acidic a liquid, the higher its pH; the more basic a liquid, the lower its pH.

Surface tension

Fig. 5.31: An effect of surface tension.

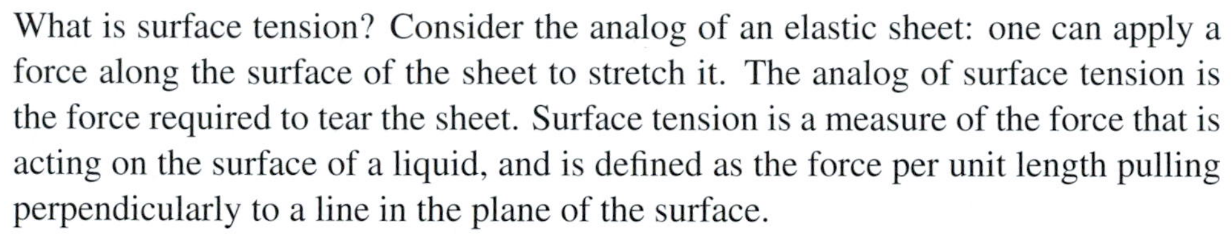

What is surface tension? Consider the analog of an elastic sheet: one can apply a force along the surface of the sheet to stretch it. The analog of surface tension is the force required to tear the sheet. Surface tension is a measure of the force that is acting on the surface of a liquid, and is defined as the force per unit length pulling perpendicularly to a line in the plane of the surface.

Compared to other liquids, water has an extremely high surface tension, which at 20°C is 7.29×10^{-2} J/m. Surface tension for a polar molecule like water is thought to arise from the dipole-dipole attraction. In a water molecule, the oxygen atom is relatively negative compared to the hydrogen atoms, which are relatively positive. On the surface of water, the oxygen atoms in water form hydrogen bonds with the hydrogen atoms and give rise to a high value for surface tension.; the water molecules on the surface form a chain of dipole bonded network.

The high surface tension of water plays a crucial role in biological processes and structures. The intermolecular hydrogen bonds attracting the surface molecules cause water to form into spherical droplets. Because of its high surface tension, water can support fairly large objects placed on its surface.

Why is water a good solvent?

The high dielectric constant and the polar nature of water — having a permanent dipole moment — are the two most important reasons that water is a good solvent and can dissolve many materials.

Consider the case of ordinary salt, that is NaCl. In air, salt forms a regular solid lattice and the Na and Cl form a stable molecule, held together by the ionic bond, and thermal motion is not strong enough to dissociate the atoms. When immersed in water, each NaCl is surrounded by water molecules, with the oxygen atom aligning itself close to the positively charged Na^+ and the hydrogen atoms being electrically attracted to negatively charged Cl^- atom. Due to effect of the water molecules on NaCl, thermal motion is strong enough to dissociate NaCl into a positively charged ion Na^+ and a negatively charged ion Cl^-. A similar dissociation occurs for many ionically bonded molecules.

The polar nature of the water molecules acts to stabilize the dissociation of ions in water. Each dissociated negative ion, say Cl^-, attracts the positive ends of the water molecules and with positive Na^+ ions similarly attracting the negative ends, as shown in Figure 5.32. The water molecules form hydrogen bonds with the ions and results in the formation of a hydrate. In effect, a hydrate is a collection of ion and water molecules held together by the hydrogen bond. The charge of the ion is screened by the water molecules and in doing so lowers the energy of the solute. The result is that the ions are bound fairly strongly to the water molecules and form a stable solution.

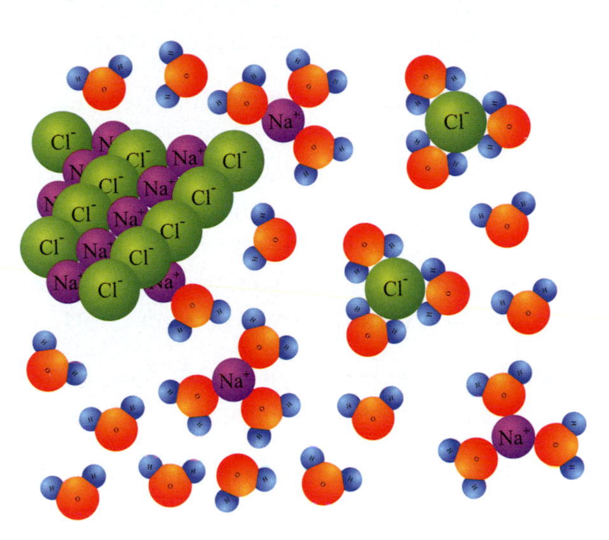

Figure 5.32: Solution of salt in water.

Water can also dissolve and hydrate molecules bound by polar covalent bonds. The positively and negatively charged poles of water attract the charged regions of the dissolving molecules. **Hydrophilic** is a term used to denote polar and ionic molecules because of their electrical attraction to water molecules. Molecules that are uncharged and unipolar such as fats and oils do not dissolve in water and are called **hydrophobic**.

5.20 Water is essential for biological processes

Water's abundance and unusual properties make it essential for life. Life most probably arose in the oceans' waters of primordial earth. Living organisms contain, by weight, about 60% to 70% water, and all processes of life depend intimately on water. The following are some of the important functions that water has for living organisms:

- Water can participate in a great variety of chemical reactions because water is a good solvent, capable of dissolving proteins, salts and sugars. Hence, water participates in many chemical reactions essential for the functioning of living cells, from photosynthesis to the production of proteins and other bio-macromolecules.

- Hydrophilic molecules such as sugars readily dissolve in water and thus water provides the necessary environment for many indispensable chemical reactions.

- Hydrophobic molecules play an important role in biology. In particular, the membranes of living cells owe their structure to the hydrophobic property of lipids that compose the cell membrane.

- In most mammals the interior of cell and the fluid that permeates the cell's environment are nearly neutral (pH of about 7.3). Small fluctuations in the pH of the fluid can cause drastic malfunctions and even the death of the organism. There are an almost limitless number of chemical reactions going on ceaselessly inside and outside the cell. Water acts as a buffer, always driving the fluid toward constant (neutral) values of pH – releasing or accepting H^+ ions in response to small changes of pH.

- Since water has a high specific heat, it can moderate the effects of temperature by accepting or releasing heat as required.

- During winter in temperate climates, since ice floats on water, it provides an insulating layer that delays the freezing of the rest of the water. Ice being less dense than water, in effect, provides a safe haven below it for many types of fishes.

- The high surface tension of water allows plants to absorb water through their roots. The absorbed water is then "pulled up" by capilliary action to heights greater than 100 m to the upper part of tall trees. Due to the strong hydrogen bonding, which forms a chainlike column of water, water can be pulled up to great heights without the chain being broken. Furthermore, the chain of water does not "tear" against the walls of the tree tubes, as would have been the case if water's surface tension were low.

5.21 The answer

Why do atoms combine? **Atoms combine to lower their energy**. This simple and unifying theme runs through all forms of bonding of atoms. Atoms lower their energy in many different ways, called bonds, due to the various types of attractive forces that exist between them. There are two irreducible sources of bonding, namely bonds arising due to the quantum mechanical properties of atoms, in particular the sharing of electrons, and bonds due to the force of electromagnetic interactions. All the bonds, including the covalent, ionic, metallic, polar and van der Waals bond, are the result of some combination of these two underlying properties of atoms.

The (permanent) bonding of atoms results in either molecules or solids. The atoms in molecules and solids have well-defined average locations and combine due to various chemical bonds — forming stable structures due to interatomic bonding. Molecules and solids are different from fluids; due to the flowing nature of fluids, atomic bonds play a very different role in determining the liquids' structure and processes.

5.22 Exercises

1. Is it true that the key mechanism in covalent bonding is electrostatic attraction?
2. Is it true that the ionization of sodium requires energy?
3. Using textbook Equation 5.1, what would be the binding energy of two atoms that are 300 pm apart?
4. Does H_2 form a polar covalent bond?
5. Is salt a 100% ionic bond?
6. Name two molecules where the atoms are 100% covalently bonded.
7. In general, would a polar covalent bond be stronger than an ionic bond?
8. Which lattice is more efficient in packing atoms into a fixed volume of space: simple cubic or face centered cubic?
9. Diamonds and graphite are both made up of carbon atoms. Then why is one very hard and the other not?
10. Why is the melting temperature of soda glass lower than that of silica glass?

6: Fluids

Q - How Does Water Flow?

Chapter Map

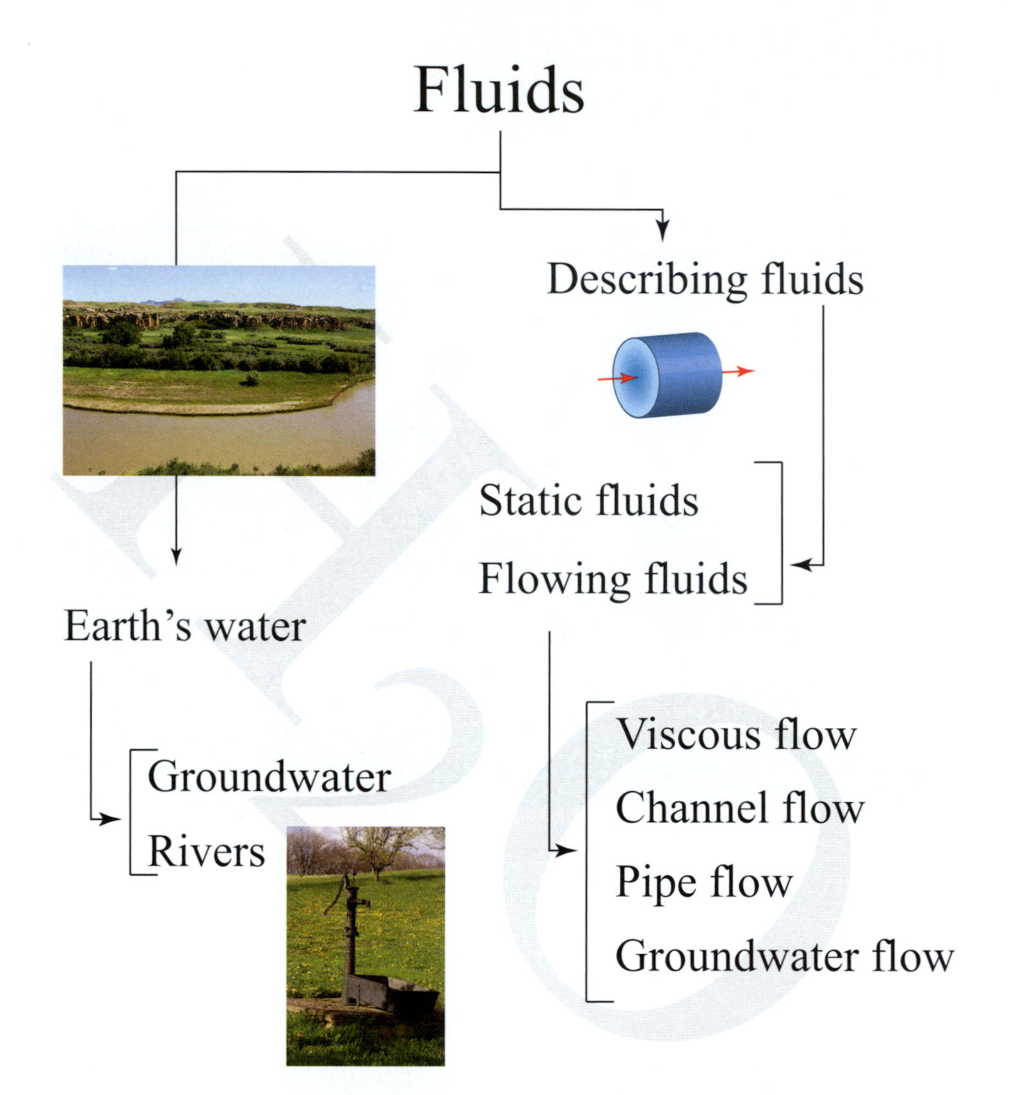
Fluids
Describing fluids
Static fluids
Flowing fluids
Earth's water
Groundwater
Rivers
Viscous flow
Channel flow
Pipe flow
Groundwater flow

6.1 The question

Water is a precious resource for life on earth and for human beings in particular. The cradles of human civilizations were all built along the banks of major rivers such as the Nile, Tigris and Euphrates, Indus and Yangtze rivers. These ancient civilizations used the flow of the river as an integral part of their societies.

Water is a substance well known to every individual and occurs on earth in all of its three states, namely solid (ice, snow), liquid (water), and water vapor. Water usually refers to its liquid phase for which it is transparent and colorless in thin layers; thick layers of water have a bluish green color.

Fig. 6.1: Almost all of earth's water is in the oceans.

Why water flows is explained by the law that all objects tend toward configurations that minimize their energy. Rivers originate in the mountains where they have high gravitational potential energy. Their flow to the oceans is an ongoing process of constantly lowering their potential energy until they reach sea level.

The question remains: **How does water flow?** How does one describe its flow? Water is a fluid and the flow of water is part of a larger subject of fluid mechanics. We briefly review some key ideas of this important field of science to understand how water does flow.

Oceans are water in a fixed volume subject to the influence of earth's rotations and energy exchange with the sun and the atmosphere. This is a vast and complex subject that is not directly linked to the flow of water and will not be discussed.

6.2 Earth's water

Water is essential for life on earth, and human beings can survive for only a few days without water. One may wonder, how much water is there on earth?

70.8% of the earth's surface is covered by water. The dry surface of the earth is not distributed evenly over globe; in the Northern Hemisphere the ratio of water to land is 1.5:1 whereas in the Southern Hemisphere the ratio is 4:1. Continents are separated by ocean basins that are connected to form a single body of water called the world ocean. The Pacific Ocean is the largest and deepest ocean basin, the Atlantic is the second largest and the Indian Ocean is deeper than the Atlantic. Earth's approximate total water supply is 1,360,000,000 km^3 and, as shown in Figure 6.2, has the following distribution:

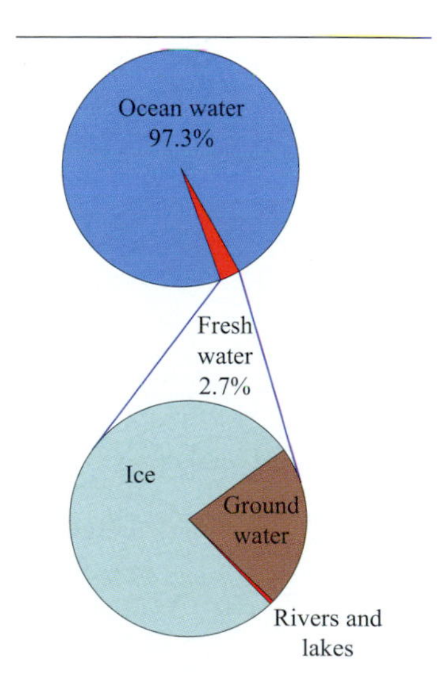

Fig. 6.2: Distribution of water on earth.

- 97.3% in oceans with a volume of 1,320,000,000 km^3 of water
- 1.8% in polar ice and glaciers with 25,000,000 km^3 of water
- 0.9 % in underground aquifers (fresh) with 13,000,000 km^3 of water
- 0.02% in lakes and rivers with 250,000 km^3 of water
- 0.001% in atmosphere (water vapor) with 13,000 km^3 of water

A very small amount of the earth's water is contained in the earth's biosphere.

Groundwater holds about 30% of the world's drinkable water. Given the increasing scarcity of water worldwide, groundwater could prove to be a crucial resource for human society.

Origin of earth's water

Much of the universe's water is thought to have been produced by the same processes that lead to star formation. When the gaseous cloud from which stars are born starts to coalesce due to gravitational attraction, strong outward blowing winds of gas and dust are created. The outflowing material impacts the surrounding gas creating shock waves that compress and heat the gas. Both hydrogen and oxygen are abundantly found in the gas, and hence water is readily produced in the warm dense gas.

The earth is about 4.54 billion years old, to an accuracy of 1%. The formation of the earth took a few hundred million years to be completed. About the time the earth was formed, the gaseous cloud forming the sun became dense enough so that fusion reaction in the core of the proto-sun was ignited. This didn't happen smoothly and, for a while, fusion most likely started in a sputtering way. Each flaring up of fusion in the proto-sun sent particles streaming out. If the earth had an atmosphere at this time, it would have been blown off, leaving the earth as a rock with neither air nor water on its surface.

After fusion in the sun had stabilized, the earth went through a phase of releasing gases from its interior in a process called degassing. There was no free oxygen in the atmosphere at this time, which was a collection of largely water vapor, methane and carbon dioxide, held to the earth by gravitational attraction. Over a relatively short time, something like 100 million years, enough material had been released to give the earth an atmosphere. Fortunately for the emergence of life, within this early period, the temperature of the earth dropped below 100°C and water vapor in the atmosphere then condensed into the oceans that we know today. The mass of water present in the oceans is about 10^{21} kg and is approximately the same as was contained in the crust when the degassing process started.

6.3 Flow of fluids

A fluid is an extended object, occupying a finite volume. In the first approximation, a fluid is a continuous distribution of mass over some specified volume. The points in the volume are specified by their position $\mathbf{x}$. The continuous distribution of mass is described by a mass density, which is mass per unit volume and is denoted, at point $\mathbf{x}$ and at time t, by $\rho(t, \mathbf{x})$; the unit of density is kg/m^3. For water, which is the only case that we will analyze in some detail, the density is constant since water is an incompressible fluid; hence, $\rho(t, \mathbf{x}) = \rho$: constant.[1]

A solid body is also approximately a continuous mass that is specified by its density. So what is the difference between a solid and a fluid?

[1]Fluids like water and oil are considered incompressible in distinction to gases which can easily be compressed. However, this does not mean that they are not compressible at all, it just means that they compress only very little.

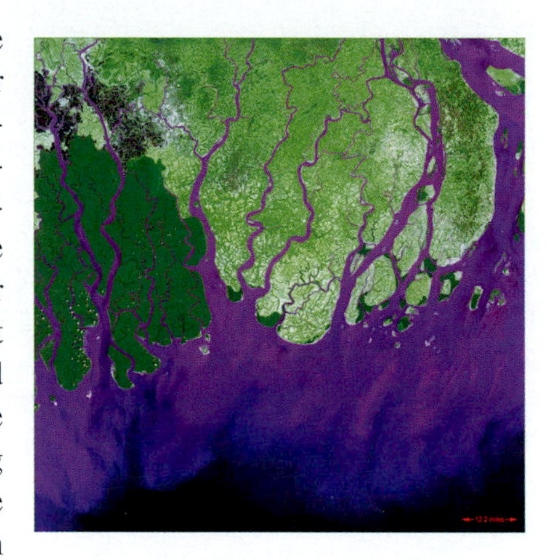

Figure 6.3: Ganges River delta. Source: USGS.

All fluids as well as solids are made from atoms. In a solid, the atoms (or molecules) that constitute it are permanently bonded to the other atoms and oscillate about their average constant position — giving rise to a relatively stable solid structure. In contrast, the atoms or molecules that make up a fluid do not have any permanent bonds with other fluid atoms or molecules. In most liquids the bonds have a duration of only a fleeting 10^{-11} to 10^{-12} seconds, after which the atoms all move around and then again form new bonds with other atoms. The atoms (or molecules) composing a liquid, unlike a solid, do not have any average position. Instead, they are in a state of incessant random motion, continually making and breaking bonds. Hence, a fluid does not have any fixed shape. Indeed, this is the reason that a fluid can take the shape of any container: the fluid atoms or molecules simply move around and adapt to container's form.

The term *fluid* refer to the common properties of liquids and gases, in particular that their atoms or molecules do not have any average position that is constant. There are, however, major differences between liquids and gases, the most important being that the density of a liquid is comparable to a solid and is about 1000 times higher than that of a gas. Atoms or molecules in a gas are virtually free from any interaction with other gas atoms or molecules, whereas the atoms or molecules in a liquid have bonds with other liquid atoms or molecules that give it the cohesion and compactness that is absent for gases. In particular, in Chapter 5 we discussed the crucial role that hydrogen bonding plays in creating the liquid state of water. We will concentrate on liquids and make only a few passing references to the behavior of a gas (some properties of gases are discussed in Chapter 10).

6.4 Describing a fluid

How can one describe the flow of a fluid? If one looks at a river, one realizes that, although the river apparently "looks" the same, each piece of the river is constantly moving. If one puts a small piece of paper on the river, one can see the paper float away. To describe the flow of a river, we need to identify, to "mark", *each* component piece of the river so that we can track how the river flows.

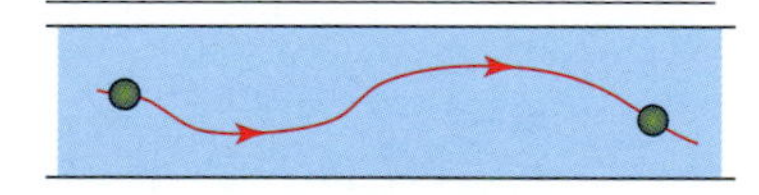

Fig. 6.4: The Lagrangian description follows the particle.

There are two ways of marking the components of a river. One way is to identify a particular element of the river and then follow how it travels from its origin (for example, in the mountains) until it reaches the sea. We can follow the particle, as function of time, by assigning a coordinate to its position at each instant, as shown in Figure 6.4. This description of motion is called the **Lagrangian description** of a physical system and is similar to how we describe the motion of particles.

Density, velocity

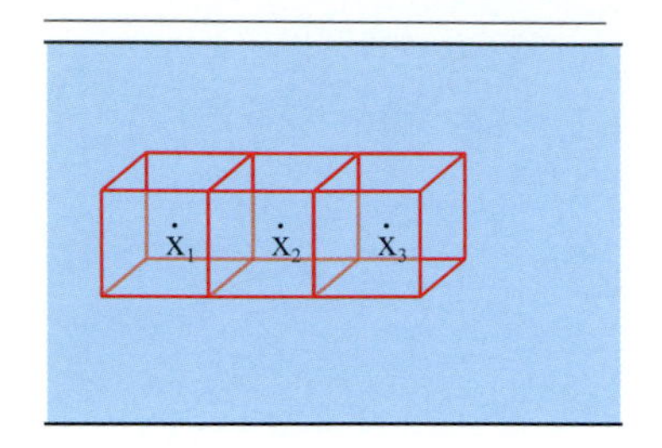

Fig. 6.5: Little cubes make up the volume of the fluid.

There is another description of a fluid, called **Eulerian description**, in which we do not follow the movement of a particular element of the fluid. Instead, one breaks up the volume occupied by the fluid into small elemental volumes that are kept *fixed* over time, as shown in Figure 6.5. In the Eulerian description one keeps track of all the elemental volumes — taking into account what flows into and out of the fixed volume element, as shown in Figure 6.6. One can consider the fixed volume to be of very small (infinitesimal) size and each volume element is labeled by **x**.

A fluid has mass density $\rho(t, \mathbf{x})$, which is the mass contained in a unit volume. The density of a typical fluid is approximately the same as that of a solid; for example, the density of water at a temperature of 4°C is 1,000 kg/m^3, whereas the density of ice at 0°C is 917 kg per meter cube. What other quantities do we need to describe the fluid? The river is in a state of constant flow and it is the *velocity* of the fluid that holds the key to its behavior. The velocity of the fluid at every point inside the fixed volume has a vector velocity, denoted by $\mathbf{v}(t, \mathbf{x})$ and which has units of m/s.

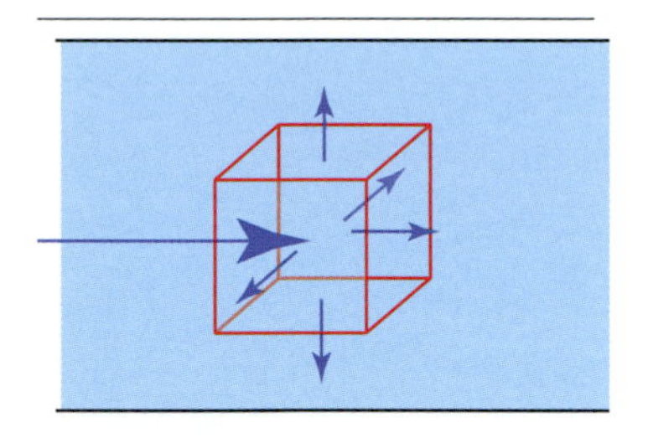

Fig. 6.6: The Eulerian description considers a fixed volume.

The velocity field $\mathbf{v}(t, \mathbf{x})$ — the word "field" being used to remind us that we are referring to the velocity of the fluid at every point of its volume — is the average of the velocities of the atoms inside the elemental volume located at **x** and taken at time t; hence, for N atoms in a volume element ΔV, the average velocity is given by

$$\mathbf{v}(t, \mathbf{x}) = \frac{1}{N} \sum_{\text{Atoms in } \Delta \text{V}} \mathbf{v}(t, \mathbf{x}_\text{A})$$

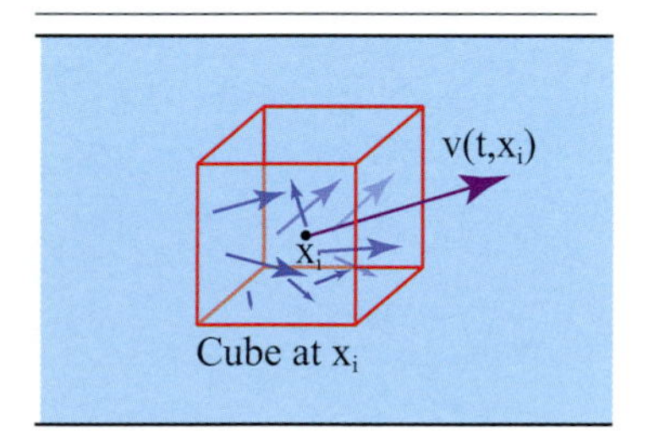

Fig. 6.7: Velocity is an average over the cube.

as illustrated in Figure 6.7.

It is important to note that the macroscopic velocity $\mathbf{v}(t, \mathbf{x})$ is given by a vector sum over the velocities of the atoms. The average velocity **v**(t,**x**) is non-zero if there is a net movement of the fluid, as in the flow of a river. The macroscopic velocity of a fluid at rest is zero, but the atoms of the fluids are nevertheless moving about randomly, without any average fixed position and with zero average velocity.

The description of a fluid consists of keeping track of the fluid's behavior over time t, in particular knowing $\mathbf{v}(t, \mathbf{x})$ — the velocity of the fluid at a fixed volume element located at **x**.

Streamline and turbulent flow

We will consider only steady flow for which velocity does not depend on time, namely $\mathbf{v}(t, \mathbf{x}) = \mathbf{v}(\mathbf{x})$. The particles comprising the fluid can be thought of as moving on trajectories defined by $d\mathbf{x}/dt = \mathbf{v}(\mathbf{x})$; on solving this equation we obtain a series of curves, called **streamlines**, such that the tangent to these curves is the velocity field, as shown in Figure 6.8.

Fluids have two forms of flow, streamline and turbulent. In streamline flow, the streamlines *never* cross whereas in turbulent flow they intersect. Turbulent flow is described by chaotic dynamics and is very complicated. We will only discuss steady streamline flow.

Pressure

The rate at which the fluid flows creates *pressure* inside the fluid. Consider the water coming out of a hose pipe; we know from experience that by increasing the "pressure" we can make the water come out with greater force. Pressure is the force per unit area that the fluid exerts on itself and anything that comes into contact with it. The unit of pressure is Pa for Pascal and is equal to one newton per meter square; that is, Pa = N/m^2 = $kg/(ms^2)$. The pressure inside a volume element located at $\mathbf{x}$ at time t is given by $p(t, \mathbf{x})$. Note that by considering the units we find that

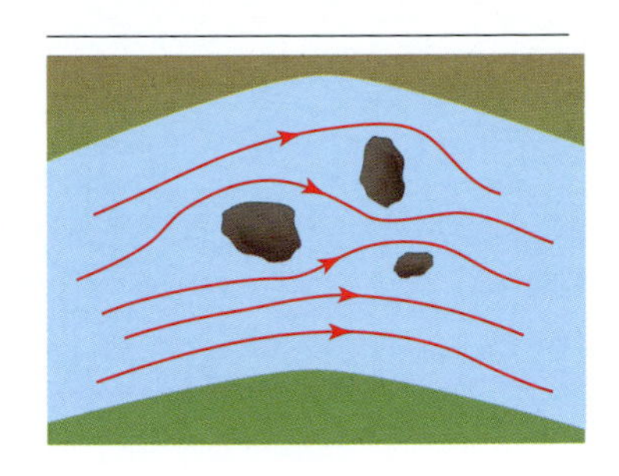

Fig. 6.8: Streamlines do not cross for a streamline flow.

$$\text{Pressure} = \frac{\text{Force}}{\text{Area}} \quad : \quad \textbf{has same units as} \quad \frac{\text{Energy}}{\text{Volume}}.$$

Pressure is *defined* as the force per unit area. As can be seen from the above equation, pressure also has the dimension of energy per unit volume and this leads to a role for pressure in fluids that is more closely related to energy. Pressure for an *incompressible* fluid at *constant temperature* has an interesting realization of being a form of the *fluid's internal energy density*. The idea that pressure is a form of energy density can be visualized by considering a column water; the water at the top pushes downwards due to the force of gravity and hence the water at the bottom has higher pressure and, equivalently, higher energy density. For an incompressible fluid like water, pressure generates the force required to hold up the column by deforming and affecting the intra- and intermolecular bonds and interactions to store the extra energy coming from the weight it is holding. Pressure energy is also stored in the modified rate at which the molecules of water move around, making and breaking bonds. Pressure is a form of internal energy that is of particular importance to the behavior of fluids.

Fluids like water are considered to be incompressible. Consequently their density is constant.

One can wonder what happens to water if one keeps increasing the pressure. Water at room temperature deforms very slightly when subjected to pressure. More precisely, at room temperature water has a compressibility factor of around 3.4×10^{-6}; a hydrostatic pressure of 6.89 kPa reduces the volume of water by about 3.4×10^{-6} of the original volume. If one keeps increasing the pressure, then at a huge pressure of 1 GPa, water undergoes a phase transition and discontinuously changes into what is called Ice IV, which is one of the forms of ice.

It can be shown that if water was exactly incompressible, then the oceans would be about 30 m higher than they are now and would, therefore, cover an extra 5 million square kilometers of the earth's surface.

Fig. 6.9: Smaller spray = more force.

How high a pressure can nature generate? Since pressure is a result of the processes and forces that constitute matter, one needs to look at nature in a more extreme environment to find the natural limit to pressure — and which is found in the interior of a star. A star is approximately a compressible gas and is in a state of equilibrium when the attractive force of gravity is balanced by the pressure generated by the compression of the gas. The star consumes all its energy by burning at higher and higher temperatures so as to generate the higher and higher pressure required to counterbalance the relentless inward force of gravitational attraction.

For stars having a mass greater than about three times the mass of our sun, gravitational attraction is so strong that no physical process, be it atomic, nuclear or subnuclear, can withstand the inward pull. Once it has exhausted all its fuel, the star can no longer generate the pressure to counterbalance gravitational attraction and undergoes gravitational collapse — becoming either a neutron star or a black hole. If the neutron star accretes enough mass, it also undergoes gravitational collapse into a black hole due to instabilities caused by pressure.

The concept of pressure, which makes its appearance in fluids, is of truly fundamental and cosmological significance. Furthermore, there is a maximum pressure force that nature can generate and sustain.

In summary, an incompressible fluid like water is described by specifying its density ρ and its velocity $\mathbf{v}(t, \mathbf{x})$ and pressure $p(t, \mathbf{x})$ for time t and at position $\mathbf{x}$ in the fluid.

6.5 Fluid flow and mass conservation

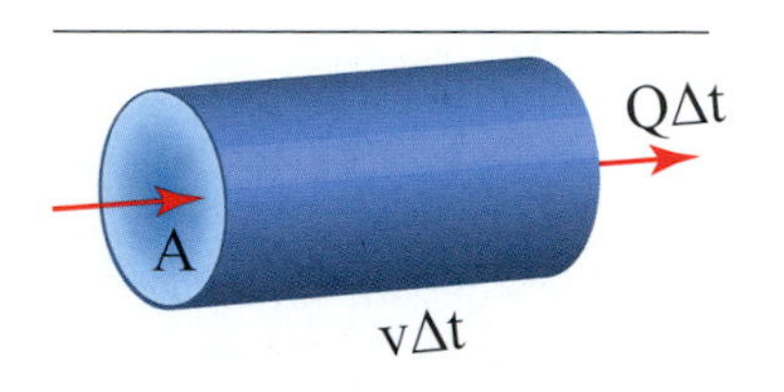

Fig. 6.10: Transport of mass.

If there are no sources or sinks that are pouring in or taking out fluid, respectively, the movement of the fluid must conserve mass.

Consider a fluid flowing through a cross-sectional area given by A. In time Δt, the volume of fluid that flows orthogonally through area A, as shown in Figure 6.10, is given by $Av\Delta t$. Let the volume of fluid flowing through the area A per unit time be denoted by Q, with units of m^3/s; the total fluid volume passing through surface A in time Δt is given by $Q \times \Delta t$.

Since there are no sources or sinks, as shown in Figure 6.10, in time Δt the volume occupied by the fluid flowing through the area A leads to the following:

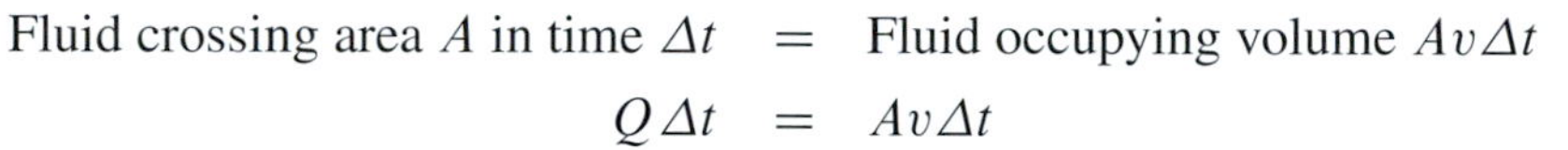

$$\begin{aligned} \text{Fluid crossing area } A \text{ in time } \Delta t &= \text{Fluid occupying volume } Av\Delta t \\ Q\Delta t &= Av\Delta t \\ \Rightarrow Q &= Av \Rightarrow v = \frac{Q}{A}. \end{aligned} \tag{6.1}$$

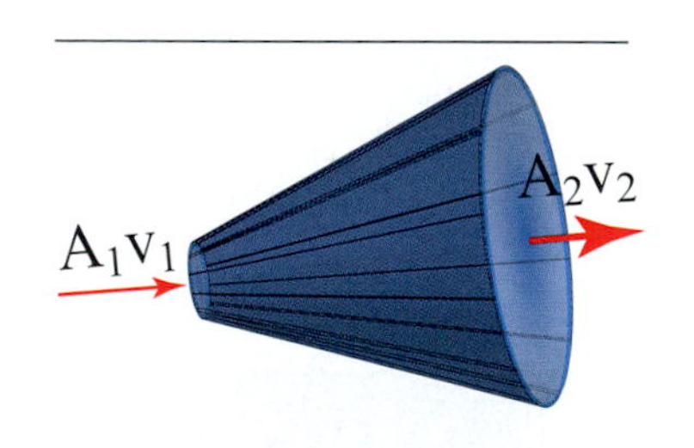

Fig. 6.11: Stream tube enclosing areas A_1, A_2. $A_1v_1 = A_2v_2$.

Consider streamline flow; a bundle of streamlines encloses an area A and let the the area increase along the stream tube, as shown in Figure 6.11. Let fluid velocity and area be v_1, A_1 at one point and v_2, A_2 at another point. There is no fluid flowing outside the stream tube since the fluid's elements flow along streamlines; hence the the fluid flowing across surface A_1 must equal the fluid flowing across another surface A_2. The conservation of fluid volume requires that

$$Q = v_1 A_1 = v_2 A_2. \tag{6.2}$$

Mass conservation requires that, for a given time interval Δt, the mass of fluid that flows through the area A must equal the mass of the fluid that occupies the volume of fluid that has passed through A. Since we are only considering fluids with constant density ρ, the mass of a fluid element is equal to $\rho \times$ volume. Hence, mass conservation of mass requires that

$$\rho Q\Delta t = \rho Av\Delta t$$

which follows from Equation 6.1. For incompressible fluids, mass and volume conservation are equivalent.

6.6 Energy and energy conservation in fluids

Fluids, similar to all natural phenomena, obey the law of energy conservation. We have to discover in what manner is this law realized in fluids. Instead of taking the route of deriving the expression for energy starting from the fundamental laws of fluid mechanics, we take the route of using our intuition for writing down the answer and then analyzing the various terms.

All objects have kinetic energy due to their movement and potential energy due to their physical location or intrinsic internal energy. The energy of a fluid at any moment is the sum of the energy of each volume element of the fluid ΔV. For simplicity, we take ΔV to be a constant; the energy of fluid is then given by

$$\Delta V \sum_i \mathcal{E}(\mathbf{x}_i), \tag{6.3}$$

where $\mathcal{E}$ is the fluid's **density of energy**, namely, the fluid energy contained in an elemental volume ΔV_i at position $\mathbf{x}_i$ as shown in Figure 6.12. The energy density consists of the kinetic energy of the fluid as well as the potential energy that includes internal energy, the most important being pressure.

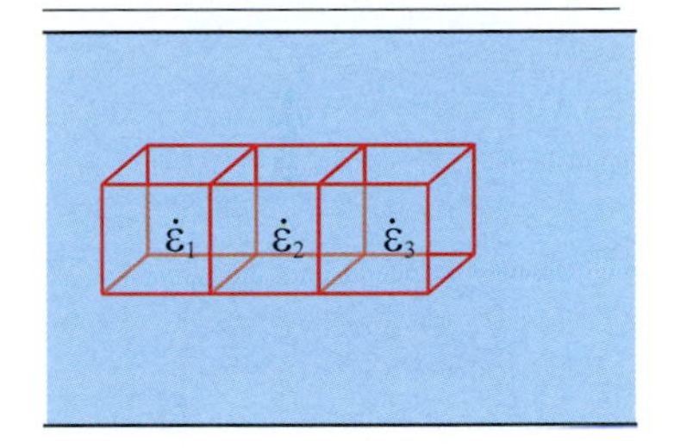

Fig. 6.12: Energy density of the fluid.

Let the fluid have no form of rotation and no viscosity; suppose for simplicity that we have steady flow so that there is no explicit dependence on time t. The energy density of the fluid consists of kinetic energy due to the movement of the fluid and potential energy due to the fluid's position in space. Recall that the kinetic energy of a body of mass m and velocity v, from the disussion in Section 3.3, is given $mv^2/2$. The fluid mass occupying a volume element ΔV is given by its mass density ρ and the fluid's kinetic energy is given by $\rho v^2/2$; hence, the energy density of a fluid occupying is given by

$$\mathcal{E} = \frac{1}{2}\rho \mathbf{v}^2(\mathbf{x}) + p(\mathbf{x}) + U(\mathbf{x}), \tag{6.4}$$

where $U(\mathbf{x})$ is potential energy density other than the pressure. Pressure is a form of potential energy that is specific to a fluid and is the reason it is separated out from the other forms of potential energy: the pressure term is a form of energy that is a measure of the fluid's *internal* energy. It costs energy $p(\mathbf{x})$ per unit volume to *inject*, into the fluid at position $\mathbf{x}$, a fluid volume element.

Energy conservation states that at two different points in the fluid, all changes in velocity and pressure must take place in such a manner so that energy is constant. Hence, for a fluid element traveling from position $\mathbf{x}_1$ to position $\mathbf{x}_2$, energy conservation yields the following equality, generally referred to as **Bernoulli's equation** (sometimes Equation 6.4 is called Bernoulli's equation as well):

$$\frac{1}{2}\rho \mathbf{v}^2(\mathbf{x}_2) + p(\mathbf{x}_2) + U(\mathbf{x}_2) = \frac{1}{2}\rho \mathbf{v}^2(\mathbf{x}_1) + p(\mathbf{x}_1) + U(\mathbf{x}_1)$$

$$\mathcal{E}_2 - \mathcal{E}_1 \equiv \Delta\mathcal{E} = \Delta\left\{\frac{1}{2}\rho \mathbf{v}^2(\mathbf{x}) + p(\mathbf{x}) + U(\mathbf{x})\right\} = 0 \tag{6.5}$$

: **Bernoulli's equation**.

Bernoulli's equation is one of the most important equations of fluid mechanics; it is a precise expression of how energy is shared — in the fluid's flow from point to point — between the fluid's velocity, pressure and potential energy $U(\mathbf{x})$.

The concept of pressure being the fluid's *internal energy density*, as in Bernoulli's equation, moves one away from the Newtonian concept of pressure being a force per unit area. As discussed in Chapter 3, the concept of energy is a very general and far-reaching concept that supersedes the concept of force, and this accounts for the broad range of application of Bernoulli's equation. All branches of fluid mechanics employ Bernoulli's equation.

6.7 Static fluid

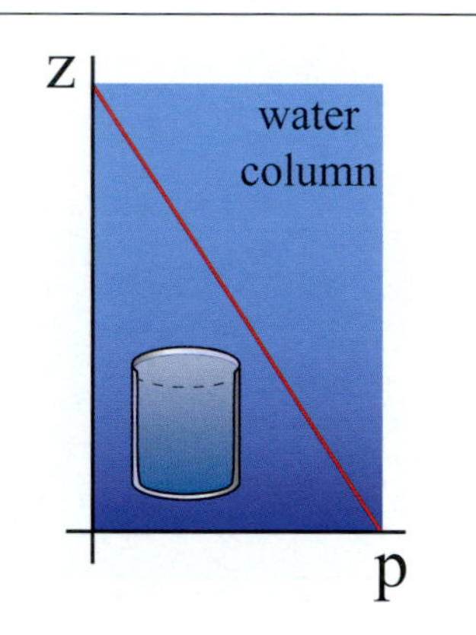

Fig. 6.13: Pressure increase in a water column.

One of the simplest applications of Bernoulli's equation is a static fluid. For making the problem more interesting consider a column of fluid under the influence of gravity; the potential energy density due to gravity, at height z, is equal to $U = g\rho z$. The velocity is zero and hence from Equation 6.5, the pressures p_1 and p_2 at two different heights z_1 and z_2, respectively, are given by

$$\begin{aligned} p_1 + g\rho z_1 &= p_2 + g\rho z_2 \\ \Rightarrow p_2 &= p_1 - g\rho(z_2 - z_1). \end{aligned} \tag{6.6}$$

Hence, for an incompressible fluid, the pressure in the column falls linearly with height, as shown in Figure 6.13.

One can do more. Bernoulli's equation is more general than we have stated — it also holds for elements of a compressible fluids (with no rotation and viscosity) that are close to each other. Consider a column of air with a density $\rho(z)$ that depends on height; we rewrite Equation 6.6 for the case of an infinitesimal separation of height; let heights z_1 and z_2 be close to each other and let the density "vary slowly" so that $\rho(z) = \rho_1 \simeq \rho_2$. Bernoulli's equation is written as follows:

$$p_1 + g\rho_1 z_1 = p_2 + g\rho_2 z_2$$

$$p_1 - p_2 = +g(\rho_2 z_2 - \rho_1 z_1) \simeq g\rho(z)(z_2 - z_1)$$

$$\frac{\Delta p}{\Delta z} \equiv \frac{p_2 - p_1}{z_2 - z_1} = -g\rho(z), \tag{6.7}$$

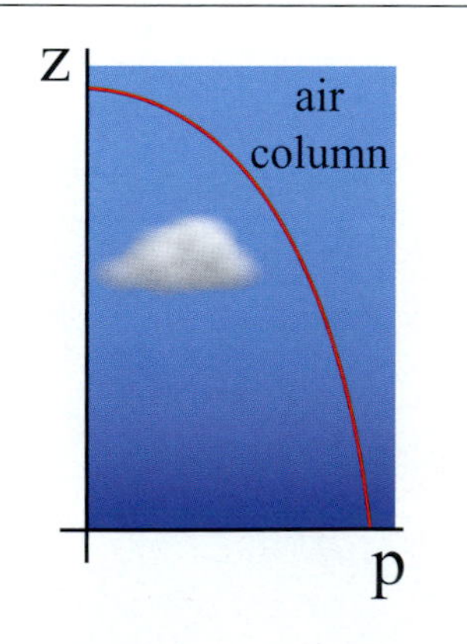

Fig. 6.14: Pressure increase in an air column.

which, after taking the limit $\Delta \to 0$, yields

$$\frac{dp}{dz} = -g\rho(z). \tag{6.8}$$

The density of air is known to fall off exponentially with height, given by

$$\rho(z) = \rho_0 \exp\{-z/h\}$$

$$\rho_0 = 1.2 \text{ kg/m}^3 \; ; \; h \simeq 8.5 \text{ km}.$$

To integrate Equation 6.8 we need to fix a boundary condition. At infinite height from the Earth's surface, the pressure is zero; hence $p(\infty) = 0$; integrating Equation 6.8 yields

$$\frac{dp}{dz} = -g\rho_0 \exp\{-z/h\}$$

$$\Rightarrow p(z) = -g\rho_0 \int_{\infty}^{z} dz' \exp\{-z'/h\} = g\rho_0 h e^{-z/h}$$

$$= \rho(z)gh.$$

The condition that the density should "vary slowly" in Bernoulli's theorem can be made more precise: Equation 6.8 holds only when z, the height of the air, is small compared to h.

Bernoulli's equation yields the dependence of pressure on height as illustrated in Figure 6.14. This result is usually derived by balancing the downward force on an element of the atmosphere due to gravity with the upward force on it due to atmospheric pressure.

6.8 Groundwater

Figure 6.15: Spring in Castellir in Northern Sardinia.

Lakes and streams on the earth's surface are forms of fresh water that also exists in the form of underground water. In Section 6.2, it was mentioned that 0.6% of the earth's water, consisting of 8.4×10^{15} m^3 of water, exists *beneath* the earth's surface. These subsurface sources of fresh water are contained in underground **aquifers**, and the water in them is called groundwater. Groundwater accounts for about 30% of the earth's total fresh (potable) water, the rest of it being mostly in the form of ice and a relatively small amount in surface water such as lakes and rivers.

An aquifer is an underground geological formation — made from gravel, sand, volcanic and igneous rocks and limestone — that is capable of storing and transmitting water. These formations are made from either permeable rock or unconsolidated sands and gravels on top of impermeable materials. When tapped by wells, aquifers can provide significant quantities of water. Aquifers may be small, only a few hectares in area, or huge, underlying thousands of square kilometers of the earth's surface. They may be only a few meters wide or they may measure hundreds of meters from top to bottom.

Porosity is a measure of the capacity of soil and rock to hold or transmit water. A typical porous material is like a tissue paper that absorbs water. An aquifer is usually made of porous material such as saturated sand, which has about 20% water, 25% gravel and 48% clay.

Fig. 6.16: A sponge is a porous material.

There is a significant amount of air trapped in the aquifer, occupying about 10% to 30% of the total interstitial volume.[2] Figure 6.16 shows the cross section of a typical porous material. Clay is highly porous containing substantial amounts of water. Nevertheless, saturated clay is impermeable to the flow of water because clay has extremely small interstitial openings that create friction, halting the flow of water.

Most groundwater originates from rain and snow that has melted and infiltrated into the ground. The amount of water that sinks into the ground depends on the type of land surface. For a porous surface, such as sand or gravel, about 40% to 50% of the rain and snowmelt seeps into the ground; the amount is 5% to 20% for less porous surfaces. The remainder of the rain and snowmelt runs off the land surface into streams or returns to the atmosphere by evaporation. Infiltration into the ground depends on the season of the year, with evaporation being greater during the warm months; during the cold months, in some regions ice hinders the water infiltration process.

Confined and unconfined aquifer

In a confined aquifer, groundwater is sandwiched by impervious layers of material (such as fine silt or clay) called **aquitards**. The surface of aquitards prevents the passage of contaminants into the aquifer. Unconfined aquifers do not have an impermeable overlying layer protecting the aquifer. Instead, the overburden consists of highly permeable material, such as sand or gravel. An unconfined aquifer has a water table, which is an interface of groundwater with air that exists in the overburden material. The typical features of an aquifer and groundwater are shown in Figure 6.17.

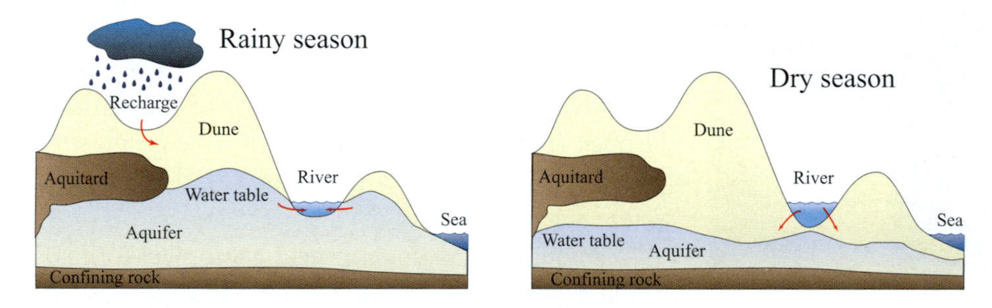

Figure 6.17: Aquifer and flow to a river.

Potentiometric surface

The movement of groundwater is due to gravity and pressure forces generated by the aquifer and aquitards. The *potentiometric surface*, shown in Figure 6.18, is the pressure on groundwater. The potentiometric pressure varies from point to point depending on the composition of the aquifer. As can be seen from Figure 6.18, if a water well is made at a position where the potentiometric pressure is below atmospheric pressure such as well A, water in the well needs to be pumped out. On the

[2]Interstice: An opening or space, especially a small or narrow one between mineral grains in a rock or within sediments or soil.

other hand, if the potentiometric pressure is higher than the atmospheric pressure as is the case for well B, water will gush out of the well on its own accord. The formation of a geyser generally appears to be triggered by a cyclic up-and-down movement of the boiling point curve within a hydrothermal system in response to changes in the potentiometric surface of the cold water that is adjacent to, and interconnected with, the hydrothermal system. A geyser's water rises many meters above the surface and is a spectacular example of the high pressure that can be generated on groundwater (see Figure 6.19).

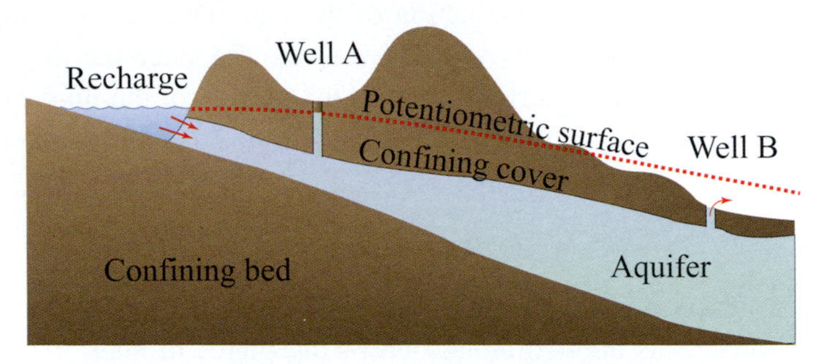

Figure 6.18: Aquifer and its potentiometric surface.

Fig. 6.19: Geyser.

Under the force of pressure and gravity, groundwater moves very slowly from recharge areas to discharge points, with flow rates in aquifers being typically measured in meters per day. Flow rates are much faster where large rock openings or crevices exist (often in limestone) and in loose soil, such as coarse gravel. Groundwater moves slowly and may take years, decades or even centuries to move long distances through some aquifers; this is the reason why there is no turbulence in its flow. However, through loose and porous topsoil, groundwater may take only few days or weeks to move through the topsoil. Groundwater typically moves along streamlines with little mixing.

Groundwater flow has the following key features:

- Groundwater flows through porous material comprising the aquifer.
- The geometry of the pathways of flow through the aquifer is highly convoluted but remains streamlined (paths not crossing each other) due to the low velocity of the groundwater.
- The role of viscosity is fundamental in groundwater flow.
- Groundwater flow is similar to the streamlined and nonturbulent flow of a viscous fluid through a very thin tube.

6.9 Viscous flow

Viscosity of a fluid accounts for energy that the fluid loses to heat energy, which is in the form of microscopic motion of atoms. The flow of groundwater depends crucially on viscosity and we briefly discuss the concept of viscosity before applying it to the study of groundwater flow.

All fluids dissipate energy through *internal friction*. If you stir water in a glass, the swirling motion gradually comes to a halt due to viscosity, the energy of the swirling being converted into heat. Of course, the rise in temperature is negligible so no one ever notices it. What is a quantitative measure of viscosity? An experimental result is that at the fluid and solid interface, all components of the velocity of the fluid are zero, with the fluid coming to a dead halt on the interface. If one observes the giant fans, say, in a car tunnel, one can see that although the fan is blowing with great speed and clearing car fumes, the dirt on the fan moves with it and is not blown away by rotation of the fan. The reason being that the dust particles *on* the fan do not move relative to the fan since the fluid has come to a complete halt at the fan's surface. Hence, the motion of the fluid relative to a solid body with which it is in contact must be zero and yields, on the fluid-solid boundary

$$\mathbf{v} = 0\,;\ v_i = 0 \text{ for all components i.} \tag{6.9}$$

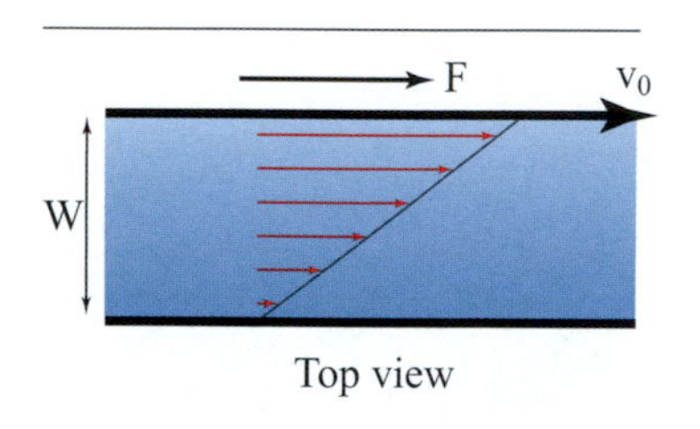

Fig. 6.20: Velocity profile of a fluid between a stationary and a moving plate.

Consider water between two parallel plates, having area A, separated by a small width W. A force F is applied along one of the plates to keep it moving at a constant velocity of v_0 with the other plate being kept stationary, as is shown in Figure 6.20. The fluid's boundary condition given in Equation 6.9 requires that the fluid's velocity at the bottom plate is zero and the fluid at the top plate has velocity v_0, the velocity of the moving plate. Since the fluid's velocity at the moving plate is nonzero, the fluid must have some *shearing* force, acting along the x-axis, to set adjacent fluid elements into motion so that the fluid velocity at the moving plate is equal to v_0.

The force F required to keep the plates moving, per unit area, is known from experiments to be proportional to the velocity gradient v_0/W, the proportionality constant η being viscosity. Hence

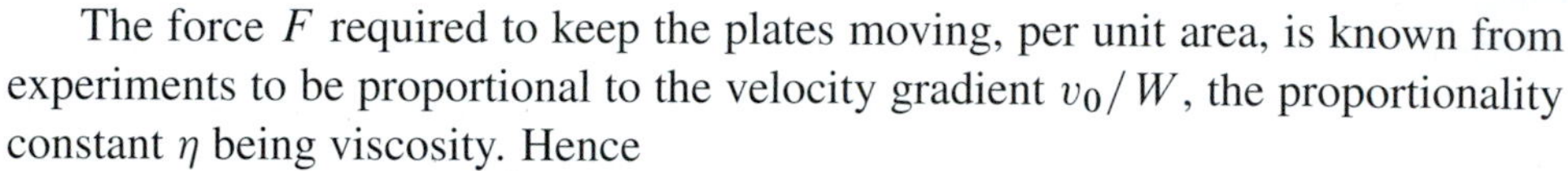

$$\frac{F}{A} = \eta \frac{v_0}{W}.$$

The unit of viscosity is Pa·s = kg/(ms).

More generally, let the top plate in Figure 6.20 move at velocity v_0 and the bottom plate move at velocity v. The definition of viscosity, for velocity along x-axis that is a function of W lying along y-axis — in the limit of small W — yields the following:

$$\frac{F}{A} = \eta \frac{v_0 - v}{W} \rightarrow \eta \frac{dv}{dy}. \tag{6.10}$$

One may wonder if this defining relation for viscosity holds for all fluids. The answer is a bit complicated. All materials can be classed as fluids and solids and a third category of materials that lie in-between the two; for the third category of materials, such as honey and tar, η becomes very large and the defining Equation 6.10 no longer holds. A modified definition of viscosity is then required and won't be discussed any further.

The viscosity of water is $\eta_w = 10^{-3}$ Pa·s, that of honey is $\eta_h = 10$ Pa·s and of tar is $\eta_t = 30{,}000$ Pa·s. With increasing temperature, viscosity decreases for liquids and increases for gases.

6.10 Flow in a channel

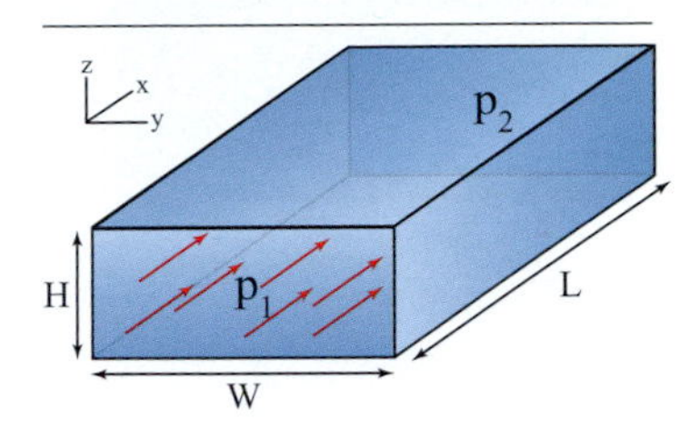

Fig. 6.21: Flow through a volume.

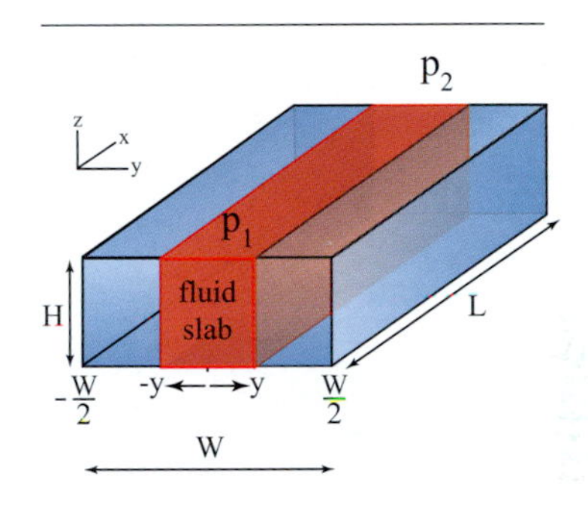

Fig. 6.22: Flow in a channel.

To get a feel for the effect of viscosity on fluid flow, consider our previous example of a fluid flowing between two parallel plates with height H and length L. Consider a flow at fixed elevation and hence the force of gravity does not play any role in the flow. Suppose the two plates, with distance W between them, are held fixed and there is pressure p_1 at one end of the fluid that is higher than the pressure p_2 at the other end; note both p_1 and p_2 are constant on the HW plane and $p_1 > p_2$ so that the flow is in the positive x-direction.

The arrangement is shown in Figure 6.21. The velocity of the flow v is entirely along the x-axis and depends only on the distance, say, from one of the plates. The net pressure force on the fluid, due to pressure, acts along the x-axis. The boundary condition on the plates due to viscosity requires that the fluid velocity be zero at the plates. Hence, viscosity creates a force in the x-direction that opposes the force due to pressure and brings the fluid velocity to a halt at the two plates; the fluid velocity is a maximum at the mid-point between the two plates.

Given that the flow should be symmetric about the mid-point between the plates, let us define $y = 0$ to be at the mid-point of the channel at a distance $W/2$ from the two fixed plates. Consider a slab of the fluid with height H and length L; the slab has a total width $2y$ centered at the mid-point and with the coordinates of the two ends being y and $-y$, as shown in Figure 6.22. The fluid volume is $2yLH$.

The magnitude of the total pressure force on the fluid volume $2yLH$ along the x-direction is

$$F = p_1 \times (2yH) - p_2 \times (2yH) = 2(p_1 - p_2)yH. \tag{6.11}$$

The area of the volume element experiencing the shearing force is $A = LH$; hence, from Equation 6.10, the shearing force in the x-direction on the fluid volume $2yLH$ comes from the shearing force on the slab's area LH at the two boundaries, namely, at y and $-y$. From the geometry of the problem, the velocity is symmetric about $y = 0$ and yields $dv(-y)/dy = -dv(y)/dy$; the shearing force at $-y$ has the opposite sign of the shearing force at y. Hence, the total shearing force is given by

$$-\left[\eta\frac{dv(y)}{dy} \times HL - \eta\frac{dv(-y)}{dy} \times HL\right] = -2\eta\frac{dv(y)}{dy} \times HL \tag{6.12}$$

The overall minus sign arises from the fact that viscosity opposes the motion of the fluid in the x-direction.

For steady velocity flow with no acceleration, the net force on each fluid element must equal zero. For the fluid to have a steady flow with constant velocity, the force along the x-direction created by pressure must be exactly equal to the opposing shearing viscous force. Hence, Equations 6.12 and 6.11 yield, for $\Delta p = p_1 - p_2$

$$\begin{aligned} -2\eta\frac{dv}{dy} \times HL &= F = 2(p_1 - p_2)yH \\ \Rightarrow \frac{dv}{dy} &= -\frac{\Delta p}{\eta L}y. \end{aligned} \tag{6.13}$$

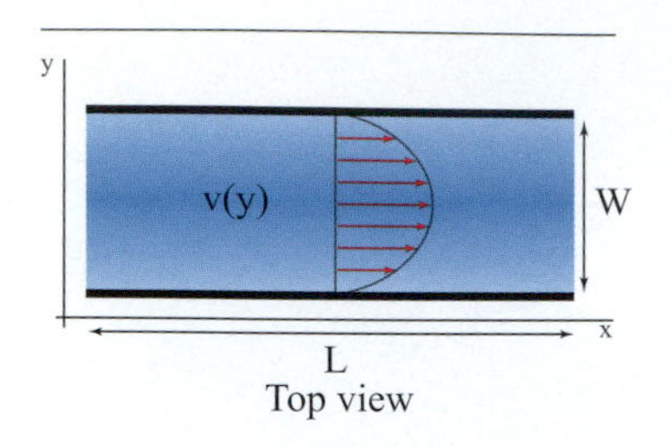

Fig. 6.23: Pressure is constant along the width W. The velocity profile in the channel is due to viscosity which requires that the velocities at the boundaries to be zero.

Integrating Equation 6.13 above yields

$$v(y) = -\frac{1}{2\eta}\frac{\Delta p}{L}\left(y^2 + C\right) \; ; \; -W/2 \leq y \leq W/2 \qquad (6.14)$$

where C is an integration constant. We impose the boundary condition that velocity of the fluid is zero at the plate located at $y = W/2$, that is, $v(W/2) = 0$; note from the symmetry of Equation 6.14 the boundary condition automatically also yields that $v(-W/2) = 0$ as is required.

The boundary condition yields $C = -W^2/4$ and hence we have the final result

$$v(y) = \frac{1}{2\eta}\frac{\Delta p}{L}\left(\frac{W^2}{4} - y^2\right) \; ; \; -W/2 \leq y \leq W/2. \qquad (6.15)$$

The velocity profile is shown in Figure 6.23.

Fluid flow

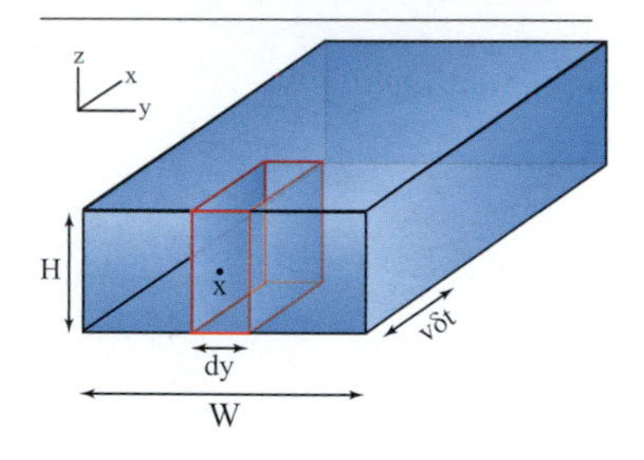

Fig. 6.24: Fluid flow in a channel.

A quantity of great interest in the flow of fluids is the rate at which fluid volume, mass or momentum or other quantities are transported across a unit surface area, namely the *fluid flow* of mass or momentum through that surface, which we obtained in Equation 6.1. We analyze the viscous flow between the parallel plates for mass conservation.

In a small time δt, the fluid in volume element $Hdy \times v\delta t$ will flow across an imaginary surface located at position x. The flow per unit height H per unit time, as shown in Figure 6.24, follows from Equation 6.1 as

$$\rho dQ = \rho\frac{Hvdy\delta t}{H\delta t} = \rho v dy.$$

Hence, flow per unit height H per unit time is given by summing over the contribution for each y and yields, from Equation 6.15

$$Q = \int_0^W dQ = \int_0^W v dy = \frac{W^3}{12\eta}\frac{\Delta p}{L} > 0, \qquad (6.16)$$

where W is the width of the channel and η is the viscosity of the fluid.

There is net positive amount of fluid mass flowing across a unit width at x since the pressure force, namely $\Delta p = p_1 - p_2 > 0$, is driving the fluid along the x-direction and sending across a mass flow equal to ρQ.

6.11 Flow in a pipe

A fluid flowing in a circular pipe, along streamlines, is called Poiseuille flow and is similar to the above simpler case. For a circular pipe of radius R, the velocity $v(r)$ for Poiseuille flow — similar to Equation 6.15 — at radial distance r from the axis of the pipe, is given by

$$v(r) = \frac{1}{4\eta}\frac{\Delta p}{L}(R^2 - r^2)$$

while $\bar{v}$, the average velocity of flow, is given by

$$\bar{v} = \frac{1}{\pi R^2}\int_0^R 2\pi r v(r)\,dr$$

$$\Rightarrow \bar{v} = \frac{1}{8\eta}\frac{\Delta p}{L}R^2. \tag{6.17}$$

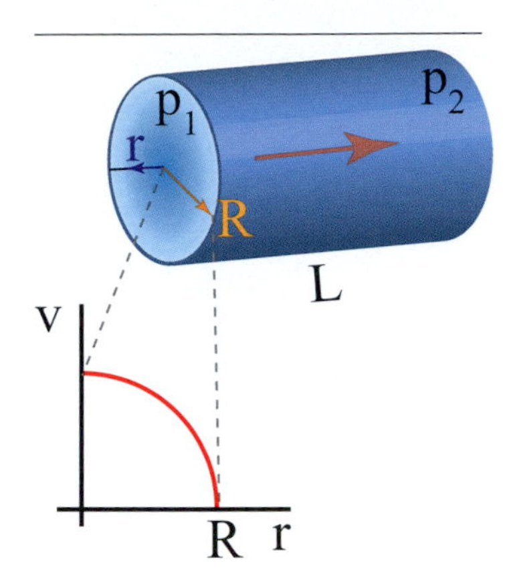

Fig. 6.25: Flow in a pipe.

The flow of mass transported across a cross section of the pipe is given by

$$Q = \int_0^R 2\pi r v(r) dr = \frac{\pi R^4}{8\eta}\frac{\Delta p}{L} \tag{6.18}$$

where recall that the flow of fluid Q is volume per unit time and $\Delta p = p_1 - p_2$. Poiseuille flow is shown in Figure 6.25. Poiseuille's formula is widely used for testing the streamline flow in round pipes and any deviation of mass discharge Q from the R^4 dependence is seen as a sign of turbulent flow.

Flow of blood

An interesting example of Poiseuille flow is the flow of blood in arteries and veins. The flow is mostly streamline, with turbulent flow being a form of illness. Many animals approximately have the same radius of the capillary artery, about 2.5×10^{-6} m or 2.5 μm. Over a length of $L \sim 1$ cm, the drop in pressure is $\Delta p \sim$3000 Pa; the viscosity of water is $\eta_w = 10^{-3}$ Pa·s. From Equation 6.17, the average velocity of blood is $\bar{v} \sim 0.02$ cm/s.

The observed speed of blood is higher than the estimate of $\bar{v} \sim 0.02$ cm/s that we obtained from Poiseuille flow. The viscosity of blood is $\eta_b = 0.0027$ Pa·s, about three times that of water, which makes the estimate even lower. The discrepancy is to be expected since red blood cells are roughly equal to the size of the artery's diameter and hence are not flowing as "smoothly" as water molecules. In fact, there is a view that blood should be treated as an object that is in-between a solid and a fluid.

The measurement of a person's blood pressure is based on the turbulent and streamlined flow of blood. The flow of blood is first fully stopped by applying pressure to an artery that runs from the heart down the elbow; the pressure is then slowly decreased. When blood just starts to flow in the artery, it is turbulent and creates a "whooshing" sound. The pressure at which this sound is first heard is the *systolic* blood pressure. The pressure is further decreased until the flow becomes streamlined and no sound is heard; this is the *diastolic* pressure, which is the pressure that blood flow exerts on the arteries.

6.12 Bernoulli's equation and viscosity

Viscosity arises due to the transfer of the fluid's macroscopic energy into the random microscopic thermal motion of the fluid's atoms and molecules. The fluid's energy consists of macroscopic and mechanical energy that is carried either by its velocity or by the pressure and potential terms. There is no term in Bernoulli's

Equation 6.5 accounting for the random thermal motion of the fluid's atoms and molecules. The fluid's energy is not conserved when viscosity is taken into account. Viscosity measures the amount of fluid energy that is transformed into microscopic heat motion. Of course, if all the energy of the fluid is taken into account, including heat energy, then one regains energy conservation.

Based on the fact that the entropy of a system must always increase, the transfer is always in one direction, with the fluid constantly losing energy to microscopic motion and never gaining from it.

Consider the energy conservation equation given in Equation 6.5. The fluid's energy conservation is violated in two distinct ways:

- The fluid always loses energy due to viscosity.
- The fluid either loses energy due to the work done by the fluid or else gains energy due to work done on the fluid.

We will be concerned only with viscous loss of energy.

The generalization of Bernoulli's equation for the case of viscosity is now addressed. Let the change of the fluid's energy density be denoted by $\mathcal{F}$. In going from point 1 to point 2, the fluid loses or gains energy and yields the following modified energy equation:

$$\mathcal{E}_2 - \mathcal{E}_1 \equiv \Delta\left\{\frac{1}{2}\rho\mathbf{v}^2(\mathbf{x}) + p(\mathbf{x}) + U(\mathbf{x})\right\} = -\mathcal{F}, \tag{6.19}$$

: Bernoulli's equation with viscosity

where the minus sign in front of $\mathcal{F}$ stems from the fact that the loss of energy due to viscosity is positive, i.e. $\mathcal{F} > 0$. For the case of energy loss due to viscosity, the pressure and potential terms in Equation 6.19 for energy ensure that the loss of energy is always positive. For frictionless reversible flow, energy loss due to viscosity is always zero.

Consider the case of a gas with pressure and velocity p_1, v_1 undergoing a sudden expansion, as shown in Figure 6.26. Its drop in velocity is so high that one can consider the final fluid velocity to be zero. The final state of the fluid is given by p_2, $v_2 \sim 0$. The modified Bernoulli's equation then states that

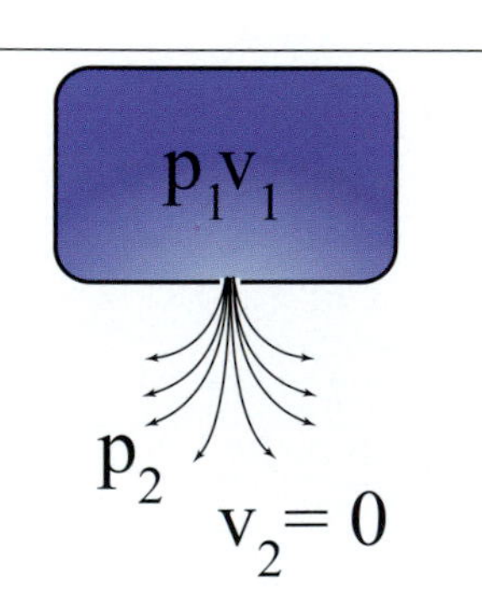

Fig. 6.26: Expansion of a gas.

$$\mathcal{F} = \frac{1}{2}\rho v_1^2 + p_1 - p_2 > 0. \tag{6.20}$$

The fluid loses energy because of the work done by the fluid in the (irreversible) expansion of the gas. Note the viscosity coefficient η does not appear on the right hand side of above equation since the process does not lose any energy due to internal friction.

For low velocity flow in a pipe, the change of velocity is negligible, i.e. $v_1 \approx v_2$. Hence, for Poiseuille flow, from Equations 6.18 and 6.19

$$\mathcal{F}_P = (p_1 - p_2) = Q \frac{8\eta}{\pi R^4} L > 0 \qquad (6.21)$$

where Q is the fluid volume discharge in the pipe. The flow in a channel constantly loses energy due to viscosity.

6.13 Groundwater flow: Darcy's equation

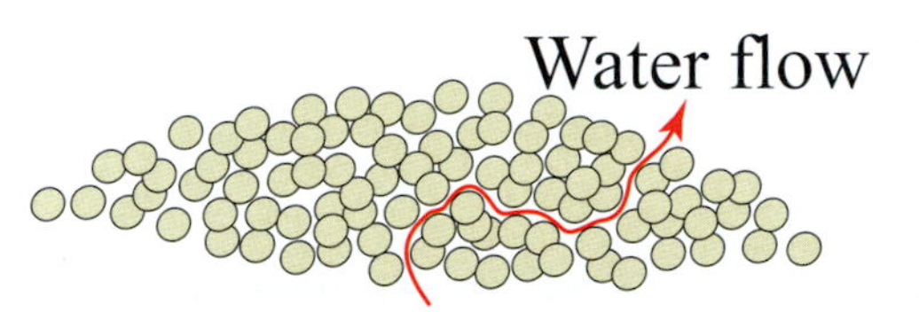

Figure 6.28: Flow through the interstices.

Groundwater flow is an example of flow through porous media such as sand and gravel. A porous medium is shown in Figure 6.27 and flow through the medium is shown in Figure 6.28. The fluid moves in the voids of the porous material and follows a very complicated and zigzag path. The equations of fluid mechanics, in particular, Bernoulli's equation for a viscous fluid given in Equation 6.19, does not refer to nature of the medium. Fluid flow through a porous medium like the groundwater aquifer has the following unique characteristics:

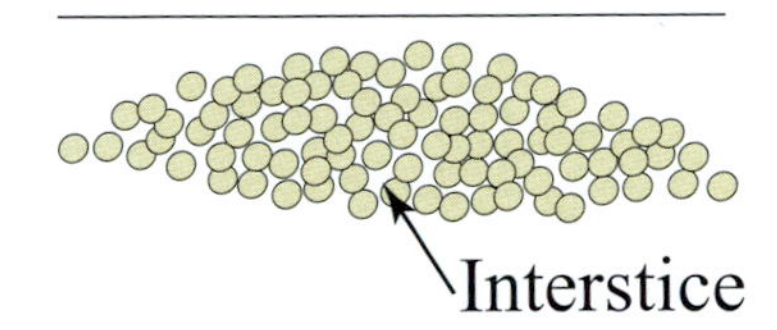

Fig. 6.27: Interstices.

- For groundwater flow, the viscosity term is much larger than for the flow of surface water.
- For most flow through the aquifer, the velocity is very small, on the order of a few meters per day. For a fluid with a typical velocity of v flowing through a porous medium with typical interstice length D, the viscosity term will dominate the velocity term if the dimensionless quantity $\rho D v/\eta << 1$. This condition is readily satisfied in the flow of groundwater through the aquifer. Hence, although the velocity changes from point to point, the velocities are so small that the v^2 term in Bernoulli's equation can be ignored.
- Groundwater flow through the aquifer is streamlined and turbulence does not occur in the flow.
- The flow of groundwater is largely determined by the interplay of viscosity, pressure and gravity.

Consider fluid flow through a pipe, with area A_p, of porous material and shown in Figure 6.29. Since the fluid takes a torturous path through the porous material, the fluid's velocity is approximated by its *average* velocity. The porous medium is characterized by its porosity ϵ, defined as follows:

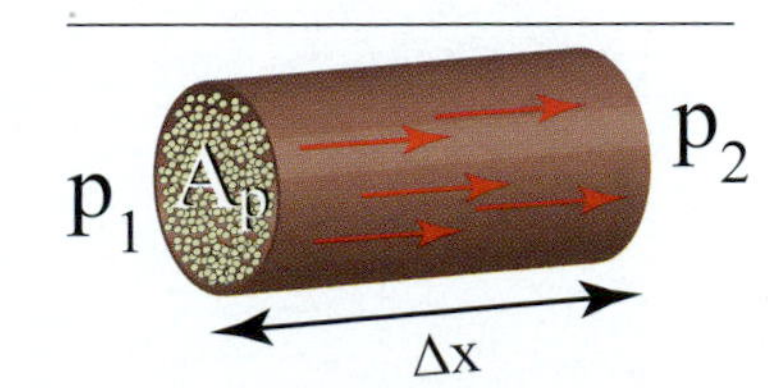

Fig. 6.29: Flow through a porous pipe.

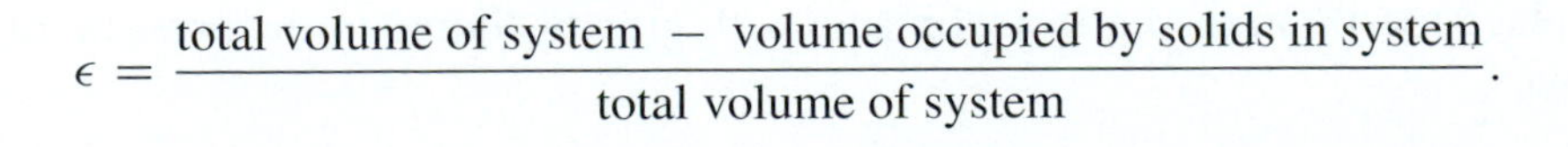

$$\epsilon = \frac{\text{total volume of system} - \text{volume occupied by solids in system}}{\text{total volume of system}}.$$

Porosity is considered high if ϵ is greater than 20% and low if it is less than 5%. The *interstitial velocity* v, which carries the fluid's kinetic energy and enters the energy equation, is defined by

$$v = \frac{1}{\epsilon} \cdot \frac{Q}{A_p}. \tag{6.22}$$

Groundwater flow can be thought of as fluid flowing through a collection of Poiseuille tubes of small radii R, as in Figure 6.30. Note from Equations 6.21 and 6.22, for Poiseuille flow of length of L along the aquifer, $\mathcal{F}_P$, the viscous loss of energy in Poiseuille flow, is given by

$$\mathcal{F}_P = Q \frac{8\eta}{\pi R^4} L = \frac{v}{(\pi R^4 / 8\epsilon A_p)} \cdot \eta L. \tag{6.23}$$

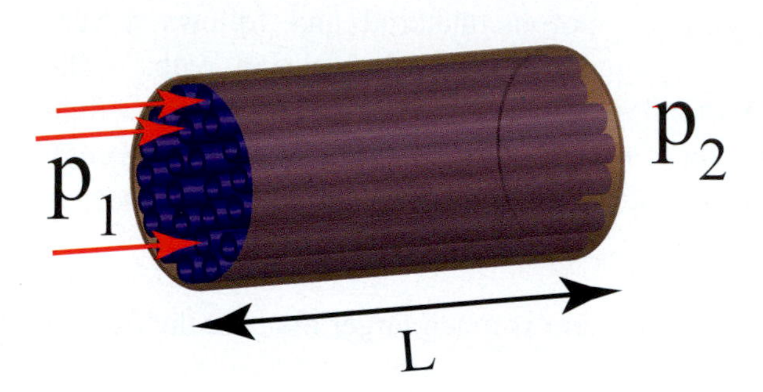

Figure 6.30: A porous pipe can be thought of as a collection of small pipes.

All the detailed and specific characteristics of the porous medium are contained in the term $\pi R^4 / 8\epsilon A_p$, which has the dimension of m^2. In analogy with Equation 6.23, Darcy (1859) made the *postulate* that groundwater flow is similarly determined by viscous energy loss $\mathcal{F}_D$ given by the following relation:

$$\mathcal{F}_D = \frac{v}{k} \cdot \eta L. \tag{6.24}$$

The constant k is the permeability of the medium, which has the unit of m^2 and is typically the square of the linear size of the interstices; k needs to be empirically determined for different types of aquifers and has the typical order of magnitude given by $10^{-12} m^2$.

For groundwater flow we can neglect the velocity term in Bernoulli's equation; the gravitational potential is given by $U = g\rho z$. Darcy's postulate for a fluid at a height z flowing along the x-direction — as shown in Figure 6.31 — yields, from Equations 6.19 and 6.24, Darcy's equation

$$\Delta(p + \rho g z) = -\frac{v}{k} \cdot \eta L \quad : \quad \text{Darcy's equation.} \tag{6.25}$$

The pressure term in Darcy's equation is the pressure on groundwater. This pressure is a result of the forces acting inside the aquifer, due to the composition and geological stresses on the permeable medium, and can be quite large. The flow takes place mostly in the horizontal plane, with gravity adding to the pressure of the flow.

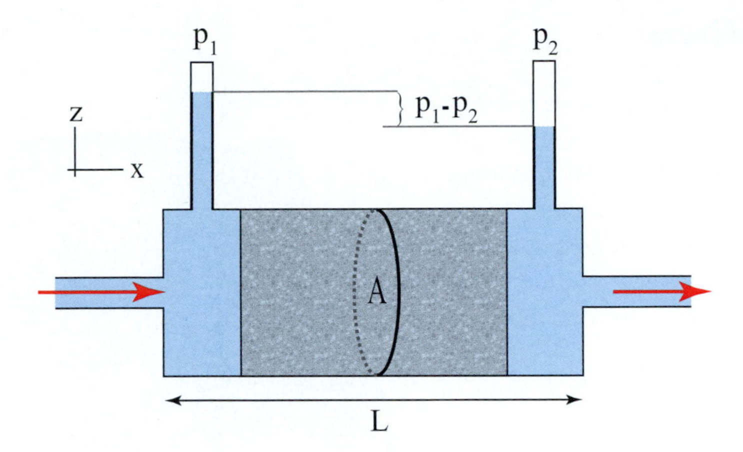

Figure 6.31: Pressure variation through a porous medium.

Optional: continuum limit of Darcy's equation

Taking the infinitesimal limit of $L \to dx$, and writing v as v_x, yields from Equation 6.25

$$\frac{\partial(p + \rho g z)}{\partial x} = -\frac{\eta}{k} v_x.$$

Writing a similar equation for flow in the y-direction, differentiating each equation with respect to x and y respectively and adding them yields

$$\frac{\partial^2(p + \rho g z)}{\partial x^2} + \frac{\partial^2(p + \rho g z)}{\partial y^2} = -\frac{\eta}{k}\left(\frac{\partial v_x}{\partial x} + \frac{\partial v_y}{\partial y}\right).$$

For an incompressible fluid, the mass of fluid entering a volume element is the amount leaving it. One can show that this leads to

$$\frac{\partial v_x}{\partial x} + \frac{\partial v_y}{\partial y} = 0.$$

Hence, one obtains

$$\frac{\partial^2(p + \rho g z)}{\partial x^2} + \frac{\partial^2(p + \rho g z)}{\partial y^2} = 0 \quad : \quad \text{Darcy's equation.}$$

The above form of Darcy's equation, which is a special case of the Laplace's equation, forms the bedrock of the study of two dimensional flows of groundwater as well as the flow of oil in oil fields. By imposing various boundary conditions on Darcy's equations, a vast range of problems for underground fluid flows have been solved. Darcy's equation can be generalized to three dimensions but solutions are more difficult to obtain.

6.14 Rivers

Rivers are used by human beings for many useful purposes, from providing water and food for consumption, for irrigation, to being a means of transportation and as a source of power. Rivers serve as a defensive shield and as a means for culture and recreation.

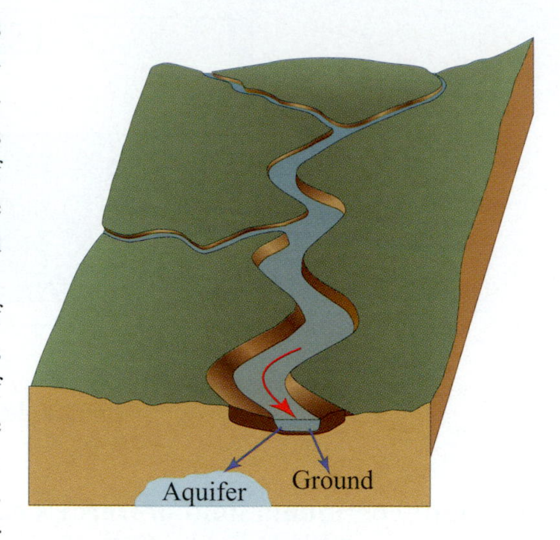

Figure 6.32: Flow of a river.

A river is the surface flow of freshwater. The source of rivers is either water from the melting of glaciers or from precipitation. Some rivers originate from groundwater. From their source, rivers flow downhill, typically terminating in a sea or in a lake; in some cases, a river can flow directly into the ground or completely dry up due to evaporation and terminate on land. The river flows in a meandering manner, with its beginning-to-end distance being about one-third of its total length.

The water in a river usually flows in an open channel made up of a stream bed, which has a varied topography, and is bounded by banks. In larger rivers there are accompanying floodplains that are flooded seasonally by the river waters overflowing their channels. River water flow also has a major component flowing through porous rocks and gravel that surround the stream bed. For many rivers in large valleys, the subsurface component of flow may be greater than the surface flow. River water also infiltrates into the soil and contributes to groundwater. See Figure 6.32.

Rivers are classified based on the stream velocity. A *youthful river* has a steep gradient and flows rapidly; its channel is eroded more deeply than laterally. A *mature river* is fed by many tributaries and has a less steep gradient and lower velocity than a youthful river. Erosion takes place more in its width than depth. An *old river* is characterized by flood plains and is wide with a low flow gradient and low erosive activity. A *rejuvenated river* is one created by plate tectonic movement.

6.15 Velocity of river flow

The most fundamental feature of a river's flow is its velocity. For example, the velocity of the river determines the extent of erosion and how much sediment it carries and deposits. For a young river, the velocity is high and this explains the deep erosion of the river beds. For an old river, being fed by many tributaries, the sediments it carries play an important role in the river's flooding and the consequent creation of alluvial soil in the river's deltaic regions. The velocity of the river plays a central role in all attempts to tame river flow and determines the design of dams, bridges, culverts and so on.

Rivers are a special case of a fluid flowing in an open channel, which is characterized by the pressure on the surface being a constant and equal to atmospheric pressure. Open channel flow can be streamlined or turbulent, steady or unsteady. River flow is streamlined on a large scale and will be treated accordingly. Of course, on a small scale, river flow is highly complex with many eddies and turbulent flows taking place midstream and on the river banks. We will not address turbulent flow.

Fig. 6.33: Amazon River.

River flow takes place due to gravity and is determined by the river's gradient, channel roughness and shape.[3] If a fluid is subjected to a constant force of gravity, it would continually accelerate and the flow cross section would decrease due to mass conservation. However, for river flow, viscosity and channel roughness oppose the fluid's motion, with the resistance force being proportional to the flow's velocity. Upstream, the velocity is low and hence gravity causes the river's flow to accelerate. The increase in velocity causes an increase in the resistance forces and leads to a force balance, leading to a *uniform flow*. The river's flow consists mostly of constant-velocity motion — characterized by no change in the flow cross section and no change in the depth of flow. Most rivers are quite wide and the river flow is, hence, considered to be at an approximately constant velocity — with the behavior of the flow on the river bed and at the banks being ignored.

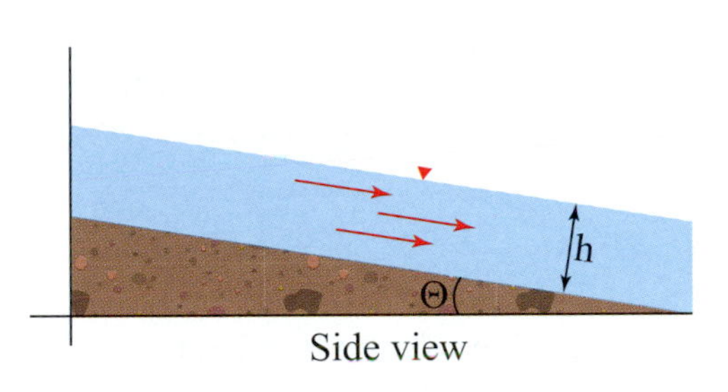

Figure 6.34: Slope of a river.

Given the complexity of a river channel and of its stream bed, any derivation of river flow from first principles is very difficult. A number of empirical rules have been discovered by observations for streamlined river flow and these form the starting point of our analysis.

Similar to the case of groundwater, where the average interstitial velocity approximately describes the flow, a river's flow is described by its *average stream velocity* v. A typical speed of a large river is a few kilometers per hour. The Amazon River has a speed of 2.4 km per hour during the dry season and increasing to 5 km per hour in the rainy season.

Consider a river with a slope of $S_0 = \tan(\Theta)$, as shown in Figure 6.34. The French hydrologist Chezy (1775) deduced from observations that the river's average stream velocity v obeys the following relation, called the **Chezy equation**

$$v = C\sqrt{R_h S_0}. \qquad (6.26)$$

The hydraulic radius R_h is the ratio of the cross-sectional area A of the stream bed divided by its perimeter P, as shown in Figure 6.35, namely $R_h = A/P$. R_h is a sort of equivalent radius, encoding the combined measure of the cross section of the river and the shape of the stream bed; for wide and shallow streams (depth $<$ 0.05 width) the R_h is the average of the depth.

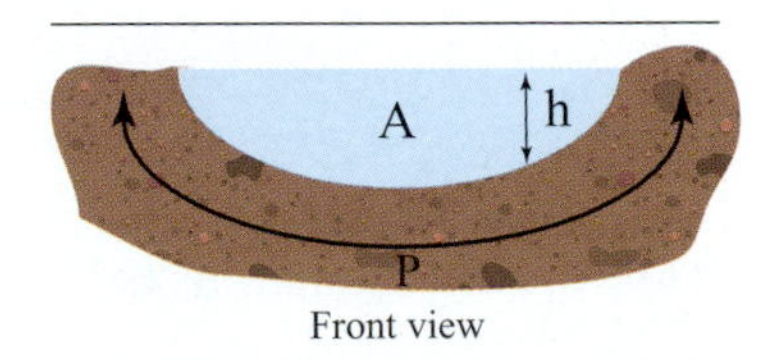

Fig. 6.35: Simplified cross section of a river.

[3]The situation is different in groundwater flow where the driving force is not only gravity but also the pressure on water generated by the aquifer.

Navier-Stokes equation

In all of our discussions, we have concentrated on the role of energy. One may wonder, is there an analog of Newton's law for fluids? The answer is "Yes". The Navier-Stokes equation is the result of applying Newton's law to fluids using the Eulerian approach. Recall from Equation 6.19 that a fluid's energy obeys

$$\mathcal{E}_2 - \mathcal{E}_1 \equiv \Delta\left\{\frac{1}{2}\rho\mathbf{v}^2(t,\mathbf{x}) + p(t,\mathbf{x}) + U(\mathbf{x})\right\} = -\mathcal{F}.$$

The frictional losses due to viscosity $\mathcal{F}$ cannot be expressed as a potential term since it does not conserve the fluid's energy; hence energy has the form

$$\mathcal{E} = \frac{1}{2}\rho\mathbf{v}^2(t,\mathbf{x}) + p(t,\mathbf{x}) + U(\mathbf{x}) - \mathcal{E}_{\text{visc}}.$$

Newtonian force is given by the gradient of the potential energy term, as in Equation 3.39. The generalization of the partial derivative $\partial/\partial x$ to three space dimensions is the gradient $\nabla = (\partial/\partial x, \partial/\partial y, \partial/\partial z)$; force per unit fluid volume is a vector given by the gradient of the potential

$$\mathbf{F}_{\text{fluid}}(t,\mathbf{x}) = -\nabla\{p(t,\mathbf{x}) + U(\mathbf{x})\} + \eta\nabla^2\mathbf{v}(t,\mathbf{x})$$

where $\eta\nabla^2\mathbf{v}(\mathbf{x})$ is known to be the force due to viscosity. Force causes the fluid volume element at $\mathbf{x}$ to accelerate. Hence, Newton's law for a fluid entering and leaving a fixed volume element at $\mathbf{x}$, is given by

$$\rho\frac{d\mathbf{v}(t,\mathbf{x})}{dt} = \mathbf{F}_{\text{fluid}}(t,\mathbf{x}) = -\nabla p(t,\mathbf{x}) - \nabla U(\mathbf{x}) + \eta\nabla^2\mathbf{v}(t,\mathbf{x}). \tag{6.27}$$

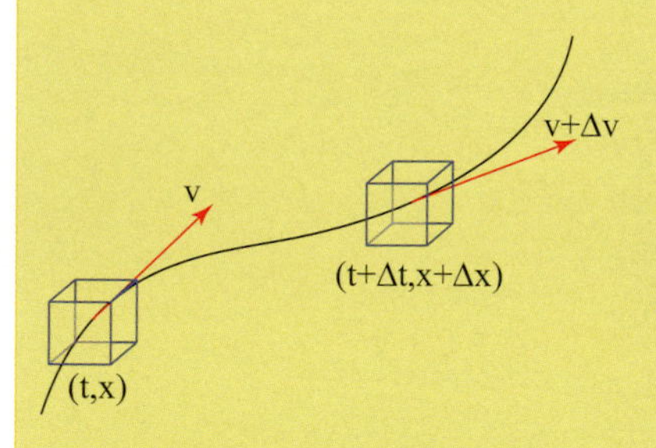

Figure 6.36: Following a fluid element.

It would seem that we are done and that we have obtained the rate of change of the fluid at volume element $\mathbf{x}$; but Newton's law tells us that the total change in the momentum per unit volume is $\rho d\mathbf{v}/dt$; if we fix our attention to the fixed volume element at $\mathbf{x}$, we will only obtain $\rho\partial\mathbf{v}/\partial t$. To obtain the total change of momentum of a fluid element at $\mathbf{x}$, we have to *follow* the fluid element as it moves out of the volume element at $\mathbf{x}$ into a neighboring element at $\mathbf{x}+\Delta\mathbf{x}$, as shown in Figure 6.36. What makes fluid mechanics nonlinear, difficult but interesting is that $\Delta\mathbf{x} = \mathbf{v}\Delta t$ and hence the neighboring element that the fluid occupies is fixed by the movement of the fluid element *itself*. The total change in the fluid's velocity, using the rule of differentiation, is given by

$$\begin{aligned}\frac{d\mathbf{v}(t,\mathbf{x})}{dt} &= \frac{\mathbf{v}(t+\Delta t,\mathbf{x}+\Delta\mathbf{x}) - \mathbf{v}(t,\mathbf{x})}{\Delta t} = \frac{\mathbf{v}(t+\Delta t,\mathbf{x}+\mathbf{v}\Delta t) - \mathbf{v}(t,\mathbf{x})}{\Delta t} \\ &= \frac{\partial\mathbf{v}(t,\mathbf{x})}{\partial t} + \mathbf{v}\cdot\nabla\mathbf{v}(t,\mathbf{x}). \end{aligned}\tag{6.28}$$

Hence, from Equations 6.27 and 6.28, we obtain the **Navier-Stokes equation**

$$\rho\left(\frac{\partial\mathbf{v}(t,\mathbf{x})}{\partial t} + \mathbf{v}\cdot\nabla\mathbf{v}(t,\mathbf{x})\right) = -\nabla p(t,\mathbf{x}) - \nabla U(\mathbf{x}) + \eta\nabla^2\mathbf{v}(t,\mathbf{x}). \tag{6.29}$$

The Irish hydrologist Manning (1890) found the following relation:

$$C = \frac{1.49}{n} R_h^{1/6}. \tag{6.30}$$

The **Manning coefficient** n has dimension of $\mathrm{m}^{1/6}$ and is a measure of the roughness of the river channel. For smooth glass, n has a value of $0.008m^{1/6}$; for hard gravel $n = 0.016\ \mathrm{m}^{1/6}$. For major rivers n is equal to $0.029\ \mathrm{m}^{1/6}$.

Combining Equations 6.26 and 6.30 yields the **Chezy-Manning relation**

$$v = \frac{Q}{A} = \frac{1.49}{n} R_h^{2/3} S_0^{1/2} \tag{6.31}$$

where Equation 6.1 has been used to define velocity in terms of Q, the rate of fluid flow. Equation 6.31 is widely used in the study of rivers and in the design of docks, bridges, culvert and levees.

6.16 Energy in river flow

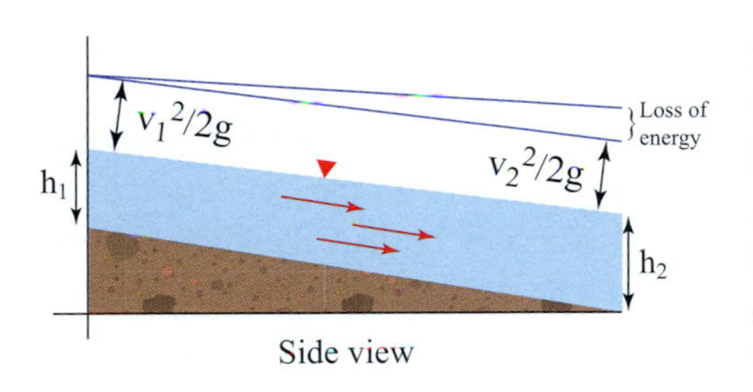

Figure 6.37: Energy loss during flow.

The concept of energy has proved useful in the study of river flow. Consider the flow of a river at an incline, flowing from a height z_1 to a lower height z_2; height is measured from the river bed, which is at zero height $z = 0$, as shown in Figure 6.37. To simplify the notation, the specific energy e is introduced which is obtained by dividing the energy density by the constant ρg; the lowercase e is used to emphasize the difference with energy per unit volume $\mathcal{E}$ defined by Equation 6.4 as follows:

$$\mathcal{E} = \frac{1}{2}\rho v^2 + p + \rho g z.$$

For a fluid at height z above the river bed, the specific energy is defined by

$$e = \frac{\mathcal{E}}{\rho g} = z + \frac{1}{2g} v^2 + \text{constant}. \tag{6.32}$$

For river flow the term p/ρ is approximately a constant and hence is dropped. Note, since ρg is a constant, specific energy e and energy density $\mathcal{E}$ are completely equivalent.

From Equation 6.1, the specific energy can be rewritten as

$$e = z + \frac{1}{2g}\left(\frac{Q}{A}\right)^2. \tag{6.33}$$

Consider a wide rectangular channel so that the river flow can be considered to be a collection of layered two-dimensional streamlined flows. The energy can

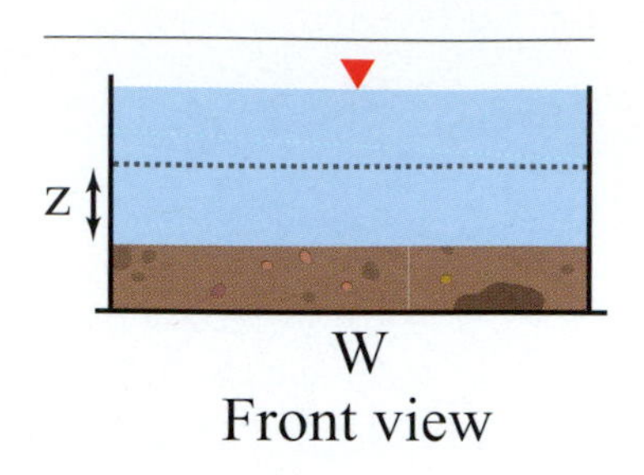

Fig. 6.38: Two-dimensional view.

be readily analyzed for this case and the results can be generalized to an arbitrary-shaped channel. If the width of the channel is W, then through cross-section area $A = zW$ the mass flow is given by $Q = qW$, where q is the fluid flow per unit width and is shown in Figure 6.38. Hence, from Equation 6.33

$$e = z + \frac{1}{2g}\left(\frac{q}{z}\right)^2. \tag{6.34}$$

Equation 6.34 shows the connection between specific energy carried by the river flow, the flow rate per unit width q and the depth of the flow z and is plotted in Figure 6.39. At each depth z, the river has streamlined flow and has an energy that, from Bernoulli's equation, is constant for that depth; hence, e is a function of depth z. It is clear by inspecting e that there is a critical depth z_c for which e is a minimum. From calculus, the minimum value of e is given for the value z_c such that

$$\frac{\partial e}{\partial z} = 0 = 1 - \frac{q^2}{gz_c^3}$$

$$z_c = \left(\frac{q^2}{g}\right)^{1/3} \Rightarrow q = \sqrt{gz_c^3}. \tag{6.35}$$

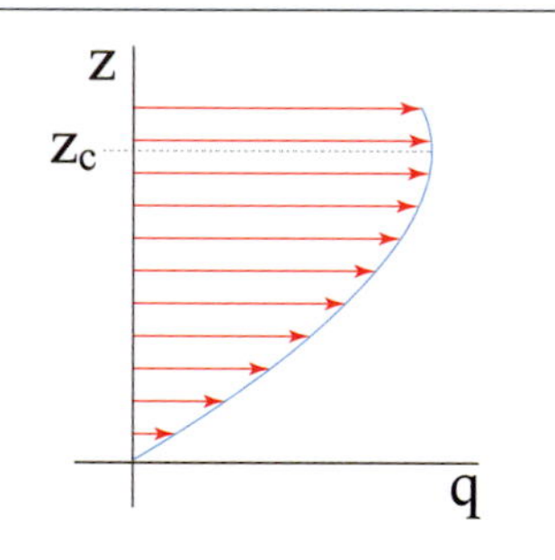

Fig. 6.39: Transport of river water versus depth for fixed energy. The fluid flow up to depth z is shown on the y-axis.

The river water flowing at depth z_c is carrying the minimum energy. Note the remarkable fact that the flow depth z_c depends only on the rate of flow q and not on many other parameters that enter the open channel flow. In fact, Equation 6.34 shows that if one can create or identify depth z_c (for which many techniques have been developed) then the depth of the channel and its flow rate can be determined.

What is the significance of the depth z_c for mass flow q? A graph of the mass flow q versus depth z, at fixed specific energy e is shown in Figure 6.39. Recall q is the amount of mass flow from the bottom of the river to depth z. At the river bed, there is no transport of water, and one has $q = 0$ for $z = 0$. As one goes to shallower and shallower depths, the amount of water being transported increases reaching a maximum at the special depth z_c. The amount of water transported then decreases further depending on the energy at zero depth. The maximum transport of a river's mass at depth z_c is a rather remarkable result that arises from the interplay of gravity and pressure.

6.17 The answer

We can now answer our question of how water flows: **water is a fluid and has two main forms of flow**.

Groundwater, forming the bulk of the earth's freshwater, flows underground through porous media called the aquifer. The flow of groundwater is driven primarily by gravity and the pressure generated in the aquifer by the geological formation of the earth's crust. The velocity of flow is so small that it can be ignored and the entire flow is determined by the interplay of pressure and gravity. Viscosity plays a significant role in groundwater flow.

Surface water in rivers flows in open channels and is driven largely by gravity. The slope of the river bed and its shape and composition are important determinants in the flow of river water; the velocity of a river is the most important feature of its flow, followed by the energy of the flow. Given the complexity of a river's flow, the velocity of a river is related to its slope, shape and composition by certain equations deduced from observations.

6.18 Exercises

1. How can an aquifer be recharged?
2. What is the difference between a Lagrangian and Eulerian description of a fluid?
3. How are the fractions Energy/Volume and Force/Area related? Which thermodynamic variable do they provide?
4. We have an incompressible fluid flowing into the opening of a conical pipe. The radius of the opening is 1 cm and the fluid velocity is 2 m/s. If the outflow from the pipe end is 1 m/s, what is the radius of the pipe end?
5. What is the difference between an aquifer and an aquitard?
6. In viscous flow, is the total energy of the system conserved?
7. Estimate how long it takes for groundwater flow through a dune of 100 m thickness.
8. If a river flows with a speed of 0.5 m/s, is 100 m wide, 10 m deep at its deepest point and has a cross section that is a segment of a circle, how much water passes every minute?
9. We have a river that is 20 m wide and 1 m deep. Its slope is a gentle 1° per kilometer. What is this river's stream velocity if the river bed is hard gravel?
10. Why is clay porous and yet impermeable when saturated?

7: Materials

Q - Why Is Steel Strong and Glass Fragile?

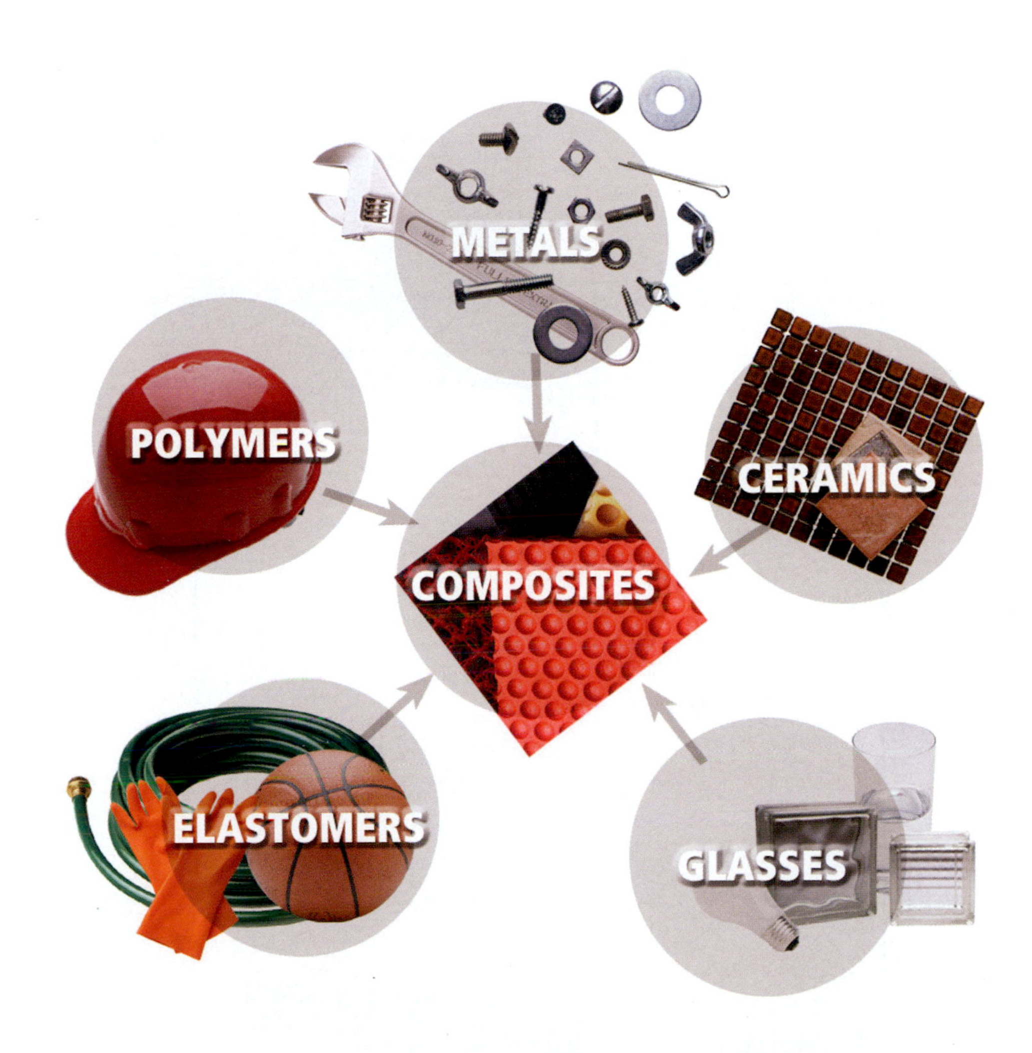

Chapter Map

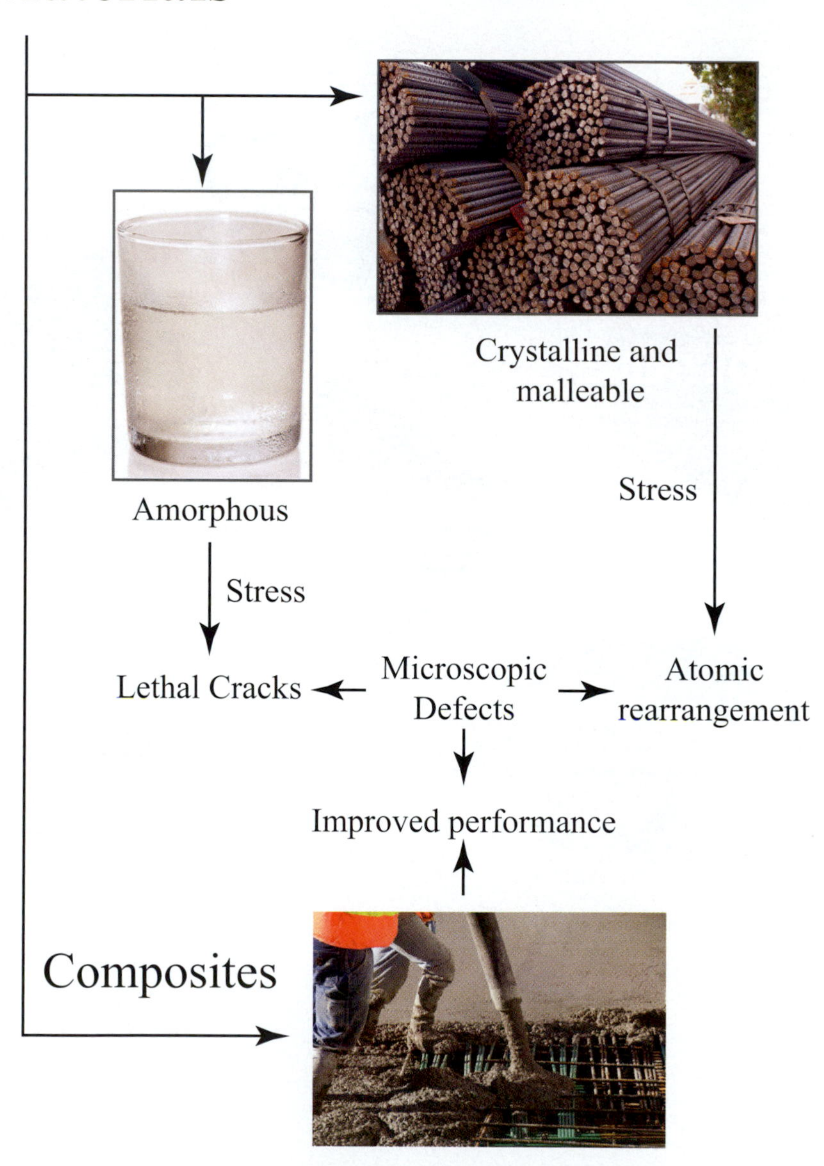

Materials
Crystalline and
malleable
Amorphous
Stress
Stress
Lethal Cracks
Microscopic
Defects
Atomic
rearrangement
Improved performance
Composites

7.1 The question

Figure 7.1: Steel bends and glass shatters. Why?

Every day we are in contact with diverse forms of materials, some man-made and others natural. Some materials are hard and others are soft; some are brittle and others malleable; some break and others bend. To strengthen iron, a blacksmith pounds it with a hammer, whereas if we were to pound a piece of glass, it would shatter. What makes glass so different from iron?

Have you ever wondered why objects have such varying and diverse properties? In particular have you ever wondered **why is steel strong and glass fragile**? This is especially odd considering the bonds involved. The bonds in steel are metallic which are generally weaker than the covalent bonds in glass (see Chapter 5). How can it be that a weaker bond gives a stronger material?

To answer these questions, we need to unravel the secrets of the materials that surround us. We need to understand precisely what we mean by "strong" and "fragile".

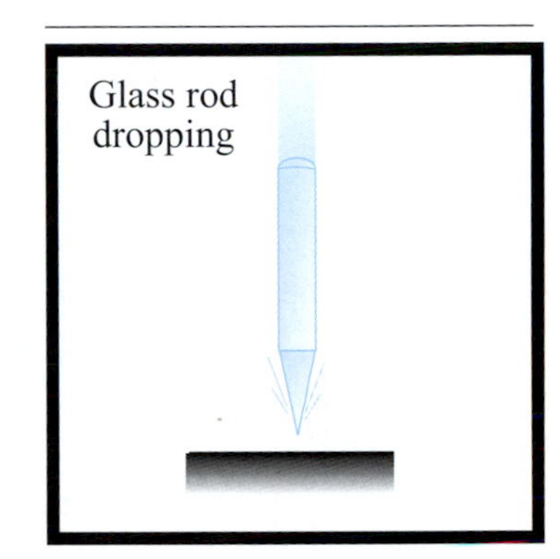

Fig. 7.2: Glass shatters on impact.

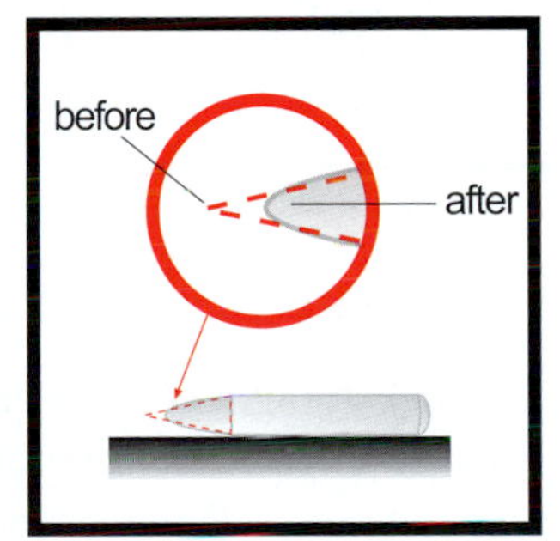

Figure 7.3: The tip of a steel rod will deform when dropped onto a hard surface.

7.2 Bulk properties of materials

First, note that if we talk about the strength or malleability of a material, we are dealing with the so-called bulk properties of a material, i.e. the properties of a large collection of atoms. As is clear from daily life, strength is a relative concept. If we need only a thin rope of a certain type of material to carry a heavy weight but for the same job need a thick rope of another type of material, then we consider the former to be stronger. Indeed, conceptually, we know that

$$\text{Strength} = \frac{\text{Maximum load carried}}{\text{Area of cross section}}.$$

In materials science, strength is generally defined in a similar way. Rather than only looking at the maximum load that can be carried by a material, one defines the more general notion of how much stress a material can withstand for characterizing

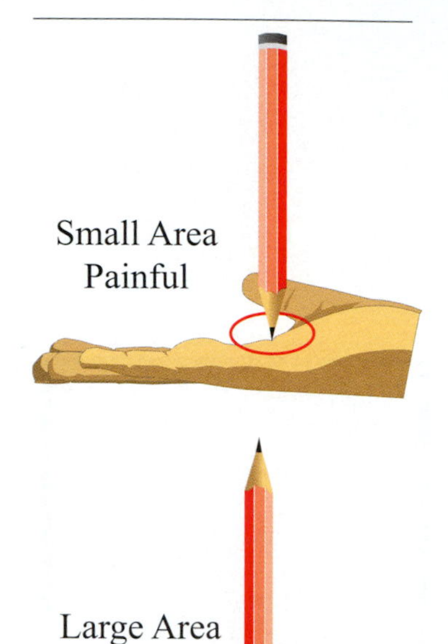

Fig. 7.4: The entire weight of the pencil is on the sharp tip.

the strength of materials. Here, **stress** refers to the application of a force per unit area and it is defined as follows:

$$\sigma = \frac{F}{A} \quad \text{Stress} = \frac{\text{Force}}{\text{Unit area}}. \tag{7.1}$$

There are many different types of stress such as compressive stress, tensile stress and shear stress.[1] The amount of stress at which a material breaks or tears is then its strength. One may wonder why stress is defined as force per unit area; the reason is illustrated in Figure 7.4. The effect of a certain amount of force applied is strongly dependent on the area over which it is distributed. Given a certain amount of force, a sharp tip of a pencil can be rather painful while the blunt end can hardly be felt. Put another way, as shown in Figure 7.5, if we hit a blunt nail with a hammer, the impact is distributed over a large area and may not be enough to drive it into the material. If we hit a pointy nail, however, the same impact affects the atoms in only a small area and the nail can push forward.

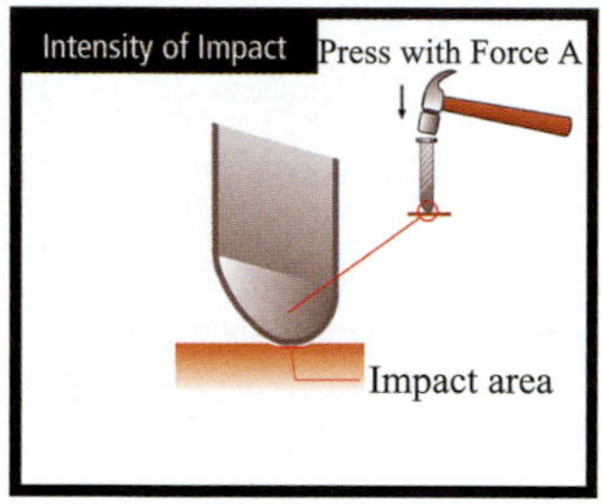

Large impact area => small stress

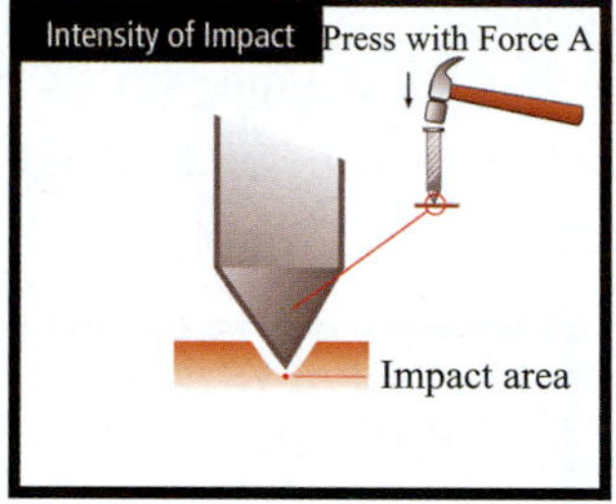

Small impact area => large stress

Figure 7.5: The effect of the impact depends on the sharpness of the nail.

Beyond the actual strength of a material, however, its response to stress when not breaking is just as important. After all, many materials will deform quite significantly before breaking. A key concept in this regard is that of strain. **Strain** is the fractional amount of deformation given a certain amount of stress as illustrated in Figure 7.6. One can then draw a stress-strain plot detailing a material's response to stress.

For many materials, the relationship between stress and strain is linear over a certain range and well described by Hooke's law (see also Chapter 3 and Figure 7.8 below). We then have

$$\text{Strain} = \frac{\text{Stress}}{\text{Young's Modulus}} \tag{7.2}$$

in the case of compression or stretching, and

$$\text{Shear Strain} = \frac{\text{Shear Stress}}{\text{Shear Modulus}} \tag{7.3}$$

[1]The definition of stress is the same as that of pressure but conceptually, there can be quite a difference. For example, the compressive stress on a material can easily be seen as the pressure on the material while the tensile stress due to pulling at both ends of a material would be more like negative pressure. Torque and shear stress are more difficult to associate with pressure.

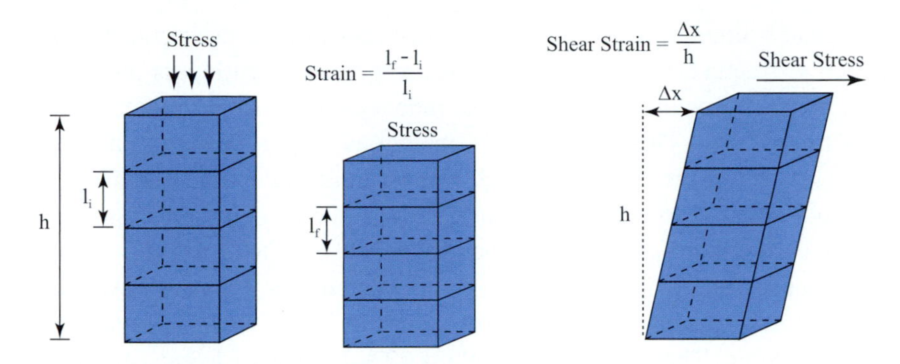

Figure 7.6: The diagram on the left shows compressive stress, and the one on the right shows shear stress.

in the case of applying a shearing force. Here **Young's modulus** is a measure of how difficult it is to compress or stretch the material, and the **shear modulus** is a measure of how difficult it is to bend the material. Strain is a dimensionless number whereas the shear and Young's moduli have dimensions of pressure, measured in units of pascal (1 pascal (Pa) = 1 N/m^2).

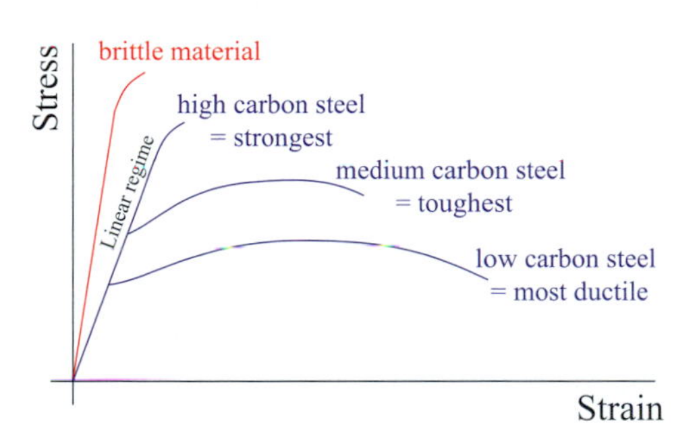

Figure 7.8: Stress-strain behavior for brittle and ductile (malleable) materials.

When a deformation follows Hooke's law, it is said to be an **elastic deformation** — the material temporarily deforms, but is restored to its original shape once the deforming force is removed. However, if a very large force is applied to, say, a knife, the steel will undergo a permanent change of shape, called a **plastic deformation**. A material is called *ductile* if it can be stretched during a plastic deformation. *Yield strength* is maximum stress for elastic deformations, with the onset of plastic deformation being caused by higher stresses.

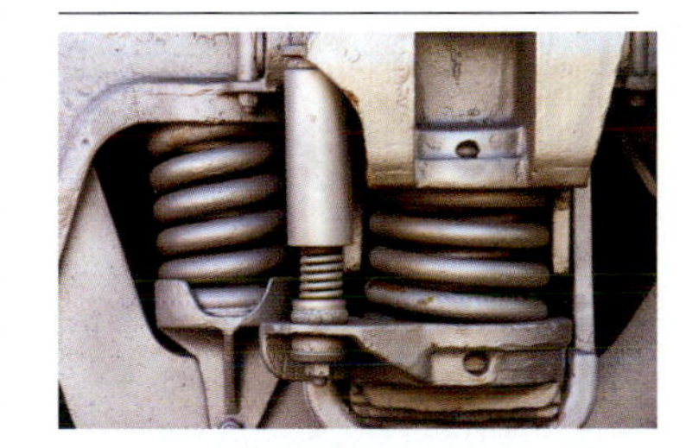

Fig. 7.7: The steel in the springs undergoes elastic deformation.

Young's modulus results from microscopic interatomic forces: if stress (force of compression or extension) is applied to a material, the atoms in the solid are displaced from their equilibrium positions, for which the energy of the solid is at its minimum. Hence, due to attractive and repulsive forces from neighboring atoms, there will be net force tending to restore the atom to its original (equilibrium) position. The stronger the force of restoration, the larger the Young's modulus.

Young's modulus for steel is about 200×10^9 Pa (= 200 GPa). This large value of Young's modulus implies that a 1-m cube of steel would shrink less than 10^{-6}m if the weight of a bus were placed on it.

Some other important concepts in the description of bulk materials are:
Hardness is a measure of a material's resistance to permanent deformation.
Brittleness is the degree to which the material fractures and breaks while being deformed.
Toughness is a measure of a material's ability to absorb energy during deformation, i.e. its ability to resist fracture.

Fig. 7.9: Bricks can be durable and strong; but due to their brittleness they can break easily.

Note that brittleness and toughness as such relate to the same concept, namely resistance to fracture, but from a different perspective. Brittleness indicates how easily a material fractures and toughness indicates how strong a material's resistance is to fracture. This difference is analogous to that between hardness and softness (softness is the degree to which a material is susceptible to deformation).

A good knife has to be hard, and not brittle, since it should not distort or fracture while cutting through something hard like, say, a piece of wood.

Although hardness and toughness seem to be similar, these are two independent properties of materials. For example, a biscuit is stiff and weak whereas steel is stiff but strong. There is a story of Pliny (23–79 AD), a Roman thinker, who was asked how could one determine if a gem were genuine diamond. His reply was that one should put the gem on an anvil and smash it with a hammer — if it does not shatter it is genuine diamond. Needless to say, Pliny must have been responsible for the destruction of many a diamond. Although diamond is the hardest naturally occurring substance known to man, this does not translate into diamond being a tough substance, since diamond is brittle and hence easily broken by a large stressing force.

In summary, *hardness* has to do with a material's resistance to permanent deformation. *Toughness*, on the other hand, is a measure of how much energy the material can absorb before the appearance of a fracture signaling the breaking of the material. If a material is brittle it will easily break, whereas if a material is tough it can withstand a large force — and undergo plastic deformation — without breaking. The difference between a brittle and tough material is illustrated in the stress-strain Figure 7.8.

Now that we have some idea on how to characterize materials, let us investigate two materials that are particularly interesting: steel and glass.

7.3 Strength of steel and glass

Above, we mentioned that strength is determined by the maximum load carried per unit area or in other words the maximum amount of stress that can be applied before a material breaks. It would then seem to be sufficient to choose a certain thickness of a rod, determine how much load it can carry and thus determine the strength of the material. However, doing so implicitly assumes that the strength of a material is independent of its thickness. Is it reasonable to make this assumption? The most straightforward way to find out is to do an experiment. The results of such an experiment are plotted in Figure 7.10 where the strengths of steel and glass are plotted versus the thickness of the rod.

The behavior of steel — namely, being independent of thickness — is perhaps not so surprising but that of glass certainly is. Below a thickness of about 25 μm, glass is stronger that steel! (Of course the exact number will depend on the types of glass and steel used.) Why would that be?

In order to find out, we need to have a closer look at steel and glass. The first thing to notice is that the atoms in steel can quite easily rearrange themselves as can be seen by bending a steel rod back and forth repeatedly as shown in Figure 7.11.

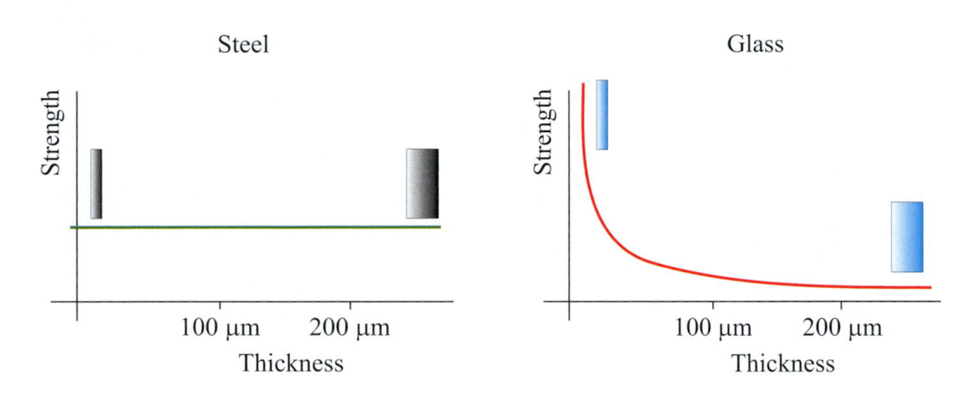

Figure 7.10: Strength versus the thickness of a rod. Glass is stronger than steel when thin enough!

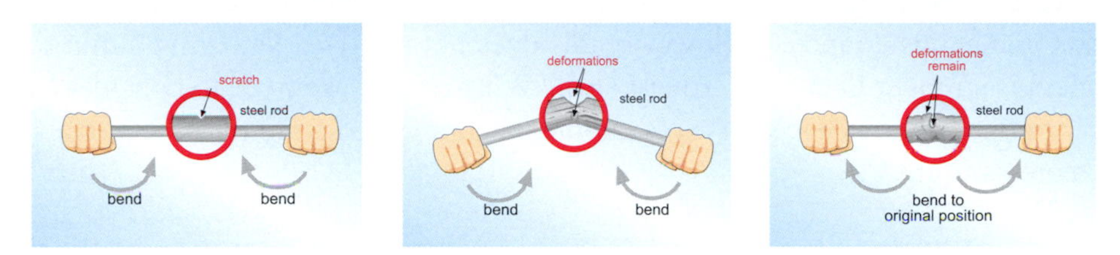

Figure 7.11: The atoms in steel can rearrange quite easily. Steel can be bent back and forth.

The same cannot be said for glass, where the atoms have very limited ability to rearrange. This can be seen by trying to bend a glass rod as shown in Figure 7.12. If one makes a small score line with, for example, a file, then usually the rod will snap right there and the break will be very clean. Indeed, if one puts the rods back together again the fit is extremely good, indicating that basically no glass atoms have moved.

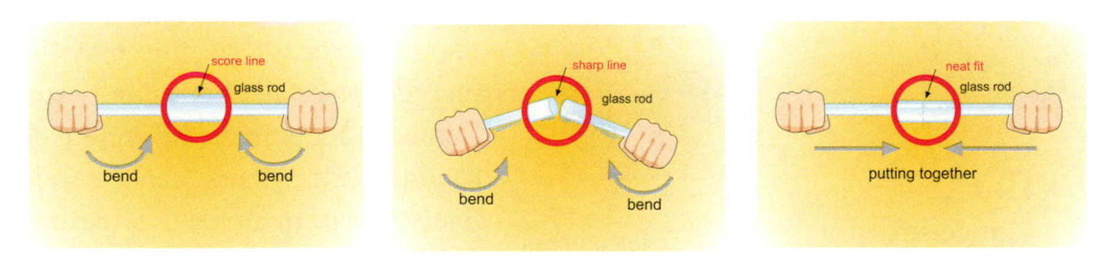

Figure 7.12: The atoms in glass cannot rearrange easily. Glass breaks.

If we look at the atomic structure, as discussed in Chapter 5, we find that glass is amorphous and has (mostly) covalent bonds between the atoms while steel is a crystalline solid whose atoms are bound by metallic bonds. Covalent bonds are directional and hence do not allow atoms to move a lot. Metallic bonds on the other hand are by and large not directional thus allowing atoms to move much more freely. This is illustrated in Figure 7.13.

Another important aspect of glass is that unlike a metal, glass is not a crystal. The atoms of glass are not arranged in an orderly fashion into a lattice, but rather are randomly distributed inside the material. Glass atoms arrange themselves in this random fashion when glass is cooled from its liquid form into a solid.

Glass can be thought of as a *liquid* with an extremely high viscosity but it is more accurate to call glass an amorphous solid.

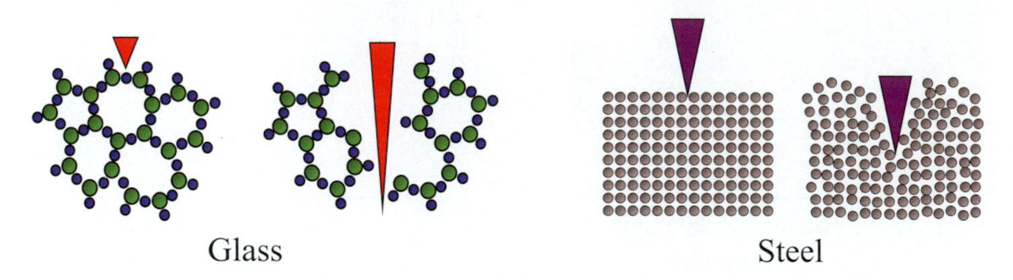

Figure 7.13: The atoms in glass cannot rearrange easily. Glass breaks.

We now understand why glass breaks (the atoms do not move) and why steel bends (the atoms do move). However, this does not really explain why a thin rod of glass is stronger than a thin rod of steel. What else do we need to consider? In order to find out, it is necessary to examine the atomic structure of the materials a bit more carefully.

7.4 Defects

Bulk materials are never perfect and what we have considered thus far holds only partially. Locations where a material is not perfect are called defects, so we now investigate these and their impact on a material's properties.

Defects are microscopic irregularities in the structure of materials. Would these defects have a significant impact on a material's properties? In order to answer that, one needs to consider what happens to stress near a defect.

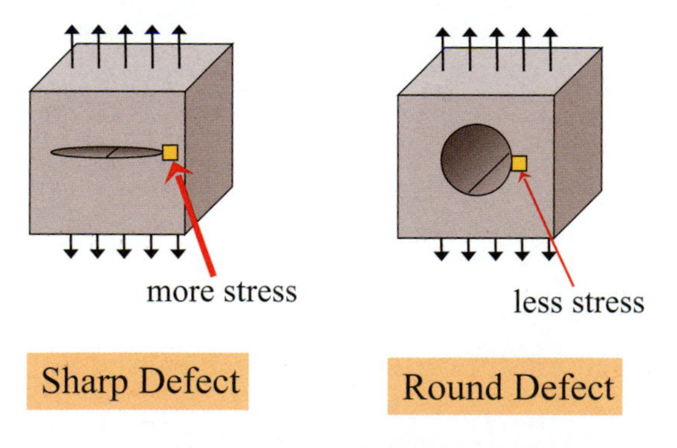

Figure 7.14: Stress is higher near a sharp edge.

Since defects are in general randomly formed during the production process of the material, they come in all shapes and sizes. Some defects can be relatively round while others can be relatively sharp. In an ideal defect-free material, an applied external stress will be evenly distributed throughout the material. What happens to those stresses if there is a defect in the material? To find out, one can drill holes of various shapes and measure how the internal stresses are redistributed. It turns out that the stress is much higher at the notch tip of a sharp hole than at the notch tip of a round hole (see Figure 7.14).

The presence of defects therefore means that stresses are distributed differently throughout the material than would be the case if the structure had no irregularities. In particular, there will be points (sharp defects) where the stresses are significantly higher.

What will happen at those points? In metals, the atoms can rearrange thus reducing stress near the defect. Consequently, in metals, the effect of an initially sharp defect is quite limited. In glass, however, atoms cannot rearrange and a sharp defect remains sharp. When an external stress is applied, and if the stress is too large, a lethal crack can be formed.

Knowing this, can we now explain why a thin glass rod is stronger than a thin metal rod? Indeed we can! In metal, defects deform to rounded shapes (atoms move) to reduce stress. As the mobility of the atoms depends on properties such as the crystalline structure, grain size, defect distribution and defect size, the absolute size of the metal rod has little impact on strength. In glass, on the other hand, a single sharp defect can be the origin of a propagating crack. The thicker the rod, the higher the probability of encountering a critical defect that triggers a lethal crack. The assumption here is of course that the probability of having a defect is constant per unit volume — something that is usually the case. Consequently, the probability of a glass rod breaking increases with its size. In other words,

- The strength of metal is related to its microstructure and mobility of atoms.
- The strength of glass is determined by its "weakest link".

A consequence of this explanation is that steel should obey scaling while glass does not. If we have a look at Figure 7.15, we see that 10 fused steel rods can support a weight that's 10 times as large as the weight supported by one rod.

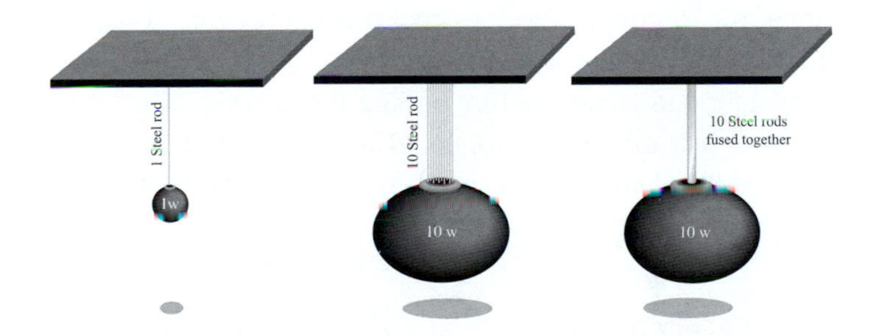

Figure 7.15: Ten fused steel rods are as strong as 10 separate rods.

Quite to the contrary, 10 fused glass rods, in general, *cannot* support 10 times the weight that one glass rod can support as illustrated in Figure 7.16.

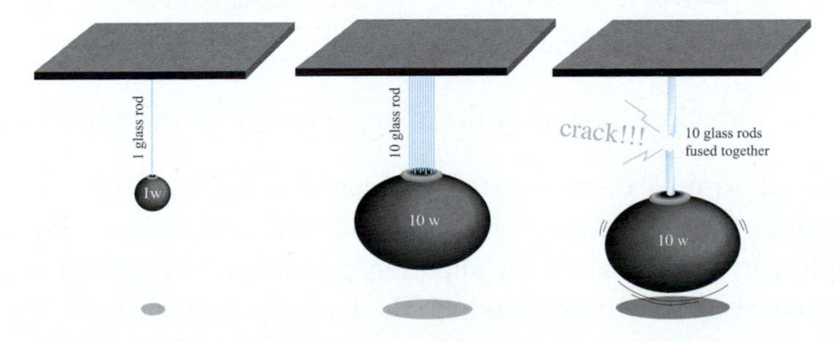

Figure 7.16: Ten fused glass rods are much weaker than 10 separate glass rods.

This is because glass suffers from the "weakest link" syndrome as shown in Figure 7.17. Although the average strength is sufficient to support 10 times the weight with 10 individual rods, when fused together, the defect in the rod with the least strength will lead to a lethal crack.

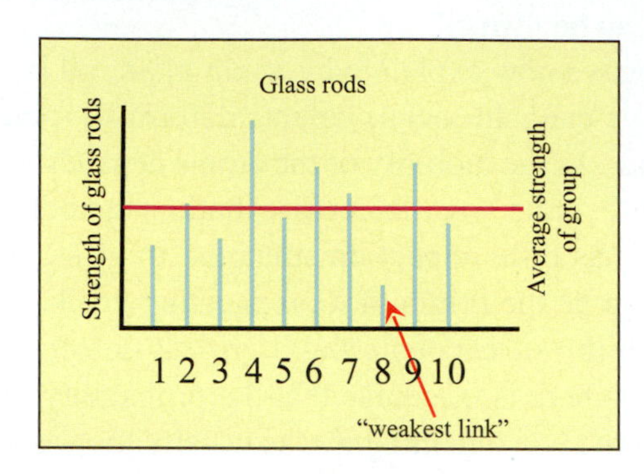

Figure 7.17: The strength of the fused rod is determined by the weakest link.

7.5 Alloys

Although we have mostly answered the chapter question, we haven't considered one issue yet. When we say "strong" with respect to a metal, we mostly talk about steel. Wouldn't the explanation given above apply just as well to iron, steel's major constituent, or other metals such as copper or aluminum? Although the discussion did not refer to any specific property of steel, the fact is that steel is much harder and stronger than iron or copper. Why is steel much stronger? It has to do with the fact that steel is an alloy. Since some key ideas put to work in alloys are also true for glass, we consider composites before ending the chapter.

One common defect of iron is a **dislocation** in which a sheet of atoms ends abruptly in the middle of the crystal, as shown in Figure 7.18. The dislocation

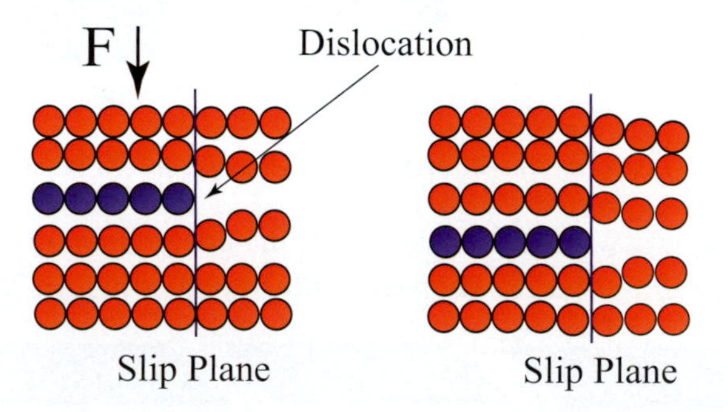

Figure 7.18: Slip plane caused by the imperfection of an extra row of atoms.

breaks up the uniformity of the crystal, and can give rise to what is called a **slip plane** caused by the iron lattice accommodating the dislocation. When a shear force is applied perpendicular to the slip plane, compared to the rest of the material, the alignment of the atoms is poorer which allows them to move more easily. Thus the strength of iron along a slip plane is considerably less than along a plane without a dislocation.

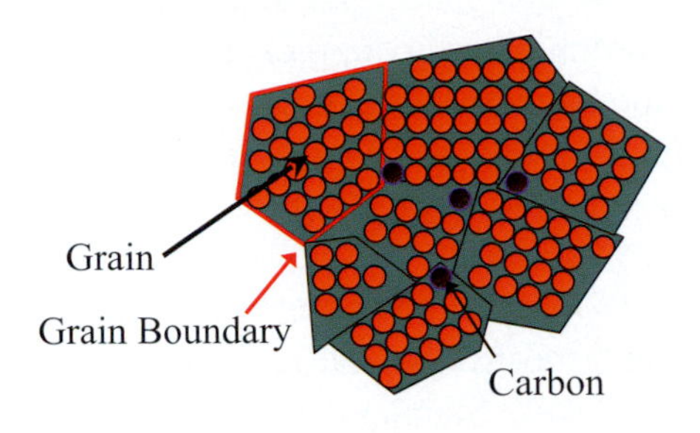

Figure 7.19: Grains, with the help of carbon, strengthen iron to yield steel.

While dislocations weaken iron under slippage, other imperfections can go the other way and actually strengthen iron by preventing slip. If iron is not a single crystal, but rather a *polycrystal* composed of many smaller crystals — called **grains**, the individual crystals will not be aligned but rather will meet at random angles, forming *grain boundaries*. Grains and grain boundaries strengthen iron against slip, since under a shearing force slip will occur along different slip planes in each grain, and since the grains are randomly aligned, the slip planes of the various grains mismatch creating large resistance to slippage.

In order to change the material properties of iron, small amounts of carbon are added to produce steel. Metallic materials that are mixtures of two or more elements are called **alloys**, and steel is therefore an iron alloy. Besides carbon, steel can contain a number of different elements such as nickel, manganese or chromium. Different amounts of carbon, as well as differences in processing of the alloy, produce steel with sometimes drastically different qualities.

We summarize below a few of the different forms of steel alloys that can be made from iron.

- Up to 0.01% by weight of carbon can dissolve (at room temperature) in iron — similar to salt dissolving in water. The carbon atoms reside in the interatomic spaces of the iron lattice, and cause localized distortions of the iron lattice. These distortions disrupt the slip planes, and so sheets of iron can no longer slide as easily under shearing forces, and hence increase the yield strength of iron besides possibly improving other properties.

- When the fraction by weight of carbon exceeds 0.01%, addition of more carbon leads to the formation of a new molecule, namely Fe_3C called **cementite**; dispersed amounts of cementite greatly increase the yield strength of steel by reducing slip. By slowly cooling the alloy from high temperature, if the amount of carbon is greater than 0.8%, it forms a material called **pearlite** — which consists of alternating layers of iron and cementite; for larger amounts of carbon, the pearlite has cementite grains dispersed throughout the material.

- Iron above a temperature of 723°C exists as a face centered cubic lattice and is called **austenite**. If austenite is suddenly cooled, a process called *quenching*, it transforms into a variety of steels depending on the final temperature. If austenite is suddenly cooled to 200°C, it forms **martensite**; this alloy has great strength and hardness — but is brittle — and the cutting edges of knives are sometimes made from it. An alloy obtained by introducing fine particles of, say, pearlite into steel is equally hard but less brittle than martensite.

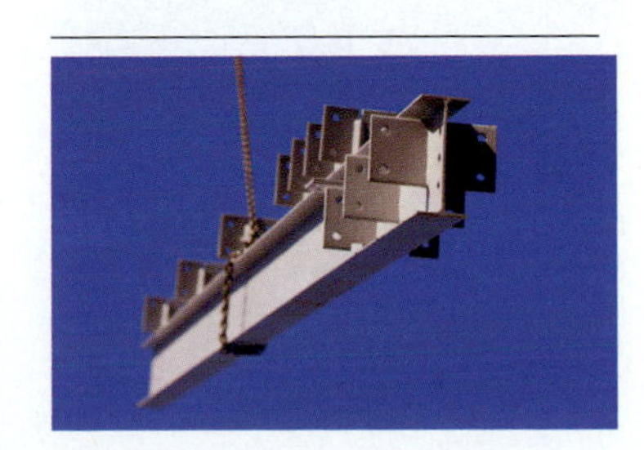

Fig. 7.20: Steel girders, strong and tough.

- One can introduce other atoms such as chromium, silicon, nickel and so on into iron; unlike carbon these atoms are too large to reside in the interatomic spaces of a iron, but instead *substitute* the Fe atoms, and cause substitutional hardening or other modifications of the properties of steel.

- To make steel corrosion resistant, at least 4% of the iron atoms of martensite steel need to be replaced by chromium, and most corrosion-resistant alloys contain about 11% chromium with some nickel. Stainless steel has a minimum of approximately 12% chromium and often some other components like nickel and manganese.

We have only scratched the surface of the great variety of alloys that can be made from iron, and each alloy has special properties. The examples above are meant to illustrate how complex and varied the properties of materials can be.

In summary, what we see is that steel is strong not only because its major constituent, iron, is a metal (indeed many metals aren't that strong at all) but more importantly because steel contains other elements, the addition of which drastically alters iron's properties. Combining various "ingredients" to obtain something new and more desirable is also at the heart of a class of materials called composites, which we have a look at now.

7.6 Composites

Fig. 7.21: Concrete needs to be reinforced with rebars to withstand the stresses in many structures.

Glass, pottery, ceramics, biscuits and so on are brittle since cracks, once formed, can easily propagate resulting in breaks. Is there a way to use the strength of glass (an excellent property) without having to incur its weakness (the fact that it easily breaks)?

Let us briefly recapitulate. To start with, glass cannot undergo plastic deformation since it cannot undergo slip. The reason being its irregular atomic structure, there is no way a whole plane of atoms can slip. If one tries to bend glass, after a generally rather small deformation, the energy goes into breaking bonds leading to the appearance of a fracture.

Once a crack appears, it propagates uncontrollably throughout the glass since there is no grain structure to stop crack propagation. Crack propagation in essence means the breakage of bonds in the glass, and this propagation takes place at the speed of sound (in glass) since the propagation of the crack takes place through thermal vibrations. The velocity of sound in solids is very high, and in glass it is about 4,800 meters per second — or about five times the speed of a typical rifle bullet.

The crack generally begins on the surface of glass, since there are always defects that weaken the chemical bonds holding glass together. If the glass is bent, stress builds up on the surface to create a crack centered on the surface defect, which then propagates at the speed of sound throughout the glass and shatters it.

Cutting Glass with Diamond: Bending glass is not the best way of cutting glass since it will break in an uncontrolled manner, given the random location of the surface defects. To cut glass, the surface of the glass is first scratched along the specified line with a hard substance like diamond. The scratch introduces a series of defects by breaking bonds on the surface, and weakens it. Hence when stress is now applied carefully, the glass can be made to crack along the scratch.

Cutting Glass with a Saw: A rapidly spinning abrasive saw — consisting of a disk mounted with a hard crystal — can slowly cut into glass without cracking it since the stress is applied to a very small area, and is used for cutting decorative glass.

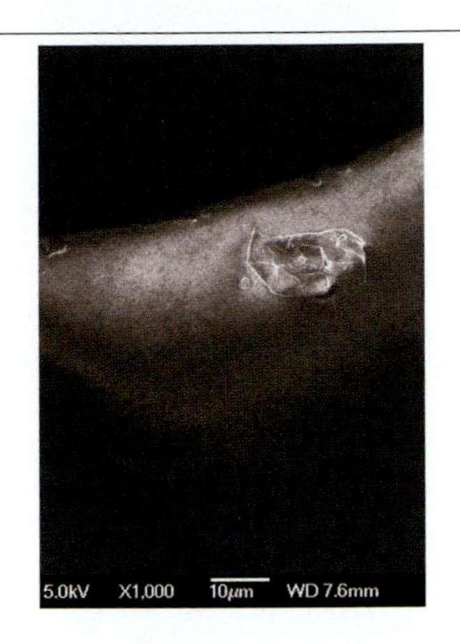

Fig. 7.22: SEM image of a glass rod with a large surface defect.

If cracks are the main culprits for the fragility of glass, then if one could stop those cracks, glass should become a lot stronger. Indeed, this is exactly what happens in many composites. One way to stop cracks is by the introduction of fiber reinforcements as shown in Figure 7.23. Fiberglass is made from this principle.

A block of homogeneous material under a uniform tension has an even distribution of stresses throughout the material. If a small crack appears, perhaps due to a local flaw, the stress in the block tends to extend the crack farther apart. This happens because the stresses which were shared out evenly before the crack developed are now concentrated at the crack tip, making it more likely for bonds to break there than anywhere else. The crack may propagate very fast, or progress in spurts as the pattern of stress in the block shifts, due perhaps to other cracks developing. Whatever way it goes, and however fast it happens, when a structure breaks, it happens as a series of developing cracks, not a simultaneous rupture of all its internal bonds. Consequently, the inclusion of crack stoppers in a material can have a drastic impact on increasing its strength.

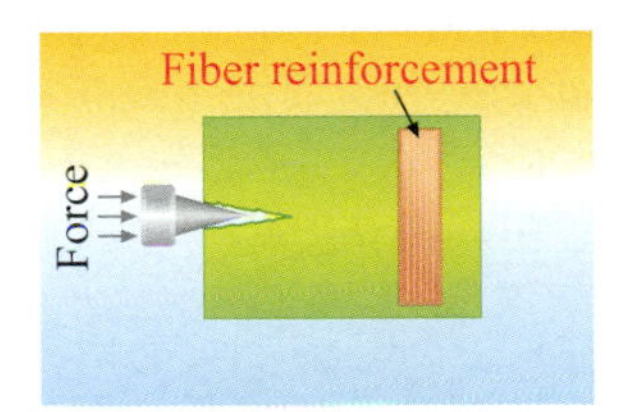

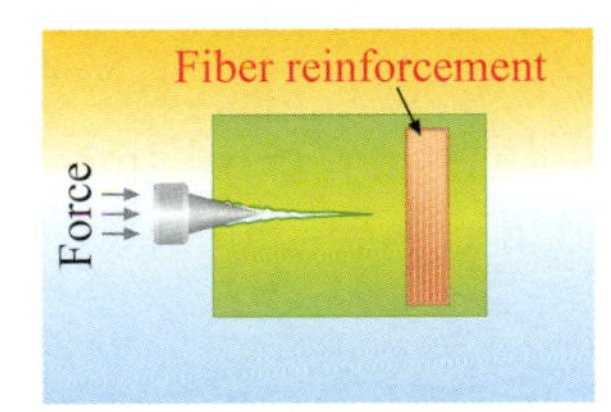

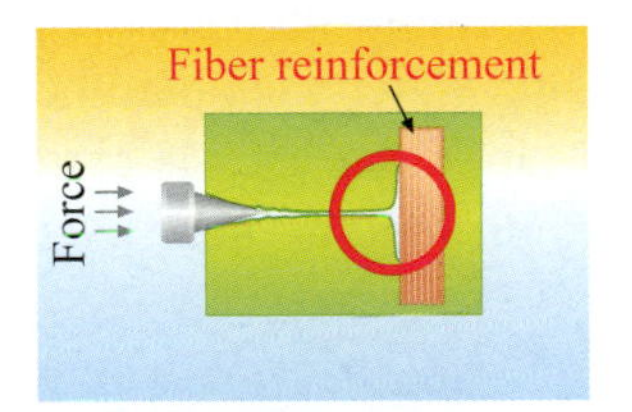

Figure 7.23: Fibers in a brittle material strengthen it by stopping cracks.

Another way to stop cracks from propagating is with the help of a bridging material such as the tiny rubber balls shown in Figure 7.24. When fibers are dispersed in a binder as in Figures 7.23 and 7.24, the properties of the composite material may be a useful blend of both fibers and binder. Even if a few fibers break in the path of a crack, they do not completely lose their contribution in resisting stress forces, which they would do if not bound in an embedding material. Composite materials can be both strong and light in comparison with noncomposite materials. The newer carbon fiber and aramid fiber composites are stronger than steel but only 20% of its weight.

One of the requirements for a successful composite is good adhesion between the constituent materials. While glass fibers bond well to resins, they do not bond to builders cement. Another requirement is that the fibers should be long. This is for two reasons. First, if they are short they may pull out of the embedding material without breaking. Second, the orientation of long fibers can be controlled more easily than short fibers. When the composite is being made, the fibers can be placed in line with the expected tensions the structure will have to meet when it is

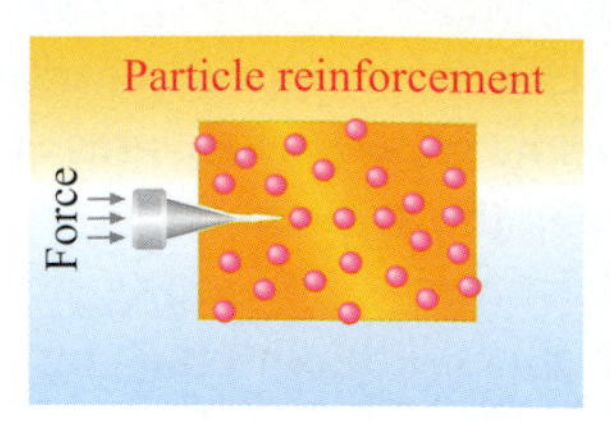

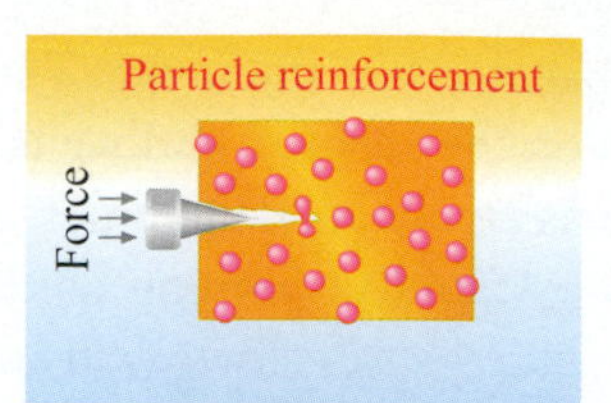

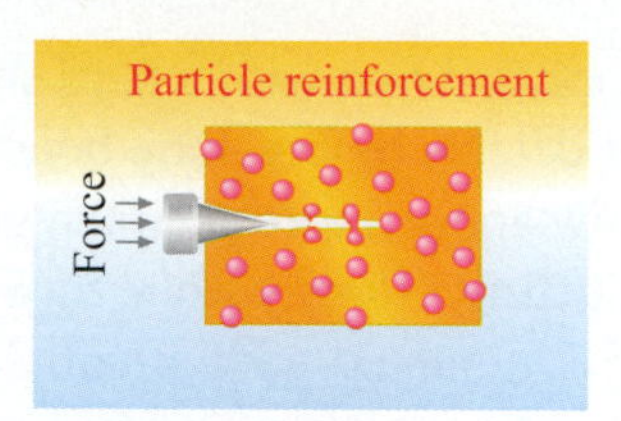

Figure 7.24: Rubbery spheres in a brittle material can be another effective way to stop cracks.

in use. A woven fiber system, such as a fabric, resists tension best along the warp and the weft. Many composites are built like plywood, with several layers, each with its own specific contribution to overall strength.

Even without the microscopic scientific understanding, composites have been in use for a long time. For example, the Egyptians often reinforced mud bricks with straw in order for them not to crack too easily under the hot sun.

Remarkable homemade composite

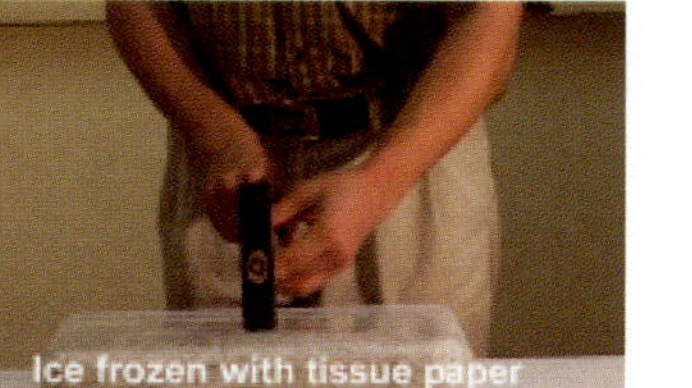

Fig. 7.25: In vain, one of the authors attempts to break tissue paper reinforced ice.

A rather remarkable composite can easily be made at home. All that is needed is some water, a roll of tissue paper and a freezer. Clearly, water, when frozen, is very brittle. An ice block can readily be smashed, even with a rather small hammer. It goes without saying that tissue paper is not exactly strong. Not only can it easily be torn to pieces it also becomes soggy and weak in water.

A composite can be made by densely filling up a cake form with tissue paper, adding water such that the tissue paper is just covered and freezing. The resulting frozen "tissue ice" is extremely hard to break. In fact a 30 x 20 x 8 cm block is very difficult to break even with a sledgehammer (see also Figure 7.25).

7.7 The answer

We have seen that defects occur in all kinds of materials, be it glass or steel. However, in malleable materials like steel, stress on sharp defects can be relieved by atomic rearrangement. In contrast, for the case of glass, stress on a sharp defect cannot be relieved and thus can turn the defect into a critical defect that will be the origin of a crack or fracture. In short, the answer to our question of why steel is strong and glass fragile is hence given by: **Steel is strong because it is both tough and tolerant to defects while glass is fragile and brittle since any of its defects can become critical.**

With the help of Figures 7.15 and 7.16, the key point of this chapter can nicely be summarized. Ten fused steel rods are 10 times as strong as a single steel rod; 10 steel rods retain their strength if they are fused together into a single thick steel rod. This is because atomic rearrangement in steel evens out the differences between defects. Consequently, the strength per unit area is not dependent on the diameter of the rod.

The situation for glass is completely different. Ten glass rods are (on average) 10 times as strong as one glass rod, but when fused together the fused rod is significantly weaker than 10 times a single rod. Since atoms cannot rearrange in glass, a single sharp defect can be critical and cause a rod to fracture, be it thick or thin. In a thin rod the probability to have such a critical defect is much smaller (in our example 10 times smaller) than in a thick rod. Consequently, the strength of glass per unit area is dependent on the diameter of the rod.

7.8 Exercises

1. Why is iron softer than steel?
2. Why do cable cars use many thin braided steel wires for carrying the cabins even though a fused rod would be simpler to manufacture?
3. What is the function of the rebars in reinforced concrete? Specifically, what type of strength do the rebars and the concrete provide?
4. If I put a mass of 100 kg on a 20-cm cube of ice, what will be the stress on the ice?
5. When putting a weight of 1000 kg on a 10-cm cube of a certain material, its height is reduced to 9.95 cm. What is Young's modulus of this material?
6. A human hair is about 100 μm thick. At this thickness, is glass or steel stronger?
7. We know that ice is brittle. Does that mean that it behaves like glass and is very strong when very thin?
8. The grain size in steel depends on a number of factors including the speed with which it is cooled. What kind of grain size (relatively large or relatively small) should I aim for if I want the steel to be very hard?
9. Can one consider wood to be an alloy?
10. Wrought iron has a tensile strength of about 300 MPa. If for small strain Hooke's law is applicable, what is its compressive strength?

8: Polymers

Q - Why Is Rubber Elastic?

Chapter Map

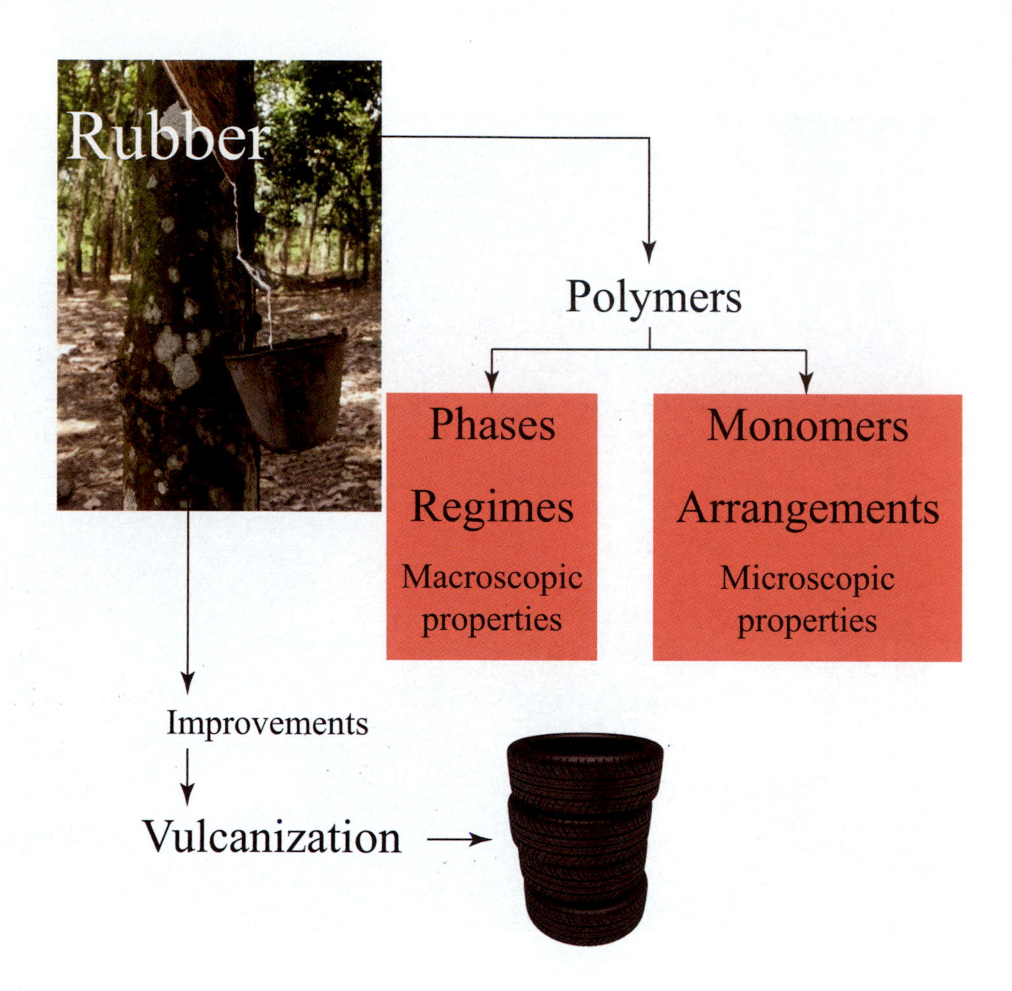
Rubber
Polymers
Phases
Regimes
Macroscopic properties
Monomers
Arrangements
Microscopic properties
Improvements
Vulcanization

8.1 The question

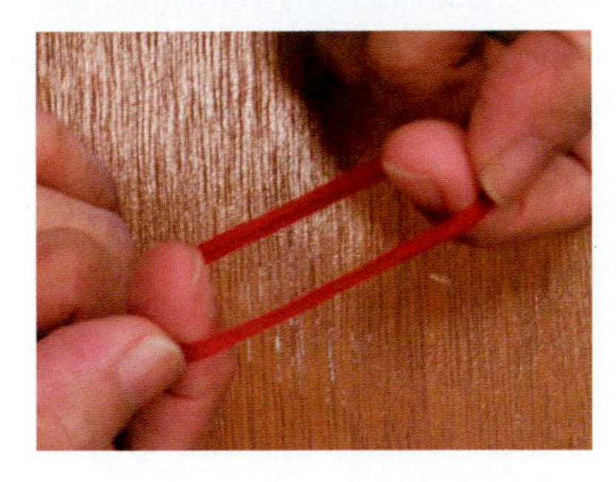

Figure 8.1: A rubber band can be stretched and relaxed many times.

Rubber bands are not only useful in daily life but also excellent representatives of a class of materials called polymers. A characteristic feature of a rubber band is its elasticity. As long as one does not overdo it, a rubber band can be stretched and released a large number of times without it changing shape or it becoming less elastic. Then again, the same is also true for a spring made from a metal. Is there any similarity? What is the difference?

In this chapter, we would like to investigate further what distinguishes polymers from other classes of materials and what the origins are of rubber's elasticity. Hence we ask: **Why is rubber elastic?**

Of course if we want to ask why rubber is elastic, then the first thing to do is to investigate how elastic it actually is. This can be done by making a stress-strain diagram. A qualitative representation of a typical stress-strain diagram for rubber is shown in Figure 8.2 and can be divided into three major parts, namely:

- When the stress is small, the corresponding strain is a linear function of the stress and well described by Hooke's law. In this region (labeled A in the figure), the deformation is fully elastic and recoverable.

- At a certain level of stress, the rubber will stretch a large amount in response to a small amount of applied stress, called necking, and this will start reducing the tensile strength of the material leading to a decline in the stress necessary to further strain the rubber. The onset of necking is visible in the graph as a hump at the end of region A. After the hump, there is a plateau (labeled B in the figure) where for equal stress the neck becomes thinner and thinner. In this stage, the deformation is nonlinear elastic and mostly nonrecoverable.

- At the end of the plateau, a steep rise (labeled C in the figure) precedes breakage.

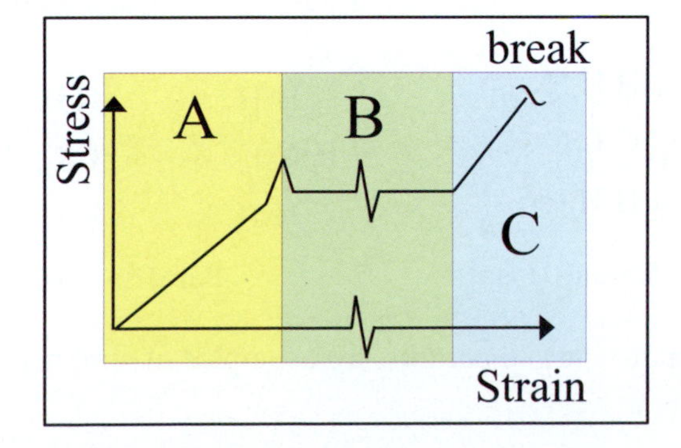

Figure 8.2: Qualitative stress-strain diagram of a polymer like rubber.

Fig. 8.3: The compressive elasticity of rubber bands reveals itself in a bouncing ball.

From daily life, it is clear that rubber is quite different from metals and glasses or ceramics. In order to understand the difference with those types of materials, we need to investigate the molecular structure of rubber. Chemical analysis shows that rubber is built up of many small chunks each containing a number of atoms. More generally, materials with such a structure are called polymers from the Greek words "poly" meaning "many" and "meros" meaning "part". The chunks themselves are called monomers. Therefore, let us now have a closer look at polymers.

8.2 Polymeric materials

Polymeric materials are made up of a very large number of polymer chains and each chain is made up of a usually large number of monomers. One may wonder how many monomers are there in a chain and it turns out that the chain length has a significant impact on the physical properties of a material. This does not, however, mean that for a given material all the chains have the same length, rather, the chains in a material have a certain average length and consequently also a certain average mass. This is quite different from simple compounds like calcium carbonate where the mass of each individual molecule is the same.

To understand why rubber is elastic, we need to understand rubber and since rubber is a polymer, we need to understand polymers. Let us therefore now consider some important aspects of polymers: polymerization (i.e. their formation from monomers), mass and size, and how the individual polymer chains are arranged in a solid.

Polymerization

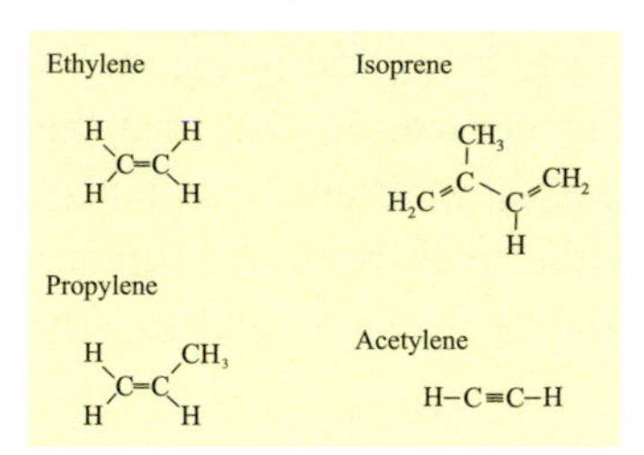

Figure 8.4: Commonly encountered monomers.

Polymers are made of long, often linear, chains of monomers, and some rather common monomers are shown in Figure 8.4. Polymerization is the process of joining the monomers by covalent bonds. If the polymers are made out of a single type of monomer, they are called **homopolymers** and if they consist of two or more types of monomers, they are called **copolymers**. Some very simple polymers, made up out of ethylene monomers, are shown in Figure 8.5. Usually, when discussing polymers, one

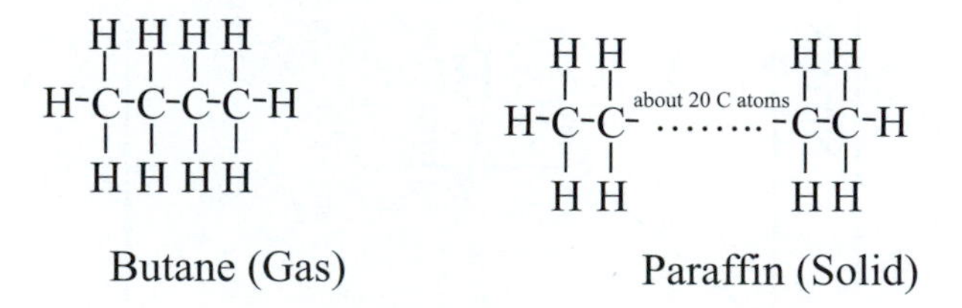

Figure 8.5: Butane gas and paraffin are examples of simple polyethylenes.

has molecules consisting of several hundred to many thousands of monomers in mind. In the process of polymerization, successive monomers generally attach themselves to a growing chain and such chains may need to be terminated by a modified monomer.

As can perhaps be imagined, there are all kinds of ways long chains can form and join together. Chains can be linear or branched, with regularly or irregularly arranged monomers.

Size and mass

The mass of a polymer is given by

$$M_{\text{polymer}} = M_{\text{monomer}} \times n \tag{8.1}$$

where n is the so-called **degree of polymerization** or in other words the number of monomers in a polymer chain. Shopping bags are often made from polyethylene chains. If for example such a chain consists of 7000 ethylene monomers, the molar mass of the chain is then 28 g/mol × 7000 = 196 kg/mol (the molar mass of ethylene is 28 g/mol). During the manufacture of a polymeric material, however, it is not really possible to exactly control the degree of polymerization for each chain and consequently the chain length will vary quite a bit following some distribution.

Figure 8.6: The path of a polymer chain resembles a necklace.

Since polymers can contain many thousands of monomers, they can be pretty long. For example, the length of an ethylene monomer is about 250 pm and hence, stretched out, a shopping bag polyethylene chain could well be 2 μm long.

In reality, however, polymer chains do not grow in a straight line. Rather, the growth process can roughly be thought of as the result of a random walk where one monomer to the next can rotate freely. This is similar to an open pearl necklace (see also Figure 8.6).

The bonds between the monomers are covalent and therefore directional. Then how could they follow a random walk? The key is that even when the direction of the bond is given, there is still (considerable) freedom as with regards to the rotation as illustrated in Figure 8.7.

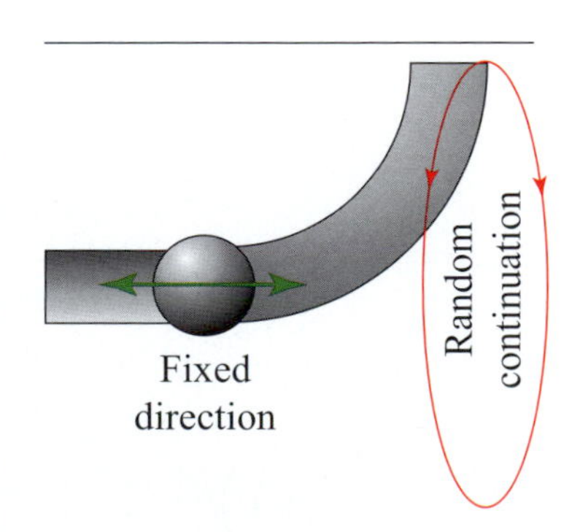

Fig. 8.7: Obtaining a random walk even if the bond direction is fixed.

Since two monomers cannot occupy the same position, the random walk must be self-avoiding. Consequently, a single chain made out of n monomers has a tendency to be rather "jumbled up" and can mathematically be thought of as an $n-$step self-avoiding random walk. Let λ be the distance between the first and last monomers. For a self-avoiding random walk, many rigorous mathematical results exist, one of them being that the average distance between the beginning and end is about

$$\lambda = \ell_0 n^{3/5}, \tag{8.2}$$

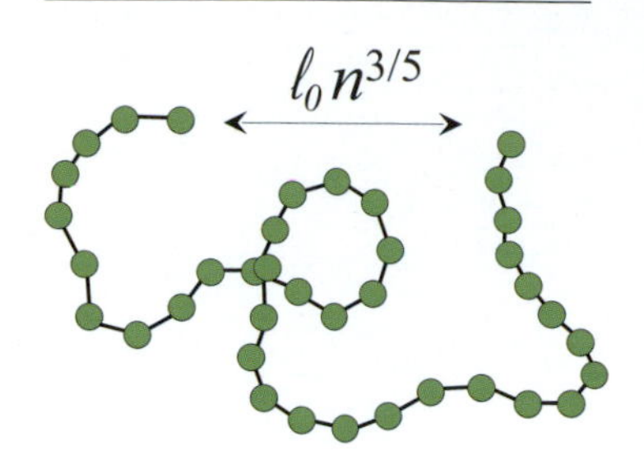

Fig. 8.8: Polymer as a self-avoiding random walk.

where ℓ_0 is the length of a single monomer and n is the number of monomers in the chain as illustrated in Figure 8.8. This tendency of single chains to be randomlike plays an essential role in their macroscopic properties when many of such chains form a chunk of polymeric material such as rubber.

The size of a polymer is often represented by the **radius of gyration** that gives the average distance from the center of mass of a type of polymer to its outer edge. Again based on the random walk approach, the radius of gyration can be found as

$$R_g = \sqrt{\frac{n\ell_0^2}{6}}. \tag{8.3}$$

The radius of gyration of the shopping bag monomer mentioned above is around $R_g \simeq 10$ nm and consequently is about 100 times smaller than its length.

Crosslinks

At room temperature, polymer chains are often in solid form bound by weak interchain van der Waals and hydrogen bonds and by strong intrachain covalent bonds. However, there can also be some interchain covalent bonds. The latter are generally referred to as **crosslinks**. The crosslinks are largely responsible for the "memory" effect in elastic polymers as illustrated in Figure 8.9. When tensile stress is applied on a rubber band in the elastic regime, the jumbled up chains may straighten a bit but the crosslinks will prevent them from sliding relative to each other. Consequently, when the stress is relieved, the crosslinks will pull the chains back toward their original shape.

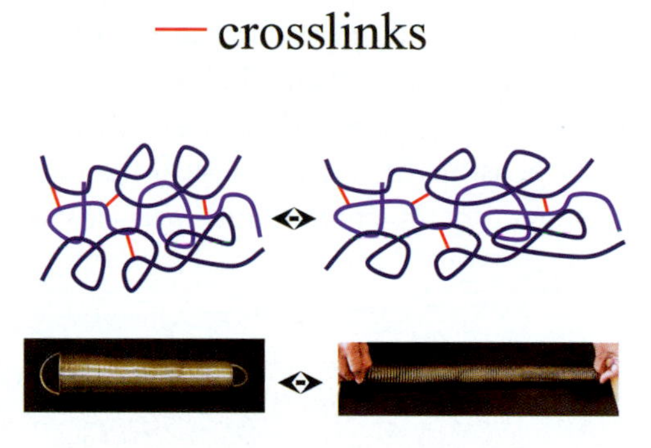

Figure 8.9: Crosslinks anchor the polymer chains to each other giving a "memory" effect.

Crystallization

Not only can chains be formed in many ways, chains can also combine in many ways to form solids. For example, polyethylene at room temperature is often a mix of crystalline and amorphous regions. In the crystalline regions, the polymer chains are somewhat regularly aligned while in the amorphous regions they are jumbled up. What is interesting is that a single chain can be part of both crystalline as well as amorphous regions, as illustrated in Figure 8.10.

Even though polymeric materials have crystalline regions, they are very different from the crystals that are formed from simple compounds such as salt or sugar.

The common plastic bags found in supermarkets and elsewhere are usually made up of polyethylene where the chains are mostly aligned. This is nicely illustrated in Figure 8.11 where the bag tears smoothly in the direction along the chains

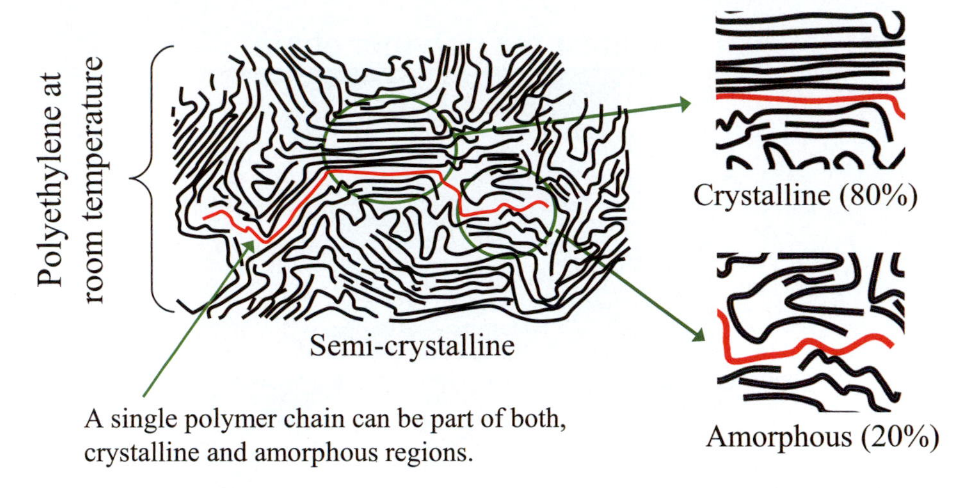

Figure 8.10: Polyethylene at room temperature can be a mix of crystalline and amorphous regions.

and roughly in the direction against the chains. In Figure 8.12 we clearly can observe drawing when pulling along the polymer chain and thinning when pulling in the direction perpendicular to the polymer chain.

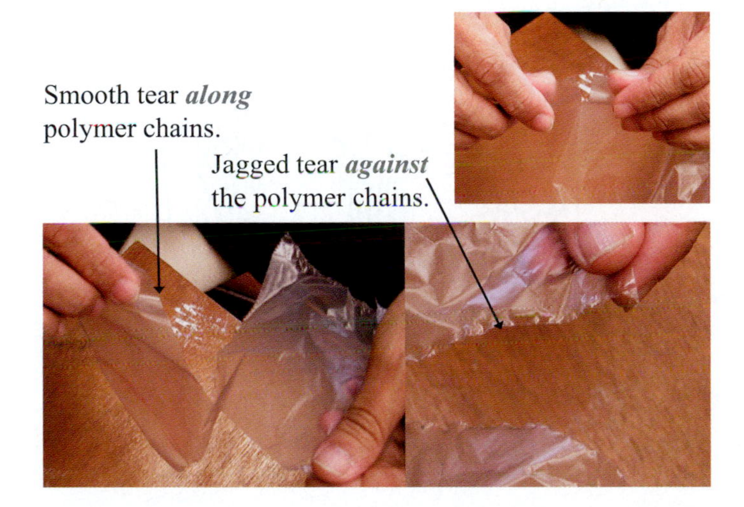

Figure 8.11: Tearing a plastic bag along and against the polymer chains.

Another type of polyethylene is the so-called ultra high molecular weight polyethylene (UHMWPE) where a single chain can consist of several hundred thousand ethylene monomers. These chains can be manufactured such that they align neatly in a crystalline arrangement yielding an extremely strong material. Indeed, fibers made from UHMWPE are about 15 times stronger than steel and three times stronger than Kevlar® and hence ideal for applications like bulletproof vests. Thus we can explain the typical stress-strain diagrams in Figures 8.2 and 8.13 as follows: In the elastic region, interchain bonds form a memory shape. In the nonlinear elastic regions, interchain bonds break and chains align little by little during the drawing process. Finally, when the chains are straightened, much stress is needed to break intrachain covalent bonds.

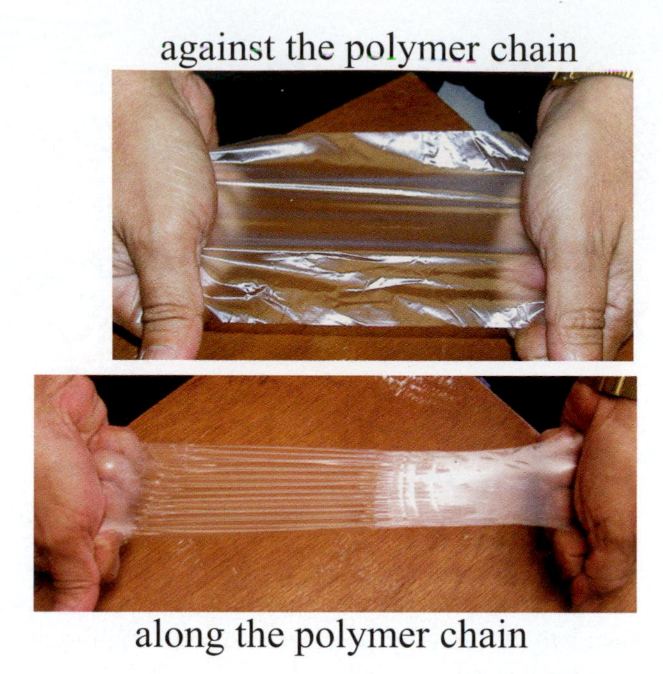

Figure 8.12: Same force pulling a plastic bag along and against the length of the polymer chains.

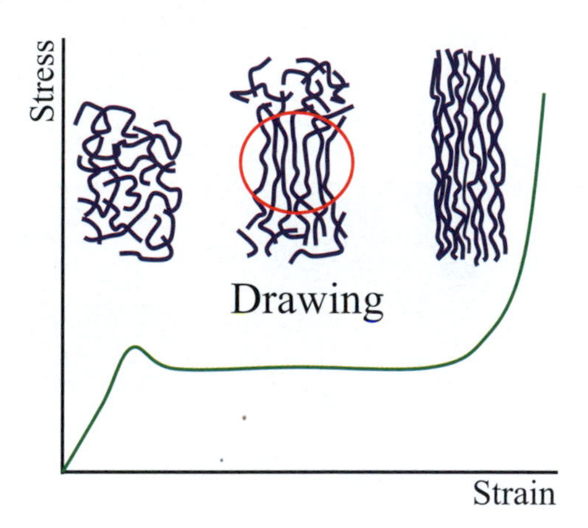

Figure 8.13: Stress-strain diagram of a typical polymer. In contrast to metals, for example, there is a nonlinear elastic regime where interchain bonds break and drawing occurs.

Silicone polymers

From the periodic table, it can be seen that silicon is in the same main group as carbon and has four valence electrons. Consequently silicon can covalently bond with up to four other atoms as is the case in silicon tetrachloride ($SiCl_4$). Just like carbon, silicon can readily form polymers. Silicon polymers are generally called silicones and are extensively used in many industries. Let us look at some examples.

Oils and greases:

Structurally simple silicon polymer chains can be obtained by linking dimethylsiloxane monomers together into polydimethylsiloxane (PDMS) at their OH groups as illustrated in Figure 8.14.

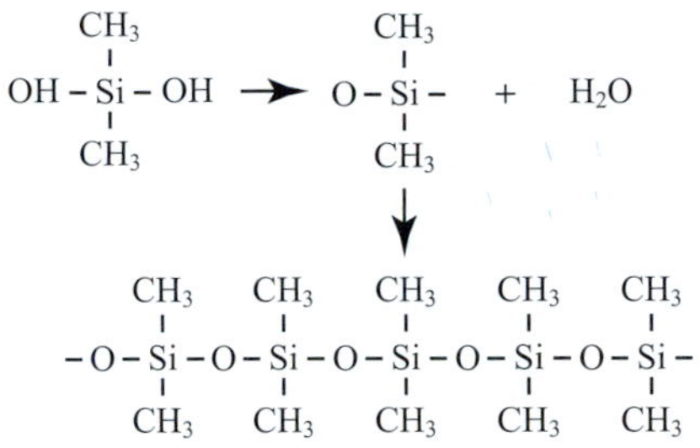

$$\mathrm{OH-\underset{CH_3}{\overset{CH_3}{Si}}-OH} \longrightarrow \mathrm{O-\underset{CH_3}{\overset{CH_3}{Si}}-} \quad + \quad \mathrm{H_2O}$$

$$\downarrow$$

$$\mathrm{-O-\underset{CH_3}{\overset{CH_3}{Si}}-O-\underset{CH_3}{\overset{CH_3}{Si}}-O-\underset{CH_3}{\overset{CH_3}{Si}}-O-\underset{CH_3}{\overset{CH_3}{Si}}-O-\underset{CH_3}{\overset{CH_3}{Si}}-}$$

Figure 8.14: Basic structure of polydimethylsiloxane (PDMS).

PDMS is a nonreactive, nontoxic, clear, oily liquid with a high viscosity and hence performs very well as a lubricant or hydraulic oil in many applications. It is also a commonly used food additive that acts as an anticaking and antifoaming agent and is listed under the label E900.

In order to control the length of the PDMS chains, the ends need to be terminated. This can, for example, be done by replacing the final OH group with CH_3.

Fig. 8.15: Most caulks are silicone based.

A nice example of a polymer based mixture with remarkable properties is **Silly Putty®** where PDMS plays a key role in its behavior.

Elastromeric sheets:

When starting out with a monomer such as $CH_3Si(OH)_3$, the third OH group can bond neighboring chains to form elastic sheets with rubberlike properties.

Sealants and caulks:

Of enormous industrial importance are also the many different types of silicone caulks and sealants.

8.3 Regimes

The properties of all materials are in one way or another temperature dependent. In the case of a generic polymer, there are five regimes when increasing the temperature from very cold to very hot as illustrated in Figures 8.16 and 8.17 where the temperature dependence of Young's modulus is shown.

1. **Glassy Regime:** [Very Cold — *Hard & Brittle*] The entangled strands cannot move and behave like an amorphous solid.

2. **Glass-Rubber Transition Regime:** [Cool — *Soft & Leathery*] The entangled strands still cannot move (much) but the atoms of the strands can.

3. **Rubbery Plateau Regime:** [Warm — *Elastic*] The entangled strands can move somewhat and the atoms of the strands can move quite freely.

4. **Rubbery Flow Regime:** [Hot — *Chewing Gum–like Flow*] The entangled strands disentangle and can move along each other.

5. **Liquid Flow Regime:** [Very hot — *Syruplike Flow*] Strands slide along each other freely.

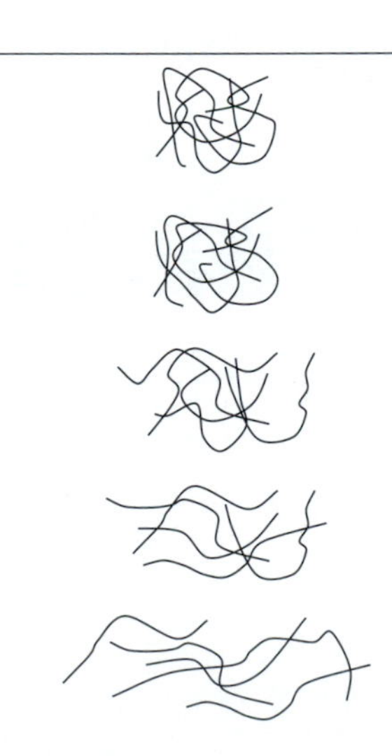

Fig. 8.16: Chains in the five regimes of a typical polymer.

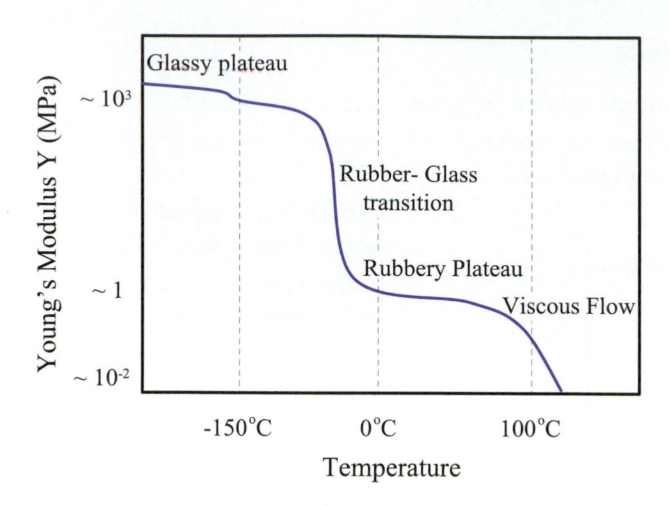

Figure 8.17: Young's modulus versus temperature.

Now that we have a basic idea of what polymers are, let us return to rubber.

8.4 Rubber

Fig. 8.18: Latex collection from a rubber tree.

Rubber originally stems from Central and South America where it has been in use since at least 1600 BC, the date of the oldest known rubber balls. The first cultivation, in the 1860s, of the rubber tree Hevea brasiliensis outside its place of origin was in Singapore Botanic Gardens. From there, it spread over southeast Asia, which is now one of the world's largest rubber-exporting regions.

An early commercial application of rubber was the erasing of pencil marks which subsequently gave it the name "rubber". The ubiquitous rubber band was invented in 1845 in England and soon rubber would become an important industrial product used in a large variety of products such as raincoats, erasers, packaging material and so on.

Rubber is a polymer made from a monomer called **isoprene** and is shown in Figure 8.19.

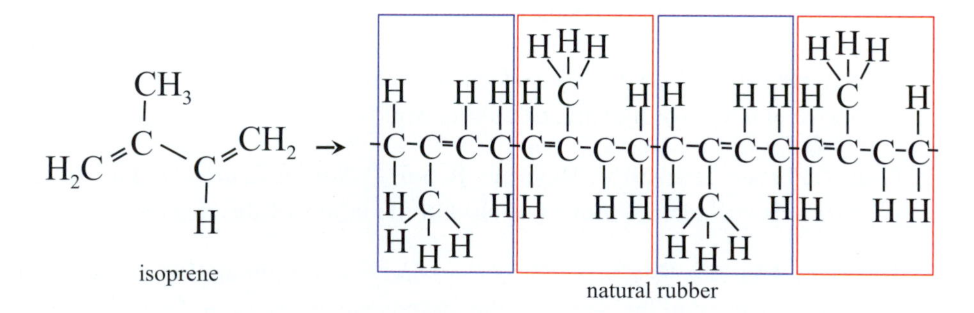

Figure 8.19: Rubber is a polymer of isoprene.

Natural rubber occurs as an emulsion called latex in the sap of some plants. For commercial purposes, it is almost exclusively obtained from the rubber tree since this tree responds to scarring by producing more latex. When the water in latex evaporates, it leaves an elastic solid which can be refined and turned into natural rubber. However, it should be noted that besides polyisoprene, latex also contains a large number of other compounds like proteins and oils.

Although refined natural rubber as such is nice, it has some serious practical drawbacks. When the weather is very cold, like in the winter, it becomes brittle and hard, while when it is too hot, like on a sunny summer day, it becomes rather soft. The question then is whether and how it would be possible to change the way rubber responds to heat. The answer to this question is that it is indeed possible to change rubber's properties in a desirable way, and a method to do this was in fact discovered millennia ago by the same Mesoamericans who have the first reported use of rubber. The modern process is generally called vulcanization (after the Roman god for fire, Vulcan) and we'll look at that now.

8.5 Vulcanization

The industrial discovery of how rubber can be improved for many practical uses is due to the American Charles Goodyear who found in 1839 that the addition of sulfur can vastly increase the heat resistance of rubber. The addition of sulfur to rubber leads to the formation of crosslinks between individual polyisoprene chains and is an example of the process of **vulcanization** which in general refers to the crosslinking of polymer chains by some means. Vulcanized rubber is shown in Figure 8.20 and the corresponding stress-strain diagram in Figure 8.21.

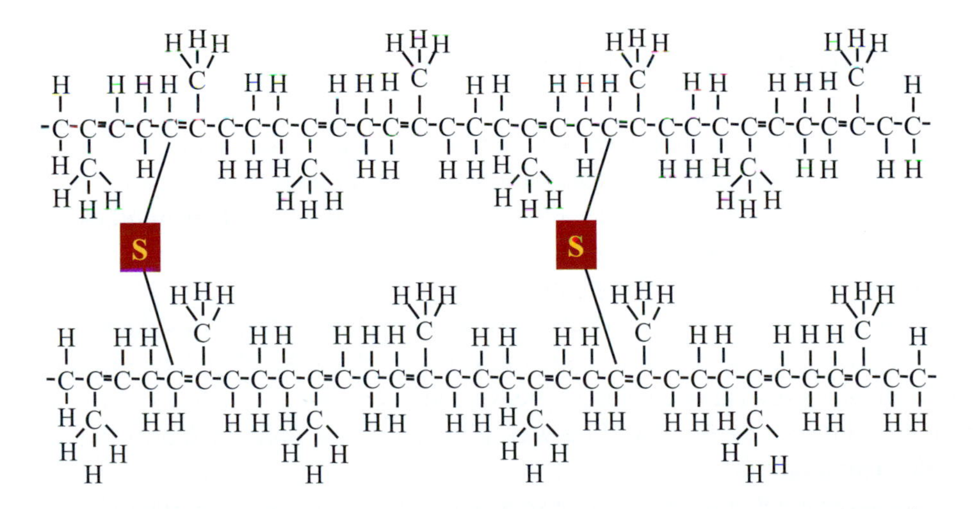

Figure 8.20: When rubber is vulcanized, crosslinks (of usually sulfur) are added to connect the chains.

As can clearly be seen in Figure 8.21, contrary to rubber with few (green line) or no (blue line) crosslinks, fully crosslinked rubber (red line) maintains a nearly constant Young's modulus over a very large temperature range. This means that fully crosslinked rubber has basically the same elastic properties from well below 0°C to over 200°C. Hence, vulcanized rubber is ideal for making items like car tires. If there is only some crosslinking like in the case of the blue line, Young's modulus drops very quickly at temperatures only slightly higher than room temperature. Consequently, when it becomes hot it is very soft. As tires get hot due to friction, this means that such rubber would be completely unsuitable for making car tires; this is the reason why only vulcanized rubber is used for making tires.

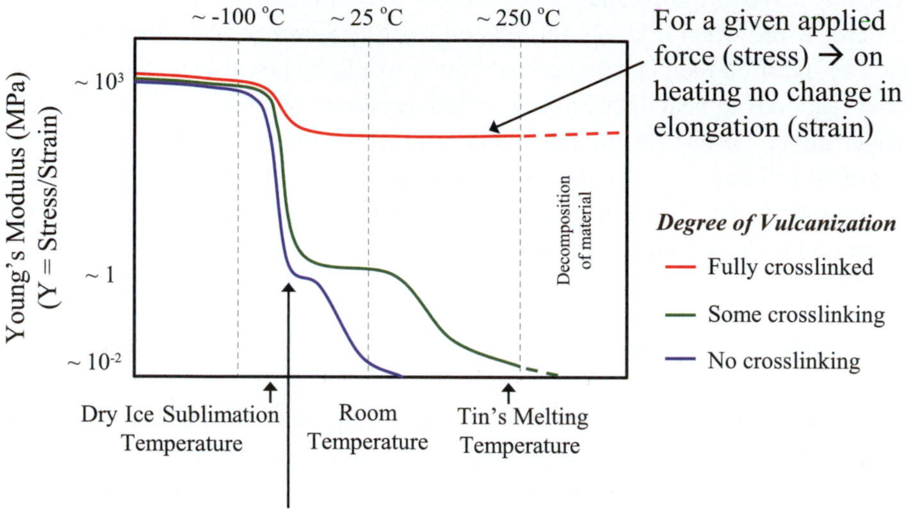

Figure 8.21: Young's modulus versus temperature including vulcanized rubber.

8.6 Thermosets and thermoplasts

Fig. 8.22: Due to their hardness, thermosets are a good material to make helmets from.

We have seen that crosslinks have a drastic impact on a polymeric material's properties and they also play a key role with regards to an important characterization of such materials: namely that as a thermoset or thermoplast. **Thermoplasts** are polymers that can flow at high enough temperatures without decomposing while **thermosets** cannot flow at high temperatures.

In thermosets, the number of crosslinks is very large. As a consequence, the polymer chains are so strongly bound to each other that even at high temperatures the thermal energy is not enough to break enough crosslinks to allow for the flow of the chains past each other. In a way, a thermoplastic solid is almost like one single giant molecule.

The large number of crosslinks also results in the polymeric material to be very hard below the glass transition temperature. Above the glass transition temperature, they generally become elastomers.

Since thermosets cannot be melted, they must be manufactured from a base material with fewer crosslinks and then be cured when molded into their final shapes. The process of curing increases the number of crosslinks, is irreversible and can be achieved, for example, by heat or chemical reaction.

In thermoplasts, crosslinks are absent or present in only small numbers. They can, therefore, be melted and be reshaped. Examples of common thermoplasts are polyethylene, polyvinylchloride (PVC) and polystyrene.

8.7 The answer

In this chapter, we asked the question why rubber is elastic. From a microscopic examination we found that rubber is made out of many monomers and therefore a member of the class of materials called polymers. Polymer chains come in many varieties but nevertheless share a number of characteristics. For example, they can have the same monomer but when polymerized into chains of different lengths may give materials with rather different properties. Another key aspect is how the various chains in a polymeric material bond to each other. If the bonds are very weak, the material may be a liquid with a high viscosity. If the bonds are strong, they form crosslinks that keep the chains anchored to each other. Now if there are many crosslinks, then the material will be very hard, but if there are some but not too many crosslinks such as is the case in vulcanized rubber, then the crosslinks will act as a memory trying to restore a material to its original shape when stretched. Thus, we find that **rubber is elastic because it is made out of crosslinked polymer chains that are in the rubbery plateau regime when near room temperature.**

8.8 Exercises

1. How many different (linear) polymers with a length of six monomers can be made if one has three different types of monomers available?

2. Even if there are no bonds at all between different chains in a polymeric material, its viscosity is likely to be rather high. Why is this so?

3. What is the radius of gyration of a polymer with a degree of polymerization of 2000 whose monomers are 0.15 nm long?

4. If the monomers in the preceding question were cubic, what would be their total volume? How would this compare to the space occupied by the polymer in the preceding question?

5. If one has a solid made up of pure polyisoprene, can it be melted?

6. Is it reasonable to describe curing and vulcanization as similar processes?

7. What happens if I heat a thermoset too much?

8. A salt crystal consists of a very regular repeating pattern of Na and Cl atoms. Then, why can't we take a NaCl pair as the monomer and consider the crystal as a polymer?

9. Do the preceding question for graphite.

10. Rubber bands are often made from polybutadiene which has a glass transition temperature of $-106°C$. Will they still be elastic when dipped into liquid nitrogen?

9: Sparks in Nature

Q - What Is Electricity?

Chapter Map

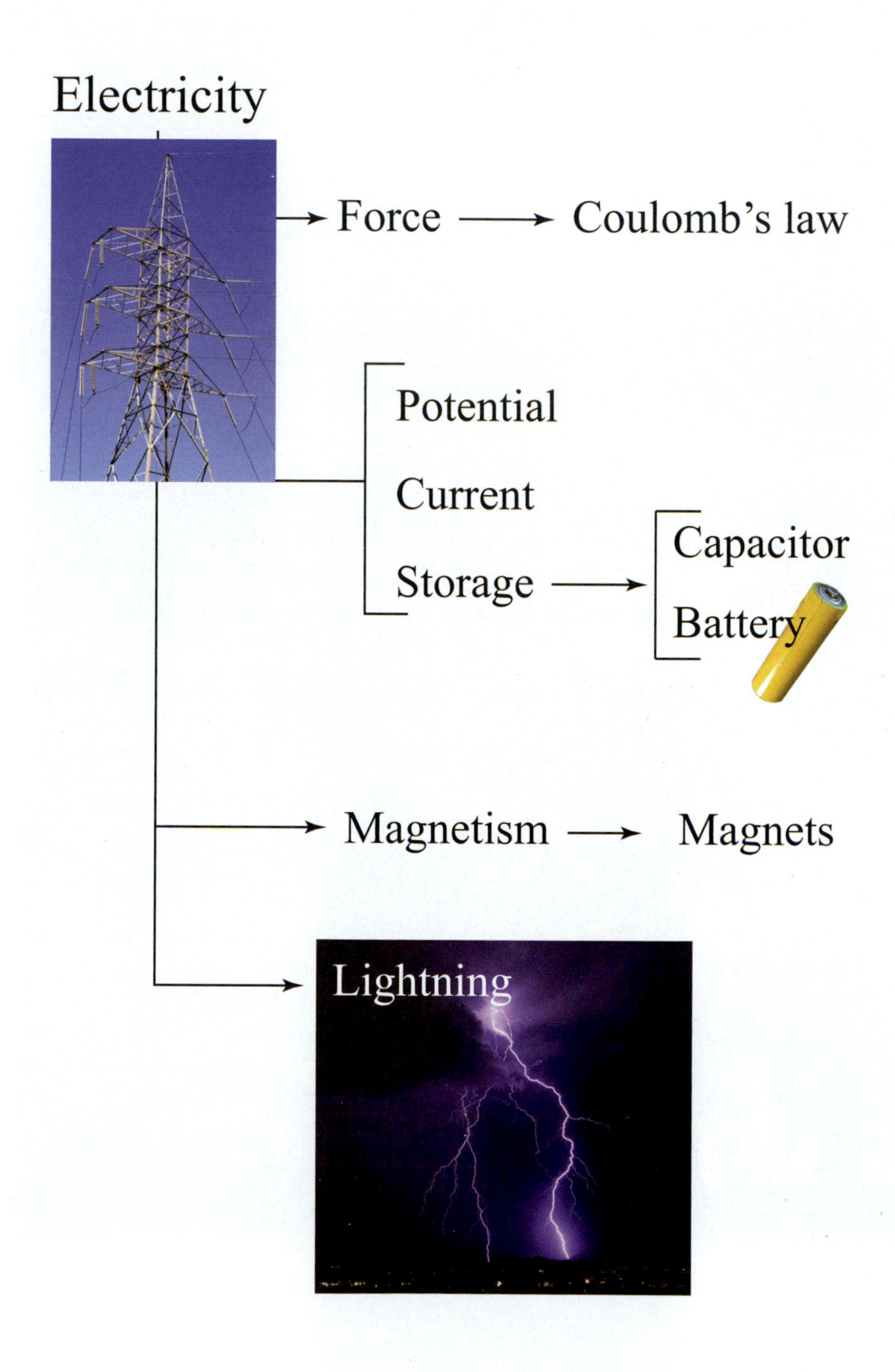

Electricity
Force
Coulomb's law
Potential
Current
Storage
Capacitor
Battery
Magnetism
Magnets
Lightning

9.1 The question

Figure 9.1: What comes out of this outlet?

Modern life is hard to imagine without electricity — it is simply everywhere. But what is it from a physical point of view? What is that thing that comes out of a wall socket? **What is electricity?**

Electricity, in one form or another, has been known to exist for millennia. Among the first well-documented electrical phenomena is static charge. The ancient Greeks had noted that if one rubs fur against amber, the amber would later attract certain light objects like hair. They even discovered that intense rubbing of the amber could "charge" it enough to create sparks. The Greek word for amber is "electron" based on the root "elek" meaning "shine" which is also the root of the modern word electricity.

Fig. 9.2: The slide leads to a buildup of static charge.

The buildup of electric charge is quite well known in drier countries with many schoolchildren experiencing how their hair stands up after a slide, as shown in Figure 9.2.

Despite the early knowledge of this interesting phenomenon, systematic investigations did not start until the seventeenth century. Let us now look at some of the basic facts that can experimentally readily be established.

9.2 Electrical phenomena

Perhaps one of the very first things to notice about electrical phenomena is that they can exert a force. How big would that force be? Is there a basic unit of electrical charge and if so what is it? Let us start by stating some experimental facts which are known today.

Electric charges and force

1. Matter has an **intrinsic property called charge** (denoted by q) in addition to the property mass.

2. **Like charges repel** each other and **opposite charges attract** as shown in Figure 9.3. This can be found out with the help of the Leyden jar, for example, and by polarizing matter as illustrated in Figure 9.9.

3. **Electric charge is absolutely conserved**. Electric charge cannot be created or destroyed. The total electric charge in the universe is a constant.

4. **Electric charge is discrete** and comes in units of the irreducible charge of the electron, which is denoted by $-e$. All *observed* charges q in nature are of the form

$$q = \pm Ne \text{ where } N : \text{integer}. \tag{9.1}$$

This "quantization" of electric charge is an experimental truth, and was known before the advent of quantum mechanics.

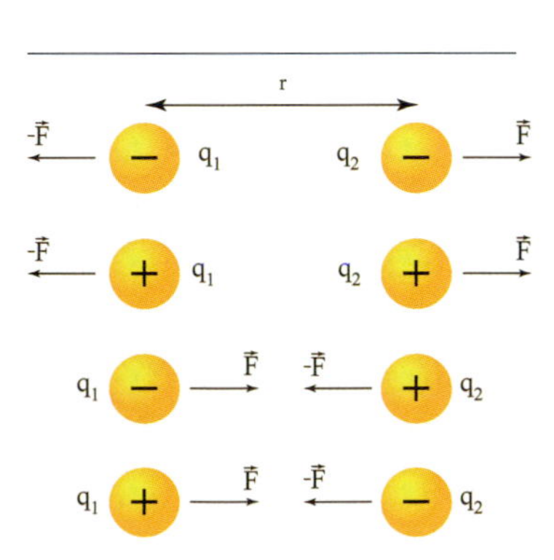

Fig. 9.3: Force between like and opposite charges.

Although only the first two items were known to him, by careful experimental observation, the French physicist **Charles Coulomb** discovered a law relating the strengths of charges and the resulting forces between them. Interestingly enough, the law he came up with looks identical to Newton's law for gravitation, but in fact has many differences from it.

Newton's law states that the magnitude of the gravitational force between two masses m_1 and m_2 separated by a distance r is given by

$$F_g = G\frac{m_1 m_2}{r^2} \quad \text{: Newton's Law} \tag{9.2}$$

where $G = 6.7 \times 10^{-11} \text{m}^3/(\text{kg s}^2)$ is the gravitational constant, and the *direction* of the force is along the line joining the two masses (see also Chapter 1).

Coulomb found that the force between two point charges q_0 and q_1 separated by distance r is given by

$$F_e = k_e\frac{q_0 q_1}{r^2} \quad \textbf{: Coulomb's Law} \tag{9.3}$$

where $k_e = 9 \times 10^9$ N m^2/C^2 is the electrostatic constant (often also called Coulomb's constant), and the *direction* of the force is along the line joining the two charges. The electrostatic constant is often expressed in terms of the electric constant $\epsilon_0 = 8.85 \times 10^{-12} \text{C}^2\text{N}^{-1}\text{m}^{-2}$ as $k_e = 1/(4\pi\epsilon_0)$. The C in k_e is the unit of charge called the **coulomb (C)**. The coulomb is defined as the amount of charge that is carried by a current of 1 A for 1 s (for more on current, see Section 9.5 below).

One may wonder as to why should force be proportional to $1/r^2$ for both the electrical and gravitational forces, and it turns out that there is no easy explanation; the answer is partly linked to the fact that the space we live in is three dimensional, and that consequently the surface area of a sphere in our space increases as r^2.

The concept of electric charge is the key to understanding the phenomena of electricity and magnetism. Unlike mass, charge is more difficult to directly experience since most matter encountered in daily life is neutral.

A good example of an electric phenomenon that can be experienced is lightning, where an electric charge moves from the clouds to the earth. An effect similar to lightning, but much lower in intensity and due to the same principles as illustrated in Figure 9.2, is the mild shock one sometimes gets in a dry climate when touching a metal object.

9.3 Electric and gravitational forces compared

Although the formulae for the gravitational and electrical forces are similar in the sense of having a common mathematical form (both being inversely proportional to the square of the distance of separation and being generated by sources), they are profoundly different as well. The key differences are the following:

The Leyden jar

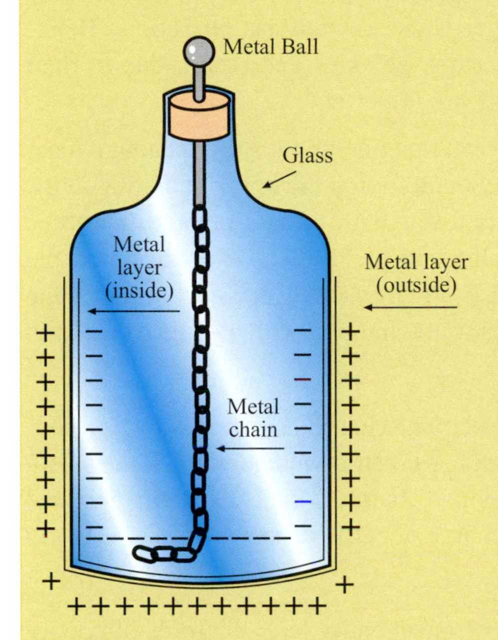

Figure 9.4: A Leyden jar

It was known by the eighteenth century that by rubbing together different materials, one could generate two types of charges, positive and negative, with like charges having the property of repelling each other while unlike charges attracting each other. It was also learned how to store these charges in devices called Leyden jars and that these stored charges could be used to create sparks.

A Leyden jar, illustrated in Figure 9.4, consists of a glass bottle or jar coated with a thin metal layer, both on the inside and outside excluding the mouth such that the inside and outside metal layers are physically separated. The inside of the jar is usually connected to the outside with the help of a chain and a metal rod held in place by an insulating cork. In modern terms, a Leyden jar is a capacitor. It was invented by the Dutch scientist Pieter van Musschenbroek in 1745 in the Dutch city of Leiden (Leyden is the old spelling of Leiden).

Originally, Leyden jars were filled with water and it was believed that the charge was stored in this water. Benjamin Franklin (see below) thought that the charge was stored in the glass but nowadays we know that the charge accumulates on the inside surfaces of the conductors.

Although many people noted the similarity between the man-made sparks and lightning, it was Franklin who first investigated the phenomena of lightning systematically by comparing hypothesis and theory with experiments. Franklin hypothesized that clouds were electrically charged, and if this were correct, then lightning would be easily understood as an electrical discharge similar to those seen in laboratories.

To test his hypothesis, Franklin flew a kite during a thunderstorm in Pennsylvania in 1752. He apparently took the precaution of standing under a shed, and holding the kite string with a dry (insulating) silk cord. (Some experimenters at that time had been struck by lightning and were killed!) When the string became wet, the electric charge accumulated by the kite flowed down to a metal key attached to the end. Franklin observed a spark jump between the key and his hand. Thus Franklin had demonstrated the electrical nature of thunderstorms!

Franklin went further: By touching the key to a Leyden jar, he charged it and then deduced that the lower part of thunderclouds was usually negative! Franklin did many other experiments, and also invented useful devices such as the lightning rod that saved many buildings. It is remarkable and impressive that despite being a prolific inventor, Franklin did not apply for any patents since he believed that having benefited from the inventions of others it would not be right to restrict his own.

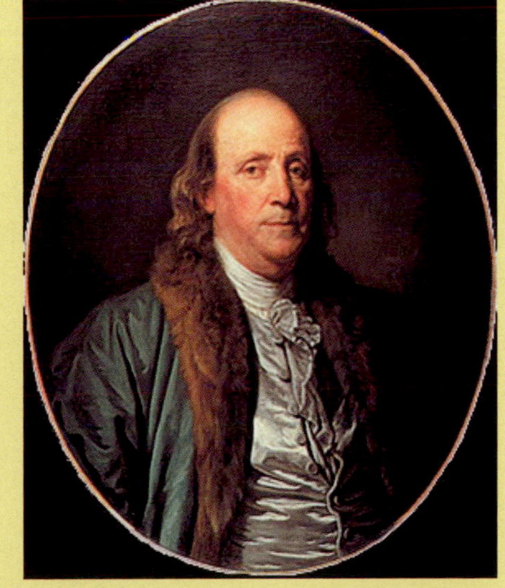

Figure 9.5: Portrait of Benjamin Franklin by Jean-Baptiste Greuze.

1. **Mass:** Mass is always a positive quantity (i.e. there is no antigravity), so gravitational potential energy always gives an attractive force between any two masses. This is the reason that one can never "shield" a system from gravity, since there is no way of canceling it out. Although gravity is an extremely weak force, its effects keep piling up; in other words, the more mass that, say, a star has the stronger is its gravitational attraction. Hence for celestial bodies, solar systems, stars, galaxies and so on, due to their enormous mass, the effects of gravity are important.

   Gravity is always attractive and one can imagine that if a large enough mass is gathered, there is no force strong enough to stop the inward pull of gravity. This is precisely what happens after a star with a mass greater than three times that of our sun has come to the end of its life. At some point, the force of the gravitational attraction is so strong that the star — under the inward pull of gravity — undergoes a gravitational collapse that results in the formation of a black hole.

2. **Electrical charge:** Unlike mass, electrical charge can be positive or negative. Contrary to gravitational forces, we can completely shield a system from electrical forces by using a negative charge to cancel a positive charge (and vice versa) since having a net charge of zero means we have effectively cancelled out the electrical forces.

3. **Universality of interaction:** Every physical entity feels the force of gravity — there are no gravitationally "neutral" entities. On the other hand, there are many fundamental entities like the neutron and photon that are electrically neutral, and do not feel electrical forces.

4. **Relative strengths:** The relative strengths are characterized by the empirical values of the two natural constants G and k_e. To compare them, we evaluate the ratio of their values for the case of an electron and proton separated by a distance r. Consider an electron having mass $m_e = 9 \times 10^{-31}$ kg and a charge of $e = -1.6 \times 10^{-19}$ C, and a proton having mass $m_p = 1.7 \times 10^{-27}$ kg and a charge $e = 1.6 \times 10^{-19}$ C.

   We then have

$$\frac{|F_{\text{coulomb}}(r)|}{|F_{\text{gravity}|}(r)} = \frac{k_e|q_e q_p|/r^2}{G m_e m_p/r^2} = \frac{k_e|q_e q_p|}{G m_e m_p}$$

$$= \frac{8.99 \times 10^9 \text{Nm}^2/\text{C}^2 \times (1.6 \times 10^{-19}\text{C})^2}{6.67 \times 10^{-11}\text{Nm}^2/\text{kg}^2 \times 9 \times 10^{-31}\text{kg} \times 1.7 \times 10^{-27}\text{kg}}$$

$$= 2.26 \times 10^{39}.$$

   We see that, for any distance of separation r of two charged bodies, the gravitational force is almost 10^{39} smaller in magnitude than the electrical force. Even if we consider bodies with much bigger masses than an electron or a proton, the effect of gravity is still extremely small. So for all cases of interest, we can ignore the effects of gravity compared to electrical forces.

One may object to the thought of ignoring gravity, considering that we always feel the force of gravity in our daily lives. This objection is correct if the mass involved is immense; for example, what we experience in daily life is the gravitational effect of the earth, which is an immense mass indeed. Compared to the gravitational effect of the earth's mass, the gravitational effect of any other massive body on the earth's surface (on, say, an object undergoing experiments in the laboratory) is negligible.

9.4 Electric potential

We study the characterization of electrical phenomena using the concept of the electric potential.

In the chapter on energy (Chapter 3), we encountered the concept of *potential energy*, an example being that of gravitational potential energy: it requires work, done against the force of gravity, to lift an object to a certain height, and the object is then said to have gained potential energy. We will now have a look at the equivalent concept for charges.

It can be proven that the electric potential is conservative (for the definition of conservative systems, see Chapter 3). In conservative systems, force and potential energy are related by $F = -dU/dx$ and applying this to Coulomb's law given in Equation 9.3, we see that the potential energy acting on an electric charge is

$$U = k_e \frac{q_0 q_1}{r}. \tag{9.4}$$

This potential energy is with regards to the system comprising the charges q_0 and q_1 (for a comparison of gravitational and electrical potential energies, see also Equation 9.6). However, it is generally useful to define the **electric potential** as the potential energy experienced by one of the charges. Thus we define the the electric potential due to a charge q_1 as follows:

$$V \equiv \frac{U}{q_0} = k_e \frac{q_1}{r} \tag{9.5}$$

such that $U = q_0 V$.

The unit of the electric potential is joule/coulomb and this is simply called the **volt**, (V), in honor of Alessandro Volta, the inventor of the modern battery, a device that creates potential differences through chemical means. Since the unit of energy is kg m^2 s^{-2} = J, the unit of the electric potential is given by kg m^2 s^{-2} C^{-1}. In other words, the unit for V in the SI system is given by

$$\text{V} = \text{joule}/\text{coulomb} \equiv \text{JC}^{-1}. \tag{9.6}$$

What is the rationale for separating out the electric potential from the potential energy of two charges? In contrast to, say, gravity, for which mass is always a positive quantity, electric charge can either be negative or positive. Hence, by separating off the electric potential V from the potential energy U, we can study V generated by a *given charge*, and then analyze how it affects other charges.

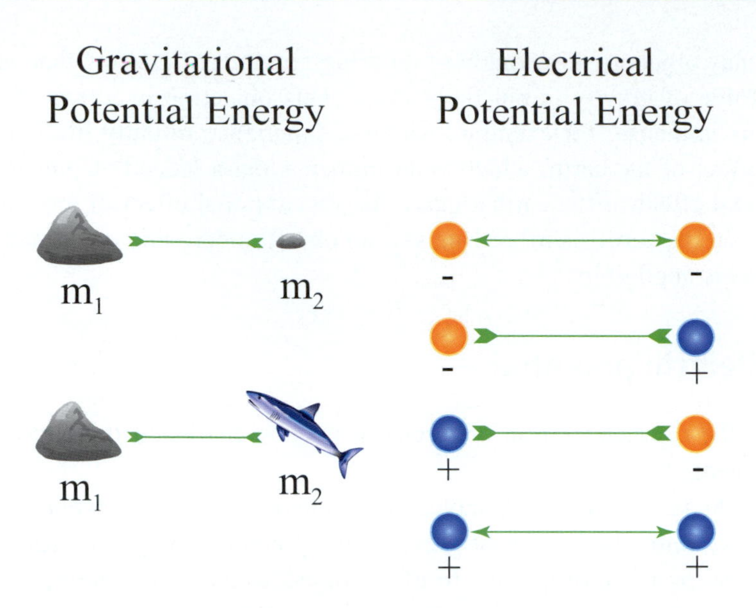

Figure 9.6: Potential energies compared.

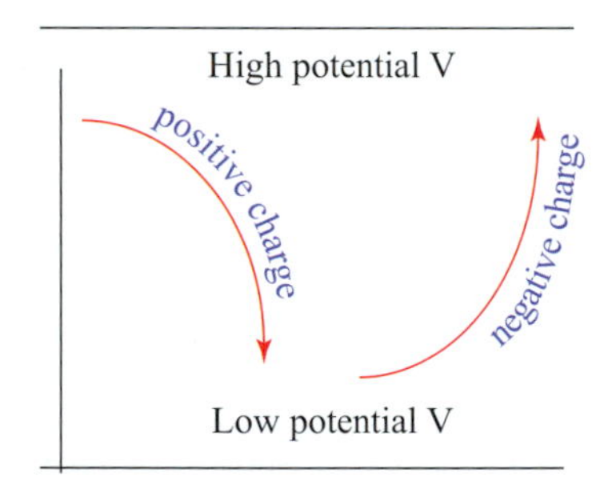

Fig. 9.7: Positive and negative charge moving in the same V-field.

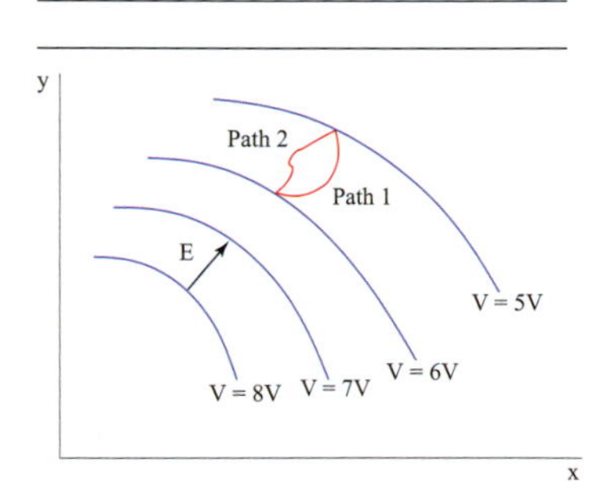

Fig. 9.8: Electric potential.

Suppose V is generated by a positive charge; a negative charge will move toward increasing values of V, whereas a positive charge will move toward decreasing values of V, as shown in Figure 9.7. A potential V that looks like a "mountain" to a positively charged particle $+q$ looks like a "crater" to a negatively charged particle $-q$.

A point with a high value of V is analogous to a point of high elevation for a body moving under the influence of gravity. Just as in gravity, a mass m can gain potential energy in going from a lower to a greater height, a positively charged particle gains potential energy in moving from a point with a lower potential to a point with a higher electric potential; however, and this is what makes electricity and magnetism so different from gravity, a negatively charged particle loses potential energy in moving from a point with a lower value of V to one with a higher value.

The fundamental irreducible charge in nature is the electron charge; by convention, the charge of the electron is taken to be negative and is denoted by $-e$. It is a natural constant, with the numerical value of e in the SI units given by

$$e = 1.6 \times 10^{-19}\,\text{C}. \tag{9.7}$$

The value of the electron charge is extremely small. In a typical lightbulb, every second over 10^{19} electron charges enter and leave the bulb's filament. Protons carry charge equal to $+e$, though the proton seldom enters the processes that we will be interested in.

The electron charge is the basis of another unit for energy. The energy gained by an electron in moving from, say, a point with $V = 5$ V in Figure 9.8 to a point with $V = 6$ V, such that the electric potential difference between the two points is 1 V, is called an electron-volt and is denoted by eV.

We consequently have

$$
\begin{aligned}
1\,\text{eV} &= e \times 1\,\text{V} \\
&= 1.6\,\text{C} \times 10^{-19} \times 1.0\,\text{J}\,\text{C}^{-1} \\
\Rightarrow 1\,\text{eV} &= 1.6 \times 10^{-19}\,\text{J}. \qquad (9.8)
\end{aligned}
$$

The electron-volt (eV) is an appropriate unit of energy to measure energies involved in atomic, chemical and (molecular) biological processes. The kinetic energy of a single molecule of nitrogen in the air at room temperature is about 1/40 eV.

Induced charges

In many substances, the outermost electrons of their constituent atoms are rather mobile as can be illustrated by induced charges and static electricity.

Static electricity is a common occurrence in drier countries and gives rise to hair standing up as in Figure 9.2. What we have here is an example of a separation of charges. Another common example is that of a balloon rubbing against a sweater. When doing so, electrons from the sweater move to the balloon thus giving it a net negative charge. The sweater, now short of electrons, will be positively charged and since opposites attract, the balloon will be attracted to the sweater as illustrated in Figure 9.9a.

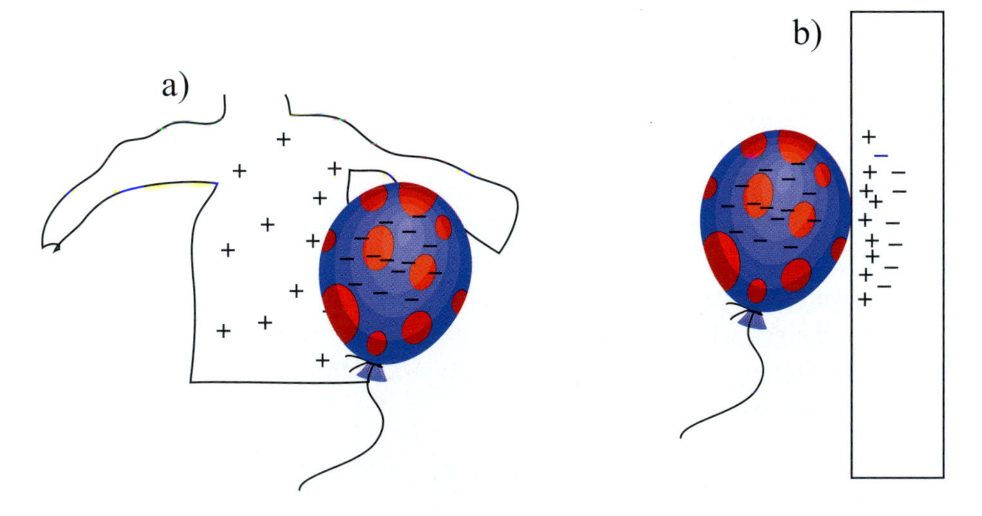

Figure 9.9: Induced charges.

When the negatively charged balloon is brought near an electrically neutral object such as a wall, the electrons in the wall will be (slightly) repelled, leaving a net positive charge as shown in Figure 9.9b. Such a charge is generally called an induced charge.

9.5 Electric current

We have seen that a positive charge will move from a higher to a lower electric potential. Subjecting charges to appropriate potential configurations gives us the means to control their movement. The movement of electric charge is called an **electric current**, and is denoted by I.

Why are we interested in electrical currents? One of the overriding considerations in our study of nature is to find new forms of energy, such as the electric potential energy. The next question that naturally arises is whether we can transform electrical energy into other forms of energy. In particular, can we move objects and power our homes and industries with this form of energy? Work adds to energy, and is one way of transforming energy from one form to another (see also Chapter 3). To do work, we must apply a force F on an object over some distance d, with the resulting work done given by the product Fd. Movement in space under the action of the electrical force is hence essential in converting electrical energy into mechanical energy. In the case of electricity, what moves are the electrical charges and they are moved by electrical forces. An electrical current is an instance of such a movement of electric charges, and is therefore an example of a means for doing electrical work.

Consider a stream of charges flowing past a fixed point. Furthermore, consider an imaginary surface S placed orthogonal (perpendicular) to the flow of charges. The flow of charge is measured by the current I, which is the amount of charge Δq that flows across the surface S per unit area and in a unit time interval Δt. Hence

$$I = \frac{\Delta q}{\Delta t}. \tag{9.9}$$

The unit for current is $\mathrm{C\,s^{-1}}$, and the SI unit of current is the ampere, denoted by A. That is

$$1 \text{ A} = 1 \text{ C/s}. \tag{9.10}$$

Electric current I in general depends on the surface through which the flow of charge is taking place, and is defined as flowing from a positive to a negative potential.

Conservation of charge is a fundamental property of nature and therefore holds for moving charges which is expressed by the laws of current conservation. If a current I flowing in a wire bifurcates into two wires with currents, say, I_1 and I_2, then charge conservation requires that

$$I = I_1 + I_2. \tag{9.11}$$

So how does one go about creating the flow of an electrical current? Since we first need to get our hands on some electrons, we start with a conductor, like a copper wire. Next we need to create an electric potential between two points in the wire that will exert a force on electric charges and make them move.

An electrical **battery** is a device that creates such an electric potential difference between its two terminals. The battery converts chemical energy into electrical energy.

Figure 9.10 shows a copper wire connected in a closed circuit with a resistor R and a battery with voltage V. A **resistor** is a device that impedes the flow of current so as to control it and plays an essential role in electronics. The electric potential, say, V_+ at the positive terminal (+), is taken to be higher than the electric potential, say, V_- at the negative terminal (−); the voltage of the battery is the potential difference given by $V = V_+ - V_-$. Due to the potential difference V, free charges in the conductor (copper wire) are set in motion and yield a current I in the circuit.

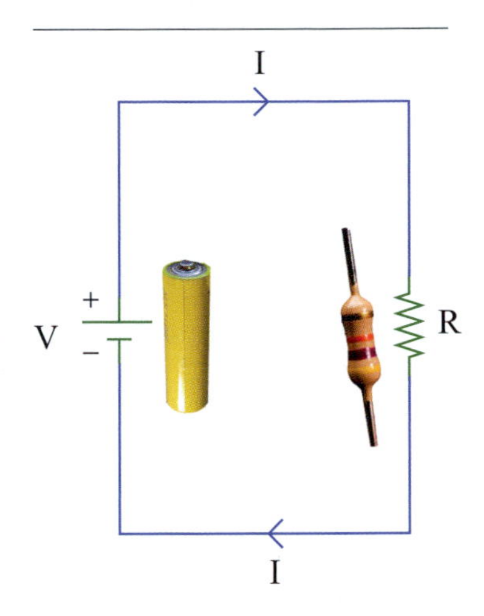

Fig. 9.10: A circuit with a resistor and a battery.

The convention in labeling the terminals of a battery means that positive charges flow from the higher potential at the positive terminal to the lower potential at the negative terminal. In conventional electric devices, it is the electrons that are moving, and since they are negatively charged, they will flow from the lower potential to the higher potential.

Consider water that is pumped to a height h, and flows back to the ground. The electrical current I can be compared to the kinetic energy of the water as it hits the ground. The battery is the analog of the pump, the potential difference V created by the battery is the analog of the height h to which the water is raised. The resistor R is analogous to the radius of the pipe, and whether it is clean, or whether there is accumulation of gravel and sand in the pipe that hinders, or even blocks, the flow of water.

What is the relation between the current I in the circuit, the resistance R and the voltage V? From the analogy of water flow, we expect that the larger the voltage V, the higher is the current I in the circuit. This is indeed the case and the current is proportional to the imposed voltage, with the constant of proportionality being the resistance. Thus one can define the resistance as

$$I = \frac{V}{R}$$

$$\Rightarrow R = \frac{V}{I}. \qquad \textbf{Ohm's law} \qquad (9.12)$$

In other words, if we create a potential difference of V in a conductor having resistance R, then a current I given by Equation 9.12 will flow in the conductor.

It is important to note that for Ohm's law, the voltage difference V is the **cause** that is responsible for the flow of current I, which is the **effect**. It should also be noted that Ohm's law is only valid for certain ranges of voltages and currents under steady state conditions, and certain materials. However, for many practical situations, it works very well.

The unit of resistance is

$$\begin{aligned} 1 \text{ ohm } (\Omega) &= 1 \text{ volt per ampere} \\ &= 1 \text{ V/A}. \end{aligned} \qquad (9.13)$$

If a battery is connected by a piece of wire, the potential difference between the two ends causes a flow of electrons. The chemical energy stored in the battery is converted to the energy of the electrons. Figure 9.11 shows charges moving in the presence of a potential and thus generating the electric current.

Materials such as plastic and wood are poor *conductors* of electricity and are termed *electrical insulators*. A measure of resistance which is independent of the size of the wire is given by the **resistivity** ρ which is related to the length l, cross-sectional area A and and resistance R of the wire by

$$\rho = \frac{\text{resistance} \times \text{area}}{\text{length}} = \frac{RA}{l}. \quad (9.14)$$

Some typical values for ρ are given in Table 9.1.

Material	ρ (ohm-meter)
Conductors	
Silver	1.59×10^{-8}
Copper	1.68×10^{-8}
Mercury	98×10^{-8}
Semiconductors	
Graphite (carbon)	$(3 - 60) \times 10^{-5}$
Germanium	$(1 - 500) \times 10^{-3}$
Silicon	$0.1 - 60$
Insulators	
Glass	$10^9 - 10^{12}$
Rubber	$10^{13} - 10^{15}$

Table 9.1: Resistivity at 20°C.

It is important to note that while the electric potential propagates at the speed of light, the *electrons* themselves flow at a much slower rate because of random collisions with the atoms composing the conductor and have a **drift velocity** of about $\sim$ 0.01 cm/s due to the presence of the external potential. Also, although there is current flow, the wire itself is electrically neutral because of the lattice of positively charged atoms!

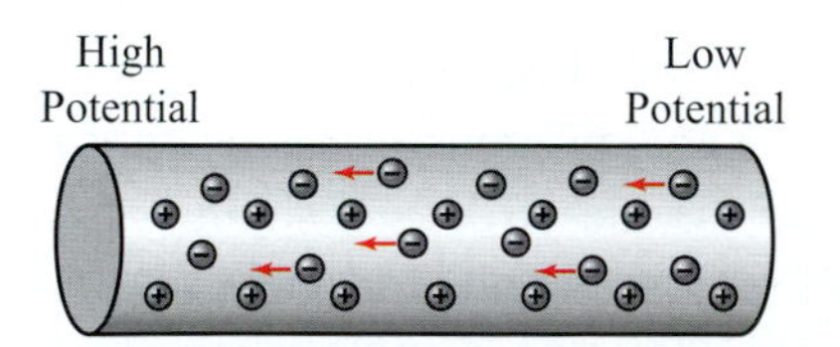

Figure 9.11: Electric current due to an electric potential.

The effect of even small current flow through the human body can be dramatic. As little as 0.015 A can cause loss of muscle control and 0.07 A can be fatal if it lasts for more than 1 s.

Conductors and insulators allow and block the flow of currents respectively, and can be understood based on principles of classical physics. Semiconductors lie in-between conductors and insulators, and their workings are based on quantum theory. Conductors have electrons that are not bound to the nuclei, and can hence move in response to an external potential difference. All parts of a conductor at equilibrium are at the same potential, since any potential difference would cause currents to flow.

One can have many resistors, which can be combined in series or in parallel (as shown in Figure 9.12), with the value of the resultant resistance being derivable from Ohm's law.

Energy is constantly lost due to the heating of the resistance R, called Ohmic heating. How much power is expended by the battery in keeping the current flowing in the circuit? In Δt time, a charge of amount Δq flows from the high electric potential to the lower potential, hence losing potential energy ΔU. The difference in the electric potential is given by the voltage of the battery, namely V, and hence using Equation 9.5 and the work-energy theorem Equation 3.4 we have

$$\Delta W = \Delta U = \Delta q V. \tag{9.15}$$

Power is defined as the rate at which work is done (i.e. $P = \Delta W/\Delta t$ as in Equation 3.10). In the case of potential energy U we have

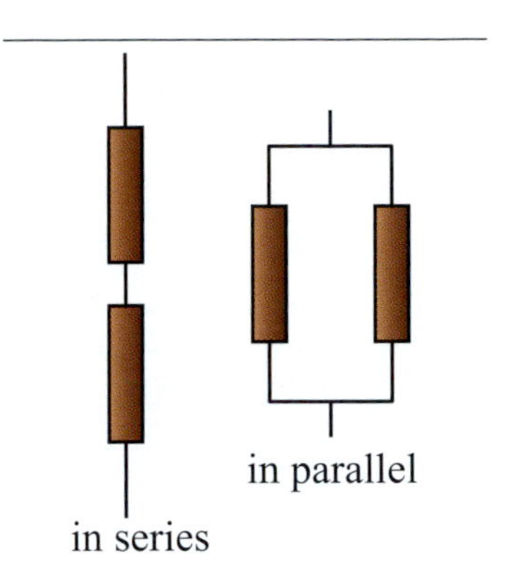

Fig. 9.12: Basic ways of combining resistors.

$$P = \frac{\Delta U}{\Delta t} \tag{9.16}$$

$$= \frac{\Delta q}{\Delta t} V \tag{9.17}$$

$$\Rightarrow P = IV \tag{9.18}$$

$$= I^2 R = \frac{V^2}{R}. \tag{9.19}$$

Note that the power loss in a circuit is proportional to I^2. In other words, if we reverse the flow of current, the sign of I will change to $-I$, but the loss due to Ohmic heating will be the same, no matter in which direction the current flows.

The loss of power of the battery due to heating is not a net loss, but rather an example of the **transformation** of energy. Lightbulbs that use a filament emanate light when the filament is heated by the flow of current, and similarly for electric ovens and so on. Hence, electrical currents serve as a vehicle for converting the energy stored by the battery into other forms of energy.

The current we have considered so far is called a DC (direct current) since its direction does not change. An AC (alternating current) is one in which the current changes direction, and hence its sign, with some (usually fixed) frequency.

9.6 Storing electrical energy

In the practical use of energy, storage for short- or long-term purposes is essential. This is true not only in the realm of technology but also in the natural world. Indeed, if plants couldn't store energy, we'd have very little to eat. Probably the most essential naturally stored energy is in the form of carbohydrates and oils. This is even so for our bodies which store energy as fat.

One way to store electric energy is with the help of batteries by reversibly transforming electrical energy into chemical energy. Another way would be to use a combined generator/pump that moves water to higher ground thus transforming electrical energy to gravitational potential energy. In both cases, electrical energy is transformed for storage. For most operations in electrical circuits, this is not exactly ideal as the stored energy must be stored and retrieved very rapidly. It is therefore beneficial to have a device that can directly store charges. One such device is the so-called **capacitor**.

Capacitor

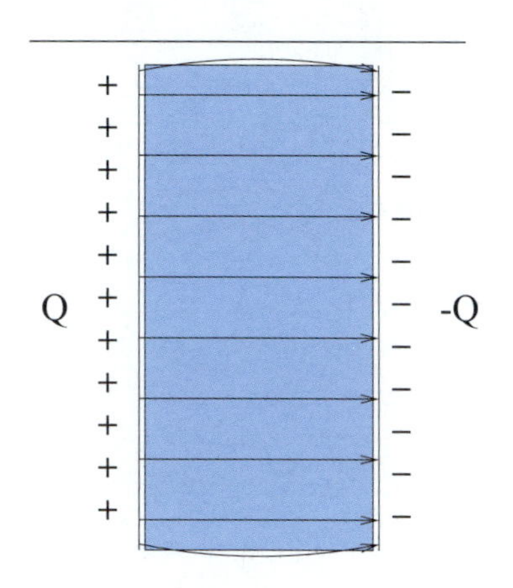

Fig. 9.13: Capacitor.

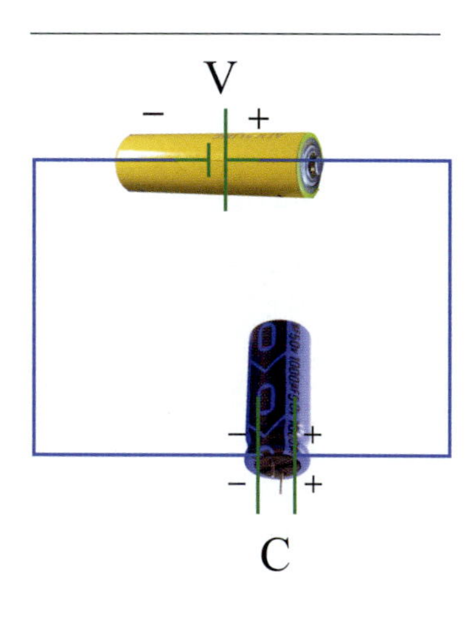

Fig. 9.14: Charging a capacitor.

How could charges directly be stored? Since most objects found in nature are electrically neutral, to obtain electrical energy it is logical to try and store positive and negative electric charges separately. However, if we bring together a large collection of positively (or negatively) charged particles, the force of electrical repulsion is very large, the charges would all fly apart, and a large amount of effort would be expended to merely keep them in place.

A more efficient arrangement is to bring together equal amounts of positive and negative charges, but separate them so that overall we are back to an electrically neutral object. This can be done by placing two conducting plates parallel as shown in Figure 9.13 and this, in essence, is the principle behind the design of all capacitors.

A capacitor can be charged by placing it in a circuit with a battery as shown in Figure 9.14. Note the capacitor is represented by two parallel lines with a gap to indicate that there is an insulator between the two conducting plates. An insulator in a DC circuit breaks the closed circuit, with no steady-state current flowing in the circuit.

The conducting plate connected to the (+) terminal of the battery will attain the potential V_+ by the flow of electrons to the battery, leaving a net charge $+Q$ on the conducting plate. And similarly for the conducting plate connected to the (−) terminal, it will have the potential V_- and charge $-Q$. The charged capacitor will therefore have a voltage difference equal to the battery, with charges $+Q$ and $-Q$ on the two respective conducting plates. After the (transient) process of charging is over, there is no more flow of charges and the current in the circuit is zero.

By charging the conducting plates with opposite charges of amount $+Q$ and $-Q$, a potential difference of amount V is created. The amount of charge a capacitor can store for this electric potential is called its **capacitance**, denoted by C (not to be confused with the SI unit of charge, namely the coulomb C). The analog of a capacitor is a water storage tank; if a certain volume of water is poured into the tank — the analog of electric charge — the increase in the height of the water is the analog of the increase in potential V while the volume of the storage tank is the analog of the capacitance C.

So what would be a good way to define the capacitance? Clearly, the capacitance C should increase when the amount of charge Q that can be put on the capacitor increases. However, for different electric potentials V, different amounts of work are needed to move a charge onto a capacitor. For example, if V is very large, it will be very hard to add a new charge. Hence, it is sensible to define the capacity as the amount of charge that can be stored on a capacitor *per volt*, and so we have

$$C = \frac{Q}{V} \quad \text{: Definition of capacitance.} \tag{9.20}$$

The SI unit of capacitance is the **farad**, denoted by F; we have

$$1 \text{ farad} = 1 \text{ coulomb per volt} \tag{9.21}$$

$$\Rightarrow 1\,\text{F} = 1\,\text{C/V}. \tag{9.22}$$

A parallel plate capacitor is composed of two conducting plates placed parallel to each other and separated by a distance d filled with an insulator. One can let air fill up the space between the conducting plates, but a dielectric material is usually placed instead of air in order to *increase* the capacitance of the capacitor.

Example 9.1: Energy stored in a capacitor

To find the electrical energy stored in a capacitor, we calculate the work required to charge it. Suppose a charge Q has been deposited on the conducting plates of the capacitor resulting in a potential difference of $V = \frac{Q}{C}$. To add more charge dQ, the work needed is the potential energy gained by the charge dQ, which is $V \times dQ$. Hence

$$dW = V \times dQ = \frac{Q}{C} dQ. \tag{9.23}$$

The work required to fully charge the capacitor can be found by integrating the charge Q from the initial value of 0 to final value of Q_f. Denoting the energy stored in the capacitor as U we obtain

$$U = \int_0^U dW \tag{9.24}$$

$$= \int_0^{Q_f} \frac{Q}{C} dQ \tag{9.25}$$

$$= \frac{Q_f^2}{2C} = \frac{1}{2} C V^2. \tag{9.26}$$

The capacitance of two parallel plates is given by

$$C = \epsilon_0 \frac{A}{d}. \tag{9.27}$$

where A is the area of the parallel plates that have air in-between the plates.

Note that capacitance C depends almost entirely on the geometrical shape of the capacitor and the material serving as the insulator. By changing the insulator from air to a dielectric material, one simply changes in U the electric constant ϵ_0 to that of the medium, namely ϵ, obtaining the general result

$$C = \epsilon \frac{A}{d}. \tag{9.28}$$

As shown in Table 9.2, a properly chosen dielectric can increase the capacitance by a few orders of magnitude.

Material	ϵ/ϵ_0
Vacuum	1
Air	1.00054
Paper	3.5
Silicon	12
Germanium	16
Water (25°C)	78.5
Titania ceramic	130
Strontium titanate	310

Table 9.2: Dielectric constants relative to that of vacuum for some common materials.

One can replace the insulating dielectric in a capacitor with a semiconductor, and create a more complex device that stores and discharges electrical energy depending on the state of the semiconductor. Another way to store electrical energy is with the help of chemical reactions. We already encountered batteries as electricity storage devices above so let us now look at them briefly.

Batteries

Figure 9.15: Typical AA batteries.

Batteries convert chemical potential energy into electrical potential energy with the help of possibly reversible chemical reactions. If the reactions are reversible, the battery is rechargeable and otherwise not. However, the basic principles are generally the same. The chemical reactions take place inside what is called a **voltaic cell** and a battery can consist of one or more of these voltaic cells in order to increase the electric potential or the capacity of the battery or both. The typical AA batteries shown in Figure 9.15 are made up of a single voltaic cell.

The basic idea of a voltaic cell is to use chemical reactions to create an electron surplus on one of its terminals while creating an electron shortage on the other. The consequence is an electric potential between the two terminals and if the terminals are connected with a circuit, a current will flow. The question now is, how can a chemical reaction give an electron surplus or shortage respectively?

The key issue is that ions (positively or negatively charged atoms) can often quite easily be obtained (for example, when salt dissolves in water). If one can then choose a chemical reaction such that the surplus or missing electrons from

Fuel cell

Similar to batteries, the idea of fuel cells is to provide electricity. Contrary to batteries, the electrodes in fuel cells are not directly affected when electricity is generated from a fuel and hence can run for rather protracted amounts of time. Some common fuels are, for example, hydrocarbons, alcohols and hydrogen.

The idea behind a fuel cell is that electrons can be separated from a fuel with the help of a catalyst. The electrons then travel through a circuit to provide electrical energy. Upon their return to the other electrode of the fuel cell, a second catalytic reaction combines the electrons with an oxidant. The physical principle underlying the energy extraction is the same as in the burning of a fuel: The reaction product has a lower energy than the reactants. For example, hydrogen can be burnt with oxygen to give a flame (heat energy) leaving water as a waste product. Similarly, in a hydrogen fuel cell, stored hydrogen is combined with oxygen (generally obtained from the air) to provide water and energy, the difference being that rather than heat, electrical energy is obtained.

Figure 9.16 shows the basic design of a proton exchange membrane fuel cell. In the cell depicted, hydrogen gas is supplied on the left hand side and air on the right hand side. At the anode, the hydrogen gas is split into positive hydrogen ions (H^+) and electrons with the help of a catalyst such as platinum. The trick now is that the proton exchange membrane (generally made up of a polymer) has been selected such that it only allows the positively charged hydrogen ions to pass. Having nowhere else to go, the electrons travel through an external circuit to the cathode thus creating a current. Finally, at the cathode, the hydrogen ions combine with oxygen from the supplied air and the arriving electrons to form water.

Hence we have the following reactions:

$$2H_2 \rightarrow 4H^+ + 4e^-$$
$$4H^+ + 4e^- + O_2 \rightarrow 2H_2O.$$

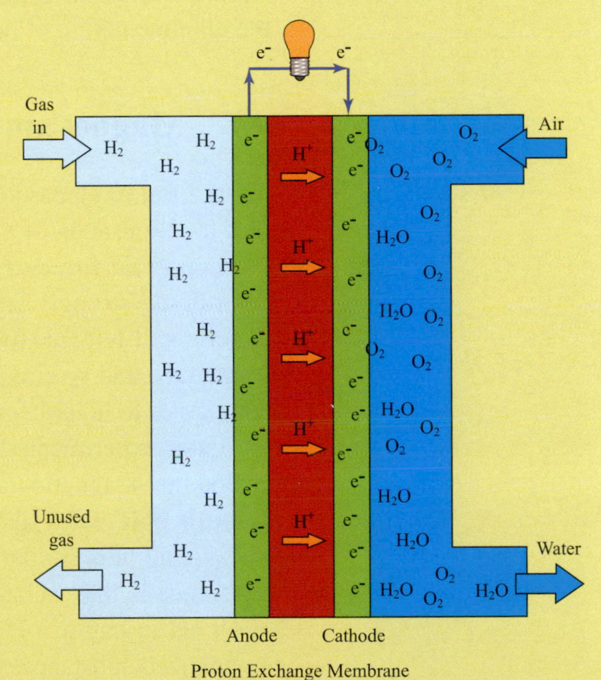

Figure 9.16: Proton exchange membrane fuel cell.

Fuel cells are attractive because they have no moving parts and very few if any environmentally problematic exhaust gases. Indeed, a hydrogen fuel cell can safely be used indoors as its only by-product is pure water. However, the engineering challenges for manufacturing cheap and reliable fuel cells are enormous. For example, the proton exchange membrane in Figure 9.16 must fulfill several apparently contradictory requirements: it must block electrons yet let protons through, it also must block both the hydrogen and oxygen gases as well as water — furthermore, its resistance to the flow of protons must be as small as possible.

The voltage over a single membrane is around 0.6 to 0.7 V, and hence for practical use a fuel cell consists of a stack of alternating layers.

the ions can be transferred to a conducting terminal, then a current between the terminals can flow. One pair of such reactions found in alkaline batteries is given by

Anode (oxidation)	$Zn(s) + 2OH^-(aq)$	$\rightarrow$	$ZnO(s) + H_2O(l) + 2e^-$
Cathode (reduction)	$MnO_2(s) + 2H_2O(l) + 2e^-$	$\rightarrow$	$Mn(OH)_2(s) + 2OH^-(aq)$
Overall reaction	$Zn(s) + MnO_2(s) + H_2O(l)$	$\rightarrow$	$ZnO(s) + Mn(OH)_2(s)$ $E_{cell} = 1.5$ V

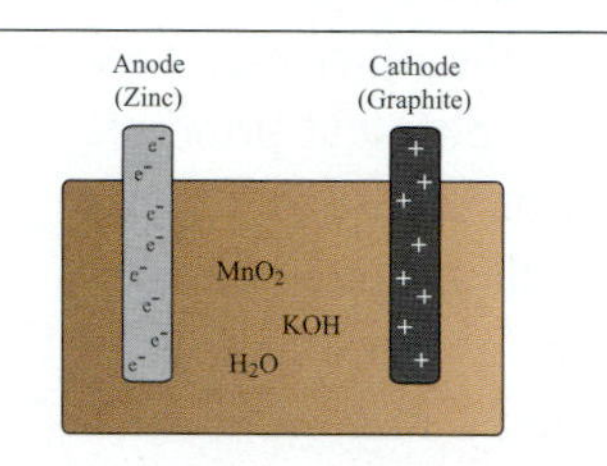

Fig. 9.17: Design of an alkaline battery.

where the letter between parentheses indicates that the compound is solid (s), aqueous (aq) or liquid (l). In order to actually put the electrons in reaction to good use, the voltaic cell needs to be designed such that they can only move from the anode to the cathode through an external circuit (i.e. a wire outside of the voltaic cell). The basic design of an alkaline battery is depicted in Figure 9.17.

The manganese dioxide is mixed in with a paste of potassium hydroxide and water that forms an electrolyte which allows for the (slow) movement of ions toward the anode and cathode respectively. Alkaline batteries get their name from the alkaline nature of potassium hydroxide.

9.7 Magnetism

In addition to electric charges and forces, it has been known for a long time, in fact for thousands of years, that analogous but apparently different forces exist: i.e. **magnetic forces**. Some naturally occurring minerals, although electrically neutral, are strongly attracted to iron. It was also discovered that these magnetic materials, if left free to rotate, always orientate themselves with the earth's north-south axis, and so already by 1000 AD the Chinese used such substances to make compasses to help them navigate.

Later experiments by William Gilbert in the sixteenth century and others revealed that magnetic materials always have two opposite poles: a **north pole and a south pole**, with like poles of materials repelling and unlike poles attracting. This situation is similar to that for the two signs of electric charges but the crucial difference is that isolated magnetic poles (magnetic monopoles) have never been observed in nature. Thus unlike the case for electric charges, one cannot create magnetized substances with an excess of one magnetic pole over another.

9.8 Magnets: storing magnetic energy

It turns out that certain materials can temporarily or permanently be magnetized. Rather contrary to the case of storing electrical energy in a capacitor where electrical charges (electrons) are moved to one of the plates, due to the absence of magnetic charges, magnetic energy is stored by creating microscopic currents at the atomic level. Creating these small currents requires energy and, in principle, the resulting magnetism can be used to later extract this energy again.

The small currents act like tiny magnets with a north and a south pole, and the magnetism of a macroscopic object is the sum of the magnetism created by the tiny currents. Because every magnet has a north and a south pole, the force between two magnets is strongly dependent on the orientation of the magnets. For magnets, we therefore do not have an elegant inverse square formula such as Coulomb's law (Equation 9.3) or Newton's universal law of gravitation (Equation 1.26).

Among the most powerful magnets available are those made from a compound of neodymium, iron, and boron with the chemical formula Nd_2Fe_14B. When brought together, even two relatively small magnets of, for example, 1 cm × 1 cm × 2 cm can lift large pieces of iron.

Noteworthy 9.1: Magnetic monopoles

The non-occurrence of magnetic monopoles in nature is somewhat of a mystery for modern theoretical physics as almost all of the unified theories of fundamental forces predict their existence. Thus understanding why they are not seen might require some deep new concepts or principles. On the other hand, the existing magnetic phenomena have been understood in terms of the electric charge.

9.9 Atmospheric electricity

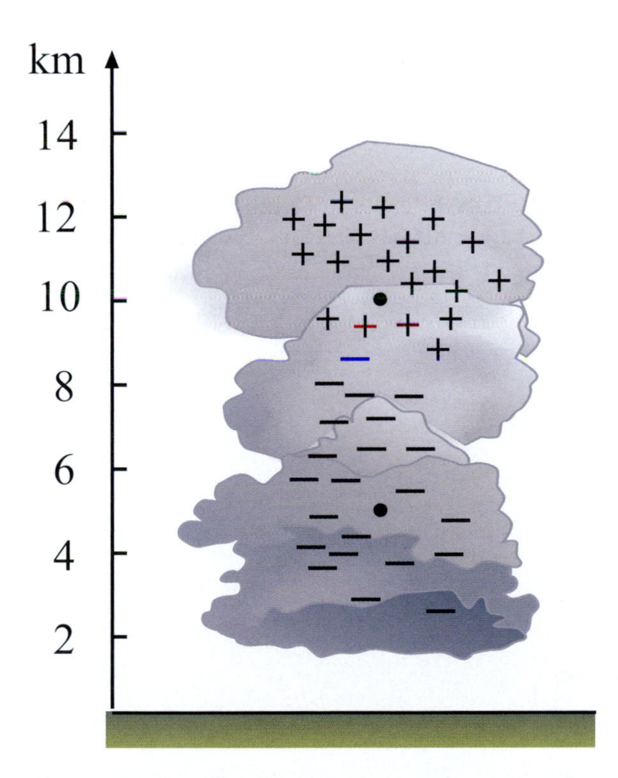

Figure 9.18: Charge distribution in a thundercloud.

A thunderstorm is made of several thunderclouds (or cells), within each of which air moves upwards during the earlier stages and then downwards during the latter stages. The activity that takes place in a thundercloud, such as the movement of air and charge buildup (see Figure 9.18), is fueled by the latent energy released when the moisture in air condenses.

For example, suppose some warm air with water vapor near ground level rises initially because it is less dense. As it reaches a higher level where the temperature is lower, it will cool, the water vapor will condense releasing heat, which then drives the air even higher. As the air inside a thundercloud rises, it sucks in more air from the bottom. The final height of the cloud may reach up to 18,000 m where the temperatures are very low.

Charge generation

The small water droplets inside a thundercloud may combine to form raindrops, which may then freeze to form ice and snow. These heavy particles start to fall and in so doing drag some of the cool air down with them in a downdraft. (Do you notice a cold wind just before it rains?)

Also, as the different particles within a thundercloud interact through collisions, they become charged. The cloud then becomes *polarized* with the heavier charged particles, which happen to be negative, moving under gravity to the base of the cloud and the lighter positively charged particles are blown by the updraft to the top of the cloud.

The net charges in the cloud may be approximated with the help of the image charge method (see below). The charges in the clouds *induce* charges on the conducting ground and the net electric field is the sum of the electric field of all the charges. For the configuration mentioned, this comes to about 2×10^4 V/m at ground level.

Fig. 9.19: Lightning bolt.

Image charge method

The electric field at ground level is not simply the sum of the electric fields of the charges in the cloud. The situation is complicated by the fact that the fields polarize the charge distribution on the earth below, so that the final electric field at the surface is modified.

There is a trick called the **image charge method** illustrated in Figure 9.20, whose justification is quite subtle, but which allows the final electric field to be calculated with ease. Basically it says that if the ground is a good conductor, then the final electric field at ground level is obtained by including the electric fields of *image charges* located below ground.

The force E experienced by a unit charge is given by

$$E = \frac{F}{q_1} = \frac{k_e q_1 q_2}{q_1 r^2} = k_e \frac{q_2}{r^2}. \tag{9.29}$$

For the situation depicted in Figure 9.20, we then have at ground level below the cloud

$$\begin{aligned} E &= 2k_e\left(\frac{q_1}{r_1^2} + \frac{q_2}{r_2^2}\right) \\ &= 2 \times 9 \times 10^9 \frac{\text{Nm}^2}{\text{C}}\left(-\frac{40\,\text{C}}{(5\,\text{km})^2} + \frac{40\,\text{C}}{(10\,\text{km})^2}\right) \\ &= -2.2 \times 10^4\ \text{V/m}. \end{aligned}$$

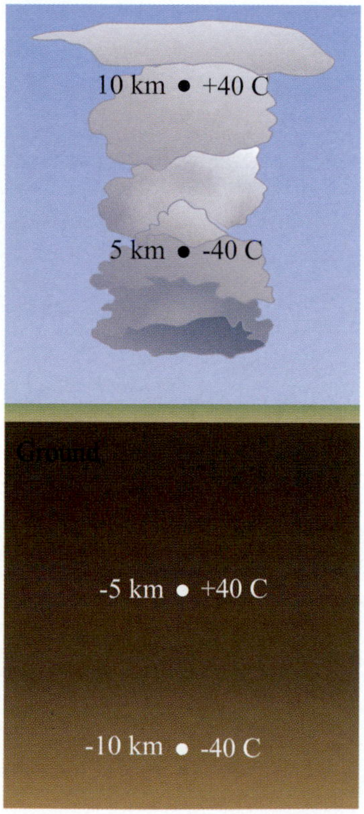

Fig. 9.20: Illustration of the image charge method.

Note the factor 2 in front of k_e which is due to the image charges below ground.

Discharge

Thus very large electric fields are generated in the vicinity of a thundercloud. The small number of ever-present free electrons in the air are accelerated by these electric fields. The collisions between these electrons and other atoms produce more free electrons leading to a chain reaction: The electrical resistance of the air breaks down and a lightning discharge is said to take place. Although dry air requires an electric field of 3×10^6 V/m to start an electrical breakdown, the presence of moisture reduces the threshold considerably.

Types of lightning

The lightning discharge can take place between two points inside a cloud, between two clouds or from the cloud to the ground; see Figure 9.21.

The last type is actually not the most common but is the most dangerous to human beings, and so we take a closer look at it.

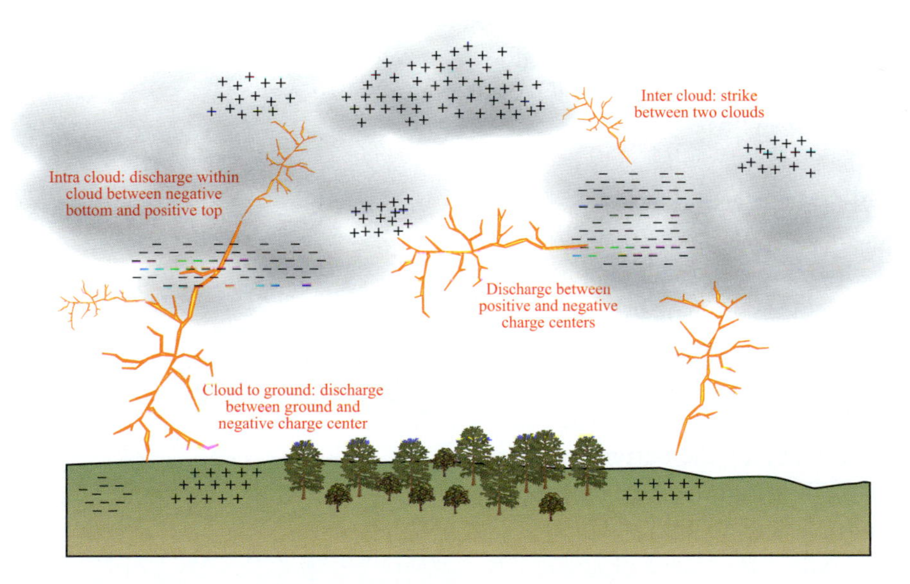

Figure 9.21: Types of lightning.

In cloud-to-ground lightning, the initial avalanche of electrons causes a downward moving **step leader** which is a channel of radius of about 5 m, proceeding in discrete steps of about 50 m with pauses of about 50 μs between the steps. This is shown in Figure 9.22.

The air in the channel becomes ionized (ion = charged atom) by the colliding electrons, and so is highly conducting. Once the leader is close to the ground (about 50 m away), as shown in Figure 9.23a, the very high potential differences between the earth and the tip of the leader cause an upward moving discharge, as shown in Figure 9.23b, which meets the leader to form a connecting path.

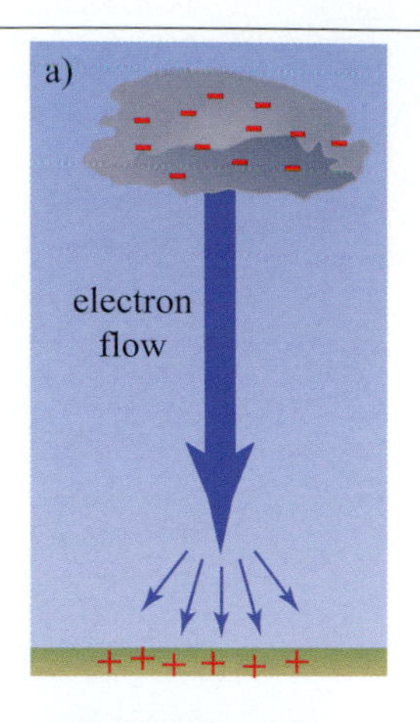

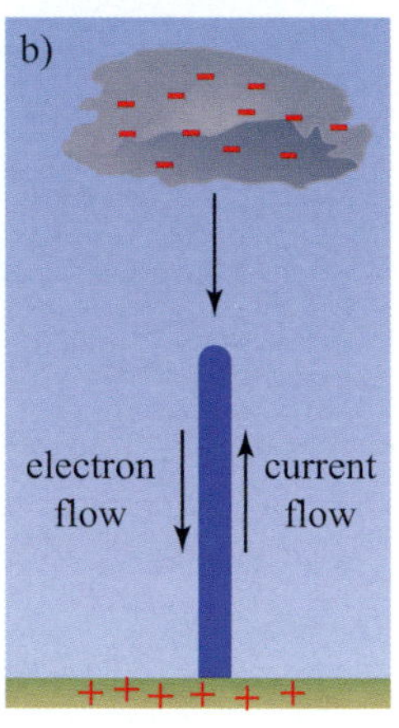

Fig. 9.23: Charge and current flows.

The conducting path then allows the electrons to leave the channel, starting with those already near the surface of the earth. Thus a *return stroke* of very bright light, moving upwards, is produced by the intense current of between 10,000 and 20,000 A which flows for about 100 μs up the channel. The electrons are then traveling at very high speeds (possibly close to the speed of light).

Accelerating charges produce electromagnetic radiation of which light is one form. Indeed the intense currents in the conducting channel produce not only the lightning flash but also radio waves.

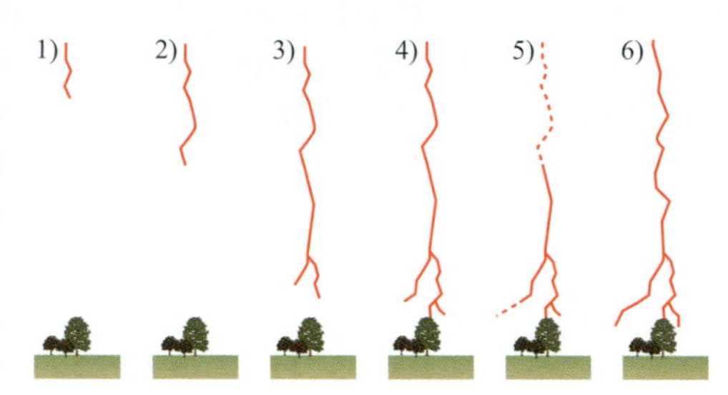

Figure 9.22: Cloud-to-ground lightning.

The high current in the channel also raises the temperature of the air to about 20,000–30,000 K and most of the energy is actually dissipated in the form of heat (which is simply random motion of the atoms), rather than as light. The high temperatures produce an intense pressure in the air which expands quickly, producing a shock wave that propagates out and decays eventually to a sound wave — thunder.

The difference in time that we notice between thunder and lightning is due to the vastly different speeds at which these two waves travel.

The global electric circuit

It is rather surprising that even in fair-weather regions of the earth, an atmospheric current flows from the ionosphere to the ground, moving positive charge to the ground. The charge carriers that make up the current are ions produced in the atmosphere by cosmic rays, ultraviolet radiation and plain collisions. With a resistance of about 200 Ω between the ionosphere and the surface of the earth, the potential difference is about 300,000 V.

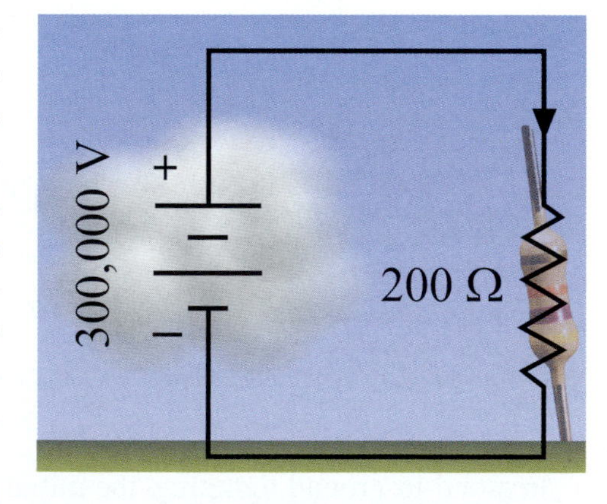

Figure 9.24: Global electric circuit.

What *maintains* this large potential difference? It is generally believed that the circuit is completed by the distribution of thunderstorms worldwide. That is, thunderstorms deliver positive charge to the ionosphere and negative charge to the ground, completing the global electric circuit as shown in Figures 9.24 and 9.25.

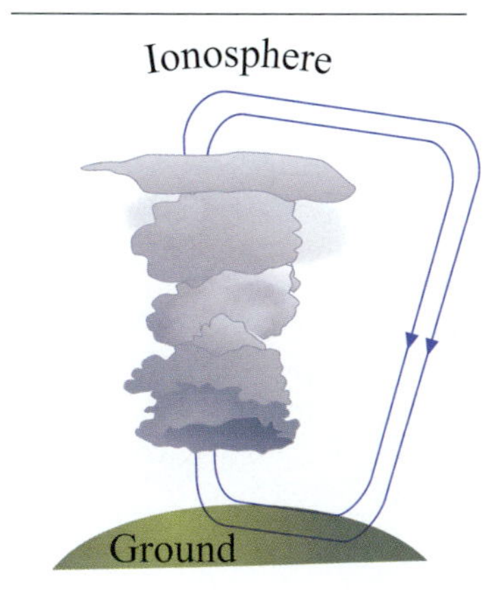

Fig. 9.25: Charge separation in a cloud.

It is estimated that there are a total of about 2,000 thunderstorms occurring at any one time on the earth. This means that on the average, each thunderstorm contributes 1 A to the global circuit.

9.10 The answer

We have seen that there are many naturally occurring electrical phenomena such as static electricity in drier countries and lightning during thunderstorms. All these electrical phenomena can be traced to the movement of electrons. Clearly when moving, electrons will have kinetic energy associated with them. What makes them special, however, is their electric charge that can have an effect over a relatively large distance and be used to do work.

In daily life, electricity is simply the availability of an electric potential for doing work. The conversion of potential to work is always done with the help of electrons since electrons can move with ease in readily available conductors such as copper. However, electrons are not the only carriers of electrical charge. For example, protons carry an equal but opposite charge and hence, in principle, protons could be used to do electrical work as well; however, the nuclei of atoms are usually not easily available. When trying to answer the question of what electricity is, we therefore should not limit ourselves to electrons and state more generally that **electricity is a form of energy involving electrical charges.**

9.11 Exercises

1. What is the effect of humidity on static electricity?
2. Explain, with the help of examples, the difference between (i) an electrically neutral object, (ii) an electrically charged object and (iii) an electrically polarized object.
3. How many electrons would it take to form 1 C of charge?
4. We have a current that flows from point A to point B. In which direction do the electrons move?
5. Does the north end of a compass needle point exactly to the geographical north pole?
6. Where is the magnetic north pole of the earth's magnetic field?
7. I have a circuit that connects the two terminals of a 6 V battery through a 100 Ω resistor. What is the current in the circuit?
8. How much power is consumed in the above question?
9. I have a circuit in which the two terminals of a battery are connected through a capacitor and a switch. When I change the switch from the on to the off position, why is there a current at first and then no current later on?
10. I have two electrically conducting plates. Plate A is at a potential of 5 V and plate B is at a potential of 8 V. If I drop a proton between the two plates, toward which plate will the proton move?

10: Odor

Q - Why Can We Smell Perfume?

Chapter Map

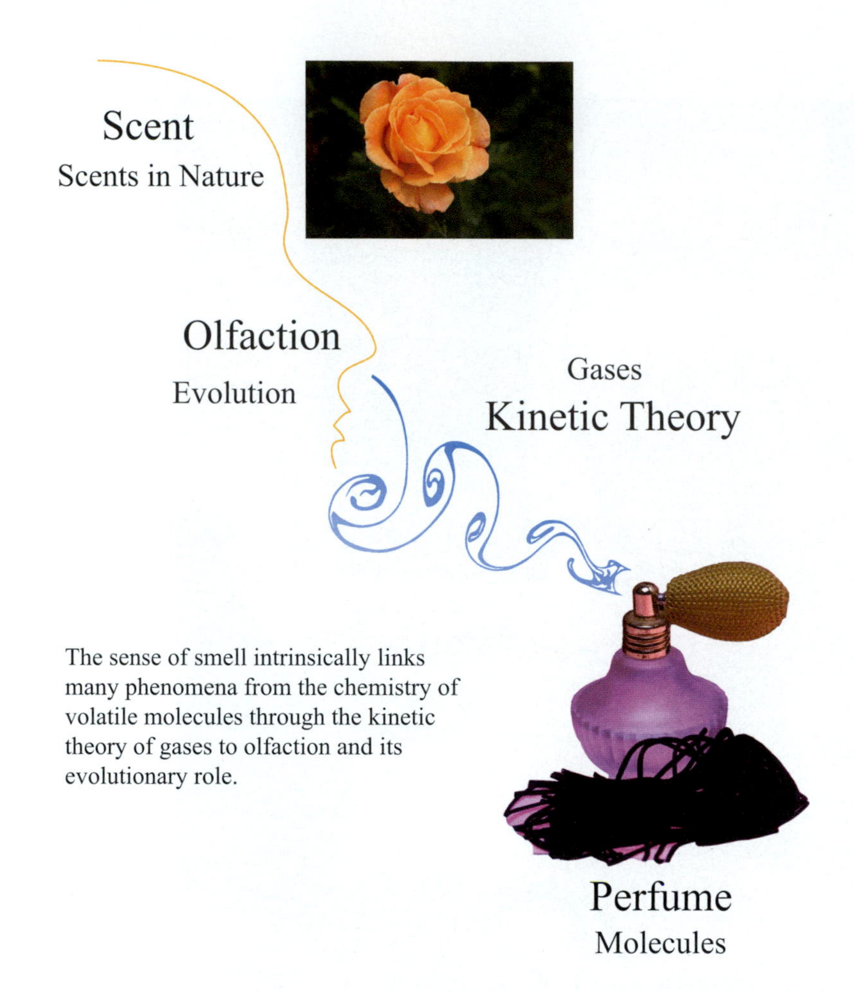

The sense of smell intrinsically links many phenomena from the chemistry of volatile molecules through the kinetic theory of gases to olfaction and its evolutionary role.

10.1 The question

Fig. 10.1: The aroma of food can make us hungry.

The smell of freshly baked bread has a strong impact on many of us. One walks past a bakery with no intention to buy food and a few minutes later finds oneself enjoying a croissant and a fragrant cup of coffee.

Indeed, smells and scents seem to influence decisions in many ways from the choice of foods, the detection of hazards as in foul smelling water to the identification of mates. Beyond the need for personal hygiene, we therefore go to great lengths not to be perceived as malodorous if not outright fragrant. Consequently, it is not surprising that perfume manufacturing is a global industry but of course we can wonder why we can perceive scents in the first place. After all, it seems reasonable that we would be able to survive just as well without the sense of smell. Hence the question for this chapter: **Why can we smell perfume?**

Figure 10.2: Scents play an important part in life!

In order to answer this question, we need to consider at least three different aspects. First, we need to know how it is physically possible to have a smell. Then we need to find out how a smell can be transported, after all we can enjoy the scent of a beautiful flower even if there are no apparent air currents. Once we know how a scent can reach us we need to be able to smell it, that is to say we need to have some kind of sensors in our bodies that can identify odors. Last, we need to consider why such a detection mechanism evolved in the first place. Clearly, our question of why we can smell perfume is far from simple so let us begin by investigating what kind of things have a smell and what do not.

10.2 Odorants

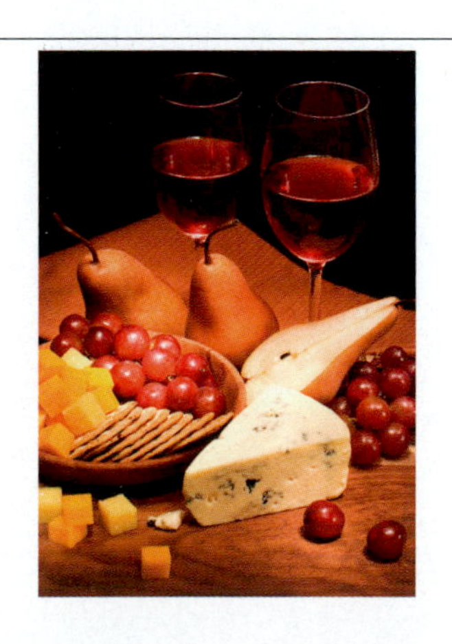

Fig. 10.3: The smell of foods like wine and cheese can easily be identified.

From daily experience we know that not all things smell. For example, if it is clean, glass doesn't have any smell and neither does a knife. This is easy to verify by lining up some glasses, cups and cutlery on a table and closing one's eyes. It will be impossible to sniff out which is which (even for a dog but only if the items are completely clean). On the other hand, generally, it will be quite easy to distinguish wine from cheese. What would be the essential difference between substances that have and do not have a fragrance? Whether or not we understand the details, it is clear that something must be transported from the substance to the nose in order to be smelled.

It is physically possible to detect odors due to the existence of volatile chemicals that evaporate slowly enough to avoid rapid depletion and fast enough to provide sufficient concentration in the air to be noticeable.

Of course, the sensitivity to smells varies greatly from species to species but the basic mechanism remains the same. A volatile chemical moves from a source like a glass of wine to a nose. Volatile chemicals that can be smelled are generally

Fig. 10.4: Despite their enormous differences, the basic mechanisms of olfaction in insects are remarkably similar to those in mammals.

called **odorants** regardless of the aesthetic value, i.e. whether they smell good or bad. The smell itself or in other words the sensation elicited by the preception of the odorant is often called an **odor**.

Most odorants are organic molecules, in other words molecules that contain both carbon and hydrogen among possible other elements, although some well-known odorants like ammonia are inorganic. Since molecules that are too heavy cannot easily move through air after evaporation, the maximum weight of an odorant molecule is believed to be around 300 dalton (a dalton [Da] is equal to the atomic mass unit u defined as one-twelfth of a carbon-12 nucleus or roughly equal to the weight of one proton). Consequently, odorants are fairly small molecules with at most around 20 carbon atoms.

Fig. 10.5: Perfume bottles can be small, as only minute quantities are required.

Probably, the total number of different odorants that a human can preceive is fewer than 1,000, and generally these odorants can be classified as belonging to one of seven primary types of olfactory stimulants. The seven **primary odorant classes** are listed in Table 10.1 with a representative chemical compound, its chemical formula, its shape and the concentration in parts per million needed for the compound to produce an equal odor intensity. As can be seen from the table, the minimum concentration required to be smelled for the class human beings are most sensitive to (putrid) is about half a million times smaller than for the class we are least sensitive to (pungent). Although the number of different odorants perceivable is not so large, since a smell can contain many different odorants in varying concentrations, the total number of odors is, as such, nearly limitless. Nevertheless, human beings can only distinguish about 10,000 different odors. Some typical odorants and their smells are shown in Table 10.2.

Now that we have some idea of what kind of molecules can elicit our sense of smell, let us investigate how those molecules can move from their source (like a perfume bottle) to our nose.

10.3 Odors spread

Fig. 10.6: A bit of food coloring dropped into a glass will slowly diffuse.

From daily experience, we know that a bottle of perfume lasts quite a while, as each application of perfume requires only a tiny amount. Furthermore, most perfumes evaporate on application and the scent is somehow carried by the surrounding air. But exactly how is it carried? How does the perfume move around in a room? In fact, although we had tentatively concluded that the perfume (or more generally an odor) does moves around, we should nevertheless ask whether this is really so or whether an alternative explanation is possible.

To find out, one can perform various simple experiments. For instance, one may open a bottle of perfume in a closed room with no perceptible movement of air, and measure how fast the smell spreads in the air. One could cover the bottle with a membrane which does not let the odorants pass through to see whether it is essential that odorants are in fact transported. One can also see if there is any preferred direction in which the smell spreads, or if it spreads in all directions. Another matter of interest would be how the strength of the perfume decreases with distance from the source.

Smell	Molecule Name	Chemical Formula	Shape	Concentration
Ethereal	Ethanol (alcohol)	C_2H_6O		800
Camphoraceous	Camphor	$C_{10}H_{16}O$		10
Musk	Muscone	$C_{16}H_{30}O$		1
Floral	Nerol	$C_{10}H_{18}O$		300
Minty	Beta-Cyclocitral	$C_{10}H_{13}O$		6
Pungent	Vinegar (acetic acid)	$C_2H_4O_2$		50,000
Putrid	Hydrogen sulfide	H_2S		0.1

Table 10.1: The seven primary odorant classes and the concentrations needed in parts per million to produce equal odor intensity.

Smell	Molecule Name	Chemical Formula	Shape
Fruity	Ethyl octanoate	$C_{10}H_{20}O_2$	
Minty	Menthol	$C_{10}H_{20}O$	
Nutty, medicinal	2,6-dimethyl pyrazine	$C_6H_8N_2$	

Table 10.2: Some typical odorants and their smells.

The results of such simple experiments lead to the following observations:

A perfume's scent:

- involves the movement of molecules
- spreads in all directions
- spreads at a finite speed
- spreads even if the air is still
- becomes weaker farther away from the source

Given these experimental facts, we can conclude that indeed, unlike, say, the case of sound waves in which there is no net physical transport of the underlying medium, for the case of scent, the perfume molecules themselves must be transported from the perfume bottle to other points in a room.

More generally, the phenomenon of a substance spreading out from a region in which it has a high concentration (for example the highly concentrated odorants in a perfume bottle) to a region where its concentration is lower (for example the air surrounding the perfume bottle) is called **diffusion**. As such diffusion can occur in many ways, the most easily notable ones being in gases and liquids. For example, if a drop of food coloring is carefully released in a glass of still water as shown in Figure 10.6, it will slowly spread out, eventually coloring the entire contents of the glass.

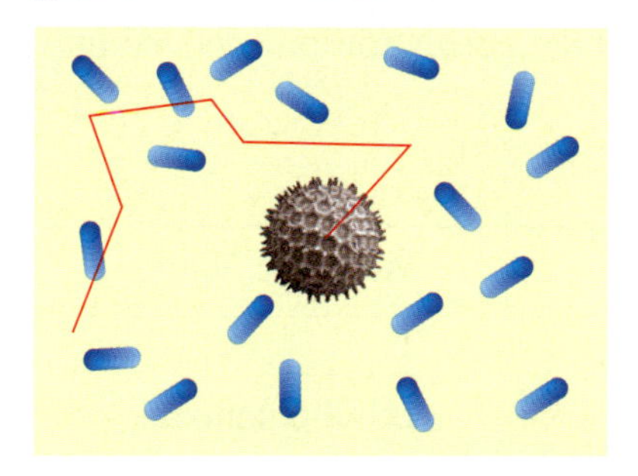

Figure 10.7: Random path of a pollen grain due to collisions with water molecules. The size of the water molecules is greatly exaggerated and their number much larger.

In order to understand why we can smell perfume, we hence need to find the mechanism that transports the perfume molecules. If the air is still how can a perfume molecule move from a perfume bottle to a person's nose? In order to answer this question, we need to investigate the behavior of air molecules that "carry" the perfume molecules in a room, and to do that it is useful to first have a quick look at a related phenomenon called **Brownian motion**.

Fig. 10.8: Robert Brown (1773-1858), British botanist and discoverer of the phenomenon now referred to as Brownian motion.

Brownian motion is named after the botanist Robert Brown who discovered in 1827 that pollen particles in water display a jiggly motion even when water is completely still. At first, he thought that perhaps the pollen particles were alive but when he repeated his observation for dust particles, he obtained the same result. He thus could exclude that the motion of the particles is due to them being alive. For a long time, the reason for this motion remained a mystery but in 1905 none other than Albert Einstein published an explanation based on the then still somewhat controversial kinetic theory of gases.

10.4 Kinetic theory of gases

We want to find out in what way and how fast can a perfume molecule move from the perfume bottle to our nose. A first assumption one could make is that a perfume molecule will move with a speed similar to that of an air molecule from the bottle to the nose. In order to find the speed of an air molecule we can investigate the behavior of a representative air molecule.

Fig. 10.9: The atoms of an ideal gas are similar to billiard balls that move and bounce without friction.

As a starting point for investigating the behavior of air molecules, and in fact that is all that we will need here, we can make some simplifying assumptions that turn out to be fairly accurate for a dilute gas such as air. The gas with simplifications representing air (or any other type of basic gas) is called an **ideal gas** and has the following defining properties:

- The gas consists of a large number of identical molecules moving with random velocities (22.4 L of the gas will contain $N_{\text{Avogadro}} = 6.022 \times 10^{23}$ molecules at room temperature and 1 atmosphere pressure).
- The gas molecules have no internal structure and can be considered hard spheres with a radius of about 10^{-10}m.
- The molecules do not interact except for brief elastic collisions with each other and with the walls of the container.
- The average distance between the molecules is much greater than their diameter.
- The position and velocity of every gas molecule is random. Physically observed properties of the gas such as pressure and temperature are the result of averaging over the random behavior of the gas molecules.

Avogadro's number is $N_{\text{A}} = 6.022 \times 10^{23}$.

The kinetic theory of gases, based on the notion that such averaging is valid, has proved to be hugely successful. One of the main results is the derivation of the **ideal gas law**, which has been experimentally verified and is given by

$$PV = nRT \qquad \textbf{Ideal Gas Law} \tag{10.1}$$

with the number of moles n given by the fraction of the number of atoms N and Avogadro's number N_A, i.e. $n = N/N_A$. The gas constant R is given by $R = 8.314$ J/(K mol).

Fig. 10.10: The fact that hot air has a larger volume is one of the principles upon which the hot air balloon is based.

For the sake of concreteness, let us consider an ideal gas at temperature T, confined in a container of volume V and consisting of N molecules. Let us further suppose that the gas is in equilibrium, which implies that there are no changes of pressure, temperature and variables taking place in the gas. Since the molecules are taken to be hard spheres, the energy of the gas consists entirely of kinetic energy.

Let the three-dimensional velocity of the nth particle be denoted by $\vec{v}_n = (v_x, v_y, v_z)$ (where the little arrow on the v indicates that it is a vector) and because there are N particles, the total kinetic energy of the gas is simply the sum of the kinetic energies of the individual particles. Hence, the kinetic energy of the gas is given by the following expression:

$$E_{\text{kinetic}} = \frac{1}{2} m \sum_{n=1}^{N} \vec{v}_n^2. \tag{10.2}$$

This is a nice expression to start with as it relates the energy of the gas with the velocity of the molecules. Clearly for moving perfume molecules around, energy would be required, so this is a good way to commence our look at the kinetic theory of gases. Next let us look at pressure, for this will give us an idea of how the molecules of the gas can "push" the perfume molecules.

Pressure

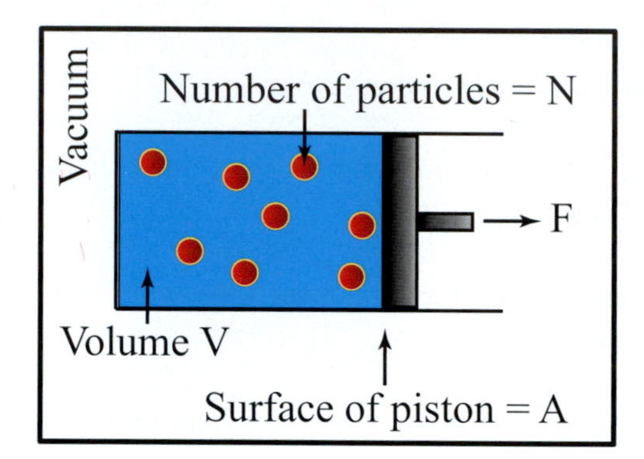

Figure 10.11: The impacts of the molecules bouncing around exert a force F on the piston.

From the atomic point of view, how does pressure arise? Even before going into the detailed mechanism, we expect that pressure should be a macroscopic manifestation of microscopic motion. In order to find out how a microscopic motion can have a macroscopic effect, consider a frictionless piston as shown in Figure 10.11, which has an area A, contains a gas in some volume V and a total number of atoms N. The piston is in equilibrium with the gas at some temperature T and our setup is such that we have a perfect vacuum outside the piston.

The atoms or molecules of the gas are constantly bombarding the piston and this will cause a force to be exerted on the piston. To keep the piston in place we need to counter this force.

How much force would we need? To figure that out, let us consider an atom traveling toward the piston with velocity $\vec{v}$. The particle hits the piston and bounces back. We assume that the energy of the atom is the same before and after bouncing off the piston in what is called an **elastic collision** where the atom does not lose any of its energy to the piston.

Assuming an elastic collision is more reasonable than it may seem at first, why would that be so? If the atom would lose energy to the piston, due to energy conservation, the piston would heat up. However, this heating up must end since if it became too hot it would be able to add energy to the atom. Consequently, the gas and the piston will tend to an equilibrium where (on average) an atom neither gains nor loses any energy in the collision with the piston. As a result, the assumption of collisions being elastic is actually quite accurate.

As shown in Figure 10.12, for a wall placed along the y-axis, an elastic collision leads to a velocity $\vec{v}\,' = (-v_x, v_y, v_z)$ which can be understood by considering for simplicity the special case of $\vec{v} = (v, 0, 0)$ (we return to the general case later). Let the velocity of the atom after the collision be $\vec{v}\,' = (v_x', 0, 0)$. Since the atom possesses only kinetic energy, we have from energy conservation

$$\frac{1}{2}mv^2 = \frac{1}{2}mv'^2 \qquad (10.3)$$

$$\Rightarrow v' = -v. \qquad (10.4)$$

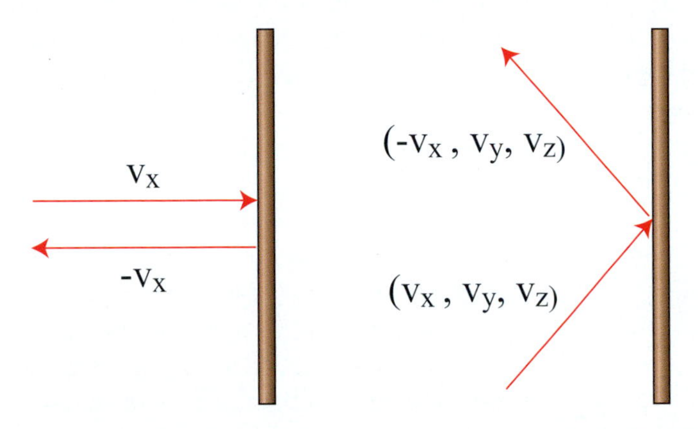

Figure 10.12: When bouncing off this wall, the velocity of an atom changes from v_x to $-v_x$.

In other words, in colliding off the piston, the particle's velocity changes from v_x to $-v_x$, and hence the momentum imparted to the piston is given by

$$mv_x - m(-v_x) = 2mv_x. \qquad (10.5)$$

During a time interval Δt, how many atoms will bounce off the piston? Atoms with velocity v_x can reach the piston during the time Δt if they are at a distance less than $x = v_x \Delta t$ from the piston. Consequently, all the atoms in a volume of size $xA = v_x \Delta t A$ will bounce off the piston. One expects half the atoms to be moving toward the piston and other half away from it. We only need to take the ones moving toward the piston to find the force on the piston, and hence we obtain

$$\text{Momentum imparted to piston during time interval } \Delta t = 2mv_x \cdot xA \cdot \frac{N}{V} \cdot \frac{1}{2} = \frac{Nmv_x^2 A}{V}\Delta t,$$

where the term xAN/V is the fraction of atoms that will reach the surface of the piston in time t. Since the force on the piston is nothing but the rate at which momentum changes on the piston due to collisions of the gas atoms, we have

$$\text{Force} = \frac{\text{Momentum imparted to piston during a time interval } \Delta t}{\Delta t}$$

$$\Rightarrow F = \frac{Nmv_x^2 A}{V}. \qquad (10.6)$$

John James Waterston

On this biographical page, rather than one of the well-known luminaries in the development of thermodynamics we would like to present the neglected pioneer J. J. Waterston. Not only was his groundbreaking work on the kinetic theory of gases rejected, he also started off with a completely wrong theory of gravitation. The quote at the bottom of the page by Lord Rayleigh should be a reminder to us all that, in addition to accepting failure, an open mind is essential for the progress of science.

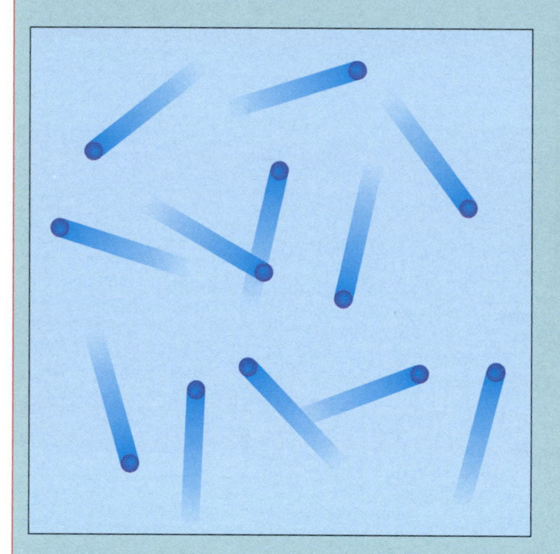

Figure 10.13: In the kinetic theory of gases, a gas consists of rapidly moving atoms or molecules.

John James Waterston was born in 1811 in Edinburgh where he attended school and later on, while he was an apprentice in civil engineering, he took classes at the University of Edinburgh. There he studied physics, mathematics, chemistry, anatomy, and surgery. Like many of his contemporaries, he was not satisfied with the notion of "action at a distance" as it appears in Newton's law of gravitation and at age 19 he published an incorrect paper trying to explain gravity in terms of colliding particles. Although this paper was completely wrong, it did give him important insights into the then still vague concepts related to particles.

After moving to London in 1833, he was posted to India in 1839 to train cadets in the East India Company. The posting allowed him to spend more time on his studies and in 1843 he published at his own cost the somewhat misleadingly entitled book *Thoughts on the Mental Functions* that tried to explain the nervous system in terms of molecular theory. In this book, he correctly derived several key features of the kinetic theory of gases like the notion that temperature and pressure are related to the kinetic energy of particles.

In 1845, Waterston submitted a more complete paper on the kinetic theory of gases to the Royal Institution that included a correct derivation of $PV/T =$ **constant** but it was rejected with disparaging comments. He then tried to convince others of his ideas but made little progress. Nevertheless, it is likely that the German physicist August Karl Kroenig was at least partially aware of Waterston's work when he published his paper proposing the kinetic theory of gases in 1856, setting off the rapid development of the theory from then on. In 1857, Waterston returned to Edinburgh where he continued his studies until his death in 1883.

In 1891, the British physicist Lord Rayleigh retrieved Waterston's manuscript from the Royal Institution's archives and arranged for its belated and posthumous publication. In the introduction Rayleigh wrote:

"The history of this paper suggests that highly speculative investigations, especially by an unknown author, are best brought before the world through some other channel than a scientific society, which naturally hesitates to admit into its printed record matter of uncertain value. Perhaps one can go further and say that a young author who believes himself capable of great things would usually do well to secure the favorable recognition of the scientific world by work whose scope is limited, and whose value is easily judged, before embarking on greater flights" (Haldane, 1928, p. 209–210).

Osmosis

The general principle of diffusion is that unconstrained molecules or atoms spread out from a region of high concentration to a region of low concentration reaching equilibrium such that the concentration is everywhere the same. For example, if we put a drop of food coloring in a glass of water, the color will slowly spread and, after some time, all the water will have an even color.

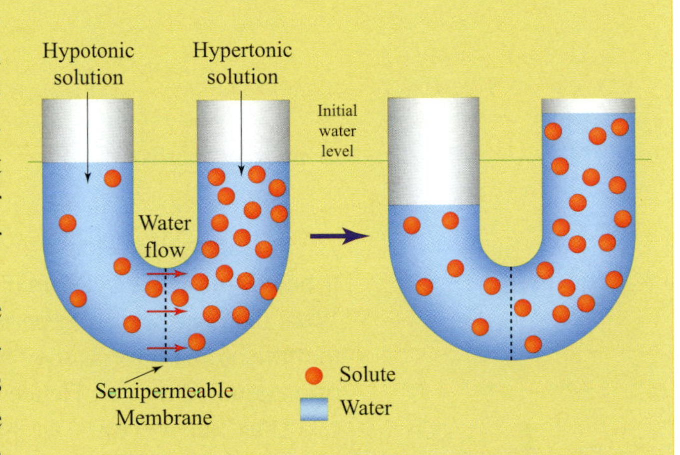

Figure 10.14: Osmosis: Water moves toward the side with a higher concentration of the solute.

A special case of diffusion occurs when we are dealing with two compartments separated by a semipermeable membrane, i.e. a membrane that blocks some molecules but not others. A good example would be sugar dissolved in water. One can choose a membrane with holes that are too small for the sugar molecules to go through but large enough for water. While usually one would then consider the concentration of the solute (i.e. sugar), in this case we need to look at the concentration of the solvent (i.e. water).

If the two compartments have the same amount of liquid, then having a higher concentration of a solute automatically implies that one has a lower concentration of the solvent. Now let us put an equal amount of liquid, one with a higher concentration of the solute than the other, in a container with two compartments separated by a membrane that only lets the solvent through as illustrated in Figure 10.14. The solute cannot pass through the membrane so nothing happens to the solute.

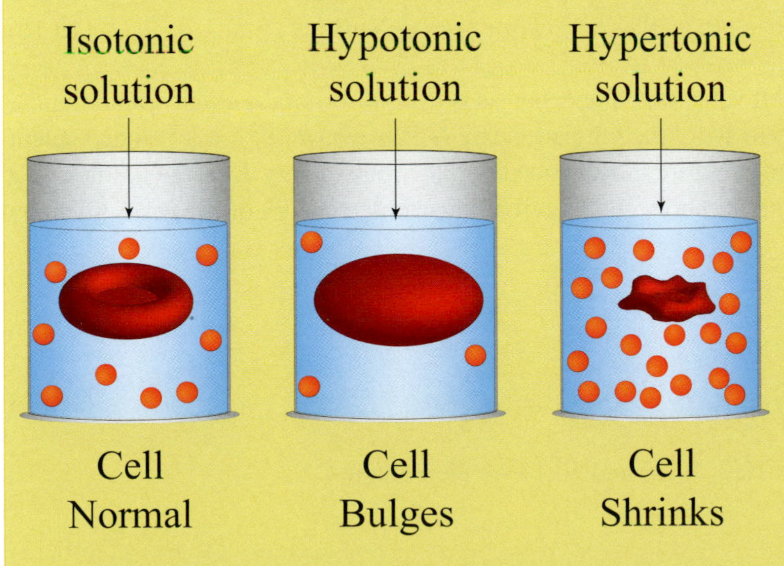

Figure 10.15: Cells depend on osmosis for survival.

The solvent (water), however, has a higher concentration on the side where there is less solute. Since the solvent can pass through the membrane, it will therefore diffuse from the side with less solute to the side with more solute. As a result, the level of the liquid rises on the side with the higher concentration of the solute. This process is called **osmosis** and is true in general, i.e. it also applies to gases and in some cases solids and gels. It is also very important for cells.

Why is osmosis important for cells? Cell membranes are semipermeable! They let water through rather easily but not many of its solvents. Although the membrane has some strength to withstand osmotic pressure, it can't be too much. As depicted in Figure 10.15, in a hypotonic environment, cells will bloat and possibly burst while in a hypertonic environment, they shrivel and shrink.

The pressure P is defined to be force per unit area, and hence the pressure on the piston due to the gas is

$$\begin{aligned} \text{Pressure} &= \text{Force per unit area} \\ &= \frac{F}{A} \qquad (10.7) \\ \Rightarrow P &= \frac{Nmv_x^2}{V} \qquad (10.8) \end{aligned}$$

or in other words

$$PV = Nmv_x^2. \qquad (10.9)$$

In reality, the velocity v of the atom is not completely fixed but randomly distributed following a certain distribution which we will not consider at this moment. What the piston really experiences is the *average* value over *all possible velocities* that the atoms have as they bounce off the piston. The average of a quantity x is often denoted by $\langle x \rangle$. Hence the average velocity is denoted as $\langle v \rangle$.

Thus far, however, we have only considered a very special set of velocities, namely, those heading straight for the piston, i.e. with $\vec{v} = (v_x, 0, 0)$. In general, the velocity of an arbitrary atom can be in any direction and has the form $\vec{v} = (v_x, v_y, v_z)$. Since all directions for the gas are equivalent, we have

$$\langle v_x^2 \rangle = \langle v_y^2 \rangle = \langle v_z^2 \rangle$$

$$\Rightarrow v^2 = \langle \vec{v}^2 \rangle = \langle v_x^2 \rangle + \langle v_y^2 \rangle + \langle v_z^2 \rangle = 3\langle v_x^2 \rangle.$$

Inserting this result into Equation 10.9, we obtain

$$PV = \frac{1}{3}N\langle mv^2 \rangle = \frac{1}{3}Nmv^2. \qquad (10.10)$$

The average of all velocities, since the gas is in equilibrium, is equal. Denoting the average of the velocity of the nth particle by $\langle \vec{v}_n^2 \rangle$, we have $\langle \vec{v}_n^2 \rangle = v^2$; independent of n. As mentioned before in Equation 10.2, the total energy E of the gas is solely composed of kinetic energy and therefore the average energy of the gas is given by

$$\begin{aligned} E &= \frac{1}{2}m \sum_{n=1}^{N} \langle \vec{v}_n^2 \rangle \\ &= N\frac{1}{2}mv^2. \qquad (10.11) \end{aligned}$$

From Equations 10.9, 10.10 and 10.11 we then obtain

$$PV = \frac{2}{3}E, \qquad (10.12)$$

which is a formula directly relating volume, pressure and energy. We now have three important physical quantities but are still missing one of the most easily noticeable quantities: temperature. How is temperature related to energy?

Temperature

We now have the means to define temperature in terms of atomic motion. This can be seen if we combine the ideal gas law of Equation 10.1 given by

$$PV = nRT = \frac{N}{N_A} RT \quad \text{Ideal Gas Law} \tag{10.13}$$

with

$$PV = \frac{2}{3} E = \frac{1}{3} N \langle mv^2 \rangle \tag{10.14}$$

which we had derived above in Equation 10.12 to obtain

$$\frac{N}{N_A} RT = \frac{1}{3} N \langle mv^2 \rangle. \tag{10.15}$$

Fig. 10.16: The functioning of a thermometer is based on the notion that it will be in thermal equilibrium with the environment.

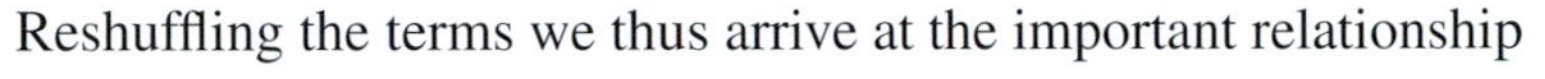

Reshuffling the terms we thus arrive at the important relationship

$$\frac{3}{2} k_B T = \frac{1}{2} \langle mv^2 \rangle \tag{10.16}$$
$$k_B = \frac{R}{N_A}$$

where the constant of proportionality k_B is so significant to have its own name. It is called the **Boltzmann constant** given by $k_B = 1.38 \times 10^{-23}$ J/K. From Equation 10.16 we see that the **temperature** is proportional to the average kinetic energy of a single atom (molecule) of the gas.

The ***Boltzmann constant*** *is given by* $k_B = \frac{R}{N_A} = 1.38 \times 10^{-23}$ *J/K.*

In other words, temperature is a measure of how fast, on the average, the atoms of a gas are moving. The faster the atoms move, the higher the temperature. The sensation of burning that we have on putting our hands into, say, a fire is because fast moving atoms from the fire impart high amounts of kinetic energy to our hands, causing atoms in our hands to move very fast, resulting in the sensation of burning.

Next, consider two gases in a cylinder separated by a frictionless piston as shown in Figure 10.17. Initially, this piston may move around depending on the relative pressures of the two gases, but eventually it will come to rest and the system is said to be in equilibrium, and the two sides will have the same temperature.

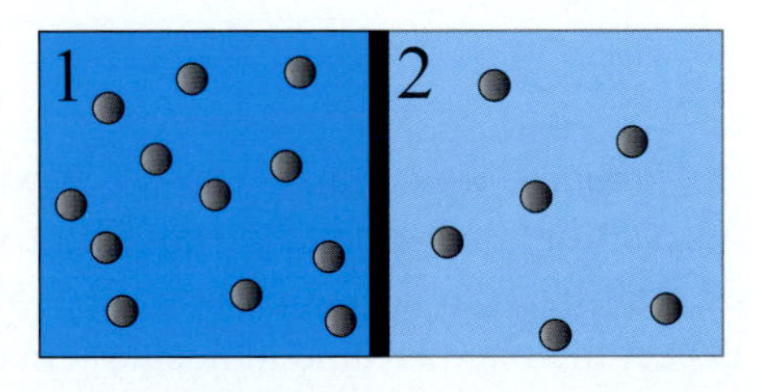

Figure 10.17: Piston separating out two gases.

Let us label the gas on the left of the cylinder with the subscript 1 and that on the right with the subscript 2. In equilibrium, the pressure exerted by both of the gases on the piston must be equal. Hence, from Equation 10.10 we have

$$\frac{N_1}{3V_1} \langle m_1 v_1^2 \rangle = P = \frac{N_2}{3V_2} \langle m_2 v_2^2 \rangle. \tag{10.17}$$

In equilibrium, the energy on both sides must also be the same so that

$$\langle \frac{1}{2}m_1 v_1^2 \rangle = \langle \frac{1}{2}m_2 v_2^2 \rangle \tag{10.18}$$

and consequently, combining this with Equation 10.17 we find that

$$\frac{N_1}{V_1} = \frac{N_2}{V_2}. \tag{10.19}$$

In other words, no matter what the gas is composed of, for example, be it nitrogen, helium and so on, **equal volumes of the various gases at the same pressure and average kinetic energy have the same number of atoms**. As shown below, same average kinetic energy means same temperature so in fact the statement implies that gases with the same pressure, temperature and volume have the same number of atoms, and it was this result which led Avogadro to postulate that 1 mol of *any* gas has the same number of atoms, given by **Avogadro's number** $N_A = 6.022 \times 10^{23}$.

Example 10.1: Speed of nitrogen

Nitrogen makes up about 80% of air and hence is a good element for obtaining an idea of how fast a typical air molecule moves at room temperature. Nitrogen has atomic number 7 (i.e. seven protons), and usually it has an equal number of neutrons so that its total atomic mass is around 14 u. One atomic mass unit is 1.66×10^{-27} kg and in air nitrogen is present as the molecule N_2, meaning that a nitrogen molecule has a mass of $14 \times 2 \times 1.66 \times 10^{-27}$ kg $= 4.64810^{-26}$ kg.
Using Equation 10.16 we then have

$$\begin{aligned} \frac{3}{2}k_B T &= \frac{1}{2}m\langle v \rangle^2 \\ \frac{3}{2} \times 1.38 \times 10^{-23}\,\frac{\text{J}}{\text{K}} \times 293\,\text{K} &= \frac{1}{2} \times 4.64810^{-26}\,\text{kg} \times \langle v \rangle^2 \end{aligned}$$

giving a speed of $\langle v \rangle = 511$ **m/s**.

Perhaps, we can now answer in what way and how fast a perfume molecule moves from the bottle to the nose. In Example 10.1, we find that the average speed of a nitrogen molecule is more than 500 m/s or 1,800 km/h. We may therefore expect that the perfume molecules emanating from the perfume bottle move with similar speeds (or perhaps a bit slower since they are heavier). If that were the case, then on opening a perfume bottle one should almost immediately smell the perfume throughout the entire room. But, of course (as a simple experiment immediately proves) this is not the case and so it is clear that something else must be going on.

What we should realize is that, given the large number of air molecules in relatively small volumes, the perfume molecule cannot go very far before colliding with other air molecules. So the question we have now is, how far and how fast can they go?

10.5 Mean free path

The average distance a molecule travels before colliding with another molecule is called the **mean free path** and often denoted with the letter l. Although typically the mean free path is much larger than the average distance between atoms, for a gas at a temperature of 300 K and one atmosphere pressure it is still very small: the approximate separation of the atoms is about 3.5×10^{-9} m, while the mean free path of the same gas molecule is about 10^{-7} m.

Let us assume that the gas molecules are hard spheres of radius r. For simplicity, we consider the motion of a single molecule and take the others as being at rest. Clearly, two molecules will collide if their centers come to within a distance of $2r$ of each other. It is easier to analyze this problem if we take the molecules at rest to be points (i.e. their radius is zero) while assigning a radius of $2r$ to the moving molecules as shown in Figure 10.18. While moving, the molecule with radius $2r$ sweeps out an area given by so-called cross-section $\sigma = \pi(2r)^2$ and in a time Δt it will sweep through a cylindrical volume of $\langle v \rangle \Delta t \sigma$ as illustrated in Figure 10.19. The number of gas molecules of radius $0r$ encountered during the time Δt is then $\tilde{n} \langle v \rangle \Delta t \sigma$ with $\tilde{n}$ the number density of the gas (i.e. $\tilde{n}$ is the number of molecules per unit volume). The mean free path l is now obtained as the distance traveled ($\langle v \rangle \Delta t$) divided by the number of molecules encountered ($\tilde{n} \langle v \rangle \Delta t \sigma$):

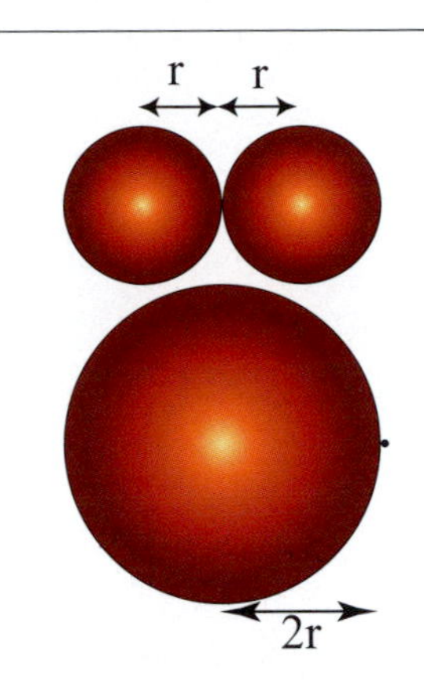

Fig. 10.18: Rather than looking at two molecules with radius r, it is easier to consider the collision of a molecule with radius $2r$ and a molecule with radius $0r$.

$$l = \frac{\langle v \rangle \Delta t}{\tilde{n} \langle v \rangle \Delta t \sigma} = \frac{1}{4\pi \tilde{n} r^2}. \qquad (10.20)$$

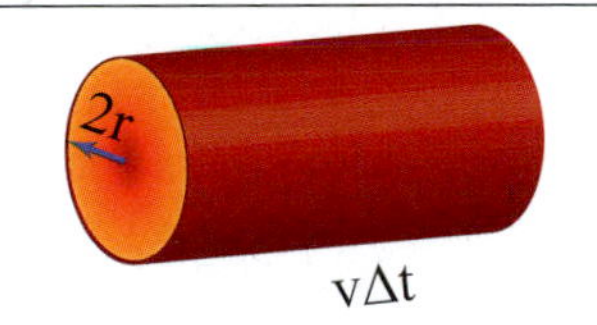

Fig. 10.19: The area swept out by a molecule of radius $2r$.

This calculation is of course only approximate as we considered only one molecule to be moving. If one makes a more complicated calculation where all the molecules move, one finds that one needs to divide by an additional factor $\sqrt{2}$ to obtain the mean free path

$$l = \frac{1}{4\sqrt{2}\pi \tilde{n} r^2}. \qquad (10.21)$$

For molecules of nitrogen gas at room temperature and pressure, the mean free path is given by

$$l = 2.25 \times 10^{-7} \text{ m}.$$

For comparison, note that the mean free path of an electron inside a conductor is given by

$$l \simeq 10^{-8} \quad \text{to} \quad 10^{-9} \text{ m} \qquad (10.22)$$

where the distance between atoms in a conductor is roughly 0.5×10^{-9} m so that the mean free path is about the distance of 10 atoms.

It is interesting to note that in interstellar space the mean free path of the H_2 atoms, which form the bulk of its matter, due to the extremely low density is a staggering

$$l = 10^{13} \text{ m},$$

or in other words just about the size of our solar system.

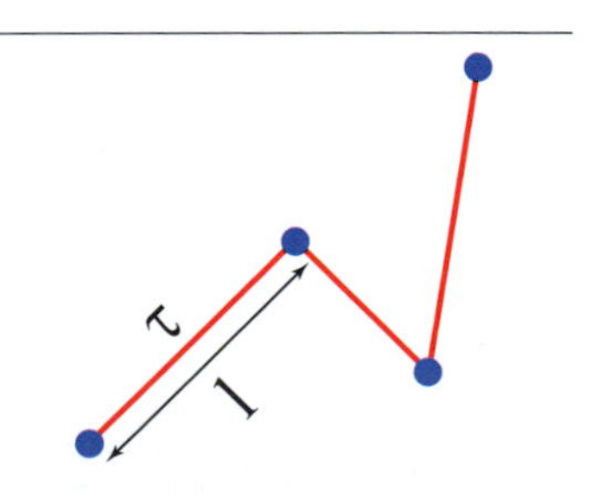

Fig. 10.20: l is the distance covered during a time interval τ.

The molecules travel the mean free path distance of l in time τ before colliding with another molecule. Hence, the average velocity of the molecule is given by

$$\langle v \rangle = \frac{l}{\tau}. \tag{10.23}$$

The number of collisions per second is given by $f = 1/\tau$ and hence

$$f = \frac{\langle v \rangle}{l}$$

yielding about 10^9 collisions per second for the atoms in air at room temperature.

Notice that the mean free path l is inversely proportional to the cross-section σ. Thus, l increases as the size of the molecules decreases. Also, since l is inversely proportional to the density, we see from the ideal gas law ($PV = nRT$) that the mean free path increases as the pressure decreases (at constant temperature and volume), or as the temperature rises (at constant pressure and volume).

The notion of a mean free path l is not limited to a gas with a single type of molecule but also valid when various types of molecules are mixed. If our single atom above, for example, were a perfume molecule and the molecules fixed in place nitrogen, then **diffusion** can be defined as the process of how a perfume molecule moves inside nitrogen.

In summary, diffusion takes place slowly because even though the molecules are moving very fast, they travel only a short distance before colliding with another molecule which changes its direction of motion.

10.6 Diffusion as a random walk

We will discuss the velocity of the perfume molecule, namely how far it can travel over a certain amount of time.

In the diffusion of perfume, there is an initial high concentration of molecules in the perfume bottle, and the molecules of perfume diffuse out into the room by randomly colliding — more than 10^9 times per second — with the molecules of air. The average distance traveled between collisions is the mean free path l and the time between collisions is τ. The high frequency of collision implies that the perfume molecule changes its direction of motion as many 10^9 times per second. The simplest way of describing the movement of the perfume molecule is therefore to consider its movement as a **random walk**.

Given that the time between collisions is τ, we model the path of the molecule by assuming that it takes a random step every τ seconds. The distance that the molecule covers in each step is the mean free path l with the *direction* of each step being *random*; we denote a single random step by

$$\mathbf{L} = \vec{n}l \tag{10.24}$$

where $\vec{n}$ is a unit vector in a random direction. As the length of a unit vector is 1, we know that $\vec{n}^2 = 1$. Hence

$$\langle \mathbf{L}^2 \rangle = l^2 \text{ since } \vec{n}^2 = 1. \tag{10.25}$$

The starting point of the molecule is the perfume bottle, which we take as the origin of the coordinate system, and suppose after time $M\tau$, during which it has taken M steps, its position is at coordinate $\mathbf{R}_M$. Since the molecule is equally likely to go in all directions, the value of its average position, after a great number of steps, should be its starting position. Thus, we have

$$\langle \mathbf{R}_M \rangle = 0 \,. \tag{10.26}$$

But if a perfume molecule on average remains at the location of the perfume bottle, how is it possible that we can smell it? The first thing to realize is that there are a great number of perfume molecules, each of which is doing its own random walk. The second thing to realize is that the actual position at any given time t will not be the average position. What we therefore would like to know is the average distance a molecule will travel away from the starting position. The position $\mathbf{R}_M$ results from the molecule taking a random step $\mathbf{L}$ from its last position $\mathbf{R}_{M-1}$, i.e.,

$$\mathbf{R}_M = \mathbf{R}_{M-1} + \mathbf{L} = \mathbf{R}_{M-1} + l\vec{n} \tag{10.27}$$

$$\mathbf{R}_0 = 0 \qquad \text{We take the perfume bottle as the origin.} \tag{10.28}$$

Since $\mathbf{R}_M$ is a coordinate, the distance from the origin is given by the root of its square. Let us therefore now first calculate the mean square of $\mathbf{R}_M$ which can be expressed as

$$\begin{aligned} \langle \mathbf{R}_M^2 \rangle &= \langle \mathbf{R}_{M-1}^2 \rangle + \langle \mathbf{L}^2 \rangle + 2\langle \mathbf{R}_{M-1} \cdot \mathbf{L} \rangle \\ &= \langle \mathbf{R}_{M-1}^2 \rangle + l^2. \end{aligned} \tag{10.29}$$

Since $\mathbf{L}$ can point in any direction we have

$$\langle \mathbf{R}_{M-1} \cdot \mathbf{L} \rangle = 0 \tag{10.30}$$

as illustrated in Figure 10.21. Another way to see that the average of the product of $\mathbf{R}_{M-1}$ and $\mathbf{L}$ equals 0 is by considering the cosine of the angle Θ between the vectors $\mathbf{L}$ and $\mathbf{R}_{M-1}$. While the vector $\mathbf{L}$ takes a random orientation, the cosine takes on a value between -1 and 1. Summing up many random values between -1 and 1 yields 0 as a result.

Since $\langle \mathbf{R}_0 \rangle = 0$ as follows from Equation 10.28, by starting from $\mathbf{R}_0$ and applying Equation 10.29 over and over again we find that

$$\langle \mathbf{R}_M^2 \rangle = l^2 M \tag{10.31}$$

where M is the number of steps taken.

The total time elapsed since the molecule started its path is $t = M\tau$ which is the time it took to reach position $\mathbf{R}_M = R(t)$; hence, using $M = t/\tau$ and Equation 10.31, we have

$$\langle \mathbf{R}^2(t) \rangle = \frac{l^2}{\tau} t. \tag{10.32}$$

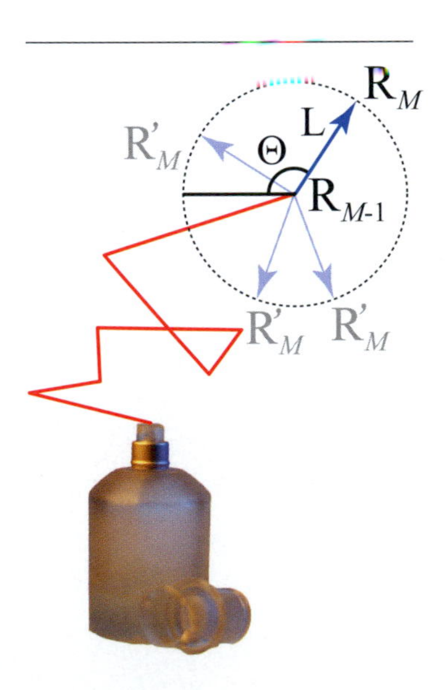

Fig. 10.21: Possible directions of L.

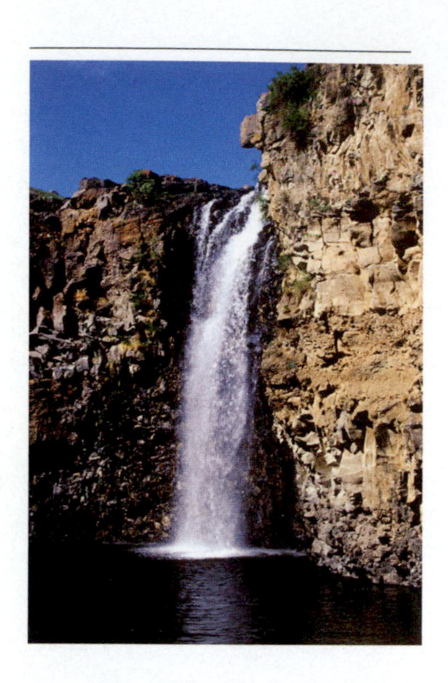

Fig. 10.22: The tendency toward equilibrium is beautifully illustrated by the flow of water.

Consequently, the distance traveled from the origin is

$$\sqrt{\langle \mathbf{R}^2(t)\rangle} \quad = \quad \sqrt{\frac{l^2}{\tau}t} \equiv \sqrt{Dt} \tag{10.33}$$

where the factor in front of the time is used to introduce the **diffusion constant** D defined as

$$D \quad = \quad \frac{l^2}{\tau} = l\langle v\rangle. \tag{10.34}$$

The right hand side is obtained with the help of the relationship $l/\tau = \langle v\rangle$ (the average velocity is the mean free path divided by the average time between collisions). Equation 10.34 is in three dimensions: what we would really like to have for the diffusion constant is the average distance in a certain direction, i.e. the diffusion constant should really be expressed in terms of $\mathbf{R}_x$. Similarly to the way we obtained Equation 10.12, we can obtain the result by multiplying with a factor 1/3 as it is reasonable to state that

$$\langle \vec{\mathbf{R}}^2(t)\rangle = \langle \mathbf{R}_x^2(t)\rangle + \langle \mathbf{R}_y^2(t)\rangle + \langle \mathbf{R}_z^2(t)\rangle = 3\langle \mathbf{R}_x^2(t)\rangle. \tag{10.35}$$

We then arrive at the correct diffusion constant

$$D = \frac{1}{3}l\langle v\rangle. \tag{10.36}$$

The important point to note from Equation 10.33 is that the root mean square distance $\sqrt{\langle R^2(t)\rangle}$ travelled by a molecule is proportional to $\sqrt{t}$, rather than t which would have been the case in the absence of collisions.

Diffusion as discussed so far explains why a perfume molecule moves throughout the room but it does not really explain why there is a tendency for the perfume molecules to be spread out evenly throughout the room. That is to say, we still do not really know why there is a tendency toward equilibrium (when just opening the perfume bottle we clearly have an nonequilibrium situation as all the perfume molecules are in or near the bottle).

The reason for this tendency toward equilibrium is in fact quite simple. If a single molecule is equally likely to go one way or another, then when we have an uneven distribution of molecules, there are many more molecules that go by chance in the direction of the smaller concentration than vice versa, thus leading to an even and uniform spreading.

10.7 Olfaction

We have seen how molecules can move from one place to another in a medium like air and we have thus gained an understanding on the transport mechanism of smell from a source like a perfume bottle to a destination like a nose. Of course, even if smell is transported, without a sense of smell to detect it we wouldn't smell anything. The technical term for the sense of smell is **olfaction** and the collection of physical components that lead to the perception of smell is called the olfactory system.

An olfactory system needs to contain some kind of a molecular detector and some processing of the information coming from the detectors for further analysis by the brain. Indeed, as shown in Figure 10.23 which depicts the olfactory system in human beings, detection and information processing are prominent features.

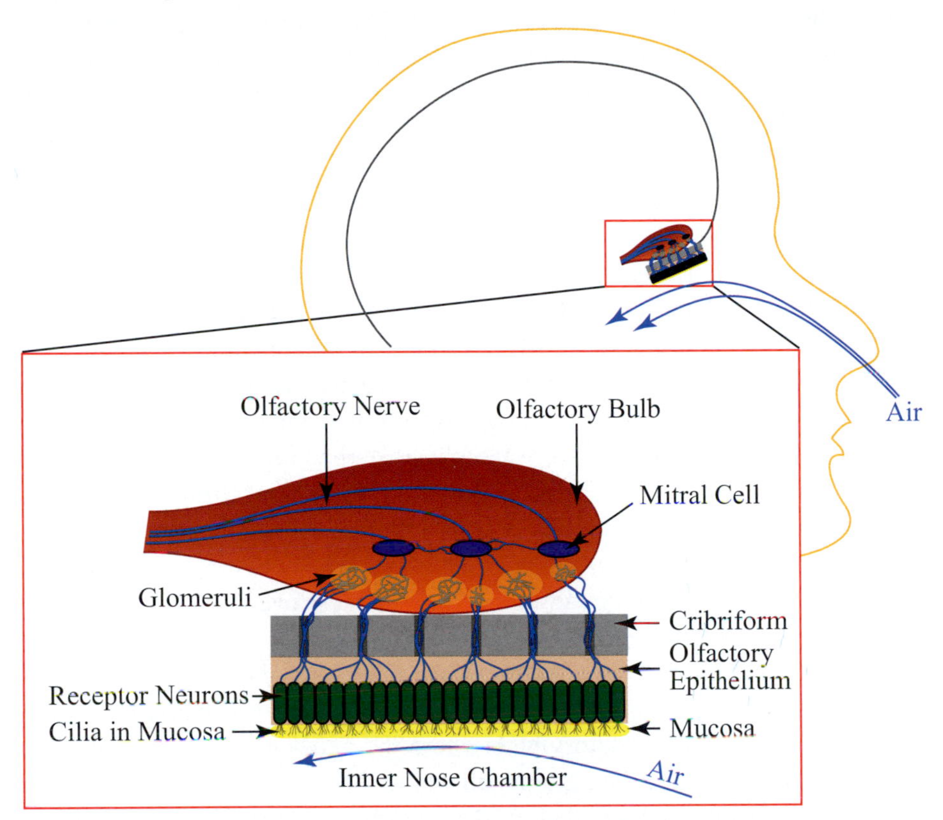

Figure 10.23: The olfactory system in human beings.

The olfactory system is located at the roof of the nasal cavities and connects olfactory receptor cells whose cilia do the actual detection of odors to the olfactory bulb which collects signals from the receptor cells and forwards the preprocessed information deeper into the brain. In a bit more detail, the process of detecting a smell roughly proceeds as follows: When an **odorant** enters the nose it comes into contact with the **aqueous mucus** into which the **cilia** of the olfactory receptor cells are immersed. It then diffuses through the mucus and reaches a cilium where it triggers a reaction that is passed on by the receptor cell. A human has about 10 million receptor cells, each having about 8 to 20 cilia. Dogs, in this respect, beat us paws down with about 100 million receptor cells, only to be left in the dust by rabbits, who have an astounding 1 billion olfactory receptor cells.

There are, in principle, several possible ways to to detect a range of different smells. One possibility is for each receptor cell to detect all the odorants that need to be smelled by an individual organism and that the quantity of the receptor cells is there solely to increase the sensitivity. Another possibility is that a receptor cell detects a family of related odorants and that receptors specialize in different

Fig. 10.24: Rabbits have the most sensitive noses among all mammals.

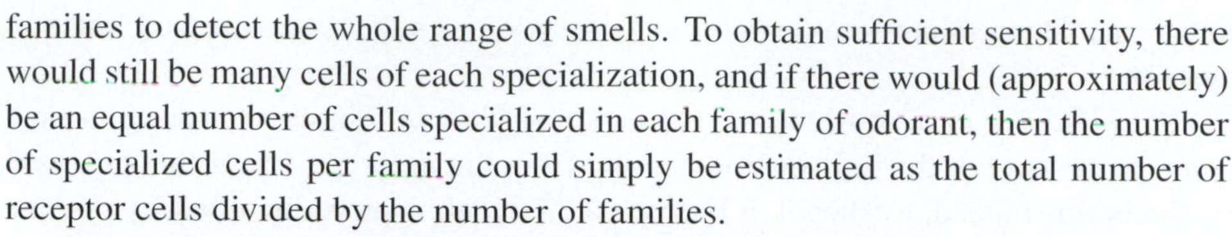

families to detect the whole range of smells. To obtain sufficient sensitivity, there would still be many cells of each specialization, and if there would (approximately) be an equal number of cells specialized in each family of odorant, then the number of specialized cells per family could simply be estimated as the total number of receptor cells divided by the number of families.

Finally, a receptor cell could be highly specialized and only detect one or a few odorants; there would then be a large number of specialized receptor cells and a smaller number of identical cells per specialization to obtain sufficient sensitivity. It turns out that it is the last possibility which nature has chosen and in total there are several hundred specializations.

Genetic analysis has revealed that there are about 1,300 receptor genes (each one of which codes for the detection of a specific chemical compound) in mice and about 350 in human beings. Since mammals have about 25,000 genes in total, this means that mice devote about 4% of their genes to the detection of odors, making it one of (if not) the largest gene family in mammals indicating the importance of olfaction to the evolutionary success for this class of organisms.

It is interesting to note that the number of receptor types in the human eye is only three — more than 100 times less! Would this imply that the eye is "less important"? Not necessarily so since the physical phenomena being detected are completely different. The nature of the electromagnetic spectrum is such that a detector can cover a range of colors so that with three types of detectors we can cover all the possible wavelengths (i.e. colors) between about 400 nm (purple) and 700 nm (red). Chemical compounds allow for much less overlap and consequently a larger number of different detectors is required.

The **receptor cells** are a type of neuron (the same kind of cell that makes up the brain) and their axons go through a plate of bone called the **cribriform** in bundles of 10 to 100 to enter the olfactory blub where the axons of the same specialization converge in a few so-called **glomeruli** where their signals are combined. The signals from different glomeruli then further converge on **mitral cells** which eventually pass them on via olfactory nerves to other parts of the brain like the olfactory cortex, hippocampus, amygdala, and hypothalamus for interpretation and further processing.

Besides the processes described above, there are many more. For example, mitral cells are interconnected and feedback from other areas of the brain is received. Not only can the smell of a tasty dish make one hungry but, vice versa, hunger can actually make the perception of a smell different! Since some of the areas which the olfactory signals are forwarded to are also involved in the retrieval and storage of memories, smells can have a strong effect on the recollection of memories. Often, when something particular is smelled, one instantly recalls the events that were originally associated with that scent. So we see that there is a neurological reason for that.

Fig. 10.25: The smell of coffee may remind us of a holiday in Italy — the process of triggering memories by smells is hardwired into our brains.

Since there are several hundred specializations for the receptor cells, how many different things can we smell? Is it equal to the number of specializations? Although that may seem reasonable at first, in fact the number is vastly greater because the smell of an object is often the combination of several odorants. Somewhat similar to how we can make a color by combining various quantities of three primary colors, smells can be made by a combination of various concentrations of odorants. With several hundred "basic"' odorants, the number of different smells is basically limitless but in practice it is believed that human beings can distinguish up to about 10,000 smells. What happens is that an odor produces a recognizable characteristic pattern of activity in the population of glomeruli similarly to how an image can be composed of many dots. In other words, odors represent topological maps (landscapes) of receptor activity patterns.

Fig. 10.26: Human beings smell pepper at concentrations of less than 1 in 1 trillion.

Now if the brain is to recognize receptor activity patterns as a kind of landscape, the arrangement of the detectors cannot be completely random. Yet, clearly, an odor as such does not have a spatial component like an image entering the eye does. It would therefore be quite inefficient to derive spatial information for the several thousand receptor cells of each type since all these detect the same chemical compound and therefore only add to the sensitivity of the detection. It would make more sense to derive the spatial information from the glomeruli since the axons of each type of receptor cell converge on only one or a few of these. Indeed, it turns out that glomeruli have a genetically predetermined location in the olfactory bulb and hence that activity pattern recognition of glomeruli can be hardwired into the brain, as this means that a given scent will lead to the same activity pattern in all the organisms of the same species.

The types of receptors are further grouped into four broad regions in the olfactory epithelium, but besides this grouping the receptor cells are arranged randomly. Not doing the "spatialization" at the receptor level saves the brain an enormous amount of processing with little downside as a multitude of receptors for each type is required to obtain any kind of sensitivity. After all, if the number of odor molecules in the air is low, only a few will be captured by the mucus on the olfactory epithelium and the location of this capture will be rather random.

Although among mammals human beings are not particularly good at "sniffing" things out (which is why we have sniffer police dogs but not sniffer police officers), our olfactory system can, nevertheless, be remarkably sensitive. For example, some people can recognize the smell of green bell pepper at concentrations of its odorant in air of less that 1 in 1 trillion!

Before we continue, it is interesting to note the somewhat special nature of the **olfactory epithelium**. It consists of three types of cells: the olfactory receptor neurons, support cells and basal cells, as well as the so-called Bowman's glands that secrete the mucus that covers the epithelium. The receptor neurons have a relatively short lifespan of about 40 days and are continuously replaced from the pool of basal cells, which function as stem cells that can specialize into the receptor cells. This process of continuous replacement of neurons from a pool of less differentiated basal cells is unique to the olfactory epithelium and occurs nowhere else in the mature nervous system.

Smell in nature

Olfaction is nature's main means of detecting environmental chemical compounds, the other one being taste which is much less sensitive. Olfaction can be found in many multicellular organisms including insects, fish and mammals.

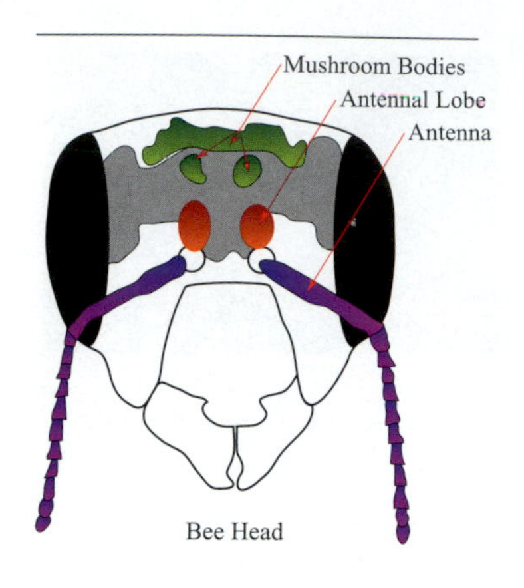

Fig. 10.27: The basic design of the olfactory system in insects is surprisingly similar to that in human beings.

Interestingly enough, the basic design of the olfactory system is very similar for all these organisms. In insects, the receptor cells are generally located in the antennae and are connected to the antennal lobe where the glomeruli are located and which functions similarly to the olfactory bulb in mammals. From the antennal lobe, signals proceed to the mushroom bodies which are parts of the insect brain associated with memory and learning. The parts of the olfactory system in a bee head are illustrated in Figure 10.27.

To many human beings, smell is mostly an aesthetic sense, but to many animals smell is a primary sense needed, for example, for the identification of food, predators and mates. Various volatile chemicals are easily produced by living organisms and the effects of those chemicals are much longer lasting than visual cues (which furthermore require a direct line of sight) or sound (which may need constant repetition). Indeed, scents are often the most efficient means of communication for an organism. In that sense, it is perhaps not surprising that social animals like bees and ants make extensive use of special types of scents called pheromones (see below for more details). The responses to such smells is innate and part of the organisms characteristic behavior.

Pheromones

Volatile chemicals can also be used to transmit messages, generally to other individuals of the same species. Such chemicals are called **pheromones** from the Greek words "pherein" (to transport) and "hormon" (to stimulate).

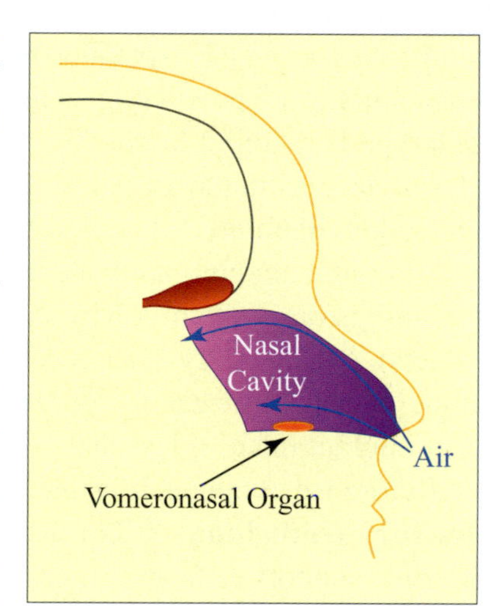

Figure 10.28: Location of the vomeronasal organ in human.

Pheromones play an important role in nature in diverse behaviors such as the marking of food trails, the signaling of alarm and attracting mates. For example, the first pheromone to be chemically characterized, called bombykol, is used by the female silkworm to attract mates. It was found in 1959 by the biochemist Adolf Butenandt (a 1939 Nobel laureate for work on sex hormones) who together with his collaborator Peter Karlson also coined the term "pheromone".

Fig. 10.29: Ants use pheromones extensively for communication.

Contrary to "regular" odors, in mammals and reptiles, pheromones are mostly detected by the so-called **vomeronasal organ** that is located between the nose and the mouth as shown in Figure 10.28. Indeed, some mammals like horses make a special facial expression called **Flehmen response** that aids supplying odors to the vomeronasal organ. Although this organ as such exists in human beings as well, it does not seem to have any function since it is not connected to the brain like in other mammals that make use of it. While the vomeronasal organ is in a sense an auxiliary olfactory system, it is important to note that in many ways it can be considered a separate sense of smell since its neurons do not connect to the same places in the brain as the olfactory nerves. Hence, it is probably more appropriate to consider the organ as a specialized chemical detector with a distinct function — in other words, as a different sense.

Evolutionary role of smell

Figure 10.30: Bees and flowers need each other.

Fig. 10.31: The olfactionary system of bees is similar to that of human beings.

Throughout their life spans, organisms face many challenges of which the needs to obtain food and to ward off predators are two key ones. As can be imagined, the ways to deal with these issues are many, and organisms evolve together with their abilities to cope with the challenges of survival and reproduction. For example, it is rather clear that plants and animals react very differently, but react they do. In a broad sense, interaction with the environment is the notion which is essential to consider, and one should therefore wonder what kinds of interaction there can be. One way to classify interactions is by their distance, while another is by their permanence. Table 10.3 shows some physical phenomena and their ranges and permanence in a general setting. However, it should be realized that such a setting usually depends strongly on whether the focus of the interaction is on detection or transmission. For example, in order for a rabbit to see an approaching eagle, it must be in the line of sight so that the rabbit can detect light being reflected from the eagle. As soon as the eagle is behind a tree, the rabbit will not be able to see it. With the exception of special cases like fluorescence, electromagnetic radiation, of which light is one form, does not "linger" and hence has very little permanence. The only type of phenomenon in Table 10.3 that generally has some permanence is the diffusion of chemical compounds (be that diffusion in water or air). It is therefore not surprising that nature has evolved a mechanism to detect volatile chemical compounds.

Indeed, the detection of chemical compounds is one of the most ancient senses and can be found in some single-cell organisms like bacteria. Having the ability to detect scents has a multitude of evolutionary advantages. For example, food can be localized from a *distance* which makes it much easier to find; likewise, danger such as the smoke from a fire can again be detected from a distance, making it easier to avoid.

Fig. 10.32: The permanence of odors is what allows search dogs to find a trail.

Of course, once one has a detection mechanism for scents, it makes much sense to not only use this passively to perceive the environment but to also use it actively to interact with the environment. Organisms that can communicate, for example, by leaving a scent that indicates readiness to mate will be at an evolutionary advantage versus organisms that have no means to communicate such matters at all. Naturally, not all organisms will evolve to use the same sense to communicate a certain type of message like readiness to mate. However, considering the simplicity with which basic scent detection can be implemented (being possible within even a single-cell organism), it is not surprising that scent detection and transmission is almost universally employed to some degree in animals.

Phenomena	Range	Permanence	Sense
Electromagnetic radiation	Long–very long	Short	Sight
Sound	Medium	Short	Hearing
Pressure differences	Very short–short	Short	Touch
Materials — chemical compounds	Medium	Medium	Taste & smell

Table 10.3: Physical phenomena, their ranges and permanence.

Playing such a central role in an organism's potential for survival, a successful design is likely evolutionarily very stable. That is to say, the basic design of a well-working olfactory system can to some degree be expected to remain intact over long periods of time. Indeed, it has been found that the design of the olfactory system of a fruit fly and that of mammals is remarkably similar. Since fruit flies and mammals diverged about 500 million years ago (which in itself is only about 500 million years after the emergence of multicellular life), this implies that our olfactory design is very old.

10.8 The answer

We are now in a position to answer the question as to why we can smell perfume. As it turns out, this answer is rather complex, involving quite a number of distinct natural phenomena. First, there is the property that volatile molecules are rather abundant in the environment. Then there is the nature of air which makes it possible for these molecules to spread out in the environment even when the air itself is not moving. The combination of the abundance of volatile molecules with their spreading out in the environment made it feasible for organisms to gain an evolutionary advantage by possessing detectors for these volatile molecules, which thus by virtue of being detectable become odorants. The evolution of a brain makes it possible to process the information from the odor receptors into the higher level perception we call smell. In other words: **We can smell perfume due to the delicate interplay of natural laws that govern the formation and evaporation of molecules, their diffusion in air, the evolution of detection mechanisms and the evolution of information processing networks**.

10.9 Exercises

Some questions on a hypothetical air molecule

Originally, the gram was defined as the mass of 1 cm^3 of water at 0°C. Although this is not exactly the definition used nowadays, it is a good way to remember that the density of water is approximately $1 g/cm^3$. For simplicity, many reasonably good calculations can be done by assuming that air consists of single hypothetical type of "air molecules" whose properties are 80% those of nitrogen and 20% those of oxygen. Given that the density of air is about 1/1000 that of water, answer the following questions:

1. Estimate the volume one single air molecule occupies in the lecture room.
2. Deduce an estimate for radius of an air molecule in the lecture room.
3. Estimate the average distance between air molecules in the lecture room.
4. Compute the mean free path of the air molecules in the lecture room.
5. Use the Avogadro's number and the definition of the mole to determine the mass of the oxygen molecule.
6. Repeat the previous question for the nitrogen molecule to estimate the mean mass of an air molecule.
7. Estimate the average velocity of an air molecule in a lecture room.
8. Using the results of Exercises 4 and 7, determine the mean time between collisions of air molecules.
9. Estimate the diffusion constant of the air molecules.

Some questions on olfaction

10. Name two key advantages of olfaction with regards to vision.
11. Is it possible to smell iron?
12. Are all odorants organic molecules?
13. We have discussed odors that spread through air. Could they spread through water? What does this mean for fish?
14. How is sensitivity to odorants achieved? In other words, what is so different in the noses of rabbits when compared to humans? After all, we do know that the design of the noses of all mammals is nearly identical.

11: Sound

Q - Why Can We Hear Music?

Chapter Map

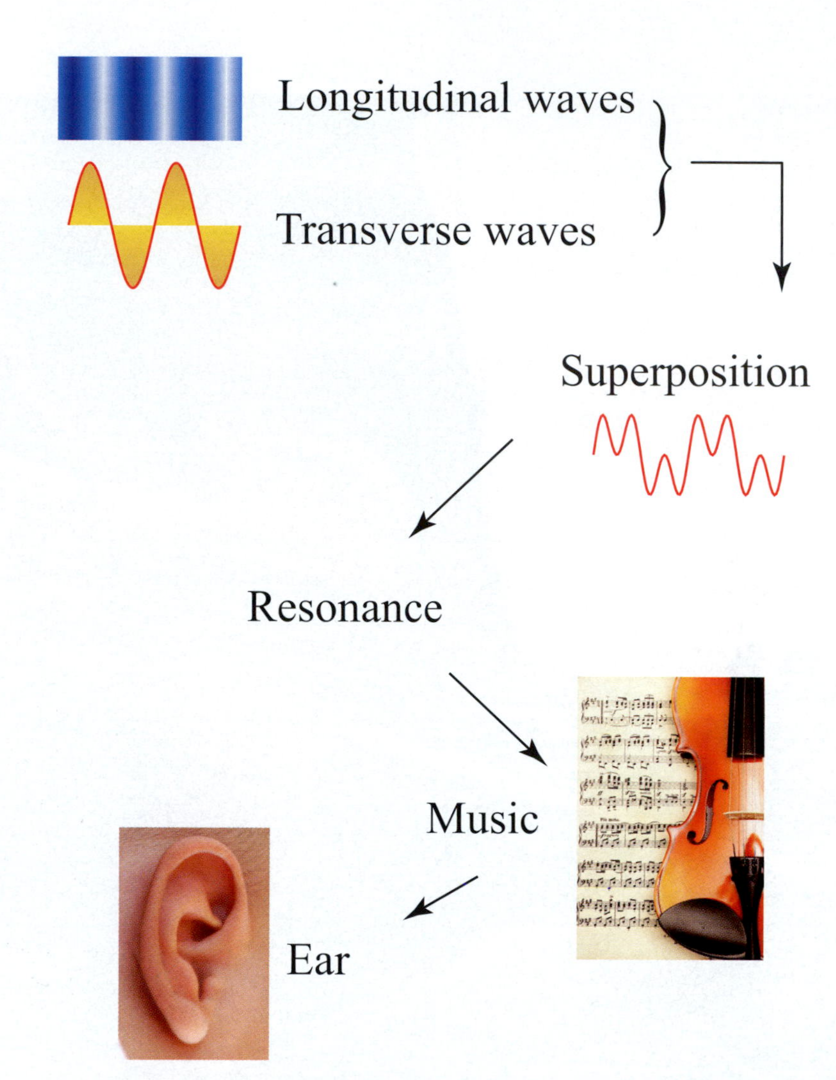

Longitudinal waves
Transverse waves
Superposition
Resonance
Music
Ear

11.1 The question

Figure 11.1: The koto is a traditional Japanese string instrument.

Music is one of the oldest cultural activities of mankind and we ask the question: **Why can we hear music?** We all know that music involves sound and that sounds can be of great variety. Besides the sounds in music, sound plays an important, if not essential, role in our lives. For example, one of the key methods of a young child to alert or communicate with its parents is through crying. Indeed, we assume that a quiet baby is a content baby. And, of course, sound is used in speech, something indispensable for human civilization and the first large-scale means of highly organized communication between human beings.

Therefore, to understand why we can hear music, we need to study sound: how it is produced, transmitted and perceived.

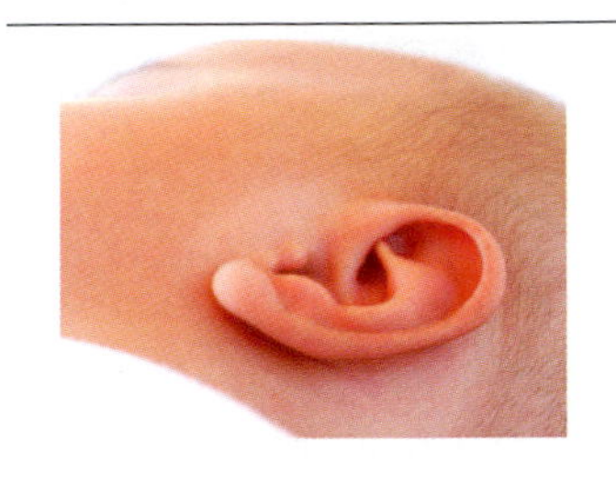

Fig. 11.2: A baby's ear.

11.2 Observations regarding sound

Let us start by enumerating a few aspects of sound that can be directly observed in daily life.

1. Sound has a **pitch** (a sound can be high or low).
2. Sound has **loudness** (sound is loud or soft).
3. If music is played in a room, one can move around in the room and continue to hear the music. Hence sound spreads throughout the room. As one moves away from the source of sound, the loudness of sound diminishes.
4. When a sound source is far away, there is a noticeable time lag between seeing an action and hearing the associated sound (e.g. the sound of a piling machine a few hundred meters away trails the visual impact significantly). Sound can thus be deduced to have a **speed**.
5. Even at what appears to be the same pitch and loudness, sounds can have different characteristics. For example, a guitar and a flute "sound" different.
6. Sound is linked to the presence of a medium, in particular, that of air; if we evacuate the air in a container surrounding a sound source, the sound disappears. But we know that one can hear sound in solids as well, as can

be seen by putting one's ear on a table that is being hit; one can hear sounds inside water as well, as one hears sounds from the surface even if one is under water. Hence sound needs a **medium** to propagate in, be it air, water or a solid, and is some kind of a "disturbance", an oscillation in or of this medium.

Two sources of sound

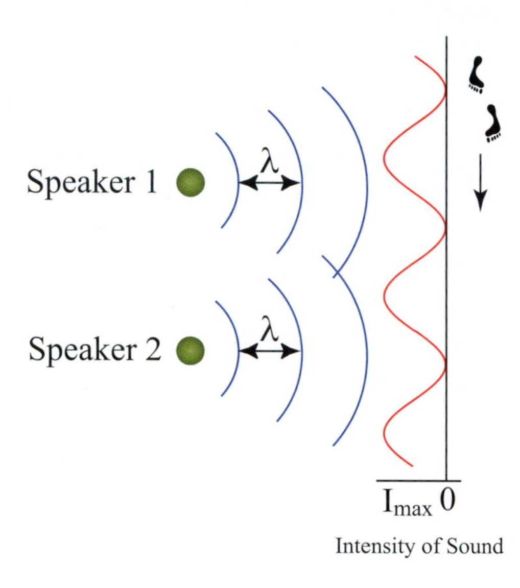

Figure 11.3: When following the footsteps the sound waxes and wanes.

A particularly important phenomenon occurs when there are two sources of sound. Consider, for example, the situation depicted in Figure 11.3 where the sound from two speakers reaches our ear. The sound we perceive is the combination of sound from both the sources. If the sound sources are identical or similar, the combined sound intensity can vary from zero to four times the volume of a single speaker, depending on the position of listener, in a phenomenon called **interference** (for more details, see later in this chapter).

If the two speakers of Figure 11.3 emit sound with slightly different wavelengths, one hears a waxing and waning of sound without even moving, called "beats".

Although sound is a wave, it is different from light since, for example, sound needs a medium and light does not.

11.3 Sound propagation in air

The medium for most of the sounds we hear in daily life is air. We consider some of air's basic physical properties to find out how sound propagates in it.

Sound is created by disturbances propagating in a medium. Sound is a form of wave motion that is the result of some kind of "oscillations" of air. In a room with perfectly "still" air, we can hold a conversation. Hence, sound does not require flow of air or wind and hence we can deduce that sound is not the result of a net movement (or transport) of air from one point to another. However, this does not mean that moving air *cannot* produce sound. It is well known that a strong breeze can create a whistling sound, and an explanation will be given near the end of this chapter.

Sound is a form of propagation of energy and momentum. But what exactly is the physical mechanism involved in the propagation of sound energy? Would it be similar to the kind of energy transport one has when throwing a stone? It is clear that this cannot be the case as a stone does not spread out like a wave. Also, if one throws a stone, energy propagates from one point to another by the motion of the stone itself; but we already know that this is not right in the case of sound since there is no net transport of air. Would there be another form of energy transport that matches the observations regarding sound? There is. This can be seen when one considers a piece of wood floating on the sea oscillating up and down as the incoming water wave goes by. Even though the incoming water wave transmits energy to the floating object — since it sets a stationary piece of wood into motion — this energy transmission does not require any net displacement of the sea water.

Fig. 11.4: Pressure inside a balloon.

An analogy to the water wave moving the log up and down is a speaker. The oscillations of the speaker's membranes compress and expand the air near the speaker; this compression and expansion then propagates through air and spreads throughout the room. To get a better notion what it means to compress and expand air, we need to understand the concept of pressure of a fixed volume of air at a given temperature — since compression and expansion generally imply a change in pressure.

Consider, for example, air that is trapped inside a balloon; if one reduces the volume of the balloon, there is a force opposing this change. We intuitively know that the force originates in the fact that the air in the balloon "resists" a reduction in its volume. More precisely, the air inside the balloon exerts a force per unit area on the (inner) surface of the balloon; the unit for the pressure P is the pascal: 1 Pa = 1 N/m^2.

For air at temperature T, by means of careful experimentation, one can find that the product of pressure and volume, i.e. PV, is constant and proportional to both the number of gas molecules, denoted by N, and the temperature T. The proportionality to the amount of gas should not come as a surprise as we can easily surmise that if we put more gas in a given volume the pressure should go up. Similarly we can easily verify that a hot gas occupies a greater volume at a constant pressure by, for example, placing a balloon in a refrigerator — the balloon will shrink. The proportionality constant between PV and NT is the Boltzmann's constant, represented by the k_B. Thus we obtain the following ideal gas law, introduced in Equation 10.1:

*The **Boltzmann constant** is given by $k_B = 1.38 \times 10^{-23}$ J/K.*

$$P = N k_B \frac{T}{V}, \qquad \textbf{Ideal Gas Law} \qquad (11.1)$$

where k_B, the Boltzmann constant, is given in Equation 10.16. The temperature T in the gas equation is not measured in the familiar Celsius scale but in the Kelvin scale, denoted by K. The conversion between the two scales is given by

$T_{\text{Kelvin}} = T_{\text{Celsius}} + 273.15.$

Kelvin: Measure of temperature which starts at absolute zero.

The theoretically coldest temperature is 0 K, also referred to as absolute zero.

11.4 Longitudinal waves

So we see that sound involves pressure variations in air. To understand this mechanism a bit better, consider a flute emitting a sound at some fixed pitch and loudness. The action of the flute can be replicated by a piston that produces a sound in the cylinder. By oscillating back and forth about its equilibrium position, as depicted in Figure 11.5, the piston creates a compression and rarefaction wave. We associate a

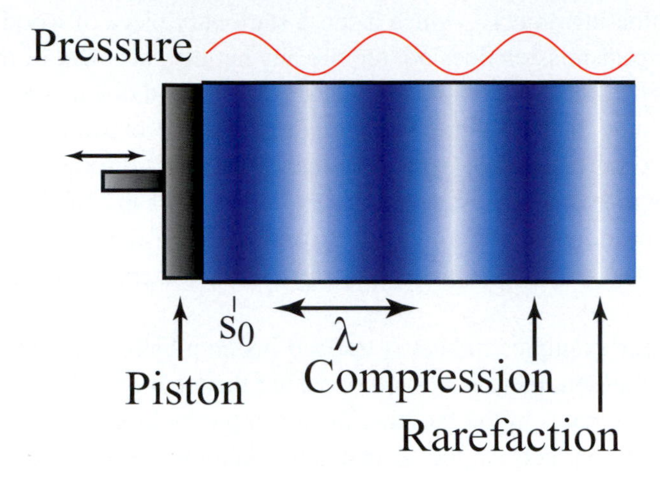

Figure 11.5: Piston generating a sound wave.

sound's pitch with how many times per second the piston oscillates back and forth, denoted by the **frequency** f with unit of hertz $= s^{-1}$ (Hz). A frequency of, say, 100 Hz means that the piston goes back and forth about its equilibrium position 100 times in 1 s. The loudness of sound is related to how far the piston departs from its equilibrium position, and the maximum distance from the equilibrium is called the **amplitude** of the oscillations.

Human hearing has a range of 20–20,000 Hz.

The simplest and probably most well-known periodic functions are the sine and cosine functions. Hence, it is natural to attempt to describe the motion of the piston and sound with (one of) these functions. Indeed, the position of the piston about its equilibrium position can be written as

$$s(t) = s_0 \cos(2\pi f t) \tag{11.2}$$

where s_0 is the amplitude (maximum displacement) of the piston's oscillations.

Suppose the ambient (constant) pressure of the air in the piston's cylinder is P_0. When the piston moves from left to right, the air next to the piston is compressed and the pressure changes to $P(t)$. This causes a *change* (increase or decrease) in the pressure of the compressed air, which we denote by

$$\Delta P(t) = P(t) - P_0.$$

Similarly, when the piston moves from right to left, the air next to the piston is expanded, resulting in a drop in the air pressure. As the piston oscillates, the compression and rarefaction of air creates a sequence of compressed and rarefied air that propagates down along the cylinder of the piston as shown in Figure 11.5, forming a sound wave. Note the important fact that there is no net transport of air: as the piston moves back and forth, the air molecules also move back and forth about their equilibrium positions.

Sound is a longitudinal wave!

If the value of s_0 is large enough we hear a sound. Hence, sound is a *pressure wave* created by the periodic (oscillatory) motion of air. Furthermore, air molecules oscillate *along* the direction of motion of the wave produced. Such waves are called **longitudinal waves**.

The change in pressure at time t follows directly from Equation 11.2 and, at some fixed point in the cylinder, as illustrated in Figure 11.6 is given by

$$\Delta P(t) = P_m \sin(2\pi f(t - t_0)). \tag{11.3}$$

where the P_m is the maximum change in pressure which is in general much less than the ambient air pressure P_0, and t_0 the starting time.

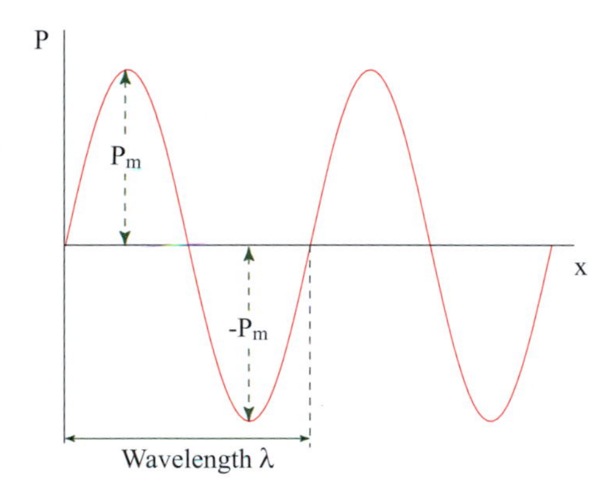

Figure 11.6: Amplitude and wavelength of a pressure wave.

The sensation of sound is caused by the pressure changes at the eardrum. One might think that the periodic nature of the pressure changes given in Equation 11.3 would result in the waxing and waning of the perceived sound. However, this is not the case because any frequency higher than 20 Hz is perceived by human beings as a continuous sound characterized by pitch f. The higher the frequency, the more shrill is the sensation of sound and hence the higher the pitch; similarly the larger P_m, the louder the volume of sound.

11.5 Transverse waves

Music is made by musical instruments, and a common type of instrument is a string instrument, for example, the violin or the guitar. How do these instruments create sound? In order to answer that question we need to know how the strings vibrate. Although the string in a string instrument is tied at both ends, for simplicity we will first study the more general case of a taut string that is free at both ends.

Wave motion in a string differs from a pressure wave in one significant manner: the oscillations of the string are perpendicular to the direction in which its wave propagates, unlike the case for sound where the medium (air molecules) oscillates in the same direction as the pressure wave. Waves created in an oscillating string

Fig. 11.8: The guitar is a typical string instrument.

are called **transverse waves** in contrast to longitudinal pressure waves. Note that the sound wave *produced* by the transverse vibrations of a string is, nevertheless, a longitudinal wave.

A transverse wave, in general, is one in which the oscillations of the medium are perpendicular to the motion of the wave. Suppose the string lies in the x-direction and its oscillation is set up in the y-direction. The position of the string is given by (x, y), where x is the position of the string along the x-direction, and the y coordinate specifies how far the string is stretched from its equilibrium position (i.e. the position when the string is not vibrating). Figure 11.7 shows such a transverse wave.

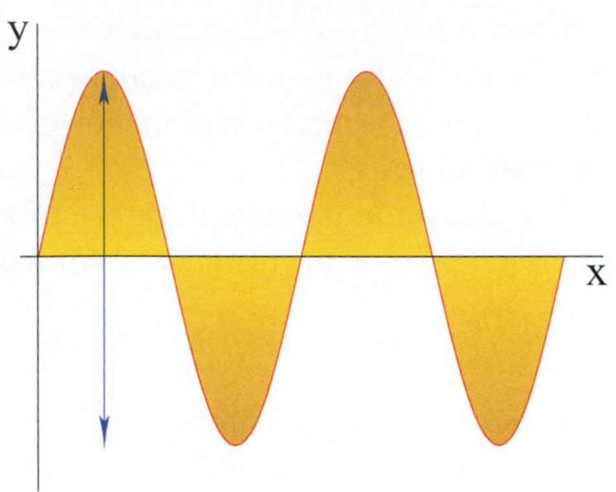

Figure 11.7: Transverse wave along the x-direction. The string vibrates in the y-direction as indicated by the blue arrow.

As illustrated in Figure 11.7, a simple wave is a periodic function of x, and hence, at $t = 0$ it can be described by

$$y(t = 0, x) = A \sin(kx) \tag{11.4}$$

where we have assumed that the amplitude is zero at the origin, that is $y(0, x) = 0$ for $x = 0$. The constant k can be viewed as a "spatial frequency" since it determines how fast the function oscillates when increasing x.

11.6 Sound velocity

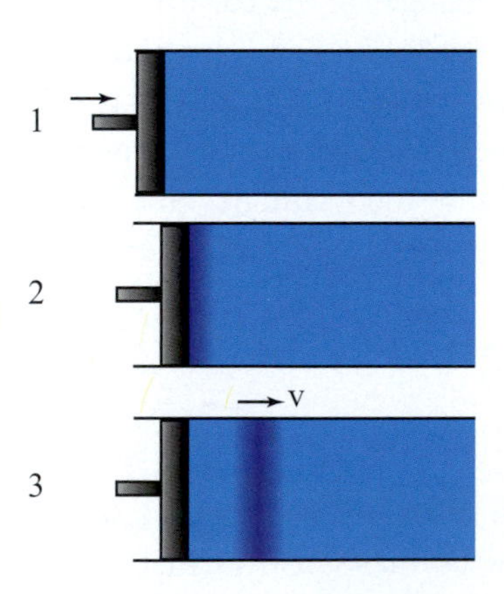

Figure 11.9: Propagating pulse.

Consider again the piston in Figure 11.5. What happens with the air that is compressed in the forward stroke? As illustrated in Figure 11.9, the region of compression travels forward with a speed v.

The velocity of propagation of the wave depends on the properties of the medium. In particular, the speed is related to how strongly a gas resists a change in volume, something that is measured with the so-called **bulk modulus** B. B is a measure of the incompressibility of a medium and is the inverse of compressibility . It is easy to imagine that the more compressible a gas is, the smaller the velocity of the pulse will be as the movement of the piston will be less resisted by the gas. Conversely, the more rigid is the medium, the higher the velocity of sound because, not being able to compress the volume medium, the energy propagating in the wave has to rapidly move to the neighboring volume elements. Hence, we expect v to be proportional to B. On the other hand, we can also expect the velocity of sound to be inversely proportional to the density ρ of the medium since the heavier the medium the greater the inertial resistance to motion. By careful measurement one can then find that

$$v = \sqrt{\frac{B}{\rho}}. \quad (11.5)$$

A string is a one-dimensional object for which the bulk modulus B is proportional to the string tension τ; the velocity of a transverse wave in a string is given by

$$v \propto \sqrt{\frac{\tau}{A\rho}}$$

where A is the cross-sectional area of the string and ρ is string's mass per unit length. For an ideal gas, when there is no inflow or outflow of energy (called an adiabatic change), it can be shown that

$$B = \gamma P, \quad (11.6)$$

where γ is the adiabatic index. Combining Equations 11.5 and 11.6 with Equation 11.1 we then obtain the useful equation

$$v = \sqrt{\frac{\gamma RT}{M}}, \quad (11.7)$$

where M is the mass of 1 mol of gas molecules. See Table 11.1 for the speed of sound in various media.

Example 11.1: Velocity of sound in neon gas

What is the velocity of sound in neon gas at 0°C?
The molecular mass of neon is $M = 20.18$ g/mol For a monoatomic gas $\gamma = 1.6$. Hence

$$v = \sqrt{\frac{\gamma RT}{M}} \quad (11.8)$$

$$= \sqrt{\frac{1.6 \times 8.314\ \mathrm{J\,K^{-1}\,mol^{-1}} \times 273\ \mathrm{K}}{20.18\ \mathrm{g\,mol^{-1}}}} = 432\ \mathrm{m/s}. \quad (11.9)$$

Now that we know the velocity of the sound wave, let us see if we can use this knowledge to find a generic expression for the pressure in the cylinder. If we were to move with the pressure wave at velocity v — say, if we were to move along the compressed portion of air — then clearly the air pressure for us would remain constant. Mathematically speaking, this means that in Equation 11.3, ΔP should

Medium	Speed $v(m/s)$
Air	331
Oxygen	316
Hydrogen	1284
Water	1402
Aluminum	5100
Iron	5130

Table 11.1: Speed of sound in various media.

not change if t_0 were replaced by $-x/v$. Hence, at points along the cylinder, we expect the generalization of Equation 11.3 to be given by

$$\begin{aligned} \Delta P(t,x) &= P_m \sin\left(2\pi f\left(t - \frac{x}{v}\right)\right) \\ &\equiv P_m \sin(\omega t - kx) \end{aligned} \tag{11.10}$$

where the angular frequency ω and the wave number k are given by

$$\omega = 2\pi f \tag{11.11}$$

$$k = \frac{2\pi f}{v}. \tag{11.12}$$

Figure 11.10 shows the variation of pressure in space.

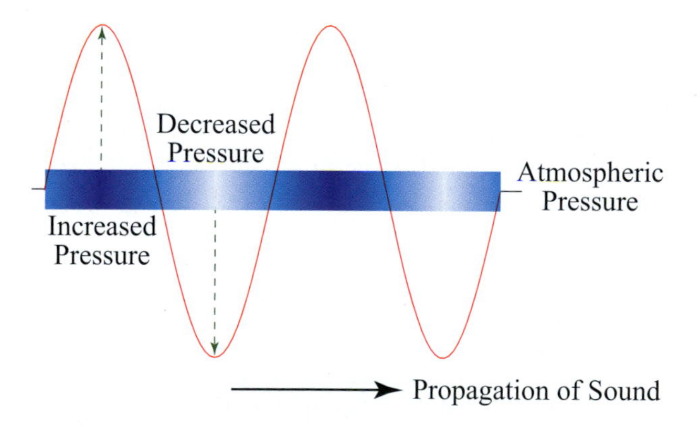

Figure 11.10: Longitudinal pressure wave.

Recall both $P(t,x)$ and P_m have the units of pressure, which is Nm^{-2}. Since the argument of a sine function is unitless, the parameter k must have a unit equal to inverse of length; this then makes kx unitless.

What is the physical significance of the quantity k? Note that since the sine function is periodic, we have

$$\sin(kx) = \sin(kx + 2\pi) = \sin[k(x + \frac{2\pi}{k})].$$

Hence when x is increased by an amount $\frac{2\pi}{k}$ the pattern of the wave repeats itself, as can be seen in Figure 11.10. Consequently we have wavelength λ given by

$$\lambda = \frac{2\pi}{k} \tag{11.13}$$

which means that k is something like a "spatial" frequency. Hence, from Equations 11.12 and 11.13, we have the important result that

$$\lambda f = v. \tag{11.14}$$

In other words, for a wave with wavelength λ arriving with a speed v at a given location, the medium oscillates with a frequency v/λ.

11.7 Sound energy and power

We have reasoned that sound is the result of pressure waves propagating in a medium. More precisely, when we hear a sound, pressure waves transmit (sound) energy to the eardrums, and it is this energy that initiates the biochemical processes inside the ear that result in the sensation of sound. So how does a pressure wave propagate this energy through air?

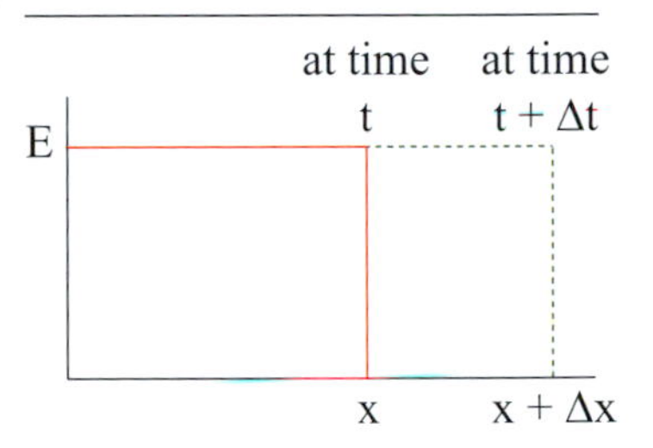

Fig. 11.11: Propagation of energy.

Figure 11.11 is a plot of energy per unit length present at different points of cylinder x. The red line shows that by time t the wave has reached up to the position x — with the wavefront of the propagating wave shown by the vertical line at x. The dotted green line shows how that the wavefront will propagate a distance Δx in time Δt, with the wavefront reaching the position $x + \Delta x$ at time $t + \Delta t$. Wave propagation from position x at time t to a new position $x + \Delta x$ at time $t + \Delta t$ is, in fact, the propagation of energy.

The piston has to constantly supply energy to the air to keep the pressure wave intact. Recall that energy is not a thing, but is an abstract and conserved quantity that, in the above process, is converted from the kinetic energy of the piston moving back and forth to sound energy.

If we consider a small volume of air in front of the piston, it will be pushed by the piston and thus gain kinetic energy. The position of this small volume of air is then given by the same equation as that for the piston, namely, $s(t) = s_0 \cos(2\pi f t)$ (Equation 11.2). More generally, by the same argument with which we obtained Equation 11.10 above, we can express the displacement of the air molecules at time t and at position x in the cylinder as

$$s(x, t) = s_0 \cos(\omega t - kx). \tag{11.15}$$

In order to obtain the kinetic energy, we need the velocity of the air (not the velocity of the wave!) which is the time derivative of Equation 11.15 given by

$$v(x, t) = \frac{\partial}{\partial t}[s_0 \cos(\omega t - kx)] = -s_0\omega \sin(\omega t - kx). \tag{11.16}$$

The kinetic energy of a small slice of air with width dx and area A such that its mass is given by $\rho A dx$ is then at $t = 0$

$$T_{\text{slice}} = \frac{1}{2}mv^2 = \frac{1}{2}\rho A dx (s_0\omega)^2 \sin^2(kx). \tag{11.17}$$

The kinetic energy of an entire wavelength is given by summing up all the slices in one wavelength; more precisely, one integrates T_{slice} over one wavelength to obtain

$$\begin{aligned} T_\lambda &= \sum_{\text{slices in } \lambda} T_{\text{slice}} = \frac{1}{2}\rho A (s_0\omega)^2 \int_0^\lambda \sin^2(kx)dx \\ &= \frac{1}{2}\rho A (s_0\omega)^2 \left[\frac{x}{2} - \frac{\sin(2kx)}{4k}\right]_0^\lambda \\ &= \frac{1}{2}\rho A (s_0\omega)^2 \left(\frac{\lambda}{2}\right). \end{aligned} \tag{11.18}$$

The oscillation of each segment of the sound wave is simple harmonic oscillation, discussed in Section 3.6. For simple harmonic motion, the average potential and kinetic energy, for one oscillation, are each equal to half the energy, as given in Equation 3.34. Similarly, for wave motion, the total potential energy is equal to the total kinetic energy over one wavelength; we thus obtain the total energy for one wavelength as

$$E_\lambda = 2T_\lambda = \frac{1}{2}\rho A (s_0\omega)^2 \lambda. \tag{11.19}$$

Here we use: $v = \lambda f$ (combining Eqs. 11.12 and 11.13) and $f = 1/(\Delta T)$ (definition of frequency) to obtain $v = \lambda/(\Delta T)$.

The power needed to keep the wave propagating is now

$$\begin{aligned} P &= \frac{\text{Energy expended in one wavelength}}{\text{Duration of one wavelength}} \\ &= \frac{E_\lambda}{\Delta T} = \frac{1}{2}\rho A (s_0\omega)^2 \frac{\lambda}{\Delta T} \\ &= \frac{1}{2}\rho A v (s_0\omega)^2. \end{aligned} \tag{11.20}$$

We have derived the important result that the amount of power needed to drive the wave is proportional to the velocity of propagation v.

Noteworthy 11.1: Whistling sound from a strong breeze

Although sound does not need the net transport of the underlying medium, it is a common experience that moving air in fact can create sound. So how do we understand this, in particular how does a strong breeze create a whistling sound?

The behavior resulting from the net movement of air is, strictly speaking, not described by waves and more properly is explained by the laws of fluid mechanics, as discussed in Chapter 6.

When a strong breeze (air moving at a high velocity) blows against a sharp object, say, a wire hanging from a building, the air has to flow around the wire; since the air has to pick up velocity it loses pressure from the conservation of energy (this fact goes under the name of Bernoulli's principle in fluid mechanics). Furthermore in going around a sharp object the velocity of the air has to change suddenly, creating net angular momentum in the air flowing around the obstacle. Conservation of angular momentum requires that the neighboring air pick up the opposite angular momentum, and this gives rise to a vortex. The net effect of the air going around a sharp obstacle is a series of vortices that seem to be emanating from the obstacle.

Since the air is blowing at a high velocity the vortices are closely spaced and this leads to a high frequency change of pressure as one vortex is followed by the next vortex. It is this high frequency change of pressure that is perceived by the human ear as a whistling sound.

For air flowing at velocity v the number of vortices created is proportional to v^4. The creation of vortices creates a drag on airplanes and becomes a major source of energy loss at high velocities; the design of airplanes, in particular their streamlined shape, is largely dictated by the need to minimize the formation of vortices and hence reduce drag.

11.8 Superposition

The superposition principle underlies combined sounds.

A characteristic feature displayed by all wave and wavelike phenomena is interference of two sources, as shown in Figure 11.3. Of course, if one can see what happens when combining two sound sources, one can also see what happens when combining three or more sound sources. It turns out that the fundamental property of amplification and weakening is not dependent on the number of wave sources, and hence one is guided to a general principle called **superposition** of which interference is an example. We define superposition more precisely below.

Waves, unlike particles, are spread over space. One may ask the basic question: what happens when two different waves meet in some region of space? What will be the resultant wave at all points of this region? Indeed, waves, and sound waves in particular, have the rather remarkable property that if there are two or more waves propagating in the same region of space, there are simple rules on how to add them, and the resultant wave is a single wave with properties derived from the constituent waves. There is no analogous property for objects like billiard balls; if two moving billiard balls meet at some point of space, they either scatter (bounce) off each other or else they cease to have their identities and coalesce according to the microscopic properties composing them. There are no rules for (classical) particles (billiard balls) analogous to the superposition of waves.

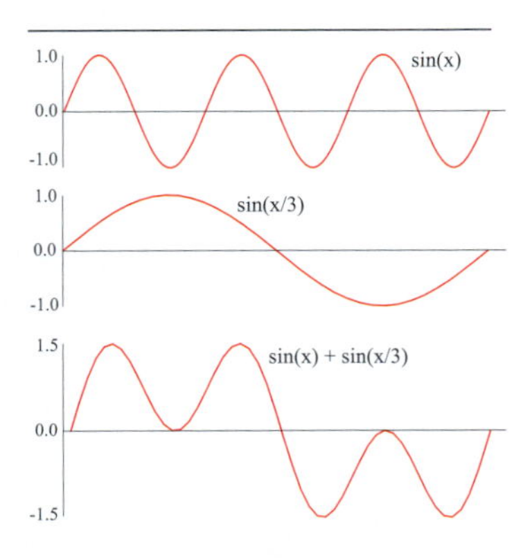

Fig. 11.12: Superposition of two waves.

Suppose two general waves are denoted $P_1(t,x)$ and $P_2(t,x)$, and the resultant wave by $P_R(t,x)$; t stands for the instant of time at which the waves are measured and x stands for the position in space at which the value of the wave is being considered. The **superposition principle** states that the resultant wave, at each instant t and at all points x, is simply the sum of the two waves. That is

$$\textbf{Resultant wave} = \text{Wave 1} + \text{Wave 2}$$

$$\Rightarrow P_R(t,x) = P_1(t,x) + P_2(t,x). \tag{11.21}$$

The superposition of $\sin(x)$ and $\sin(3x)$ is depicted in Figure 11.12.

Noise-canceling headphones

Figure 11.13: Some headphones have noise-cancellation electronics.

On long flights or even in trains or buses, rumbling ambient noises can be rather annoying. Not only can the noise prevent one from getting some sleep, its characteristic of being fairly loud at lower frequencies also makes it difficult to listen to music or the soundtrack of movies with standard headphones.

The question that naturally arises is whether there is a remedy to this uncomfortable situation. A first idea might be to try to dampen external sounds as much as possible by using, for example, a headphone with plenty of insulation that fully encloses the ears. Although this approach works reasonably well for the higher tones, it doesn't work very well for lower tones. Indeed it doesn't work exactly for those tones that are the most irritating.

As so often, knowledge of science can come to the rescue. In our initial investigation of sound, we have already encountered the phenomenon of interference of sounds which we used to conclude that sound must be a wave. We then found a general formula for interference in the superposition principle. This tells us that theoretically, any sound can be canceled out if it is combined with its own mirror image. This is illustrated in Figure 11.14 where wave A and its mirror-image wave B are combined to yield a straight line (i.e. no sound) at the bottom. When two waves cancel each other out, the interference is called destructive interference.

High-quality noise-canceling headphones use exactly this approach. A (or several) small microphone(s) is (are) built into the earcup that picks up the ambient (low-frequency) sounds and feeds this signal into a sound processor and amplifier which drives the speaker in the earcup such that the generated sound is the mirror image of the ambient noise. Then, when the sound of the speaker and the ambient noise arrive at the ear, they will cancel each other out resulting in silence. Of course, in the real world, one has to deal with some practical limitations and the cancellation will not be perfect. Nevertheless, a substantial reduction is possible.

If one would like to listen to music or a soundtrack, the superposition principle comes to the rescue a second time. Without having to give up any of the cancellation benefits, one can just add the "wanted" sound to the signal that drives the speakers in the earcup for a much improved listening experience!

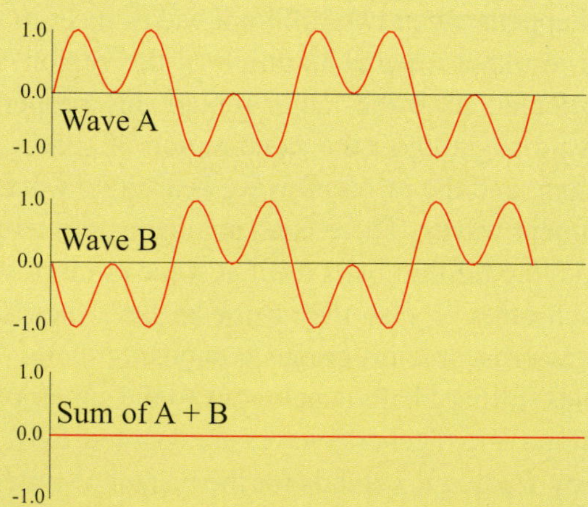

Figure 11.14: Destructive interference. If waves A and B are added up they cancel each other out.

The generalization to the superposition of an arbitrary number of waves is straightforward, and we can superpose waves that have different frequencies and amplitudes.

11.9 Intensity

We already know that sound can be loud or soft, but we would also like to know how much sound energy is being deposited at the eardrum. Intensity, denoted by $I(x)$, is a measure of the transport of energy and is defined to be the amount of energy per unit time that is crossing a unit area at point x, or equivalently, is equal to the power P flowing through unit area at x.

From daily experience, we know that if sound has a very high intensity, it is painful to the ear, signifying a large influx of energy.

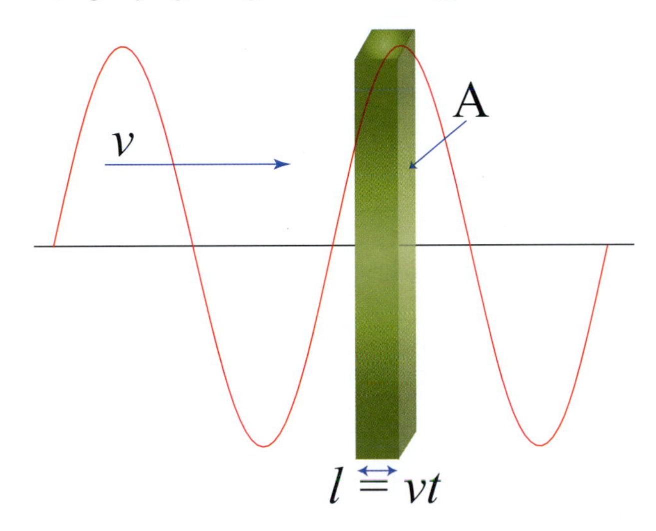

Figure 11.15: Intensity of a wave.

Figure 11.15 shows energy flowing through an area A. The energy of the wave contained in the box indicated in Figure 11.15 flows through surface A in time t.

We hence have

$$\begin{aligned} I(x) &= \frac{\text{Power}}{\text{Unit area}} = \frac{\text{Power}}{A} \\ &= \frac{1}{2}\rho v(s_0\omega)^2 \end{aligned} \tag{11.22}$$

where we have used Equation 11.20 to obtain the last line.

Note that the intensity I is proportional to s_0^2; in other words, the intensity I is proportional to the square of the maximum displacement of the piston generating the pressure wave. Therefore, if we have two waves, we find

$$\begin{aligned} I(x) = P_R^2(t,x) &= (P_1(t,x) + P_2(t,x))^2 \\ &= P_1^2(t,x) + P_2^2(t,x) + 2P_1(t,x)P_2(t,x) \\ &\neq P_1^2(t,x) + P_2^2(t,x). \end{aligned}$$

Note it is the cross-term $P_1(t,x)P_2(t,x)$ in the expression for the intensity I that creates the constructive and destructive interference of two (or more) waves. The loudness that the human ear perceives, however, is not the intensity I, but rather the logarithm of the intensity. The sound level for a sound of intensity I is defined by the units of decibel β and is given by

$$\beta = 10 \log\left(\frac{I}{I_0}\right), \tag{11.23}$$

where I_0 is the reference intensity and is taken to be the intensity at the threshold of human hearing and is equal to $I_0 = 1.00 \times 10^{-12}$ W/m^2. The human ear's sensitivity for loudness ranges from 0 dB to 150 dB (though above 120 dB would be painful). Normal speech corresponds to an intensity of 50 dB. Due to the logarithmic nature of hearing, the human ear can detect an enormous range of intensities having a staggering factor of more than one trillion from softest to loudest.

11.10 Beats

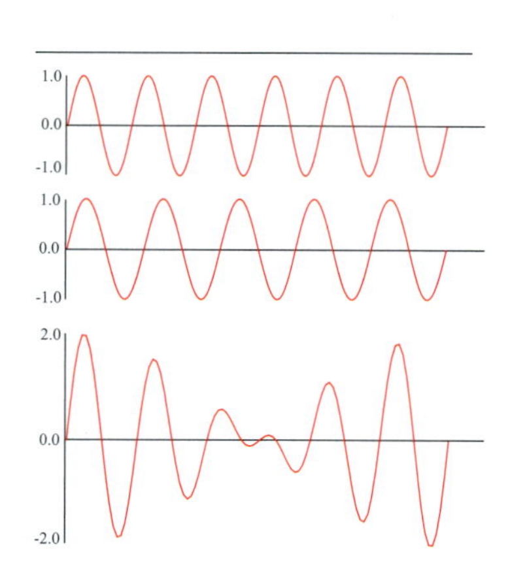

Fig. 11.16: Beating sound waves.

What happens when we superpose two waves having the same amplitude but different frequencies? We have experienced such phenomena many times in daily life when two sounds with different pitch (which is a nontechnical term for frequency) are heard by us at the same time. Let two waves have different frequencies f_1 and f_2. For simplicity, let us assume that we are interested in the oscillations of the wave at the same point in space (which is the case for the ear receiving two different sound waves), and hence examine the dependence of the wave only on time.

Using Equation 11.21 and the fact that two generic waves can be described by cosine functions like $P_1(t) = A\cos(2\pi f_1 t)$ and $P_2(t) = A\cos(2\pi f_2 t)$, we have for the resultant wave

$$\begin{aligned} P(t) &= P_1(t) + P_2(t) \\ &= A\cos(2\pi f_1 t) + A\cos(2\pi f_2 t) \\ &= 2A\cos\left(2\pi\frac{f_1 - f_2}{2}t\right)\cos\left(2\pi\frac{f_1 + f_2}{2}t\right). \end{aligned} \tag{11.24}$$

The wave generated can therefore be thought of as a wave with frequency $(f_1 + f_2)/2$ with its amplitude being modulated by the factor $\cos(2\pi\frac{f_1 - f_2}{2}t)$. Since the modulating factor goes from its maximum value of 1 to its minimum value of 0, the sound that results periodically goes through a maximum and a minimum and is called **beats**.

Consider the special case of $f_1 \simeq f_2$; then since the first term $\cos(2\pi \frac{f_1 - f_2}{2} t)$ is almost a constant, the sound seems to have a frequency of $(f_1 + f_2)/2$. However, the amplitude will slowly vary from 0 to $2A$. The phenomenon of beats is well known in music; we will show later how it can be used for tuning the frequencies of various instruments.

11.11 Standing waves

To hear a sound we must have a source for that sound, for example, the vibrating string of a musical instrument. Above, we have seen how such a vibration can be described but thus far we have not considered the ends of the string. Before investigating the realistic scenario where both ends are fixed, let us first look at what happens when one end is fixed.

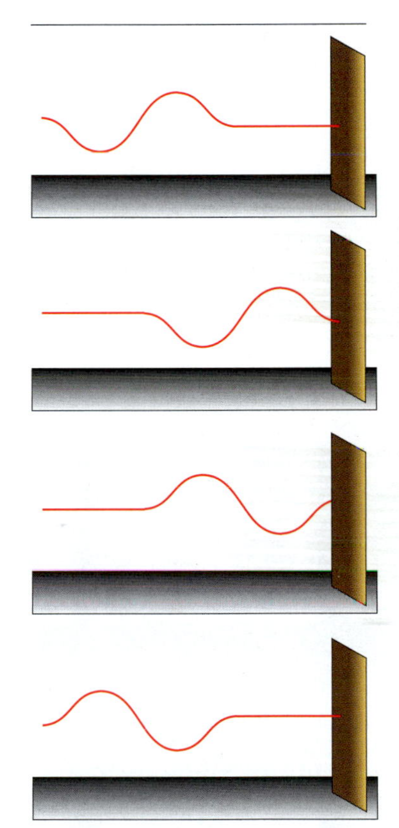

Fig. 11.17: Wave reflection at wall.

Consider a string that is tied at one end, as shown in Figure 11.17. A moving wave can simply be described by subtracting vt from the sine function's argument: i.e as $y(t, x) = \sin[k(x - vt)]$. After the wave reflects off the wall we have two waves in the string, namely one wave moving from the open end to the wall and a second reflected wave moving from the wall toward the open end. The superposition of the incident and reflected waves yields a so-called standing wave.

To see how the mathematics works out, consider two waves with the direction of propagation of the second wave in the opposite direction from the first wave:

$$\begin{aligned} y_{\text{standing}}(t, x) &= A \sin[k(x - vt)] + A \sin[k(x + vt)] \\ &= [2A \sin(kx)] \cos(kvt) \\ &= [2A \sin\left(2\pi \frac{x}{\lambda}\right)] \cos(2\pi f t). \end{aligned} \tag{11.25}$$

From Equation 11.25, it follows that the space and time dependences of the standing wave have *factorized*; in other words, the time dependence of the standing wave is given by $\cos(2\pi f t)$ and space dependence by $\sin(2\pi x/\lambda)$. The points for which the displacement of the wave is always zero (i.e. $y(t, x) = 0$) are called nodes of the wave. To find all the nodes we look for the points x for which $\sin(2\pi x/\lambda) = 0$; since the sine function $\sin(\pi n) = 0$ for all $n = 1, 2, 3...\infty$ one obtains that the nodes, at points x_{node}, are given by

$$\begin{aligned} \sin(2\pi \frac{x_{\text{node}}}{\lambda}) &= 0 \\ 2\pi \frac{x_{\text{node}}}{\lambda} &= \pi n \; ; \; n = 1, 2, 3...\infty \\ \Rightarrow x_{\text{node}} &= 0, \frac{\lambda}{2}, \lambda, \frac{3\lambda}{2}, 2\lambda, ... \end{aligned} \tag{11.26}$$

In other words, the **nodes** of the wave are fixed, and do not propagate (which explains the name: standing wave). An **antinode** is a point where the amplitude of oscillation of the standing wave is a maximum. A standing wave can be created with any wavelength λ.

Note that, similar to a propagating wave, each point of the medium oscillates about its equilibrium position. An example of a standing wave is shown in Figure 11.18.

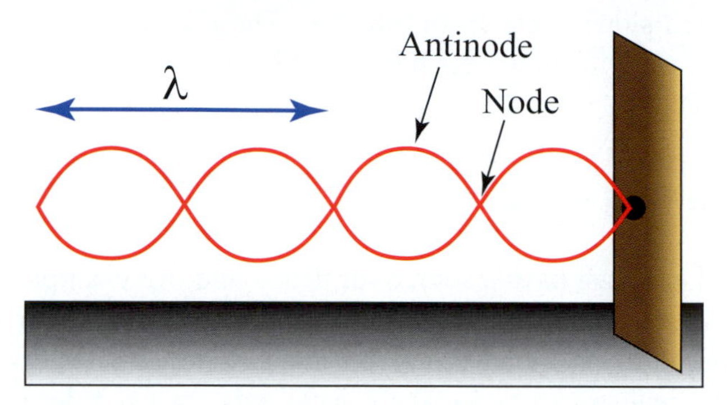

Figure 11.18: Creation of a standing wave by superposing two waves, having the same wavelength and velocity, but traveling in opposite directions. The diagram shows the configuration of the standing wave at two different moments.

11.12 Resonance

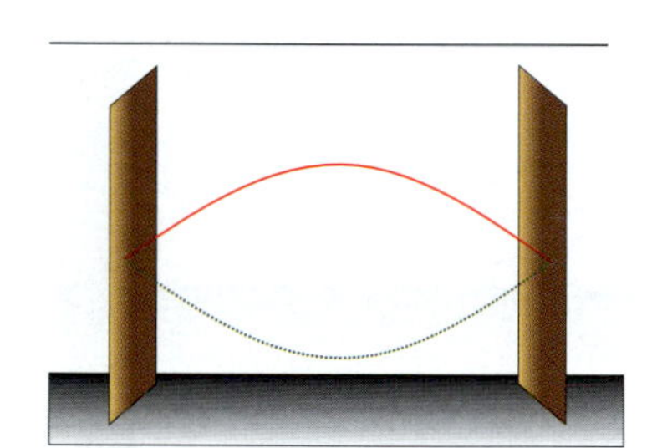

Fig. 11.19: First harmonic.

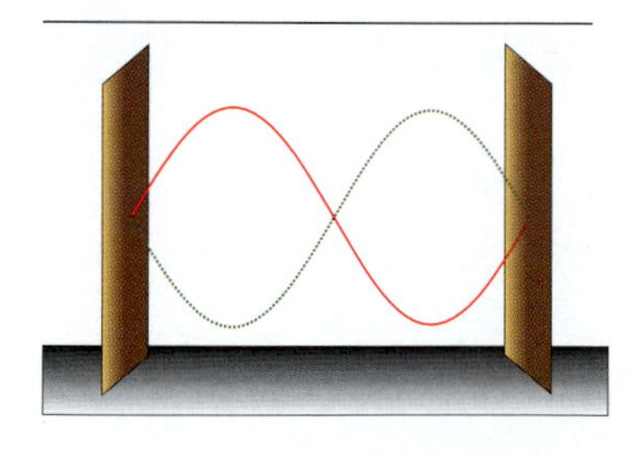

Fig. 11.20: Second harmonic.

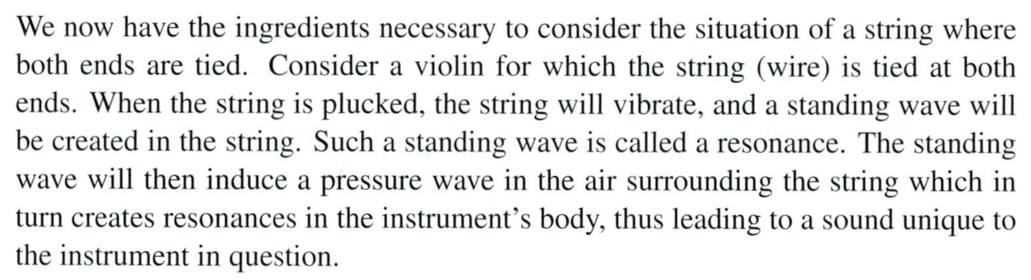

We now have the ingredients necessary to consider the situation of a string where both ends are tied. Consider a violin for which the string (wire) is tied at both ends. When the string is plucked, the string will vibrate, and a standing wave will be created in the string. Such a standing wave is called a resonance. The standing wave will then induce a pressure wave in the air surrounding the string which in turn creates resonances in the instrument's body, thus leading to a sound unique to the instrument in question.

The transverse vibrations in the string are given by Equation 11.25, namely

$$y_{\text{resonance}}(t, x) = [2A \sin\left(2\pi \frac{x}{\lambda}\right)] \cos(2\pi f t). \tag{11.27}$$

But we are not done yet; we have to account for the fact that both the ends of the string are always fixed. Suppose the length of the violin string is L, with one fixed end at $x = 0$ and the other end fixed at $x = L$. The displacement of the string at both ends must be zero. We hence must have

$$y_{\text{resonance}}(t, 0) = 0 = y_{\text{resonance}}(t, L) \tag{11.28}$$

$$\Rightarrow \sin\left(2\pi \frac{L}{\lambda}\right) = 0 = n\pi \tag{11.29}$$

$$\Rightarrow \lambda = \frac{2L}{n}, \quad n = 1, 2, 3\ldots. \tag{11.30}$$

Unlike the case for arbitrary standing waves, the violin string can only have a discrete number of wavelengths as given in Equation 11.30. The reason for this result is easy to understand. A node can be located at a number of points for a standing wave, but with one node always being located at the boundary point where the traveling wave is reflected. For a resonant wave, in addition there must be a node located at the other end point of the string as well. Only an integral, n number of half-wavelengths, $\lambda/2$, can fit into the length L of the violin string as standing waves, as can be seen from Figure 11.21. All other wavelengths that are allowed for standing waves are eliminated by destructive interference (destructive interference occurs when one wave cancels out another — this can happen, for example, when the nodes of one wave are at exactly the spot of the antinodes of the other wave).

Since wavelength λ is related to frequency f by Equations 11.12 and 11.13, we have from Equation 11.30

$$f = \frac{v}{\lambda} = n\frac{v}{2L}. \tag{11.31}$$

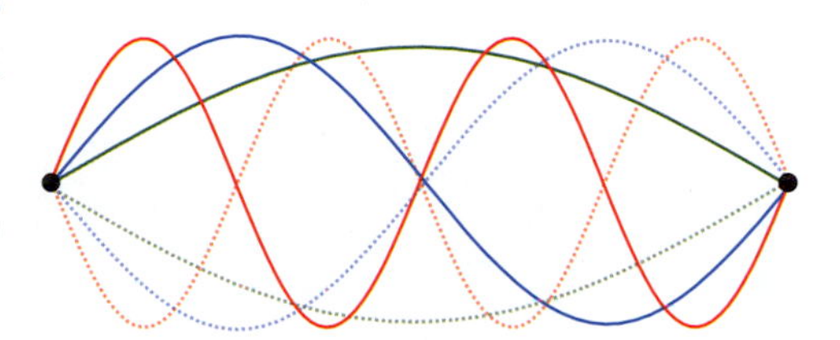

Figure 11.21: Resonant waves in a violin. Note the resonant wave in green has wavelength of $\lambda = 2L$ and is the longest wavelength that can be accommodated within the violin; the resonant wave in blue has wavelength of $\lambda = L$, and that in red has wavelength $\lambda = L/2$.

The frequencies given above are called **resonant** or harmonic frequencies. Figure 11.21 shows a few of the resonances for a fixed length L.

The lowest frequency possible in the string, i.e. the one with wavelength $2L$, is called the fundamental mode. The wavelengths of the other harmonics are then L, $L/2$, $L/3$, etc. In other words, in terms of the length of the string, the harmonics after the fundamental are given by the sequence of numbers 1, 1/2, 1/3, etc. and the sum of these numbers $(1 + 1/2 + 1/3 + \cdots)$ is called the harmonic series in mathematics (see Section 2.4 for some more details).

Resonance can also take place for a medium that is more complex. Consider a circular metal ring used in some forms of music. If one were to strike the ring, there would be oscillations set up, with an integral number of wavelengths around the circumference of the ring as depicted in Figure 11.22. If the radius of the ring is r, the allowed wavelengths for oscillations are then given by

$$\begin{aligned} n\lambda_{\text{Ring}} &= 2\pi r \\ n &= 0, 1, 2, \ldots . \end{aligned} \tag{11.32}$$

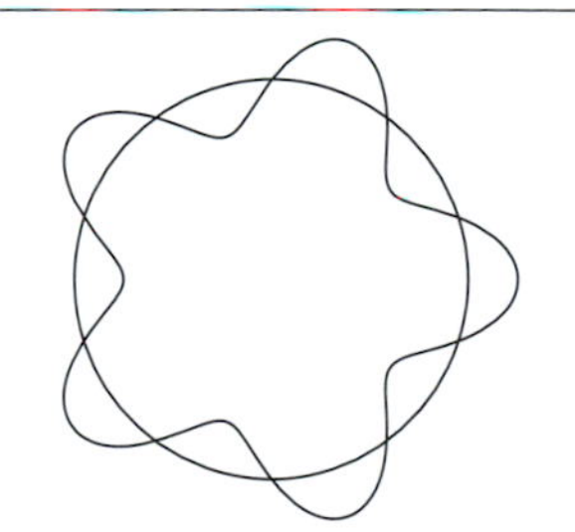

Fig. 11.22: Resonance wavelength on a circle.

The reason that for a circular ring the resonant wavelengths can take only integral multiples of $2\pi r$ is due to periodicity of the medium for a ring; for an open string this periodicity is not required.

11.13 Musical instruments

To proceed further, we need to analyze what a musical instrument is. The two largest classes of (mechanical) instruments are wind and string instruments. Wind instruments are essentially similar to the example of the piston, and therefore directly create the longitudinal sound waves. String instruments, on the other hand, first create transverse waves in a string whose vibrations in turn create the longitudinal sound waves.

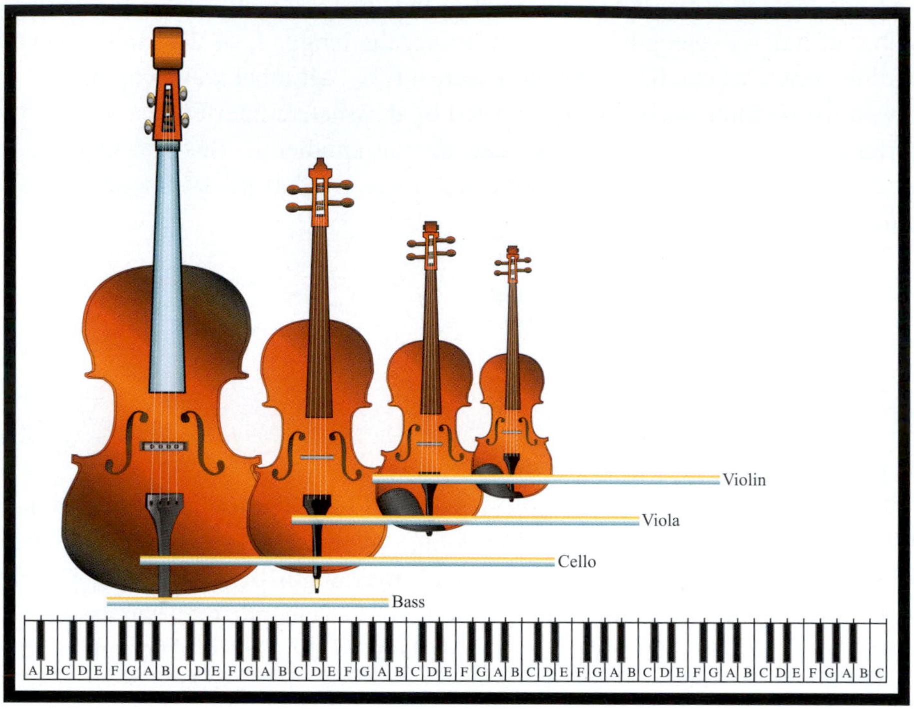

Figure 11.23: Pitch depends on the length of the string.

Specific musical instruments

We are now in a position to understand the acoustic aspect of musical instruments. Musical instruments in general are designed to produce a specific set of frequencies of sound.

When a musical instrument is played, a resonant frequency is excited that for stringed instruments results from the transverse oscillations of the string, while for wind instruments the resonant frequency is a pressure wave directly created in the instrument. The advantage of having a resonant wave is that it can sustain a large amplitude of oscillations that push back and forth like the piston that we considered, and thus audible sound waves are produced over a long period of time.

Tuning

The phenomenon of beats is well known in music, and may be used in the following manner for tuning the frequencies of various instruments. First, a tone of a standard frequency (often 440 Hz) is created with a special tuning fork. Then, on the instrument, the note that should have this frequency is played and one listens to whether there is a beat. If so, the instrument needs tuning since having a beat means that the frequencies of the tuning fork and the instrument are not the same. One can then adjust the pitch of the instrument (e.g. by changing the tension of the string on a violin) until the beat stops (or becomes very slow). No beats mean that both the tuning fork and the musical instrument have the same frequency of oscillation.

Harmonics

If a stringed instrument has a length of L, then its resonant frequencies are given by Equation 11.31

$$f = \frac{v}{\lambda} = n\frac{v}{2L} \quad : \text{String Instrument.} \tag{11.33}$$

The resonant frequencies for wind instruments are the following:

$$f = \frac{v}{\lambda} = n\frac{v}{4L} \quad : \text{Open at one end (odd } n \text{ only)}$$

$$f = \frac{v}{\lambda} = n\frac{v}{2L} \quad : \text{Open at both ends.}$$

Stringed and wind instruments are designed to generate different frequencies of sound. The bass and bass saxophone have the largest length, and hence generate the lowest frequencies (bass sound); at the other extreme, a violin and soprano saxophone have the shortest length and generate the highest frequencies.

The **amplitudes** produced by an instrument are different for the various harmonics. The distribution of the amplitudes as a function of frequency is quite complicated, and was only achieved through a history of trial and error leading to an aesthetically pleasing result.

11.14 The ear

It goes without saying, sounds and the means to create sounds are of little meaning if they cannot be perceived. The sense of hearing is called **audition** and in mammals, like human beings, the detection of sound takes place in the ear. The design of the ear is shown in Figure 11.24 and consists of three basic parts: the outer, the middle and the inner ear.

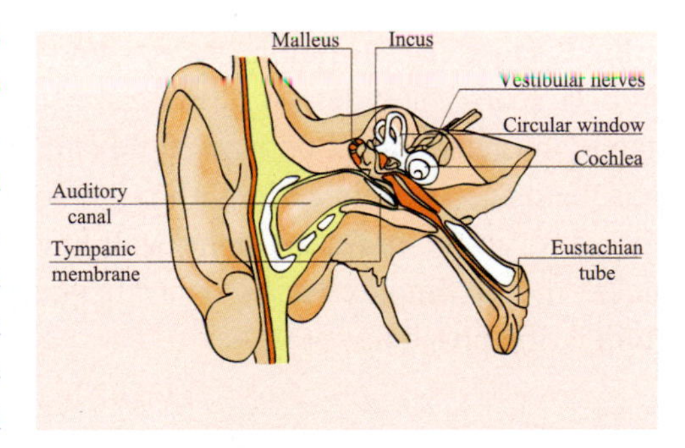

Figure 11.24: Illustration of the human ear.

The outer ear includes the **pinna** (the externally visible part), the ear canal and the **tympanic membrane** (eardrum). The main function of the pinna is to collect sounds and can vary drastically between mammal species. Elephants, for example, have huge ears that also serve as radiators to cool blood, while seals have no external ear parts at all. The ear canal channels sounds to the tympanic membrane which transforms the air pressure oscillations into mechanical vibrations.

The middle ear consists of the air-filled cavity behind the eardrum and three bones called the **malleus** (hammer), **incus** (anvil) and **stapes** (stirrup). The function of the middle ear is to transmit the vibrations of the eardrum to the inner ear. First, the mechanical vibration is taken up by the malleus which passes it on to the incus which in turn passes it on to the stapes. The middle ear is filled with air, and for the eardrum to vibrate as freely as possible it is important that the average pressure in the cavity equals that of the outside environment. This is achieved with the help of the Eustachian tube which links the middle-ear cavity to the back of the pharynx (the area behind the mouth). Under normal circumstances, the Eustachian tube is collapsed and hence closed to the flow of air. When swallowing or when exposed to a positive pressure in the inner ear, it will briefly open to allow pressure equalization to take place. However, it does not as readily open when exposed to negative pressure in the inner ear and hence, when descending rapidly, as is the case of a landing airplane, a temporary discomfort may ensue as many travelers experience.

The inner ear consists of the **cochlea** (the organ with the receptor cells that inform the brain of a sound) and the vestibular apparatus that senses motion and gravity. The inner ear is filled with a fluid, and it is the motion of this fluid that is detected by the receptor cells. More specifically, vibrations of the stapes cause vibrations in the fluid of the cochlea. The vibrations in the fluid are then picked up by the receptor cells with the help of tiny hairs that are moved back and forth by the fluid.

The human ear can hear a range of frequencies from about 20–20,000 Hz. Any typical sound one hears is "composed of" not a single frequency — as in the case of sound with a given pitch — but of many frequencies.

Evolutionary role of sound

Sound detection is rather old. Indeed the basic design of the ear is similar in all animals from fish to human beings and thus covering a time span of about 400 million years. The design of the ear is therefore, evolutionarily speaking, rather stable and hence must convey an advantage over nonhearing organisms. For human beings, communications through sounds evolved into speech and therefore play a very special role. It may well have been that it was the ability to communicate more complex issues using language that made the emergence of human civilization possible.

11.15 The answer

Now we can answer the chapter question of why we can hear music. There are several key aspects to this. First, we can hear music because we live in an environment (air) that allows for the transmission of pressure waves in the form of sound. Second, sound waves can be created rather easily by a variety of means in a rather controlled way. Third, we have a specialized sophisticated organ to detect sound. However, sound alone does not music make! The final key aspect is that we also have a central nervous system (brain) that can *interpret* the various auditory inputs to a larger whole. For example, the recognition of a melody requires the temporal correlation of sounds. In short, we can hear music because **nature allows for the creation and transmission of vibrations, we have an auditory system and a brain to interpret the meaning of a sound**.

11.16 Exercises

1. A 60 Hz vibration produces waves in air that propagate at 340 m/s. What is the
   (a) frequency?
   (b) wavelength?
   (c) length in terms of seconds of one period?

2. It is said that human beings can typically hear frequencies in the range of 20–20,000 Hz. What wavelengths do these correspond to in air?

3. Explain the meaning of the well-known English idiom "Keep your ear to the ground". In some Western movies, one sees the American Indian scouts putting their ear to the ground. Explain the physics reason behind that action.

4. Do sound waves bend toward or away from the ground on a hot and sunny day? Explain why and suggest how you might test the phenomenon.

5. In many science fiction movies, explosions taking place in outer space are seen and heard at the same time. What errors in physics are there in those scenes?

6. (a) Give a few examples of animals that use sound reflection (echoing) to navigate or hunt for food. What frequencies do they use? Can humans hear those sounds?
   (b) Whales and elephants use sound for communication. What frequencies do they use?

7. If someone drops a coin or a spoon on the floor, you can tell without even looking what the object was. Explain why.

8. The same fundamental note played on a violin sounds different from that played on the piano or some other instrument. Explain in physical terms what is happening.

9. Why does one hear a sharp sound when a whip is cracked?

10. Some insects such as mosquitoes and bees produce a buzzing sound. How do they do that and what frequencies do they produce?

11. If there is a large explosion, one finds that from some distance one can hear the tremor well before the sound of the explosion. Why would this be so?

12. If we have a fundamental frequency of 400 Hz, what is the frequency of the third harmonic?

13. We have a string instrument and want to increase the pitch of a string. Give at least two ways to do so.

14. We have two nearby sound sources. One has a frequency of 440 Hz and the other a frequency of 446 Hz. What will be the frequency of their beat?

12: Nature's Solar Cells

Q - Why Are Leaves Green?

Chapter Map

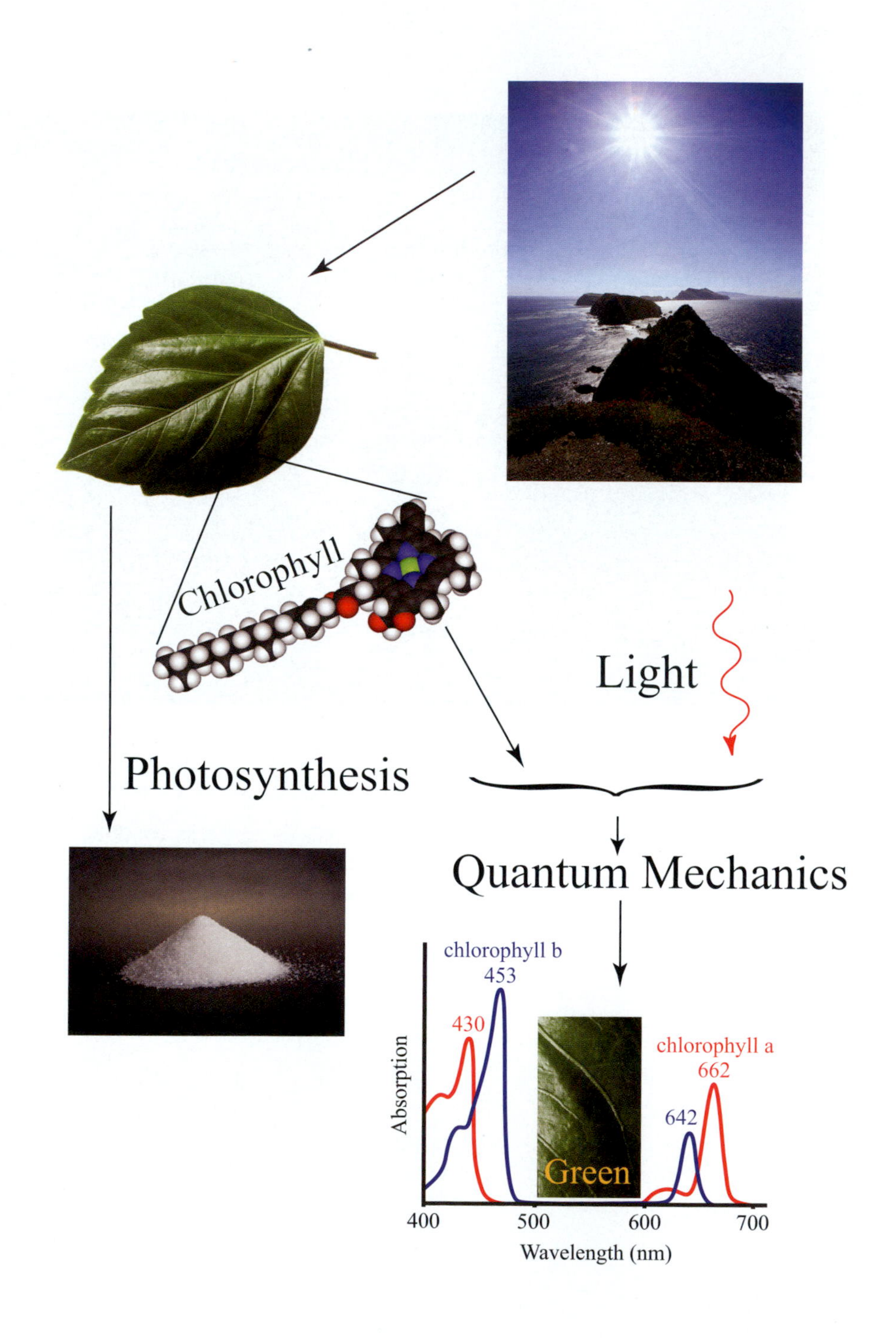

Chlorophyll
Light
Photosynthesis
Quantum Mechanics
chlorophyll b
453
430
chlorophyll a
662
642
Green
Absorption
400
500
600
700
Wavelength (nm)

12.1 The question

Figure 12.1: A sunlit green leaf.

Much of life on earth is ultimately powered by the sun. Some organisms like plants and certain bacteria can directly convert the energy contained in sunlight into food while others feed on them, possibly through a complex predator-prey network. For example, carnivores like lions and tigers do not eat plants directly but consume plant-eating animals like deer and gnus.

The mechanism of capturing sunlight is therefore one of the most important processes for life as we know it. For human beings, the most visible place where the capture of sunlight takes place is in leaves of plants that we perceive with the color green. If leaves are so important and their color is green, then there might very well be a good reason for the leaves to have that color. Therefore, let us ask: **Why are leaves green?**

In order to investigate this question, we first need to find out a bit more about color and about leaves.

12.2 Color

It has been known for a long time that sunlight can be split into a rainbow of colors with the help of a prism. Why this is so, however, was not understood until much later. As such, light is **electromagnetic radiation** and colors are nothing but light of different wavelengths as illustrated in Figure 12.2. The wavelengths of electromagnetic waves span an enormous range: from the very short like gamma rays that are less than a nanometer long to the very long like radio waves that have a wavelength of several kilometers.

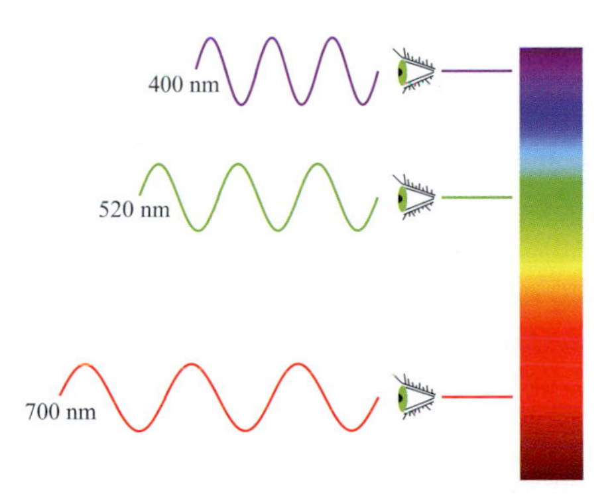

Figure 12.2: Colors are light of different wavelengths.

When light impinges on matter, it can be absorbed, reflected or transmitted (or a combination of these). Substances that absorb visible light are generally referred to as **pigments** (visible light mostly ranges from about 400 nm to 700 nm). Not surprisingly, different pigments absorb light of different wavelengths and the color of a pigment is the result of the light most strongly reflected.

12.3 Leaves

Leaves are specialized plant organs that convert light energy into chemical energy by a process called **photosynthesis**. Since photosynthesis is the main function of leaves, it may well be that the color of leaves has something to do with this process. Let us therefore find out exactly where and how photosynthesis takes place. The first thing to do is to look at how leaves are constructed. The basic anatomical structure of a leaf blade is shown in Figure 12.3.

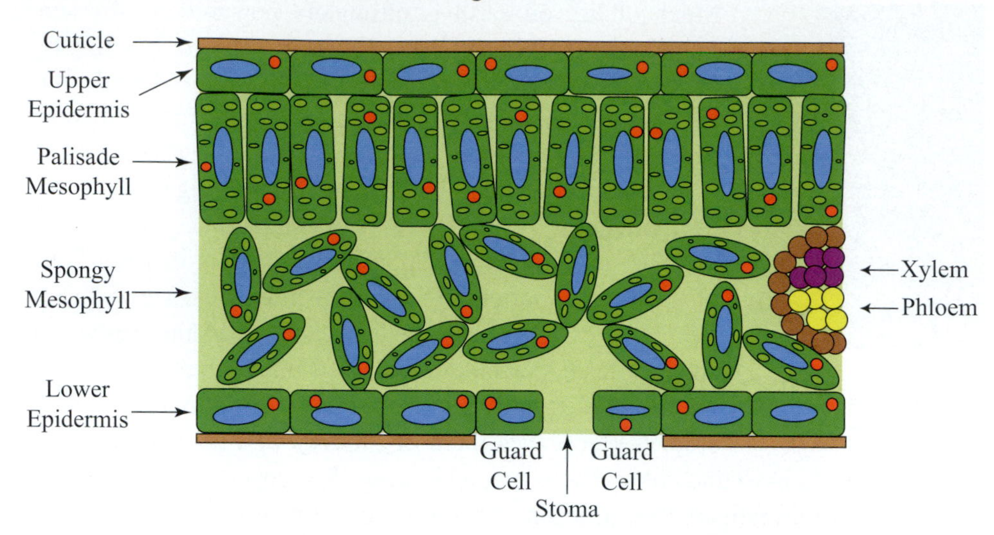

Figure 12.3: Basic anatomy of a leaf.

The upper and lower surfaces consist of the so-called **epidermis**, a tissue that among other functions separates the inside and the outside of the leaf as well as regulates gas and water exchange. The epidermis is therefore a multicellular analog of the cell wall and thus illustrates nicely the recurring theme of the construction of ever more complicated structures from simpler building blocks. The interior of a leaf blade consists of the palisade and spongy mesophyll as well as a vascular bundle that contains two types of veins: the xylem, which transports water from the roots to the cells in the leaf, and the phloem, which moves glucose — the main product of photosynthesis — containing sap to the rest of the plant.

The epidermis is usually coated in a waxy transparent cuticle to prevent water loss. In many leaves, the upper and lower epidermis are somewhat different with the lower epidermis having the majority of the so-called **stomata**, small pores that can open and close to regulate the exchange of gases and water.

The mesophyll layer is generally divided into two sublayers: the palisade mesophyll layer consisting of tightly packed, vertically elongated cells and the sponge mesophyll layer consisting of more rounded, less tightly packed cells. Careful experimentation shows that photosynthesis takes place in both of these layers in organelles called **chloroplasts**.

Although, the shapes and detailed structure of leaves can vary greatly in, for example, whether or not the mesophyll has two sublayers, how many cells thick a sublayer is, whether or not the leaf is flat, they all share in common that photosynthesis takes place in chloroplasts. Furthermore, despite the great variety of plants, these chloroplasts are virtually identical throughout. Let us therefore have a closer look at chloroplasts next to find out where and how photosynthesis takes place in a chloroplast.

Chloroplasts

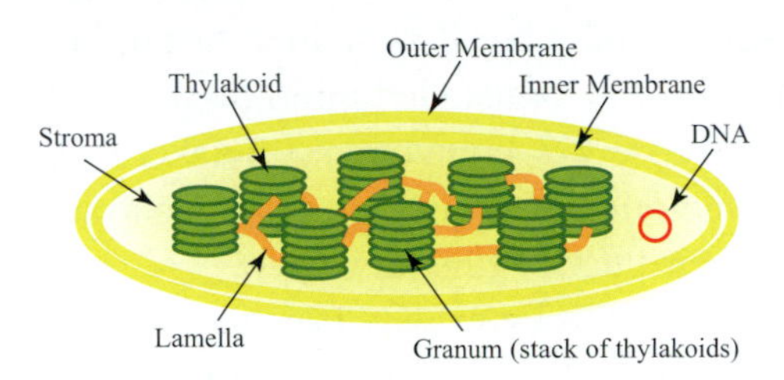

Figure 12.4: Basic structure of a chloroplast.

As mentioned above, chloroplasts are the organelles in which photosynthesis takes place. They are usually flat disks with a diameter of about 2 to 10 μm and a thickness of about 1 μm. The interior of the chloroplast, termed **stroma**, is surrounded by an inner and an outer phospholipid membrane. The stroma contains stacked **thylakoids** as well as small pieces of circular DNA and even some ribosomes. This basic structure is illustrated in Figure 12.4. Thus here we see an example of genetic information outside of the nucleus (in eukaryotes, most of the DNA is found in the nucleus; see also Chapter 17 on DNA).

The number of chloroplasts is in general quite large. A square millimeter of a leaf can contain more than half a million of them.

If one looks at the overall structure of a chloroplast, it almost appears as if it were an independent single-cell organism. Indeed, the prevailing theory is that they originate from endosymbiotic cyanobacteria.

Although we have moved closer to the site where photosynthesis takes place, we're not quite there yet, as it turns out that the actual photosynthesis takes place only in the thylakoids. Therefore, let us continue our investigation by considering thylakoids.

Thylakoids

If one carefully studies a thylakoid one finds that it consists of two main parts: the membrane that embeds the various proteins that carry out photosynthesis (we're finally there!) and its interior called the lumen. The structure of a thylakoid is illustrated in Figure 12.5.

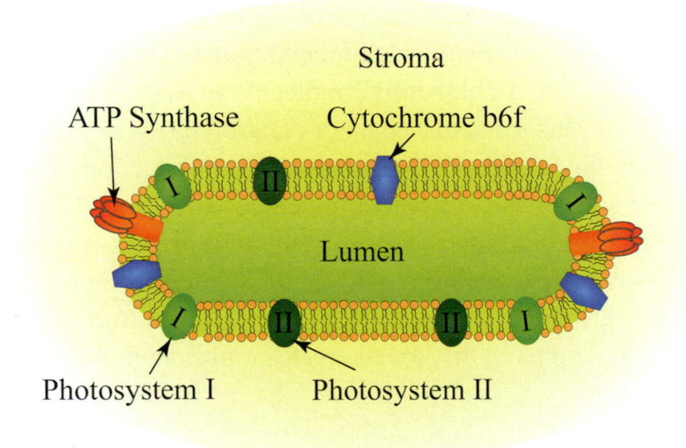

Figure 12.5: Basic structure of a thylakoid.

Interestingly enough, the membrane of the thylakoid consists of a lipid bilayer that displays some reminiscence of bacteria as well as the inner chloroplast bilayers. Therefore, just like chloroplasts, thylakoids may be of symbiotic origin. In other words, it may well be that mesophyll cells' key organelle (chloroplast) is the remnant of an ancient symbiotic bacterium whose key component (thylakoid) in turn is an even more ancient remnant of a symbiotic bacterium. However, contrary to chloroplasts, thylakoids have no DNA in their interior (the lumen).

Before continuing, let us briefly recap: we want to know why leaves are green. It is likely that the color of a leaf has to do with its function, photosynthesis. Of course, we cannot be sure of that but it is a good starting point. Hence, we've drilled down to where photosynthesis takes place. Now that we've arrived at that point, let us consider the process in somewhat more detail.

12.4 Photosynthesis

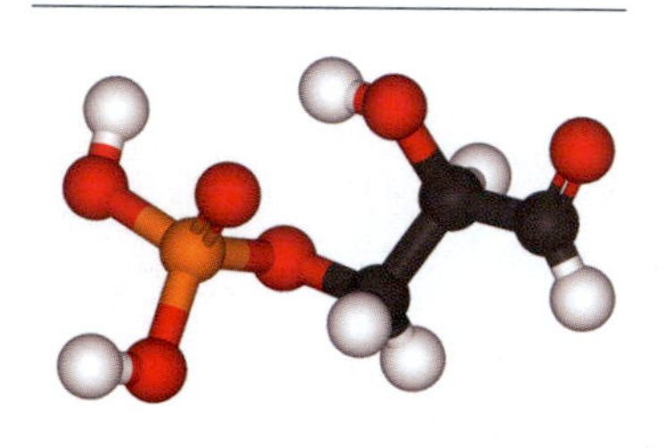

Fig. 12.6: Triose phosphate, the end product of photosynthesis. Triose phosphate can readily be transformed into sugars.

Photosynthesis is the capture of light energy and its conversion into chemical energy in the form of bonds that can release this energy when broken. The direct end product of photosynthesis is the molecule triose phosphate shown in Figure 12.6 which can either serve as a food source directly inside the cell or be further processed to produce glucose that can be transported to other cells. For the moment, taking into consideration only the resulting glucose production, photosynthesis can be summarized in the chemical equation

$$6CO_2 + 12H_2O + \text{light} \rightarrow \quad C_6H_{12}O_6 + 6O_2 + 6H_2O. \qquad (12.1)$$

All in all, photosynthesis is a rather complex process divided into two distinct stages. In the first stage, taking place in the thylakoid membrane, light is used to store energy in certain molecules, and in the second stage, generally taking place in the stroma, this stored energy is released to fixate carbon dioxide in triose phosphate as illustrated in Figure 12.7. This stage is often called **carbon fixation cycle** or **Calvin cycle**.

In somewhat more detail, the first stage of photosynthesis in which the actual capture of light takes place in the thylakoid membrane is illustrated in Figure 12.8. The first stage begins with a **chlorophyll** molecule giving up one electron when absorbing some light. This leads to a chain of electron movements in molecules in the membrane of the thylakoid and eventually to the production of NADPH from NADP when a second type of chlorophyll molecule absorbs a photon. As the first molecule is one electron short after the absorption of a photon, it needs to regain it in some way. This is done by splitting a water molecule into an oxygen atom and hydrogen and then stripping the hydrogen atom of its electron leaving a proton. The resulting proton surplus in the lumen creates a proton gradient that is used in the synthesis of ATP:

NADPH is a key molecule used in processes such as sugar and lipid synthesis.

$$\begin{aligned} 2H_2O + 2NADP^+ + 2ADP + 2P_i + \text{light} \rightarrow \\ 2NADPH + 2H^+ + 2ATP + O_2 \end{aligned} \qquad (12.2)$$

where P_i is so-called inorganic phosphate, a stable ion of phosphoric acid, H_3PO_4.

The second stage of photosynthesis (called the Calvin cycle or the dark reaction) takes place in the stroma of the chloroplast and begins with the capture of a carbon dioxide molecule. The NADPH from the first step is then used to manufacture triose phosphate ($C_3H_5O_3\text{-}PO_3^{2-}$):

$$3CO_2 + 9ATP + 6NADPH \; + \; water \rightarrow \\ C_3H_5O_3\text{-}PO_3^{2-} \; + \; 9ADP + 8P_i + 6NADP^+. \quad (12.3)$$

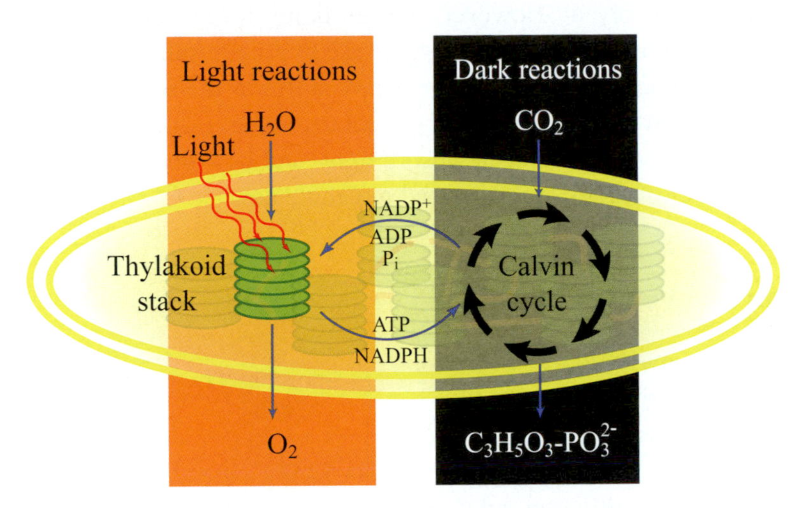

Figure 12.7: Schematic representation of photosynthesis. The dark reactions take place in the stroma of the chloroplast and produce the triose phosphate which is then exported to the cytosol of the cell where it is used to produce sugars and other organic molecules.

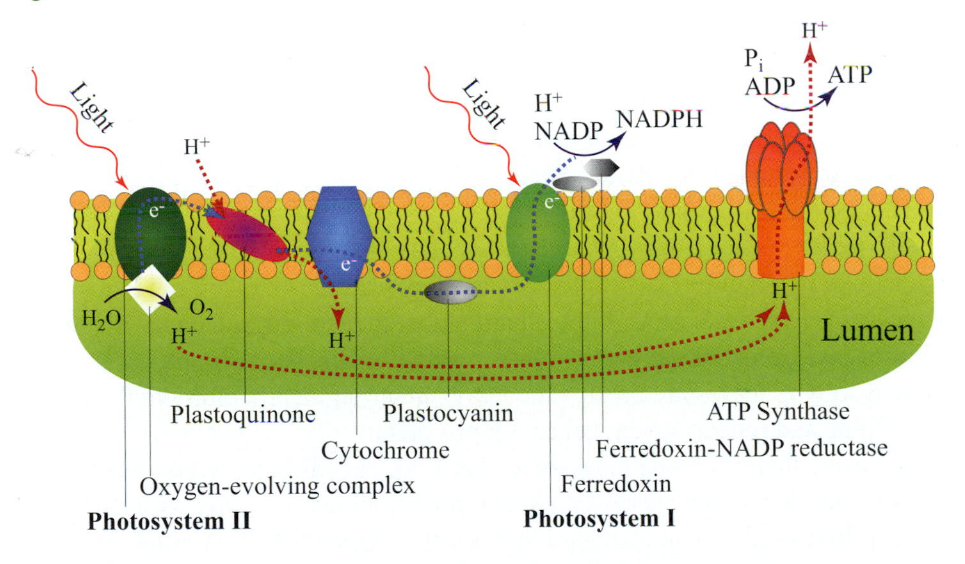

Figure 12.8: The main processes in the photosystems of thylakoid.

Much of the triose phosphate is exported to the cytosol of the plant cell containing the chloroplast where it is used to manufacture sugars, fatty acids and amino acids. Most of the triose phosphate that remains in the stroma is used to produce starch which can be used, for example, at night when the light is insufficient to support the energetic needs of the plant.

In bacteria, photosynthesis takes place in much the same way as in plants except for that they don't have chloroplasts. Cyanobacteria, for example, have membranes which are very similar to those of thylakoids in plants. Cyanobacteria produce oxygen and may to a large extent have been responsible for the early atmospheric buildup of oxygen. Since multicellular life as we know it needs large quantities of environmental oxygen, without cyanobacteria multicellular life might indeed never have evolved. However, oxygen is not necessarily a by-product of photosynthesis. Some bacteria, for example, use hydrogen sulfide instead of water to obtain electrons and protons and hence leave sulfur as the by-product.

Now that we have a rough idea of how photosynthesis works, we need to ask: Why does the chlorophyll molecule absorb light, and how does it do that? Let us therefore have a closer look at this special molecule now.

Chlorophyll

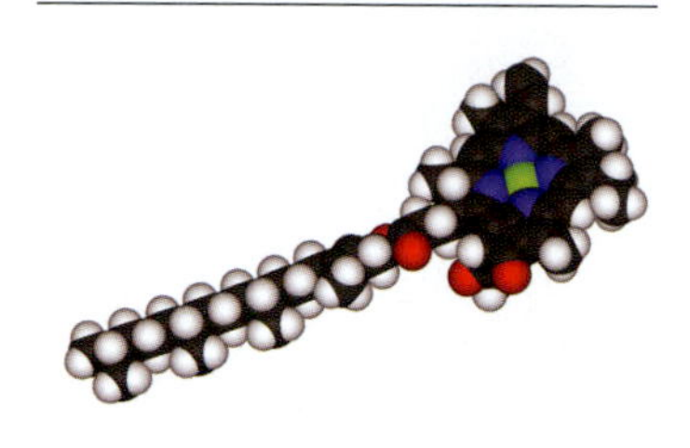

Fig. 12.9: Chlorophyll a.

As we have seen above, in the process of photosynthesis, light is absorbed by chlorophyll molecules (a chlorophyll molecule is shown in Figure 12.9). Since light consists of different colors, each with a different wavelength, it is then natural to ask which wavelengths are absorbed by chlorophyll and which are not, namely, what is the so-called **absorption spectrum**. In an absorption spectrum, light of all colors is shone on an object and it is subsequently measured which wavelengths are reflected (for more details, see Section 12.5 below). The more light is reflected, the less is absorbed and vice versa. The absorption spectrum for chlorophyll is shown in Figure 12.10, and it can be seen that red and blue light is absorbed the most strongly while green light is hardly absorbed at all. Indeed, as illustrated in Figure 12.11 most of the green light is reflected. This is a first and basic explanation as to why leaves are green. The leaves main pigment, chlorophyll, absorbs red and blue and reflects green, hence it looks green!

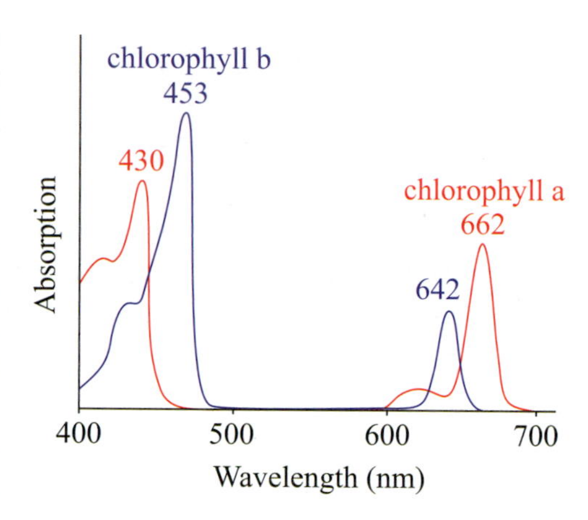

Figure 12.10: Absorption spectra of chlorophyll a and b.

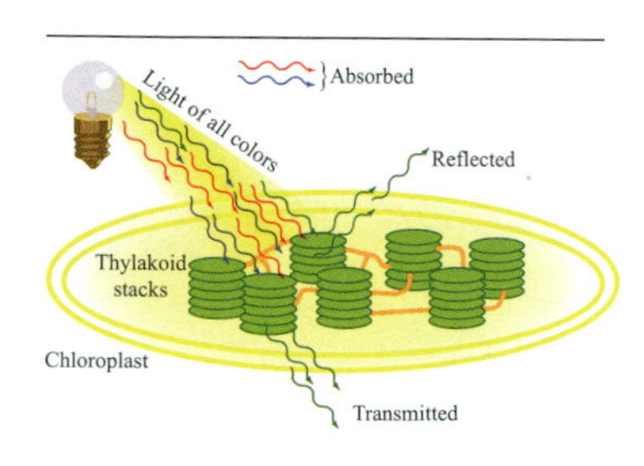

Fig. 12.11: Light absorption in a chloroplast.

However, there is something somewhat odd. From basic experiments with prisms we know that the energy contained in light varies smoothly with the wavelength. Why would a plant disregard the energy in green light? After all, we know for sure that green light is there since it is reflected from the leaf. Also by investigating the spectrum of the sun, we know that there is ample green in sunlight.

Well, since we know that absorption of light is essential for life on earth and furthermore that this absorption is puzzling, let us investigate some more absorption spectra to see whether this can give us some insights.

12.5 Absorption spectra

Absorption spectra are measured with the help of so-called spectrometers as illustrated in Figure 12.12.

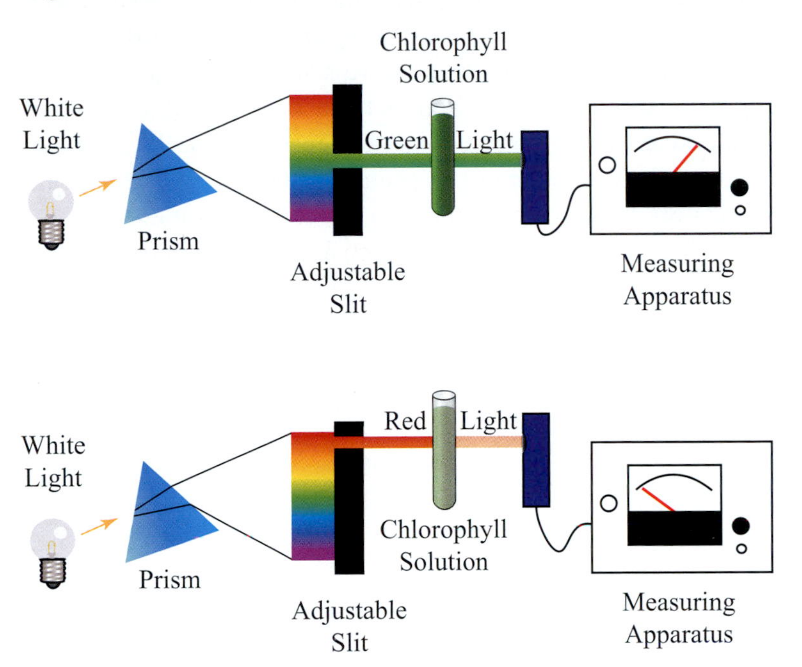

Figure 12.12: Basic design of a spectrometer.

As mentioned in Section 12.4 above, absorption spectra are obtained by shining light on an object and by measuring the transmitted or, in some cases, reflected light.

Before one can measure what is absorbed, however, one needs to know which colors emanate from the light source and what intensity those colors have. Therefore, let us now look at the color spectra of some possible light sources. Since plants use sunlight, it appears to be a good idea to start with sunlight. The color spectrum of sunlight is shown in Figure 12.13. What is remarkable is the large number of black lines! Where would those come from? We'll get back to that later (see Chapter 23), but for now it is clear that the sun is not an ideal light source containing all wavelengths.

Figure 12.13: The spectrum of the sun. Note the many black lines.

Let us therefore consider another common light source: the lightbulb. The spectrum of a lightbulb is shown in Figure 12.14. A lightbulb works by running an electric current through a thin metal wire. The current heats up the wire and the wire then starts to glow with a spectrum that corresponds to the temperature of the wire. Indeed, a lightbulb is a reasonably good approximation of the so-called **black body radiation**. In principle, a black body is an object that does not reflect

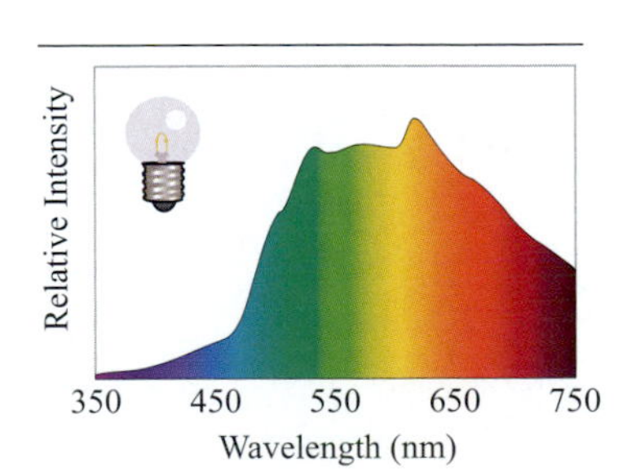

Fig. 12.14: The spectrum of an incandescent lightbulb.

any incoming radiation, i.e. it absorbs light of any color impinging on it. Any radiation, part of which is visible light, emitted from this object is purely due to its temperature. For practical purposes, however, the spectrum of any sufficiently hot object will closely resemble that of a black body as the reflected component will only be a small fraction of the total light coming from that object. This is why completely different materials like burning charcoal (i.e. carbon), melting iron or melting glass (i.e. silicon dioxide) all have a very similar orange-yellow color. Charcoal burns at around 1350 K, iron melts at 1811 K and silica glass melts at 1650 K —- all quite similar temperatures and hence they produce similar colors.

The fact that every object emits a spectrum of radiation with a peak in intensity depending only on the object's temperature and not its molecular constituents (here surface reflection is neglected) was discovered in the nineteenth century. Figure 12.15 show this "universal" spectrum.

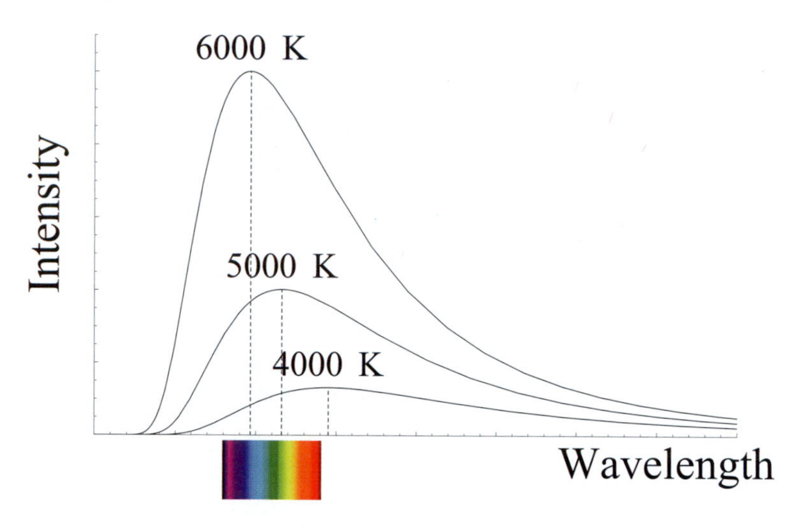

Figure 12.15: The intensity of the radiation emitted from a so-called black body depends on the temperature of this black body.

For a long time the intensity distribution was a complete mystery. None of the theories known during the late nineteenth century could even remotely explain this shape. Before delving deeper into this, let us first continue our investigation into various absorption spectra by considering one more type of light source: the fluorescent lightbulb. An illustration of its spectrum is shown in Figure 12.16. As can clearly be seen, in this case, there are some spikes.

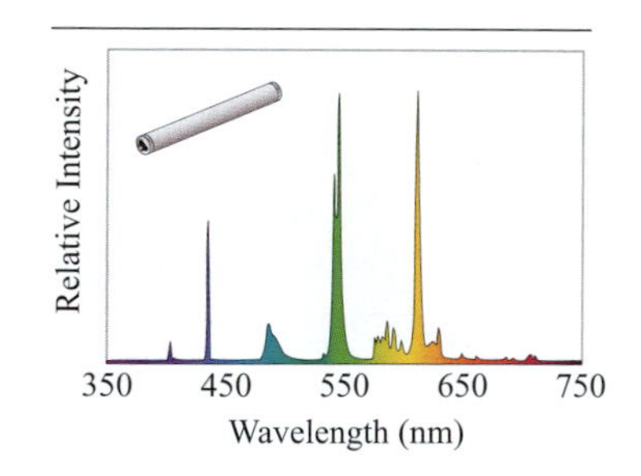

Fig. 12.16: The spectrum of a warm white fluorescent lightbulb.

Now that is odd! Sunlight has black lines, an incandescent lightbulb has a black body spectrum, and fluorescent light really mostly consists of only a few spikes. What is going on here? Do all these phenomena belong together? What we see is that an incandescent lightbulb is the best light source among the three examples discussed above, so let us use this source to shine light on some more materials to see whether we can observe something interesting.

We could choose some basic materials like steel or stones, but to start off it is often best in science to begin with the simplest available. The simplest element is hydrogen and consequently a good choice. However, since hydrogen is in a gaseous form at room temperature, rather than considering the light reflected off it (this would be a very small amount of light), it is more straightforward to consider the light passing through it. The resulting absorption spectrum is shown in Figure 12.17 — it shows a number of clearly visible black lines.

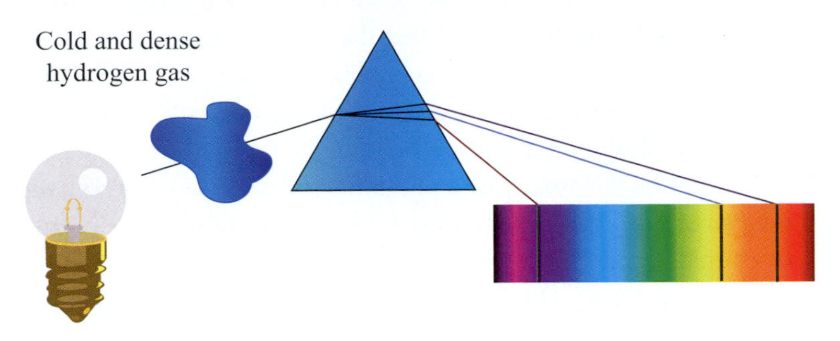

Figure 12.17: The absorption spectrum of hydrogen shows a small number of clearly visible black lines.

At this point, let us briefly sit back and reflect on the experimentally provided facts. We have found that the color of hot objects is in general mainly determined by their temperature and described by a curve which cannot be explained with the help of nineteenth century theory. We further have found that light passing through hydrogen gas is absorbed at very specific wavelengths only. Last, we have found the reverse of absorbing specific wavelengths in the intensity peaks of the fluorescent light emission spectrum.

It was known at the end of the nineteenth century from the invention of the radio and other research that oscillating electric currents generate electromagnetic radiation of which light an example. Hence it did not seem particularly farfetched to try to explain the black body spectrum by considering a black body as consisting of many small "atomic radio transmitters", i.e. small oscillators. Doing this alone, however, does not yield satisfactory results. What was missing?

The breakthrough came in the year 1900 when the German physicist Max Planck discovered that the spectrum of a black body radiation can theoretically be derived by assuming that black body consists of tiny *quantized* oscillators such that the energy of each of these oscillators is an integer multiple of a constant times the oscillation frequency of this oscillator. In other words, his assumption was that

$$E_{\mathrm{osc}} = nh\nu \tag{12.4}$$

with E_{osc}, the energy of the oscillator, n an integer, h a constant nowadays called the **Planck constant** and ν the frequency of the oscillator. Thus, unbeknown to him, Max Planck ushered in one of the greatest scientific revolutions in the history of mankind: the advent of quantum physics. We're not quite there yet, however.

The Planck constant h is given by $h = 6.626 \times 10^{-34}$Js.

The next important breakthrough came when Albert Einstein explained the so-called photoelectric effect in 1905 by assuming that light itself consists of tiny packets nowadays called **photons** (see also Figure 12.25). In a sense, photons are like particles with a definite amount of energy depending on the wavelength of the light. A single photon, of wavelength λ, has energy equal to

$$E = h\nu = \frac{hc}{\lambda} \qquad \textbf{Planck relation} \qquad (12.5)$$

where c is the speed of light and λ the wavelength. This relation is generally referred to at the **Planck relation**. (Note that the relation between frequency ν and wavelength λ is given by the general relationship: $\lambda\nu = c$, as in Equation 11.14.) The direct consequence of Einstein's insight is that the intensity of a light source is not given by some kind of strength of its light wave but by the number of photons emitted. In other words, the energy of a photon is reflected in its wavelength while the intensity of a ray of light is a function of the number of photons in the ray.

This sounds odd! From optics we clearly know that light behaves like a wave. Indeed, the English scientist Thomas Young performed his famous "double-slit" experiment, around the year 1800, where he showed that light passing through two nearby slits produces a characteristic interferences pattern of bright and dark stripes on a screen (see also Chapter 13). Then Einstein introduced particle-like photons to explain the photoelectric effect. Isn't a particle like a small ball or bullet — nothing remotely close to a wave? One of the most fundamental insights of quantum mechanics is that all matter has both wave as well as particle properties — so photons, representing what in general one would call a wave, can behave like particles, while vice versa electrons, representing what in general one would consider to be a very small ball or bullet, can behave like a wave (see also p. 273). The details of this so-called particle-wave duality are discussed in Chapter 24.

What we know now is that light comes in little packets called photons, each carrying a well-defined amount of energy depending on its wavelength. What we don't know yet is how this is connected to the curious black lines in the absorption spectrum of hydrogen.

At the end of the nineteenth century, there was another mystery we haven't mentioned here. In 1909, Ernest Rutherford clearly established the atomic nature of matter by showing that a positive nucleus is surrounded by electrons. The picture then was that the electrons circle the nucleus in a fashion similar to the motion of the planets as illustrated in Figure 12.18.

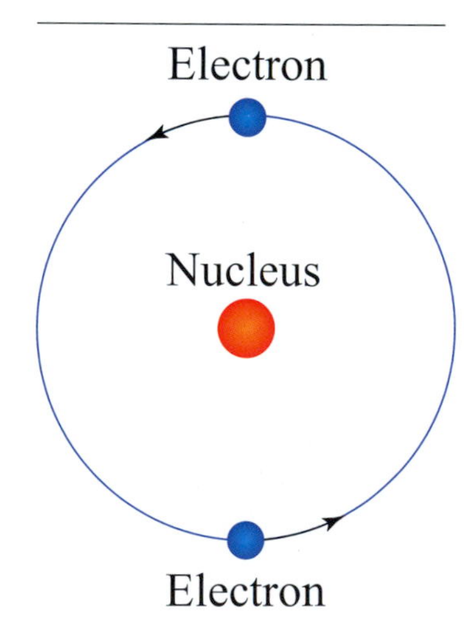

Fig. 12.18: Model of an atom as a planetary system.

An atom as a planetary system is a nice idea, but from the outset it was clear that such a model was not particularly realistic. From experiments with electricity, it was well known in 1897 that charged particles flying in a circular orbit should radiate, thus losing energy. Indeed, it was calculated that atoms would rapidly collapse in the planetary model. But, as is evident from our own existence, atoms are, in general, exceedingly stable.

In 1913, the Danish physicist Niels Bohr mediated the problem of the atomic collapse by conjecturing that electrons can only move in well-defined orbits and that a change in orbit is associated with the emission of a photon when the orbits decreases and with the absorption of a photon when the orbit increases, as illustrated in Figures 12.19 and 12.20. One can then plot the allowed energy levels like in Figure 12.21, which shows the energy levels of hydrogen. From this figure, it is also possible to see the analogy of lifting a stone up. The higher energy levels have a higher potential energy, and this energy can be tapped into just like in the case of water flowing through turbines in a dam.

The difference in the energy of the electron when jumping from one orbit to another needs to be exactly equal to the energy of the photon absorbed or emitted thus guaranteeing the conservation of energy — one of the most important notions in all of science. Since the energy of a photon is a direct function of its wavelength as per Equation 12.5, absorption and emission occur only at very specific wavelengths.

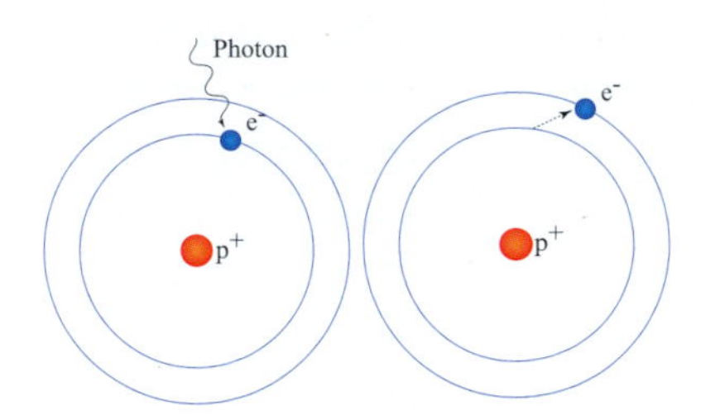

Figure 12.19: When absorbing a photon, the electron jumps to a higher orbit.

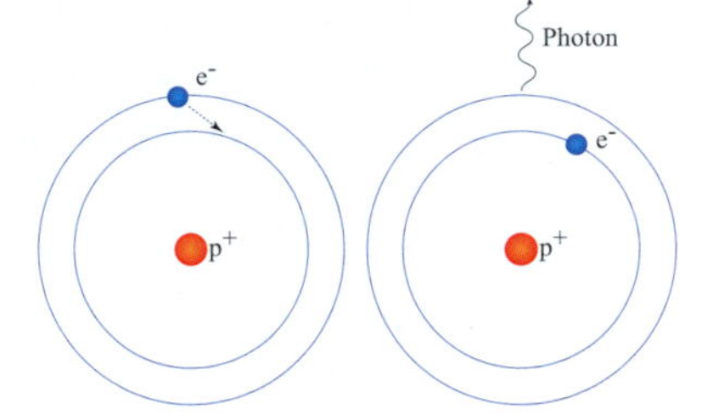

Figure 12.20: When emitting a photon, the electron jumps to a lower orbit.

We can now understand the black lines in the absorption spectrum of hydrogen. Only those photons with exactly the right wavelength to make an electron jump from one orbit to another are absorbed. All the other photons will simply pass through.

Of course, one can ask what happens to an electron that has jumped to a higher orbit (such a higher orbit is usually called an excited orbit). In Bohr's model, electrons prefer to be in as low an orbit as possible under the constraint that each orbit can only contain a certain number of electrons as determined by the Pauli exclusion principle (see also Section 4.7). For example, the lowest orbit can hold two electrons while the next higher orbit can hold eight electrons. Therefore, an element like lithium which in total has three electrons per atom in its usual state will have two electrons in the lowest orbit and one electron in the next orbit. If by absorption of a photon an electron has jumped to a higher orbit, the electron will after a short time return to its preferred orbit by emitting a photon.

But then, if the absorbed light is reemitted shortly after, why do we nevertheless see the black lines? The reason for this is that the measuring apparatus which records the spectrum only takes in light from one direction – the direction where the light source is relative to the apparatus with the hydrogen in-between. When light is absorbed and reemitted, the reemission is in *all* directions and hence the fraction of reemitted light in the original direction is quite small, giving rise to the black lines.

By experimentation with other elements, it turns out that each element has a characteristic absorption spectrum, and from this one can infer that the energy differences between orbits are also element dependent.

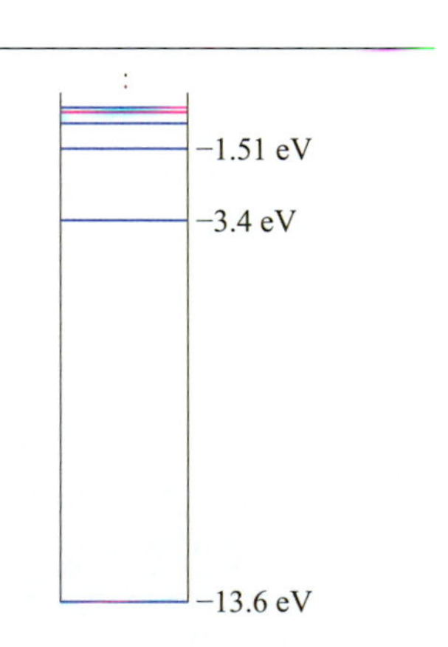

Fig. 12.21: The energy levels of hydrogen.

So now we can also explain why the spectrum of the sun has many black lines: they correspond to the various elements in outer regions of its atmosphere. Indeed, we can even explain the spikes in the spectrum of fluorescent light now. In fluo-

Louis de Broglie

Figure 12.22: Portrait of Louis de Broglie.

Louis de Broglie was born in France on August 15, 1892. As his family name is derived from the Italian name Broglia, it has a somewhat unusual pronunciation sounding similar to the English letter combination "broy". Although his early interest was in the humanities, having obtained a degree in history in 1910, de Broglie soon switched over to physics and mathematics which gained him a degree in 1913. After that he was conscripted into the wireless section of the military, a post he kept until the end of World War I in 1918.

After the war, Louis de Broglie resumed his studies and specialized in theoretical physics. He was especially interested in the then-novel phenomena involving quanta and he obtained his doctorate degree in 1924 at the Faculty of Sciences of Paris University for the thesis "Researches on the quantum theory."

De Broglie's thesis contained results that were essential ingredients for what would become a theory that transformed all of science and with it the world we live in: quantum mechanics. The key result was the prediction of the wave nature of the electron, spectacularly verified by Davisson and Germer in 1927 when they experimentally discovered electron diffraction. In 1929, he was awarded the Nobel Prize "for his discovery of the wave nature of electrons".

To this very date, it remains puzzling as to why nature displays both particle-like and wavelike properties. However, experimentally, this duality has been confirmed countless times: Light, usually a wave, can be particles (photons) as in the photoelectric effect while electrons, protons etc., usually thought of as particles, can be waves. For more details, see p. 273.

Although one may think that the wave nature of electrons is rarely encountered in daily life, one particular outcome is quite well known: the scanning electron microscope (SEM). Figure 12.23 shows an SEM image of fly legs. The reason why SEM images have much higher resolutions than optical microscopes is that the wavelength of the electron is much shorter than that of visible light. The wavelength of an electron in a typical SEM can be about 5 pm (5×10^{-12} m), roughly 100,000 times shorter than green light!

Figure 12.23: SEM image of fly legs.

Even though de Broglie's greatest work was in physics, he kept his interests in the humanities and wrote about the philosophy of science as well as the value of modern scientific discoveries.

Louis de Broglie lived a long and fruitful life. He died on March 19, 1987, at the age of 94.

The particle-wave duality

Basic interference and refraction experiments readily show that light behaves like a wave. Although Newton had proposed the corpuscular nature of light, the failure to explain wavelike phenomena with this picture meant that the idea was largely abandoned. Atoms, on the other hand, clearly seemed to be particles that to some extent behave just like bullets or tiny balls.

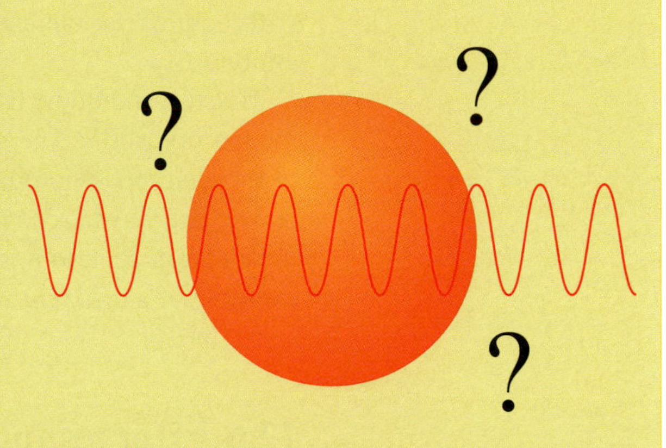

Figure 12.24: Is light made of particles or waves?

It appears that in classical (i.e. nonquantum) physics waves are waves and particles are particles. It has been proved impossible to make bullets behave like a wave or waves like little bullets. Modern experiments, however, show that without doubt, light can behave as if it were made up of particles (see the photoelectric effect below) and electrons can behave as if they were waves (see the scanning electron microscope image on p. 272). How can this be resolved?

With the advent of quantum mechanics, it has become clear that all fundamental physical phenomena have both properties at the same time, and that it depends on the circumstances which aspect is the most clearly recognizable. Thus we can say that light consists of photons, the particle that makes up electromagnetic radiation, even though in daily life light as well as other electromagnetic phenomena like radio-"waves" clearly behave like waves.

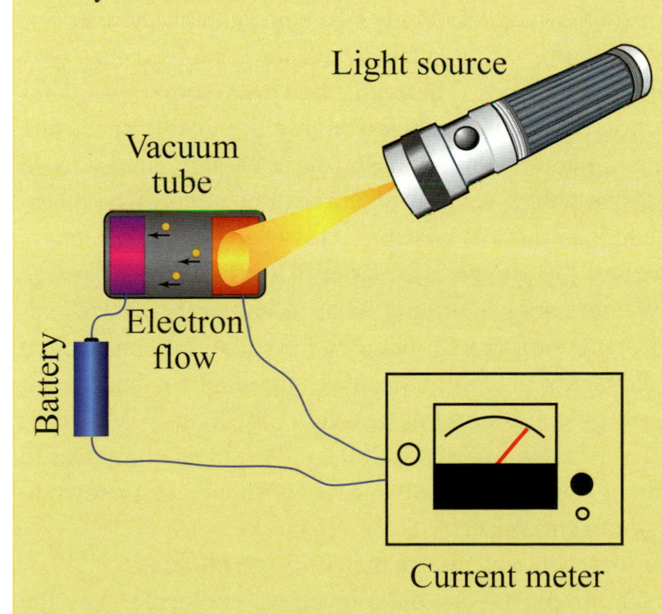

Figure 12.25: The photoelectric effect reveals the particle aspect of light.

The particle nature of light can be observed in the photoelectric effect. In this effect, light strikes a charged plate and liberates electrons if their energy is sufficiently raised by the impinging light to leave the surface. Experimentally, it can be seen that rather than the intensity, the wavelength of the light determines whether a current flows or not. In the wave picture, this is impossible to explain. If, however, light is taken to consist of "light quanta" (i.e. photons), with an energy proportional to their wavelength, then the experimental observation is readily understood as was first shown by Einstein in 1905. For this Einstein received the Nobel Prize in 1921 (remarkably Einstein did not receive the Nobel Prize for his special and general theories of relativity).

The photoelectric effect is employed in the photo-sensors that are encountered in daily life, for example, in switches that automatically turn on a porch light when it gets dark in the evening.

rescent light, the electrons were initially not pushed to a higher orbit by incoming light but by the supplied electric current. When those electrons return to their favored orbit, they emit a photon at certain exactly determined wavelengths creating the spikes. Since light sources that emit only one or two wavelengths are somewhat uncomfortable for human beings, most fluorescent lights are engineered such that with help of various physical principles a broader spectrum besides the spikes is emitted.

Last, it should be noted that jumping to a higher orbit is similar to a ball being moved up a hill. The act requires energy, namely, exactly the energy contained in the absorbed photon, and this "stored" energy can be recovered. As mentioned above, one way to recover the stored energy is by reemitting a photon, thus bringing the electron back to its preferred orbit. However, this is not the only way! A good example of a different way is photosynthesis.

12.6 Harvesting sunlight

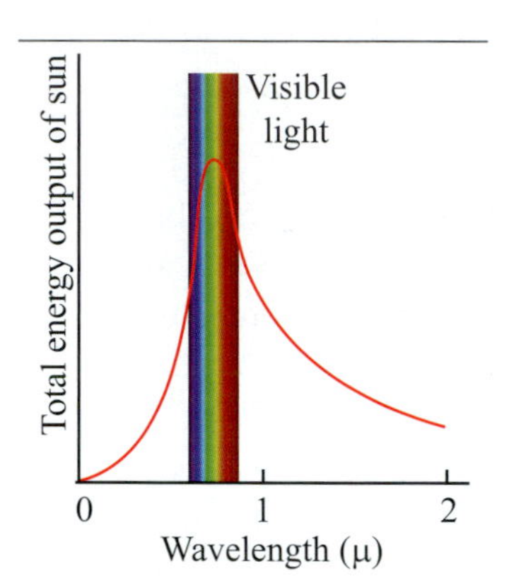

Fig. 12.26: The sun has the highest energy output at visible wavelengths.

So, all in all, we have seen that the light energy coming from the sun consists of small packets in the form of photons. A photon can only be absorbed as a whole giving up all of its energy or none. It is a matter of take it or leave it. Therefore, if a molecule wants to harvest the energy in a photon, it must do so by a mechanism that can exactly handle this energy with no leftovers. Not surprisingly, therefore, molecules can in general only absorb photons with certain very specific wavelengths. And this is exactly what happens for chlorophyll! Chlorophyll can only absorb red and blue light.

Why does chlorophyll absorb only red and blue? To some extent, there is certainly a lot of evolutionary freedom. A molecule could have evolved that only absorbs green light but not blue and red. There are, however, some considerations that narrow down the possible frequencies to roughly the visible spectrum. If the wavelength is too long, a photon simply doesn't have enough energy to do many useful things. Indeed, one would want to obtain as much energy as possible and that means a wavelength as short as possible. However, if the wavelength becomes too short, the energy of the photon is so large that it can easily damage organic molecules. Photons with energies around what to human beings is visible light have exactly the desirable property in that they have about the maximum amount of energy without damaging organic molecules important for human existence. This can also be seen by considering ultraviolet (UV) light. UV light is just beyond blue and with a wavelength of about 200 to 400 nm as compared to blue with a wavelength of about 450 nm. Excessive UV exposure is well known to substantially increase the risk of skin cancer.

Let us now have a closer look at how leaves actually absorb photons.

As mentioned previously, in plants the photoreceptive molecule, i.e. the molecule which absorbs photons, is chlorophyll. In isolation, a chlorophyll molecule brought to an excited state by the absorption of a photon will rapidly return to the ground state by reemitting a photon, as can be seen in the fluorescence of chlorophyll illustrated in Figure 12.27.

In order not to lose the energy gained from the absorption of a photon to fluorescence, chlorophyll molecules are organized into so-called photosystems where the vast majority of chlorophyll molecules act as an antenna complex. In the antenna complex, chlorophyll molecules are arranged in a two-dimensional array embedded in the membrane of the thylakoid consisting of 200 to 300 individual chlorophyll molecules together with other molecules. There are two types of photosystems and each thylakoid has many of these pairs.

In a photosystem, when a photon arrives, an electron in one of its chlorophyll molecules in the antenna complex is elevated to a higher energy state. The energy captured is then passed on to neighboring chlorophyll molecules by resonance energy transfer via a number of "hops" until it reaches the reaction center where it can be used for useful work. This process is schematically represented in Figure 12.28. In the reaction center, a special pair of chlorophyll molecules passes the exitation to a primary electron receptor from where the energy is used to do useful work.

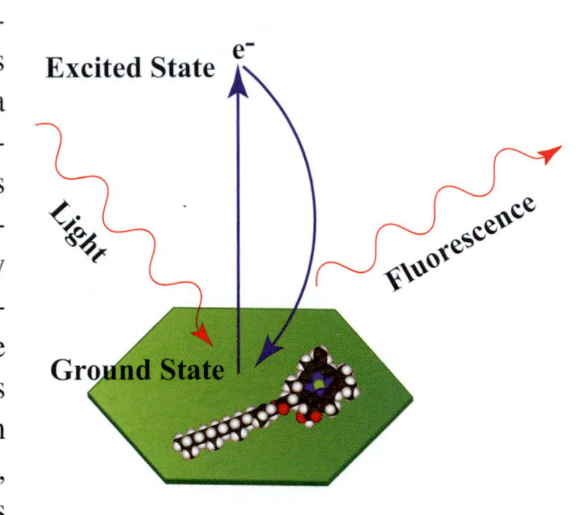

Figure 12.27: Fluorescence of a single chlorophyll molecule.

Some details of photosynthesis are illustrated in Figure 12.29. Photosystem I is the more basic of the two photosystems and the only system employed by green sulfur and purple sulfur bacteria. Its primary product is NADPH. Photosystem II is similar to photosystem I and works in conjunction with it. In cyanobacteria and chloroplasts, photosystem II sends its electron to photosystem I which therefore is its final electron acceptor.

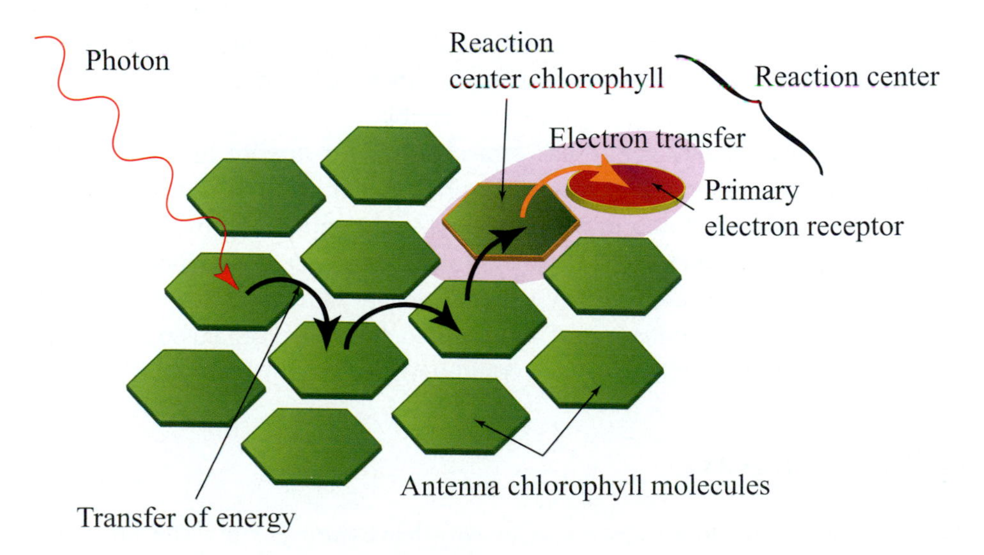

Figure 12.28: Schematic illustration of the photosystem.

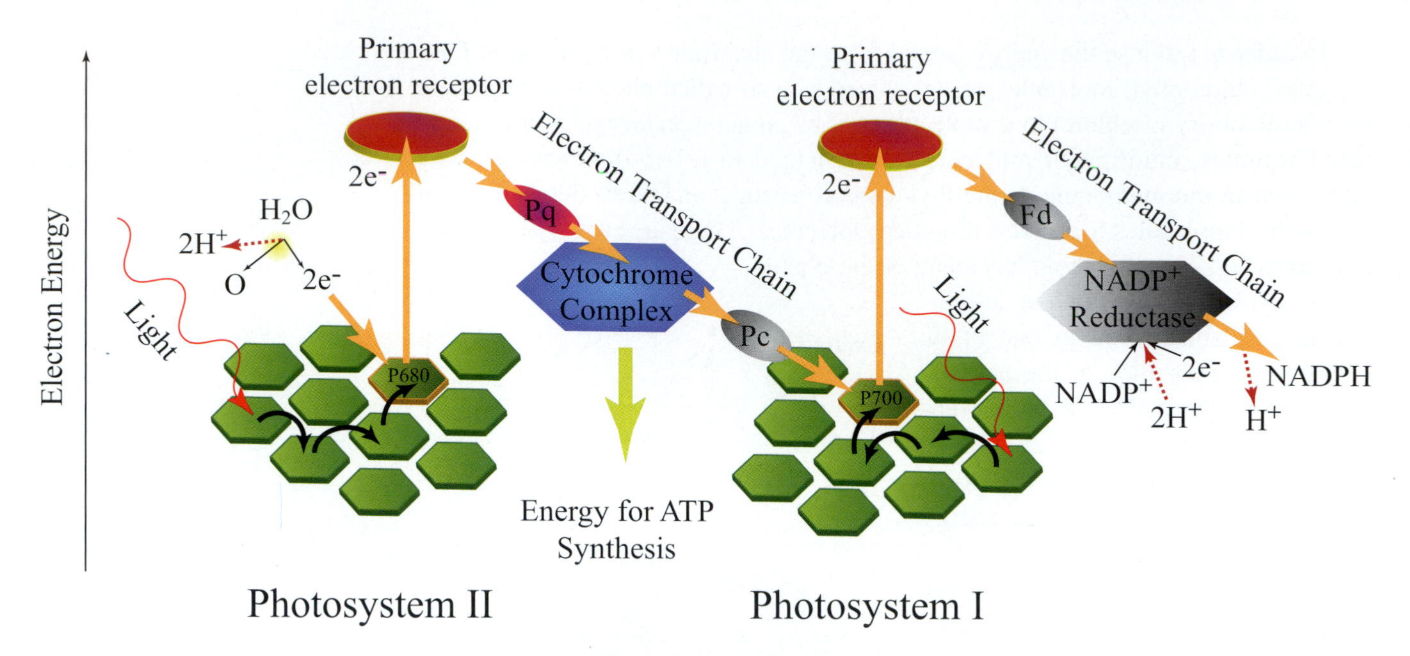

Figure 12.29: Some details of photosynthesis.

When photosystem II gives up an electron it must somehow regain it. As mentioned in Section 12.4, this is done by splitting water and taking the hydrogen molecule's electron with oxygen as the "waste" product.

12.7 Bioluminescence

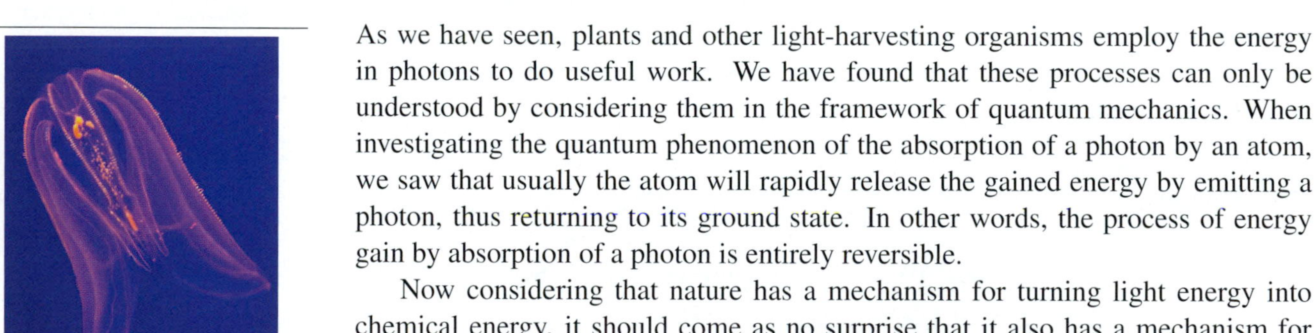

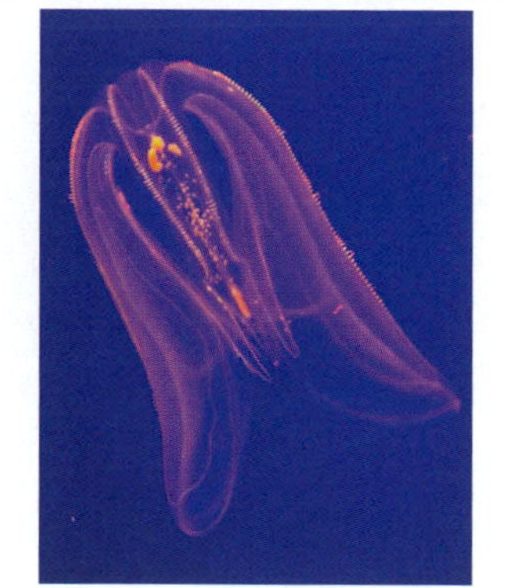

Fig. 12.30: Luminescent comb jelly.

As we have seen, plants and other light-harvesting organisms employ the energy in photons to do useful work. We have found that these processes can only be understood by considering them in the framework of quantum mechanics. When investigating the quantum phenomenon of the absorption of a photon by an atom, we saw that usually the atom will rapidly release the gained energy by emitting a photon, thus returning to its ground state. In other words, the process of energy gain by absorption of a photon is entirely reversible.

Now considering that nature has a mechanism for turning light energy into chemical energy, it should come as no surprise that it also has a mechanism for turning chemical energy into light. In general, light emitted by living organisms as, for example, in Figure 12.30 is called bioluminescence, a phenomenon rather more common than usually directly experienced by human beings. Among the species notable for the emission of light are such diverse organisms as squids, mantis shrimp, fireflies, jellyfish, and certain fungi.

12.8 The answer

We can now formulate an answer as to why leaves are green. The absorption of light is a quantum mechanical process that can only occur when the energy of a photon exactly matches an energy level change in the absorbing molecule.

The energy levels of the main light-absorbing pigments (chlorophyll) are such that only red and blue light match these energy levels. Hence, red and blue light are absorbed while the energy in green light does not match the available energy levels in chlorophyll, and is consequently either reflected or transmitted. This reflected green light is the light we see when observing leaves, and that is why leaves are green.

All in all, we can state: **The harvesting of energy from light is a fundamentally quantum mechanical process in leaves. Leaves are green because the chlorophyll molecule harvesting light can only absorb red and blue light, reflecting the green light that we see.**

12.9 Exercises

1. On an inhabitable planet orbiting a significantly hotter star than our sun, would the leaves more likely be blue or red?
2. In which part of the chloroplast does the Calvin cycle occur?
3. How many photons does it take to split one water molecule?
4. What would the color be of fluorescent chlorophyll molecules?
5. If an atom would have three different excited levels, how many different wavelengths can the absorbed or emitted photons have?
6. Why are there two different lines in Figure 12.10?
7. We claim that the wavelength of a photon must exactly match the energy difference between the states of an atom or molecule. Then why are the peaks in Figure 12.10 so broad? Shouldn't they be very sharp?
8. Is it conceivable to have a photosynthesis-like process in leaves that are transparent of visible light (that is, the visible light cannot be used as an energy source)?
9. What color would leaves have if they could also absorb green light?
10. What color would leaves have if they could only absorb green light?

13: Vision

Q - Why Can We See Sunlight?

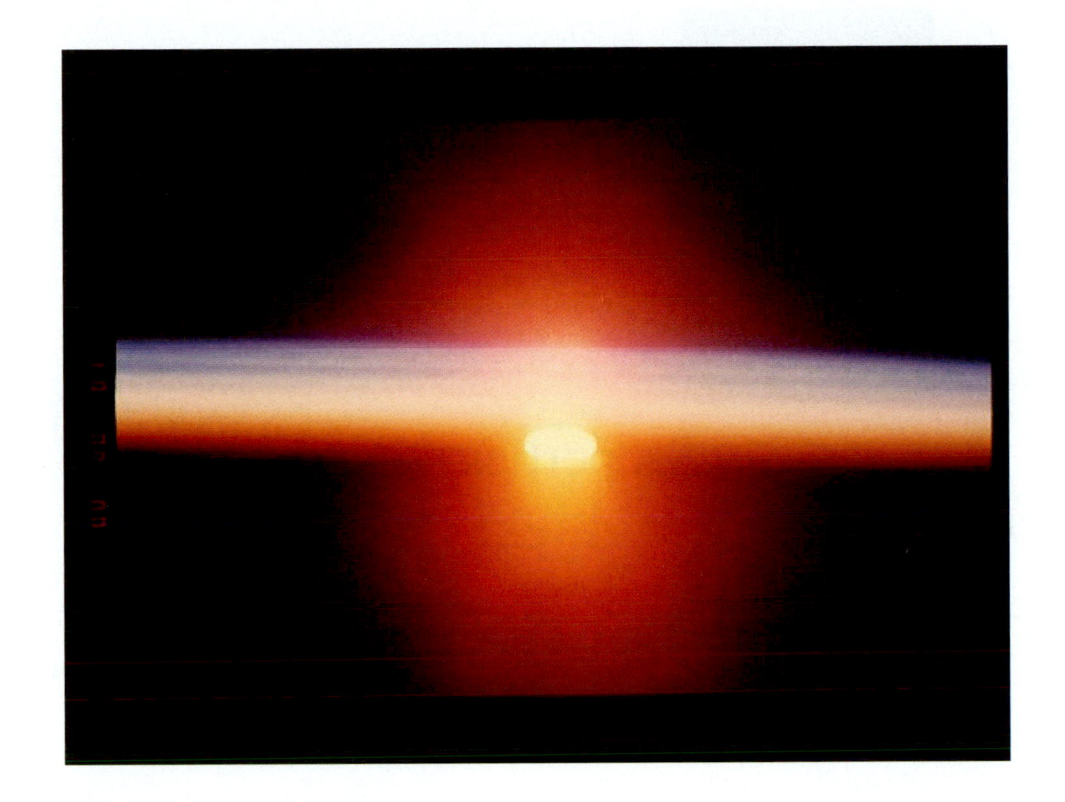

Chapter Map

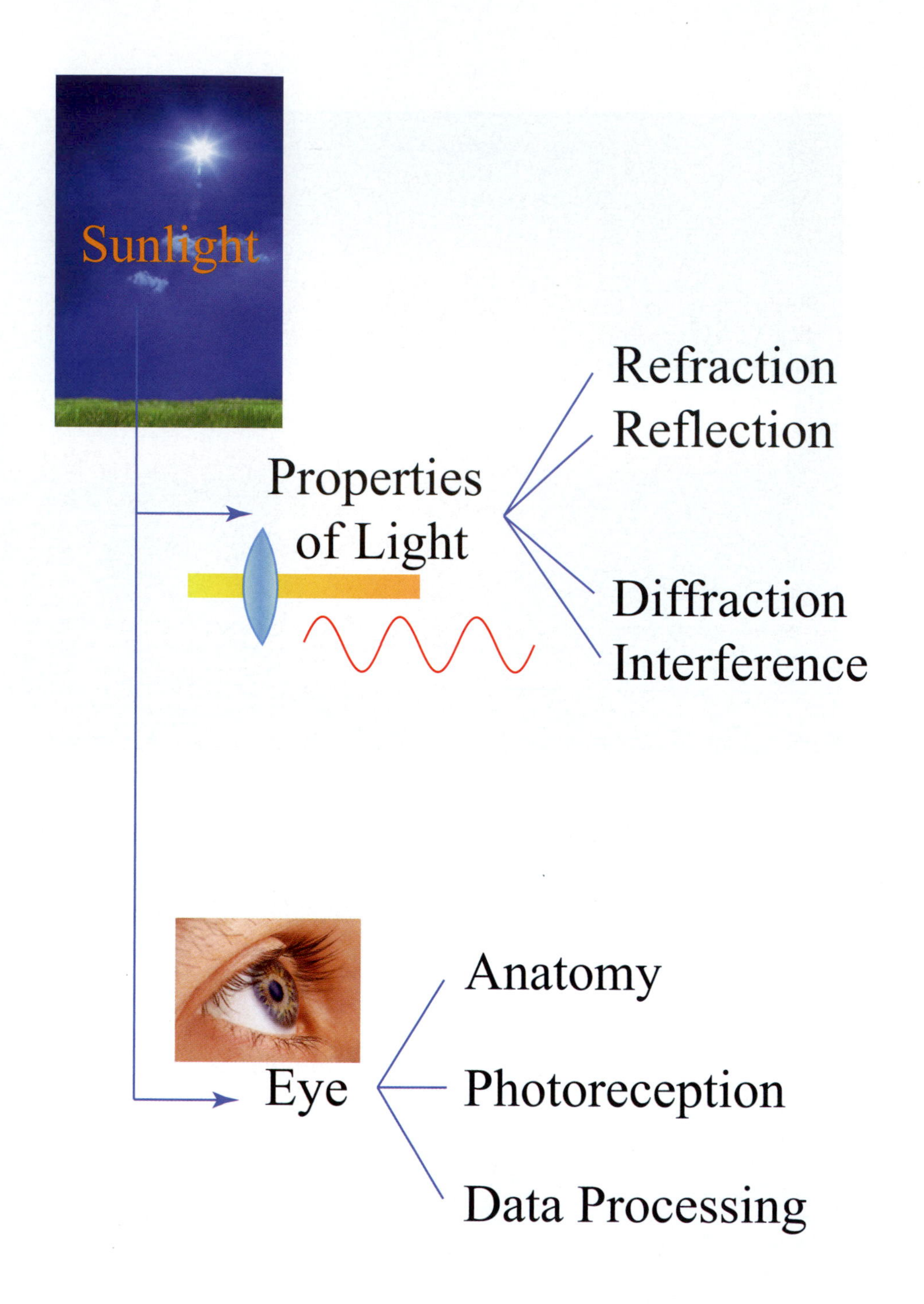
Sunlight
Properties of Light
Refraction
Reflection
Diffraction
Interference
Eye
Anatomy
Photoreception
Data Processing

13.1 The question

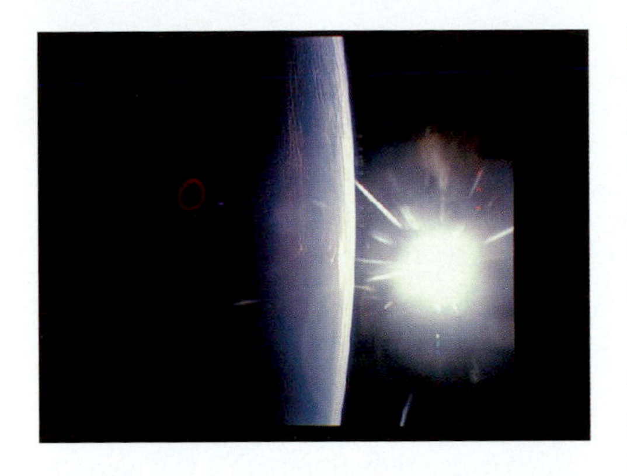

It is easy to get distracted by the shimmering reflections of the evening sun in a calm mountain lake. Indeed the things we see (or don't see!) have a strong impact on our emotions, and vision plays an essential part in how our society is organized.

However, vision doesn't come cheap; our eyes occupy a prominent place in our heads and are directly connected to the brain which needs to carry out a rather enormous amount of processing to make sense of all the incoming data — a feat well beyond early twenty-first century computer technology.

Fig. 13.1: The cat's eye is very similar to a human's but is dichromatic.

Virtually all the naturally occurring light originates from the sun, and the question for this chapter is: **"Why can we see sunlight?"**

In order to answer this question, we need to consider a) what is sunlight, b) how light propagates from a source to a destination and c) how light can be detected by living organisms. Even when detected, though, without some form of information processing, the data obtained from the light detectors will be useless so we need to "see" what is minimally required to have some form of vision. Last, one may wonder why the sense of vision evolved in the first place.

More generally speaking, light is a phenomenon that has fascinated human minds for many millennia, but a satisfactory explanation of its properties and underlying principles has been possible only during the last century. Besides being essential for vision (the eye "sees" objects by receiving light that has bounced off or passed through objects), light has almost limitless applications in daily life.

13.2 Colors

We know just by looking at the various signs about town at night that light comes in many colors, and one can show that white light is actually a combination of all colors. In a famous experiment performed by Newton, he took a prism and shone white light on the prism; the resulting refracted wave, received on a screen, was seen to consist of all the colors of the rainbow, as illustrated in Figure 13.3. One can show that the different colors of light are associated with different wavelengths, ranging from about 400 nm (violet) to about 700 nm (red). As the wavelength of light decreases, the human eye perceives it as light changing its color from red to violet. In other words, each distinct color has its own unique wavelength, and white light is a combination of waves of different wavelengths.

Fig. 13.2: The wavelength of light determines its color.

Of course, if one has the colors with wavelengths ranging from 400 nm to 700 nm one can wonder what goes on for different wavelengths and whether light for such wavelengths exists. It turns out that it does but, for historical reasons, it is generally referred to by the more generic term **radiation** or more precisely EM radiation where "EM" stands for electromagnetic. Light in the range from 400 nm to 700 nm is then often called "visible light". Shown in Figure 13.4 is a section of the different wavelengths of radiation.

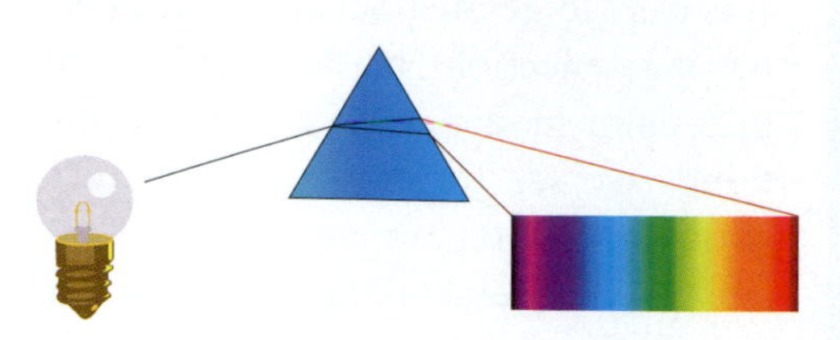

Figure 13.3: The wavelengths of light emitted by a white electric bulb's filament can be resolved into many colors by a prism.

The wavelengths of radiation span a vast range, from extremely long wavelengths exceeding kilometers to very short wavelengths smaller than the size of a proton. From left to right, in Figure 13.4, we have radio waves which have a wavelength of about 500 m, followed by short radio waves, and so on down to around a 1-m wavelength for TV waves. We then have microwave radiation with a wavelength around 1 cm, followed by infrared radiation down to 10^{-6} m. Radiation of wavelength between 10^{-7} m and 10^{-9} m is invisible and is called ultraviolet radiation. Radiation of wavelength from 10^{-9} m to 10^{-12} m is called x-rays, and radiation shorter than 10^{-12} m all the way to a vanishingly small wavelength is called **gamma rays**.

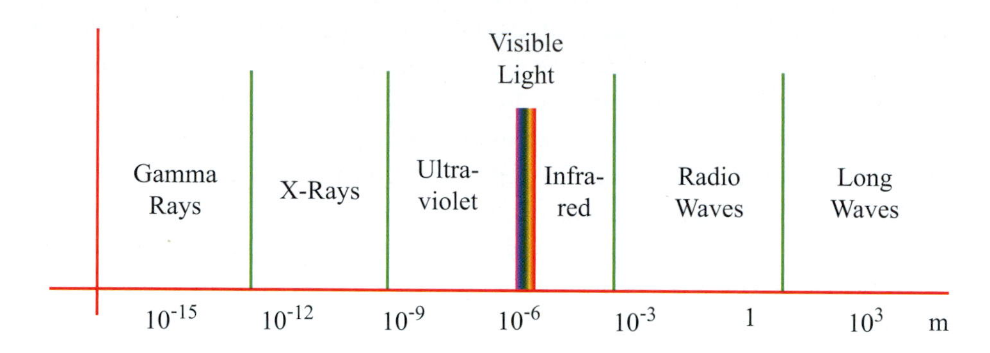

Figure 13.4: The spectrum of visible and invisible light.

One can easily see from the list of the different forms of radiation that radiation has unique properties that have been put to society's use in varied manners, from x-rays in medical science to TV, radios, mobile phones, microwaves, communications and so on.

13.3 What is sunlight?

Sunlight is an example of "white-like" light, i.e. a combination of different wavelengths. Much of the light from the sun is in the visible range but some is invisible, such as infrared light ("below red") and ultraviolet light ("above violet") as shown in Figure 13.5.

Clearly standing in sunlight, one can experience warmth. Indeed, with the help of a lens, one can light a fire when the sun is shining brightly. As there will be no fire when it is cloudy, it is clear that the energy for lighting the fire does not come from the lens but the sun. Indeed, any bright enough source of light can start a fire with the help of a lens.

One can ask about the energy carried by each particular wavelength that comprises sunlight. This is shown in the left panel of Figure 13.5. Note how the peak in the energy distribution is in the middle of the visible spectrum. Is this a coincidence? Probably not!

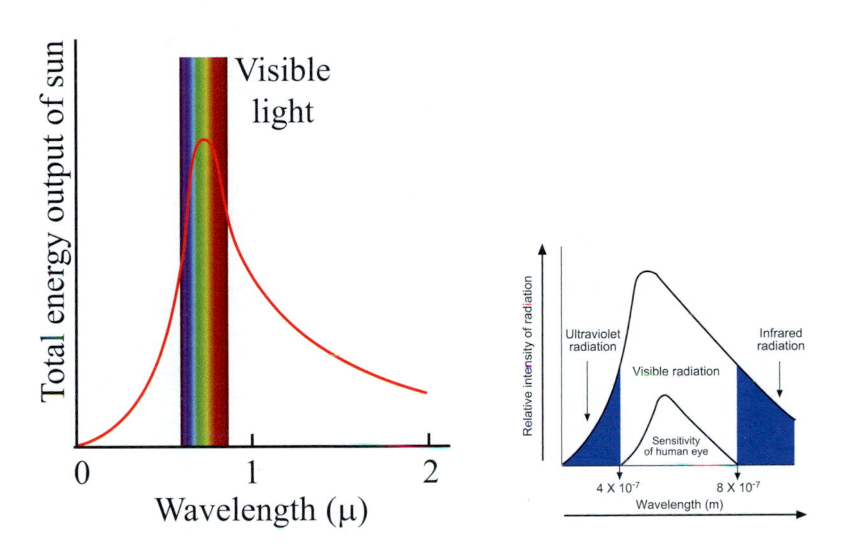

Figure 13.5: The electromagnetic spectrum of vision and the sensitivity of the human eye.

A common expression for a very hot object is "red hot" — why the adjective "red"? In fact the hotter an object is, the more it radiates at shorter wavelengths. So "blue hot" would be very hot indeed!

The sunlight that we experience on the surface of the earth is only a small portion of the solar radiation incident on the earth; a large portion of that radiation is being diminished by our atmosphere, by the processes of reflection, scattering and absorption. This is rather fortunate as exposure to many of sunlight's components can pose a health risk; excessive exposure of the human skin to ultraviolet light, for example, increases the risk of skin cancer.

The stratosphere (upper atmosphere) has a layer of ozone which is a strong absorber of the biologically harmful short wavelength ultraviolet radiation having wavelength below 280 nm. It is for this reason that there is a great deal of concern at the rate at which pollutants such as fluorocarbons deplete the ozone layer, allowing the ultraviolet radiation to penetrate.

The troposphere (lower atmosphere) also plays a useful role in diminishing harmful radiation. At this level, clouds (which are nothing but water and ice particles) and suspended matter such as dust and smoke reflect and scatter radiation. Indeed the blue color of the sky (and the reddish color of sunrises and sunsets) is because that scattering depends on wavelength.

What happens to all the energy that the sun provides the earth? A substantial portion (about 25%) of all frequencies of light that reach the earth is reflected back into space. Of all frequencies of light incident on the earth, 54.4% is in the infrared, visible and ultraviolet. Half of what gets through drives the weather, while only a tiny 0.15% of the incident light is used in photosynthesis by plants. Figure 13.6 shows some of the energy fluxes on earth.

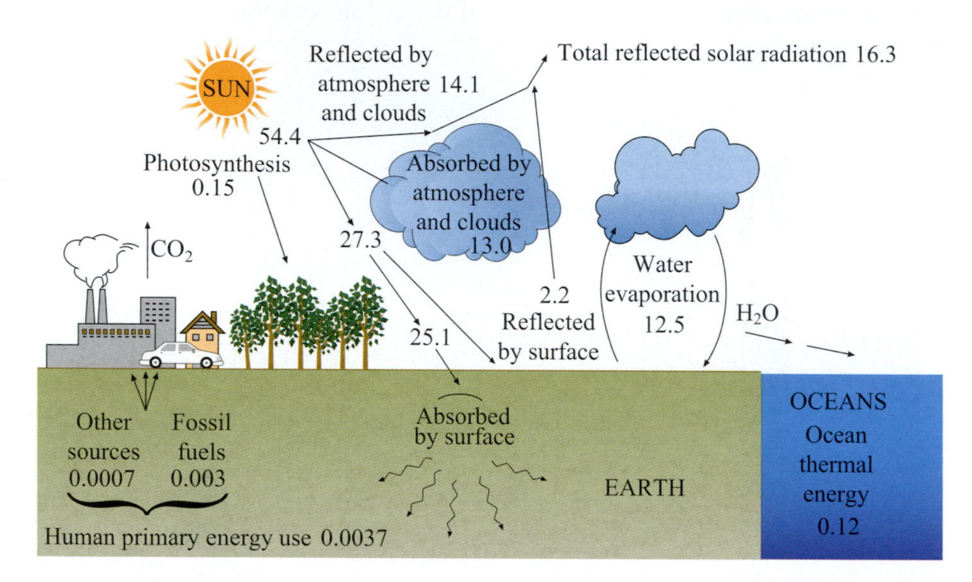

Figure 13.6: Areas where the sun's energy ends up. The numbers indicate percentages in terms of infrared, visible and ultraviolet wavelength of sunlight that is incident on the earth.

Hence, we see that light is a form of energy whose energy content is determined by its wavelength. However, without an organ to detect light, there would be no vision, so let us now first have a brief look at the anatomy of the human eye.

13.4 Basic anatomy of the human eye

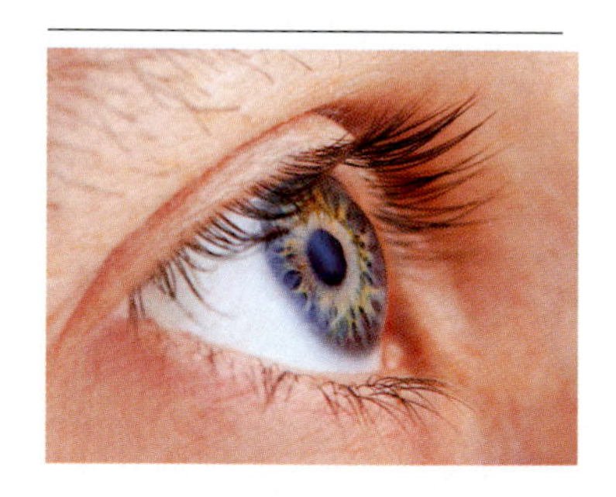

Fig. 13.7: A human eye.

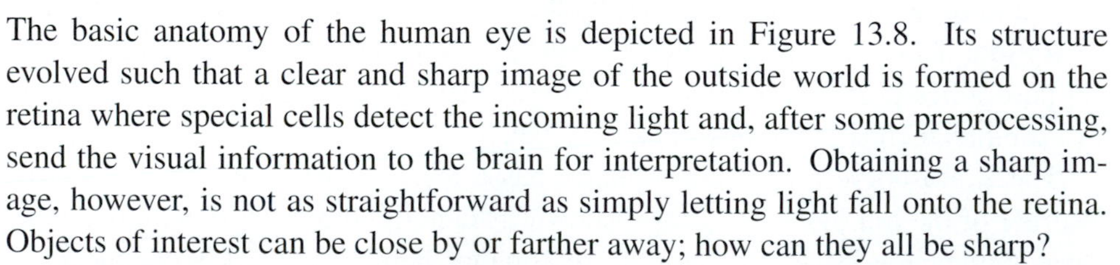

The basic anatomy of the human eye is depicted in Figure 13.8. Its structure evolved such that a clear and sharp image of the outside world is formed on the retina where special cells detect the incoming light and, after some preprocessing, send the visual information to the brain for interpretation. Obtaining a sharp image, however, is not as straightforward as simply letting light fall onto the retina. Objects of interest can be close by or farther away; how can they all be sharp?

To find out, let us briefly investigate the anatomy of the eye. It turns out that it consists of three main layers called tunics (in biology, a **tunic** is a covering or enclosing membrane or tissue), namely:

1. **Fibrous tunic:** The **fibrous tunic** is the outer layer of the eyeball and consists of the cornea and the sclera. The **cornea** is the transparent tissue that covers the iris and pupil, while the **sclera** is a whitish dense fibrous coat that protects the eye and helps maintain its (basically spherical) shape. The cornea is continuous with the sclera.

2. **Vascular tunic:** The **vascular tunic** consists of the iris, the ciliary body and the choroid. The **iris** controls the amount of light that enters the eye and consists of involuntary circular and radial muscles. The **ciliary body** controls

the curvature and thickness of the lens thus changing the focal point. The **choroid** contains blood vessels that supply the eye with oxygen and nutrients and that remove waste products while its black pigmentation prevents internal reflection. The choroid is continuous with the ciliary body and the iris.

3. **Nervous tunic:** The **nervous tunic** is the innermost sensory layer of which the **retina** is part.

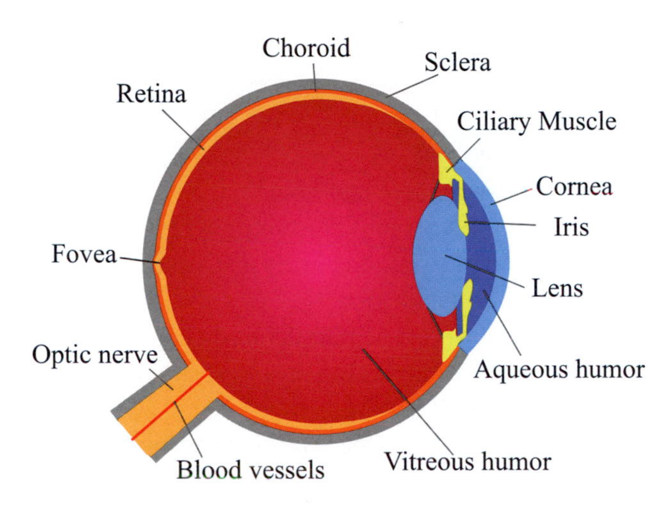

Figure 13.8: Cross section of the human eye.

Light first enters the eye by passing through the cornea. It then travels through the **aqueous humor** toward the pupil and subsequently through the **lens**. Although light is refracted in both the cornea and aqueous humor, the main control over the refraction of the incoming light (and hence its image on the retina) is through changing the shape of the lens. After passing through the lens, light travels through the **vitreous humor**, a jelly-like transparent substance, before impinging on the retina where the photoreceptors are located. The center of the retina is called the **fovea** and is its most sensitive region. When we want to look at something, we turn our head such that the fovea is on a direct line to that object and our eyes maximize the focus on this area.

From this, we can see that in order to understand why we can see sunlight, we need to know not only about light detection but also about how light propagates and how light is refracted by different materials and shapes. Let us therefore investigate these topics now.

13.5 Propagation of light

One of the first things noticeable about light is that it appears to reflect and refract since we can see objects around us when the light is on and we cannot see anything when the light is off. This suggests that light is well described by the notion of a ray perhaps consisting of small "light particles". However, a set of wonderful experiments shows that light can give interference patterns — similar to those that can be observed in water waves in a pond (for details see p. 292). This then suggests that light is a wave. Is light a wave or particles? Let us start by having a closer

Fig. 13.9: In this setting light appears very much to be ray-like.

"look" at easily verifiable properties of light.

Light:

- propagates
- refracts
- diffracts
- has color
- becomes fainter farther from the source
- carries energy
- behaves like a ray

If we put a light source in a box with a small hole and observe the light coming out of the hole in a dark room, we see that the light moves in straight lines within a homogeneous medium like air. With the help of detailed analysis employing various lenses, this can be verified and one can conclude that **light is a ray**. Depending on the source or the lenses, light can also converge, be parallel or diverge, and this is represented by a set of converging, parallel or diverging rays as shown in Figure 13.10.

If we consider light as a ray, does that imply that there is a propagation speed? As such this is not clear from the outset since when one turns on a light switch, the light immediately appears. If light has a finite speed, it surely must be very fast. But how fast? The first reasonably successful measurement of the speed of light was carried out by the French physicist Armand H. L. Fizeau in 1849. He shone a narrow beam of light on the teeth of a rotating cogwheel and placed a mirror behind it as illustrated in Figure 13.11.

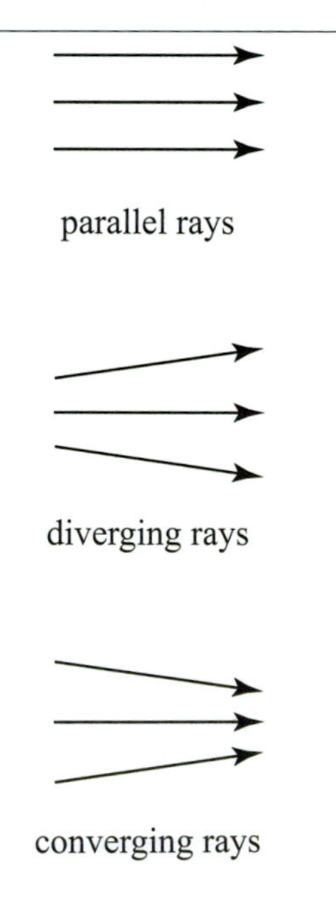

Fig. 13.10: Light can be considered as consisting of converging, parallel or diverging rays.

The rotating cogwheel breaks up the light into separate pulses. The idea is to then slowly speed up rotation and observe how much light returns through the gaps between the cogs. When the cogwheel rotates very slowly, light will go through a gap and then return through the same gap. Then at some speed, the disk moves fast enough such that the tooth following the gap is exactly blocking the returning light thus giving a dark image. If the cogwheel is spun even faster, then the returning light will hit the next gap and hence one has a bright image again. We can now calculate the speed as follows: If the number of turns per second is f, then it will take $1/f$ seconds for one turn and if the cogwheel has N teeth and consequently N gaps, it will take $1/(Nf)$ seconds to move from gap to gap. In other words, the time between gaps Δt is given by

$$\Delta t = \frac{1}{Nf}. \tag{13.1}$$

If the light pulse travels back and forth exactly once during this time interval, we have for the speed of light c equal to $2d/\Delta t$. That is

$$\Delta t = \frac{2d}{c}. \tag{13.2}$$

Combining Equations 13.1 and 13.2 we then obtain for the speed of light

$$c = 2dNf. \tag{13.3}$$

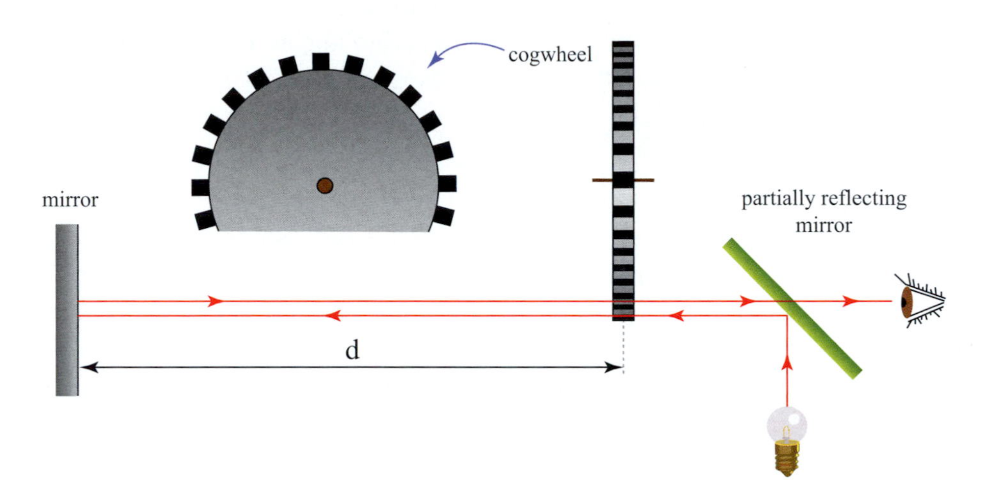

Figure 13.11: Schematic illustration of Fizeau's apparatus to measure the speed of light.

In Fizeau's experiment, the cogwheel had $N = 720$ teeth, the distance to mirror was $d = 8633$ m and $f = 25.2$ is the number of turns per second. Inserting these values into Equation 13.3, the speed of light is calculated to be 3.13×10^8 m/s which is very close to the actual value of 2.9979×10^8 m/s.

$$\textbf{Speed of light in vacuum: } \mathbf{c = 2.9979 \times 10^8 m/s}. \tag{13.4}$$

The following question naturally arises: how do light rays propagate in the presence of material bodies? In particular, two well-known cases are the **reflection** of light off polished surfaces (mirrors) as well as the **refraction** of light in passing from one medium to another.

Reflection

Consider the reflection of light off a mirror. As shown in Figure 13.12, the light ray starts at point A, is incident at point O of the mirror at an angle Θ_i with respect to the perpendicular of the mirror surface, and then is reflected through point B, making a reflected angle Θ_r. It is an experimental fact that the angles are equal, i.e.

$$\Theta_r = \Theta_i. \tag{13.5}$$

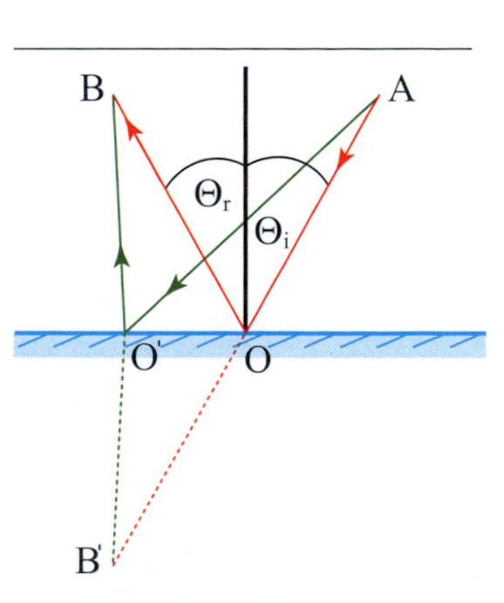

Fig. 13.12: Reflection of a ray of light off a mirror.

What does this mean (geometrically speaking)? If we look at the distance covered by the ray going from point A to B through O, we find that the distance is minimal when $\Theta_r = \Theta_i$. In order to illustrate this, let us consider the green path in Figure 13.12. Consider a reflection in the mirror of the paths from A to B; the reflection does not change the distances of the paths, and in particular OB = OB' and O'B = O'B'. Hence, path AOB is as long as path AOB', and likewise AO'B = AO'B'. If $\Theta_r = \Theta_i$ (i.e. the case of the red line) a little bit of geometry shows that the path AOB' is a straight line. A straight line is the shortest distance between two points and hence all other paths between those two points are longer. From this argument, we can conclude that when reflected, light takes the shortest path touching the mirror from point A to point B.

Refraction

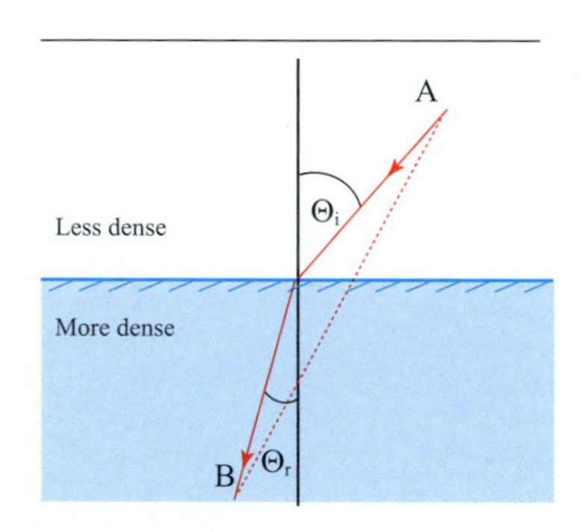

Fig. 13.13: In refraction, light entering a denser medium bends toward the normal.

Refraction is the bending, that is, the change of direction in the path of a ray of light when it passes from one medium to another (for exampple, from air to water). How strongly the light is deflected is related to the so-called **index of refraction** n. The index of refraction is related to the speed of light in a medium and defined as

$$\textbf{index of refraction} \qquad n \equiv \frac{\textbf{speed of light in vacuum}}{\textbf{speed of light in medium}} \tag{13.6}$$

which of course implies that $n = 1$ for vacuum. Intuitively, we expect that light will travel slower in a denser medium compared to a less dense medium. The speed of light in a medium with refractive index n is given by

$$\frac{c}{n}: \text{ Velocity of light in medium with refractive index } n. \tag{13.7}$$

Light leaving a less dense medium and entering a more dense medium bends toward the normal, whereas light leaving a more dense medium and entering a less dense medium bends away from the normal.

For example, a ray of light propagating through air hitting a piece of glass is refracted as shown in Figure 13.13.

Table 13.1 gives the indices of refraction for some typical transparent materials.

Medium	n
Vacuum	1.0000
Air	1.0003
Water	1.33
Plexiglas	1.51
Crown glass	1.52
Quartz	1.54
Diamond	2.42

Table 13.1: Indices of refraction for yellowish orange light.

When carefully measuring the relation of the angles made by light in the two media, one can find that

$$n_1 \sin \Theta_1 = n_2 \sin \Theta_2. \qquad \textbf{Snell's law} \qquad (13.8)$$

Snell's law is an experimental fact. Above, we saw that for reflection, the light chooses the shortest path. In the case of refraction, however, the path is bent as can be seen in Figure 13.13 (one can also easily see this by inserting one's hand into water, which will then look bent) and we can conclude that the light does not choose the shortest path. Light continues to travel in a straight path within a medium, but is bent when leaving one medium and entering the other medium.

How is light bent in travelling from one to the other medium? Which path would light be choosing then?

The easiest way to arrive at an answer is to first consider the following analogy: the velocity of light is higher in air than in glass, just like a man running on land is faster than a man swimming. Suppose a lifeguard starting from his chair is trying to reach a buoy floating in a swimming pool. What path should the lifeguard follow so that he reaches the buoy in the shortest possible time? If the lifeguard goes in a straight line, he will spend a certain fraction of time running on the ground, and the remaining fraction swimming. Since he swims slower than he runs, he should try running for a longer time on the ground before diving into the pool. On the other hand, if he runs too far on the ground, he will end up swimming more than he should. The exact point where he should dive into the pool for minimizing the time he takes to reach the buoy is the analogy of how light rays should bend as they leave one medium and enter the other.

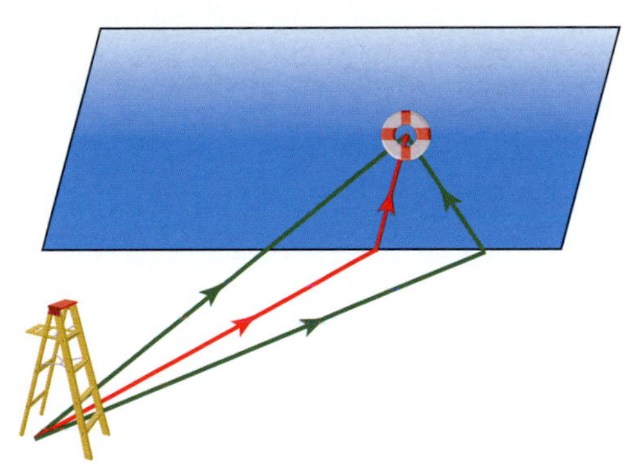

Figure 13.14: Lifeguard choosing optimum path (red), with green path being nonoptimal.

Perhaps, one could surmise that light does the same as our lifeguard and takes the *quickest* path, rather than the shortest, in going from one medium to the next. One way to find out whether this is true is to simply calculate which path would be the quickest when going from a point A in one medium to a point B in the other medium and then see whether the answer matches up with Snell's law. Doing so, one indeed finds this to be the case.

Lenses

The fact that light refracts when moving from one medium to the next can be used to construct lenses with varying properties depending on their shape. The lens in our eyes has a somewhat oval shape, so let us consider this type of lens first.

The function of an eye is to detect light such that the brain can form an image. In many animals, the actual detection takes place in the retina but there clearly is a problem. Objects of interest can be nearby or far away — how can the image be sharp in either case? One option is with the help of a lens as depicted in Figure 13.15.

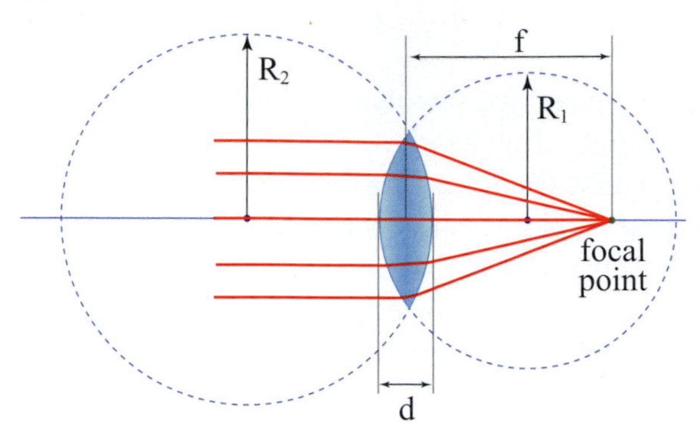

Figure 13.15: Biconvex lens.

In most basic lenses, the lens surfaces are shaped such that they lie on spheres of radius R_1 and R_2. The line connecting the centers is called the axis. Often R_1 and R_2 are identical such that a lens is symmetrical with respect to a plane perpendicular to the axis. If a parallel beam of light passes through this kind of lens, its rays will converge on the focal point f. The focal point is located on the axis, and its distance from the center of the lens can be calculated with the help of the so-called **lensmaker's equation** given by

$$\frac{1}{f} = (n-1)\left[\frac{1}{R_1} - \frac{1}{R_2} + \frac{(n-1)d}{nR_1R_2}\right] \quad \text{Lensmaker's equation} \qquad (13.9)$$

where n is the index of refraction of the lens material and d is the thickness of the lens.

If one is at the focal point behind a lens then much of the incoming light that originates some distance away (such that those rays are approximately parallel) will end up at a point. For vision, that would not be all too useful since that would mean that a somewhat distant object is only perceived as a dot. How could this be changed? One thing one could do is to see what happens if one is not exactly at the focal point. Let us place a screen a bit behind the focal point as in Figure 13.16; in this case we obtain a scaled yet sharp image.

It is notable that the image is upside down. Would this also be the case in our eyes? Indeed it is! The image on the retina is inverted and it is our brain that subsequently flips it to make it upright. If S_1 and S_2 are relatively large when compared to the thickness of the lens, the following relationship holds:

$$\frac{1}{f} = \frac{1}{S_1} + \frac{1}{S_2}. \qquad (13.10)$$

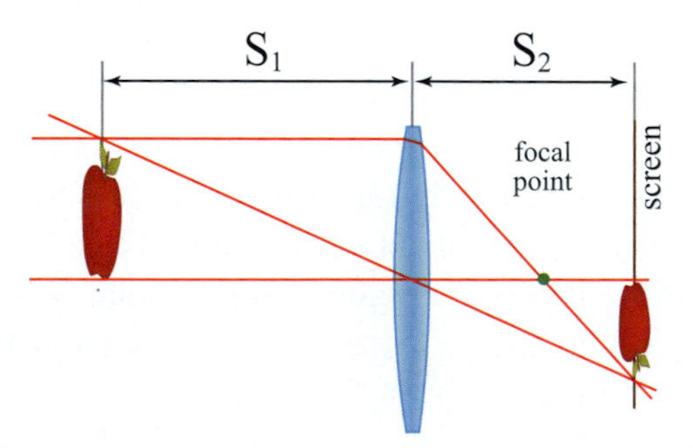

Figure 13.16: Image formation with a biconvex lens.

Considering the fact that the eye has a fixed size and that the distance between the lens in the eye and the retina is fixed, this means that we get a nicely sharp image only for objects at one certain distance. The rest would be more or less blurry. Although better than no vision, this is clearly not an ideal situation. How could this be remedied? One way would be to move the lens backward and forward (roughly this is what happens in a camera). Another way is to change the shape of the lens — that is, to change R_1 and or R_2 in Figure 13.15 — and this is exactly what happens in mammalian and many other animals' eyes.

Another common function of lenses is magnification, which is illustrated in Figure 13.17. The magnification M of a (thin) lens is well described by the formula

$$M = -\frac{S_2}{S_1} = \frac{f}{f - S_1} \tag{13.11}$$

where in the second equality Equation 13.10 was used. Some other common lenses are shown in Figure 13.18.

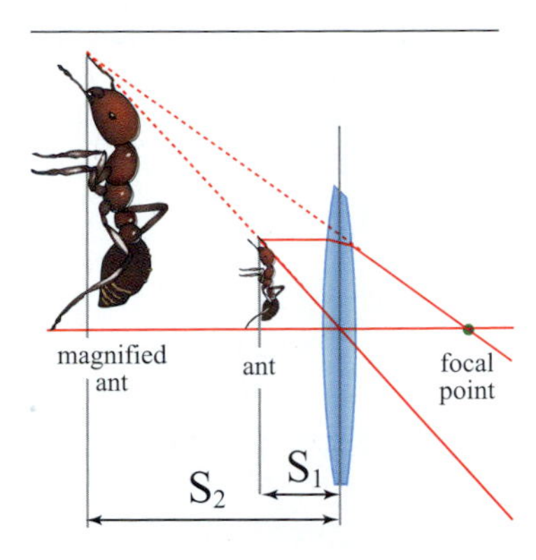

Fig. 13.17: Magnification with the help of a biconvex lens.

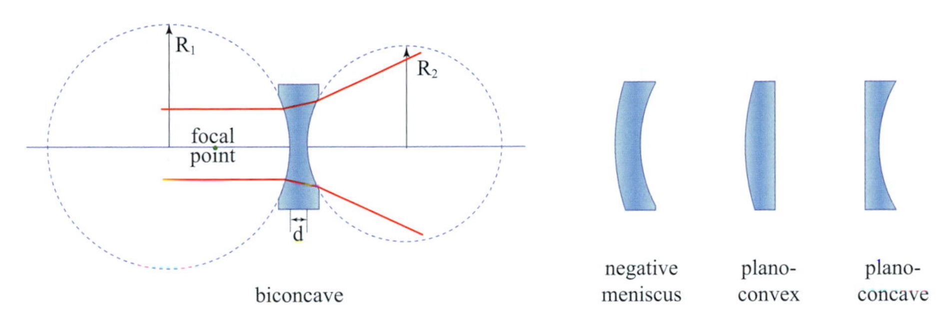

Figure 13.18: Some other common lens shapes. A biconcave lens, a negative meniscus lens where the inner radius is smaller than the outer radius, a plano-convex lens and a plano-concave lens.

Total internal reflection

If light travels from a denser to a less dense medium, as in the case of light traveling from water into air, given the right circumstances we can have the phenomenon of **total internal reflection** in which light is completely reflected back into the more dense medium, as shown in Figure 13.19.

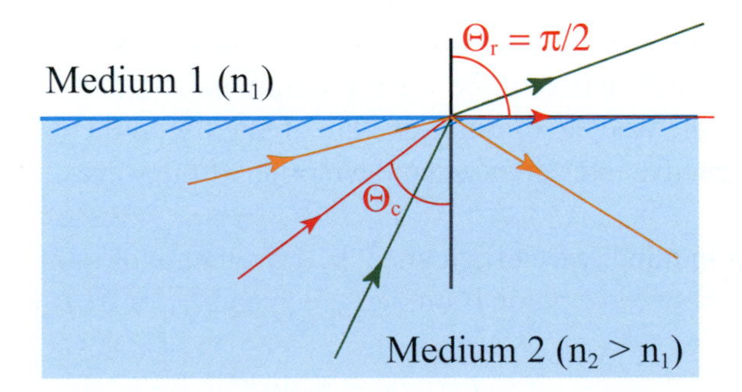

Figure 13.19: The orange ray displays total internal reflection.

This comes about in the following way: When a light ray hits the interface at an angle Θ_1, the ray is refracted away from the normal (recall that the ray goes from a medium with a higher index of refraction to a medium with a lower index of refraction). At some angle of incidence $\Theta_1 = \Theta_C$, the refracted beam will be

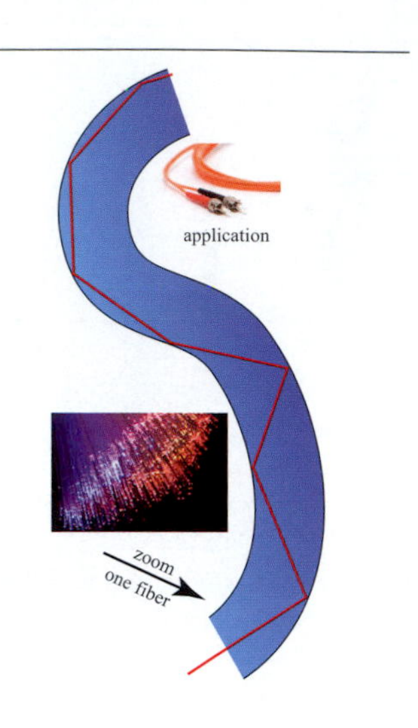

Fig. 13.20: Fiber optics are important in communication.

parallel to the interface giving $\Theta_2 = \pi/2$. For angles exceeding Θ_C, the refracted ray will then remain in the originating medium and thus be reflected at the interface. Given the indices of refraction of the two media, we can use Snell's law to calculate the critical angle Θ_C

$$\begin{aligned} n_2 \sin \Theta_C &= n_1 \sin \frac{\pi}{2} = n_1 \\ \sin \Theta_C &= \frac{n_1}{n_2}. \end{aligned} \tag{13.12}$$

One important application of this phenomenon is the field of **fiber optics**. Fiber-optic cables are made out of dense optically transparent fibers, and light propagates inside these light pipes due to total internal reflection with almost no loss of energy (see Figure 13.20). Fiber optics have many uses, from transmitting high volumes of telecommunication signals to medical laporoscopic and microsurgery procedures.

Interference

Thus far, we have found that describing light as a ray is in excellent accordance with experiments involving reflection and refraction. Indeed, general rules like Snell's law allow one to make a great many verifiable predictions. It would seem that all is well; but, in 1801, Thomas Young carried out an experiment that would become one of the most famous physics experiments of all time.

Put a bright light source in front of an opaque screen which has two slits. These slits let through light, and act as two sources of light. The fascinating outcome of this experiment is that the screen will show a so-called interference pattern of dark and bright lines with intensities depicted in Figures 13.21 and 13.22. Interference patterns are the salient property of waves and are well known to occur for sound and water waves (see also Chapter 11). Hence, the surprising (in "light" of the results on reflection and refraction) conclusion from the double-slit experiment of Figure 13.21 is that **light is a wave**.

From observations, one can find that two waves which are in phase (that is, the troughs and crests of the two waves are at the same place) show constructive interference where the motion of the wave is amplified and that waves that are out of phase (that is, the troughs of one wave are aligned with the crests of the other wave) show destructive interference and the motion of the waves cancel each other out.

Keeping this in mind, we can calculate the wavelength of light: The difference δ in path lengths between a point P on the screen and S_1 or S_2 is given by

$$\delta = r_2 - r_1 = d \sin \Theta \tag{13.13}$$

where d is the distance between the slits. For a bright fringe, we must have constructive interference and hence the path difference δ must be a multiple of the lights wavelength. Therefore, we have

$$\delta = d \sin \Theta_{\text{bright fringe}} = m\lambda \quad (m = 0, \pm 1, \pm 2, \ldots). \tag{13.14}$$

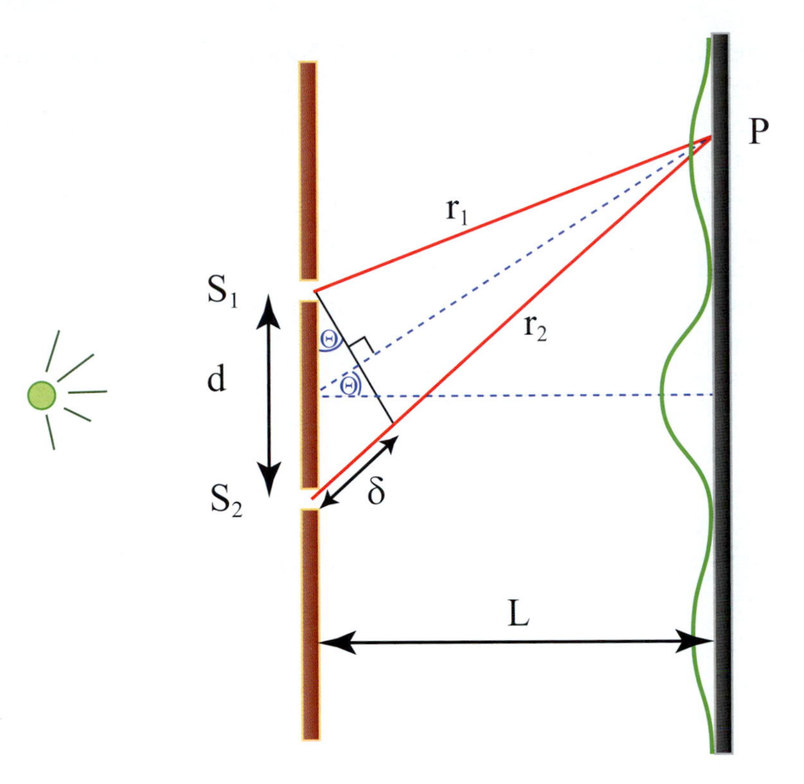

Figure 13.21: Double-slit experiment. S_1 and S_2 are two slits separated by distance d and that act as sources of light, with r_1 and r_2 being their distance from point P on the screen and $r_2 = r_1 + \delta$. The distance between the slits and the screen is L. The midpoint between the two slits makes an angle Θ with the point P.

Similarly for dark fringes we must have destructive interference and hence the path difference δ must be a multiple of the wavelength plus one half wavelength.

$$\delta = d \sin \Theta_{\text{dark fringe}} = (m + \frac{1}{2})\lambda \quad (m = 0, \pm 1, \pm 2, \ldots). \tag{13.15}$$

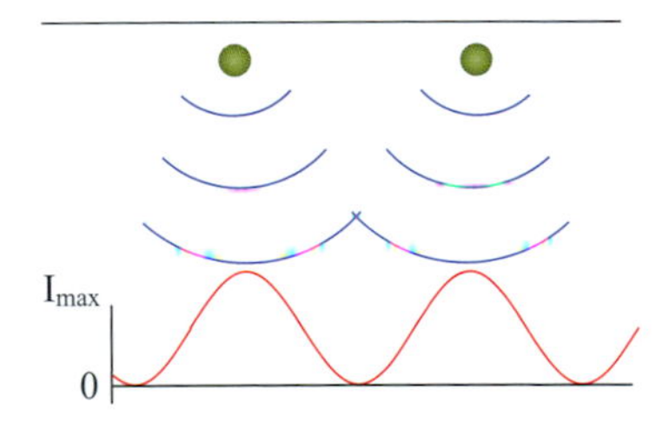

Fig. 13.22: Interference.

In principle, the wavelength of light now follows by solving Equation 13.14 for λ and measuring the angle $\Theta_{\text{bright fringe}}$ (d is an experimentally given quantity). The problem is that this angle is very hard to measure. Fortunately, however, we can determine it in a different way. If $L \gg d$ and $d \gg \lambda$ then the position P of the fringe on the screen can be approximated as $P = L \sin \Theta_{\text{bright fringe}}$. Rewriting this we obtain $\sin \Theta_{\text{bright fringe}} = P/L$ which we can substitute into Equation 13.14. As both P and L are easy to measure we can now determine the wavelength as:

$$\lambda = \frac{Pd}{L}. \tag{13.16}$$

Noteworthy 13.1: Light: a wave or a particle?

The law of reflection and refraction that holds for light can be derived by considering light to be made of small particles that Newton called "corpuscules". For example, the law of reflection holds for billiard balls bouncing off a wall and is a consequence of the conservation of momentum.

The famous experiment of Thomas Young, however, showed the remarkable result that light behaves like a wave.

In the eighteenth century, there were two conflicting views on the nature of light. One opinion, held by Newton, was that light is made out of small pointlike objects and was called the corpuscular theory of light. The other, held by Huygens and his followers, was that light is a wave. Young's experiments on the interference of light conclusively proved that light is wavelike. However, the laws of reflection and refraction follow from the corpuscular theory of light.

Question: So is light a particle or a wave? What is the resolution of the conflicting views on the nature of light?

The answer is that the two descriptions of light are approximations valid in different domains. Only the advent of quantum theory makes it possible to provide a unified model, in which the two approximations are mutually consistent (see also Chapters 23 and 24). It turns out that light is both a particle and a wave in a complementary sense: if it behaves like a particle it does not behave like a wave and vice versa.

Example 13.1: Calculating the wavelength

Consider a double-slit experiment with the screen at a distance of 1 m from the slits, a spacing between the slits of 10^{-5} m and the first bright fringe being 5×10^{-2} m from the center line. Filling these quantities into Equation 13.16 we obtain

$$\lambda = \frac{(5 \times 10^{-2}\,\text{m})(10^{-5}\,\text{m})}{1\,\text{m}} = 5 \times 10^{-7}\,\text{m}. \qquad (13.17)$$

Diffraction

Consider light passing through a circular aperture, impinging on a screen. The pattern of intensity at the screen shows a sharp maximum at the point of the screen closest to the aperture, and then has a series of maxima (separated by minima) of lower and lower intensity. See Figure 13.23. This phenomenon is called **diffraction**, and is unlike the interference pattern due to a double-slit experiment where all the maxima have the same intensity. However, diffraction can be obtained from the interference of light by considering the circular aperture as the limit of infinitely many concentric slits and clearly shows that light has the characteristics of a wave.

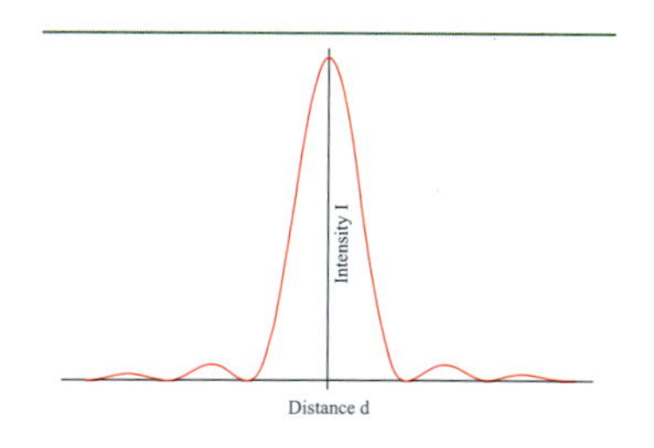

Fig. 13.23: Diffraction pattern.

13.6 Detection of light

Now that we have seen how light can propagate from a source to a destination, we need to consider how it can actually be detected by, for example, the human eye.

Light detection begins with light-sensitive proteins called **photoreceptors**. The photoreceptors are part of larger assemblies forming either an organelle in a single-cell organism or an essential component of specialized light-detection cells.

Certain bacteria, such as the photosynthetic cyanobacteria, have an organelle called the **eyespot apparatus** that helps them swim toward or away from light. In a sense, therefore, an eyespot apparatus, as the name suggests, is the simplest possible eye.

In more complex organisms, light detection takes place in specialized cells. For example, the so-called **rod cells** are quite sensitive but cannot resolve color while the so-called **cone cells** are tuned to a certain color (more precisely a range of similar colors) but not so sensitive.

A simple illustration of a rod cell is shown in Figure 13.24 while a cone cell is shown in Figure 13.25.

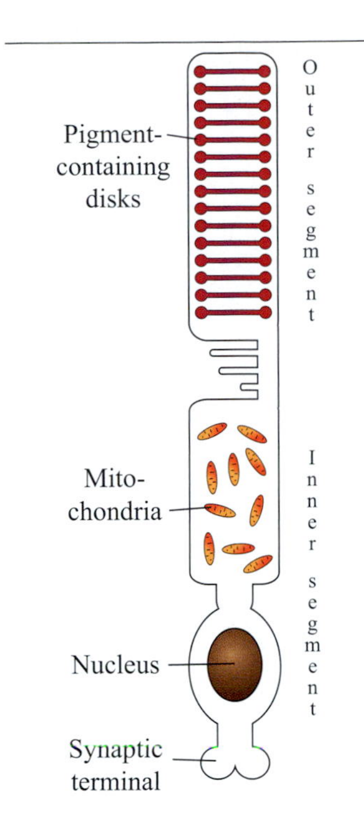

Fig. 13.24: A human rod cell.

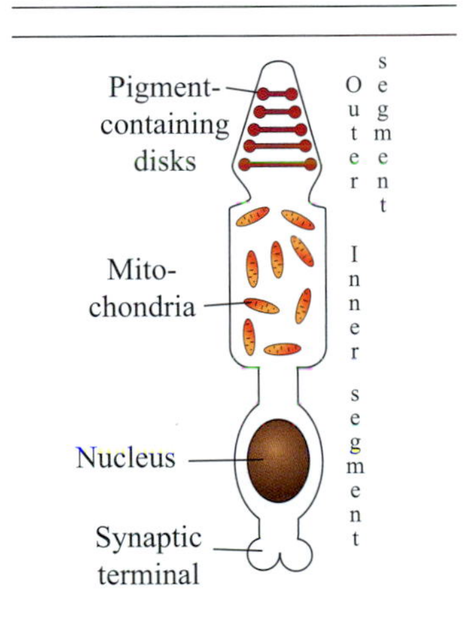

Fig. 13.25: A human cone cell.

Spatial information

In order to obtain spatial information from light, a large number of photosensitive cells are placed side by side inside a membrane called the **retina** (see Figure 13.8). As can be imagined, processing the spatial information of millions of cells requires a significant amount of resources and hence it is beneficial to implement some compression scheme. It turns out that the retina consists of several layers and the arrangement of the cells in these layers plays an essential role in the compression. In human beings, the rods and cones are located in the last layer (and consequently light needs to pass through previous layers before it can be detected). When light impinges on a rod or cone, a signal is sent to a layer consisting of bipolar and horizontal cells. After processing, the signal is then passed on to ganglion cells whose axons connect to the brain. The number of ganglion cells, totaling about 1.25 million in human beings, is much smaller than the number of rods and cones that number about 100 million and 5 million, respectively. Around the fovea things are a bit different since most of the light-sensitive cells are cones and the ratio between cones and ganglion cells is around 2:1.

However, it is not just that ganglion cells collect the inputs from a number of cones or rods and concentrate the signal to improve sensitivity, as happens in the case of odor detection (see Chapter 10). What really happens is a form of edge detection. When neighboring cones or rods send a signal at the same time, only the ones on the edge will cause their corresponding ganglion cell to trigger. Hence, the retina only sends outlines to the brain thus drastically reducing the bandwidth required of the connection from the retina to the brain.

Let us have a look at this computational process in a bit more detail as illustrated in Figure 13.26. Most photoreceptors are wired in a configuration such that they have a center-surround receptive field. That is to say, they result in a strong signal to the brain when the amount of light impinging on them is very different from that impinging on neighboring cells, otherwise the signal is weak or there is no signal. For the center-surround receptive field, two different cases are possible: on-center (see the top row in Figure 13.26) or off-center (see the bottom row).

As such, photoreceptor cells will always output a signal when sufficient light strikes them. The signal is then passed on to bipolar or horizontal cells as shown in Figure 13.26, distinguished by two types of bipolar cells. When receiving light, the neighboring cells of an on-center cell send a signal to the horizontal cells which then inhibit the bipolar cells, and only a weak signal results if the on-center cell also receives light, or no signal if the on-center cell receives no light. Vice versa in the case of the off-center cells.

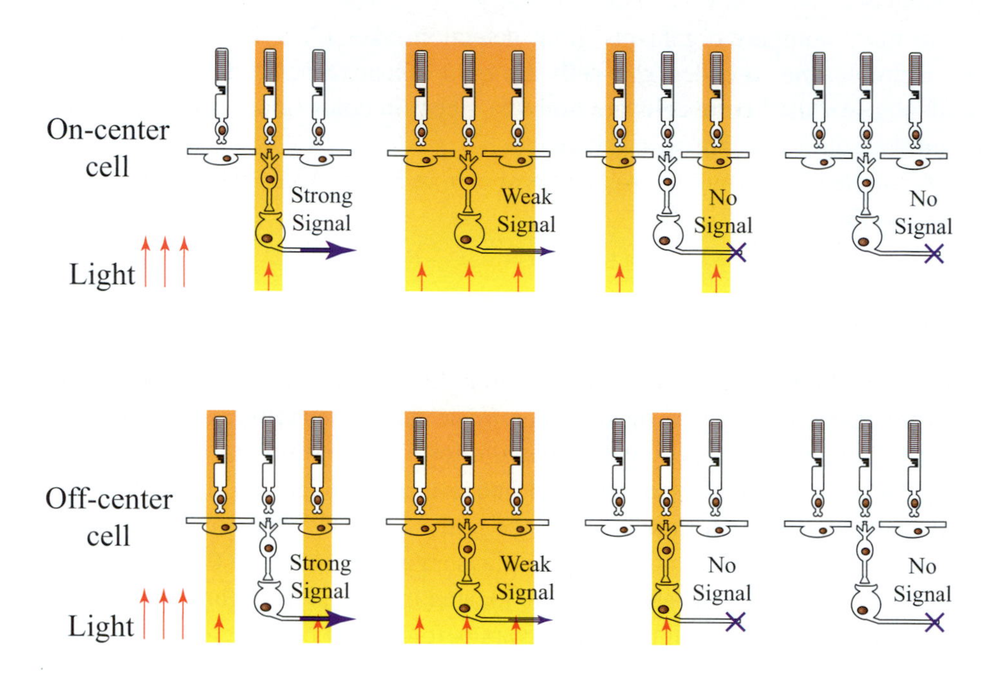

Figure 13.26: Information compression in the retina. Rather than pixel-for-pixel images, our brain mostly receives contours and outlines.

Color vision

Color vision is achieved by having cone cells that are sensitive to slightly different wavelengths. Most mammals have two types of cone cells and are dichromatic. Human beings and some other primates, though, have three different cones with peak sensitivities around 430 nm, 540 nm and 570 nm, respectively, and are therefore trichromatic. However, it doesn't end there since many animals such as birds are tetrachromatic (they have four different cones) and some crustaceans like the mantis shrimp are thought to have a stunning 16 different receptor types.

13.7 Evolutionary role of vision

The light from the sun is a source of energy exploited early on in the evolution of life on earth. However, especially in a medium like the water in the oceans, the light intensity is not everywhere the same (for example, it becomes darker at greater depths) and an organism which can position itself at an optimal place for photosynthesis will be at an advantage. Indeed, conceptually, the notion of photosynthesis is closely related to light detection — as one can argue that an ongoing conversion of light into chemical energy is analogous to a signal that light is there.

It is therefore rather conceivable that light detection first evolved in photosynthetic bacteria.

When multicellular life emerged, having the ability to detect more than the direction of the incoming light would have been a huge advantage — in particular, some form of spatial resolution to recognize predator and prey. But since light contains more information than just being there or not at a certain point in that it has a wavelength (color), making use of that information can convey additional advantages. Hence, the emergence of color vision occurred soon after or perhaps around the same time the first true eyes evolved.

Fig. 13.27: The human eye is optimized for daytime vision and is trichromatic.

Fossils of what appear to be in essence modern eyes date back about 530 million years to the so-called Cambrian explosion (see also Chapter 21) when all of the current multicellular body plans emerged. Multicellular life itself is probably about 1 billion years old, but due to scant paleontological evidence the status of vision during the period before the Cambrian explosion is not so clear. What is clear, however, is that once it emerged, vision was, from an evolutionary point of view, hugely successful since the vast majority of all animals have one type of eye or another.

13.8 The answer

We can now answer the question as to why we can see sunlight. First, we can see it because it is a wave phenomenon that can propagate even if there is no medium. Hence, it can move through space from the sun to the earth. Second, it is a form of energy, and by using this energy, detection is possible. Third, we have specialized cells that can tell the presence or absence of light and sometimes its color. Fourth, we have an organ (the eye) that preprocesses the visual data so that an efficient stream of visual data can be sent to the brain. Last, we have a brain that can interpret those visual data.

13.9 Exercises

1. Galileo was one of the first to systematically measure the speed of light. His attempt involved two observers positioned in two towers that were about 10 km apart. The idea was that the first observer opens a shutter in a lantern and then as soon as the second observer sees the light from the first lantern, opens his shutter. Galileo would then measure the time it takes from opening the first shutter to seeing the light from the second lantern arrive at the first lantern. Unfortunately for Galileo, this experiment turned out to be inconclusive. Why is that so? Would it have been sensible to choose towers that were farther apart?

2. In Young's double-slit experiment, if we know that the distance from the center line to the first bright fringe is 5 cm, what is the distance from the third to the fourth bright fringe?

3. What would it be like if our eyes evolved such that we could see deep ultraviolet light?
4. If vision is so advantageous, why wouldn't plants have it?
5. Why is it important for visual preprocessing to take place in the retina?
6. Which elementary particle is associated with the microwaves in the eponymous kitchen appliance?
7. Microwaves can cook food (and hence your brain as well). Yet we can see inside through a metal sheet with small holes. Why can the light pass through but apparently not the microwaves?
8. Why would it have been difficult for Fizeau to do his experiment inside a laboratory?
9. What is the difference between a biconcave lens and a biconvex lens? The lens in our eye is closest to which one of the two?
10. Design a symmetric biconvex lens with a focal length of 2 m.

14: Biopolymers

Q - Why Does Life Use Polymers?

Chapter Map

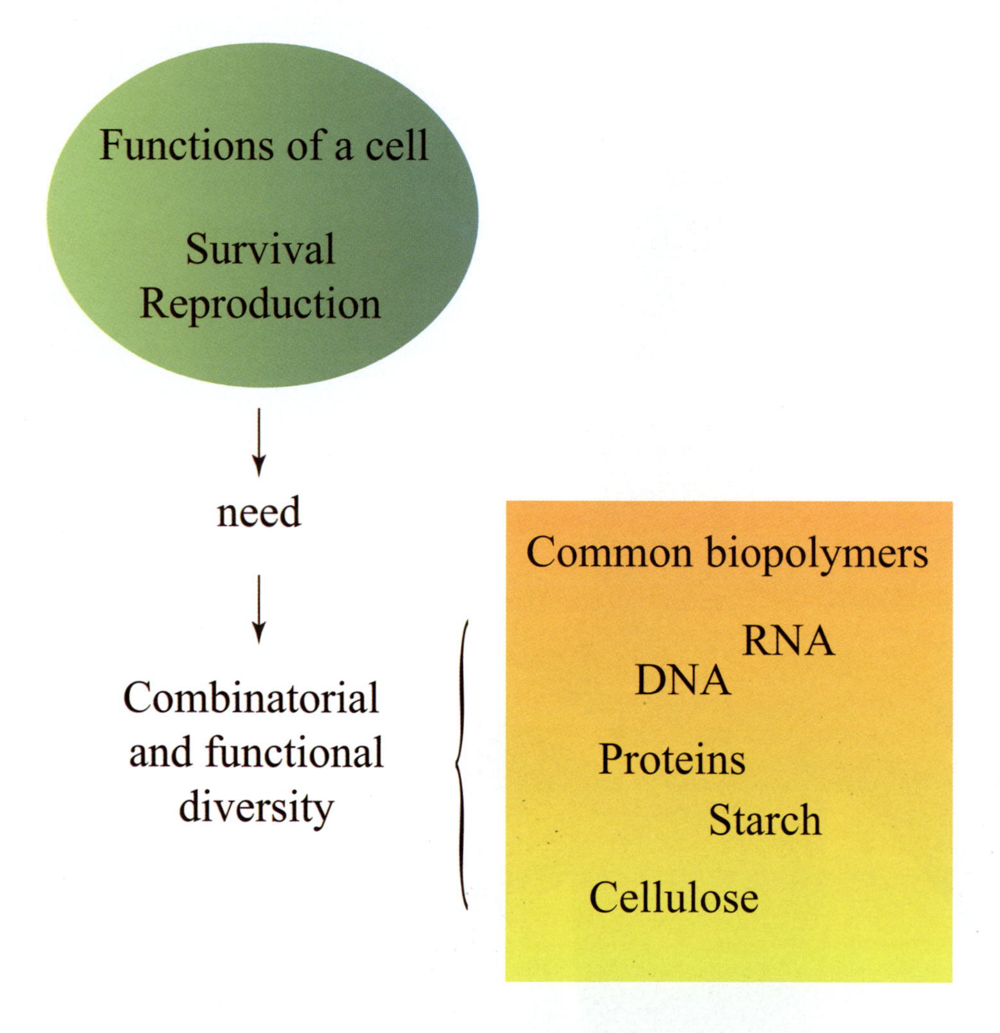

Functions of a cell
Survival
Reproduction
need
Combinatorial
and functional
diversity
Common biopolymers
RNA
DNA
Proteins
Starch
Cellulose

14.1 The question

When we look at cells, or life in general, we see that nature makes extensive use of polymers. Indeed, DNA, RNA, proteins, starch and cellulose, for example, are all polymers. Such polymers are essential for life on earth as all modern cells use one single (basically universal) genetic apparatus involving DNA, RNA and proteins.

In contrast, the materials humankind has used for engineering purposes and machinery, since the beginning of civilization, have been much simpler like copper and iron. Notwithstanding their simplicity, the enormous importance of acquiring the skills to handle those materials is reflected by naming entire episodes of humanity's development after them — e.g., the Bronze Age (with bronze being a mixture of copper and tin). Without the invention of steel, a "simple" mix of iron and some other molecules like carbon, the modern Industrial Revolution would not have been possible. Then, if relatively simple materials like steel are so useful, and presuming that, in general, nature is pretty clever in its selection of materials, we need to ask: **Why does life use polymers?**

14.2 Common biopolymers

Before we can zoom into our question, we first need to get some idea of what biopolymers are and where we can find them. Given in Table 14.1 are parts of a cell where the common biopolymers can be found.

	Animal Cell	Plant Cell	Monomer
RNA	Nucleus Cytoplasm	Nucleus Cytoplasm	Nucleotide
DNA	Nucleus Mitochondrion	Nucleus Chloroplast	Nucleotide
Protein	Nucleus Cytoplasm	Nucleus Cytoplasm	Amino Acid
Starch	—	Chloroplasts	Glucose
Cellulose	—	Cell Walls	Glucose

Table 14.1: Locations, in a cell, where biopolymers can be found.

As can be seen from the table, the number of different types of monomers is rather small. Let us now have a closer look at the structure of these biopolymers.

Starch

Starch is a biopolymer that serves as an energy storage. Consequently, it must be built of something that can deliver energy to a cell. And, indeed, the monomers that form the basic units for starch are glucose molecules which carry energy that can be utilized by the cell. A glucose monomer consists of six carbon, six oxygen

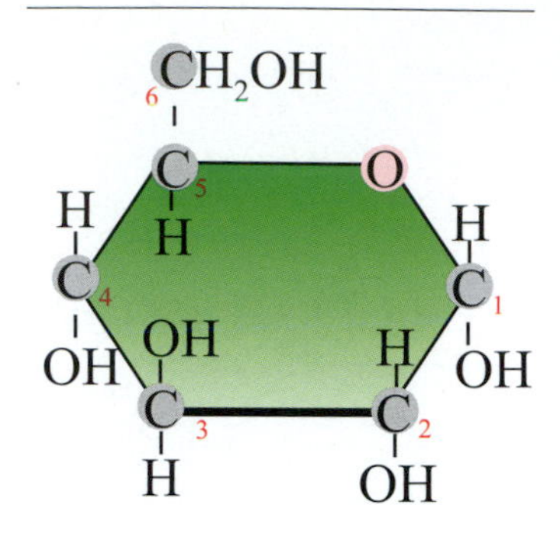

Fig. 14.1: Glucose monomer.

and 12 hydrogen molecules, as shown in Figure 14.1. In general, starch consists of two components: *amylose* which is a linear chain of glucose molecules and *amylopectin* which is a branched chain of glucose molecules (although this also has many linearly chained glucose molecules just like amylose). The carbon atoms are numbered as indicated and amylose is formed by linking the carbon$_1$ atom of one glucose monomer with the carbon$_4$ atom of another glucose monomer as depicted in Figure 14.2. In the process of the linkage, one water molecule is freed.

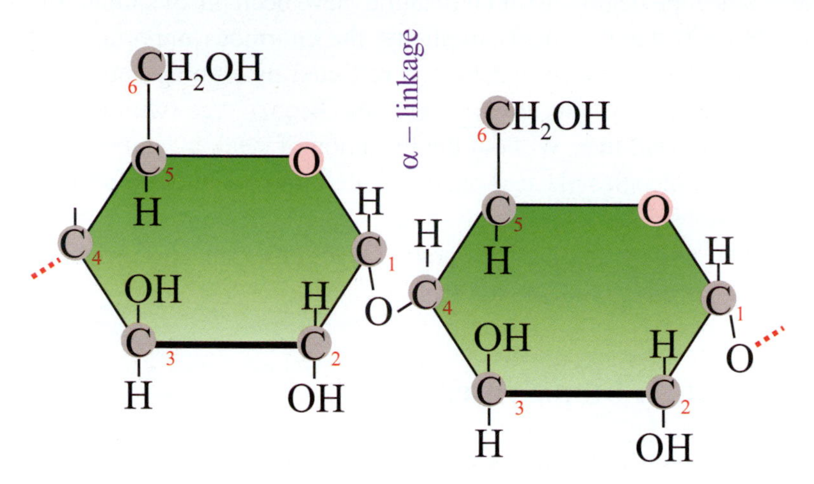

Figure 14.2: Glucose monomers as linked in this graph form amylose, a component of starch.

Cellulose

Fig. 14.3: Cellulose is a major component of wood.

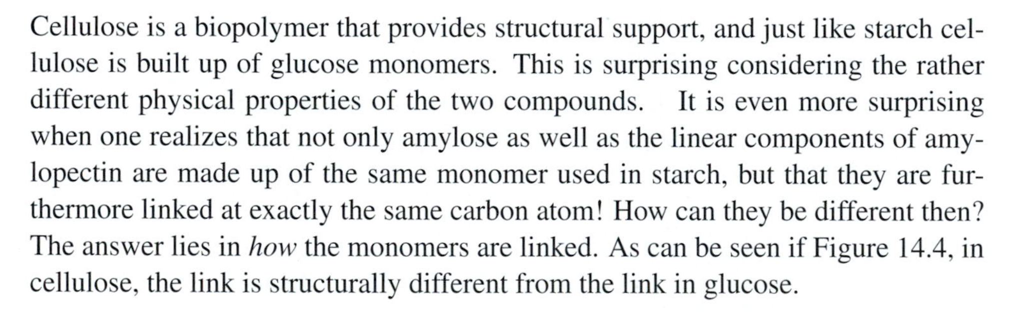

Cellulose is a biopolymer that provides structural support, and just like starch cellulose is built up of glucose monomers. This is surprising considering the rather different physical properties of the two compounds. It is even more surprising when one realizes that not only amylose as well as the linear components of amylopectin are made up of the same monomer used in starch, but that they are furthermore linked at exactly the same carbon atom! How can they be different then? The answer lies in *how* the monomers are linked. As can be seen if Figure 14.4, in cellulose, the link is structurally different from the link in glucose.

Proteins

Proteins are biopolymers built up of monomers called amino acids; there are 20 fundamental amino acids that are used to manufacture the bulk of proteins. Proteins fulfill myriad functions in the cell, for example, acting as catalysts for reactions (in which case they are enzymes) or providing structural support (for example, as part of the cytoskeleton). While starch and cellulose are based on a single monomer, in human beings and most other life-forms there are 20 distinct amino acids.

Besides the standard 20 amino acids, there are other nonstandard amino acids, including those found in proteins (both in prokaryotes and eukaryotes) and those that occur as intermediates in metabolic pathways that occur in living organisms. Nonstandard amino acids consist of amino acids that have been chemically mod-

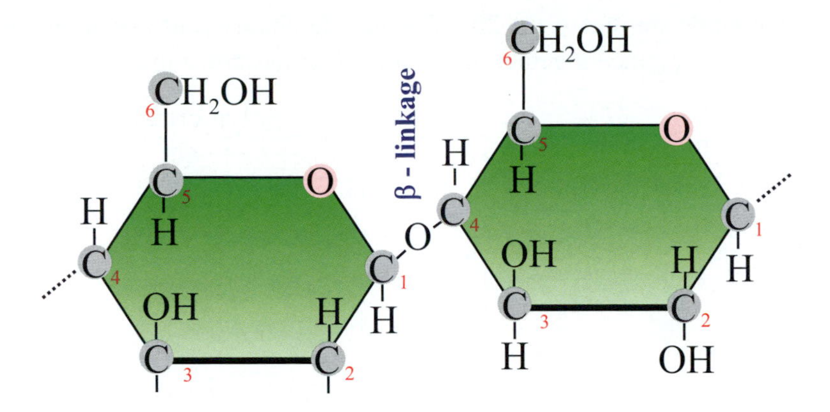

Figure 14.4: Glucose monomers as linked in this graph form cellulose.

ified on having been incorporated into a protein as well as those amino acids that occur in living organisms but are not found in proteins. For example, collagen is the most abundant protein — by mass — in vertebrates and contains two nonstandard amino acids, namely 4-hydroxyproline and 5-hydroxylysine.

It is difficult to estimate the number of nonstandard amino acids but the numbers are thought to be not larger than the standard 20 amino acids.

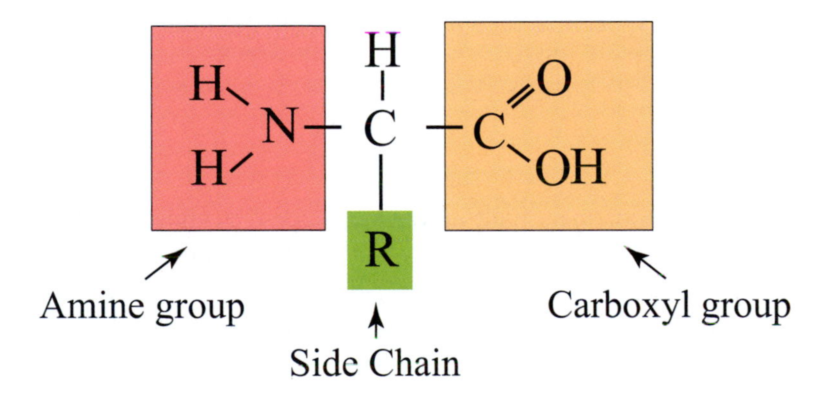

Figure 14.5: Structure of an amino acid monomer. One of the 20 possible side chains is shown in Figure 14.6.

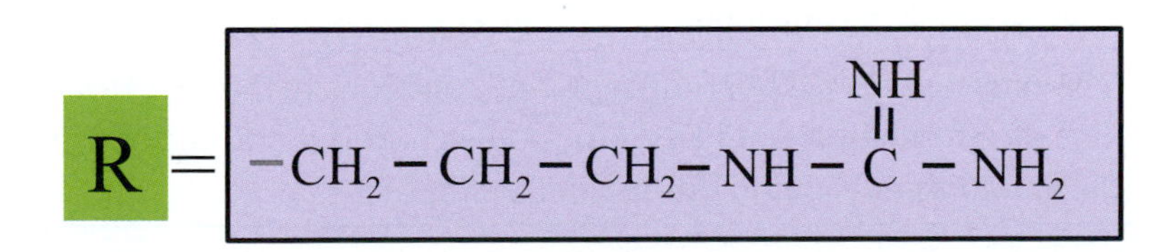

Figure 14.6: The side chain arginine.

As can be seen from Figure 14.5, the amino acid monomer consists of three parts: an amine group consisting of one nitrogen and two hydrogen atoms, a side chain and a carboxyl group. What is essential to note is that the amine group and the carboxyl group are identical for all amino acids, and that the differences in the amino acids are due to the 20 different side chains. Individual amino acid monomers are then joined together in peptide bonds as shown in Figure 14.7.

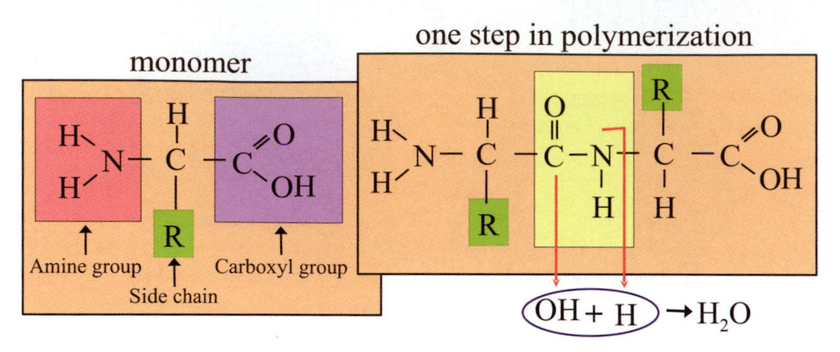

Figure 14.7: The amino acids in proteins are linked by the peptide bonds shown in the yellow box.

As can be seen, the peptide bond links the amine group of one monomer with the carboxyl group of the next monomer. This is independent of the side chain and identical for all amino acids. It can hardly be overstated how brilliant an arrangement this is as it implies that any amino acid can be linked to any other amino acid in an arbitrary sequence. This is somewhat identical to how carriages are constructed in trains. Imagine how inconvenient it would be if only certain carriages could be combined but not others.

Nucleic acids

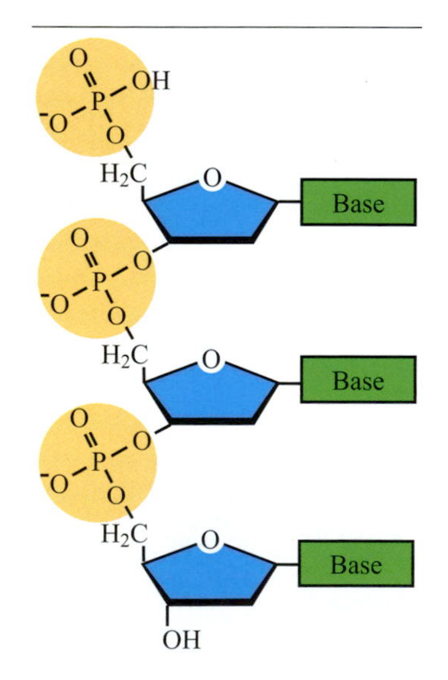

Fig. 14.8: Nucleotides can be strung together to form polynucleotides.

Nucleic acids are biopolymers built up of nucleotides, as shown in Figure 14.8. There are two types of nucleic acids, RNA and DNA, that fulfill different roles as summarized in Table 14.2. Each of these types of nucleic acids is built up out of four different nucleotides. Three of those nucleotides are the same in DNA and RNA while the remaining one is slightly different.

DNA	Deoxyribonucleic acid (DNA) performs one main function: it stores genetic information. It is generally, but not always, double stranded and undergoes stages of folding.
RNA	Ribonucleic acid (RNA) fulfills myriad functions in the cell, from transport of information from DNA to protein factories and for transporting the amino acids required for protein synthesis.

Table 14.2: The two types of nucleic acids.

The structure of a nucleic acid monomer is shown is Figure 14.9 where it should be noted that the five-carbon sugar is different for DNA and RNA as illustrated in Figure 14.10. In DNA, the oxygen atom attached to the 2′ carbon atom is missing. Just as in the case of proteins, nucleic acids are composed of three distinct and well-defined parts. There is a five-carbon sugar to which first one or more phosphate groups consisting of a phosphorus atom and three oxygen atoms is attached and second an amine base of varying composition.

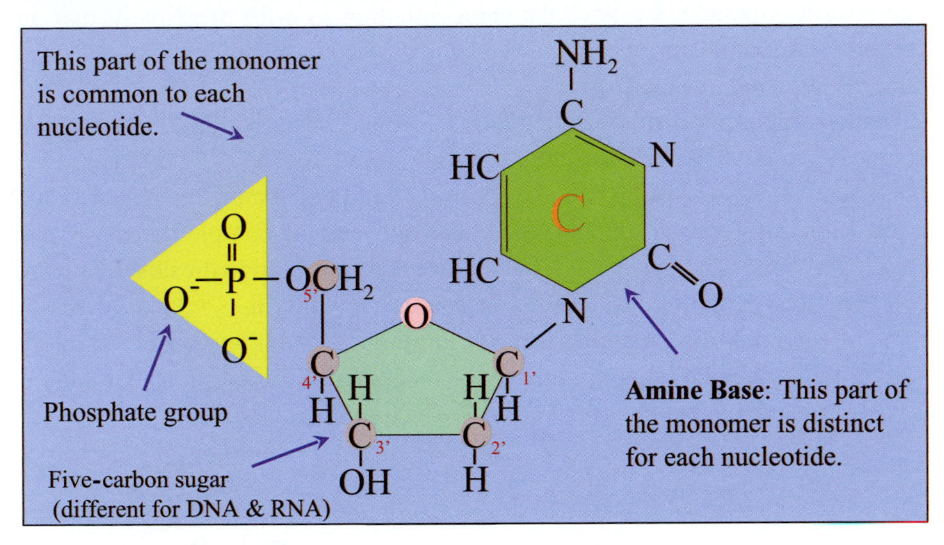

Figure 14.9: The structure of a nucleic acid. Depicted here is the case where the five-carbon sugar is deoxyribose as it occurs in DNA. In RNA, the five-carbon sugar has an extra oxygen atom at the 2′ carbon (below the plane).

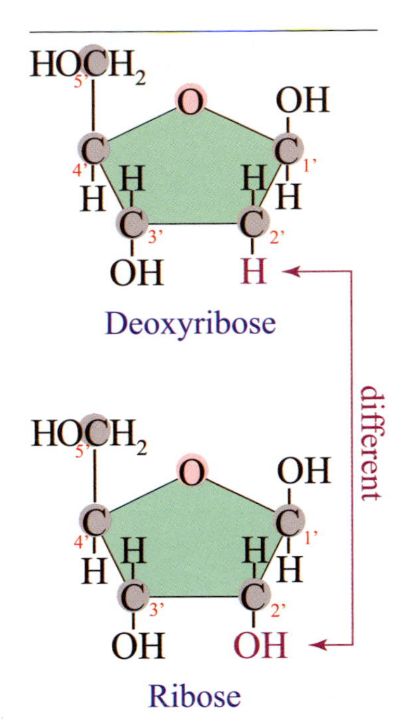

Fig. 14.10: The difference between ribose and deoxyribose.

The five different amine bases found in DNA and RNA are shown in Figure 14.11.

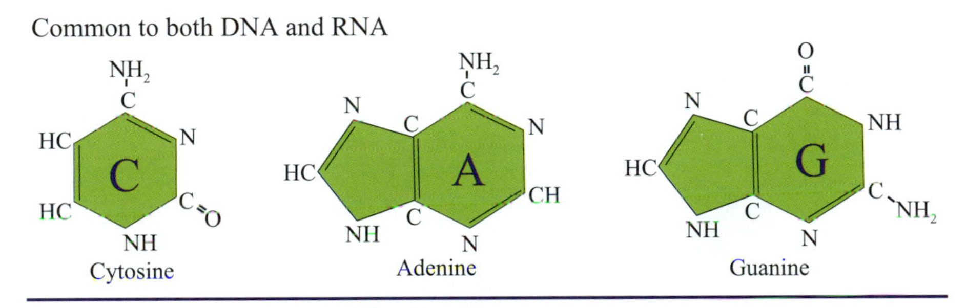

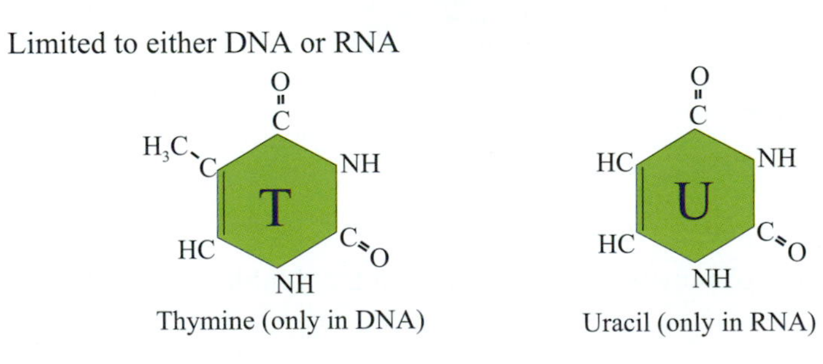

Figure 14.11: The amino bases that distinguish the different nucleic acids. The amino bases are also classified as having single or double rings.

In the formation of RNA and DNA polymers, the five-carbon sugar and the phosphate group are always the same while the differences lie in the amine bases.

Since the monomers are linked by a covalent bond between the phosphate group of one monomer and the carbon sugar of the next group (regardless of which amine base it contains), we have the same situation as with proteins in that any sequence of monomers is possible. An example of how the nucleotides are strung together is depicted in Figure 14.8.

Another important difference between DNA and RNA is that DNA generally is double stranded while RNA usually is single stranded. The two strands in DNA are complementary consisting exclusively of A-T and G-C pairs, and hence as such have the same information content. The bonds between the complementary strands are hydrogen bonds and hence are relatively weak as opposed to the covalent bonds which link up the monomers. As a consequence of this, it is relatively easy to separate the strands (for example, when reading out a gene) but hard to break the DNA strand (so that the sequence and hence the stored information is maintained). An illustration of the double-strand DNA is given in Figure 14.12.

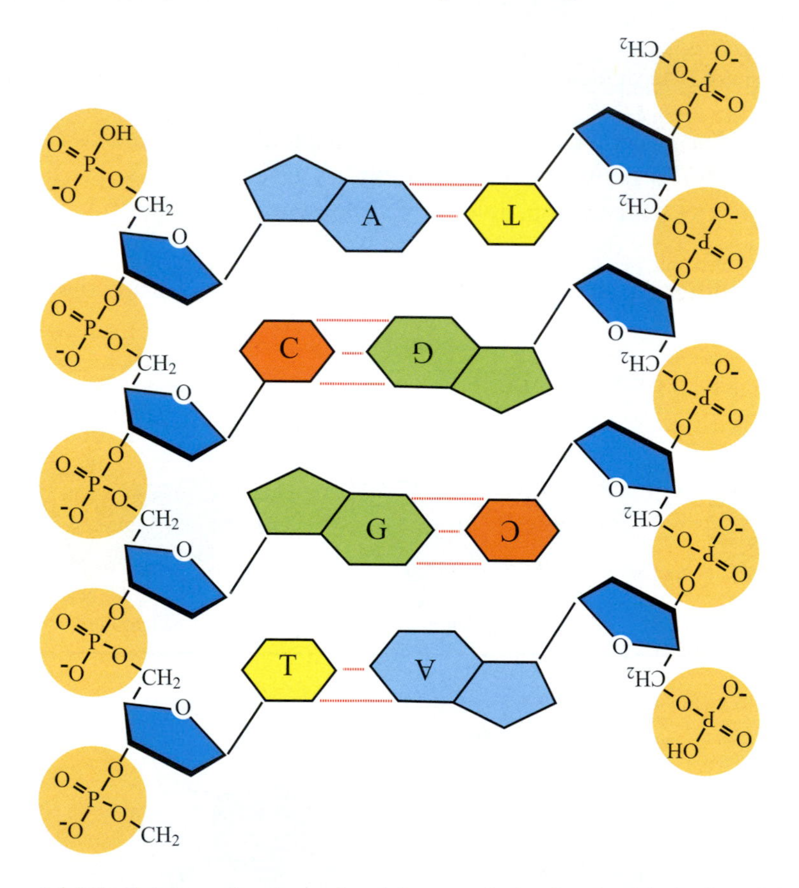

Figure 14.12: Pairing of polynucleotide strands in the double helix of DNA.

14.3 Functions of a cell

Now that we have some basic idea of the functions and structures of the common polymers found in cells, we can continue our investigation into why life makes such abundant use of them.

The prime functions of an organism, in general, are survival and reproduction notwithstanding specific implementations of multicellular life-forms — like in ant colonies or slime molds. In the case of single-cell organisms like bacteria, one can also describe the central function of the cell as growth and division. What are some of the key processes that are required? A short list is given in Table 14.3.

- Absorption of raw materials
- Processing of raw materials
- Processing of information

Table 14.3: Some key processes occurring in a cell.

Such processes require work and hence energy, which will be discussed in more detail in Section 14.6. For the moment, let us leave energy out of the picture and consider the resources. Clearly, only elements that occur in the earth's crust and lower atmosphere in reasonable quantities are suitable to serve as supplies. With the help of those elements, key processes and requirements such as those listed in Tables 14.3 and 14.4 must be carried out.

Versatility	There are many functions going on inside a cell and hence the (processed) materials need to display significant diversity.
Ease of Use	Despite allowing for diversity, the processed materials need to be easy to manufacture.
Specificity	Usually, each (macro-)molecule has one (or a few) specific function(s) in a cell. For example, enzymes are proteins which catalyze specific biochemical reactions in cells.

Table 14.4: Some key requirements regarding the processes occurring in a cell.

There are basically two ways to attack the problem of constructing a cell which can carry out the above key processes and which fulfills the key requirements.

The first approach is similar to how we human beings construct machines: take up certain materials and mold them into shapes that fulfill the desired functions or assign a certain material or compound to a certain function. While it is perhaps conceptually not entirely impossible to follow this route for living systems, there are serious drawbacks that seem to form insurmountable obstacles. First, a mapping of compounds to function would not be possible as a general mechanism since the number of simple compounds is very limited. Second, if the shaping and molding approach as in a steam engine constructed out of steel were to be followed, where would the machinery to build the machine come from? Not only does there appear to be a chicken-and-egg problem, the energies and environment seem unlikely to be available in a natural environment.

A second approach would be to take some sort of building blocks that can be assembled into a large number of different structures. The building blocks themselves should either be readily available elements or molecules, or somewhat more complicated structures built up from readily available elements or molecules. This is indeed the path nature has chosen. Table 14.5 gives an overview of the various monomers that build up the main classes of biopolymers and also shows the elements that in turn build up the monomers. It is remarkable that only six elements are used.

	No. of Monomers	No. of Elements	No. of Functions
RNA	4	5 (C,H,O,P,N)	Very large
DNA	4	5 (C,H,O,P,N)	Small
Starch	1	3 (C,H,O)	Very small
Cellulose	1	3 (C,H,O)	Very small
Proteins	20 (most organisms)	5 (C,H,O,S,N)	Extremely large

Table 14.5: The number of monomers and their constituting elements.

14.4 Diversity

Fig. 14.13: How many different strips can one make?

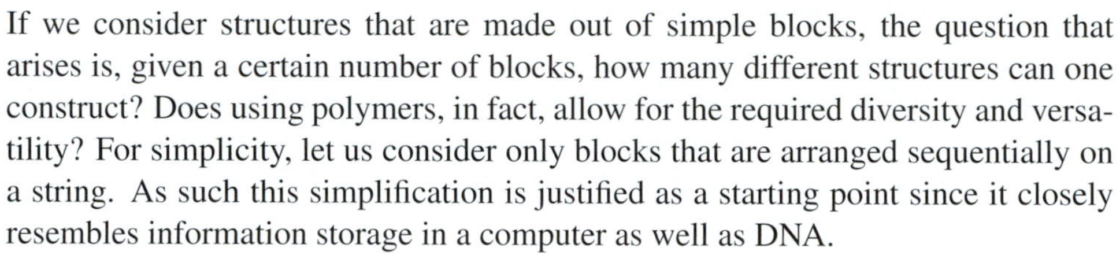

If we consider structures that are made out of simple blocks, the question that arises is, given a certain number of blocks, how many different structures can one construct? Does using polymers, in fact, allow for the required diversity and versatility? For simplicity, let us consider only blocks that are arranged sequentially on a string. As such this simplification is justified as a starting point since it closely resembles information storage in a computer as well as DNA.

In the very simplest case, we only have a single type of block and we can make only one type of structure of a given length. However, when some kind of computation or process logic is required, at the very least one needs two types of blocks. In computing, those blocks are called bits that assume the values zero and one. It should be noted that having two types of blocks or a block that can assume different values or attributes are conceptually the same thing. In other words, it does not matter whether I say that I have one block with the number 1 painted on it and another block with the number 0 painted on it or a single block that can display either a 1 or a 0. Similarly, we could also think of having a block painted in blue or red, respectively, or a single "chameleon" block that can display either color.

A string of a given length obeys the following relationship:

$$\text{Number of combinations} = (\text{Types of blocks})^{\text{Length of String}}. \tag{14.1}$$

So in the case of a binary string (i.e. a string built up of 1s and 0s) of length 3, we have the following $2^3 = 8$ combinations: (000, 001,010, 011, 100, 101, 110, 111). A binary string of length 8 has 256 combinations and a binary string of length 16 has 65,536 combinations. Of course, in practical applications, not all of these different combinations may have different meanings. For example, if there is no natural beginning or end to a string, then the combinations 011 and 110 (011 read backwards) may mean the same. In other scenarios, complementary strings may have the same meaning (i.e. strings where one type of block matches with another type of block) so that 000 means the same as 111, or so that 101 means the same as 010.

If one has four types of building blocks, as is the case for nucleotides, the number of combinations increases much more rapidly with length. For example, an RNA strand of three nucleotides can already yield $4^3 = 64$ combinations and a strand of eight nucleotides a whopping $4^8 = 65{,}536$ combinations. Knowing that RNA strands are generally quite a bit longer than eight nucleotides, one can easily imagine the huge number of different RNA strands possible in cells.

However, the already enormous possible diversity of RNA (or for that matter DNA) strands of a given length pales when compared to the possibilities in proteins. As proteins are made up of 20 types of amino acids, a super-short string of only three amino acids already gives $20^3 = 8{,}000$ possibilities. A string of eight amino acids would give $20^8 = 25{,}600{,}000{,}000$ possibilities!

Hence we can clearly see that nature has found a process that for all intents and purposes allows for a limitless number of combinations. Having such a large pool of possibilities translates into a very large number of possible functions, exactly what is needed for complex processes like those found in living cells.

14.5 Transcription and translation

Now that we have some ideas as to *why* nature is using biopolymers, let us briefly investigate *how* it does so. After all, there's not much point in having the world's greatest building material if one cannot use it. In order to obtain some insights into this matter we need to consider two aspects: first, the process of using the materials, and second, the energy to run the process. In this section, we discuss the processes of transcription and translation while we'll consider energy in the next section.

Fig. 14.14: A musical score transcribes a notion completely different from its representation. Like DNA to life? Or more like DNA to RNA?

In basically all modern cells, the main classes of biopolymers are associated with certain types of functions. In other words, there is a rather strict logical organization at a coarse-grained level, even though, rather unsurprisingly, considering the complexity of a cell, the organization into different roles is less strict when considering a more fine-grained level. DNA stores genetic information, RNA transports and communicates information while proteins are the workhorses and factories.

There are several notable points with regards to this organization. First, it is universal. That is to say, it is the same for a bacterium, plant or human being. This alone is a good indication of how successful the scheme is. However, that is not all; the organization is also evolutionary very old and probably dates back more than 3 billion years. Considering that the earth is about 4.4 billion years old and that it has been inhabitable for about 4 billion years, we find that nature has conserved the apparatus for at least three times as long as it took to develop it. From this, one can conclude that not only does the organization work extremely well but it may also hint at the possibility that it is the consequence of some biological natural laws.

Of course, the question that arises is: If DNA is used for information storage, then how is this information processed? After all, DNA stores information for proteins as well, but DNA consists of four types of nucleotides yet proteins consist of 20 types of amino acids. Clearly there cannot be a direct one-to-one correspondence between nucleotides and amino acids. Also, one should wonder whether the information is directly processed on the DNA or indirectly by transporting the information to some kind of protein factory. Interestingly enough, in this respect, human technology and nature follow similar paths by choosing the indirect path as this is more efficient in a general setting.

What happens in a cell is that information is first transcribed into messenger RNA (mRNA) that transports the information (encoded in the sequence of nucleotides) to a complicated organelle made out of RNA strands and proteins called ribosomes that uses blocks of three successive nucleotides on the mRNA strand to carry out the protein assembly. Three successive nucleotides form informational blocks called **codons** and these codons map to certain amino acids or indicate the beginning or end of the assembly process.

Since there are four different nucleotides to choose from and a codon consists of a succession of any three of those nucleotides, we can conclude that a single codon can in principle encode for $4^3 = 64$ items. With 20 amino acids and a start and stop instruction, a codon hence only needs to be able to code for 22 items and consequently there are $64 - 22 = 42$ extra combinations. Rather than having a complicated machinery to exclude unnecessary or unwanted combinations, nature employs them to build in redundancy. One way to do that is to simply ignore the third nucleotide for certain combinations of the first two nucleotides. An example is the sequence of two nucleotides whose amine bases are cytosine (C). All the four combinations (CCU, CCC, CCA, CCG) encode for the amino acid proline.

A schematic representation of the transcription and translation process is depicted in Figure 14.15.

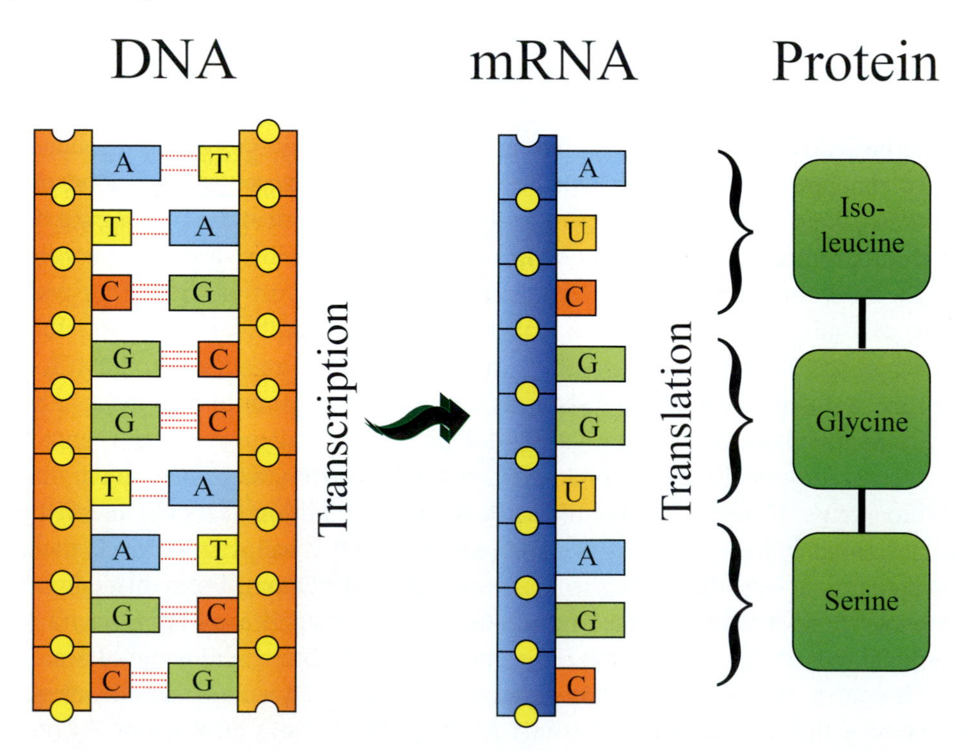

Figure 14.15: Pairing of polynucleotide strands in the double helix of DNA. In DNA, A pairs with T while G pairs with C. (Note that in RNA T is replaced by U and that the pair is A-U instead.)

14.6 Energy in a cell

Let us now have a bit closer look at where the energy comes from for the many processes that occur in a cell. Externally, forms of energy that may be available to a cell that is a primary producer of food (i.e. it does not depend on other life-forms as a food supply) are, for example, sunlight, lightning, heat, and chemicals. As discussed in Chapter 3, energy can be transformed from one type into another. In the case of cells, energy is stored chemically. For example, in a plant, energy from the sun is stored by assembling water and carbon dioxide into sugars like glucose or fructose and then further into starch. Of course, being able to store energy alone is not enough. The reaction must also be reversible but it should not reverse spontaneously.

Hence, in general a reaction looks as shown in Figure 14.16. There is a small hump that needs to be overcome in order to either store or release energy. Without the hump, the energy storage would be easier because the activation energy would be lower. However, there would be nothing stopping the reversal of the reaction. With the hump, when needed, at the cost of a small activation energy, the stored energy can be released on command. This is a bit like storing a ball on the top of a mountain. If the mountain top would be perfectly round, the ball wouldn't stay there but roll right down again. If, however, we dig a small hole on the top of the mountain, then the ball will no longer spontaneously roll down the hill once we've put it up there. Then, when we want to use the stored potential energy in the ball, we can give it a nudge to lift it over the edge of the hole and the ball will roll down. At the bottom of the mountain, we can then use its kinetic energy to do work.

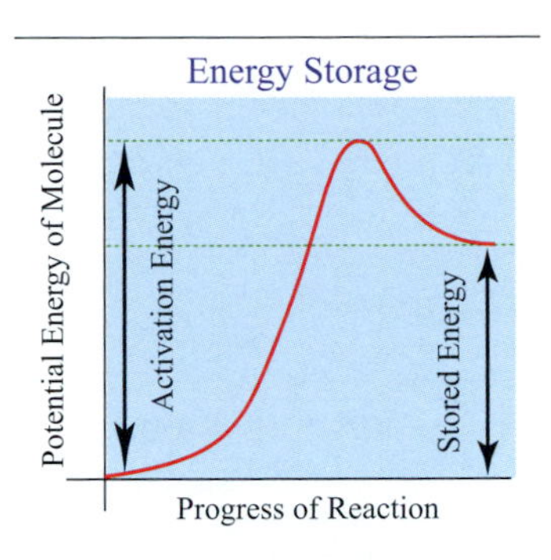

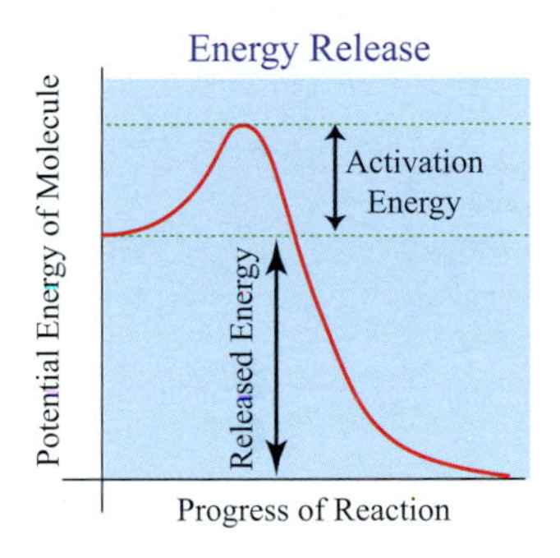

Fig. 14.16: To store or release energy, an activation energy is necessary.

ATP

A consequence of the requirement that the reaction that stores energy does not reverse spontaneously is that the potential energy required at the start of the reaction is larger than the actual energy stored. This can be seen by the hump in the figure. Then, when it is time to release the stored energy, a small amount of activation energy gets the energy-releasing reaction going.

Starch or glucose are nutrients. In principle, one could imagine that a cell uses the energy from those nutrients directly to power the myriad processes occurring in it. However, this is not the case. Instead, cells temporarily store the energy from nutrients in so-called carrier molecules, the most common of that is the nucleic acid adenosine-triphosphate (**ATP**). ATP is synthesized from ADP (adenosine-diphosphate) in a reaction that costs energy (this energy is taken from a nutrient like glucose). As the reaction is reversible (i.e. ATP can turn into ADP setting energy free), the process of ATP synthesis is suitable for storing energy. The conversion of ATP into ADP yields about 11-13 kcal/mol of usable energy to the cell.

The process of supplying energy is one of coupled reactions. That is to say, desired reactions are carried out together with the conversion of ATP to ADP. All in all, this is somewhat similar to our use of electricity. It is generated elsewhere (efficient power plants), transported to where it is needed and then used to make possible a desired reaction/process (e.g. the emission of photons from a filament in a lightbulb).

The question that naturally arises is, why would nature go through so much trouble and not directly use the energy stored in, for example, glucose? The reason for this is one of efficiency.

Enzymes

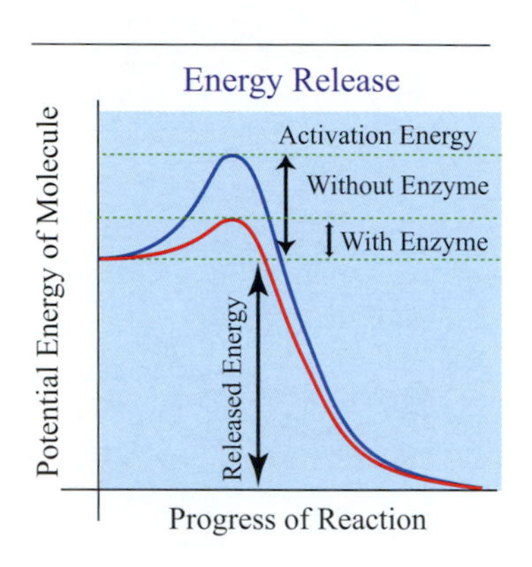

Fig. 14.17: Enzymes lower the activation energy.

We mentioned above that certain chemical reactions need an activation energy to overcome a hump. In order to minimize unwanted spontaneous reactions, this hump should be high by default but low when the reaction is desired. Hence it is highly desirable to have some kind of a catalyst. Indeed, **enzymes** are such catalysts. They can speed up chemical reactions in a cell enormously.

Is there a way one can visualize the working of an enzyme? Let us again consider the example of the ball on the mountain. In order to prevent it from rolling down spontaneously, we had dug a small hole to store it in. In order to get the ball rolling down the hill we need to push it over the rim of the hole, and this will require a certain amount of energy. Now, if instead we had constructed a little gate at one side of the hole, then we'd only need to open the gate to get the ball rolling. Surely that would require a lot less energy, and indeed that is exactly what an enzyme accomplishes. It lowers the barrier for a reaction. Best of all, having a gate, if properly constructed, does not affect the probability of the ball starting to roll down due to a random event (perhaps by being hit by a large hailstone). Hence, it is in principle possible to have both a safe storage of energy as well as relatively easy access to this energy.

14.7 The answer

We are now in a position to give a possible answer as to why polymers are essential to life. While without having general *a priori* laws of biology akin to the fundamental laws of physics, it will be impossible to answer our question with certainty, we can nevertheless conjecture with some confidence that it is the unique combination of easy processing with diversity and versatility that leads nature to the use of polymers. Indeed, we may state that life needs polymers because **polymers are the only class of materials that can both be easily processed and allow for the diversity necessary for life.**

14.8 Exercises

1. How many different polymers with a length of five monomers can be made if one has 10 different types of monomers available?

2. Which is the only major class of biopolymers that contains sulfur in their base structure?

3. What is the difference between transcription and translation?

4. Both cellulose as well as starch consist of glucose monomers. Then why are they so different?
5. Speculate as to why herbivores can eat grass and properly digest it while human beings cannot.
6. How many different side chains do the amino acids found in human beings have?
7. What is the key difference between the DNA of a cyanobacterium and that of a human being?
8. What is the type of bond that holds the pair A-T together in a strand of DNA?
9. Name two organelles that have their own DNA.
10. Name two key differences between DNA and RNA.

15: Proteins

Q - Who Does All the Work?

Chapter Map

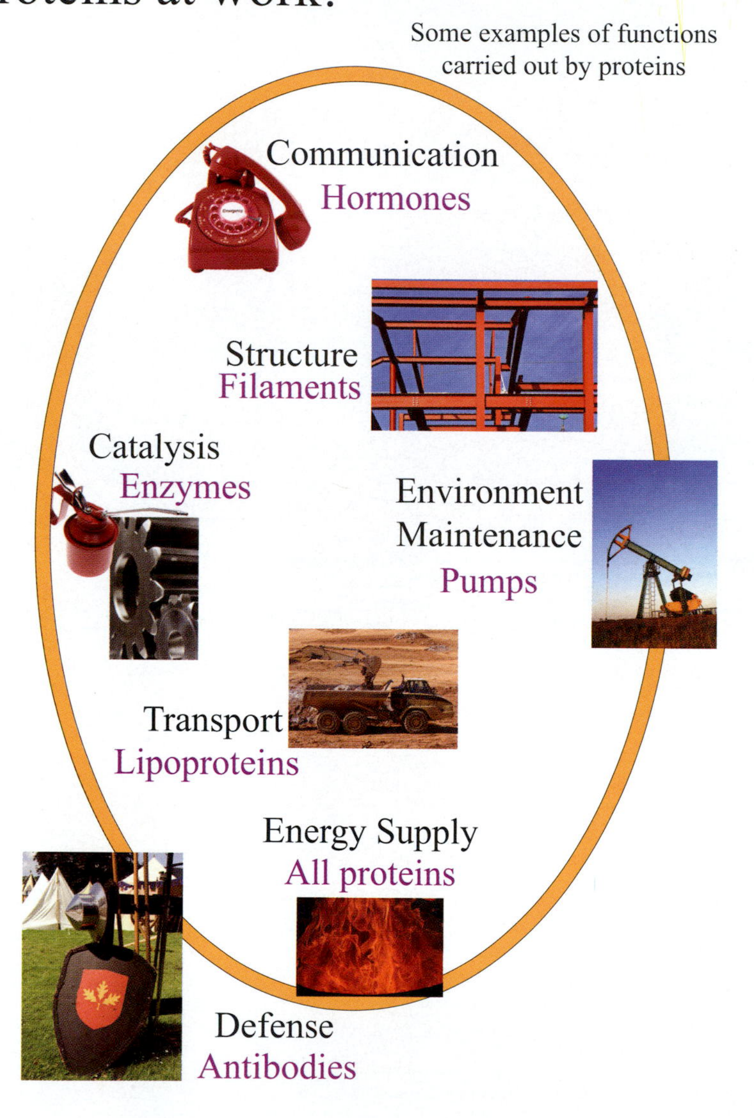

Proteins at work!
Some examples of functions carried out by proteins
Communication
Hormones
Structure
Filaments
Catalysis
Enzymes
Environment Maintenance
Pumps
Transport
Lipoproteins
Energy Supply
All proteins
Defense
Antibodies

15.1 The question

Figure 15.1: In cells, the work never stops.

Fig. 15.2: Is a cell like a refinery? What keeps all the processes going?

Being alive is not easy! In order for cells to live and reproduce, myriad functions need to be carried out. To name a few: There needs to be a cell membrane or wall to separate the inside from the outside. This membrane needs to be produced to allow for growth and to be equipped with mechanisms that allow desirable substances to be brought into the cell and waste products to be moved out of the cell. There needs to be metabolism to process food and materials, there needs to be repair of cell damage, growth and reproduction.

All these tasks need to be carried out. Considering the complexity of modern life, one cannot expect that they are all handled by a single class of cellular constituents. However, in view of the fact that cells seem to designate certain types of molecules for certain jobs, for example, DNA for information storage and RNA for information transport, it makes sense to ask: **Who does all the work?**

In a way we can consider an organism, be it single- or multicellular, like a kind of a house with walls that separate the inside from the outside, wires that connect its various rooms and inhabitants who carry out various activities.

If we want to know who is doing all the work in this house, we need to first know how work can be done at all. Of course, if we say that, we should wonder what "work" actually is. In this context one could say that work results in, from the viewpoint of the cell, useful outcomes — be they, for example, the manufacture of needed materials or locomotion. All of these outcomes have one thing in common: They require chemical reactions to take place.

While some chemical reactions take place quite easily at the temperature an organism would find itself in, very often these desired reactions occur only at a very slow pace or not at all. Living systems therefore make extensive use of catalysts in the form of enzymes that speed up and enable the machinery of the cell. To start our exploration into who does all the work, let us therefore now look at enzymes in a bit more detail.

15.2 Enzymes

Enzymes speed up chemical reactions.

The key function of **enzymes** is to speed up specific chemical reactions without being consumed themselves — although they may change while speeding up a reaction, this change is generally reversible. Besides speeding up reactions, enzymes have a second crucial characteristic for living organisms: specificity. By and large, certain enzymes catalyze only certain specific reactions. After all, there is a huge number of processes going on simultaneously in a cell and the reaction rates

Enzymes are often very specific as to which reactions they speed up.

for most of these will need to be different. Therefore, with enzymes catalyzing only a specific reaction, cells can control the processes taking place by adjusting the concentrations of the enzymes that catalyze these processes.

Consequently, the number of different enzymes in a cell is rather large. Most enzymes are named by adding the suffix "-ase" to the name of the substrate being catalyzed. For example, lactase is an enzyme that splits lactose into two sugar monomers (galactose and glucose) and DNA polymerase is an enzyme that catalyzes the polymerization of DNA from nucleotide monomers. As can be seen from these two examples, enzymes can catalyze reactions of two broad classes: anabolic and catabolic reactions. In **anabolic reactions**, two or more substrates are brought together while in **catabolic reactions** a single substrate is split into two or more products. Table 15.1 lists a few key enzymes and their functions.

Enzyme	Type	Function
Lactase	Catabolic	Splits lactose into sugar monomers
Amylase	Catabolic	Splits starch into sugar monomers
DNA Polymerase	Anabolic	Assembles DNA
Peptidase	Catabolic	Breaks up proteins
Lipase	Catabolic	Breaks down fats

Table 15.1: Examples of key enzymes found in cells.

While all enzymes are proteins, not all biochemical catalysts are enzymes. For example, RNA can have catalytic functions as well. This is an important fact in the context of the origin of life since enzymes are necessary for enzyme production, leaving one with a chicken-and-egg problem unless other catalysts can be brought into the picture.

Most of the molecules in our environment are rather stable — we don't see spontaneous explosions all around us — and hence it is clear that there is some barrier stopping them from reacting. Consequently, in order for molecules to react this barrier needs to be overcome. By observation in a laboratory, it can easily be found that the effect of an enzyme on the speed of a reaction can be enormous, often in the order of millions of times. But how is that possible? Wouldn't that require an enormous amount of energy? After all, from thermodynamics, we know that nothing comes for free and from chemistry we know that chemical reactions that do not spontaneously occur have an activation energy that needs to be overcome.

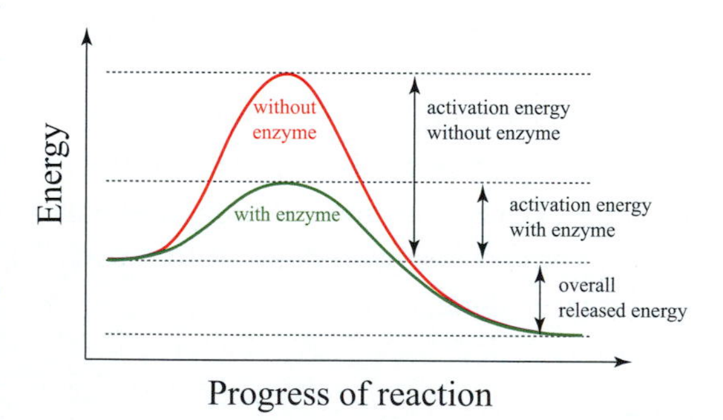

Figure 15.3: Enzymes lower the activation energy.

The **activation energy** is the barrier that needs to be overcome and indeed all enzymes work by lowering the activation energy as shown in Figure 15.3. As mentioned above, from the fact that the difference in speed between a catalyzed and uncatalyzed reaction can be enormous, one might be tempted to conclude that the activation energy must be very high. This, however, is not usually the case. When dealing with complex molecules, the reaction rate is determined not only by the energy barrier that needs to be overcome, but also by how likely the reactants are aligned such that a reaction can take place — something that may happen only very rarely. For less precise alignments, either (much) more energy is required or the reaction doesn't take place at all. Hence, besides lowering the activation energy, enzymes speed up reaction rates by bringing substrates together in exactly the right way as illustrated in Figure 15.4.

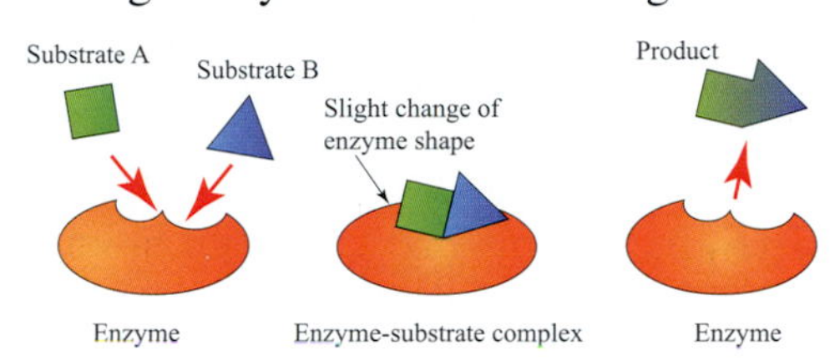

Figure 15.4: Schematic enzyme action.

Still, even if the activation energy is lowered, a barrier remains. How is that overcome? For some reactions, an energy supply might be necessary, but in many cases the constant random motion of the molecules due to thermal motion will be sufficient. In other words, if the energy barrier is lowered enough by an enzyme, thermal random motion imparts enough energy to the molecule for it to cross the barrier and the reaction takes place.

In general, the scientific field of enzyme reactions is called **enzyme kinetics** and it is an essential part in the study of cellular chemical reactions. Let us now have a look at the enzyme kinetics of a typical catabolic reaction: the cutting of a polysaccharide chain.

Figure 15.5 shows schematically how the enzyme lysozyme cuts a polysaccharide chain into smaller pieces.

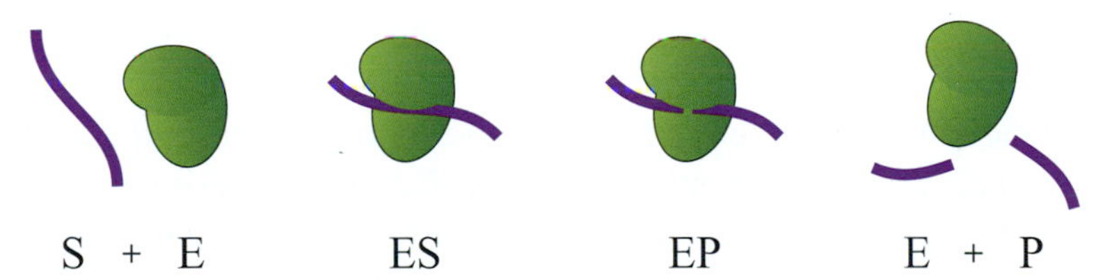

Figure 15.5: Schematic representation of lysozyme cutting a polysaccharide chain. S+E: Substrate and Enzyme, ES: Enzyme-Substrate complex, EP: Enzyme-Product complex, E+P: Enzyme and Product.

If we denote the enzyme as E, the substrate as S and the product as P, we can write the reaction quite generally as

$$\mathrm{E} + \mathrm{S} \underset{k_{-1}}{\overset{k_1}{\rightleftarrows}} \mathrm{ES} \overset{k_{\text{cat}}}{\rightarrow} \mathrm{E} + \mathrm{P} \tag{15.1}$$

where k_1 is the rate at which the enzyme and the substrate form enzyme-substrate complexes ES, and k_{cat} the so-called **turnover number** that gives the number of substrate molecules that one enzyme molecule can process per second.

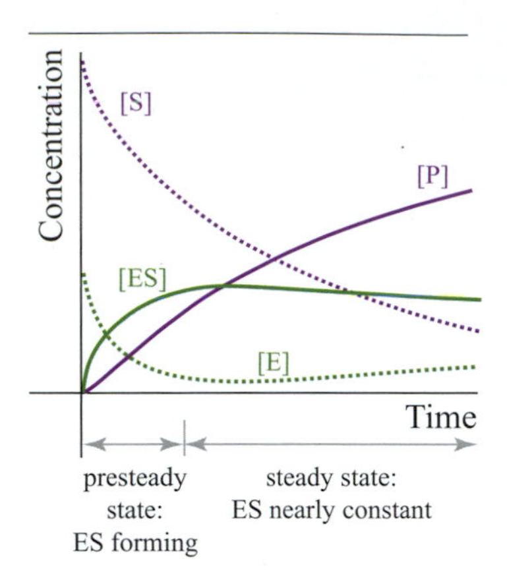

Fig. 15.6: Steady state of the ES complex.

What we are generally interested in is what steady state will be obtained if we have a given total concentration of the enzyme $[E_T]$ and the substrate [S] (the square brackets indicate concentration).

Intuitively, we would expect that the concentration of the enzyme-substrate complex [ES] increases rapidly at first and that it then levels off at a point determined by the turnover number as is shown in Figure 15.6. Since [ES] is (by definition) constant at the steady state we have that the rate of [ES] breakdown equals the rate of [ES] formation. In other words

$$k_{-1}[\mathrm{ES}] + k_{cat}[\mathrm{ES}] = k_1[\mathrm{E}][\mathrm{S}] \tag{15.2}$$

where [E] is the concentration of the *free* enzyme (i.e. those enzyme molecules that are not bound to a substrate). As the total concentration of the enzyme $[E_T]$ is taken fixed for the moment we have $[E] + [ES] = [E_T]$ that can be used to rewrite Equation 15.2 as

$$[\mathrm{ES}] = \frac{k_1}{k_{-1} + k_{\mathrm{cat}}}[\mathrm{E}][\mathrm{S}] = \frac{k_1}{k_{-1} + k_{\mathrm{cat}}}([\mathrm{E_T}] - [\mathrm{ES}])[\mathrm{S}]. \tag{15.3}$$

It is now convenient to introduce the Michaelis constant as $K_m = (k_{-1} + k_{\mathrm{cat}})/k_1$ and express the rate at that products are made as

$$V = k_{\mathrm{cat}}[\mathrm{ES}] \tag{15.4}$$

where V stands for velocity — a somewhat different meaning from standard mechanics. Inserting K_m and V into Equation 15.3 we obtain the so-called **Michaelis–Menten equation**

$$V = \frac{k_{\mathrm{cat}}[\mathrm{E_T}][\mathrm{S}]}{K_m + [\mathrm{S}]} \qquad \textbf{Michaelis–Menten Equation}. \tag{15.5}$$

The above equation is a good description of the kinetics of many enzymes in situations where the substrate concentration is relatively large when compared to that of the enzyme. The Michaelis–Menten equation is based on applying concepts from basic chemistry (mass-action law) and physics (diffusion). While this is reasonable in many cases, often it is not directly applicable to living organisms as the interior of a cell is a crowded place.

Although enzymes can range in size from roughly 60 to about 2500 amino acid monomers, generally only a few of their atoms are involved in the actual catalysis, and the location of those atoms is called the **active site**.

A great trick in nature's repertoire is that processes are controlled not only by enzyme concentrations but also at a second level by smaller molecules that affect the functioning of an enzyme as illustrated in Figure 15.7.

That is to say, there are so-called **inhibitors** that reduce or even suppress the effectiveness of an enzyme and there are **activators** that enable or increase an enzyme's functioning. From a computational point of view, what is interesting is that the inhibitors and activators are often direct or indirect products of the catalyzed reaction thus forming a feedback mechanism.

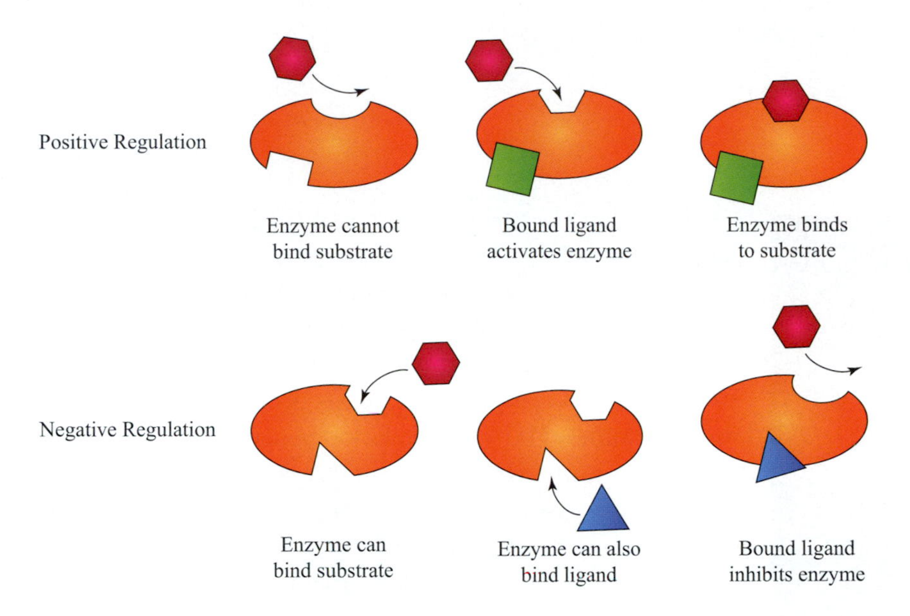

Figure 15.7: Small molecules can activate or inhibit the functioning of an enzyme as part of intricate feedback control mechanisms.

So now we have seen that enzymes play a key role in making many reactions possible. In a way we could say, "OK, done, we know who does all the work — it's the enzymes!" and that's true enough. But it's a little bit like saying that the tools do all the work. Therefore, let us probe a bit further. If we talk about tools, we often think about building. In order to build any structure of some intricacy, besides the walls, windows etc., one needs a skeleton and a means to move things around. Similarly, a cell needs various structural materials that we consider next.

15.3 Structural proteins

Among the most important activities that need to be carried out by a cell are growth, division, motion and maintenance of cellular structure. Although these activities are rather diverse, they share some common characteristics. First, they need materials that are easy to assemble and disassemble, that are neither too rigid nor too flexible, that are rod or filament like, and these materials need to allow for a rather diverse range of specific functions. These requirements immediately exclude mono-atomic or simple structures (like steel); DNA and RNA are unsuitable for various reasons among which are the complexity of their assembly process and their lack of structural rigidity. Lipids clearly lack structural stability in rod- or filament-like arrangements and so the only major class of biomolecules with the right properties is proteins. Some of the functions of the cytoskeleton are summarized in Table 15.2.

It is not surprising then that eukaryotic cells make extensive use of structural proteins. They have three main types of protein assemblies that, together, form the bulk of the **cytoskeleton**. These main types are the microtubules that play a key role in the division of a cell, the microfilaments that are essential for motion, and the intermediate filaments that provide structural support.

Jöns Jakob Berzelius

Figure 15.8: Portrait of Jöns Jakob Berzelius who coined the term “protein”.

Jöns Jakob Berzelius was one of the founding fathers of modern chemistry. He was born on August 20, 1779, in Väversunda, Sweden. Having lost both parents during his early childhood, he grew up with relatives. After studying medicine at Uppsala University from 1796 until 1802, and working for several years, Berzelius became a full professor in Stockholm in 1807 where he remained until his death in 1848.

Berzelius spent great efforts to analyze substances and counts among his greatest achievements the definite confirmation of the **law of constant proportions**, the law that states that inorganic substances are composed of different elements in fixed proportions by weight. Putting the law of constant proportions on a firm footing was of enormous importance since at that time the distinction between compound and mixture was not entirely clear.

One of the techniques that Berzelius perfected was electrolysis that aided him greatly in his quest to accurately describe the constituents of compounds (he was the first to electrolyze salts). He discovered no fewer than three new elements — cerium, selenium and thorium — and showed that silicon is not a compound as was the thought at that time but an element.

With his systematic and careful approach, Berzelius determined the atomic weight of nearly all the elements known to him quite precisely, an essential piece of information in the development of the periodic table somewhat later by Mendeleyev. Having to deal with so many elements and compounds, he simplified the existing chemical notation by introducing the modern chemical notation of using the first one or two letters of the Latin name of an element to represent it with the proportions given by a superscript (he wrote H^2O rather than H_2O). Furthermore, he invented the mercury cathode and introduced words such as organic chemistry, isomerism, allotropy, catalysis, protein, polymer, polymerization, halogen, and electronegative.

	1818	1873
Gold	☉	Au
Mercury	☿	Hg
Lead	♄	Pb

Figure 15.9: Old and new symbols.

Another major contribution was his notion of atomic “duality”. His thinking was that since electrical potentials can split compounds, atoms must be charged and that a chemical combination of atoms neutralizes this charge. Berzelius noted that this worked well for inorganic compounds but not organic compounds — a mystery he was never able to solve.

On a more practical level, Berzelius’ work was characterized by a highly systematic approach and great care for precision that would later become the hallmark of science in general.

Egg white

Figure 15.10: Egg white: raw, whipped and fried.

An unfertilized egg consists of three main parts: the egg shell, the egg white and the egg yolk. Egg white consists of about 88% water and 12% solids of which 92% is proteins. It is often referred to as **albumen** from the latin word "albus" meaning white. Egg white protects the yolk and provides water and proteins for a developing embryo. It makes up about 2/3 of the weight of an egg (note that the egg yolk contains about an equal ratio of proteins by weight but also a significant amount of fat, cholesterol, vitamins and minerals that occur only in small quantities in egg white).

Egg white is an excellent substance to illustrate some of the fascinating physical properties of polymers. In a raw egg white, the chains of amino acids that constitute the proteins are folded in specific ways forming compact molecules that are suspended in water thus forming a gel-like liquid. When egg white is heated up, the protein molecules start to unfold in a process called **denaturation** that leads to long strands of amino acids that can easily entangle with each other and form new (weak) bonds as illustrated in Figure 15.11. Thus, the egg white in a boiled or fried egg becomes a soft solid.

More generally, proteins and nucleic acids can denature due to a number of different factors such as being exposed to strong acids or bases. Since the function of a protein is strongly related to the way it is folded, denatured proteins cannot carry out their task and may lead to the death of an organism. In the case of frying egg white, denaturation is irreversible, but under many circumstances denaturation is reversible and proteins can regain their structure even after being denatured.

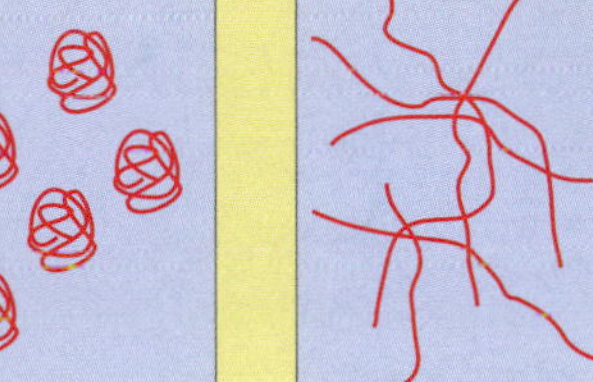

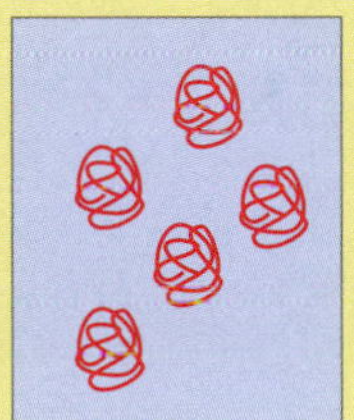

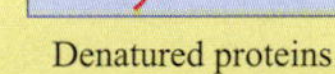

Figure 15.11: Denaturation of proteins.

When egg white is beaten with a whisk, denaturation takes place as well but additionally air is mixed in, leading to a volume increase of up to around eight times. The stability of the foam depends on the strength of the bonds between unraveled protein strands. A common way to strengthen those bonds is by whisking the eggs in a copper bowl. The minute amounts of copper that are released during the preparation of the foam are enough to make a noticeable difference. Hence there is indeed a scientific reason to the recommendation by many chef's schools to use copper bowls for the whisking of egg whites.

Egg white contains more than 40 different proteins, the most common of which are shown in Figure 15.12, and all the amino acids are necessary for human nutrition. The combination of valuable physical properties with high nutritional value make egg white one of the most common cooking ingredients in the world.

Ovalbumin	54%
Ovotransferrin	12%
Ovomucoid	11%
Globulins	8%
Lysozyme	3.5%
Ovomucin	1.5%

Figure 15.12: Most common proteins in egg white.

Since cells themselves as well as their environments constantly change, the cytoskeletal system is highly dynamic. Let us now consider the three main types of cytoskeletal elements in some more detail.

Separation of chromosomes in mitosis
Control of vesicle traffic
Cell movement
Connection of cells to each other

Table 15.2: Some of the functions of the cytoskeleton.

Microtubules

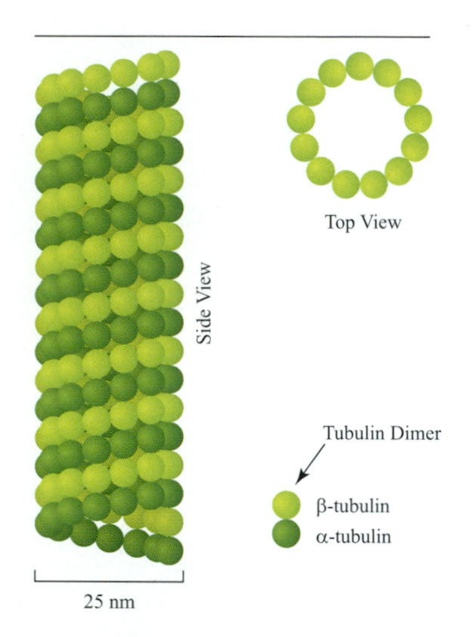

Fig. 15.14: Schematic representation of a microtubule.

Microtubules can self-assemble. Yet this self-assembly can precisely be controlled by the cell.

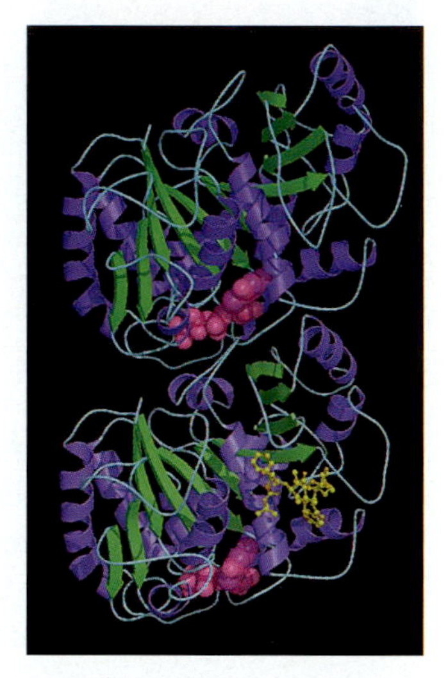

Figure 15.13: Tubulin dimer. Guanine nucleotide in pink.

Microtubules play a key role in cell division where they are used in the segregation of chromosomes. As illustrated in Figure 15.13, microtubules are built up of protein dimers (that is, protein molecules composed of two subunits that in this case are nearly identical) called tubulin that can self-assemble into hollow tubes of about 25 nm in diameter as shown in Figure 15.14.

What is rather remarkable about the self-assembly process is that it can precisely be controlled by the cell. That may sound weird: On the one hand, the tubule assembles by itself but on the other hand, this process can be controlled. What does that mean and how is that possible? What self-assembly means here is that the cell only needs to create the right conditions for the tubulin dimers to polymerize into tubules. There is no need for some specific mechanism like in the case of, for example, the copying of DNA to put the dimers into the right place. If there are the right conditions and tubulin dimers are present, microtubules will form. Control of the assembly process is then obtained by manipulating the conditions. To understand that, let us consider the dynamics of a microtubule as illustrated Figure 15.15.

In order for tubulin dimers to polymerize into microtubules, its subunits need to bind to the nucleotide GTP (guanosine triphosphate) that is also the G in DNA and RNA. In free dimers (i.e. not part of a tubule), GTP remains as is. When a dimer is part of a tubule, however, the GTP nucleotide bound to the so-called β subunit gradually hydrolyzes into GDP (guanosine diphosphate) while releasing some energy. The GTP bound to the α unit, on the other hand, remains as is. What is essential now is that the kinetics of a dimer with two GTP nucleotides is different from that of a dimer with a GTP and GDP nucleotide. GTP-bound tubulin does not readily fall off from the microtubule while GDP bound tubulin does, but only if it is at the end and not when it is in the middle.

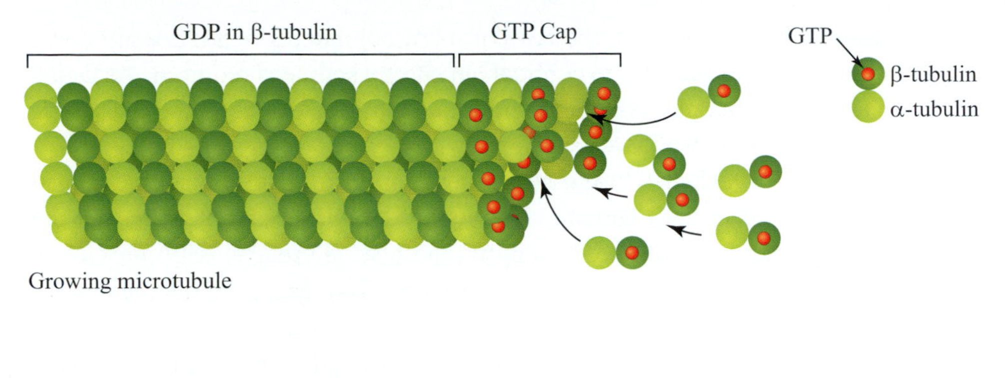

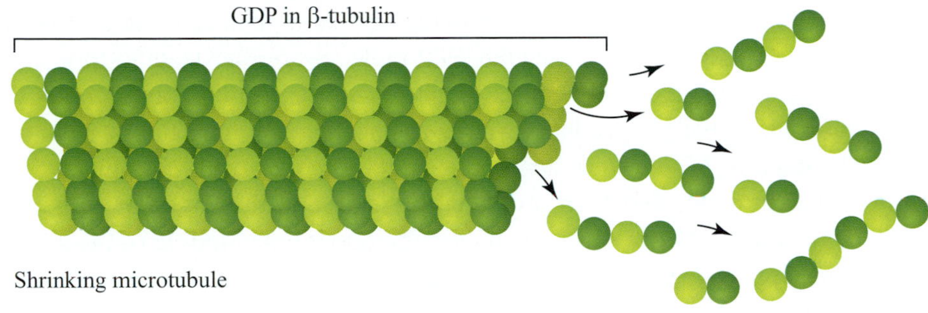

Figure 15.15: Depending on the presence of a GTP cap, microtubules can grow or shrink.

Since the tubulin dimers attach to each other in a specific way, the tubule can grow in only one way. Consequently, when a "fresh" dimer with GTP on both subunits attaches to a tubule, this dimer will not fall off for a while. Now if another fresh dimer attaches to the front before the previous dimer's GTP hydrolyzes into GDP becoming prone to fall off, it will stay put and be part of the tubule until it becomes a dimer at the other end of the tube where it can fall off. For a free tubule there is consequently a kind of a race between attaching dimers in the front and hydrolization in the front and detachment in the back.

Microfilaments

For structural support and locomotion, microfilaments such as the one depicted in Figure 15.16 are used. Figure 15.17 shows actin (as well as tubulin) in a stained cell. Just like microtubules, microfilaments grow from basic units and only in one direction. In this case, however, the basic units are monomers of the globular protein molecule **actin** shown in Figure 15.18, and the resulting filaments with a diameter of about 7 nm are not hollow. Microfilaments provide much of the needed support for a cell to deal with the frequent stretching and compression that is especially common in multicellular organisms.

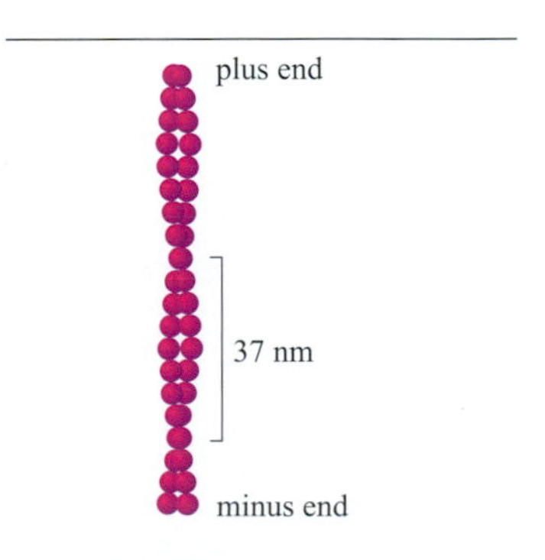

Fig. 15.16: Schematic representation of a microfilament.

Besides providing structural support, microfilaments have a host of other functions, but perhaps the most noticeable one is that they underlie motion in many cell types. Indeed, microfilaments are a key component of muscular contraction in that myosin "walks" along an actin filament. How this works is schematically depicted in Figure 15.19.

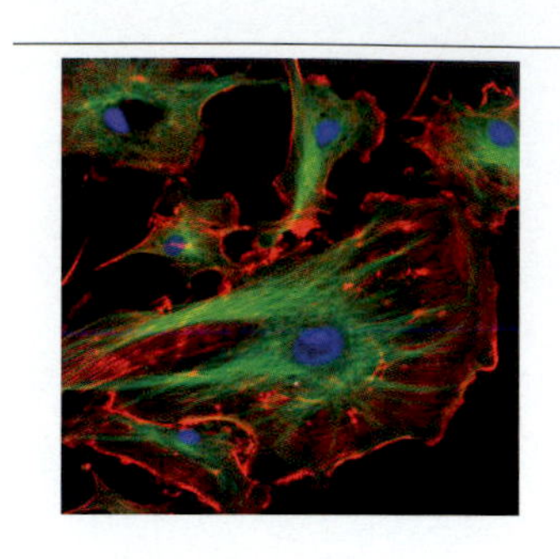

Fig. 15.17: Stained cell. Blue: nucleus; green: tubulin; red: actin.

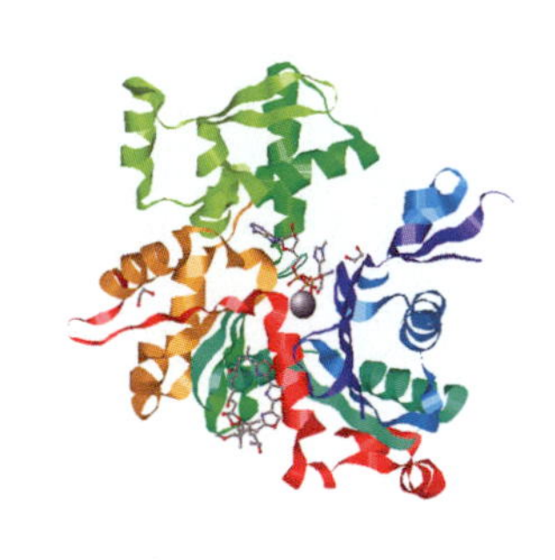

Fig. 15.18: Actin monomer.

Of course, as we know from the laws of thermodynamics, work doesn't come free. Consequently, for muscles to contract, an energy source is required. As usual, this is ATP which splits into ADP and phosphate while releasing energy. The five stages of one step in the walk are as follows:

Stage 1 (Head attached): In the beginning, a myosin head is firmly attached to an actin molecule in the filament. This is the default arrangement in that the myosin head will be without energy supply (and also the reason why muscles in dead organisms are stiff, a phenomenon call rigor mortis). Stage 2 (Head released): An ATP molecule binds to a special region of the myosin head. This causes a small change in the myosin head that reduces the strength with which it is bound to the actin molecule allowing it to slide to the next actin molecule. Stage 3 (Head moves): The myosin head encloses the ATP molecule further leading to a shape change of the head that causes it to move along the filament by about 5 nm. ATP is now hydrolized but the products ADP and phosphate remain tightly bound to the head. Stage 4 (Bind to new actin molecule): The myosin head binds weakly to the new actin molecule which results in the release of the phosphate produced by the hydrolysis of ATP in stage 3 and the subsequent tight binding of the myosin head to the actin filament. Stage 5 (Force generated): This in turn causes the head to change back to its original shape while exerting a force on the acting filament resulting in the filament sliding relatively to the myosin filament. At the same time, the head releases its ADP thus returning to the initial state — ready for a new cycle.

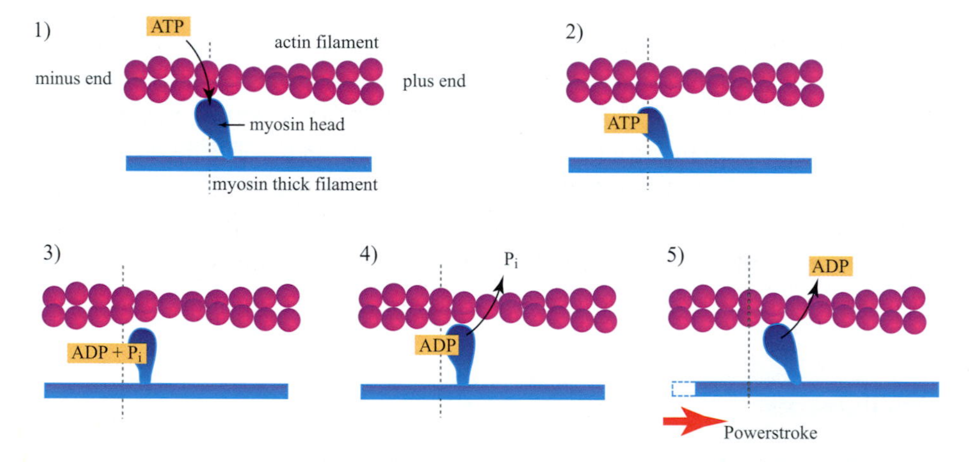

Figure 15.19: Muscles contract by sliding bundles of myosin against microfilaments. P_i is so-called inorganic phosphate, a stable ion on phosphoric acid, H_3PO_4.

Intermediate filaments

Intermediate filaments have a diameter of around 10 nm and, therefore, their size is in between microtubules and microfilaments. Depending on the cell type, they can be made from different proteins but all share a common overall structure of protein dimers entwined into coiled rods as shown in Figure 15.20

Intermediate filaments are rather abundant in the skin and nerve cells of mammals. An important family of intermediate filaments is formed by the **keratins**.

Keratin filaments can be linked together by disulfide bonds into networks that can survive the death of the cells that they are a part of, leading to tough protective layers such as in nails, claws, scales and the outer part of the skin. They are also the basic constituent of hair.

Another large family of intermediate filaments is formed by the so-called **neurofilaments** that occur in high concentrations along the axons of vertebrate neurons where they determine the axon's diameter. Since the diameter of the axon directly affects the speed with which electrical signals can travel in the axon, these filaments have a strong impact on the overall functioning of the neural network.

Thus far, we have considered structural components inside a cell. However, structural components can also be outside of cells. Indeed, the most abundant protein in animals, making up about 25% of the protein total, is a structural component that can be found both inside and outside of a cell. It is a protein called **collagen**, so let us have a look at that a bit more closely.

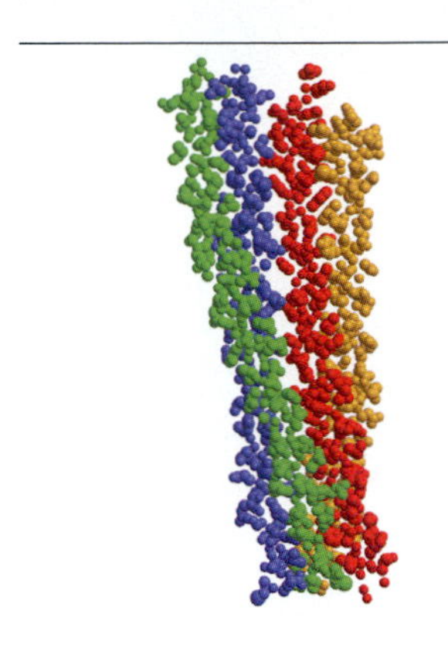

Fig. 15.20: Schematic representation of an intermediate filament building block.

Collagens

Figure 15.21: Collagen is a key constituent of, for example, cartilage, bones and teeth. It also provides toughness to skin.

The word "collagen" is derived from the Greek words "kolla" for glue and "genes" for born. Hence, literally, collagen means glue maker and indeed, collagens are among the earliest known glues that mankind used, perhaps as early as 8,000 years ago. Collagen glues can be obtained by boiling skin in water, for example.

Just like the filaments discussed above, collagens are made of building blocks. The resulting polymers have great tensile strength and are the main constituents of cartilage, tendons, bones and teeth. It is also part of the skin where together with the keratin polymers discussed above they provide the skin its characteristic toughness. As collagen provides tensile strength, its degradation in the skin leads to wrinkles, a common sign of old age. As such, collagen is a kind of fibrous protein (note that fibrous proteins are only found in animals).

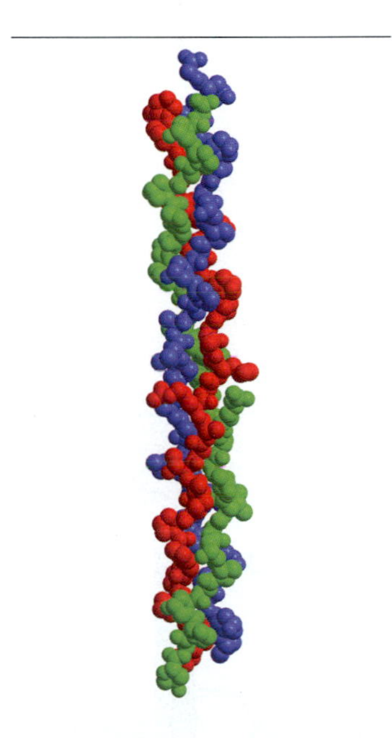

Fig. 15.22: Tropocollagen.

The basic building block for collagen is the tropocollagen molecule that consists of three collagen α chains that are twisted together into a helix as illustrated in Figure 15.22. Each chain in the tropocollagen molecule is about 1000 amino acids long and the total length of the tropocollagen molecule is about 300 nm with a diameter of about 1.5 nm. It should be noted that the collagen chain is a characteristic amino acid sequence not to be confused with the collagen fibers that are the final fibers resulting from three specific hierarchical steps of combining those chains. A special property of these tropocollagen molecules is that they can self-assemble into collagen fibrils of between 10 nm and 300 nm diameter that in turn are bundled into the final collagen fibers that can be up to around 3 μm thick.

All in all, one could argue that the microtubules are a kind of bone for a cell, while actin fulfills the role of the muscles. Intermediate filaments can be considered similar to ligaments. For a cell to have strength, shape and the ability to move, all these cytoskelatal components need to work together.

Figure 15.23 illustrates the complementary nature of the three types of structural proteins.

By varying the relative abundances, a wide range of materials with vastly different physical properties can be constructed by a process of self-assembly. What we see here is that enzymes are not necessary for the assembly of the filaments and hence that not all work — even work involving proteins — needs enzymes. It's good that we didn't stop our investigations too early. This fact may also be relevant in relation to the origin of life.

Excellent! We have found that proteins are important materials in the construction of our "cellular house".

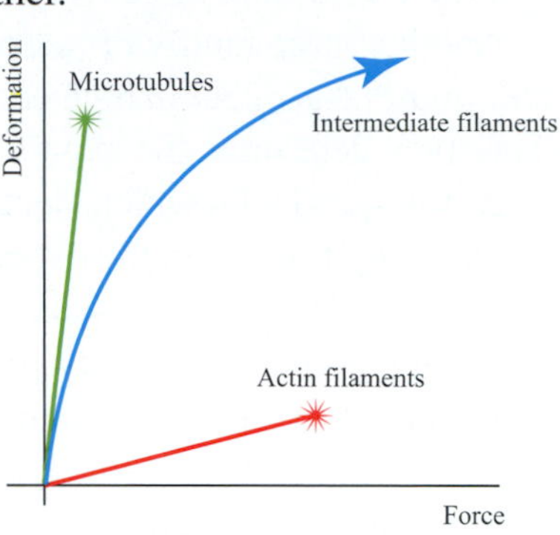

Figure 15.23: Mechanical properties of structural proteins.

All houses have doorbells which is an example of the wider need for having a communication system. There must be a way to send signals from one part of a cell to another part of the cell or, in a multicellular organism, from one cell to another. In the next section, we investigate some of these signals.

15.4 Sending messages

In multicellular organisms, including plants, signals need to be sent from one part of the body to others. The chemical messengers for these signals are called **hormones**. Although not all hormones, notably steroid hormones like estrogen and testosterone, are proteins, many are. Perhaps the best known example of a protein hormone is **insulin**.

Fig. 15.24: Hormones are messengers. Many hormones are proteins.

Insulin plays a key role in the metabolism of carbohydrates. In mammals, it is produced in the pancreas and some of its functions are listed in Table 15.3.

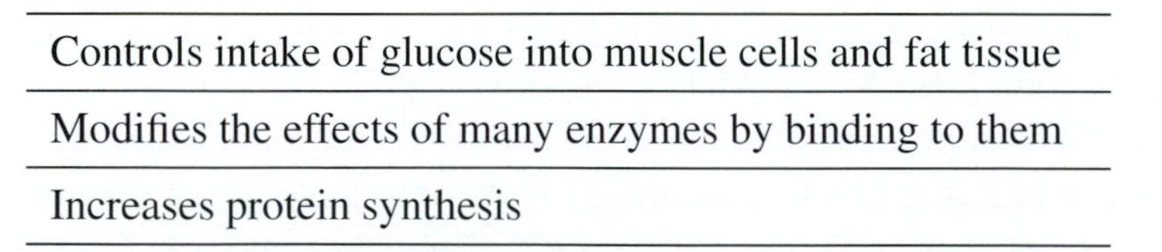

Functions of insulin
Controls intake of glucose into muscle cells and fat tissue
Modifies the effects of many enzymes by binding to them
Increases protein synthesis

Table 15.3: Some of the functions of insulin.

In order for many organs and tissues to function properly, blood glucose levels need to be maintained in a very narrow range (in a sense this is a bit similar to domestic energy supply — if the voltage swings too much, our appliances will break). This is no easy task since food intake is generally very irregular, that is, there is food intake during a meal or when prey is caught but between meals there is no food intake for many hours or perhaps even days. Furthermore, the actual total quantity of glucose in the blood of a human is surprisingly small at about 5 g or roughly 1 tsp of sugar.

How, then, can the glucose level be maintained in such a narrow range? There must be some kind of a feedback mechanism. Indeed there is but rather than just having a single agent, there are in fact two! First there is insulin that decreases blood glucose levels and then there are several hormones that increase the blood glucose level of which the protein **glucagon** (just like insulin produced by the pancreas) is the most prominent one.

Figure 15.25: Glucose levels are stabilized by two opposing "forces".

When after a meal, for example, glucose levels increase in the blood, insulin is released from the pancreas into the blood. As a result, muscle cells and fat tissue increase their intake of glucose thus lowering the glucose level in the blood. When there is too little glucose in the blood, the pancreas releases glucagon that causes the liver to release glucose. It is almost a bit like the benign tug of war depicted in Figure 15.25.

Running a house, including the signaling system, requires a lot of energy, but what if there is a power outage and all the emergency stocks have been used up? That is to say, there is no food coming from the outside world and there is no fat left to consume. As a last resort, an organism can use up proteins for energy supplies.

15.5 Energy

The assembly of proteins from their atomic constituents (carbon, nitrogen etc.) requires a significant energetic effort. First, the groups that form the amino acids need to be collected, next the amino acids have to be assembled and then these amino acids need to be strung together to form the protein. Clearly, it will be beneficial to an organism if it can skip some of these steps efficiently. Therefore, in general, dietary proteins are only split into amino acids and not further. This process works quite well since virtually all life on earth uses the same 20 amino acids and hence these building blocks are universal to life as we know it. Indeed, some organisms (humans being among them) cannot even produce all the required amino acids from their atomic constituents and need to obtain certain amino acids from food. Such amino acids are called essential amino acids.

Proteins	*4*
Carbohydrates	*4*
Alcohols	*7*
Fats	*9*

Energy contained in 1 g of the major food categories (in calories).

In other words, dietary proteins are generally used to make new proteins by breaking the ingested proteins up into amino acids in the digestive process, transporting these to cells and processing them there.

However, the fact that this process can save energy can also be viewed in a different light. It indicates that amino acids store a significant amount of energy that in principle can be liberated if required. This is exactly what happens when organisms run short of carbohydrates and fats. In that case, the amino group (NH_3) is stripped from the amino acid. Since the resulting compounds are all part of the regular energy pathway of the cell used for carbohydrates and fats, these resulting compounds can therefore readily be used.

Hence, besides their many other functions, proteins are not only tools but also the energy source to carry out the job.

Sometimes one can read about so-called "sick" buildings, where the environment inside is less than optimal. Similarly, in an organism, its fluids need to be maintained in just the right state. Let us therefore consider some aspects of a "healthy" cellular fluid.

15.6 Fluid and acid/base balances

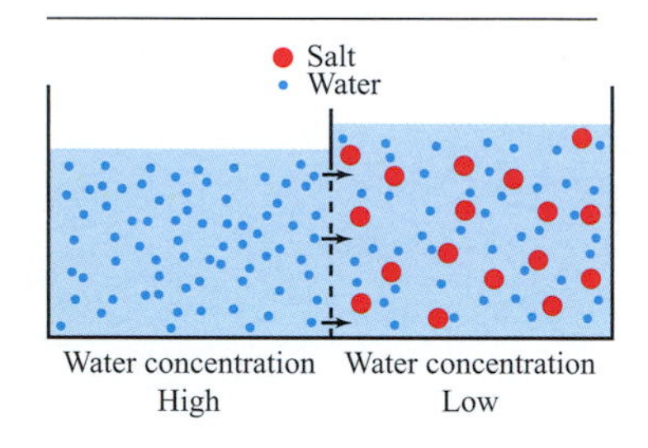

Fig. 15.26: Water cannot actively be transported by cell membranes. It is regulated with the help of osmosis.

Due to the limited amount of resources (food, energy) efficiency is essential to life. In general, efficiency and the wide range of operational parameters necessary for survival are conflicting tendencies. For example, organisms may need to endure rather large temperature variations even though the chemical reactions underlying survival may only be at their maximum efficiency in a very narrow range of temperatures. Organisms therefore need to finely balance these tendencies.

An important ingredient of this balance is the maintenance of stable internal conditions, a process called **homeostasis**. Water is an essential part of life as we know it. Even so, water cannot be actively transported by cell membranes and hence water transport in and out of cells is determined by **osmosis** (see also p. 219). The key to maintaining the right amounts of fluids in the various cells of a body is then to ensure that the concentrations of ions and proteins in the fluids that surround the cells or the blood stream are adequate.

While it is quite obvious that a lack of water can be life threatening, the consequence of having water in the body transported in and out of cells by osmotic pressure also means that too much water can be life threatening since this would result in very low concentrations of ions and proteins in the blood. For example, a condition called hyperhydration (also known as water poisoning or water intoxication) can occur if extremely large amounts of water are consumed in a very short time such that the excessive water cannot be removed by the body as urine and perspiration quickly enough. Although the exact amount of water to cause hyperhydration will vary from person to person, the consumption of 5 to 8 L of water within one to one and a half hours can be lethal.

Fig. 15.27: Drinking too much water in a short period of time can be lethal.

Besides ion and protein concentration, another essential aspect of the fluids in our body is their pH or in other words the acid-base balance since proteins are, in general, very sensitive to the pH value in both their conformation and their function. The largest contributor to the pH value comes from H^+ ions — i.e. protons — that are the result of metabolism. While excessive H^+ can be removed from the body after suitable processing by urination or respiration, these processes are rather slow and hence an interim buffering mechanism is necessary to maintain the pH at a stable level. One of the buffering mechanisms involves proteins. When the pH increases, some proteins can release H^+ ions from one of their amino acid's carboxyl groups thus lowering the pH. Conversely, when the pH decreases, the amino group of a terminal amino acid of some proteins can accept an H^+ ion to increase the pH.

Thus far we have basically considered activities that are carried out in a certain area or a certain location. However, especially in the case of multicellular organisms, the specialization of cells often requires the transport of materials and finished products. In this area too, proteins play an important role (though it should be mentioned that not all materials and products are transported by proteins).

15.7 Transport

When we consider the transport of materials and products, we need to look at transport in the blood as well as transport through the cell membrane. Proteins that transport materials are called **carrier proteins** or **transport proteins**. First let us consider transport through the membrane.

The basic building blocks of cell membranes are phospholipids, which form a bilayer separating the inside and outside of a cell. By itself, such a bilayer is highly permeable to small nonpolar molecules such as O_2 and CO_2 and somewhat permeable to small uncharged polar molecules such as water and urea. However, the bilayer is virtually impermeable to any kind of charged molecules or ions like Na^+ or K^+. Since many of the nutrients and other molecules that a cell needs are large, polar and/or charged, the question immediately arises as to how such molecules can be moved across the cell membrane.

One answer lies in special proteins that can transport molecules from one side of the membrane to the other. It turns out that there are two key mechanisms: one that does not need any energy called **facilitated diffusion** and another that does need energy called **active transport**.

Examples of facilitated diffusion are given in Figure 15.28 where two types of proteins are shown. The first type, called a channel protein, forms an aqueous pore through which specific molecules can diffuse along the concentration gradient. The second type, called a carrier protein, undergoes conformal changes to assist in the transport of a molecule through the membrane.

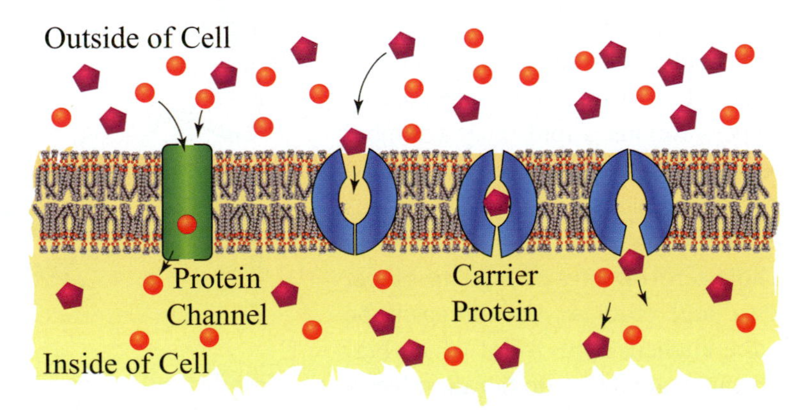

Figure 15.28: Facilitated diffusion.

In active transport, energy is supplied to the protein, again called a carrier protein due to its conformation to the molecule to be transported, and transport can occur against a concentration or electrostatic gradient. An example of active transport is shown in Figure 15.29 where the so-called **sodium-potassium pump** is depicted. The sodium-potassium pump moves sodium ions (Na^+) from inside the cell to the outside and potassium ions (K^+) from the outside of the cell to the inside, in both cases against the concentration gradient. At the energetic expense of using up one ATP molecule, at first three (Na^+) ions are moved out and then two (K^+) ions are moved in. The six steps of this process are shown from left to right in Figure 15.29.

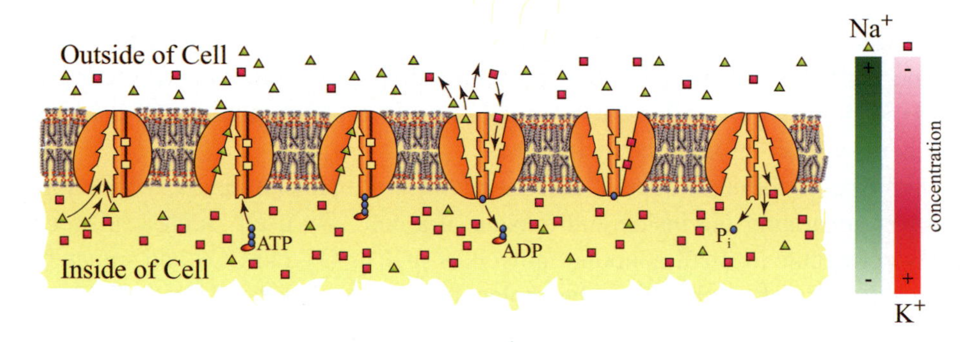

Figure 15.29: Active transport: sodium-potassium pump.

Another fundamentally different way to transport molecules from one side of the membrane to the other in eukaryotes is by a process called **endocytosis**. In endocytosis, molecule(s) are engulfed with a part of the cell membrane to first form a bud that is then separated as a vesicle as shown in Figure 15.30. There are three basic mechanisms of endocytosis: phagocytosis, where solid nutrients but also pathogens like bacteria and viruses are engulfed that will be further broken down inside the cell; pinocytosis, where liquid and dissolved molecules are engulfed that do not need to substantially be broken down; and receptor mediated endocytosis, where the vesicle will have a protein coat.

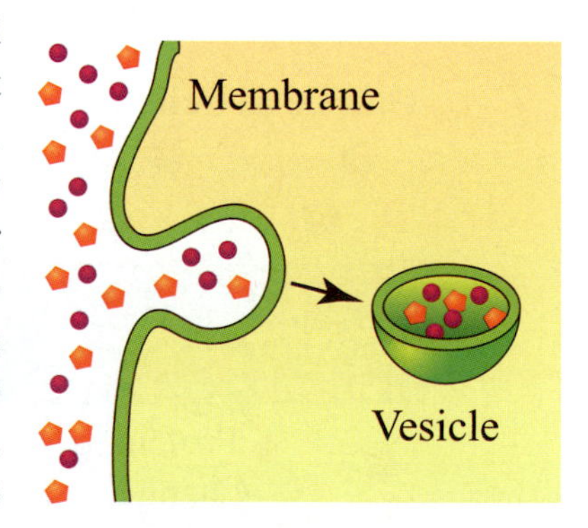

Figure 15.30: Endocytosis.

In multicellular organisms, nutrients and materials are mostly transported from one part of the body to another through the bloodstream. One important class of materials is formed by the many types of lipids. Lipids play a number of key roles in the body. For example, cholesterol is an essential part of all cell membranes in animals. However, if an organism tries to transport cholesterol directly through the blood, that creates a huge problem. As can easily be verified in the kitchen, lipids — of which butter would be an example — do not dissolve in water; this is one of the reasons we need soap to do the dishes! But if they are essential, they still need to be transported. How could that be done?

In order for lipids to be transported in the bloodstream, they require the help of transporter lipoproteins. Lipoproteins contain both lipids and proteins. In order to transport lipids like cholesterol through the blood, lipoproteins form small vessels with the charged part of the protein facing outwards making the vessel water soluble and the lipids on the inside. Figure 15.31 shows the so-called low-density lipoprotein complex often abbreviated as LDL. LDL is the main means of transporting cholesterol from the liver through the blood to cells in the body. Depending on the amount of lipids it carries, LDL is around 22 nm in size, a smaller variant called high-density lipoprotein (HDL) measures around 8 to 12 nm but rather than carrying cholesterol away from the liver, it carries cholesterol to the liver. The correct ratio of HDL and LDL is important for our health.

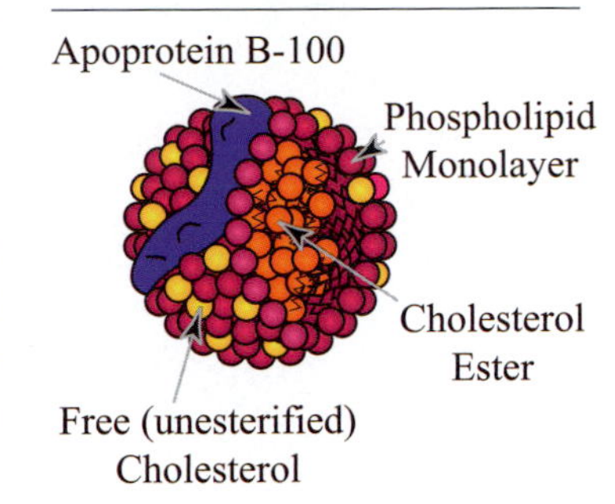

Fig. 15.31: Low-density lipoprotein (LDL) consisting of a phospholipid monolayer combined with an apolipoprotein encasing triglycerides and cholesterol.

Another essential ingredient of multicellular animal life is oxygen. Only a relatively small number of cells have direct access to oxygen and consequently all other cells need to have oxygen transported to them. This is done with the help of hemoglobin, a complex protein that in humans consists of four globular protein subunits each with a heme unit containing one iron atom. Since each iron atom in this configuration can bind to one oxygen molecule (O_2), one hemoglobin molecule can transport as many as eight oxygen atoms a time. Figure 15.32 shows a schematic representation of the hemoglobin molecule.

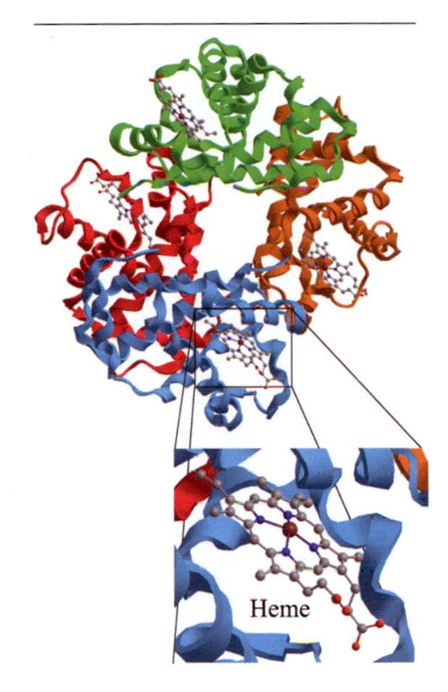

Fig. 15.32: Hemoglobin consists of four globular protein subunits each with a heme containing one iron atom.

15.8 Versatility

As the previous sections in this chapter illustrate, proteins have a great number of rather diverse functions. Indeed bacteria can have thousands of different proteins; for example, *Escherichia coli* (*E. coli*) have around 2,400 proteins with an average length of about 320 amino acids while human beings have tens of thousands of different proteins. One could therefore think that nature makes pretty good use of the possible amino acid sequences. But is this so?

To consider the total number of possible proteins, let us first go back and look at the possible number combinations one can obtain when throwing dice. What we do is the following: We throw the dice and then read off the faces from left to right as if they were numbers. For example, if we throw two dice and the one on the left ends up with a "3" and the one on the right with a "5", then we'd note it as the number "35". How many different numbers can two dice make in total? Well if the first die is a "1", then it can combine with any of six faces of the second die giving a total of six combinations starting with a "1" (11, 12, 13, 14, 15, 16). Similarly, if the first die is a "2", we can have 21, 22, 23, 24, 25, 26 giving another six combinations. Thus we can continue for each of the faces of the first die and find that the total number of different numbers is 36 or 6^2. In other words, we have the number of faces to the power of the number of dice. So if we had thee dice, we could make 6^3 different numbers.

Now back to proteins: if we want to know the total number of possible different amino acid sequences, we have to take the number of different amino acids and raise them to the power of the length of the sequence (i.e. the type of amino acid corresponds to the face on the dice and the length of the amino acid sequence to the number of dice). Consequently, if we have a very small protein consisting of a sequence of 50 amino acids we have 20^{50} different possible combinations and if we take an average *E. coli* protein we have 20^{320} different possible combinations.

Fig. 15.33: Two dice can give 36 different combinations (when ordered: i.e. 2-1 and 1-2 are not the same).

These are huge numbers! $20^{50} \approx 10^{65}$ is 1 million times 1 billion times the number of atoms in the earth (the earth contains approximately 10^{50} atoms) and $20^{320} \approx 10^{416}$ is far greater than the number of atoms in the entire universe that is somewhere around 10^{78}.

How did we come up with that number of atoms in the universe? Mostly, the universe consists of hydrogen so lets only consider that to get an estimate. Roughly, 1 kg of hydrogen has 10^{26} atoms, a typical star has a mass of about 10^{30} kg, a galaxy has about 10^{11} stars and the universe has about 10^{11} galaxies. Hence the total number of atoms in the universe is about: $10^{26} \times 10^{30} \times 10^{11} \times 5 \times 10^{11} \approx 10^{78}$. Note that this is only a back-of-the-envelope type of calculation and there easily could be 100 times more atoms but the key is the order of magnitude to get some idea about what is going on.

Of course, the total number of all possible proteins is even greater as we need to sum up all the possibilities for each length of the amino acid chain, but we'll neglect that here for simplicity. Taking into consideration that the muscle protein titin is a staggering 26,926 amino acids long, a length that allows for $20^{26,926}$ or $\approx 10^{35,031}$ combinations, it seems to be clear that the number of proteins in existence today is almost zero when compared to the total number of possible proteins.

To verify this, let us do some more back-of-the-envelope calculations. If we assume that there are around 100 million species and that each species has 10 thousand different proteins and that the proteins of each species are completely different from all other species, we find that the total number of different proteins is 10^{12} (of course in reality, the average number of proteins per species is quite a bit smaller and in general species share many proteins, but this calculation should give us a good upper bound). Even if we forget the very large proteins and just consider the average protein length of the *E. coli* bacterium, this means that the fraction of proteins in existence is smaller than $10^{12}/10^{416} = 10^{-404}$. This fraction is so small that for practical purposes one could just as well say it's zero.

But isn't it unreasonable to just consider the species in existence today? Shouldn't we also take all the extinct species into account? As strange as that may sound, it really doesn't make that much of a difference. Let us again grossly overstate the number of different proteins that ever existed to obtain an upper bound. Life on earth is approximately 4 billion years old. If there were 100 million species right from the outset and all species are completely replaced every 100 years with new species that have all different proteins while each species has again 10 thousand proteins, we have $10^8 \times 4 \times 10^9 \times 10^{-2} \times 10^4 = 4 \times 10^{19}$ as the total number of different proteins that ever existed (it should be stressed that the true number will be far smaller than that). So the fraction of all the proteins that ever existed to all the possible proteins of average length is then smaller than 10^{-397}, or in other words a number with 397 zeros after the decimal point — again practically zero.

To turn this around, we can conclude that the way proteins are made, there is a virtually infinite pool of possibilities. With proteins, nature will never run out of things to try!

Protein structure

What we have described above covers only the actual amino acid sequence. This is referred to as the **primary structure** but in general the function of a protein strongly depends also on higher order structures that result from specific ways in which the sequence folds. The **secondary structure** is formed by localized substructures that can occur multiple times over the length of the chain. Often these substructures are alpha-helices and beta-sheets, examples of which are clearly visible in the zoom-up in Figure 15.34. Then the substructures are spatially arranged to form the **tertiary structure**. Last, in the **quaternary structure**, several amino chains, each having a tertiary structure, are combined to form a single large protein. In the tubulin dimer in Figure 15.34, all four structural elements are present: The primary structure is the sequence of amino acids, the secondary structure is the many alpha-helices and beta-sheets, the tertiary structure is the globular molecule that forms one part of the dimer and the quaternary structure, the combination of the globular molecules into the dimer.

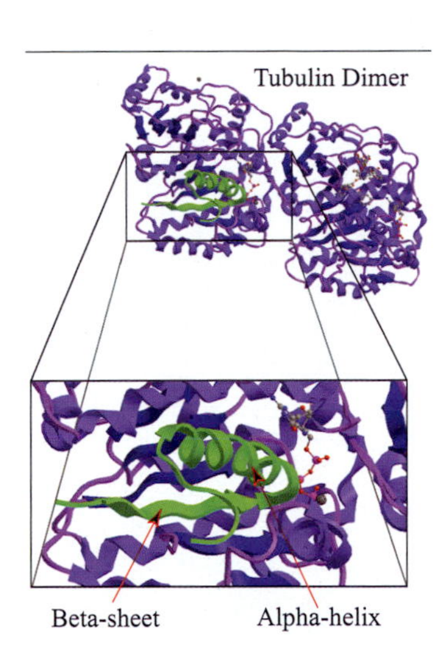

Fig. 15.34: Alpha-helices and Beta-sheets in a tubulin dimer.

The versatility and great flexibility of the protein's structure is exemplified by its role in the body's immune system that defends it against harmful and toxic substances that abound in the environment.

15.9 Antibodies

The number of different harmful bacteria and viruses is enormous. More formally, intruding substances that cause an immune response are called **antigens**. The question then is, what is a good way to deal with them? Clearly, before any harmful intruder can be neutralized, it has to be recognized. In principle, this recognition can be done by either a powerful cell or protein that can recognize many different intruders — and can then take care of them. Or fighting antigens can be done by more specialized but simpler molecules that then signal more powerful cells or molecules to take action if they cannot do so themselves. In light of the very large number of possibly harmful bacteria and viruses, it would appear to be useful to have small detection molecules specifically tuned to certain intruders and then specialized "destroyer" cells that can take on most if not all recognized intruders. This indeed is the solution nature has chosen as a part of the immune system. The specialized detection molecules are the antibodies.

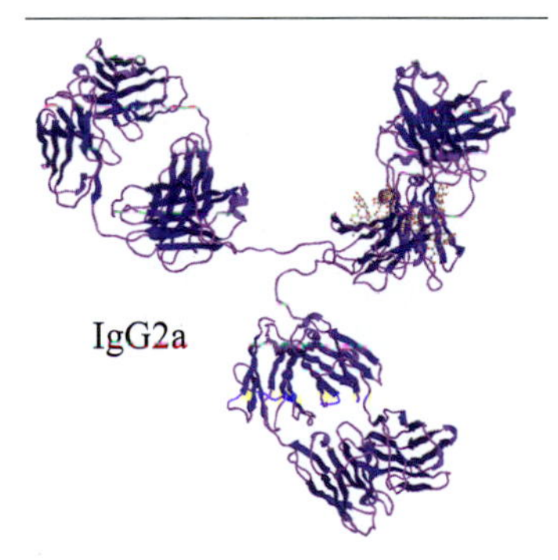

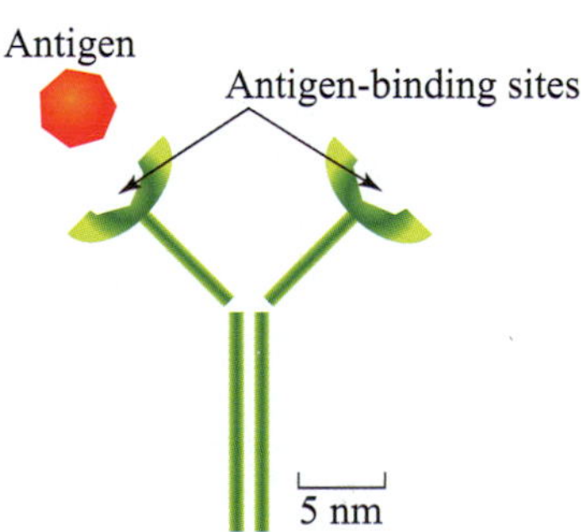

Fig. 15.35: Top: IgG2a monoclonal antibody. Bottom: Schematic representation of typical antibody.

The body's immune system consists of a) white blood cells such as macrophages, T- and B-lymphocytes, b) bone marrow, where the white blood cells are produced and c) the thymus gland, where T-lymphocytes undergo maturation. The T-lymphocytes' primary role is the detection of antigens and that of B-lymphocytes is the production of protein antibodies that tag an antigen. Macrophages' primary role is to destroy the intruding antigen.

Antibodies are Y-shaped proteins that have two sites called paratopes that recognize specific antigens, as illustrated in Figure 15.35. Conceptually, the situation is similar to that of a lock and a key. The antigen can be considered a lock; the protein's primary and secondary structures create a shape whose function is to complement the shape of the antigen and fit it, like a key fits a lock. Antibodies float freely in the bloodstream and when they encounter a lock, thus recognizing an antigen to be destroyed, they bind to it, tagging the identified intruder for destruction by the immune system.

Antigens are detected by the T-lymphocytes that, in turn, encounter and digest the bacteria or viruses. The T-lymphocytes then communicate information gained to the thymus gland, which then decides what specific B-lymphocytes are to be produced. The B-lymphocytes then synthesize protein antibodies. The primary structure of the proteins is the sequence of amino acids chosen by the B-lymphocytes — based on the information that it has received, via the thymus gland, regarding the intruding antigen. The protein then folds to form the secondary structure with many alpha-helices and beta-sheets tailored for the antigen in question. The folded protein undergoes further folding to form its tertiary structure, which is further modified to match the antigen. The final folded structure of the protein can be said to be an induced structure, tailored to lock-on and tag the antigen in question.

15.10 The answer

Of course if we ask "Who does all the work?" we have to be a bit careful with what we mean by "all". Clearly, in a strict sense, no single type of molecule will do *all* the work. However, as we have seen in this chapter, when it comes to what can be called the heavy lifting, we see that the main workhorses of the cells are the proteins. Considering the many functions of proteins, both structural as well as organizational, it is perhaps not all too surprising that about 50% or more of the dry mass of most organisms consists of proteins. With this we are now able to formulate an answer to the chapter question "Who does all the work?" Very roughly one can say that the task division in the cell is divided into three parts: DNA stores long-term information, RNA transfers and controls this information and proteins carry out the key tasks in the cell. Or in short: **The workhorses of the cells are the proteins.**

15.11 Exercises

1. What are the main elements that make up proteins?

2. Give an estimate of the atomic weight of a protein with an average length.

3. What is an alpha-helix and a beta-sheet?

4. Is there an enzyme that assists in the assembly of microtubules?

5. By what means can the growth of a microtubule be reversed?

6. If there were only 12 different amino acids, how many different proteins with a length of six amino acids would be possible?

7. If we proclaim that we need as many different possible proteins as there are atoms in the universe, then how long would a protein need to be? (For

simplicity assume that there are 20 different amino acids and that all proteins have equal length.)

8. Repeat Exercise 7 but allow the proteins to be of different lengths.

9. If the average chicken egg has a weight of 58 g and its shell weighs about 6 g, how much protein does a chicken need to produce for its eggs in a year? (Chicken usually lay about five eggs a week.)

10. With so many different possible proteins, the need for classification is clear. Can you think of some?

11. Give an example from daily life where denaturation can be observed.

16: RNA

Q - Jack of All Trades, Master of None?

Chapter Map

RNA

Assists with

Replication

Information storage

Catalysis

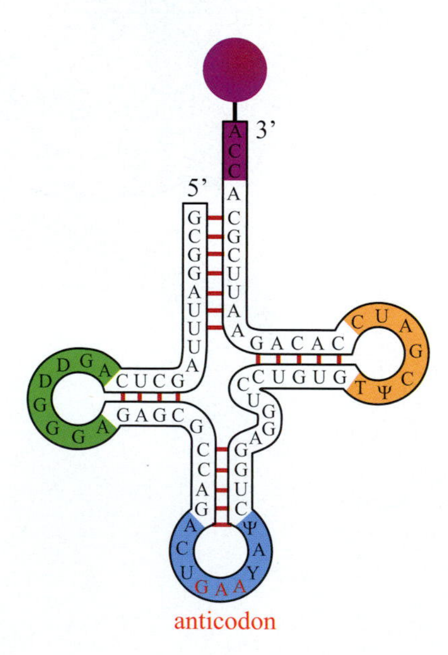

16.1 The question

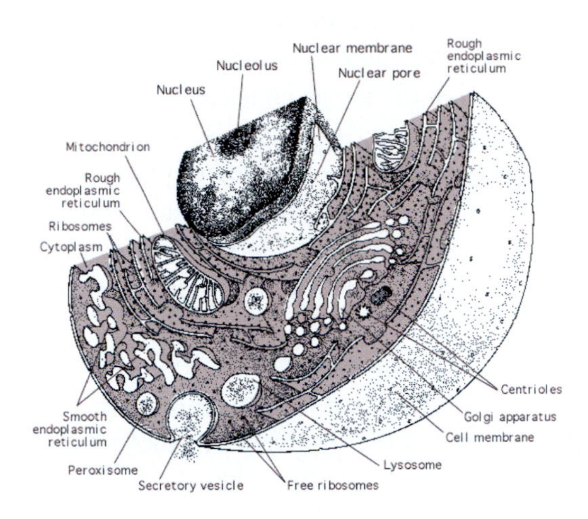

Figure 16.1: Modern cells are highly complex structures. Source: NHGRI.

Modern cells, be they a bacterium or part of a large animal or plant, are by any standards very sophisticated as illustrated in Figure 16.1. They consist of many highly adapted interdependent molecules.

For example, proteins are manufactured with the help of highly specialized molecular complexes called ribosomes, but ribosomes themselves contain proteins. How is this possible? Is this another one of life's chicken-and-egg problems? To find out, we need to have a closer look at how protein production actually works and take it from there. It turns out that special molecules of ribonucleic acid (RNA) bring information about the structure of the protein from DNA to the ribosome and that different RNA molecules bring the building blocks (amino acids) from which the ribosome produces the protein. Initially, therefore, RNA appears in a somewhat auxiliary role. Important but not glorious.

Some further inspection shows that RNA is quite flexible, and involved in a rather large number of processes in the cell. Considering the hard work done by proteins and the center-stage role of DNA in storing genes, does this make RNA a jack of all trades? Perhaps an evolutionary relic soon to be replaced?

To probe further, let us contemplate some essential aspects which one would expect any cell, even the earliest ones, to have.

If we look at early cells, we know that there must have been some kind of information storage to make heredity possible, some kind of (self-) replication to allow for offspring and some kind of catalysis to speed up otherwise tardily slow chemical reactions. Due to its complicated nature, interdependency between different types of molecular groups is almost certainly something that has evolved as a specialization to specific requirements — like being the most efficient catalyst. Consequently, we may surmise that a single class of molecules, which is at least adequate for all of the three necessary functions, should have been the first to originate.

If we take a hint from modern cells we find that there are three types of polymers to consider: DNA, RNA and proteins. As argued in Chapter 14, only polymers are considered to be suitable for functions like heredity in living systems. Now clearly, one has to be careful when trying to draw conclusions about pre-modern cells by starting from modern cells. However, since the modern genetic apparatus in essence dates back some 3 billion years and since evolutionary processes do have a clear path even when their details are unknown, it is reasonable to take the modern cell as a starting point and work back from there.

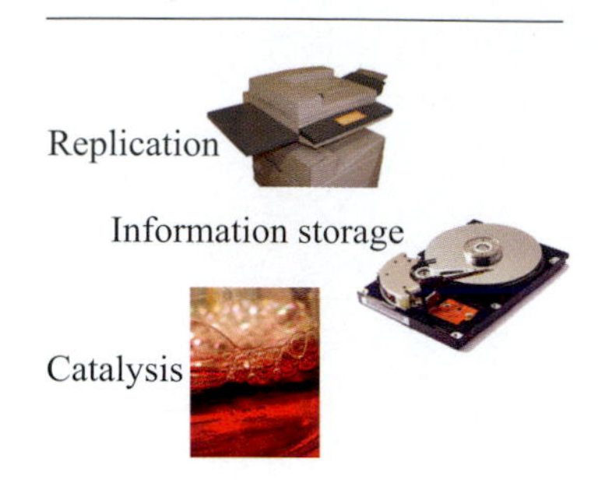

Fig. 16.2: Key requirements for life.

Among DNA, RNA and proteins, RNA is the only type of molecule that is known to be good, albeit perhaps not always excellent, with regards to all three must-haves of early cells: information storage, self-replication and catalysis. DNA, while excelling in information storage, is at best marginal in catalysis (not enough flexibility in its 3-D structure) and self-replication (as a consequence of not being able to catalyze replication). Similarly, proteins excel in catalysis but are poor choices for information storage (no base pairing and hence hard to copy) and self-replication (most 3-D structures cannot be reversibly unfolded). Hence among the three modern types of molecules, RNA seems to be good at many things but outstanding at nothing. Hence, the chapter question: **Is RNA a jack of all trades and a master of none?**

In order to answer this question, we need to understand what RNA can do. Therefore, let us now look at RNA in the context of each of the three functions that must have been available to some degree in early cells: replication, catalysis and information storage. See Figure 16.2.

16.2 RNA and replication

A - T (DNA)
A - U (RNA)

G - C (DNA)
G - C (RNA)

Fig. 16.3: Nucleotide base pairs are A-T, G-C for DNA, and A-U, G-C for RNA.

In modern cells, genetic information is stored in the DNA and much of the catalytic work is carried out by proteins. Although both DNA and proteins are polymers built up of monomers that can be in any order, they are otherwise rather dissimilar. DNA has only four distinct monomers while proteins have 20 in human beings. Furthermore, the structure of the nucleotide monomers that form DNA is very different from that of the amino acid monomers that form proteins: The nucleotides consist of a phosphate groups, a five-carbon sugar and a base while an amino acid consists of an amino group, a carbon atom with a side chain and a carboxyl group.

It is therefore necessary to employ some sort of a mechanism that translates the information about a protein that is stored in DNA as a nucleotide sequence to the correct amino acid sequence found in a protein. It turns out that this is achieved with the genetic code discussed in Section 18.10 in the context of information processing where groups of three successive nucleotides encode for a specific amino acid.

Contrary to the case of proteins, DNA and RNA consist of very similar building blocks, with almost identical monomers. Both DNA and RNA are built up of nucleotides containing the bases adenine (A), guanine (G) and cytosine (C), while DNA further uses thymine (T) and RNA uracil (U), the unmethylated form of thymine (a methyl group has the formula CH_3 and is named after methane that has the formula CH_4, see also p. 305 for the structure of the amino bases). Thus both DNA and RNA use four distinct nucleotides.

The sugar in the DNA nucleotides lacks an oxygen atom in the five-carbon sugar ribose — which the RNA nucleotides do have — and is, hence, called deoxyribose. A special property essential for the process of information storage and processing is that these nucleotides can pair up in predetermined and generally fixed ways. The base adenine (A) can pair up with the base thymine (T) or uracil (U) while the base guanine (G) can pair up with cytosine (C). Since the pairs are formed between the bases of these nucleotides, such pairs are generally referred to as **base pairs**. See Figure 16.3.

The bonds between the base pairs are hydrogen bonds and as a consequence quite weak. That means that they can fairly easily be broken, if so required.

Let us now have a bit closer look at how RNA assists in the process of obtaining a protein from a sequence of nucleotides in DNA. When a gene encoding a protein needs to be expressed, first the relevant sequence of nucleotides is copied onto a strand of so-called **messenger RNA (mRNA)** with the help of RNA polymerase such that the mRNA strand is exactly complementary to the DNA being copied. This process is called **transcription**. Now DNA has two complementary strands, so how how does RNA polymerase know which one to copy? RNA polymerase only works in one direction (reading the DNA strand from the 3' to the 5' direction) and thus it is predetermined which of the two strands is copied. If this were not so, then all copyable regions would need to be symmetric, clearly an undesirable situation.

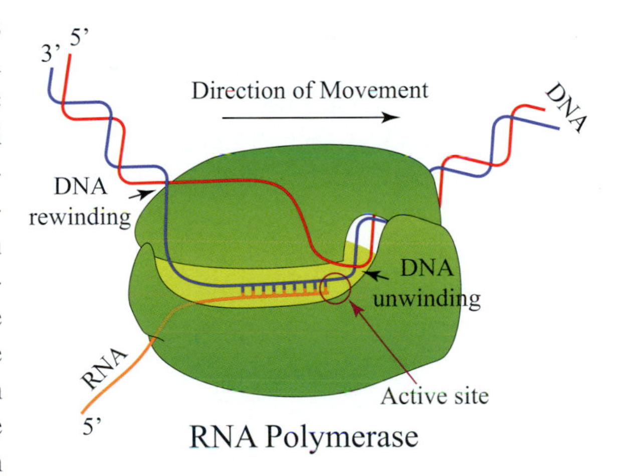

Figure 16.4: Schematic representation of transcription.

At first, in what usually is referred to as initiation, RNA polymerase binds to a specific region of the DNA identified by a certain sequence of nucleotides (e.g. CAAT in eukaryotes). After stabilization with the help of a small protein called factor sigma, the RNA polymerase separates the double-stranded DNA to form a bubble so that it can pair the first nucleotide monomer with the beginning of the DNA sequence to be copied. It then moves down the DNA, pairing one nucleotide after another with the DNA while attaching that nucleotide covalently to the growing RNA sequence; this process is generally called elongation, as illustrated in Figure 16.4. Rather than leaving the growing RNA sequence paired with the DNA, RNA polymerase detaches the newly formed RNA trailing a few nucleotides behind the attachment site such that the DNA can rewind. Finally when the RNA polymerase reaches a stop signal in the DNA, termination occurs. In eukaryotes, the resulting RNA strand is then post-processed before being translated into a protein by a ribosome, while in prokaryotes the mRNA can be translated immediately.

It should come as little surprise that quite some energy is necessary to carry out the process described in the previous paragraph. How is this energy supplied?

Rather than having separate energy and nucleotide sources, RNA polymerase uses energy-rich triphosphate versions of the nucleotides, i.e. ATP, UTP, CTP, GTP. When these triphosphate nucleotides arrive at the RNA polymerase, the energy stored in the phosphate-phosphate bonds is released by splitting the triphosphate nucleotides into phospate groups and the needed RNA nucleotide monomer.

16.3 RNA and catalysis

The catalytic activity of a molecule is strongly dependent on its three-dimensional structure since the reaction rate of substrates only increases significantly if the substrates are brought together in exactly the right way. This then requires specific chemical properties from the catalyst to bind first the substrates and second the correct spatial arrangement so that the substrates are spatially suitably oriented for the reaction to proceed at a fast pace. Furthermore, in many cases, an enzyme needs to be able to (partially) change shape to, for example, first capture the substrates, then with a slightly different conformation catalyze the reaction, and last change shape one more time to release the product — quite akin to the workings of a tool like a pair of pliers.

In the same way, in order for RNA to be able to catalyze a large range of different reactions it needs to be able to assume many different shapes. Catalytic RNA molecules are often referred to as **ribozymes** by combination of the words **ribo**nucleic acid and **enzyme**.

Although RNA as such is single stranded, the fact that it is made up of nucleotides that allow for complementary base pairing means that an RNA sequence can have some of its sections pair up with others of its sections thus creating intricate three-dimensional structures. Furthermore, it turns out that some short structural elements are used quite frequently, as illustrated in Figure 16.5.

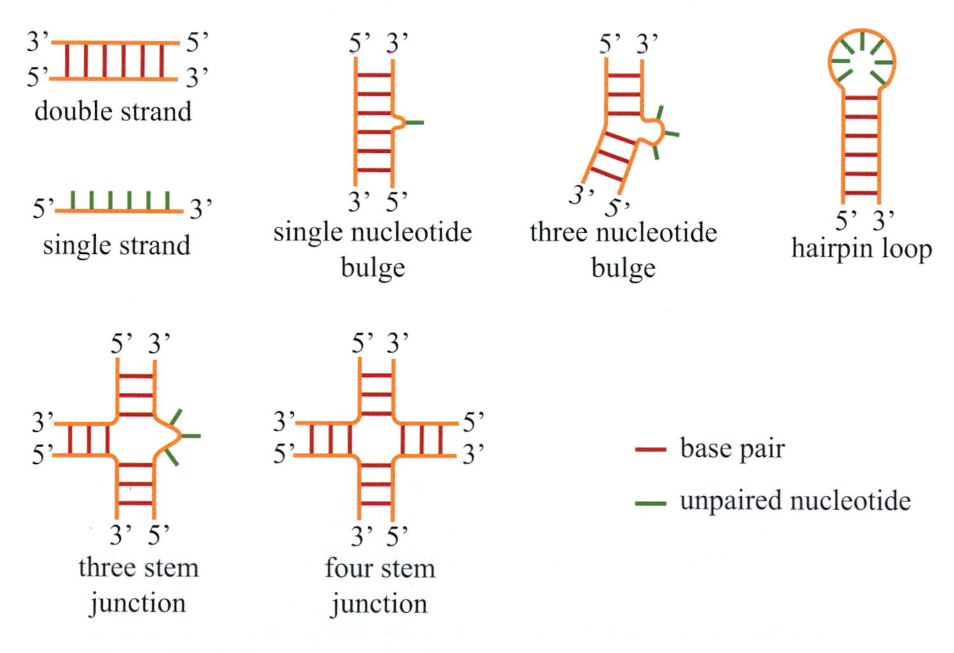

Figure 16.5: Some common elementary RNA secondary structures.

In modern cells ribozymes are quite rare with the exception of the ribosome, essential for life on earth, that contains relatively large ribosomal RNA molecules generally abbreviated as **rRNA**.

In ribosomes, the messenger RNA is translated into proteins according to the genetic code that matches sequences of three nucleotides to certain amino acids by stringing these amino acids together with covalent bonds.

Although ribosomes also contain about 35% proteins besides rRNA (three rRNA molecules in prokaryotes and four rRNA molecules in eukaryotes), the catalytic activity needed for joining amino acids into a protein is carried out by the rRNA.

In bacteria, only one type of RNA polymerase is used, but in eukaryotes there are three types: RNA polymerase I is responsible for three of the four rRNA molecules by transcribing a precursor rRNA that is then modified into the three types of rRNA while RNA polymerase III directly transcribes the fourth rRNA molecule. On the other hand, mRNA molecules are transcribed by RNA polymerase II that is similar to the RNA polymerase of bacteria.

Just like in the case of enzymes, ribozymes often have metal atoms to assist them in their function. Due to the relative rarity of ribozymes versus enzymes in modern cells, one could suspect that ribozymes are not as versatile. However, experiments have shown that ribozymes can catalyze a great number of reactions. The key difference with enzymes is that ribozymes appear to have in general lower maximum reaction speeds. It is therefore quite conceivable that many reactions that are now catalyzed by enzymes were once catalyzed by ribozymes during the earlier stages of evolution.

What we therefore see with regards to the catalytic activity of RNA is that while perhaps often simply adequate, in a case where it really counts, namely protein production, RNA is very capable.

16.4 RNA and information storage

In order to be an efficient information-storage medium, a molecule is necessary that is made of a variety of monomers (minimally two) that can be strung together in arbitrary sequences. As mentioned above, out of the main molecules found in biological systems, information storage seems to require the use of molecules such as DNA, RNA or proteins.

However, the ability to store information is of little use unless it can also efficiently be accessed, or in other words read out. A key advantage of DNA and RNA over proteins in this regard is the mechanism of base pairing. Due to base pairs, exact complementary copies of sections, or of entire RNA or DNA molecules, can be made. This is not readily possible with proteins. A second issue is that of conformation. In general, the three-dimensional structure of a complicated RNA or protein molecule does not allow access to all (or most) of its molecules since the overall structure cannot reversibly be unwound. Of course, only small sections could be used for information storage but that would be inefficient.

Consequently, proteins are not particularly suitable for information storage. RNA, on the other hand, is suitable due to it allowing for base pairing, but its folding would imply some limitations. In this sense, it is perhaps not unexpected that all information in modern life is stored in DNA.

There is, however, good reason to believe that RNA may have fulfilled the role of information-storage medium earlier on during evolution.

So what are the reasons for assuming that RNA preceded DNA? As per its name, RNA contains the five-carbon sugar ribose that can relatively easily be

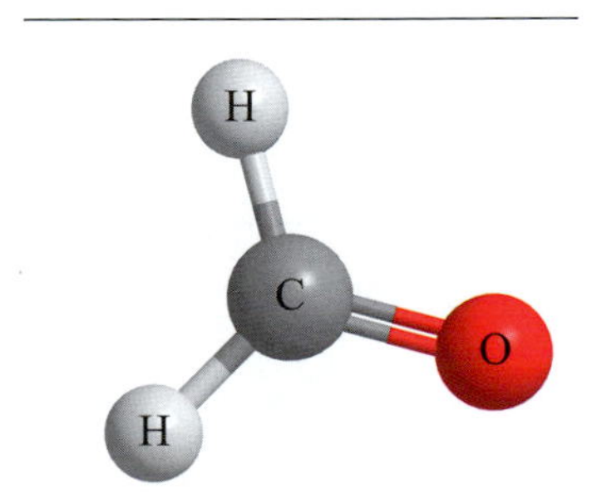

Fig. 16.6: Formaldehyde can form ribose in an environment that may have been present on the young earth.

formed from formaldehyde (H_2CO, see Figure 16.6). The deoxyribose in DNA on the other hand is harder to obtain. In modern cells, it is manufactured from ribose with the help of an enzyme. However, chemically, the deoxyribose-phosphate backbone of DNA is significantly more stable than RNA and hence it is understandable that DNA would be the preferred infomation-storage molecules. What this does not imply is that RNA is unsuitable — rather that it is not as suitable.

Well then, perhaps RNA is not so good in information storage and not a master at this. Although that may be true to some degree, one should nevertheless be somewhat careful. The genome of many viruses is made up of RNA only. Indeed, viruses with both RNA and DNA are rather rare. Even though it is arguable whether viruses are life-forms or not, it is clear that information is being transported. Therefore, at least under certain circumstances, RNA can be a good information carrier.

In the context of viruses, it is notable that viral genomes can be single-stranded or double stranded, be they made up of DNA or RNA. Hence in viruses we find examples of single stranded DNA.

16.5 RNA and computation

Computational processes in general contain instructions that need to be translated and control statements that influence the sequence of how these instructions are executed. For example, in a simple program we may want to sum the numbers between 1 and 10 and display the (interim) results on a monitor on the condition that the result is larger than 20. In order to do this in, for example, the programming language C++, we could write the program

Fig. 16.7: The keyboard, a common means to enter a program.

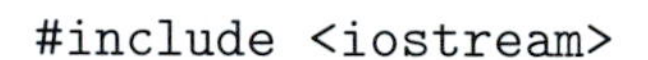

```
#include <iostream>

int main()
{
   int i;
   int sum = 0;
   for(i=1;i<=10;i++){
     sum = sum + i;
     if(sum > 20) std::cout << sum << std::endl;
   }
   return 0;
}
```

The first line indicates that a set of instructions defining input and output needs to be used. The line `int main()` indicates the start of the code that needs to be processed while the following two lines, `int i;` and `int sum = 0;`, indicate that we have an integer variable with the name `i` and also an integer variable with the name `sum` that is initialized to 0. The next line `for(i=1;i<=10;i++)` starts a loop that is executed while the condition `i<=10` is fulfilled (the symbols <= means "smaller equal"). In more detail, `for(i=1;i<=10;i++)` means that our integer variable `i` starts at 1, after which the commands between the curly brackets are executed, and then `i` is increased by 1 (this is indicated by `i++`). The part `i<=10` in `for(i=1;i<=10;i++)` means that the program should proceed to the next instruction `return 0;` when the integer `i` is larger than 10. The line `if(sum > 20) std::cout << sum << std::endl;` is a conditional statement and is part of the loop. The part `if(sum>20)` checks whether the value of the variable `sum` is larger than 20. If so, the part `std::cout << sum << std::endl;` is executed printing the value of `sum` on the monitor. If `sum` is smaller than or equal to 20, nothing happens. Finally when `i<=10` is no longer true `return 0;` is executed which marks the end of the program.

In computers, a large collection of similar constructs eventually amounts to what in a certain way can be construed to yield unrecognizable outcomes such as a word processor.

Translation

One of the key processes in a cell, the translation of the information stored in messenger RNA (mRNA) into a protein, has a distinctly computational flavor to it. After an mRNA strand has been transcribed, and if necessary processed, it proceeds to the ribosome where it is translated into a protein. This procedure is called **translation** since specific sequences of nucleotides need to be "translated" into a corresponding amino acid. After all, in human beings there are five times as many different amino acids than there are nucleotides. To be more specific, three successive nucleotides, referred to a a **codon**, represent a certain amino acid.

Since there are four different types of nucleotides, there are in total $4^3 = 64$ possible different codons. The table that matches these codons — with amino acids as well as instructions start and stop — is called the genetic code and shown in Figure 18.14.

The question, of course, is how is the genetic code being processed? One option would be for the ribosome to recognize the codons and then grab a suitable amino acid. This is, however, not quite how nature does it. In nature, amino acids destined for protein production are attached to specialized RNA molecules called **transfer RNA (tRNA)**. The tRNA molecules, an example of which is shown in Figure 16.8, has a complementary sequence of bases called an anticodon on one side, which exactly matches a codon. Since a specific tRNA molecule can only have one kind of amino acid attached to it, if the ribosome finds a tRNA whose anticodon matches the codon of the mRNA currently being processed, then the ribosome

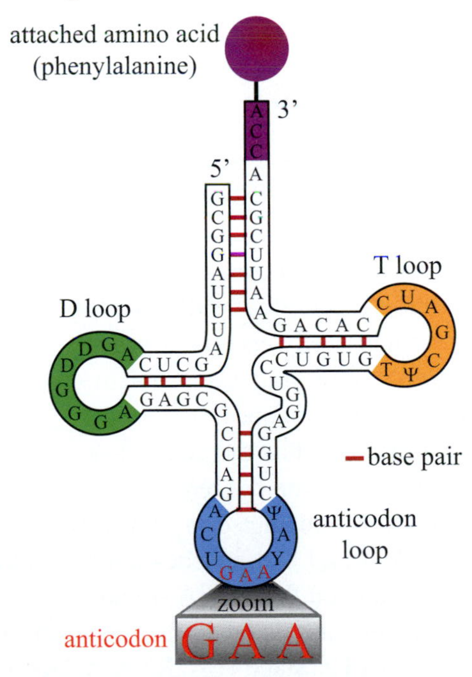

Figure 16.8: The tRNA molecule for the amino acid phenylalanine.

has the correct amino acid for the protein to be manufactured. In a computational framework, this is a straightforward process and can easily be expressed in a computer language such as C++ in that the code could be, for example,

```
if(!current_anticodon == current_codon) useAminoacid();
```

where `current_anticodon` and `current_codon` are binary variables and `useAminoacid()` a function. Since there are significantly more codons than different types of amino acids, even taking account start and stop signals, many amino acids can be represented by more than one codon. Of course, it would be possible that nature simply doesn't use all the possible 64 codons but this is not the case. All codons are used. Since base pairing needs an exact match between codon and anticodon, some amino acids are therefore carried by more than one type of tRNA.

A notable feature of tRNA is that after its synthesis, some bases are chemically modified. For example, in Figure 16.8, the bases 'D, Y, Ψ' in the thereafter named D loop, are modifications of uracil.

Since being read as such does not destroy an mRNA strand, it can in principle be read over and over again. When that happens, enormous amplification can occur and a single gene can yield a huge quantity of protein.

In contrast, rRNA (after suitable processing) is the final product when being transcribed and consequently in order to manufacture the 10 million or so ribosomes a mammalian cell needs, multiple copies of the necessary rRNA genes exist in the genome.

Protein assembly

In Chapter 15, we have seen that proteins have an enormous number of different functions but we haven't considered yet how they are actually made. What is the cellular mechanism that produces proteins? In the chapter on information (Chapter 18 and briefly in the section above), it is discussed how only four different nucleotides code for 20 different amino acids. This is a typical coding problem that one could easily encounter in, for example, computer science or engineering, and nature's solution to the problem is summarized in Figure 18.14 that displays the standard genetic code.

Once one has the code, there must be a machinery that processes this code and produces the desired end result: a protein. A schematic representation of this machinery is depicted in Figure 16.9 where it is shown how a ribosome reads an mRNA strand and attaches amino acid monomers — according to the currently read codon on the mRNA strand — to a growing protein.

In some more detail, the four main stages of the assembly process are given in Figure 16.10. Stage 1: A tRNA molecule arrives at the ribosome carrying an amino acid. The amino acid carried corresponds to the tRNA's anticodon as specified by the genetic code. Stage 2: The tRNA molecule with the anticodon matching the codon currently being read by the ribosome attaches to the ribosome. Stage 3: The ribosome attaches the amino acid brought in by the tRNA to the growing amino acid chain. Step 4: The ribosome releases the tRNA molecule.

But wait a moment, if ribosomes contain proteins and ribosomes make proteins including the ones for new ribosomes, then where do the first ribosomes come from? The answer to that question is that they are already there. When cells divide, not only is the DNA copied, essential components are divided between the two cells as well. So roughly, each daughter cell will receive half the parent's ribosomes and basically half of all other constituents besides DNA as well. Notably, the mitochondria that are the power plants of cells have their own DNA and are not encoded in the DNA that forms the chromosomes, hence if ribosomes and mitochondria were not in both daughter cells, they could not be produced and the daughter cell lacking them would quickly die.

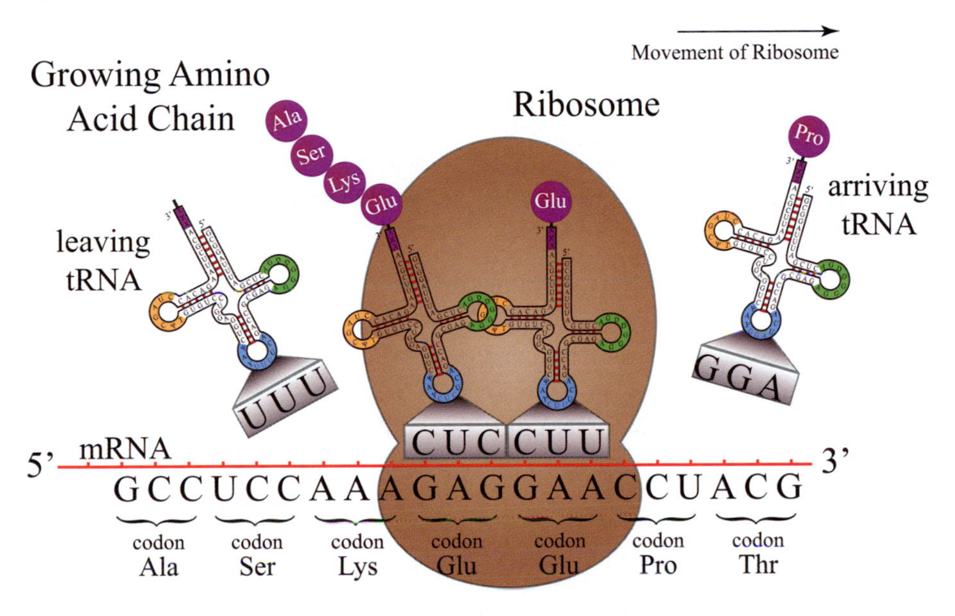

Figure 16.9: Ribosomes assemble proteins.

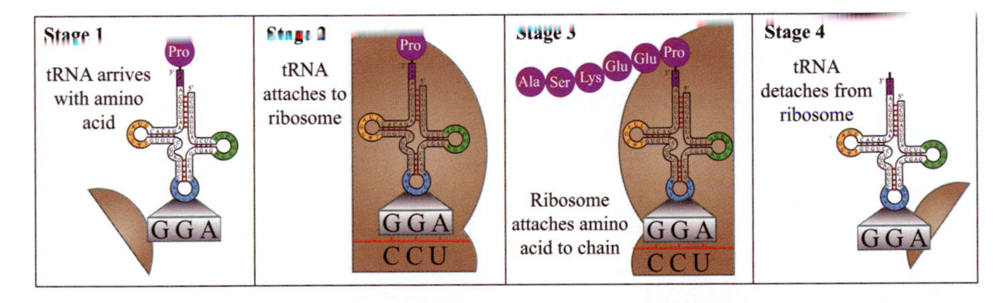

Figure 16.10: Stages in the assembly of proteins.

Control

In computational processes, control is essential, especially if constant adaptation to changing environments is necessary. Due to their versatility, RNA molecules are involved in a number of regulatory activities such as RNA processing, modification and editing. Let us now have a look at some interesting classes of RNA molecules.

snRNA

The so-called **small nuclear RNA (snRNA)** is a class of RNA molecules involved in RNA splicing and the regulation of transcription factors.

They are also involved in the regulation of telomeres. Telomeres are highly repetitive sequences of DNA that can be found at the terminal ends of chromosomes, as shown in Figure 17.14. In general, telomeres become shorter as a cell divides since the DNA polymerase responsible for replication cannot proceed all the way to the end of a strand. By having telomeres, none of the essential genetic material will be missed at the end of a chromosome when a cell divides. However, the telomeres do become shorter and are therefore thought to play a role in aging. A subclass of snRNA, the small nucleolar RNA (snoRNA) plays an important role in the chemical modification of certain types of RNA molecules like rRNA and tRNA.

snRNA are generally about 150 nucleotides long and form complexes with proteins called small nuclear ribonucleoproteins (snRNP) to fulfill their function.

eRNA

In general the uses of promoters and inhibitors lead to a kind of on-off regulation that resembles a digital process. It appears that by interfering with the transcription apparatus, **efference RNA (eRNA)** allows for a somewhat analog fine grain control.

tmRNA

Common to all bacteria, but thus far not found in eukaryotes, is a class of RNA molecules that have both tRNA as well as mRNA regions. Their main purpose is to deal with ribosomes where the production of a protein has become stuck. Unfinished proteins, besides possibly being entirely useless, could also be damaging to a cell as their function is unpredictable. It is therefore important to identify stuck ribosomes and tag the incomplete protein for destruction. This is done by the **transfer-messenger RNA (tmRNA)**.

siRNA

One of the mechanisms of regulating gene expression is through RNA interference (RNAi). In this process, small 20- to 25-nucleotide long, double-stranded RNA molecules called **small interfering RNA (siRNA)** play a key role. For example, if a certain type of mRNA strand has already been transcribed and needs to be stopped from being translated into a protein by a ribosome, then the so-called RNA induced silencing complex (RISC) can be used as follows: One of the strands of siRNA called the guide strand is incorporated into the RISC complex that subsequently binds to the complementary regions on an mRNA strand. The RISC complex then destroys the mRNA thus silencing it.

This mechanism is also efficient to inhibit the formation of viral proteins and part of an ancient immune response used by most plants and animals. In this case, the siRNA locks onto the viral RNA and blocks its translation thus preventing the production of viral proteins essential for the virus' reproduction.

miRNA

Similarly to siRNA, **micro RNA (miRNA)** plays an important role in gene regulation by base pairing to mRNA strands. However, they are single-stranded rather than the double-stranded siRNAs. miRNAs are encoded in RNA genes that are significantly longer than the actual miRNA they code for. When transcribed, an miRNA-coding RNA gene first yields a primary transcript (pre-miRNA) that is processed in the nucleus to a roughly 70-nucleotide long pre-miRNA. Finally, this pre-miRNA is then processed into miRNA in the cytoplasm.

Thus, once again, we find that RNA does an excellent job in yet another category of function.

16.6 RNA genes

In modern cells, information is stored in DNA. Those parts of the DNA that encode the information for the production of proteins are called genes. However, besides proteins, as we have seen in the sections above, there are many essential RNA molecules. Not surprisingly, the information for these molecules is also stored in the DNA in regions that are called RNA genes. The resulting RNA molecules, for example, the catalyst rRNA or the transport molecule tRNA, are then called **non-coding RNA (ncRNA)** since they do not code for proteins.

16.7 Overview

The main RNA functions with regard to protein production are summarized in Table 16.1.

Molecule	Abbreviation	Function
transfer RNA	tRNA	brings amino acid monomers to ribosome
messenger RNA	mRNA	brings instruction for protein to ribosome
ribosomal RNA	rRNA	catalyzes the joining of amino acids

Table 16.1: Key RNA types for protein manufacture. tRNA and rRNA are non-coding while mRNA is coding.

Some of the other noncoding RNA not involved in protein synthesis are listed in Table 16.2.

Given names such as small and micro, one might be tempted to assume that that all noncoding RNA is relatively small. This is not the case, however. For

Molecule	Abbreviation	Function
small nuclear RNA	snRNA	regulatory functions in eukaryotic nuclei
efference RNA	eRNA	gene regulation
transfer-messenger RNA	tmRNA	identifies faulty ribosomal activity in bacteria
small interfering RNA	siRNA	regulates gene expression and combats viruses
micro RNA	miRNA	control gene expression

Table 16.2: Key noncoding RNA types.

example, in female mammals, one of the two X chromosomes is inactivated (so as to have the same number of active X chromosomes as males — namely, one) by an RNA gene named Xist that is 18,000 base pairs long. Nevertheless, the majority of ncRNA are rather small.

16.8 The answer

We started this chapter by wondering whether RNA is a jack of all trades. In its modern meaning, jack of all trades has a rather negative connotation implying mediocrity. When we look at RNA molecules and their many functions, it becomes clear that RNA is far from mediocre in the many things it does. Indeed, it appears to be very efficient in most, if not all, of the functions it performs — be this replication, catalysis, information storage or control. RNA is far from mediocre and in many senses a master. Then if so, why are there DNAs and proteins? The answer to that is likely not so much the inadequacy of RNA but the excellence of DNA and proteins in their current use — they are the grand masters, so to speak. Of course, it is very well possible that RNA, DNA and proteins emerged simultaneously as specializations of a currently unknown class of polymers. Be that as it is, from what we know today, RNA is a good candidate for the precursor molecule of the modern cell.

It has been said that the original "jack of all trades" expression was somewhat different: "Jack of all trades, master of none, though ofttimes better than master of one". And in this form, the figure of speech may very well be true. Based on what we know today, it is quite feasible to imagine life with only RNA out of the trio RNA, DNA and proteins. Neither proteins nor DNA by themselves could fulfill the three essential functions of replication, catalysis and information storage. So we now can answer our chapter question somewhat tongue in cheek by **RNA, master of all trades, much better than grand master of one!**

16.9 Exercises

1. Give two reasons why RNA might have preceded DNA in the evolution of life on earth.
2. What is the anticodon corresponding to the amino acid lysine?
3. If different parts of RNA can form base pairs leading to intricate structures, why is RNA generally not called double stranded?
4. Can RNA polymerase read DNA in both directions?
5. How is the energy supplied for protein production?
6. How is the energy for mRNA production supplied?
7. What is the difference between a ribosome and a ribozyme?
8. The catalytic activity of the ribosome is carried out by which type of biopolymer?
9. Why is it so that RNA can be a reasonably efficient catalyst?
10. What is "noncoding RNA"?

17: DNA

Q - What Determines the Structure of DNA?

Chapter Map

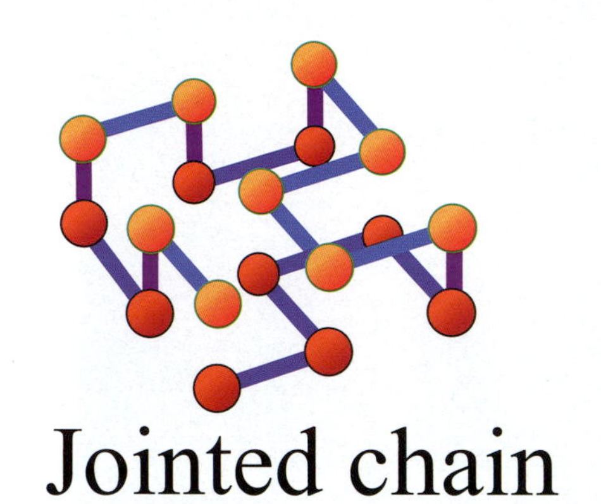

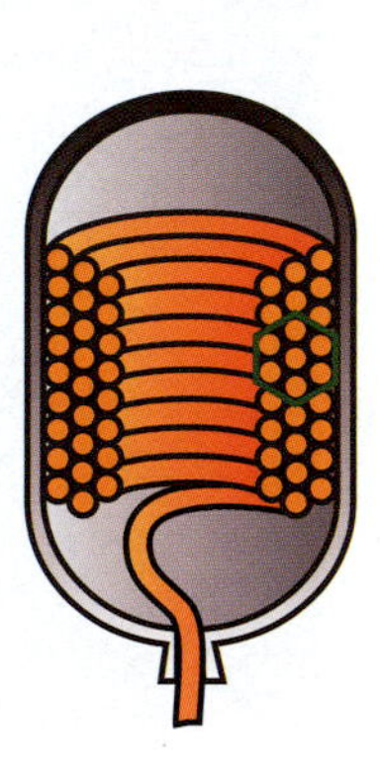

17.1 The question

DNA (deoxyribonucleic acid) and other biopolymers constitute the molecular foundation of life. We explore the question: **What are the underlying physical processes that determine DNA's structure?** The probabilistic model of the DNA, based on the random walk, as well as the deterministic approach of viewing the DNA as a rigid beam both provide precise explanations on how the structure of the DNA is determined.

The primary function of the DNA is the storage of genetic information, which in turn requires transcription and translation by RNA biopolymers. The functioning of DNA and proteins are closely linked to their structure; we discuss a few cases how their structure is formed and briefly touch on how "structure determines function".

17.2 DNA and random walk

The DNA is double helix; each strand of the helix is a biopolymer made from four distinct monomers, called nucleotides and labeled by A, C, G and T; the monomers are separated by a distance of about 0.3 nm. Knowing the sequence of the monomers in one strand automatically fixes the other strand by a rule of complementation, leading to the double helix. A monomer in one helical strand together with its complement in the other strand is called one base pair.

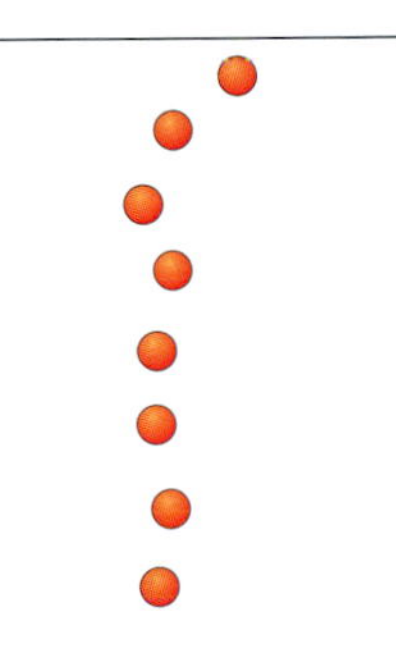

Fig. 17.1: A typical configuration for the DNA monomers.

The DNA is described by specifying the position of each monomer; for the discussion of this chapter, the specific properties that differentiate the four monomers from each other do not come into play and hence they are treated as being identical; the DNA is completely specified by the position of each monomer, with a typical configuration shown in Figure 17.1. All biopolymers — namely the DNA, RNA and proteins — share many properties and the study of DNA leads to general results that are applicable to all biopolymers.

DNA molecules that are enclosed inside the nucleus of a cell are compact objects and their structure is described by "beam theory" that is discussed in Section 17.10. Consider, instead, viruses and bacteria whose cell membranes have ruptured and which are surrounded by the debris of their DNA. In many cases, we are not interested in the detailed structure of the debris DNA but in the average size of the "blob" that comes out — or other similar coarse-grained properties.

In particular, in many cases the end-to-end distance of the DNA is all that we are interested in. Furthermore, experiments with a single DNA molecule or a protein that measures the force required to extend the molecule do not depend on specific details of the biopolymers; these experiments find their explanation in the coarse-grained description of the biopolymers.

A DNA molecule is usually in a solvent at some finite temperature T. The monomers of the DNA collide more than 10^{10} per second with the atoms and molecules of their environment. Instead of trying to follow the movement of each and every monomer, a DNA molecule — as a net result of all the random collisions as well as due to their interactions — is described by assuming that each monomer of the DNA is undergoing a *random walk*.

In Section 8.2 we found that the structure of polymers, in particular their average size, can be predicted by using the concept of a (self-avoiding) random walk. We investigate whether a similar approach to the biopolymers yields any insight into their coarse-grained properties.

Consider a large collection of the same DNA, with only the number of its monomers being specified; since the monomers have a separation of about 0.3 nm, in effect the total length, given by L, of the DNA is taken to be specified and fixed. The various possible allowed configurations of the DNA are considered as one of its possible random outcomes.

From the point of view of random walk theory, the various observed properties of the DNA are obtained by averaging over all the allowed configurations of the random walk. In particular, random walk theory is used for obtaining the *average* properties of DNA, such as its average size, its moment of gyration and many more detailed and specific properties. For DNA with a length L that is reasonably long, the DNA's observed average size is well described by a random walk.

17.3 Review of random walk

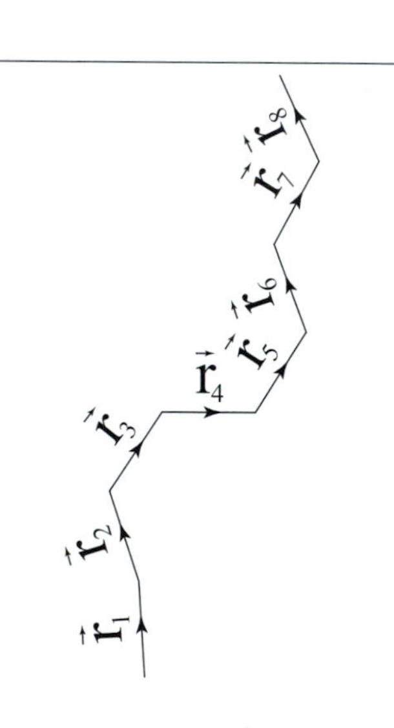

Fig. 17.2: A curve (path) overlapping the positions of the monomers.

Random walk is studied in some detail so as to describe the behavior of the DNA. We rederive the result of the random walk given in Equation 10.31 with the purpose of generalizing it to the case of the DNA. Consider an M step random walk with "mean free path" given by ℓ. As discussed in Section 10.6, the mean free path ℓ is the average distance that a particle travels before undergoing a collision that causes a random change in its direction.

Let the first monomer of the DNA be located at the origin. The mean free path of the random walk is *not* the separation of the monomers but rather a property of the entire DNA molecule; we will show later that ℓ for the DNA is given in terms of the correlation length of the DNA's monomers, called the *persistence length*. The conformation (shape) of the DNA molecule is the result of an M-step random walk that starts at one end of the DNA, follows the layout of the monomers and ends at the other end. Since each step has a length of ℓ and the length of the DNA is given by L, the number of steps is given by $M = L/\ell$ and is shown in Figure 17.2.

The position of the first random step is indicated by a vector going from the first monomer and ending at position $\mathbf{r}_1$; similarly, the distance from the i-th to the $i + 1$-th random step is denoted by $\mathbf{r}_i$ all the way to the last step $\mathbf{r}_M$ as shown in Figure 17.2. The positions of the DNA monomers define the path that the random walk follows from its origin to its final position $\mathbf{R}_M$.

Each random step has a fixed size, namely, that of the mean free path; hence

$$\mathbf{r}_i^2 = \ell^2. \tag{17.1}$$

Each step of the random walk is assumed to be independent. Hence, Equation 10.30 is equivalent to

$$E[\mathbf{r}_i \cdot \mathbf{r}_j] = \begin{cases} \ell^2 & i = j \\ 0, & i \neq j. \end{cases} \tag{17.2}$$

The final position of a particle doing a random walk is given by a vector sum of all its successive earlier positions and we obtain

$$\mathbf{R}_M \equiv \sum_{i=1}^{M} \mathbf{r}_i . \tag{17.3}$$

There is no preferred direction for the random walk and, hence, the average position is always the starting point. For a random walk starting from the origin, we have

$$E[\mathbf{R}_M] = 0. \tag{17.4}$$

After M steps, the average of the square of the distance of the particle from its starting point, from Equations 17.1 and 17.2 is given by

$$\begin{aligned} E[\mathbf{R}_M^2] &= \sum_{i,j=1}^{M} E[\mathbf{r}_i \cdot \mathbf{r}_j] = \sum_{i=1}^{M} E[\mathbf{r}_i^2] = \sum_{i=1}^{M} \ell^2 \\ \Rightarrow E[\mathbf{R}_M^2] &= \ell^2 M. \end{aligned} \tag{17.5}$$

It can be shown that for the number of steps $M >> 1$ the probability distribution for the random walk to be at some final position R is given by

$$P(R, M) = \sqrt{\frac{1}{2\pi\ell^2 M}} e^{-\frac{1}{2\ell^2 M} R^2}$$

as shown in Figure 17.3.

For a random walk in three dimensions, the probability distribution to be at a final position $\mathbf{R} = (x, y, z)$ is simply the product of the distribution for a random walk in each of the three dimensions and is given by

$$P(\mathbf{R}, M) = \left[\sqrt{\frac{1}{2\pi\ell^2 M}}\right]^3 e^{-\frac{1}{2\ell^2 M}\mathbf{R}^2} \ ; \ \mathbf{R} = (x, y, z). \tag{17.6}$$

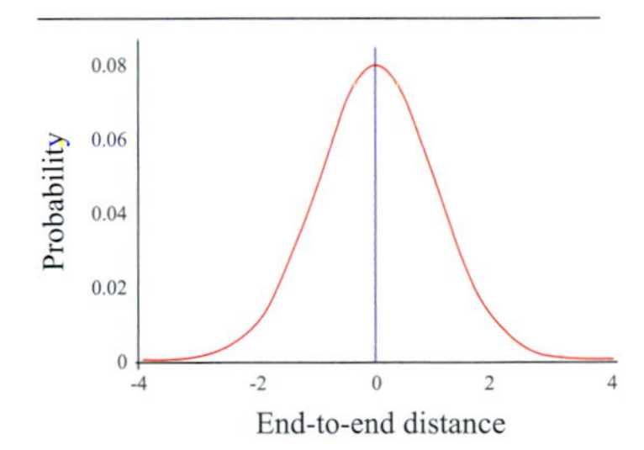

Fig. 17.3: End-to-end distance of a DNA molecule; the x-axis is given by $R/(\ell\sqrt{M})$.

From Figure 17.3 it can be seen that the random walk has the highest probability of ending up at position $\mathbf{R} = 0$. In other words, the particle is driven toward its starting point not by any external force, but rather by the very fact that the largest number of paths available to it consist of paths that start and end at the same point. This force is sometimes called the "entropic" force to remind us of the statistical preference of the particle to be in a state that is the most likely.

17.4 DNA random walk; persistence length ξ

For describing the structure of the DNA, the random walk reviewed above has to be generalized. In the earlier discussion in Section 10.6 on diffusion as well as the condition given in Equation 17.2, each random step is *independent* of the previous one. In contrast, DNA is described by a random walk that is *correlated*, namely a random step is influenced by the previous step(s) taken. The reason that DNA has

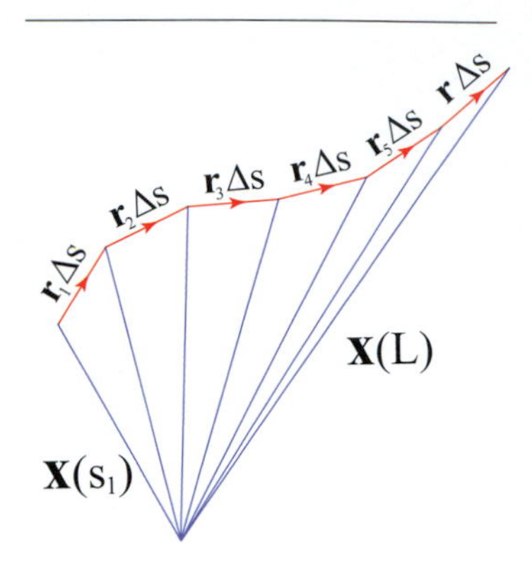

Fig. 17.4: The position along the curve is denoted by $\mathbf{x}(s)$.

a correlated random walk is because its monomers are covalently bonded and have a rigidity that tends to keep the neighboring monomers aligned with each other.

For simplifying the analysis the DNA molecule is approximated by a *continuous curve*, with the coordinates of the curve specified by $\mathbf{x}(s)$; the position on the curve is specified by the parameter s lying in the range $0 \leq s \leq L$, as shown in Figure 17.4. The length of segment of the curve from point s on the DNA to another point $s + \Delta s$ is given by the difference between the location of the two ends of the segment, namely $\Delta\mathbf{x}(s) = \mathbf{x}(s + \Delta s) - \mathbf{x}(s)$.

Let $s_i = i\Delta s$, with $i = 1, 2, ...I$ and $\Delta s = L/I$. The position vectors $\mathbf{x}(s_i)$ are chosen so that the steps of the random walk are given by the following

$$\mathbf{x}(s_{i+1}) - \mathbf{x}(s_i) \equiv \Delta\mathbf{x}(s_i) = \mathbf{r}_i \ \Delta s$$

and are shown in Figure 17.4. There are many ways of choosing the parametrization of the curve $\mathbf{x}(s)$ by the variable s; to make the choice of the parametrization unique, one imposes the condition that the length of the "tangent vector" $\mathbf{r}_i$ be unity, that is

$$\mathbf{r}_i \cdot \mathbf{r}_i = 1.$$

The total length of the DNA is given by generalizing Equation 17.3 to

$$\mathbf{R} \equiv \sum_{i=0}^{I-1} \Delta\mathbf{x}(s_i). \tag{17.7}$$

The correlation of the DNA can be shown to be given by

$$E[\Delta\mathbf{x}(s) \cdot \Delta\mathbf{x}(s')] = e^{-|s-s'|/\xi_P}(\Delta s)^2 \tag{17.8}$$

where ξ_P is called the *correlation* or *persistence* length.

As the name indicates the persistence length of the DNA is a measure of the average distance over which the segments of the DNA's curve move together.

The expectation value of the distance squared, from the first to last step of the random walk, similar to Equation 17.5 and for $L >> \xi_P$, is given by

$$\begin{aligned} E[\mathbf{R}^2] &= \sum_{i,j=1}^{I} E[\Delta\mathbf{x}(s_i) \cdot \Delta\mathbf{x}(s_j)] \\ &= \sum_{i,j=1}^{I} (\Delta s)^2 e^{-|s_i - s_j|/\xi_P} \\ &\simeq \int_0^L ds \int_0^L ds' e^{-|s-s'|/\xi_P} \\ \Rightarrow E[\mathbf{R}^2] &\simeq 2L\xi_p. \end{aligned} \tag{17.9}$$

Recall from Equation 17.5 that a random walk of M steps with step size ℓ and *no correlation* is given by $E[\mathbf{R}_M^2] = \ell^2 M$. From Equations 17.9 and 17.5, the correlated DNA random walk is equivalent to an uncorrelated random walk with the following identification of the parameters for the two cases:

$$\ell = 2\xi_P \; ; \; M = \frac{L}{\ell} = \frac{L}{2\xi_P} \tag{17.10}$$

The distance ℓ is called the Kuhn length in physical biology.

17.5 DNA: a freely jointed chain

The significance of the persistence length can be understood by considering a concrete example. Suppose each step is equal to half the persistence length; that is, $\Delta s = \xi/2$. Let the first step $\Delta\mathbf{x}_1$ be fixed; the next step $\Delta\mathbf{x}_2$ has step size $s_2 - s_1 = \xi/2$. From Equation 17.8, the correlation of $\Delta\mathbf{x}_2$ with the first step $\Delta\mathbf{x}_1$ given by $E[\Delta\mathbf{x}_1 \cdot \Delta\mathbf{x}_2] \sim e^{-(s_2-s_1)/\xi_P} = e^{-1/2} \sim 0.60$, which means that 60% of the time the two steps are aligned. The third step $\Delta\mathbf{x}_3$ of step size $s_3 - s_1 = \xi$ has the following correlation with the first step $E[\Delta\mathbf{x}_1 \cdot \Delta\mathbf{x}_3] \sim e^{-1} \sim 0.37$, which means that about 40% of the time it points in the same direction as $\mathbf{r}_1$.

The fourth step of step size $s_4 - s_1 = 3\xi/2$ has the following correlation with the first step $E[\Delta\mathbf{x}_1 \cdot \Delta\mathbf{x}_4] \sim e^{-3/2} \sim 0.22$ showing that it is correlated with $\Delta\mathbf{x}_1$ only 20% of the time and hence has a less significant correlation with the first step. And by the fifth step, no significant correlation is left.

Hence, the DNA random walk can be approximately thought of $\Delta\mathbf{x}_1$, $\Delta\mathbf{x}_2$ and $\Delta\mathbf{x}_3$ being mostly in the *same* direction. The steps after $\Delta\mathbf{x}_4$ are decorrelated with the first step $\Delta\mathbf{x}_1$ and point in random directions.

The physical interpretation of the **persistence length** is that it is the distance over which the DNA remains roughly straight. The DNA polymer behaves like a *rigid rod* of length equal to the persistence length ξ_P. The rigid rods are linked to other rigid rods by free joints since each of the individual rigid pieces can point in any direction. This picture gives rise to the **freely jointed chain** model of the DNA, as shown in Figure 17.5. The length of the rigid rod is equal to ξ_P. One can think of building up the entire DNA polymer by attaching rigid pieces in a "chain", one by one, with each rigid piece being added in a random direction; the number of rigid rods required to make the DNA is given by L/ξ_P.

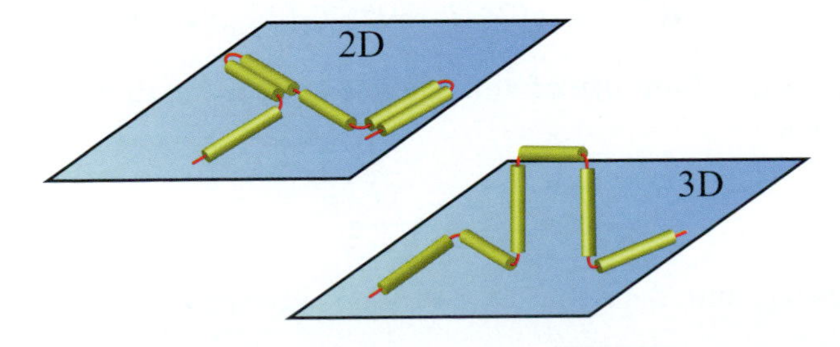

Figure 17.5: Random walk, in the freely jointed chain model, in two and three dimensions. The length of the rigid rods is equal to the persistence length ξ.

Random walk theory can be used to predict the stretching of the DNA when subjected to a stretching force; Figure 17.6 shows the result of such a computation. The response of a DNA to a loading force can be studied in detail and can be used to test the results of the random walk theory.

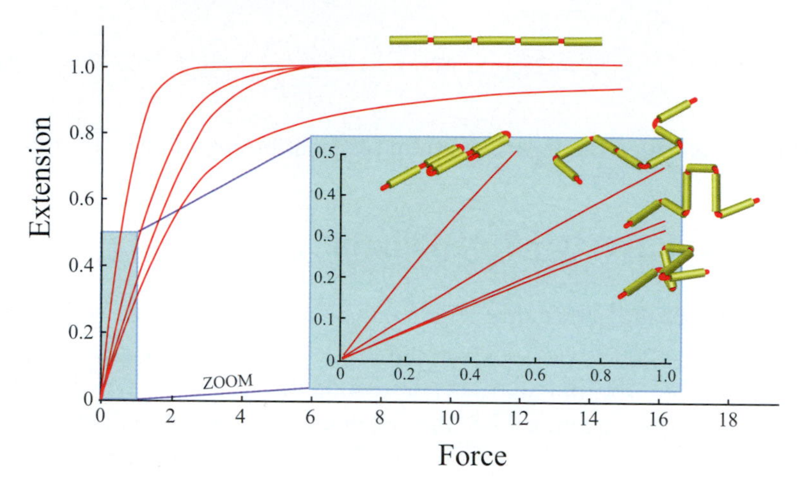

Figure 17.6: The extension versus force graphs for a freely jointed chain for one, two and three dimensions pointing only in orthogonal directions; also shown is the case for a rigid rod random walk in three dimensions pointing in any direction. The inset shows the configurations for the chains for the four cases.

17.6 Size of viral and bacterial DNA

The random walk model for the structure of the DNA is valid as long as the length of the DNA is much larger than its persistence, that is $L >> \xi_P$. A typical example is the lambda bacteriophage virus that has a DNA with length of 16.6 μm and a persistence length, at room temperature, of $\xi_P \simeq 50$ nm; the persistence length is clearly much smaller than the length of the DNA.

The length of DNA can be expressed in terms of the number of base pairs N_{bp}. The average distance between base pairs is about 0.34 nm and hence the length of the DNA is given by $L \approx 0.34\ N_{bp}$ nm. One measure of the size of DNA is the square root of the average squared size of the DNA; hence, from Equation 17.5, DNA size is given by

$$\sqrt{< \mathbf{R}^2 >} = \sqrt{2\xi_p L} = 0.82\sqrt{N_{bp}\xi_P} \text{ nm}.$$

Another measure of the size of viral and bacterial is the radius of gyration R_g, defined as follows:

$$\sqrt{< R_g^2 >} = \sqrt{\frac{L\xi_P}{3}}.$$

Since $L \simeq 0.34 N_{bp}$ nm, the radius of gyration is given by

$$\sqrt{< R_g^2 >} = 0.34\sqrt{N_{bp}\xi_P} \text{ nm}. \tag{17.11}$$

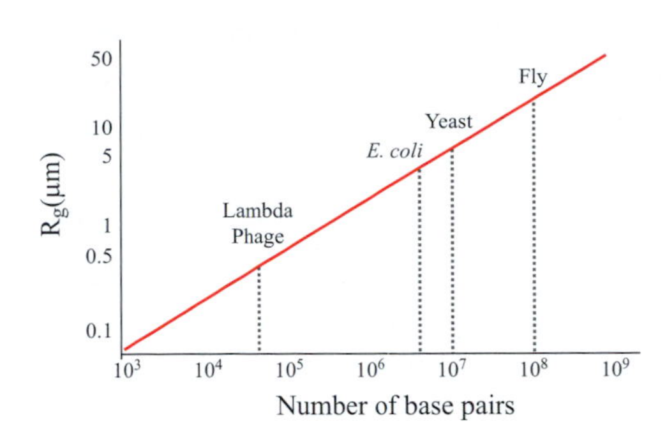

Figure 17.7: Radius of gyration versus genome size. Note the horizontal axis has a logarithmic scale.

Figure 17.7 shows the radius of gyration R_g for various DNAs with varying number of base pairs.

Consider the virus T2 phage whose DNA has escaped from its capsid. The viral DNA has about N_{bp} =150,000 base pairs. The DNA length is given by $L = 0.34N_{bp} \sim$ 51,000 nm and the persistence length is $\xi_P \sim$ 50 nm. Hence, from Equation 17.11, $< R_g^2 >^{1/2} \sim 2\,\mu$m is the radius of gyration of DNA. The crude random walk model gives the correct first approximation for the observed structure of the exploded DNA. The most important reason for the discrepancy is that the DNA seems to be constrained.

17.7 Proteins: compact random walk

One of the key ideas driving structural biology is that structure determines function; in particular, protein folding and unfolding is thought to determine the functional behavior and properties of proteins. Proteins are biopolymers with 20 types of amino acids as their monomers (see Chapter 15). Globular proteins, in equilibrium, form compact structures with the monomers folded upon themselves. Proteins in the unfolded state are described by a free random walk, similar to the one that describes the DNA; we will discuss the structure of only the folded globular proteins.

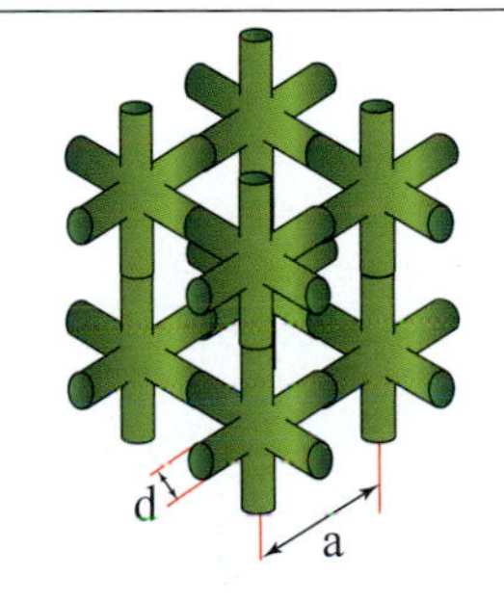

Fig. 17.8: Three-dimensional lattice with links.

One may wonder if random walk theory can explain any coarse grained and average properties of folded proteins. The random walk considered for the DNA and RNA were open structures, with monomer sites being surrounded by the solvent. In contrast, proteins are explained by a self-avoiding volume-filling random walk taking place on a three-dimensional cubic lattice; shown in Figure 17.8 is a lattice with spacing a and links having a diameter of size d.

The protein monomers occupy all the lattice sites in a fixed volume. A typical configuration of a compact random walk is shown in Figure 17.9; each link corresponds to a step taken by the monomers is filling up the lattice. This model of the protein is very crude, but — due to its simplicity — it yields precise and quantitative results on protein folding.

A site not occupied by the protein monomers is going to be occupied by the solvent. It is empirically observed that the solvent typically makes contact with the monomers only at the surface of the protein; to achieve this, the protein tries to exclude the solvent by filling a volume with its monomers and letting the solvent come in contact with only the monomers on the surface. This structure of the protein is described by a compact random walk that is defined in the following manner: For a protein having N^3 monomers, fix an $N \times N \times N$ cubic lattice. The amino acid monomers are only allowed in the nearest neighboring lattice sites that are not occupied; a typical example of a compact random walk is shown

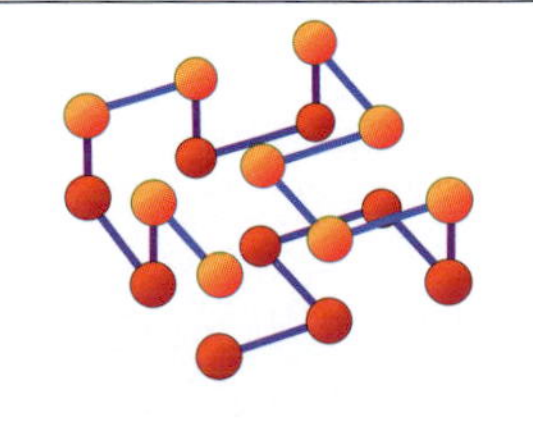

Fig. 17.9: Compact polymer in three dimensions.

in Figure 17.9, with the links showing how the monomers were placed. For a compact random walk, the monomers visit *every* site on a fixed volume of the three-dimensional lattice. The solvent is pushed out to the surface by the protein monomers, which occupy all the lattice sites in the given volume.

The average size of an unfolded protein or a DNA — described by a simple random walk and taken to be equal to $\sqrt{\langle \mathbf{R}^2 \rangle}$ — goes as $L^{1/2}$. The self-avoiding random walk, which is appropriate for RNA, has a larger average size that goes as $L^{3/5}$. In contrast, a compact random walk is volume filling with no holes or gaps. In N number of steps, the protein monomers cover the cubic lattice and hence N is proportional to the volume of the protein. Since the number of steps N is proportional to the length of the protein, we see that, since the protein's linear size is proportional to the cube root of its volume, the average size of the protein scales as $N^{1/3} \sim L^{1/3}$.

Since the random walk is volume filling, the protein's mass is proportional to volume; hence, the compact random walk predicts that mass is proportional to N, the number of monomers. Experiments show that the protein mass depends approximately linearly on the number of monomers, as expected.

17.8 Secondary protein structures

A striking property of globular proteins is the importance of patterns formed by the protein monomers. It is assumed that the compact random walk drives the formation of secondary structures. The folded state of the protein, in this model, is given by a configuration that is a unique state with the lowest energy.

For example, a 2×3 lattice has only four topologically distinct patterns as shown in Figure 17.10a; all other ways of covering the lattice are related to the four patterns by rotations and reflections. Figure 17.10b shows the distribution

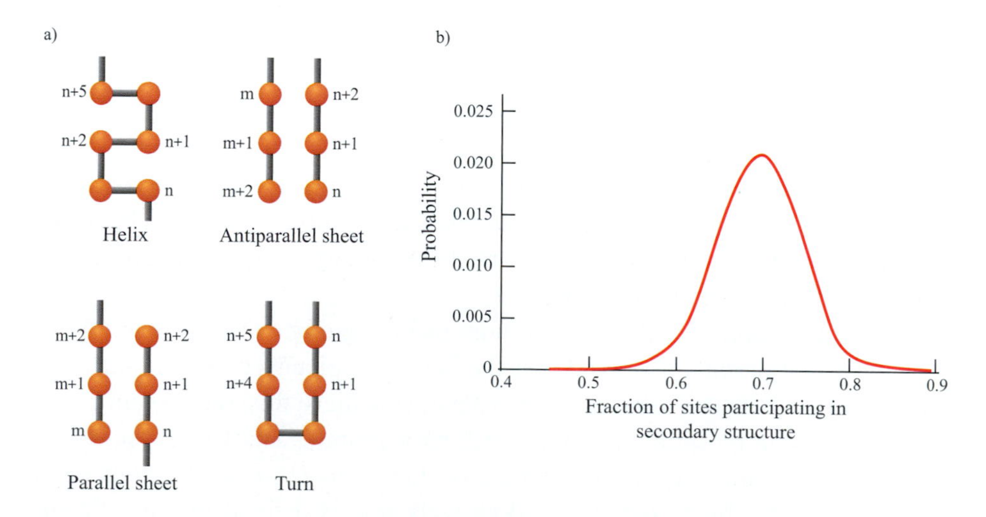

Figure 17.10: a) All the distinct secondary structures in lattice models of proteins for a 2×3 lattice. b) The probability distribution for the occurrence of the protein's secondary structures.

of secondary structures for a 20×20 lattice (consisting of 400 monomers); it is seen that 70% of all the monomers in the compact random walk model belong to a secondary protein structure.

To find the folded configuration for a protein, one has to enumerate all the possible folded configurations of a protein — a virtually impossible task for even a relatively small 100-monomer protein.

Proteins are made of monomers that consist of 20 distinct amino acids, of which 10 are hydrophilic and 10 are hydrophobic.

- Monomers that are hydrophilic readily form a hydrogen (H) bond with the solvent containing water.
- Monomers that are hydrophobic do not form an H-bond with the solvent containing water.

The compact random walk model takes the view that the hydrophobic and hydrophilic nature of the monomers guides the the protein toward its folded state. The collapsed folded state of the protein is induced by hydrophobicity and results in the formation of secondary structures.[1]

Since hydrophobicity is a key property driving the formation of secondary structures, we simplify the problem by making a drastic reduction in the number of protein configurations by classifying all monomers into two groups, namely H (hydrophobic) and P (hydrophilic). To illustrate how the truncated model works, consider the simple example of a 2×3 lattice with three topologically distinct secondary structures shown in Figure 17.11a; the numbers on the lattice sites indicate the sequences of steps — and the links denote the path — taken by the compact random walk. There are six lattice sites which either H or P can occupy, leading to $2^6 = 64$ distinct configurations.

To account for the force of hydrophobicity, we assign an energy to each configuration by the following rule: The energy ϵ of the protein is increased if a hydrophobic monomer H is in contact either with the solvent or with a hydrophilic monomer P.[2] Figure 17.11b shows the energy for different configurations; the orange dots are the hydrophobic monomers and the green dots are the hydrophilic monomers; the dashed line indicates interaction of the hydrophobic monomer with the solvent. The graph shows that the configuration with the shape of an upside-down U, namely Π, is a unique state with the lowest energy cost of 2ϵ; hence, configuration Π is the shape of the folded protein.

17.9 RNA: self-avoiding random walk

Consider the case when the RNA is a single-stranded biopolymer.[3] RNA has three monomers A, C, G in common with DNA; with the fourth monomer being U and

[1] There is another view that the formation of secondary structures is the result of the manner in which hydrogen bonds are formed.

[2] A more refined model can be made that assigns a different energy cost depending on whether the H monomer comes into contact with the solvent of the P monomer.

[3] The double-stranded RNA is not analyzed.

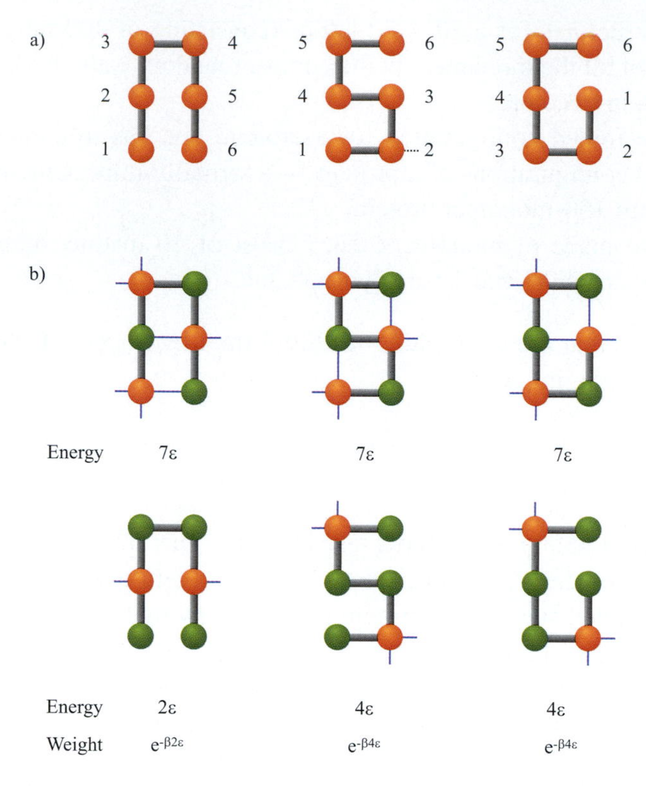

Figure 17.11: Secondary structure in lattice models of proteins. Gray lines show the path taken by the monomers in forming the protein. a) Three topologically distinct structures for a protein. b) The orange dots are hydrophobic and the green dots are hydrophilic. The lines emanating from the orange dots indicate the energy cost of placing the hydrophobic monomers at that lattice site.

not T. Similar to proteins, the RNA monomers are either hydrophilic or hydrophobic. Suppose the RNA has N_{bp} nucleotide monomers. The RNA has the following three phases, depending on the bonding of the monomers with the solvent:

- The deoxidized phase in which it has no bonds with the solvent. In this phase, the structure of the RNA can be described by a self-avoiding random walk; its average size given by $\lambda N_{bp}^{3/5}$, with λ being proportional to the persistence length of the RNA. See Figure 8.8.

- The molten globule phase. The tRNA is an example of globular RNA, characterized by well-defined tertiary interactions.

- The completely oxidized phase in which all the base pairs making up the RNA are H-bonded with the solvent. The fully bonded phase of the RNA is volume filling similar to globular proteins. In this phase the RNA is described by a compact random walk. Assuming each monomer has a volume of about 1 nm^3, the RNA monomers occupy a volume equal to N_{bp} nm^3, and hence we have that the average linear size of the RNA is proportional to $N_{bp}^{1/3}$.

One may ask, why is RNA described by a self-avoiding or a volume-filling walk whereas for DNA an unconstrained and simple random walk is quite an accurate description? The answer partly lies in the length of the biopolymer. The DNA is a very long polymer that is free to move in all directions and hence the likelihood of the base pair arriving at an occupied site is negligible. The monomers of the deoxidized RNA are also free to move but since the RNA tends to be much shorter than the DNA its structure is more compact, leading to the self-avoiding random walk. For the completely oxidized phase, RNA's binding to the solvent makes it cluster together and yields a volume-filling random walk.

17.10 DNA: bending a rigid beam

We have seen in Section 17.5 that DNA can be thought of as a collection of rigid rods linked by a flexible chain. One can go further and try and understand how rigid are the rigid rods and their effect on the dynamics and structure of DNA.

We will study two properties of DNA that result from the bending of the rigid rods:

- DNA often forms loops with sizes that are small compared to the persistence length ξ_P. One needs to know the probability of DNA forming loops, which is the result of the competition between the energy required to bend DNA versus the number of configurations available for such loops.
- In many cases, DNA is packed within a capsid that has a size much smaller than the persistence length ξ_P. For packing purposes the rigid rods need to be bent into circular loops and one needs to know how much energy is required to bend DNA into a loop.

For these and many other applications in physical cell biology, we need to study how does one describe the bending of rigid beams and, in particular, how much energy is stored in the bent beam.

Consider an iron rod, with cross-sectional area A, that is bent under the action of F, an external force. Figure 17.12 shows a section of the iron rod that is bent in an arc; there is a neutral axis that does not undergo deformation, with the material above the axis being stretched and the material below being compressed. Let the unstretched length of the iron rod be L and let the material above the neutral axis be elongated by an amount ΔL. Hooke's law states that, within the elastic limit, such that the force applied F does not permanently deform the material, the force per unit area required to stretch the material is proportional to the fractional extension in the length of the material. In other words

$$\frac{F}{A} = Y\frac{\Delta L}{L}. \tag{17.12}$$

The constant of proportionality Y is called the Young's modulus for the material; it has dimensions of force per unit area given by kg $\mathrm{m}^{-1}\mathrm{s}^{-2}$. See also Section 7.2.

As shown in Figure 17.12, the neutral axis of length L subtends an angle θ with the origin. Consider the extension of the beam at a distance z above the neutral

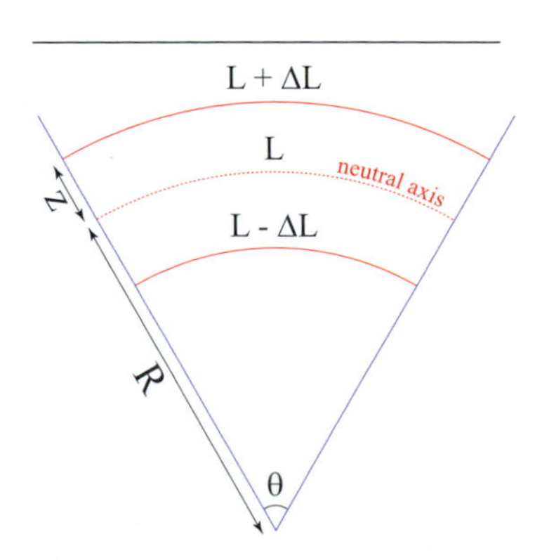

Fig. 17.12: Stretching a rod.

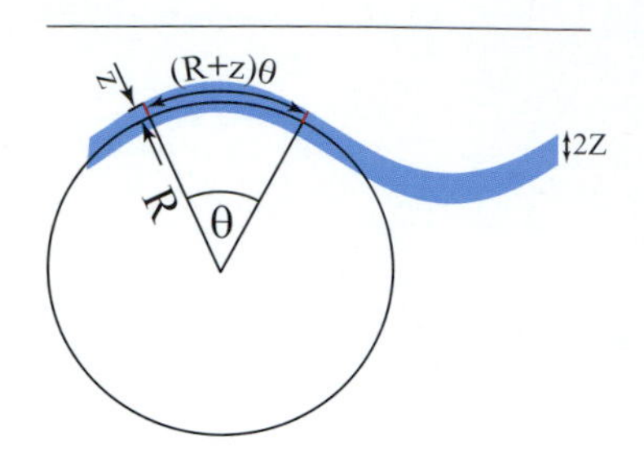

Fig. 17.13: Curvature of a beam.

axis. The unstretched length L is given by the neutral axis, namely $L = R\theta$; the stretched length is given by $L + \Delta L = (R + z)\theta$ and hence $\Delta L = z\theta$. The fractional extension is given by

$$\frac{\Delta L}{L} = \frac{z\theta}{R\theta} = \frac{z}{R}. \tag{17.13}$$

The strain energy, per unit volume, at point z is given by

$$E = \frac{1}{2}Y\left(\frac{\Delta L}{L}\right)^2 = \frac{1}{2}Y\left(\frac{z}{R}\right)^2. \tag{17.14}$$

Consider the special case of bending the beam into a circular loop of radius R, partly shown in Figure 17.13. The energy stored in the circular loop is given by integrating the strain energy over the volume of the iron rod; assuming the rod has a constant cross section, one obtains the total strain energy required to bend the beam as follows:

$$\begin{aligned} E_b &= \frac{1}{2}YL\int dA\left(\frac{z}{R}\right)^2 = \frac{1}{2}Y\frac{L}{R^2}\int z^2\,dA \\ \Rightarrow E_b &= \frac{1}{2}Y\frac{IL}{R^2} \end{aligned} \tag{17.15}$$

where the geometric moment of the iron rod is given by

$$I = \int z^2\,dA.$$

For a beam with an exactly circular cross section, we have that $dA = zdzd\phi$, where z is taken to be the radial coordinate. For a beam with radius d, the geometric moment is the following:

$$I = 2\pi\int_0^d z^3\,dz = \frac{\pi}{2}d^4.$$

A beam bent into a perfectly circular loop has $L = 2\pi R$; hence, the strain energy of a loop is given by

$$E_b = \frac{\pi YI}{R}. \tag{17.16}$$

For a DNA molecule, the typical value of Y is given by about 10^8 Pa; this is approximately Young's modulus for teflon, which is 5×10^8 Pa. In contrast, Young's modulus for steel is about 10^{11} Pa and rubber has a value from 10^7 to 10^8 Pa.

17.11 Looping of DNA: the *lac* operon

A gene is a segment (sequence) of DNA that has a well-defined function. The genome for a particular species is the collection of all of its genes. The gene provides a code that is usually expressed by being copied by a (messenger) mRNA, which is then processed by a ribosome for synthesizing a protein. There are, however, other gene segments in DNA — called *operons* — whose function is to *control* the expression of some other gene. The structure and function of DNA are closely related. One of the ways that an operon controls gene expression is by forming DNA loops: the formation of loops blocks the expression of some genes and promotes the expression of some other genes. Figure 17.14 shows some DNA loops that are involved in controlling gene expression.

The *E. coli* bacterium is one of the most widely studied organisms and has led to a number of important results. We discuss a remarkable operon of the *E. coli*, namely the *lac* operon, discovered by Jacob and Monod and shown in Figure 17.15; see Section 18.7. The *lac* operon is one of the most famous operons and its importance in molecular biology has been compared to the hydrogen atom's seminal influence on quantum mechanics.

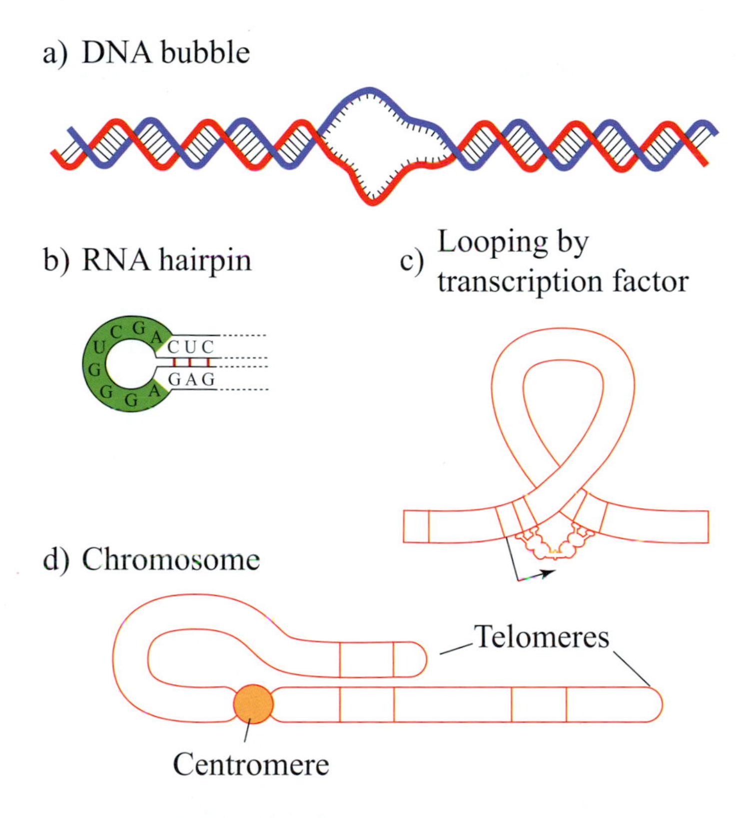

Figure 17.14: Loops in polynucleotides.

As is the case with most bacteria, the *E. coli* obtains its energy by using a set of proteins that allow for the uptake (into the cell) and digestion of glucose. There are, however, circumstances when the *E. coli* finds itself in an environment that has no glucose but which is rich in lactose. The *E. coli*, after a finite time interval, switches off the production of the mRNAs that produce glucose-digesting proteins and switches on the mRNAs that produce proteins for the uptake and digestion of sugar lactose. We do not discuss the mechanism of how the *lac* operon is switched on, but rather focus on a more specific and narrow question of *how* the operon represses the production of the mRNAs required for the digestion of glucose.

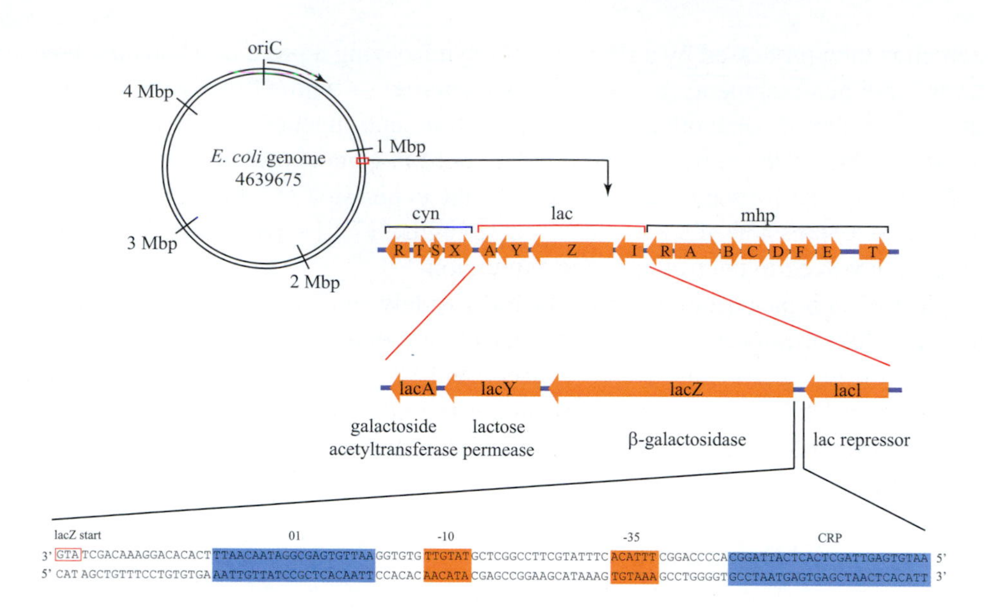

Figure 17.15: The *lac* operon gene in the *E. coli* genome.

When the *lac* operon is switched on, there appears a *lac* represser — which is a single protein that binds two distinct sites on DNA thus forming a loop, as shown in Figure 17.14c. Once the loop is formed, the protein that copies the relevant DNA gene into mRNAs — required for producing the glucose digesting proteins — is no longer able to bind to DNA and hence the mRNAs are no longer produced.

The *lac* represser has three possible binding DNA sites, labeled by O_1, O_2 and O_3. To form the represser DNA loop, the protein located at O_1 needs to bind to either O_2 or O_3. The separation of O_2 and O_3 from O_1 is 401 and 92 base pairs, respectively, and is shown in Figure 17.16.

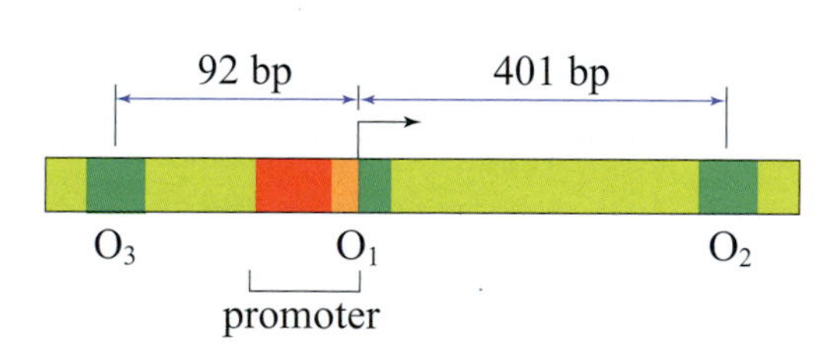

Figure 17.16: Position of the *lac* operators.

Depending on which sites are linked, the looped portion of DNA can be either 401 or 92 base pairs in length. Since a separation of 92 base pairs is much less than the persistence length ξ_P — which is equal to about 150 base pairs — creating a loop of a DNA segment involves substantial bending of DNA.

The loop can form only if the site O_1 is in physical proximity of either O_2 or O_3. Suppose the minimum distance of separation is d for the protein to bind the two sites of O_1 with either O_2 or O_3. From the random walk picture of the conformation of a DNA strand, we need to know the *probability* that the relevant binding sites are within a distance of d. Let N_{bp} be the number of base pairs between O_1 and either O_2 or O_3. One considers a random walk starting at the repressing DNA site O_1 and finds the probability that the walk returns to the same site after N steps. The number of steps taken is given by $N = L/\xi_P$, where the length of the walk is given by $L = 0.34 N_{bp}$ nm.

From Equation 17.6, the probability of being a distance $\mathbf{R}$ from the starting point is given by

$$P(\mathbf{R}, N) = \left[\sqrt{\frac{1}{2\pi\ell^2 N}}\right]^3 e^{-\frac{1}{2\ell^2 N}\mathbf{R}^2}$$

$$\text{where} \quad \mathbf{R} = (x, y, z).$$

Since $\mathbf{R}^2 = d^2$ is small, we can take the probability to be given by

$$P(\mathbf{R}, N) \simeq \left[\sqrt{\frac{1}{2\pi\ell^2 N}}\right]^3.$$

Hence, the probability that the two sites are near each other is given by

$$p_d \simeq \text{const } N^{-3/2}.$$

Since $N = L/\xi_P \approx 0.34\ N_{bp}$ nm, the probability is given by

$$p_d \simeq \text{const } N_{bp}^{-3/2}. \tag{17.17}$$

17.12 Energetics of DNA looping

Consider DNA to be in a (liquid) solvent with particles at some temperature T; due to the random thermal motion of the liquid molecules, a DNA molecule is constantly buffeted by random collisions with the molecules of the liquid. Any restrictions in the movement of DNA, such as the formation of a loop, would decrease the number of random configurations that are accessible to DNA. The entropy of DNA is determined by the number of configurations available to it, and hence creating a loop would decrease the entropy of DNA. The formation of the loop entails bending DNA that costs energy. It is the *interplay* of energy and entropy, something that appears time and again in natural processes, that is the hallmark of (molecular) biology.

The bending energy for creating a perfectly circular loop is given from Equation 17.16 by

$$E_{\text{loop}} = \frac{\pi Y I}{R}.$$

The energy required for bending DNA needs to be supplied by the thermal energy available in the cell, which is in the form of the random motion of the solvent molecules. The only length scale in DNA dynamics is the persistence length ξ_P and, in equilibrium, DNA can spontaneously form a loop with a radius of curvature that is approximately equal to the persistence length. Hence, for looping to take place, the thermal energy $k_B T$ needs to be comparable to the bending energy for a loop of radius ξ_P and yields[4]

$$k_B T \simeq \frac{\pi Y I}{\xi_P} \quad \Rightarrow \quad \xi_P \simeq \frac{\pi Y I}{k_B T}$$

$$\Rightarrow E_{\text{loop}} \simeq \frac{\xi_P k_B T}{R}. \tag{17.18}$$

[4]A more rigorous derivation yields the same intuitive result.

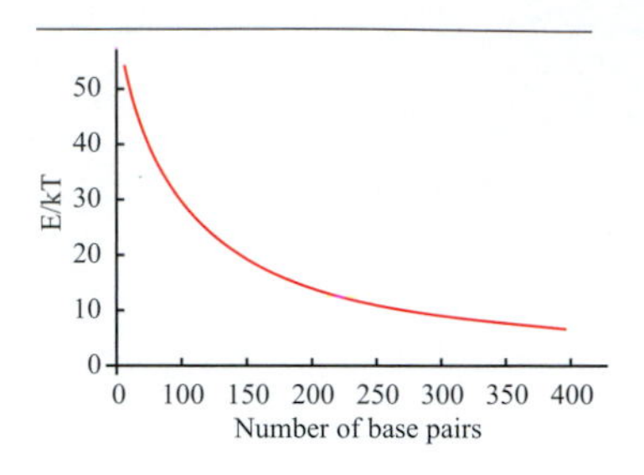

Fig. 17.17: Energy of DNA loops versus the number of base pairs.

Note we have replaced the properties of DNA, encoded in YI, by thermal energy $k_B T$. The energy of DNA loops versus the number of base pairs is shown in Figure 17.17.

A useful way of writing the energy required to bend DNA into a loop is in terms of the number of base pairs N_{bp} that are in the loop. Since $2\pi R = L = 0.34 N_{bp}$ and using $\xi_P = 50$ nm, we have the following:

$$\frac{E_{\text{loop}}}{k_B T} = \frac{2\pi}{N_{bp}}\left(\frac{\xi_P}{0.34}\right) \approx \frac{3000}{N_{bp}}. \tag{17.19}$$

Competition between bending energy and thermal motion

We introduced the concept of free energy F of a system at temperature T in Equation 3.43 of Section 3.8, which encodes the amount of energy E available — from the system's total energy — for doing useful (mechanical) work. The entropy S is a measure of the amount of energy that has to remain as thermal (random) energy. For a process, at constant T, that causes a small change in the free energy of the system, Equation 3.47 yields

$$\Delta F = \Delta E - T\Delta S. \tag{17.20}$$

For the case of the formation of a DNA loop, the change in the energy of DNA is given by the energy required for bending DNA and is given by

$$\Delta E = E_{\text{loop}} \approx \frac{3000}{N_{bp}} k_B T. \tag{17.21}$$

Entropy is a measure of how likely a particular configuration is and depends on the probability of a given configuration to occur. One expects that the entropy S_{AB} for two systems, which are weakly coupled, should be equal to the sum of the entropy of the individual systems, namely equal to $S_A + S_B$. In other words, for two systems specified by the number of configurations Ω_A and Ω_B, one expects that

$$\begin{aligned} &S_{AB} = S_A + S_B \\ &\Rightarrow S_{AB} \propto \ln(\Omega_A \Omega_B) \\ &\therefore \;\; S = k_B \ln(\Omega) \end{aligned}$$

where Ω is the number of configurations available to system S and Boltzmann's constant k_B is the proportionality factor. Entropy is further discussed in Section 18.6.

The number of configurations for a state is proportional to the probability of the system being in that state; that is, $\Omega \propto p$. Hence, the change in the entropy ΔS for DNA looping is proportional to the logarithm of the probability that the repressive receptor sites are in close proximity of each other and yields, from Equation 17.17,

$$\Delta S = k_B \ln(p_d) + \text{constant} = k_B[\ln(\text{const} \times N_{bp}^{-3/2})] = k_B[-\frac{3}{2}\ln N_{bp} + \text{const}].$$

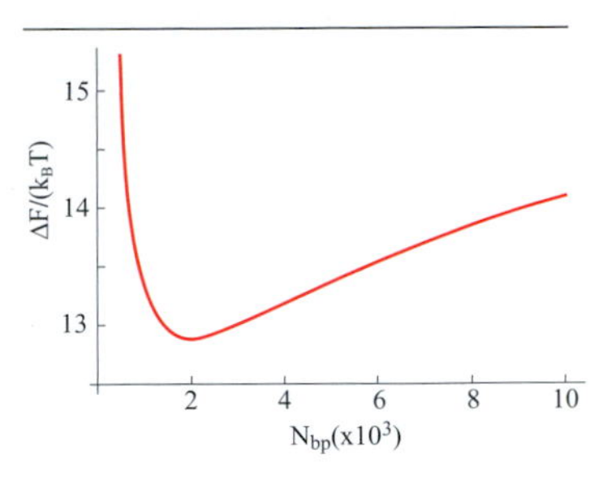

Fig. 17.18: Free energy in loops.

Collecting our results given in Equations 17.20 and 17.21, the total change in DNA's free energy due to the formation of a loop is given by

$$\Delta F \approx k_B T \left(3000 \frac{1}{N_{bp}} + \frac{3}{2} \ln N_{bp} + \text{const} \right).$$

The change in the free energy ΔF — due to the formation of the loop — shows that a large loop lowers the energy of bending, which is proportional to $1/N_{bp}$; on the other hand, a large loop increases the entropy of the system since it is proportional to $\ln N_{bp}$. The system has the highest likelihood of being in a configuration for which ΔF is a minimum.

Figure 17.18 shows the dependence of the change of free energy as a function of N_{bp}; remarkably, there is a minimum for ΔF for $N_{bp} = 2000$ that can be obtained by solving the following:

$$\frac{\partial(\Delta F)}{\partial N_{bp}} = 0 = -3000 \frac{1}{N_{bp}^2} + \frac{3}{2} \frac{1}{N_{bp}}$$

$$\Rightarrow N_{bp} = 2000.$$

It follows, from the minimum of ΔF, that the highest probability is for DNA to form a loop that contains about 2000 base pairs; the free energy cost of bending DNA rises sharply for smaller loops, as seen in Figure 17.18, and results in DNA preferring to have loops much larger than the persistence length ξ_P.

The probability for the formation of a DNA loop is *proportional* to the J-factor which is given by the following:

$$J = \mathcal{N} e^{-\frac{\Delta F}{k_B T}} \tag{17.22}$$

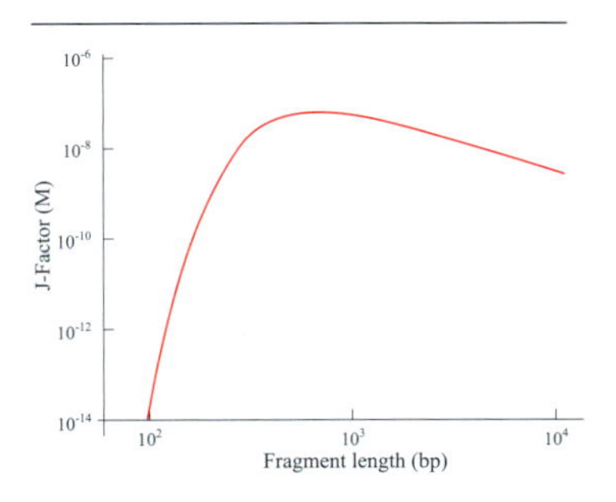

Fig. 17.19: J-factor as a function of DNA length.

where $\mathcal{N}$ is a factor coming from the number of configurations that correspond to a specific value of ΔF.

Figure 17.19 shows that the J-factor for the formation of DNA loops. For large fragments, with $N_{bp} >> 2000$, the cost of bending DNA is negligible and the curve is dominated by the random walk result given by $\Delta F \sim 3(\ln N_{bp})/2$; the short loops are dominated by the energy required for bending DNA given by $\Delta F \sim 1/N_{bp}$. The intermediate-size loops show the competition between bending energy and random walk.

Figure 17.19 shows there is a small but nonzero probability proportional to 10^{-14} that a DNA loop of 92 base pairs will form *spontaneously* and a much higher probability, proportional to 10^{-10}, that a DNA loop will have 401 base pairs. When the *lac* operon is switched on, a protein permanently locks the spontaneously formed DNA loop. The looped DNA, in turn, disallows the expression of the gene that produces the glucose-digesting proteins.

In summary, the mechanism by which *lac* operon switches off the uptake and digestion of glucose hinges on the spontaneous formation of DNA loops. The formation of loops, in turn, results from the competition between the decrease in the free energy by *increasing* loop size and hence reducing bending energy versus decreasing the free energy by *decreasing* the loop size and thus decreasing the number of configurations available to DNA. DNA minimizes its free energy by fixing the loop size to be $N_{bp} = 2{,}000$ base pairs.

17.13 Packing of DNA

The packing of a DNA into a capsid is of great interest in the study of viruses and bacteria. A capsid is a membrane that contains the genome of a virus. Consider the attack of the lambda phage virus on a healthy *E. coli* cell. The following steps are completed within 20–30 minutes, and on its completion the virus goes on to attack some other host:

- The lambda phage takes about a minute to puncture the cell membrane of the *E. coli* bacterium and inject its approximately 50,000-bp genome into the host cell.

- Over the next 10–20 minutes, the viral genome utilizes the molecular machinery of the *E. coli* cell to replicate itself and produce the proteins that are encoded in the viral genome.

- Once a sufficient concentration of viral proteins has been achieved, capsids are spontaneously assembled, together with ATP-driven packing motors.

- The packing motors rapidly wind the viral genome and pack it into the capsid.

- On completing the packing of the virons (a viron is an individual virus), tails are added to the viral capsid and the host cell is burst open, releasing the mature virons — carried inside capsids — and leaving behind the decimated *E. coli* bacterium.

Structural biology has made a lot of progress in understanding how the host cell membrane is punctured by the virus, how the capsid is assembled and the nature of the ATP-driven packing motor. We leave these questions and instead examine the following:

- What is a suitable measure of the degree of DNA packing inside the capsid?

- What are the salient features of the structure of the capsid?

- How is the virus genome packed into the capsid?

The length of a DNA molecule is typically in excess of 10 μm and it is packed into a capsid with a linear dimension of roughly 50 nm. The size of the capsid is typically equal to or less than the 50-nm persistence length ξ_P and hence the packing of DNA entails spending a large amount of energy for bending and packing DNA into the capsid. The packing energy required is so large that thermal energy can never supply enough energy. Instead, there are special molecular motors with their own supply of energy that carry out the work of packing DNA.

The degree of compression for a DNA packed into a capsid is given by the ratio of the volume occupied by DNA compared to the total volume of the capsid. All viral capsids, up to a factor of two, have the same size. The capsid of the famed lambda phage can be thought of as a sphere with a radius of 27 nm and encapsulating a 48,500-bp genome. A single base pair typically occupies a volume of 1 nm^3. The packing ratio for the *lambda phage* is given by

$$\frac{N_{bp}\ \text{nm}^3}{\frac{4}{3}\pi(27)^3\ \text{nm}^3} \approx \frac{5\times 10^4}{4\times(27)^3} \approx 0.6 \quad : \quad \text{Lambda phage.}$$

The viral DNA practically occupies the entire volume and is akin to crystalline densities. On the other extreme is the volume occupied by eukaryotic DNA, which is contained in a nucleus that is enclosed within a special membrane inside the cell. DNA has an approximate length of 10^9 base pairs contained in a nucleus with a radius of about 5μm. Hence, DNA has a low packing ratio given by

$$\frac{10^9\ \text{nm}^3}{4\times(5000)^3\ \text{nm}^3} \approx 0.002 \quad : \quad \text{Eukaryotic DNA.}$$

17.14 DNA: bending energy

We consider the packing of viral DNA since this is relatively well understood. For the purpose of simplifying the calculation, it is assumed that the viral capsid has the shape of a cylinder with height H. Given the high packing ratio of a viral bacteriophage, the viral DNA inside the capsid is highly ordered. Experimental studies show that the viral genome is packed in a concentric arrangement that, to a very good approximation, is concentric circles. The circular DNA strands are separated by a distance of d_s, with each strand having six neighbors, as shown in Figure 17.20.

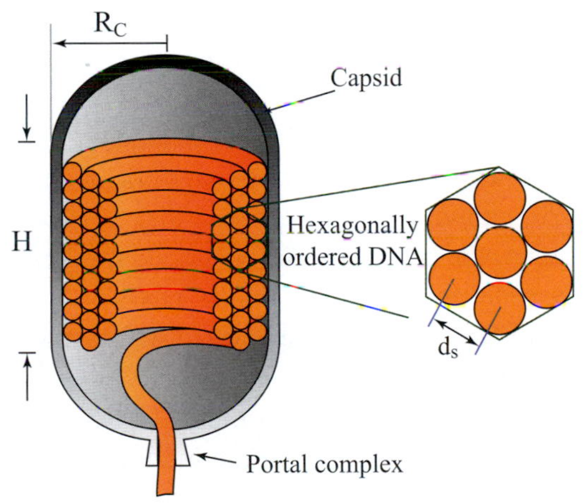

Figure 17.20: Viral DNA in a capsid. The height of the cylinder H = 100 nm and the radius R_c = 20 nm.

There are the following three competing forces in the packing of viral DNA:

- Entropy effects that drive DNA toward being as spread out as possible.
- The energy of bending DNA into a radius of curvature far smaller than the persistence length.
- Electrostatic energy because both DNA and the solvent are charged.

The entropic effects are 10 times smaller than the other two forces and will be ignored. DNA bending energy is lowered if the radius of curvature is as large as possible. However, this entails that large sections of the wound-up DNA are close to each other, which increases their electrostatic energy. Lowering the electrostatic energy drives DNA loops far from each other into regions of the capsid where their radius of curvature is small, thus increasing their bending energy. It is the competition between lowering either the bending energy E_b or the interaction energy E_c that fixes the average distance of separation d_s between the parallel DNA strands.

The volume occupied by the viral DNA is about Ld_s^2, where L is the length of the packed DNA. For viral DNA, the capsid is almost completely filled up by the tightly packed DNA. Since the volume V_{capsid} of most viral capsids is approximately equal, we have

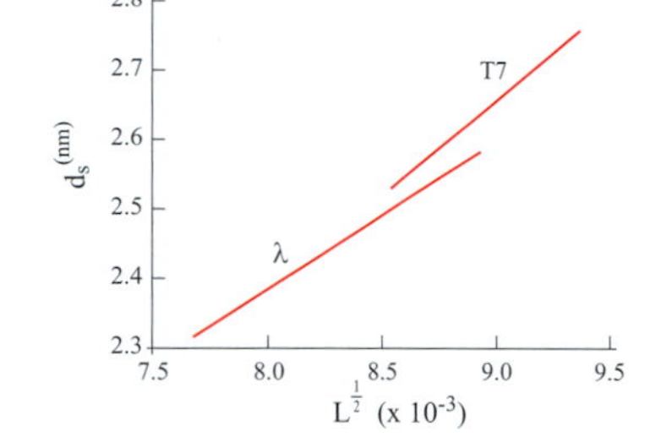

Fig. 17.21: DNA spacing as a function of packaged length.

$$V_{\text{capsid}} \simeq Ld_s^2$$

$$\Rightarrow d_s \approx \sqrt{V_{\text{capsid}}/L} \simeq L^{-1/2}.$$

For a limited range of genome length, experiments show the above relation is approximately correct; see Figure 17.21.

The free energy of the packed DNA, with length L and distance d_s, is given by

$$F_{\text{tot}}(d_s, L) = E_{\text{b}}(d_s, L) + E_{\text{e}}(d_s, L). \tag{17.23}$$

17.15 DNA: circular loops and electrostatic energy

We estimate the strain energy associated with the packing of viral DNA. We make a simplified model of the capsid as a cylinder, neglecting its conical apex. The cylinder has a radius of $R_c \sim 20$ nm and a height of about $H \sim 100$ nm. As shown in Figure 17.20, although DNA is packed into a helix (not to be confused with DNA double helix), the change in the height of the helix in one rotation is much smaller than the radius of curvature. Hence, DNA is considered to be packed into circular loops of radius R_i, with spacing d_s and stacked into a *cylinder* of height H and radius of R_c.

Let $n(R_i)$ be the number of loops with radius of R_i; the total strain energy is given by the sum of the strain energies of the individual loops and hence, from Equation 17.18,

$$E_{\text{b}} = \xi_P k_B T \sum_{i=1}^{M} \frac{n(R_i)}{R_i} \tag{17.24}$$

where M is fixed by the requirement that the total length L of the loops is equal to the length of DNA, namely,

$$\sum_{i=1}^{M} n(R_i) R_i = L.$$

By a series of approximations, it can be shown that Equation 17.24 yields the following:

$$E_\mathrm{b} = \frac{\xi_P k_B T H}{2d_s^2} \ln\left(1 - \frac{d_s^2 L}{\pi H R_c^2}\right) \tag{17.25}$$

$$F_b = -\frac{\partial E_\mathrm{b}(d_s, L)}{\partial L} = \frac{(\xi_P k_B T / 2\pi R_c^2)}{1 - (d_s^2 L / \pi H R_c^2)}$$

where L is the total length of the viral DNA, R_c is the radius of the cylindrical capsid with height H and d_s is the distance between the loops.

DNA: electrostatic energy

There are electric charges both on DNA as well as in the solvent in which DNA is immersed. Instead of doing a first principle calculation of the effect of the charge distribution, one can instead find the effective potential between the strands of DNA. One is led to the conclusion that parallel strands of DNA interact through a pair potential per unit length given by $v(d_s)$. For N strands of DNA packed into a hexagonal arrangement with six nearest neighbors as shown in Figure 17.20, the electrostatic energy is given by

$$E_c = 3NHv(d_s)$$

where the prefactor of 3 is from the six nearest neighbors, divided by 2 to avoid overcounting. Experiments yield the following result:

$$E_c = \sqrt{3}(c^2 + cd_s)F_0 L\, e^{-\frac{d_s}{c}} \tag{17.26}$$

where c is the concentration of ions in the solution.

The form of the pair potential $v(d_s) \simeq \exp(-d_s/c)$ can be understood to be the result of electrostatic screening. Electrostatic attraction is between unlike charges and repulsion is between like charges. Since the solution has free charges, electrostatic forces lead to each positive charge being surrounded by negative charges and vice versa, thus reducing the range of the interaction and leading to an exponentially falling potential.

The total electrostatic energy of packing E_c depends on the concentration of ions in the solution. If one alters the concentration of ions, one could create a barrier to the ability of the virus to inject its DNA into the host cell.

Collecting results given in Equations 17.25 and 17.26 yields the following total energy of the packaged DNA:

$$F_\mathrm{tot} = \frac{\xi_P k_B T H}{2d_s^2} \ln\left(1 - \frac{d_s^2 L}{\pi H R_c^2}\right) + \sqrt{3}(c^2 + cd_s)F_0 L\, e^{-\frac{d_s}{c}}. \tag{17.27}$$

From Equation 17.27 we can compute the distance d_s for which the packed DNA is in equilibrium as well as the force F required to pack DNA.

We hence have the following:

$$\frac{\partial F_{\text{tot}}(d_s, L)}{\partial d_s} = 0 \quad : \quad \text{Fixes } d_s$$

$$F = -\frac{dF_{\text{tot}}(d_s(L), L)}{dL} \quad : \quad \text{Determines the force.} \tag{17.28}$$

Figure 17.22 shows experimental results on the rate of packing for bacteriophage $\Phi 29$ and the force required to complete the packing.

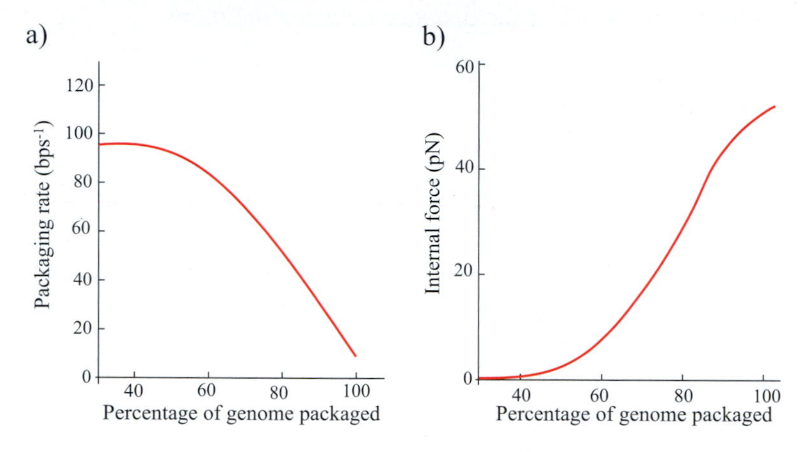

Figure 17.22: DNA packaging in a bacteriophage. Source: Rob Phillips.

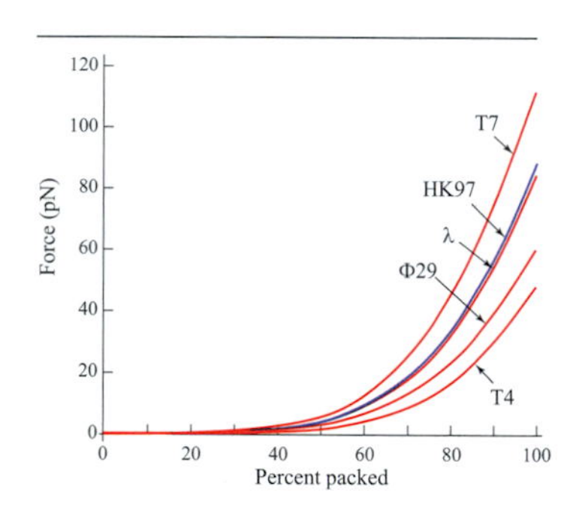

Fig. 17.23: Packing forces for different viruses.

Figure 17.22a shows that the rate of packing slows down as more and more viral DNA is packed into the capsid; Figure 17.22b shows the force F, given by Equation 17.28, required to pack DNA. The rate of packing slows down due to the increase in the internal force exerted on the packed DNA and, hence, Figure 17.22a follows from Figure 17.22b.

Figure 17.23 shows different bacteriophages and the packing force required. The case of $\Phi 29$ has been verified by experiments, with the other curves representing the predictions made by Equation 17.27 above.

Note that for T7, F_{tot} in Equation 17.27 yields that a maximum force of over 100 pN is required to pack DNA compared to the maximum force for, say, the T4 virus of 50 pN. This suggests that either the T7 phage will be unable to complete DNA packing into the capsid, or else that the T7 phage is equipped with more powerful molecular motors compared to the T4 phage.

17.16 The answer

The structure of DNA and other biopolymers is one of the central subjects of molecular biology. We used concepts from the theory of random walk and from the mechanics of beam bending to quantitatively explain some of the structural features of DNA, RNA and proteins.

Random walk theory was used to explain a few coarse features of the biopolymers, in particular its average size. The simple random walk was found to be appropriate for describing DNA. The globular folded phase of protein was analyzed using the concept of a compact random walk that quantitatively accounts for the protein's observed secondary structures. The deoxidized phase of the RNA was best described by a self-avoiding random walk and the oxidized RNA phase by a volume-filling random walk.

The idea of persistence length of DNA, which emerges from the random walk model, has far-reaching consequences and, in particular, leads to the freely jointed rigid rod model of DNA. The length of the rigid rods are equal to DNA's persistence length and the bending of DNA leads us to the study of beam theory. The formation of DNA loops is a result of the competition between entropy and DNA's strain energy of bending and which, in turn, partly explains the function of the *lac* operon.

The packing of viral and other DNA into capsids, which have dimensions smaller than the persistence length, is explained using concepts of beam theory. The structure of the packaged DNA results from the competition of strain energy with electrostatic energy.

We can now answer our question: **The structure of DNA is determined by the interplay between energy, entropy and biological constraints.**

17.17 Exercises

1. Would it have been possible to make DNA a binary code?

2. I stand at a point x and randomly take 10,000 steps of 1 m forward or backward. What should I expect my average position to be?

3. Suppose I stand at a point x and then randomly take 10,000 steps of 1 m forward or backward. On average, what is the distance I would be from the origin?

4. What type of distribution is depicted in Figure 17.3 (think, for example, of Bernoulli, uniform and normal distributions)?

5. If I start the random walk shown in Figure 17.3 in the center of a petri dish, how large does the petri dish need to be so that probability to bump into the rim of the petri dish is smaller than 0.27% if we take 10,000 steps of 1 m?

6. What is the persistence length of DNA?

7. Is energy required to package DNA in a viral capsid?

8. I discovered a new bacterium with a genome slightly smaller than that of *E. coli*. How many base pairs will it roughly have?

9. Since there are bacteriophages, would there also be "viro"phages?

10. In the *lac* operon as shown in Figure 17.16, what does the promoter promote? Is it the production of lactose, glucose or something else?

11. What is the packaging ratio of the lambda phage? If I'd have a "super-size" phage that has twice the volume of the lambda phage but the same amount of packed DNA, what would its packaging ratio be?

18: Information

Q - Does DNA Compute?

Chapter Map

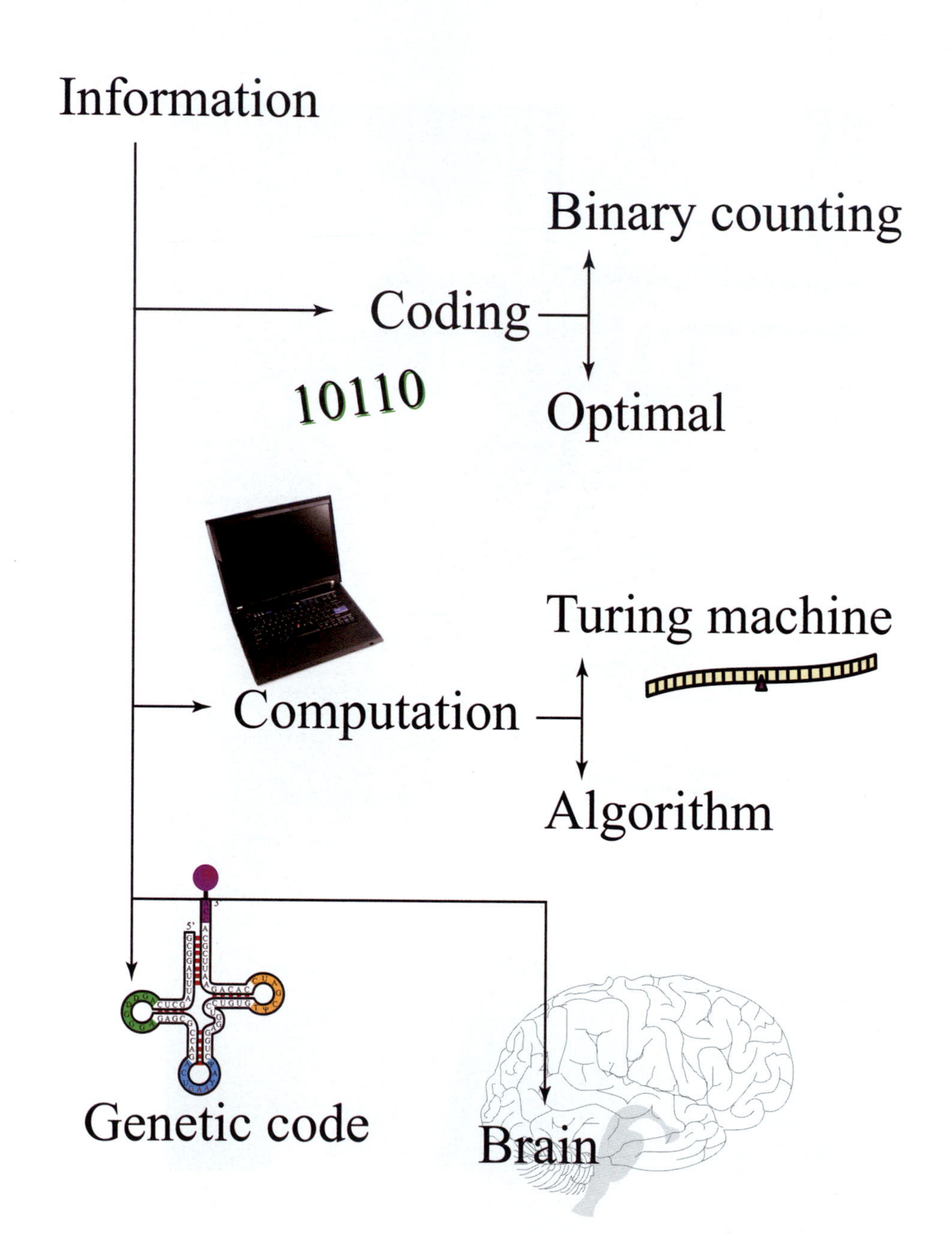

Information
Binary counting
Coding
10110
Optimal
Turing machine
Computation
Algorithm
Genetic code
Brain

18.1 The question

Figure 18.1: Grandmother with daughter and grandchild.

"She looks just like her mother!" Affection for the faces of little children is almost universal in human beings, and one of the most common observations is that children often resemble their parents or sometimes grandparents in appearance. But it's not just appearance that's similar in child and parent. Although it may sound somewhat silly, children are also of the same species and share virtually all biological functionality with their parents.

Therefore, even without any deep scientific analysis, it is clear that somehow information is transmitted from parent to child. Now, if we want to understand this process properly, we need to know not only about how it works but also about what information actually is and how it can be measured. Indeed, we cannot really talk about information transmission until we have gained a better understanding of information itself. We will therefore start this chapter by investigating the notion of information.

Even for the next step, we still cannot trivially consider how information is transmitted from parent to child since we need at least some rudimentary form of information processing. We say "at least" here as this is obvious from the daily life observation. Of course in reality, we can expect the form of information processing involved in reproduction to be rather complicated. But what is "information processing"? To answer that we need to understand the basics of computer science and hence consider algorithms and the Turing machine, a generic kind of computer.

When we have an idea of what information is and how in a generic sense it can be processed, we can return to our observation that children often resemble (one of) their parents. Information about life is stored in DNA and hence information can be passed from parent to child with the help of DNA. However, the process of passing on information, just like many actual functions essential for life, is carried out by proteins and enzymes, which have a composition very different from DNA. So how does information flow from DNA such that the proteins and enzymes are produced which are essential for life and reproduction? Indeed, we can ask the main question of this chapter: **Does DNA compute?**

18.2 Measuring information

It is almost always a good idea to start off with things as simple as possible! If we want to discuss information, we need to have some kind of tokens to represent this information. In principle, we could have any number and kind of tokens but the simplest would be to have only two possible tokens (e.g. a green and a red ball, a triangle and a square, an empty and a full glass) or one token with two possible values (e.g. a coin with head and tail, an egg that is broken or not, a light that is yellow or blue).

In information theory, one often uses the fictious figures Alice and Bob as the sender and receiver of a message, respectively.

In information theory, a token which can assume two values is called a bit, and the most convenient token is that of a number which can assume the values 0 and 1. Messages that need to be transmitted are then first encoded into a series of zeros and ones — called a string — and the question is then, how many bits are actually necessary to communicate this string from one point to another?

Another way to pose this question is as follows: If Alice wants to send a message to Bob, how many bits does she actually need to transmit when sending a digital file by e-mail (here we are talking just about the bits needed for the file and not the bits that one will need for things like the e-mail address, the subject and so on)?

If we create a string by repeatedly flipping a coin assigning, for example, a 1 to heads and a 0 to tails, then the string we obtain is a completely random sequence of zeros and ones with a length equal to the number of coin flips. If we want to transmit this, the number of bits that need to be sent is exactly equal to the length of the string.

As such, that is probably not all too surprising. After all, if we have a string of 1,000 bits, we would certainly expect that we need to send those 1,000 bits if we want to transmit the information to our friend. However, in a more general setting, is it always true that one needs to send all the bits one has in order to transmit all the information contained in them?

The answer is a clear no! If, rather than a string with a random sequence of zeros and ones, we have a string where zeros and ones alternate, instead of sending the 1,000 bits, we'd be much better off to only send a zero followed by a one and then a command stating "repeat 500 times". Hence we can see that the number of bits that needs to be sent depends on the patterns found in the string.

Now we can take this a step further and state that in reality a 1,000-bit string with a random sequence of ones and zeros contains much more information than a string where a zero and a one are repeated 500 times. Indeed, we can see this in daily life, too. A manual with 1000 different pages certainly contains a lot more information than a manual with two pages repeated 500 times. Of course, this has nothing to do with what is the meaning of the message and whether the information is actually useful — information theory does not concern itself with meaning and usefulness.

Thus we see the essential connection between the quantity of information and the number of bits which need to be transmitted. Indeed, one can now define the following:

The information content of a message is defined as the minimal number of bits needed to transmit that message.

Of course, we must now ask: "How do we find the minimal number of bits needed to transmit the message?" In order to answer that question we should have a look at something called optimal coding. Before doing so, however, we need to know about binary counting so we'll consider that first before continuing.

18.3 Binary counting

The binary system is very useful for the study of information.

We are all used to the familiar decimal system where we have 10 basic numbers (0 to 9) whose actual value in a given number depends on their position in the number. For example, if we have the number 253, the 2 really stands for 200 and so on. As trivial as this may seem, the brilliance of this system is that the value of the *entire* number can simply be obtained by adding up successive digits each multiplied by the correct value according to their positions, i.e. $253 = 2\times100+5\times10+3\times1$. Of course successive values are simply obtained by adding zeros. Sure, we all know this but it wasn't always like that. For example, to obtain the Roman numeral IV (our number 4) we take V (the number 5) and subtract I (the number 1). Roman VI is then our number 6. Another point is that additions are really easy in the decimal system. For example $15+6+3 = 24$ which we can calculate by adding $5+6+3 = 14$ and then carrying over the 1. Try this in Roman $XV + VI + III = XXIV$!

So it's clear that the decimal system is handy, and not surprisingly it's now used all over the world. What is more surprising is that the Romans didn't use it even though it had long before been invented in India. (Sometimes, great ideas travel surprisingly slowly.)

The decimal system is a number system with base 10. That means that the value of the rightmost digit in an integer is 10^0, the next one 10^1, then 10^2 and so on. As mentioned above, doing so is convenient but as such this is not necessary. In fact, one can use any base one wishes. For example, if we use base 6 and have the number 312, this means that we have $3 \times 6^2 + 1 \times 6^1 + 2 \times 6^0$ which would be 122 in our decimal system. Still a good system but not as handy as the decimal system.

However, one other base besides 10 is extremely handy and useful. It is the base 2, or binary, system. It is special because it is the simplest logically possible number system as it contains only two symbols: a 0 and a 1.

Wouldn't it be possible to have a number system with just one symbol and then encode numbers by either having or not having that symbol? Not really, as "not having the symbol" is logically equivalent to having a "not" symbol, thus leaving one again with two distinct symbols in total. How about a system with just one symbol where the number of times the symbol is repeated represents the size of the number? In fact this is how children learn how to add 3 + 4: make two fists, count 3 fingers by straightening them, then count another 4 fingers by straightening 4 more fingers, then return to the beginning and count all the straightened-out fingers to obtain 7. Wouldn't that be a workable system? The answer to that is "no". First, there is a practical limit. Writing large numbers would require a lot of space. Just try to imagine adding 12,309 and 3,765 with the help of fingers. But more serious is the following conceptual problem: We started our discussion by talking about the transmission of information. If we want to transmit a sequence of numbers or even only a single number, we need to know the beginning and the end of the number. Hence, either we introduce a start/end symbol which brings us back to having two symbols in total or we have to demarcate the start/end by having no symbol. But as mentioned above, no symbol is logically equivalent to having a symbol indicating "no symbol" bringing us back to the minimum of two symbols.

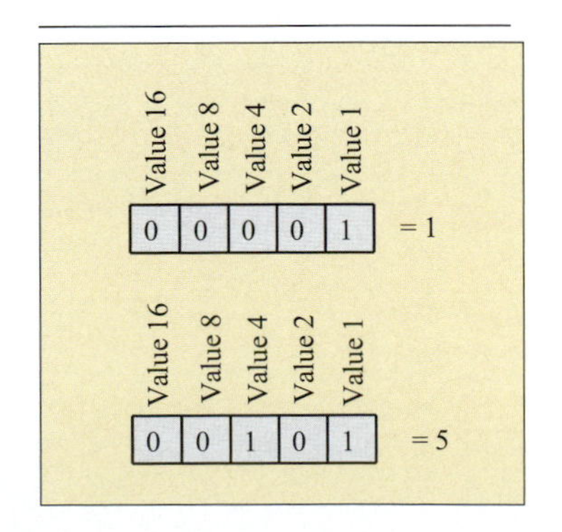

Fig. 18.2: Examples of binary numbers.

Therefore, if information needs to be transmitted, be it from person to person directly, over the Internet or from parent to child with the help of DNA, we can't do with fewer than two symbols. And if information is not transmitted, it cannot be known and hence perhaps it would be fair to say that then it isn't information at all.

Consequently, when dealing with information, the binary system is a natural choice to use as it is the simplest possible system that can be employed.

18.4 Optimal coding

Let us assume again that we have a string of randomly chosen bits — for example, 1 million bits. But, contrary to before when we used a regular coin flip to determine whether a bit is a 0 or a 1, this time, let us assume that we have a very special coin which on average gives a head (corresponding to a 1) only once every 1,000 flips. In other words, the probability to get a head is 1/1,000 and hence the probability of obtaining a tail is 999/1,000.

Clearly, on average, a string generated with this special coin would have 1,000 ones and 999,000 zeros. As the ones are the result of a coin toss, they will be spread throughout the string in a random fashion. Indeed, two ones could even succeed each other (even though the probability of this occurring is rather low).

Still, if we inspect the string, we mostly see zeros and although we don't have the patterns of alternating zeros and ones as above, we should expect that this fact can somehow be used to reduce the number of bits transmitted and hence that the information content of this string is less than that of a random string where zeros and ones occur with equal probability.

How can Alice transmit a string more efficiently than sending the 1 million bits? Well, since the number of ones is small, she could do the following: First, Alice tells Bob that the starting point is a string with 1 million zeros, then she sends him the *positions* of the ones. Would this really be better?

To find out, let us see how many bits are necessary to transmit one position. Since the string is 1 million bits long, the farthest position would be 1 million. In binary 1 million can be encoded by 20 bits (note: $2^{20} = 1{,}048{,}576$) and hence Alice would need to send $1{,}000 \times 20 = 20{,}000$ bits to send *all* the positions.

So, how many bits would she need to send in total with this scheme? Since 2^{20} is quite a bit larger than 1 million, some of the numbers between 1 million and 2^{20} can be used to encode some special meaning which Alice and Bob have agreed upon. For example, 20 successive ones could mean: "This is a string with all zeros and the next 20 bits indicate the length of the string, followed by the end code". Nineteen ones and then a zero then could mean "end, we're done sending positions of ones". Such special sequences of ones and zeros are generally referred to as control sequences. Therefore, with a scheme like this, Alice could transmit the full information in this string of 1 million bits with only 20,060 bits — quite an improvement!

Indeed, what we also can see is that the control bits are only a tiny fraction (60 of 20,060) and that we hence can basically ignore them for our considereations.

Sending the positions was a pretty good idea. But, can Alice do even better?

Yes, she can! With the current scheme, the positions can be sent in any order, but of course that is not necessary at all. It would seem to be rather natural to agree that the positions are sent in order. If they are sent in order, then, rather than sending the entire position, Alice can do with only sending the *distance* from the current bit with a value 1 to the next bit with a value 1.

How many bits would she need for that distance? Probability theory teaches us that it will be extremely unlikely that the distance is larger than 4,000 positions and hence we only need to be able to count binary from 1 to 4,000 which can be done with a mere 12 bits ($2^{12} = 4{,}096$). Again, as 4,096 is larger than 4,000, there are quite some numbers left which can be used as control sequences. One of them could, for example, indicate the very rare event that the distance is larger than 4,000 positions. Ignoring control sequences, Alice can therefore send her message now with only 12,000 bits (1,000 distances between positions each being coded by 12 bits). A huge improvement, indeed, from 1,000,000 down to 12,000.

Of course, we should ask ourselves: What is the theoretically smallest number of bits that one would on average need to send this kind of a string of 1 million bits generated from a special coin where the probabilities of obtaining a 1 or 0 are unequal? More generally, this could be extended to a die with, for example, eight sides, each of which has a different probability of occurring.

To proceed, let us consider a die with the probabilities of throwing a side depicted in Table 18.1. If we throw this die 1,000 times, how many bits do we need

$$
\begin{aligned}
1 &\Rightarrow \tfrac{1}{2}\\
2 &\Rightarrow \tfrac{1}{4}\\
3 &\Rightarrow \tfrac{1}{8}\\
4 &\Rightarrow \tfrac{1}{16}\\
5 &\Rightarrow \tfrac{1}{32}\\
6 &\Rightarrow \tfrac{1}{64}\\
7 &\Rightarrow \tfrac{1}{128}\\
8 &\Rightarrow \tfrac{1}{128}
\end{aligned}
$$

Note that the probabilities in Table 18.1 add up to 1.

Table 18.1: How should one code for transmission of the rolls of an eight-sided die with the above probabilities?

to transmit on average to send the outcomes from Alice to Bob? Since there are eight different sides, the simplest would be to use three bits per throw since if we have three bits we can count from zero to seven giving eight possible values. So 1,000 throws would require 3,000 bits. Again, we have to ask whether we can do better than that. Is there a better way to *code* the throws?

Well intuitively, one would say that it would be good to code it in such a way that sides which occur often (i.e. side 1 or 2) require as few bits as possible while allowing some more bits for less likely outcomes (i.e. sides 7 or 8). Clearly, the smallest possible number of bits for transmission is a single bit. Hence, let us

1	$\Rightarrow$	0
2	$\Rightarrow$	10
3	$\Rightarrow$	110
4	$\Rightarrow$	1110
5	$\Rightarrow$	11110
6	$\Rightarrow$	111110
7	$\Rightarrow$	1111110
8	$\Rightarrow$	1111111

Table 18.2: How should one code for transmission of the rolls of a die with the above probabilities?

start by assigning a 0 to face 1 of the die. Since we are transmitting the outcomes sequentially, if the next outcome is not face 1, we cannot simply assign a single bit with value 1 to face 2 because then we couldn't distinguish the other possible outcomes. Hence, let us assign the code 10 to face 2. Of course, now we still need to code for the remaining faces, and following the same spirit we can make the assignments shown in Table 18.2.

Such a code would certainly work. A bit with a 0 indicates that the next bit encodes a new outcome as does the case when there are seven 1s on a row. An example is shown in Figure 18.3.

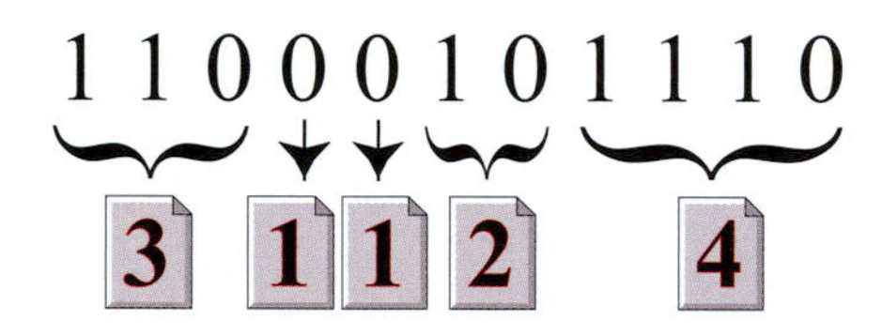

Figure 18.3: Example of how the successive outcomes 3-1-1-2-4 are encoded for transmission.

Having this code, how many bits would we need, on average, to transmit 1,000 outcomes? We can solve this as follows (assuming we throw the die 1,000 times): On average we would throw a one 500 times (see Table 18.1), and to transmit these outcomes we need 500 bits (see Table 18.2). We would also throw a three 125 times which would need 375 bits to transmit, and so on for the other possible outcomes. What we find by adding up the average number of bits necessary for all the faces is that the 1,000 outcomes (on average) need 1,984 bits to be transmitted.

Now if we look at the above a bit closer, we realize that what we actually did is to multiply the probability of obtaining an outcome by the number of bits required to transmit that outcome times the total number of throws. However, since we are dealing with averages anyway, there's really no need to multiply by the total number of throws. Indeed, it is much more convenient to look at how many bits on average need to be transmitted per throw as shown in Table 18.3. In the above example, that would be 1.984 bits per throw. Now, that's excellent because that's clearly a lot better than sending three bits per throw.

Having the code, we can calculate how many bits on average we need to transmit a sequence of throws. But how can we get a code like that in the first place?

1	$\Rightarrow$	$\frac{1}{2}$	$\times 1$	$= \frac{1}{2}$
2	$\Rightarrow$	$\frac{1}{4}$	$\times 2$	$= \frac{1}{2}$
3	$\Rightarrow$	$\frac{1}{8}$	$\times 3$	$= \frac{3}{8}$
4	$\Rightarrow$	$\frac{1}{16}$	$\times 4$	$= \frac{1}{4}$
5	$\Rightarrow$	$\frac{1}{32}$	$\times 5$	$= \frac{5}{32}$
6	$\Rightarrow$	$\frac{1}{64}$	$\times 6$	$= \frac{6}{64}$
7	$\Rightarrow$	$\frac{1}{128}$	$\times 7$	$= \frac{7}{128}$
8	$\Rightarrow$	$\frac{1}{128}$	$\times 7$	$= \frac{7}{128}$
			Total	1.984

Table 18.3: Average number of bits per outcome.

Intuitively, we knew that it would be a good starting point to code things such that likely outcomes require fewer bits than unlikely outcomes. If we inspect Table 18.2 carefully, we see that the number of bits in the code is exactly equal to minus one times the base two logarithm of the probability! For example, let us consider the outcome 8:

$$\text{Number of bits} = -\log_2\left(\frac{1}{128}\right) = \log_2(128) = \log_2\left(2^7\right) = 7.$$

In more general terms, we see that the number of bits n_i that need to be transmitted for a outcome i which occurs with a probability p_i is given by

$$n_i(\text{outcome } i) = -\log_2(p_i). \tag{18.1}$$

Now of course the outcome i only occurs with the probability p_i, so following Table 18.3 above we see that on average outcome i requires

$$\langle n_i \rangle = p_i n_i(\text{outcome } i) = p_i \times -\log_2(p_i) = -p_i \log_2(p_i). \tag{18.2}$$

Consequently, summing up all the possible outcomes we obtain that the average number of bits necessary to transmit the result of *each* outcome is given by the formula

$$\langle n \rangle = -\sum_{i=1}^{8} p_i \log_2(p_i). \tag{18.3}$$

For a message consisting of N throws of the die, the total number of bits that would be required to send this message, from Equation 18.3, is then

$$N \langle n \rangle = -N \sum_{i=1}^{8} p_i \log_2(p_i) \; ; \; \textbf{Total number of bits}. \tag{18.4}$$

English text consists of 26 letters and the occurrence frequency of each letter is known; one needs to include punctuation marks and spaces giving a total of, say, 30 basic symbols. Hence, if an English text message contains N symbols, then the total number of bits required for transmitting the message is given by an expression very similar to the one given in Equation 18.4.

Equation 18.3 is Shannon's famous formula, and interestingly it can be shown that this formula is valid for so-called **optimal coding** which is the coding that yields the smallest average number of bits when transmitting a string given the frequencies of its symbols. Equation 18.3 can therefore also be interpreted as a formal definition of information!

At this point, we can look at Equation 18.3 in a different light. If the probability to obtain a certain outcome is very large, then the surprise to obtain this outcome is very small and indeed the information gained is minimal. In the extreme case that the probability of an outcome is 1, we already know what is going to happen, and hence the actual event will provide us with no new information. If we plug $p_i = 1$ into Equation 18.3, we indeed obtain $\langle n \rangle = 0$ as $\log(1) = 0$. On the other hand, if the probability of an outcome is very small, then the surprise and hence the information gained is very large. Plugging a small p_i into Equation 18.3 indeed gives a big number for $\langle n \rangle$. Thus we see that there is a deep connection between, surprise, information content and the number of bits that need to be transmitted.

In summary, if an outcome is very unlikely, the surprise to have obtained it is very large and hence the information gained likewise. In an optimal coding scheme, the consequence is that an outcome with a lot of information requires a large number of bits for its transmission.

We now have an idea of how information can be defined and how it can be quantified. We have also seen how essential it is to have some kind of "coding". However, to perform this kind of coding, we need to be able to process the information and hence we need to carry out some kind of computations. Therefore, we will continue our investigation by examining the notion of computation closer in Section 18.7. Before doing so, let us briefly digress into several related topics.

18.5 Dealing with errors

For either practical and/or conceptual reasons, it is rather common in physics and mathematics to consider idealized systems. For example, friction is often ignored and atoms are sometimes describes as little billiard balls in order to simplify calculations. Indeed, some calculations may only be possible under certain assumptions that simplify the system. When simplifications are made, the idea is that the essence of the problem is only affected a little bit by these simplifications and that the main characteristics are maintained. In such cases, errors are thus added deliberately but we believe that we can ignore them.

In other cases, the effects of errors may not be negligible and one then must account for them. For example, in our deliberations in Section 18.4 above, we tacitly assumed that every bit Alice sends out is actually received by Bob. In real life, however, we know that transmission lines are less than perfect and hence that errors frequently occur.

In biological systems, the thermal noise at the cellular level is enormous. Indeed, it is probably fair to say that the cellular machinery is trying to carry out its functions in the equivalent of a thunderstorm. Even with the best of machineries, errors in such an environment will be unavoidable.

When there are errors, the first choice one needs to make is whether one should try to ignore or correct them. Sometimes, ignoring an error may be significantly less troublesome than fixing it. For example, if we hung up a picture in our room 1 cm higher than planned, we may just want to leave it that way rather than to drill a new hole and to fix and paint the old hole. If we receive junk mail, we may be much better off to simply dump it in the trash rather than getting into an enervating protracted argument with the mailer.

Closely related to this is the severity of the error. If an error is not too severe, it is much more likely that we can ignore it. However, if an error causes some serious problems, then we would certainly be inclined to correct it. For example, take the binary string for 32 given by 100000 and assume a 1-bit error occurs. If the error is in the first bit, the string changes to 000000 or, in decimal, to 0 — a radical change. However, if the error is in the last bit, we obtain 100001 or, in decimal, 33 — a much smaller change.

It should be noted that errors are not necessarily bad. Evolution, for example, would come to a grinding halt if all errors would be eliminated. Since the term "error" has negative connotations, the word "mutation" is more commonly used in this context. Yet in principle, it's exactly the same thing.

Therefore, we see that for a real-world system, two aspects need to be taken into account. First, when possible, most effort should be directed at avoiding severe errors and second, there should be a mechanism for error correction.

Of course, errors cannot be corrected unless there is also a mechanism for error detection. One of the simplest schemes for error detection in binary strings is that of adding a so-called parity bit. In this scheme, a bit with value 1 is added to the string if the number of ones in the string is even. Otherwise a bit with value 0 is added. With the help of a parity bit, single errors and uneven multiple errors can always be detected but not if there are an even number of multiple errors. Often, parity is good enough, but of course sometimes more effort is warranted.

When an error is detected, and correction is desirable, with a parity bit scheme the receiver needs to get back to the transmitter to ask for a resending of the string. This may not always be practical, and consequently methods have been developed to not only detect but also correct an error. One of the simplest effective ways to do so is by redundancy. If, for example, a string is sent three times in a row, then any single error can be corrected by means of a majority vote.

Another way to introduce redundancy is to create a code where multiple-bit combinations signify the same. In other words, if Alice would want to tell Bob which side of a coin she threw, they could agree that two or several different similar strings refer to the same side. For example, if they use a string of 5 bits, they could agree that a majority of 0s means a head and a majority of 1s a tail. If Alice then transmits five 0s to indicate that she threw a head, any two errors would not lead Bob to get the wrong message. Strings like 00000, 01010, 00100 etc. would all mean the same.

18.6 Information and the laws of physics

At first sight it would seem that information theory and physics are rather distinct subjects. Indeed, the laws of physics describe things like the speed of light, the motion of the planets or the principles of electricity. Although not formally part of the physical description, it should be noted, however, that none of these laws would be sensible without the notion of information, be it explicit or not. Especially when considered in the light of the special theory of relativity, it is clear that physical laws cannot do without a transfer of information. The motion of a planet can only be affected by a sun if there is information about the existence of the sun at the location of the planet.

In quantum theory, the link with information is much more direct as we can only "know" about the state of a system when we perform a measurement.

However, a striking link between physics and information theory comes from the unexpected field of statistical mechanics. We briefly review the concept of entropy as this provides a remarkable link to Shannon's formula for information.

Entropy and Boltzmann

The **entropy** S is a measure of the disorder of a system. In all natural processes, entropy remains constant or increases, as discussed in Section 3.8. For a system in equilibrium, its large-scale properties can be described by a few quantities such as its energy, temperature, volume and so on; these quantities define the macrostate of the system. Every **macrostate** corresponds to a large number of microscopic states of the system, called its microstates. Every physical system is made out of atoms (and molecules) and a classical **microstate** specifies the position and velocity of every atom.

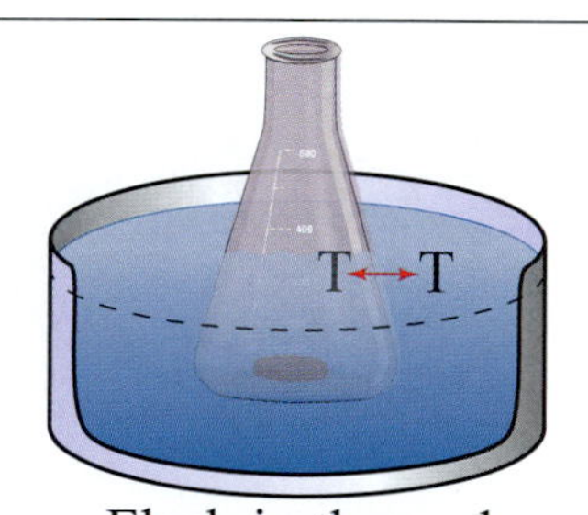

Flask in thermal equilibrium

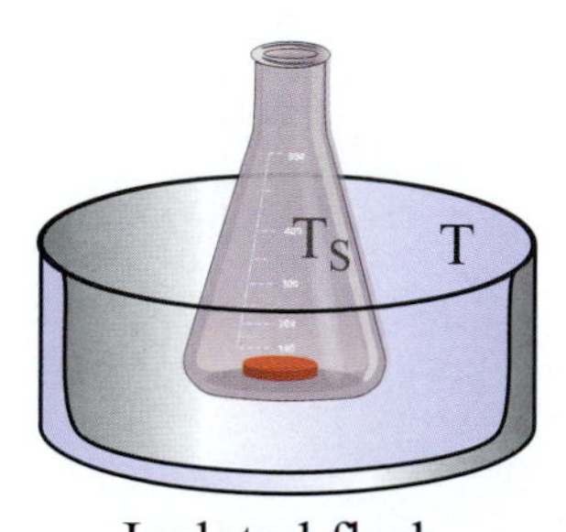

Isolated flask

Fig. 18.4: For an isolated flask there is no heat exchange.

For example, consider a gas with constant energy E and occupying a given volume V; the atoms composing the gas can be at many different positions and moving in different directions; every *specific* configuration and velocity of *all* the atoms in a gas is *one* microstate of the gas. The number of states $\Omega(E, V)$ — corresponding to the macrostate E, V — is equal to the *total* number of microstates. For a system in equilibrium, the Austrian physicist **Ludwig Boltzmann** related entropy of a system to the microstates of the system by the following equation:

$$S(E, V) = k_B \ln \Omega(E, V). \tag{18.5}$$

Boltzmann's formula for entropy holds for a perfectly isolated system in equilibrium. The only perfectly isolated system is the universe itself. If a system is not perfectly isolated and is in contact with the environment (a large system, sometimes called a heat bath), then for equilibrium the temperature of both the system and environment must be the same (see Figure 18.4). Let the system's microstates be denoted by $i = 1, 2, ...\Omega$. Let p_i be the probability for the system to be in the ith microstate and let E_i be its energy; the probability is given by the famous **Boltzmann distribution:**

$$p_i = \frac{1}{Z} e^{-E_i/k_B T} \; ; \; \sum_{i=1}^{\Omega} p_i = 1 \tag{18.6}$$

$$\Rightarrow \; Z = \sum_{i=1}^{\Omega} e^{-E_i/k_B T}, \tag{18.7}$$

where Z is the system's so-called **partition function** and one of the key constructs of statistical mechanics.

It can be shown that the (average) energy and entropy for the system is given by

$$E = \sum_{i=1}^{\Omega} p_i E_i \tag{18.8}$$

$$S(E, V) = -k_B \sum_{i=1}^{\Omega} p_i \ln p_i. \tag{18.9}$$

Equation 18.9 is of great generality and holds for systems in equilibrium as well as for systems that are far from equilibrium. For a perfectly isolated system in equilibrium, it is one of the *axioms* of statistical physics that *all* the microstates of such a system are equally likely; hence $p_i = 1/\Omega$ and this yields

$$S(E, V) = -k_B \sum_{i=1}^{\Omega} \frac{1}{\Omega} \ln\left(\frac{1}{\Omega}\right) = k_B \ln \Omega(E, V).$$

The entropy of an ideal monatomic gas — at temperature T and having N atoms each with mass m — is given by

$$S(E, V) = N k_B \left(\frac{5}{2} + \ln\left(\frac{V}{N}\right) + \frac{3}{2} \ln\left(\frac{m k_B T}{2\pi\hbar^2}\right)\right). \tag{18.10}$$

For a system at temperature T and obeying the Boltzmann distribution given in Equation 18.6, the definition of entropy given in Equation 18.9 leads to the following:

$$\begin{aligned} S(E, V) &= -k_B \sum_{i=1}^{\Omega} p_i \ln p_i = -k_B \sum_{i=1}^{\Omega} p_i \left(-\frac{E_i}{k_B T} - \ln Z \right) \\ &= \frac{E}{T} + k_B \ln Z \end{aligned}$$

where E is the (average) energy of the system. The free energy of the system F, introduced earlier in Section 3.8 as the amount energy of a system available for doing useful work, is obtained from

$$Z = e^{-F/k_B T}.$$

Hence, in terms of the macrostate of the system, entropy is given as

$$TS = E - F$$

and was presented earlier in Equation 3.43.

Shannon and entropy

What is surprising is that Equation 18.9 for entropy has exactly the same form as Shannon's formula for information (Equation 18.3):

$$S = -k_B \sum_{i=1}^{\Omega} p_i \ln p_i \text{ , Entropy Equation 18.9}$$

$$\text{Total number of bits} = -N \sum_{i=1}^{\mathcal{A}} p_i \log_2(p_i) \text{ , Information Equation 18.4.}$$

Note for entropy, k_B is the Boltzmann constant and Ω the number of possible different microscopic configurations (microstates). For the total number of bits required for transmitting a message, N stands for the total number of symbols in the message and $\mathcal{A}$ is the number of basic symbols used in the text of the message.

Similar to information as discussed after Equation 18.4, we can now understand why entropy is often considered the measure of "ignorance". When the entropy is large, the probability of the system being in one specific microstate or another is very small, and consequently our ability to "guess" its state is also very small. The less we can guess the state, the larger our ignorance; therefore, the larger the entropy, the larger our ignorance.

18.7 Algorithms

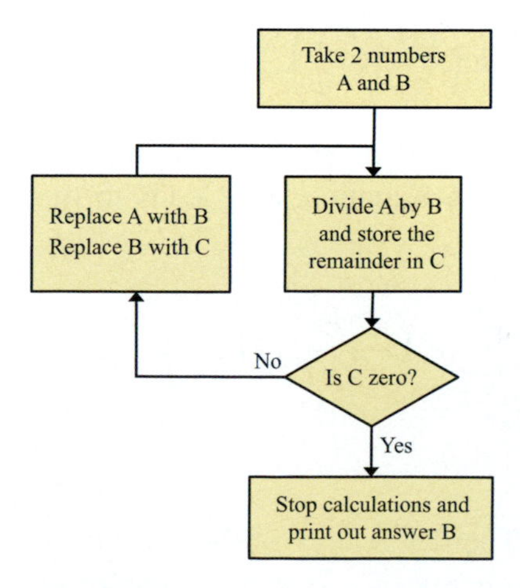

Figure 18.5: Flowchart for Euclid's algorithm.

Nowadays, computers and microprocessors are everywhere, from desktop or laptop computers to washing machines, from refrigerators to music-playing postcards. Indeed, computers and what they do, "computing", are so ubiquitous that we hardly pay attention to them. Not only that, they have become such an integrated part of our daily lives that many perceive the concepts and ideas related to computing as completely natural. Of course, as such, this is a good thing. However, it is even better to realize that it wasn't always like this.

Until not long ago (that is, the early twentieth century), computers were human beings who did numerical calculations for things like science, engineering or accounting. They did so by systematically carrying out a given number of calculatory steps. Such a collection of systematically carried-out steps to solve a certain problem is called an **algorithm**.

The word "algorithm" is probably a distortion of the name of the influential Persian mathematician **Al-Khwarizmi**. Interestingly enough, not only was his name given to this central notion in computer science, as he had written a treatise about algebraic methods, its Arabic title was distorted into the word "algebra" and thus his legacy also lives forth as a central part of mathematics. Al-Khwarizmi was the link between Indian mathematics and European mathematics since he used the Indian number 0 that was popularized in Europe when Fibonacci translated his work in the early thirteenth century.

One of the earliest and still best-known algorithms is **Euclid's algorithm**. It is used to find the greatest common divisor and works like this: Given two integer numbers a and b, obtain the remainder c of the division of a by b. If the remainder is not 0, relabel b to a and c to b and repeat the step of the previous sentence. Do so until the remainder is $c = 0$. When this is the case, the greatest common divisor of the original numbers is b.

The flowchart for Euclid's algorithm is depicted in Figure 18.5, while an example is given in Figure 18.6.

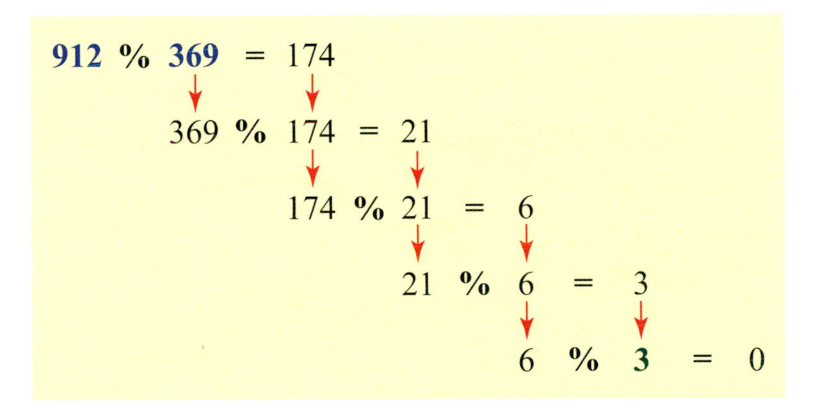

Figure 18.6: Example of Euclid's algorithm. The greatest common divisor of 912 and 369 is 3. (Note that % denotes the modulo operation as in for example the computer language C. The modulo operation gives the remainder of a division.)

Let us now have a look at an extensively studied "natural algorithm". Some bacteria like *Escherichia coli* (usually abbreviated as *E. coli*) prefer to metabolize glucose (as this is energetically favorable). However, if glucose is not available while lactose is, it can produce the enzymes necessary for catabolizing lactose (see also Section 17.11).

The algorithm employed by the bacterium can be described as in Figure 18.7.

At the center of the actual mechanism in the bacterium is the so-called ***lac* operon**. An **operon** is a strand of DNA that generally includes a promoter (a sequence of nucleotides that enables the structural genes to be transcribed), an operator (a sequence of nucleotides that regulates the activity of the structural genes) and one or more structural genes that are transcribed into mRNA to later be translated

IF	lactose present
AND	glucose not present
AND	cell can actually synthesize active LacZ and LacY
THEN	transcribe *lacZYA* from *lacP*

Figure 18.7: The algorithm implemented by the *lac* operon.

into proteins or enzymes.

In Section 17.11 we discussed how the *lac* operon *switches off* the production of proteins required for digesting glucose. We now discuss another function of the lac operon, namely how it *switches on* the production of lactose-digesting proteins.

Regardless of whether there is glucose or not, the *lacI* gene is always expressed and leads to the production of a repressor protein that then binds to the operator of the *lac* operon. When the repressor is bound to the operator, the transcription of the structural genes is severely impaired and proceeds only at a very low level. When, however, lactose is present, one of its metabolites called allolactose will bind to the repressor, changing its shape. With this changed shape, the repressor can no longer bind to the operator, and consequently the transcription of the structural genes increases.

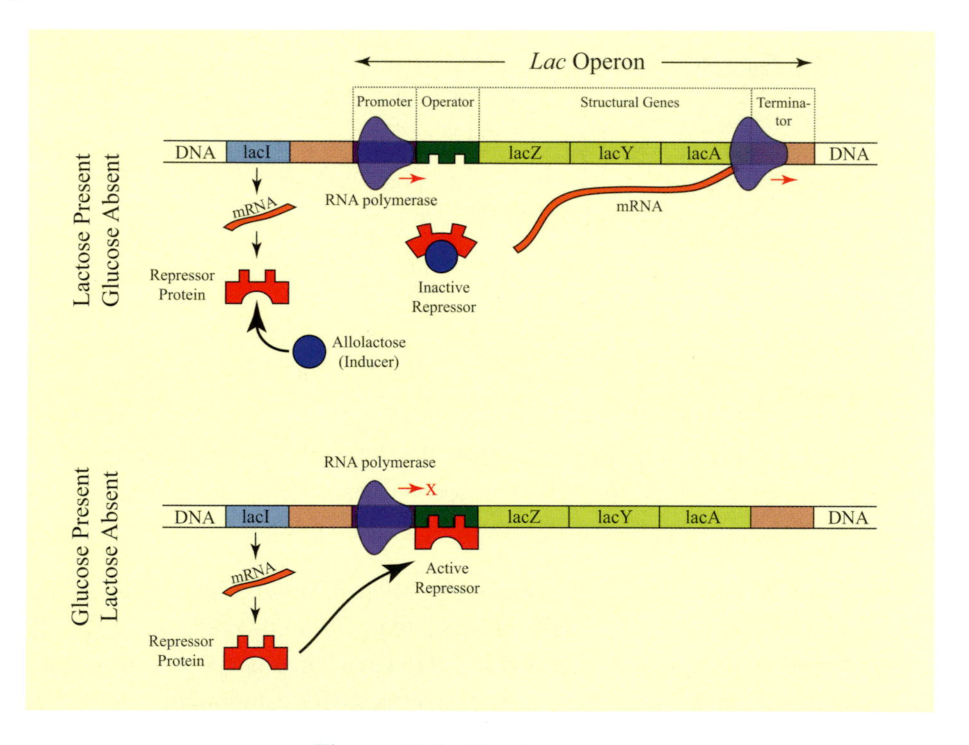

Figure 18.8: The *lac* operon.

Hence, when lactose is present, *lacZ*, *lacY* and *lacA* are expressed, but if glucose is also present they shouldn't be expressed to a large extent as glucose is preferred. So the question is, how is the distinction between absence and presence of glucose made? It turns out that a molecule called cAMP can drastically increase the transcription of the *lac* operon when it is bound to the promoter. Now when glucose is present, cAMP concentrations are low and hence transcription of the operon is low as well. But when glucose is absent, cAMP concentrations increase and along with it the production of lactose-catabolizing enzymes also increases.

We therefore see that the algorithm implemented is far from trivial. It is not a matter of a single if-then scenario.

18.8 Computation and Turing machines

Thus far, we have seen that certain kinds of problems, like determining the greatest common divisor, can be solved by applying an algorithm in a general process called **computation**. The earliest computers were human beings, but by their very nature algorithms are repetitive, and for more complicated problems they need to be applied over and over again. Not only is this boring, due to human nature, but human computers are also rather error prone.

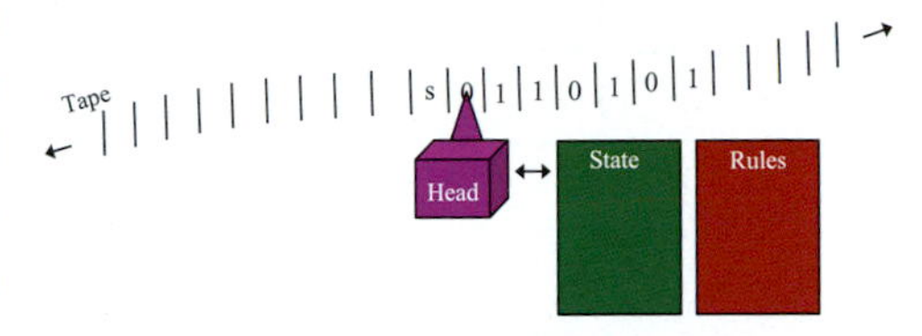

Figure 18.9: The basic design of the Turing machine.

Computation can be defined as finding a solution to a problem by means of an algorithm.

It was therefore realized early on that it would be of enormous benefit if a machine could be built that carries out the calculations of an algorithm. But of course, to build one machine for each algorithm is rather impractical, and one can therefore ask: Is it possible to have a machine that can carry out the calculations of various algorithms? And if such a machine exists, what would be the most general machine? We all know the answer to the first part of the question: Yes, there are machines that can carry out many types of calculations and apply many different types of algorithms, the prime example being the desktop computer. While the desktop computer is astoundingly versatile in its application, conceptually it's far from simple and, hence, is rather difficult to analyze scientifically. For answering the second part of the question, we have the notion of the so-called **Turing machine** that we discuss in some more detail now.

The design of the Turing machine is shown in Figure 18.9. As can be seen, as such, it's a very simple device but it will become clear below that its behavior is anything but simple. The Turing machine consists of the following five parts:

1. An **infinite tape** with readable and writable fields on it.
2. Each field has one of a finite number of **symbols** on it.
3. There is a **read-write head** that is positioned somewhere on the tape.
4. The machine is in one of a finite number of **states**.
5. There is a set of **rules** that determines what the machine will do given a certain symbol read from the tape and the state that the machine is in.

An example of a simple Turing machine for a binary counter is shown in Figure 18.10.

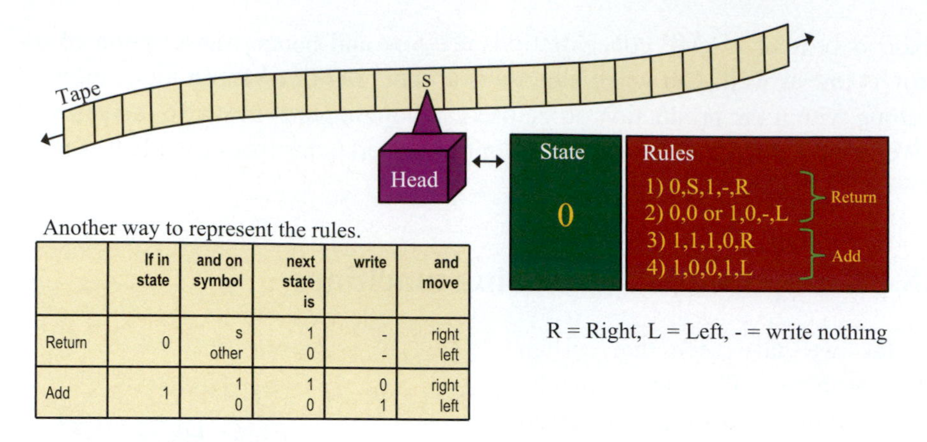

	If in state	and on symbol	next state is	write	and move
Return	0	s other	1 0	- -	right left
Add	1	1 0	1 0	0 1	right left

Rules are defined in this format: (current state, current symbol, new state, new symbol, left/right)

Figure 18.10: Turing machine for a binary counter.

The rules are defined in the following way: (current state, current symbol, new state, new symbol, direction of motion). A dash "-" means "do nothing", "R" stands for "move right" and "L" for "move left". Therefore rule 1 given by (0,S,1,-,R) means the following: If the current state is "0" and the symbol on the tape is "S", then change the state to "1", do nothing (i.e. do not write a new symbol to the tape) and move one step to the right.

Our Turing machine now has the state "1" and is pointing on the tape to symbol "0". To continue, we then look up the rule corresponding to this combination. As can be seen in the figure, that is rule 4. We then have the situation depicted in Figure 18.11.

Letting the machine run, it will create the following sequence of symbols: (0, 1, 01, 11, 001, 101, 011, 111, 0001,...), that is indeed 0, 1, 2, 3, 4, 5, 6, 7, 8 ... counted in binary. It should be noted that with the given rule set, the binary numbers need to read right to left rather than the other way around as is usually the case.

It is indeed rather surprising to see how a machine with such a simple design

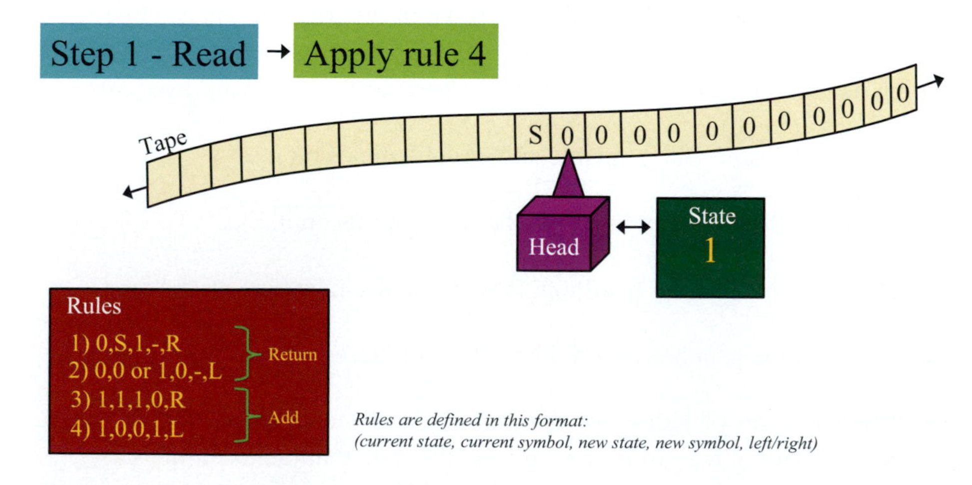

Figure 18.11: Turing machine for a binary counter after a few steps.

and such restricted movement can do a nontrivial task like counting binary with a set of only four rules. What is truly fascinating, however, is the fact that the Turing machine is not restricted to mathematically simple computations (binary counting, though nontrivial, is nevertheless mathematically a very simple operation). It can do a lot more, as we will see in the next section.

At this point, the striking similarity between the conceptual Turing machine and the genetic apparatus should be mentioned. In the genetic apparatus too, we have a tape (for example, mRNA), a read head (for example, ribosome), and a set of rules (for example, the genetic code discussed below). Parallels for other aspects of the machine like writing to the tape and having a state also exist in more indirect ways.

18.9 Universal computing

A universal Turing machine can in principle compute anything for which an algorithm exists.

Of course, a disadvantage of a Turing machine like the binary counter described in the previous section is its specialization. It can only do one thing. For multiplication, one would need a new set of rules with associated symbols and states. A multiplication Turing machine is a lot more complicated than the binary counter, but it can be done. All in all, this is not a very satisfactory state of affairs and one should therefore wonder whether it is possible to design a Turing machine that can fulfill a multitude of operations.

The fact that such a Turing machine can be designed should not be all too surprising. After all, one can imagine designing a machine that simply combines sets of rules corresponding to certain tasks and using an extra set of rules with associated symbols and states to discriminate between the tasks.

Now that we have a machine not only for one but also for many tasks, the next logical step is naturally to wonder whether there is a machine that can do *all* tasks. The stunning answer is "yes", namely the **universal Turing machine**. We do need to be a little bit careful, though, as to the meaning of "all tasks". Clearly, a universal Turing machine cannot break the laws of nature or revive the dead. With all tasks, the following is meant: all calculations or computations for which an algorithm exists.

The celebrated Church-Turing thesis proves exactly this, namely that a suitably designed Turing machine can carry out any computation for which an algorithm exists.

The thoughts in the previous paragraph should immediately remind us of the personal computer, and indeed the personal computer is an example of a universal Turing machine. So far so good; everyone who has used a computer knows that it can do a large number of things, but doesn't the Church-Turing thesis imply that the same can also be said for a tape with a read-write head, some memory and a set of rules? Yes, it does! Stronger still, a consequence is that **all universal Turing machines are equivalent in what they *can* compute**.

Therefore, one could implement an e-mail program on a tape ... but would this be practical? Almost certainly not! The notion of a universal Turing machine is a conceptual one; it tells you what in principle can be done, not whether this is practical or not. For practical purposes, specific designs that fulfill requirements like being fast enough will be chosen. In fact, except for research purposes, the authors are not aware of any large-scale direct use of Turing machines à la Figure 18.9, even though *universal* Turing machines (computers and chips with similar functionality) are ubiquitious.

A universal Turing machine consisting of a tape with read-write head, a state and some rules is a conceptually simple device. As an actual machine, it lacks two aspects that are rather common in the natural world: parallelism and direct interaction of units. That is to say, in the natural world we have many units like cells that temporally evolve at the same time (that is, in parallel) and whose temporal evolution is strongly influenced if not determined by its interaction with other units.

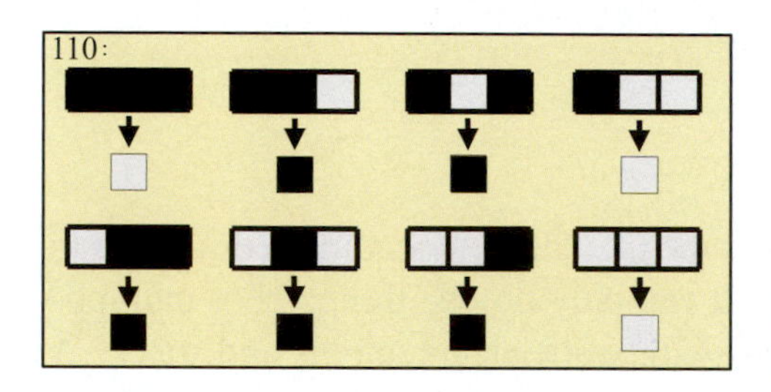

Figure 18.12: Rule 110.

One could then ask: What is the simplest system with interacting units that is capable of universal computing? Rather surprisingly, it can be shown that the elementary cellular automaton with rule 110 shown in Figure 18.12 is a universal Turing machine. It only evolves toward one direction when starting with a single black cell (namely the left hand side; the right hand side is always white) and has some regular-looking patterns on the left edge as can be seen in Figure 18.13 (more details on the what cellular automata are can be found in Section 20.4).

The reason why computations can be done is that rule 110 also generates moving structures that can interact, thus fulfilling the role of a logic gate.[1] The question, of course, has to be how those structures can be made use of in a program. Since rule 110 determines the evolution of the cellular automaton entirely from step 1 onwards, to program the automaton one needs to choose the colors of the cells in the initial row. In this case, one would not start with a single black cell but with a suitably chosen arrangement of black and white cells.

It should be stressed that the difference between rule 110 and a Turing machine with a tape is only one of implementation. In terms of theoretical computational power, all universal Turing machines are equal. This does not, however, imply that they are all equally suitable as paradigms for analyzing given classes of systems like cells, solids or microprocessors.

The genetic apparatus can be considered as a species of universal Turning machine.

What is important in the context of natural systems is that rule 110 shows that a given set of interacting units like cells *can* be capable of complex and complicated computations even if the evolution rule and the interactions are rather simple.

So what does this basically mean? If you have some tokens, some rules and some memory, you can in principle construct a machine that can implement any algorithm.

Wouldn't reproduction or indeed life itself be some kind of an algorithm? If so, clearly, DNA could be regarded as a kind of memory, but where are the rules? In fact, there are rules at many levels (for example, rules for the sequence of amino

[1]A logic gate is a construct that takes several inputs and produces a single logic output. For example, in an "and" logic gate, if both the inputs evaluate to a "true", then the output will be "true".

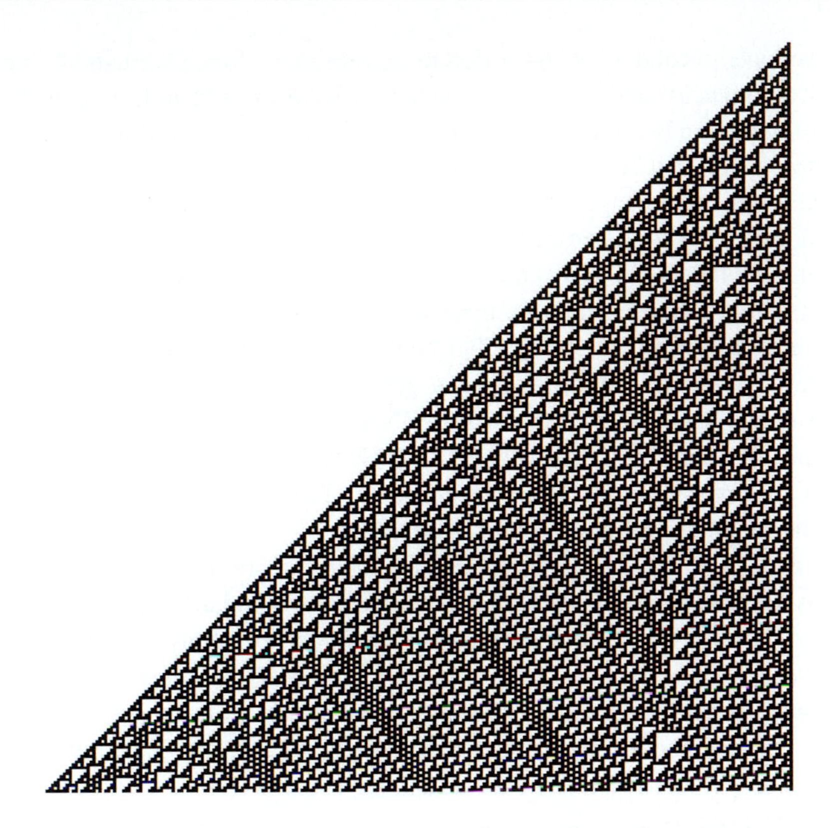

Figure 18.13: Rule 110 can be used as a universal computer.

acids in proteins, rules for cell division and so on), but in the next section we will consider one of the most basic sets of rules, namely, the set of rules that governs how a sequence of nucleic acids in DNA is translated into a sequence of amino acids in proteins.

18.10 Genetic code

Already around the time that Watson and Crick revealed their model of DNA in 1953, it was known that there must be some kind of nontrivial rule set involved in the translation from DNA to protein. DNA consists of four different types of nucleotides, and proteins usually consist of 20 different types of amino acids. Clearly, there cannot be a one-to-one correspondence between nucleotides and amino acids. A (short) sequence of nucleotides must hence combine into words called **codons** with the nucleotide types the letters of its alphabet.

How long are these words? Are they all of the same length? From the outset, we know that combinations of two nucleotides to specify one amino acid would still be insufficient since this would maximally only allow for $4^2 = 16$ (the number of different types of nucleotides to the power of number of nucleotides in the combination) correspondences. Therefore, at the very least, one needs three nucleotide codons or variable-length codons whose maximum length exceeds two nucleotides.

Life has chosen to use fixed-length codons, consisting of three nucleotides, and hence we have in total $4^3 = 64$ different possibilities. Since we usually have 20 different amino acids and it is reasonable to account for a start and a stop codon, we therefore have nearly three times as many possibilities as are minimally required. Would this mean that some three-nucleotide combinations are never used? After all, in the English language, many of the possible letter combinations do not make valid words. Or perhaps, some amino acids are coded for by different words? That turns out to be the case and this redundancy is likely not accidental.

It turns out, however, that the three-nucleotide combinations in DNA are not directly translated into amino acids but go through an intermediate step. First, the sequence of nucleotides that describe a gene in DNA is transcribed (copied) to an RNA strand, the so-called messenger RNA or mRNA, and then this mRNA is translated (decoded) into amino acids in the ribosomes that read the mRNA three letters at the time. DNA and RNA are made up of the same nucleotides except for DNA's thymine being replaced by uracil in RNA.

The table with the standard genetic code as it occurs in, for example, human beings is shown in Figure 18.14. It is called the "standard" genetic code as it applies to the vast majority of current life-forms and probably hasn't changed in the last several billion years. As can be seen, some amino acids like leucine are coded for by as many as six different codons while others, like tryptophan, are only coded for by one codon. As there is no apparent pattern, the question that arises is whether this code is the result of a random event or is it the end result of an evolutionary process?

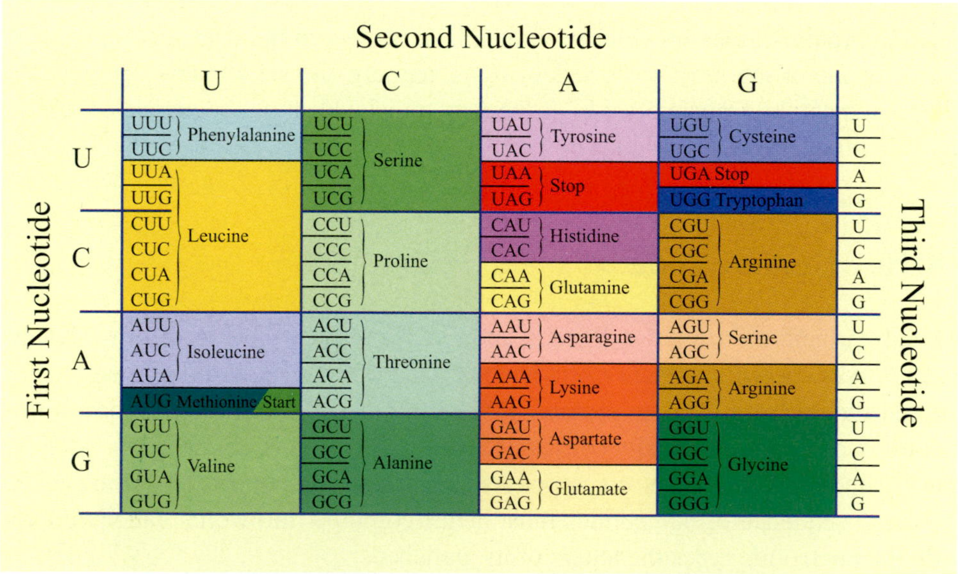

First Nucleotide	Second Nucleotide: U	C	A	G	Third Nucleotide
U	UUU Phenylalanine	UCU Serine	UAU Tyrosine	UGU Cysteine	U
	UUC	UCC	UAC	UGC	C
	UUA Leucine	UCA	UAA Stop	UGA Stop	A
	UUG	UCG	UAG	UGG Tryptophan	G
C	CUU	CCU Proline	CAU Histidine	CGU Arginine	U
	CUC	CCC	CAC	CGC	C
	CUA	CCA	CAA Glutamine	CGA	A
	CUG	CCG	CAG	CGG	G
A	AUU Isoleucine	ACU Threonine	AAU Asparagine	AGU Serine	U
	AUC	ACC	AAC	AGC	C
	AUA	ACA	AAA Lysine	AGA Arginine	A
	AUG Methionine Start	ACG	AAG	AGG	G
G	GUU Valine	GCU Alanine	GAU Aspartate	GGU Glycine	U
	GUC	GCC	GAC	GGC	C
	GUA	GCA	GAA Glutamate	GGA	A
	GUG	GCG	GAG	GGG	G

Figure 18.14: The standard genetic code.

The answer to this question may lie in error analysis. As we have seen in Section 18.5 above, errors can be dealt with by building redundancy into a system. Furthermore, errors that cause severe problems need to be suppressed more strongly than errors that cause more minor damage. Indeed, in an evolutionary context, some errors are necessary. But the probability of an error (mutation) being beneficial is inversely related to the impact of the error on the system.

We need to ask what types of errors need to be considered as "grave" and what types of errors as possibly "benign". One characteristic to consider is how the error affects the hydrophobicity of the resulting protein. A change of a hydrophilic to a hydrophobic amino acid can have drastic consequences, as in the case of sickle cell disease. In sickle cell disease, a one-nucleotide mutation resulting in a change of a certain GAG codon to a GUG codon — that is, a single change in a single GAG codon with other GAG codons in the same gene not necessarily having this mutation — results in normal hemoglobin to turn into sickle cell hemoglobin.

Indeed, for the standard genetic code, single-nucleotide errors often lead to amino acids that are similar as with regards to hydrophobicity, and consequently the mutation may not be harmful in many instances. As a result, it is rather reasonable to assume that the genetic code is the product of an evolutionary process and highly optimized as with regards to its function.

18.11 Brain

Thus far we have mainly looked at what information is, how it can be described and its relation to the laws of physics. We have also seen how information can be processed with the help of the genetic code. However, thus far we have not considered large-scale information processing. In human beings, as is the case for animals in general, the central processing unit is the **brain**. Let us therefore have a brief look at it.

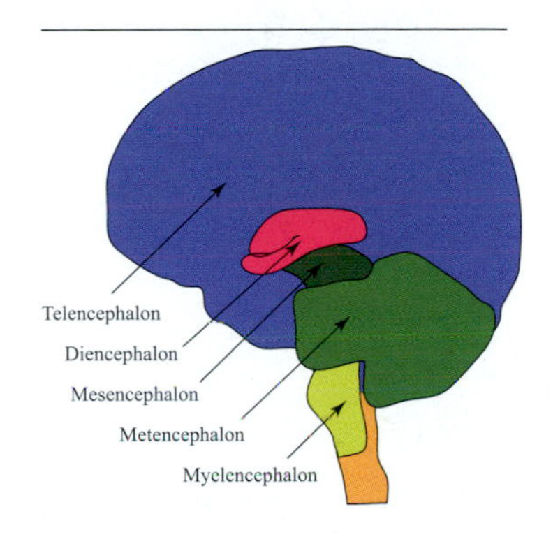

Fig. 18.15: The brain is often considered to consist of five major divisions as labeled.

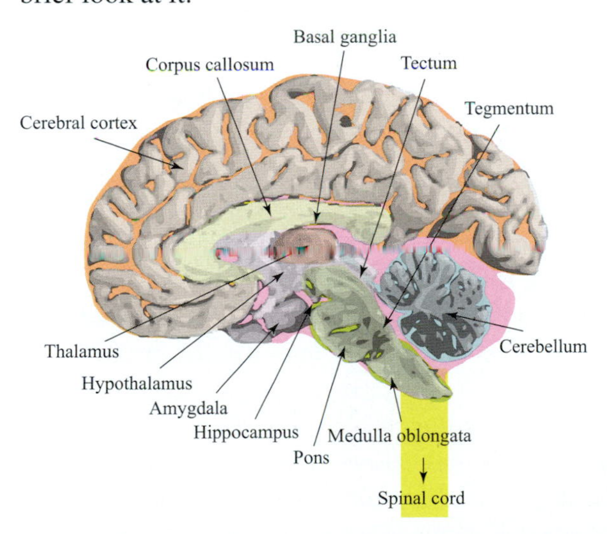

Figure 18.16: Approximate regions of the human brain.

As such, brains are extensions of the spinal cord that started to evolve more than 500 million years ago and, more obviously, after the emergence of multicellular life about 1 billion years ago. Although the variations can be quite large, the structures of vertebrate brains are all quite similar (in contrast to, for example, octopuses and crustaceans) and consist of two halves called hemispheres. Roughly, a human brain weighs about 1.3 to 1.4 kg and can be considered as having five major divisions as illustrated in Figure 18.15: telencephalon, diencephalon, mesencephalon, metencephalon and myelencephalon. Sometimes, the telencephalon and diencephalon are grouped together as the **forebrain**, and the metencephalon and myelencephalon are the **hindbrain**. The mesencephalon is also referred to as the **midbrain**.

Hargobind Khorana

Figure 18.17: Hargobind Khorana.

Hargobind Khorana was awarded the Nobel Prize in Physiology or Medicine (shared with Robert W. Holley and Marshall Warren Nirenberg) in 1968 for his work on the interpretation of the genetic code and its function in protein synthesis. Codons are the codes in DNA for synthesizing proteins. The fact that codons consist of three DNA nucleotide bases was first demonstrated by Francis Crick and Sydney Brenner in 1961, the same year that pioneering experiments on the codon were carried out by Marshall Nirenberg and Heinrich J. Matthaei. Subsequent work by Hargobind Khorana identified the rest of the code, and shortly thereafter Robert W. Holley determined the structure of transfer RNA, the adapter molecule that facilitates translation. With this, Khorana and his team established how the information required for protein synthesis is encoded in DNA, namely is spelled out in codons that are three-letter words: a set of three nucleotides constitute a single codon, which is the code for a specific amino acid. Khorana made another contribution to genetics in 1970 when he and his research team were able to synthesize the first artificial copy of a yeast gene.

Khorana was born near the village of Multan in British India. Although poor, his father was dedicated to educating his children and his was practically the only literate family in the village inhabited by about 100 people. He was home schooled by his father; he later finished his high school in Multan, where Ratan Lal, one of his teachers, influenced him greatly. Khorana earned his BSc in 1943 and MSc (in 1945) from Punjab University at Lahore under the supervision of Mahan Singh. He earned his PhD from the university in 1948 and Roger J. S. Beer was his supervisor.

Khorana spent two years (1950–1952) at the University of Cambridge where he started his research work on proteins and nucleic acids. In 1952 he went to the University of British Columbia, Vancouver, and in 1960 moved to the University of Wisconsin-Madison. In 1970 Khorana became the Alfred Sloan Professor of Biology and Chemistry at the Massachusetts Institute of Technology where he worked until retiring in 2007 and at present is Professor Emeritus at MIT.

Khorana's current work is on the molecular mechanisms of visual transduction in the vertebrate photoreceptor cells and explores the molecular mechanisms underlying the cell-signaling pathways of vision in vertebrates. The system has evolved to sense single photons and at the other extreme cope with millions of photons in strong light. The biochemistry in the rod cell ultimately causes closing/opening of the conductance channels in the plasma membrane. This results in hyperpolarization of the cell, activation of the synapses to the subsequent sets of cells in the retina and, eventually, in a signal to the brain. Khorana's studies are concerned primarily with the structure and function of rhodopsin, a light-sensitive protein found in the retina of the vertebrate eye, in particular, the mutations in rhodopsin that are associated with retinitis pigmentosa, which causes night blindness.

Khorana's research has always taken an experimental approach and has been interdisciplinary, including chemistry, biochemistry, molecular biology and biophysics.

Cerebral cortex	Language, thought, reasoning, voluntary motor control
Corpus callosum	Communication between brain halves
Basal ganglia	Movement coordination
Hippocampus	Memory formation
Amygdala	Information processing related to long-term memory
Thalamus	Information relay
Hypothalamus	Body temperature, emotions, hunger, thirst
Tectum	Sensory-information processing
Tegmentum	Motor functions, awareness
Pons	Breathing, arousal, sensory-information relay
Cerebellum	Movement, balance, posture
Medulla	Breathing and heart-rate control

Table 18.4: Key functions of major brain regions.

The five major divisions can further be subdivided as illustrated in Figure 18.16. For the **telencephalon** we have the **cortex**, **basal ganglia**, **hippocampus** and **amygdala**. For the **diencephalon** we have the **thalamus** and **hypothalamus**. For the **mesencephalon** we have **tectum** and **tegmentum**. For the **metencephalon** we have the **pons** and the **cerebellum**. For the **myelencephalon** we have the **medulla**.

Some of the key roles of the various brain parts are listed in Table 18.4.

Information processing in the brain is carried out with the help of neurons and to some degree glia cells. Glia cells mostly function to support the role of neurons by, for example, holding them in place and supplying nutrients and oxygen. However, they also may influence the signal transmission between neurons and may therefore play an important role in the overall processing taking place in the brain. It is notable that the number of glia cells is far larger than that of neurons (roughly by a factor 10). There are many different types of neurons and their location is not exclusive to the brain as they can also be found in, for example, the spinal cord or nerves. However, all neurons are involved in signaling — they are information messengers. The basic structure of a neuron is illustrated in Figure 18.18.

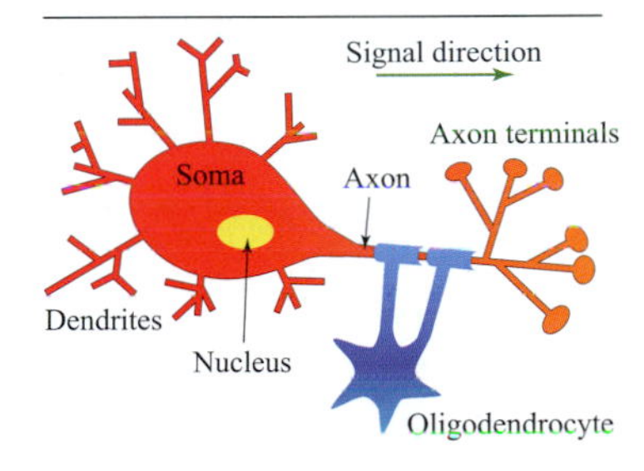

Fig. 18.18: Structure of a typical neuron in the brain.

Neurons consist of three different parts: the cell body (called **soma**), an axon with axon terminals at the end and dendrites. The **axon** looks a bit like a (possibly rather long) tail and is used to bring the outgoing signals of the cell to the axon terminals where the signal is transmitted to other neurons. The **dendrites** look like the branches of a tree and are attached to the soma. They receive the incoming signals from other neurons. In general, the connections between neurons are many to many. That is to say, a neuron will receive inputs at its dendrites from several different neurons while sending its output via the axon terminals to several neurons as well. The resulting neural network is rather complex, and its structure also depends on the region of the brain, even though the neurons themselves are not particularly complex cells when compared to most other cells in the body (for a simple neural network model, see Section 20.5). The signal traveling in the axon is an electrical impulse that makes it very fast but also susceptible to leakage. A special type of glia cells called **oligodendrocyte** has extensions that form insulating myelin sheets around the axon to prevent such electrical leakage (myelin is an

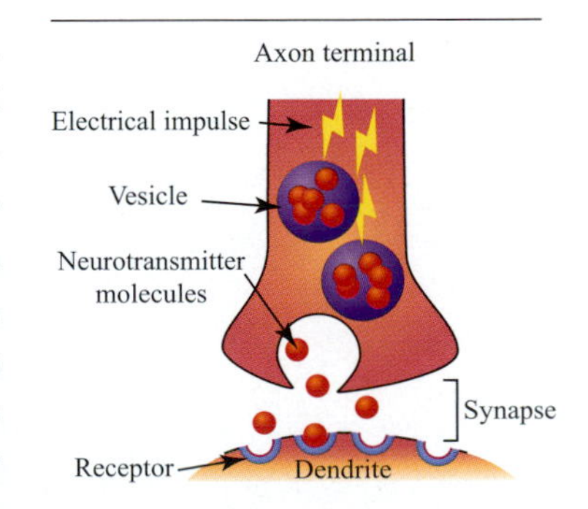

Fig. 18.19: Structure of a synapse.

	Ant Colonies	Neurons in Brains
Number of units	High	High
Robustness	High	High
Connectivity	Local	Local
Memory	Short term	Both: short and long term
Stability of connections	Weak	High
Global interactions	Trails	Brain waves
Complex dynamics	Exist	Common

Table 18.5: Comparing ant colonies and brains.

electrically insulating material made up of about 80% lipids and 20% proteins). The total length of myelinated nerve fibers in the brain is thought to be around 150,000–180,000 km!

The area where an axon terminal meets a dendrite is called a **synapse** and the communication is carried out with the help of molecules called **neurotransmitters**. The basic design of a synapse is illustrated in Figure 18.19. The human brain contains about 100 billion neurons and trillions of glia cells. The number of synapses in the cortex is even larger, estimated to at 0.15 quadrillion!

Ants and brains

As such, when considering ants, one may not readily associate them with brains. Although ants do have brains, they are very small indeed, roughly ranging from 10,000 to 100,000 neurons. The interesting comparison is not between the brains of ants and the brains of mammals but the comparison between the ant colony and the neuron collective that makes up a mammalian brain. Neurons by themselves are neither intelligent nor capable of carrying out complex tasks such as motor coordination. What turns a "skull-full" of neurons into a brain is the intricate network of connections. Similarly, in many ways, a single ant is not an ant at all. Ants can only be what they are as members of a collective.

Table 18.5 shows some of the key similarities and differences between brains and ant colonies.

Individually, ants do not gather, store or process much information, but through interactions the collective response of the colony can nevertheless be quite complex. It is almost as if the ant colony is a single organism. Of course, there are also many differences between ant colonies and brains, the most significant perhaps being the lack of a long-term memory and the weakness of the interaction between ants. Ants communicate with each other with the help of (volatile) chemical markers that they deposit in the environment — for example, when marking a trail. As a consequence, the interaction between ants is not very strong, and furthermore, due to the evaporation of communication molecules, not very long lasting. Neurons, on the other hand, are wired in a network with extremely long-lasting synaptic connections that facilitate strong interactions between linked neurons.

What we can see, however, is that information can be processed in a variety of ways, not necessarily in a centralized fashion. It may be fascinating to contemplate the consequences of considering human beings as ants and human society as an ant colony. Can human society, as a whole, be viewed as an organism with its own intelligence?

18.12 The answer

We started out this chapter by noticing how much children resemble their parents. Without knowing anything about the exact mechanisms, by logical deduction we concluded first, that there must be some information transfer and second that the information must be processed somehow. Since the common idea is that DNA stores genetic information, we explored whether DNA can also compute.

To get a grip on the question, we first ventured into investigating what information actually is and discovered the deep and subtle links between entropy, the number of tokens needed to transmit a message and the notion of information. When studying computation, we discovered that the personal computer as well as the genetic apparatus can both be considered "species" of the universal Turing machine, and hence in principle capable of carrying out any type of computation for which an algorithm exists.

While the parallels between the abstract notion of a universal Turing machine and the biological machinery that produces proteins are rather striking, DNA by itself is but part of this machinery. It seems, therefore, fair to state that the answer to our question "Does DNA compute?" should be no. However, what is more important here is that **DNA does not compute but, rather, is part of a computational process**. This computational process involves measurable and quantifiable information and algorithms that are executed.

18.13 Exercises

1. Write the number 2,500 in binary.
2. How many bits does one need to encode a position in a string of 200,000 symbols?
3. Suppose you are given a five-sided biased die with a probability of 1/8 to obtain a 1 or 2, and a probability of 1/4 to obtain a 3, 4 or 5. What is an optimal code for transmitting the throws of this die?
4. Given the fact that the four types of nucleotides in mRNA only need to code for 20 amino acids and furthermore a stop sign, is it possible to design a code that is shorter than the standard genetic code? If so, how could this be done?
5. Can a personal computer be considered a universal Turing machine?
6. Are there any amino acids that are effectively only coded for by two nucleotides? (That is, the third nucleotide can be ignored.)
7. Design a Turing machine that can write the word "Life" on its tape.

8. What is the function of allolactose?
9. Why was a base 2 logarithm chosen in Equation 18.3?
10. If I flip a coin six times, how likely is it to obtain exactly three heads and three tails?

19: Nanoworld

Q - Why Can a Gecko Climb a Wall?

Chapter Map

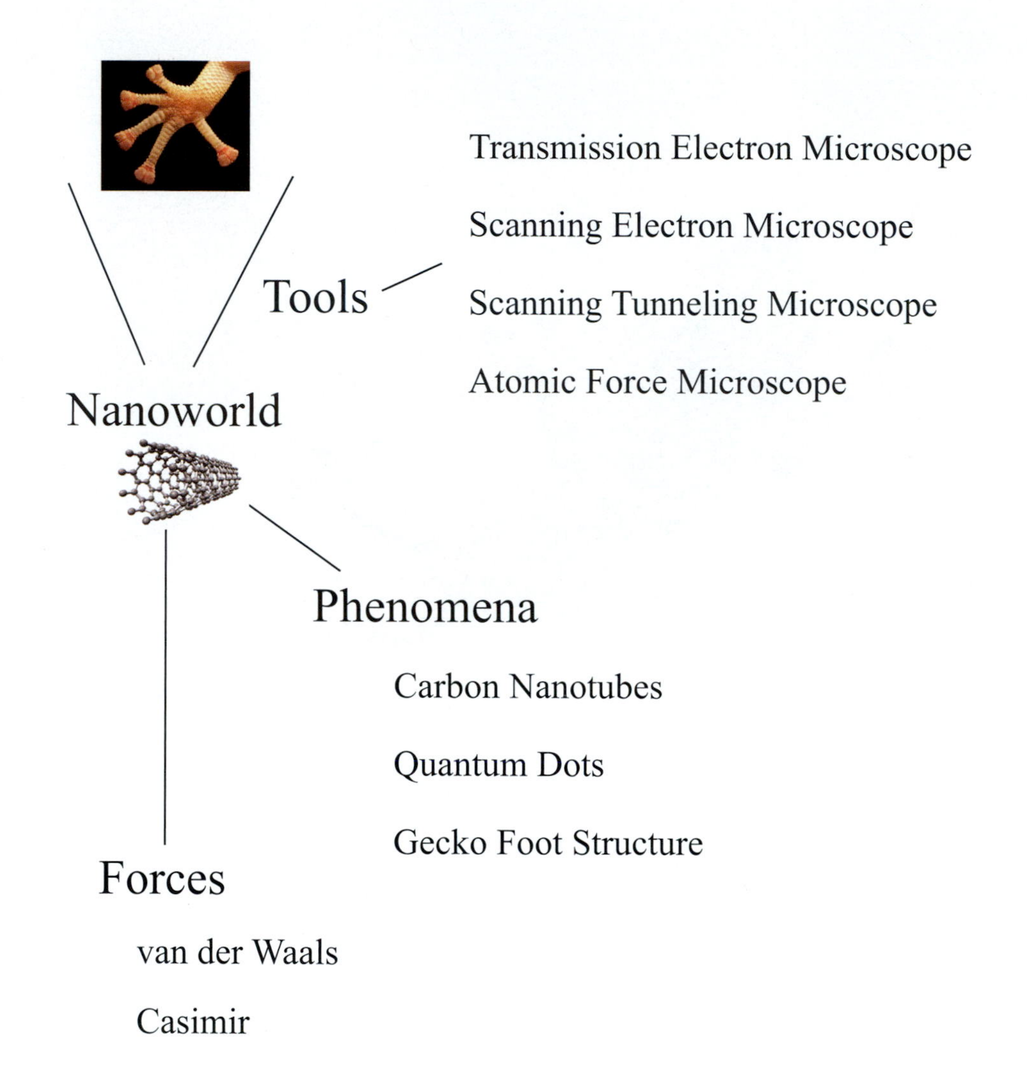

Tools
Transmission Electron Microscope
Scanning Electron Microscope
Scanning Tunneling Microscope
Atomic Force Microscope
Nanoworld
Phenomena
Carbon Nanotubes
Quantum Dots
Gecko Foot Structure
Forces
van der Waals
Casimir

19.1 The question

Figure 19.1: Geckos can climb on all kinds of surfaces.

Fig. 19.2: Geckos can easily climb walls.

The gecko belongs to the lizard family and inhabits the tropical and subtropical areas of the world. Geckos weigh about 50 to 150 g. They are nature's supreme climbers, racing on smooth surfaces at a meter per second. What makes geckos so exceptional is that they have specialized toe pads that enable them to climb vertical walls and walk upside down on ceilings as shown in Figures 19.1 and 19.2. It has been shown that the gecko has the same adhesion to a great variety of surfaces, including surfaces that are under water or in a vacuum.

If one examines the trail of a gecko one sees it is completely dry; the gecko does not use any kind of "glue" or suction pads for climbing walls and ceilings.

The question is: What is the force that allows the gecko to walk on walls and ceilings without falling down? Since no chemicals or fluids are involved, the force must be due to the physical interaction of the gecko's feet with the atoms and molecules of the wall. As discussed in Chapter 5, a solid material consists of bonded atoms and molecules that have a separation of about 0.1–10 nm (*nanometer*), where 1 nm = 10^{-9} m.

We explore the nature of the force holding the gecko to surfaces. We will discover that the force being used by the gecko is generated by nanometer-scale interactions of the gecko's feet with the material of the wall. The study of the gecko's remarkable abilities leads us to an interesting realm of phenomena, namely, the nanoworld — a scale that is indeed very small.

19.2 How small is the nanoworld?

The term *nanoworld* refers to phenomena (processes and structures) that have a characteristic size from 1–100 nm. The term *mesoscopic* is also used for the nanoworld and points to phenomena that have a scale between the microscopic world of atoms and the macroscopic world of everyday life.[1] A nanoscale object is a large collection of atoms and molecules, containing anywhere from 10–10^6 atoms or molecules.

All living systems carry out all of their cellular operations essentially using nanomachines and nanoprocesses. Multicellular life is a manifestation of a multitude of nanostructures and processes at the cellular level. Large collections of atoms and molecules are involved in the organization, production and operation of all living systems. For example, DNA is a molecular nanoscale structure that is the storehouse of genetic information; the process of protein synthesis and the storing of biological energy in ATP are both nanoscale processes.

[1]The term *scale* is used for indicating the characteristic physical size of a phenomenon.

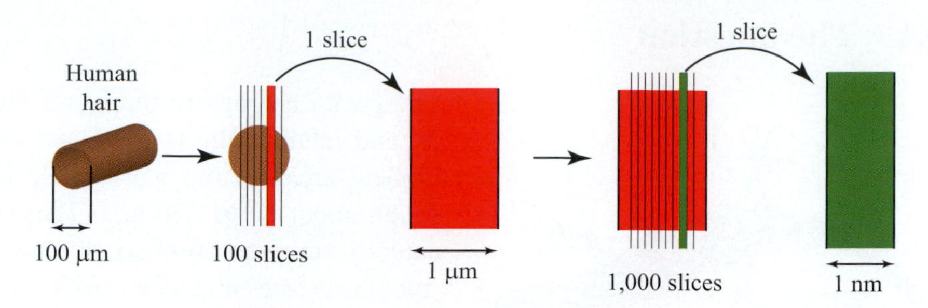

Figure 19.3: The scale of the nanoworld.

Note that the typical minimum size of a "large" object that we can directly perceive using our five senses is around 0.1 mm = 10^{-4} m. To get a feel for the size of a nanometer, consider slicing a human hair. A strand of human hair has a diameter of 10^{-4}m. Slicing the hair into 100 equal strands yields a strand of width 10^{-6} m; take a resulting single strand and again slice it 1,000 times; the final strand has a width of 1 nm, as shown in Figure 19.3. The typical separation of atoms in a gas is about 10^{-7} m = 100 nm and that of two atoms in a solid or a liquid is about 1 nm. The size of an atom is typically about 10^{-10} m = 0.1 nm.

19.3 Electron microscopes: exploring the nanoworld

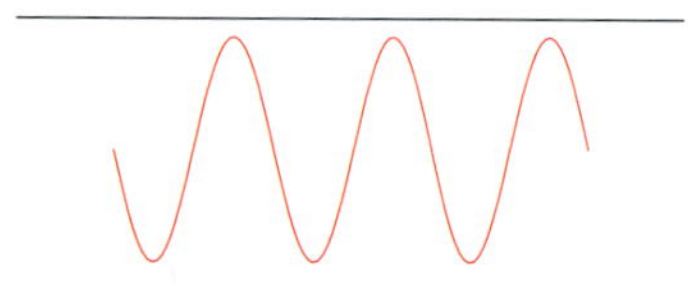

Fig. 19.4: Wave for the electron.

Science is an empirical discipline founded on experimental observations of nature. Nanotools, which are experimental instruments that allow us to look at objects that have a size of about 1–100 nm, are essential for studying the nanoworld. It is primarily due to breakthroughs made in the experimental tools for studying mesoscopic phenomena that the realm of the nanoworld has become accessible to scientific studies. To answer our questions regarding the nanoworld, in particular to understand how the gecko walks on a ceiling, we need apply our nanotools on the foot of the gecko.

Optical microscopes are inadequate for studying the nanoworld since the smallest object that can be viewed using visible light has a size of about 10^{-7}m, which is many times larger than the typical scale of nanophenomena. Using x-rays or shorter wavelength radiation to study nanostructures is not very useful since the these tend to damage the material being studied.

Since optical microscopes are inadequate for studying nanoscale phenomena, what other means do we have for creating images of the nanoworld? The counterintuitive world of quantum mechanics, discussed in Chapter 24, comes to the rescue. An electron, when it is observed, is a particle and has a definite position. What is the state of the electron when it is *not being observed*? Quantum mechanics gives a rather counterintuitive answer: The unobserved electron is in a transempirical state that has the likelihood of being at many points *simultaneously*; the probability of finding the electron at different points is given by the Schrödinger wave function.

The extent to which the transempirical state of the electron is spread out in space is given by the electron's de Broglie wavelength λ_e as shown in Figure 19.4. For an electron having a momentum $p = m_e v$, the de Broglie wavelength is given by

$$\lambda_e = \frac{h}{p} = \frac{h}{m_e v}.$$

For an electron with velocity $v = 6 \times 10^6$ m/s, the de Broglie wavelength is $\lambda_e = 0.12$ nm. For an electron moving at a velocity much less than that of light, the kinetic energy is given by

$$\text{Kinetic Energy} = \frac{1}{2} m_e v^2.$$

The energy of a photon (light) with wavelength λ, for velocity of light denoted by c, is given by

$$E = \frac{hc}{\lambda} = \frac{1240}{\lambda} \text{eV nm}.$$

The mass of the electron is 0.511 MeV/c^2. For an electron with kinetic energy equal to 1 eV, the associated de Broglie wavelength is 1.23 nm, about 1,000 times smaller than the wavelength of a 1-eV light particle (photon) having a wavelength of 1240 nm $= 12.4 \times 10^{-7}$m, which is a bit outside the visible spectrum of light.

We have the remarkable result that electrons can behave like a (transempirical) wave. The wavelike behavior of the electrons can be used to form images, just like the wave properties of light are used for making an optical microscope. As we saw above, 1-eV electrons can form images of distances that are about 1,000 times smaller than a light particle with the same energy.

This range of fairly low energies for electrons is not damaging to most materials and provides a nondestructive probe for imaging nanometer phenomena. One can use a stream of electrons, just like one uses a stream of light (particles), to scatter off the material being examined and form images similar to the optical microscope.

The electron microscope was invented in the 1930s and could only be focused on samples very close to the electron source and hence was not very useful. It was further developed during World War II but it was only by the 1980s that many practical problems of focusing were solved, leading to the current-day transmission electron microscopes.

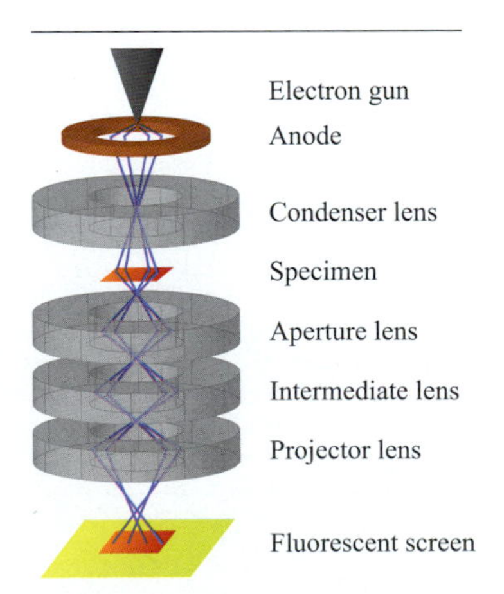

Fig. 19.5: Design of a transmission electron microscope.

Transmission electron microscope

The transmission electron microscope (TEM) shines electrons on a specimen and reconstructs the image of it by collecting the electrons that go through it, as shown in Figure 19.5. The TEM is a very useful device for the nanoworld and can resolve distances of up to 10^{-11}m; the best resolution is equal to 5×10^{-11}m. The TEM can penetrate through a material of a size up to 100 nm and a layer-by-layer image can be built up. Samples thicker than 100 nm are damaged due to heating caused by the electrons.

Scanning electron microscope

The scanning electron microscope (SEM) is similar to the TEM, except that the electron beam in this case scans only the surface of the material, much like how

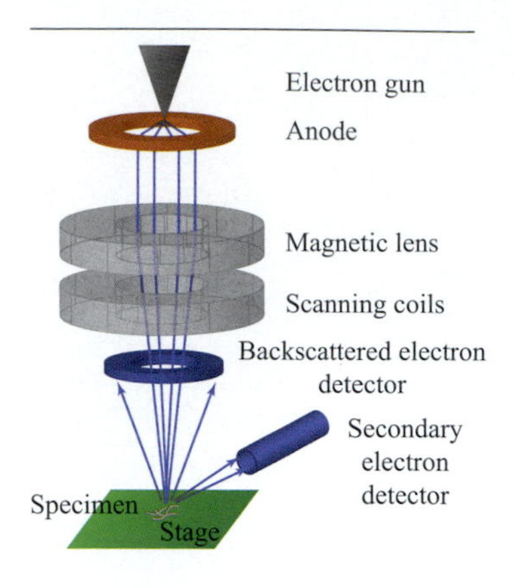

Fig. 19.6: Design of a scanning electron microscope.

the electron gun scans the TV screen. Figure 19.6 is a schematic representation of the SEM. The SEM usually has a resolution of 1–2 nm and can magnify up to 250,000 times.

The SEM can be used only for samples that have a conducting surface. The sample's surface is made conducting by coating it with a thin layer of a heavy metal, usually gold. The power of resolution of the SEM is usually lower than the TEM, but its depth of focus is several orders of magnitude greater. Unlike the TEM, the SEM is confined to studying only the surface of a sample.

As the surface of the object is scanned with the electron beam point by point, secondary electrons are set free. The intensity of this secondary radiation is dependent on the angle of inclination of the object's surface. The secondary electrons are collected by a detector that sits at an angle above the object. The signal is enhanced electronically. The magnification can be varied smoothly and the image appears a little later on a viewing screen.

Electrons having energy from 1 eV to, say, 300 eV have wavelengths from a few nanometers down to 0.01 nm. The typical energy of electrons used in the SEM is from 5–30 keV and for the TEM from 100–300 keV. The reason that much higher energies than 1 eV are used for the electrons — even though the de Broglie wavelength of a 1 keV electron is 1.23 nm — is because the problem of *focusing* the electrons is a competing factor in the resolution of these microscopes. Tightly collimating many electrons into a sharp focus is not easy since the electrons, unlike the photons in a beam of light, strongly repel each other due to electrostatic repulsion, especially since they are at a short distance from each other. Hence, in practice, the resolution that can be achieved by an electron microscope is much lower than the de Broglie wavelength of the electrons that are being used; namely, only objects much larger than the de Broglie wavelength of the electrons being used can be imaged.

19.4 STM/AFM stylus system

Many nanotools use a physical probe, called a stylus, to scan out the surface (and in some cases the bulk) of a sample using the stylus-sample interaction. Usually, the vertical position of the stylus is recorded as a function of the horizontal position — over the sample — of the stylus.

The main *advantage* of the stylus system is that the accuracy of the reading depends on the probe's sharpness, which can be as small as 2–3 nm; the STM/AFM scanning systems have another major advantage, namely that the sample can be shaped and modified by the scanning device. The main *disadvantage* of scanning microscopy is that the scanning process is slow and, moreover, it is difficult to obtain a magnified image of the sample. Furthermore, physical contact of the stylus with the sample may damage both.

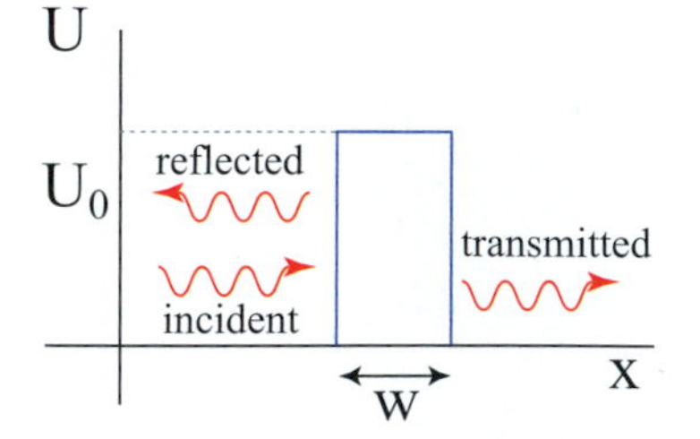

Fig. 19.7: Particle with energy less than U_0 tunneling across the potential barrier.

Scanning tunneling microscope (STM)

The scanning tunneling microscope (STM) works on the principle of quantum tunneling. Classical physics forbids a particle from crossing an energy barrier if the particle has an energy less than the barrier. In quantum physics, the transempirical state of the electron is spread out over an extent given by the de Broglie wavelength. There is, hence, a finite, albeit small, probability — which depends on the height U_0 and width W of the barrier as shown in Figure 19.7 — that an electron can cross a classically forbidden barrier in a phenomenon called *tunneling* as discussed in Section 24.7.

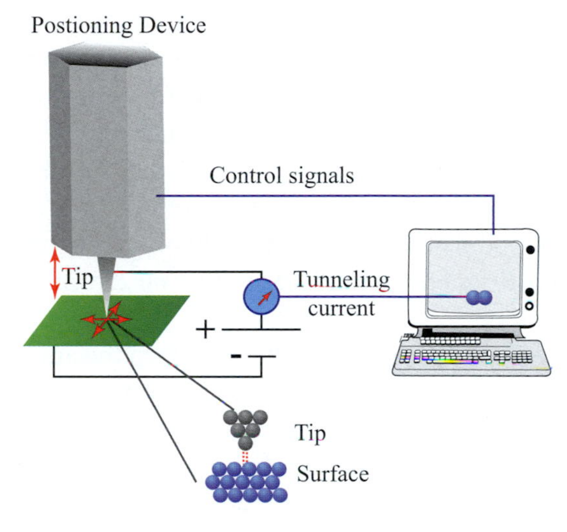

Figure 19.8: Design of the scanning tunneling microscope.

The direction of current flow depends on the bias that is fixed for an experiment. If the stylus tip has a negative bias and the sample is positive, electrons flow from the stylus to the sample. The current flows from the sample to the stylus for the reverse bias. For the STM, the width W of the tunneling barrier is the height of the stylus from the sample. By keeping the tunneling current a constant by varying the height of the stylus, one can map out the surface of the sample — as shown in Figure 19.8. A typical distance of the stylus from the sample is in the range from about 0.4–0.7 nm. The STM has a 0.1-nm lateral resolution and 0.01-nm depth resolution. The lateral resolution is limited by the sharpness of the stylus and the vertical resolution is limited by the precision in measuring the tunneling current.

Atomic force microscope (AFM)

The STM has the limitation that it can only be used on a sample that conducts at least a small current. The AFM, developed in 1986, avoids this limitation by measuring deflections of the stylus that are designed for responding to various physical properties; the stylus in general responds to atomic forces of interaction, primarily the magnetic and electrostatic forces. The properties are used for the stylus-sample interaction and a single stylus can scan for many different properties.

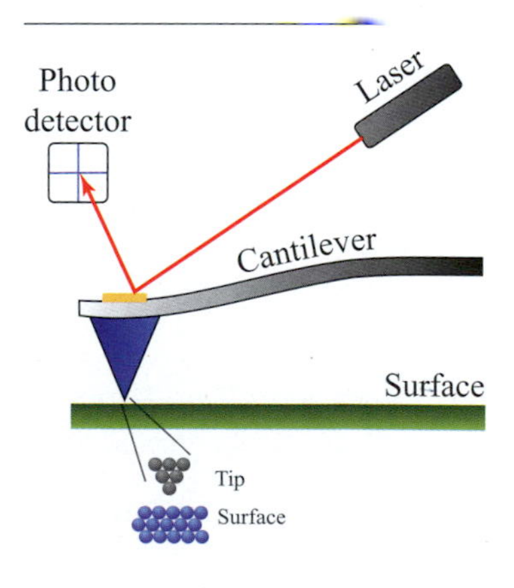

Fig. 19.9: AFM cantilever.

The stylus is attached to a cantilever, as shown in Figure 19.9. The cantilever is first aligned approximately close to the position of the sample being scanned. The stylus is then lowered and a feedback system is used to bring the stylus within a few nanometers of the sample surface. The sample is then moved under the stylus. A laser measures the deflection of the cantilever as the stylus scans the surface, as shown in Figure 19.9. A computer records the position of the cantilever and converts it into an image. High-resolution AFM has an accuracy comparable to that of the STM and TEM.

For example, the AFM can measure the van der Waals force, discussed in Section 5.7. The stylus is brought to a distance of 2–20 nm above the sample surface. The cantilever is deflected due to the attractive van der Waals force and the deflection is measured as the surface is scanned. Other applications of the AFM are to measure changes across the sample's surface due to changes in temperature, heat capacity and chemical reactivity.

19.5 How "big" is the nanoworld?

One would like to have a physical intuition of how much room there is in the nanoworld. For this purpose, we recount a lecture given by Richard Feynman in 1959 to illustrate the depth of the nanoworld. This seminal lecture, entitled *There is Plenty of Room at the Bottom*, is considered to be one of the starting points of nanoscience and nanotechnology. To illustrate that *the nanoworld is a staggeringly small world*, Feynman showed that — using nanotechnology — the entire 24 volumes of the *Encyclopedia Britannica* could be written on top of a pinhead.

The essential idea is that if one magnifies a pinhead 25,000 times then the area of the pinhead, having a diameter of approximately 0.0015 m, can accommodate the text of the entire encyclopedia. Conversely, if one could miniaturize the writing of the encyclopedia by 25,000, then it would fit onto the head of a pin. As shown in Figure 19.10, a typical page contains 80 lines with about 120 letters per line. What is the minimum size of a miniaturized alphabet? The human eye, using a magnifying glass, can resolve objects of a minimum diameter of about 0.0002 m, which upon 25,000-fold miniaturization will have a diameter of about 8.5×10^{-9}m — a distance of about 32 atoms for an ordinary metal. Hence, each miniature letter will contain about 1,000 atoms, enough atoms for storing (writing) as well as for retrieving (reading) text.

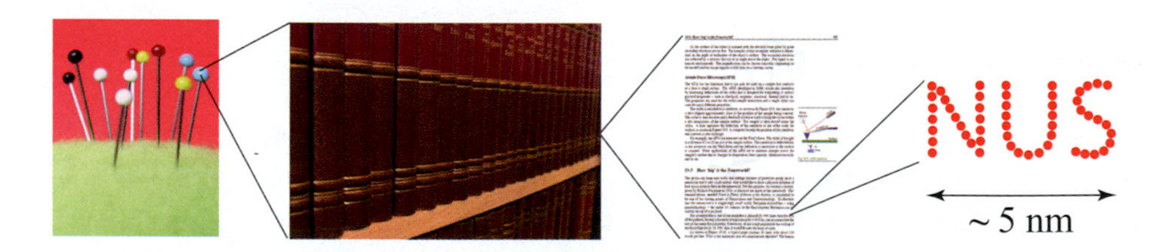

Figure 19.10: With nanosized letters, an entire encyclopedia could be written on a pinhead!

Partition the pinhead into rectangles with a side of about 10^{-8} m. One could then write — with room to spare — the entire encyclopedia, atom by atom, on a pinhead. Feynman then shows that if the *volume* of a pinhead is used for storing information, then 25 million books (each equal to the size of a volume of the encyclopedia) — approximately equal to all the books ever written — can be stored in a cube with a side equal to 10^{-4}m: the size of a speck of dust. Feynman then observes that in biological systems, DNA is doing precisely what we are imagining for the encyclopedia, namely storing tons and tons of information at the nanoscale.

The creation of nanotools like the STM and SEM have made the vision of Feynman into a practical and realistic exercise.

19.6 Gecko, van der Waals and Casimir forces

Nanoforces operate at the nanometer or smaller scale. All chemical bonds between atoms, discussed in Section 5.2, are nanoforces operating at the nanometer scale.

We now discuss two nanoforces, namely the van der Waals and Casimir forces, since many research teams have experimentally determined that it is, indeed, the van der Waals and Casimir forces that provide the mechanism for the gecko's amazing ability of walking on walls. It has been argued that the van der Waals and Casimir forces are not two separate distinct forces, but rather the expression of the *same* force taken in different limits. Essentially, the underlying force has a long distance and macroscopic manifestation as the Casimir force and in the short range limit it appears as the van der Waals force between polarizable materials.

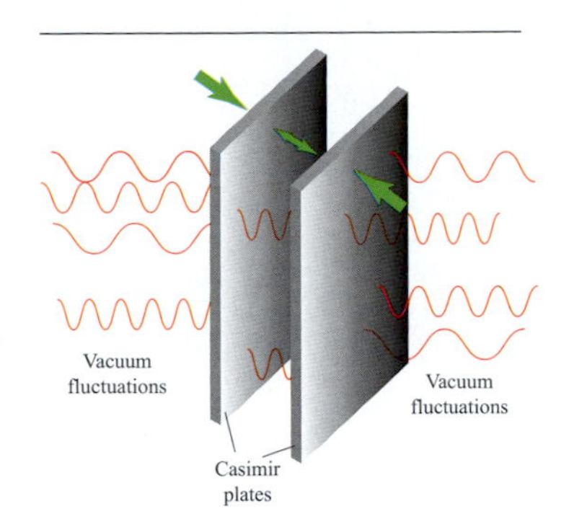

Fig. 19.11: Casimir effect.

One can easily verify that the gecko is using the van der Waals force. Teflon® is one of the few materials for which the van der Waals force is negligible, and this is the reason that almost nothing sticks to it. Sure enough, the gecko cannot hang on to Teflon® — confirming the view that the gecko uses the van der Waals force.

The van der Waals force is briefly discussed in Section 5.2, and we concentrate on understanding the Casimir force. If two parallel and neutral metallic plates are placed at a distance, say z, then there is a net attractive force between them, called the Casimir force. The result is rather puzzling since the plates are perfectly neutral and there is no induced polarization in either plate. The explanation of the Casimir force lies in the behavior of the electromagnetic field. Consider the case of no parallel plates; when the field is quantized, the energy of the vacuum state — namely, the quantum state of the field that has the lowest energy — is determined by contributions from all wavelengths of the field.

When the metallic plates are introduced, the boundary condition on the vacuum state is changed. The change in the geometry of the vacuum state is shown schematically in Figure 19.11. The electric field on the plate must be zero, and this condition removes certain wavelengths for the quantum field that were previously allowed.

The Casimir force does not depend on electric charge and neither does it depend on gravity. It is the result of the quantum properties in the *momentum* of the electromagnetic field. Transempirical photons that populate the vacuum "bounce" off the metallic plates, imparting momentum to the plate and hence creating a force due to radiation pressure.

Parallel conducting plates, with area A, have a net attractive Casimir force per unit area given by[2]

$$\frac{F_C}{A} = -\frac{\pi^2 \hbar c}{240} \frac{1}{z^4} \quad : \quad \textbf{Casimir force/area} \qquad (19.1)$$

where $\hbar$ is Planck's constant (divided by 2π) and c is the speed of light.

[2] What makes the Casimir force remarkable is that the vacuum energy of the quantum electromagnetic field is infinite for both cases, namely in the presence and in the absence of the metallic plates; however, the *change* in the energy of the vacuum state when a boundary is introduced — given in Equation 19.1 — is *finite*. The change in the energy leads to the Casimir force. The van der Waals force is also the result of the quantum fluctuations of the electromagnetic field since the presence of atoms changes the boundary condition for the quantum vacuum.

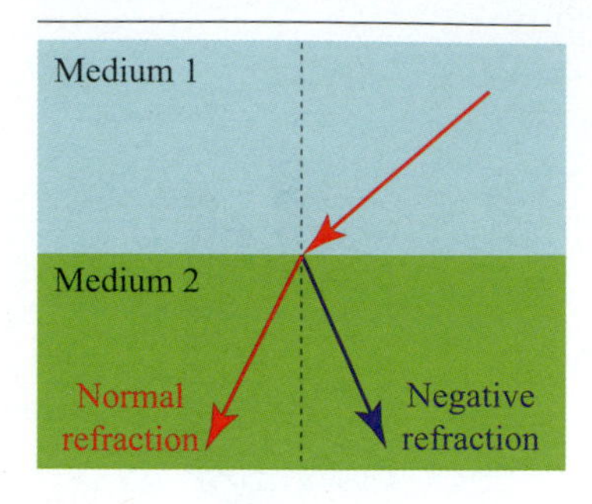

Fig. 19.12: Negative refraction.

The Casimir force is a surprise since one would not expect neutral conductors to have any (attractive) force between them. The Casimir force/area is negligible for macroscopic distances, but becomes important at nanometer distances, having the value of 100 kPa (1 atmosphere) for z =10 nm.

The Casimir force per unit area is an important cause of friction in the nanoworld since it causes surfaces that are a nanometer apart to stick to each other. Nanomachines can run smoother and with less or zero friction if one can manipulate the Casimir force. The Casimir force is *repulsive* for some special materials that have a negative refractive index and are called left-handed materials. The behavior of light refracting through such a material is shown in Figure 19.12, and a frictionless nanomachine could be created by coating its component parts with this material. However, these materials usually have quite a complicated structure and coating nanomaterials with these may, in practice, be difficult.

If one interposes a thin layer of the left-handed material between the parallel conducting plates, then it can be shown there is a net *repulsion* between the conducting plates due to the presence of the material. An even more surprising possibility is that of *quantum levitation*. One conducting plate could hover over the other at the distance where the repulsive Casimir force balances the weight of the plate; the plate would then levitate. The same conducting plates would not levitate if not for the left-handed material interposed between them.

The van der Waals force is attractive; the Casimir interaction becomes strong at nanometer distances and is attractive for most types of matter; the attractive forces are additive. The network of gecko hairs is at a nanometer distance from a typical surface and form intermolecular bonds with the surface by means of van der Waals and Casimir forces. When millions of gecko hairs make contact with a surface, they collectively create a powerful bond that is a 1,000 times stronger than the force geckos needs to hang on to a wall.

Gecko: walking on the ceiling

Fig. 19.13: Foot of a gecko.

The scanning electron microscope has been employed to physically examine the gecko's feet. On looking at the gecko, one finds that it neither emits any glue nor does it have suction pads that allow it to walk on walls. Instead, it is found, as shown in Figure 19.14, that the hair on gecko's toepads are of nanometer size.

Geckos climb vertically and even walk on inverted surfaces with ease — using millions of micron-scale adhesive foot hairs on each toe. A single gecko foot hair (called seta) is only as long as two diameters of a human hair, that is, 10^{-4}m long. Each seta ends in up to 1,000 even tinier tips that are only 2×10^{-7}m wide. The end point of a seta is too small to see using optical microscopes since their size is smaller than the wavelength of visible light. It is these nanosize hairs that are responsible for the attractive forces that hold geckos to surfaces.

Each foot hair splits into hundreds of spatula shaped tips measuring only 200 nm in diameter and permitting intimate contact with rough and smooth surfaces alike. A van der Waals mechanism implies that the remarkable adhesive properties of gecko setae are merely a result of the size and shape of the setae spatulae, and are not strongly affected by surface chemistry. Theory predicts greater adhesive

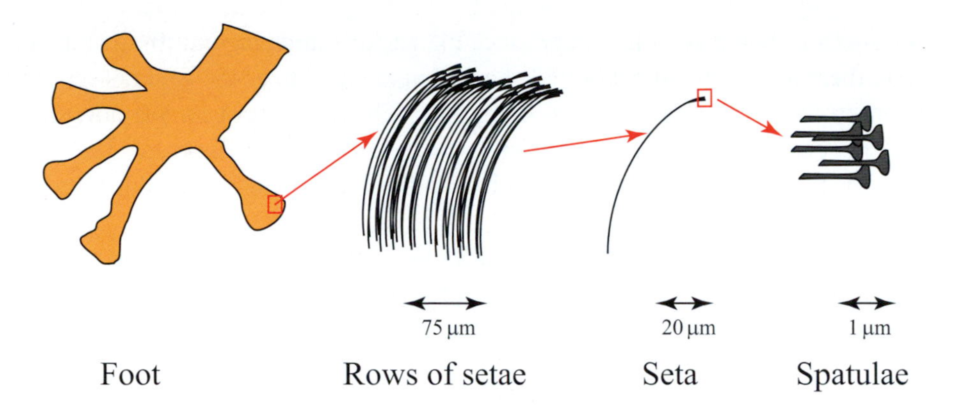

Figure 19.14: Zoom onto the foot of a gecko.

forces simply by subdividing setae and, hence, increasing the surface density of the gecko's setae. Geckos' adhesive microstructure requires minimal attachment force, leaves no residue, is directional, detaches without measurable forces and is self-cleaning; it works underwater, in a vacuum, and on nearly every surface material and profile.

The self-cleaning effect is a result of the fact that dirt has slightly less attraction for the gecko's hairs than for the surface on which the animal is walking. This results in a net autocleaning effect. The stickiness is unlike conventional adhesives, which either adhere weakly and detach with ease or are strong and hard to remove. By contrast, gecko hairs adhere strongly and detach easily, as shown in Figure 19.15.

The seta adhesion is so strong that a single seta can lift the weight of an ant having a weight of 200 μN = 20 mg. A million setae can easily fit onto the area of a dime and could lift the weight of a child (20 kg). The maximum possible force of 2,000,000 setae on a gecko = 2,000,000 × 200 μN = 400 N = 40,788 g (on earth), or about 41 kg! The weight of a Tokay gecko is approximately 50–150 g.

A gecko can hold up its entire body weight with only a single toe. If the adhesive is so strong, how does it get its feet off? The seta spatula peels off like tape. If a gecko increases the angle that the seta makes with the surface beyond 30°, it just pops off, as shown in Figure 19.15. At angles of over 90°, the hairs easily become detached. Because the hairs are solid structures, they are not damaged in this process and can be used repeatedly.

There are about 850 species of gecko, and there are other animals that have evolved geckolike setae. Each species has a different type of seta. Some species have much simpler hair without the 1,000 tips of the gecko.

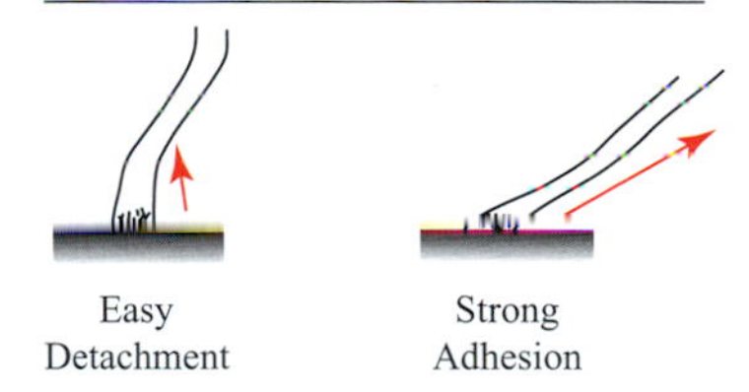

Fig. 19.15: Strong adhesion in one direction, easy to remove in the other.

19.7 Carbon nanotubes: artificial nano "hair"

We saw that the gecko can walk on walls and ceilings due to the tip of the gecko's seta (foot hair) having a radius of about 2×10^{-7}m. We also saw that the van der Waals and Casimir forces used by the gecko depend only on a nanoscale contact of the gecko's seta with the surface in question and do not depend on the the material making up either the seta or the surface. Any nanosized artificial "hair" in contact

Fig. 19.16: Master of nanotechnology.

with a surface should be able to reproduce the gecko's amazing climbing abilities.

Are there any artificial nanosized "hair" that are as thin as the gecko's seta and that can provide adhesion similar to the gecko? The answer is yes. We now study carbon nanotubes and silicon nanowires, which are wirelike structures that have nanometer radius, that will turn out to have adhesion properties similar to gecko hairs.

To start with, we study carbon nanotubes in some detail. It is a prime example of a nanostructure and, in particular, has great potential for many applications — including being used for making a geckolike adhesive.

In our discussion of graphite in Section 5.12, we saw that graphite has a layered structure shown in Figure 5.15. Each carbon atom is made from a nucleus of six protons, six neutrons and six electrons. Two of the electrons occupy the *s* shell and four electrons are in the *p* shell; the *p* shell allows eight electronic states, hence there are four electrons and four vacancies in the outer *p*-shell. In graphite, each carbon atom forms three covalent bonds with other carbon atoms in its plane, forming a six-sided hexagon, as shown in Figure 19.17.[3]

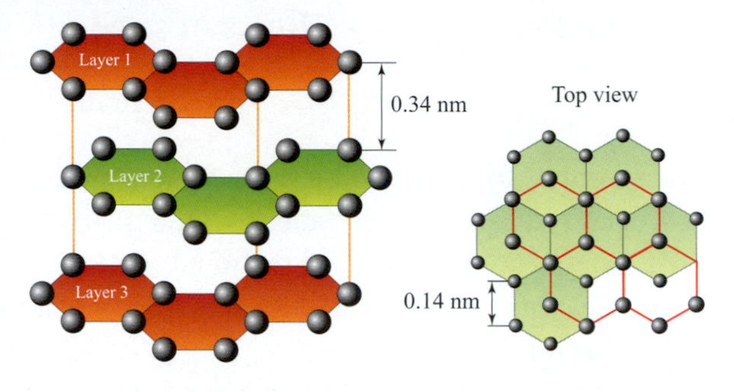

Figure 19.17: Structure of graphite.

The remaining electron from each atom is shared with all the other carbon atoms in the plane in the following manner: An electron from one carbon atom being shared with the neighboring atom can jump to a third neighboring carbon atom and so another, and in doing so can travel throughout the plane. Such an electron is said to be delocalized; one can see that all the electrons that are not covalently bonded are delocalized in a given plane. The delocalized electrons in a given plane have a (weak) van der Waals interaction with the delocalized electrons in the two adjacent planes. Hence, the interplanar force between the carbon atoms is much weaker than the intraplanar bonding. The interlayer distance is 0.34 nm and the interatomic distance in a plane is 0.14 nm, as shown in Figure 19.17.

To form a carbon nanotube, a single layer of graphite is peeled off and rolled into an SWNT (single-walled nanotube); the carbon atoms on the edges of the single layer are bonded to form a hollow tube. An SWNT has a typical diameter of 1.2–1.4 nm with the smallest being 0.5 nm; there is no theoretical limit on the length of an SWNT and as of 2009, millimeter lengths have been fabricated.

There are numerous varieties of SWNTs since there are many ways of bonding the carbon atoms on the edge of the graphite layer to form a tube; one can take any two carbon atoms located in a layer and bond them by rolling the plane.

[3]In diamond, all four electrons in the outer shell of each carbon atom form four covalent bonds with four other carbon atoms, leading to a lattice structure for diamond that results in the hardest known naturally occurring solid.

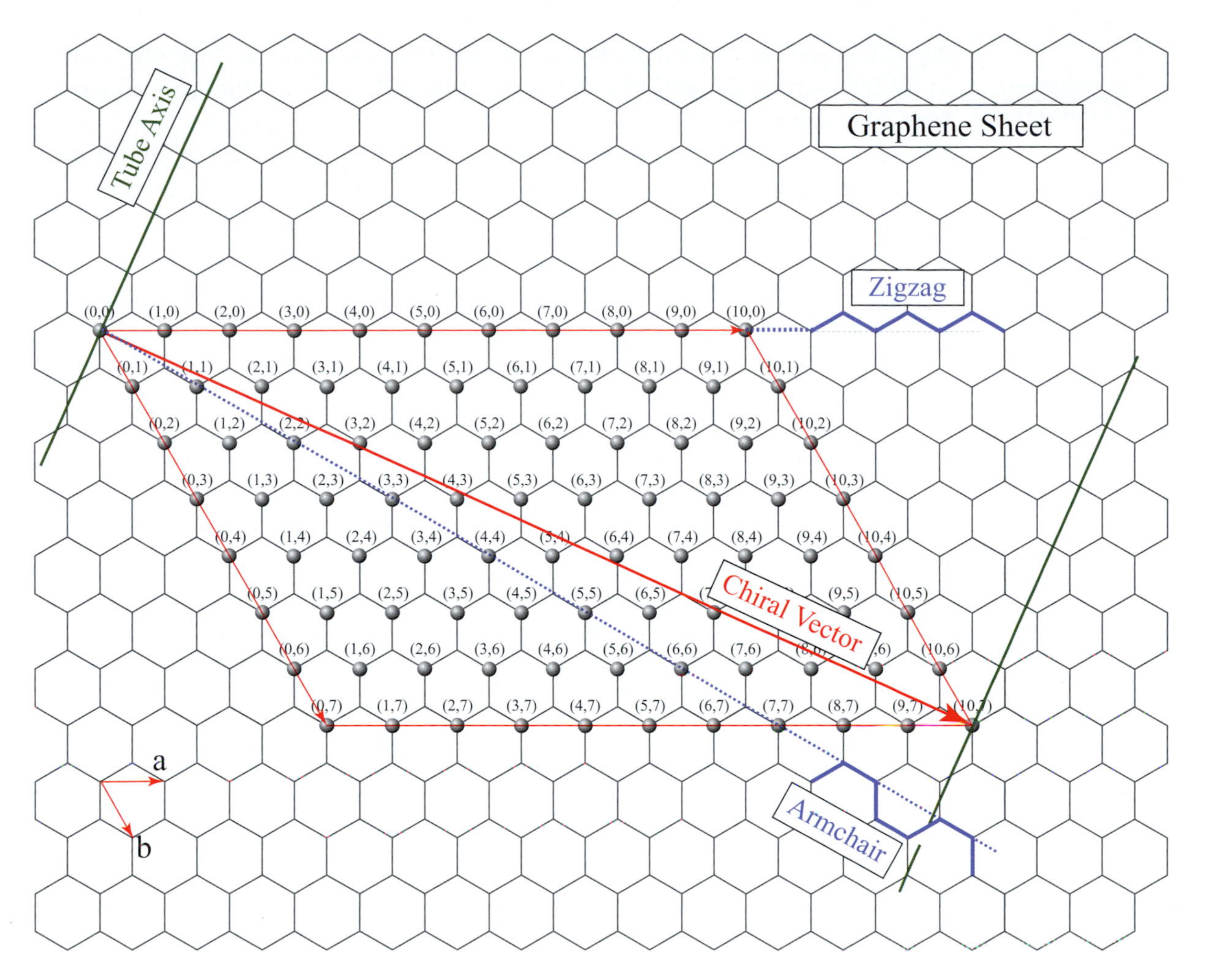

Figure 19.18: Rolling vector used to describe carbon nanotubes. In this figure, a (10,7) vector is depicted.

The vector joining two carbon atoms in graphite, called a rolling vector, is given by

$$\mathbf{R} = m\mathbf{a} + n\mathbf{b} = (m, n).$$

The distance between any two carbon atoms is specified by a vector joining them; in particular, graphite is a honeycomb two-dimensional lattice that is specified by two basis vectors **a** and **b**, as shown in Figure 19.18.

The carbon nanotube has three typical patterns depending on how it is formed: namely (a) the *armchair* specified by the rolling vector (m, m), (b) *zigzag* given by the rolling vector $(m, 0)$ and (c) the general *chiral* carbon nanotube with the arbitrary rolling vector (m, n). See Figure 19.18.

There is an SWNT for each choice of the rolling vector (m, n). As seen in Figures 19.18 and 19.19, the rolling vector specifies in which direction the graphite sheet is rolled as well as determines the radius of the SWNT.

Carbon nanotubes are formed by a process of self-assembly, in which the bonds between the atoms form spontaneously — by carbon arc discharge, chemical vapor deposition and laser-assisted catalytic growth. The product of self-assembly results in SWNTs, of which about two-thirds are semiconductors and the remaining one-third are conductors.

One can do fabrication of carbon nanotubes so that many SWNTs of different radii are formed concentrically, with many SWNTs (having certain fixed radii) being located in the hollow region of an outer SWNT to form an MWNT (multiwalled nanotube), as shown in Figure 19.20. MWNTs are interesting in their own right, having a greater tensile strength than the SWNT. However, the nanoscale properties of the SWNT are better understood by theory and hence are widely studied.

Carbon nanotubes have many remarkable properties, a few of which are listed below.

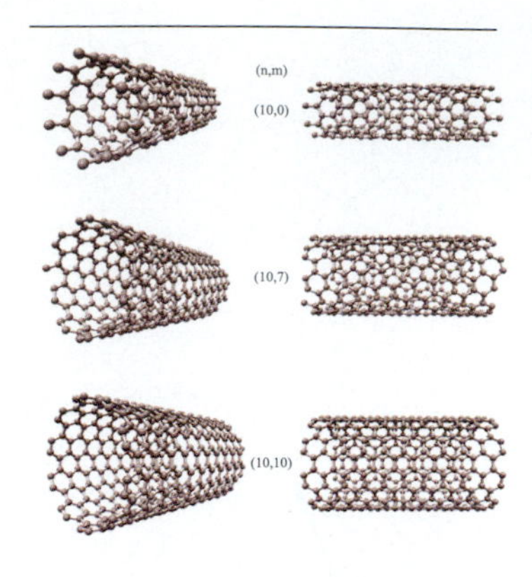

Fig. 19.19: Singlewall carbon nanotubes.

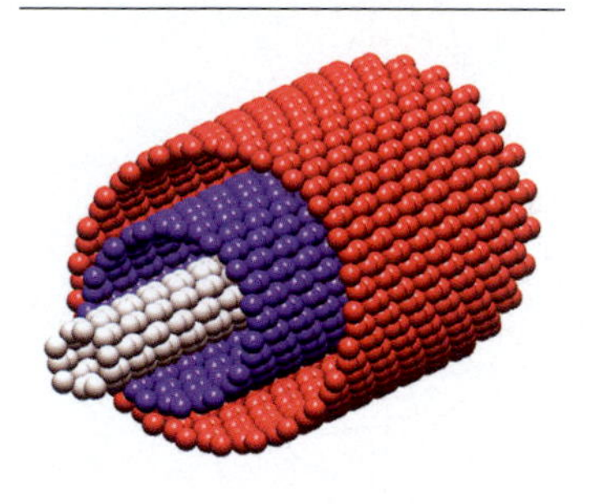

Fig. 19.20: Multiwall carbon nanotube.

- Depending on the choice of its rolling vector, a carbon nanotube is either a conductor of electrical currents or a semiconductor. Carbon nanotubes that are conductors have very high electrical conductivity. It is estimated that carbon nanotubes can carry currents up to 10^9 A/cm^2, whereas copper wires melt for currents of 10^6 A/cm^2. High currents do not generate large amounts of heat for carbon nanotubes. For these reasons, it has been suggested, to minimize the loss of energy by electrical currents due to heating, all electrical transmission wires should be made from carbon nanotubes.

- Carbon nanotubes are excellent conductors of heat. The measured thermal conductivity of carbon nanotubes is 3 W/(km K). This is almost two times that of diamond 2 W/(km K) — currently the best known thermal conductor.

- Carbon nanotubes are extremely flexible and bounce back to their original shape after being deformed. They can be twisted, flattened and bent into a circle and around corners without breaking. They are stronger and lighter than carbon fiber materials and are used for making polymer composite fibers.

- Tensile strength is the force per unit area that breaks, or permanently deforms, a material. Carbon nanotubes have great tensile strength, which has been measured to be 150 GPa and which is 375 times greater than that of steel's 0.4 GPa.

- Carbon nanotubes are much harder than steel, having a Young's modulus of 1,000 GPa, while at the same time having only one-sixth the density of steel and therefore being much lighter. Carbon nanotubes are almost as hard as the hardest known material, namely diamond, that has a Young's modulus of 1,220 GPa. Unlike steel and diamond, carbon nanotubes are not brittle but, instead, are highly flexible and malleable.

19.8 Carbon nanotubes: artificial gecko adhesive

Adhesives as strong as the gecko hairs could be formed by using arrays of multi-walled carbon nanotubes (MWNTs). The tangled mass of MWNTs is similar to the

Leo Esaki

Leo Esaki, also known as Leona Esaki, was born in Osaka, Japan, in 1925. Esaki completed work for a BS in physics in 1947 and received his PhD in 1959, both from the University of Tokyo. Esaki is an IBM Fellow and has been engaged in semiconductor research at the IBM Thomas J. Watson Research Center, Yorktown Heights, New York, since 1960.

Prior to joining IBM, he worked at the Sony Corporation, where his research on heavily doped Ge and Si resulted in the discovery of the Esaki tunnel diode; this device constitutes the first quantum electron device. Since 1969, Esaki has, with his colleagues, pioneered "designed semiconductor quantum structures" such as man-made superlattices, exploring a new quantum regime in the frontier of semiconductor physics.

Leo Esaki shared the Nobel Prize in Physics in 1973 with Ivar Giaever and Brian David Josephson for his discovery of the phenomenon of electron tunneling. He is known for his invention of the Esaki diode, which exploited that phenomenon. This research was done when he was with Tokyo Tsushin Kogyo (now known as Sony). He also contributed as a pioneer of semiconductor superlattice while he was with IBM.

Sumio Iijima

Although carbon nanotubes had been observed prior to his "discovery", Iijima's 1991 paper on carbon nanotubes generated unprecedented interest in the carbon nanostructures and has since fueled intense research in the area of nanotechnology. Iijima was awarded the Benjamin Franklin Medal in Physics in 2002, "for the discovery and elucidation of the atomic structure and helical character of multi-wall and single-wall carbon nanotubes, which have had an enormous impact on the rapidly growing condensed matter and materials science field of nanoscale science and electronics".

Sumio Iijima is a Japanese physicist. Born in Saitama Prefecture in 1939, Iijima graduated with a Bachelor of Engineering degree in 1963 from the University of Electro-Communications, Tokyo. He received a Master's degree in 1965 and completed his PhD in solid-state physics in 1968, both at Tohoku University in Sendai.

Between 1970 and 1982 Iijima was a researcher at Arizona State University and studied crystalline materials using high-resolution electron microscopy . He visited the University of Cambridge during 1979 to perform studies on carbon materials. He worked for the Research Development Corporation of Japan from 1982 to 1987, studying ultra-fine particles, after which he joined NEC Corporation where he is still employed. His seminal work on carbon nanotubes was carried out in 1991 while working with NEC.

He has been a professor at Meijo University since 1999, is the Director of the Research Center for Advanced Carbon Materials at the National Institute of Advanced Industrial Science and Technology, is a Senior Research Fellow of NEC Corporation and is the Dean of the Sungkyunkwan University Advanced Institute of Nanotechnology (Seoul, South Korea).

hierarchical structure of a gecko's foot, with a branching of MWNTs of different of diameters.

A gecko adhesive does not require being pressed into a surface — the fibers engage by being dragged parallel to the surface with minimal force into the plane. A gecko adhesive exhibits "frictional adhesion" where the fibers push off the surface if there is a force that is perpendicular to the surface, giving automatic release.

Researchers have grown MWNT nanohairs on top of microhairs, as illustrated in Figure 19.21a. The microhairs are grown on a polymer substrate, on which they are anchored. When pressed onto a glass surface, the tangled portion of the nanotubes takes the shape of the glass surface and comes into close contact with the surface. The nanohairs align themselves along the surface, thus dramatically increasing the proximity between the nanotubes and the surface and, hence, maximizing the van der Waals and Casimir forces that operate at nanometer separation. Attractive forces between the ends of the nanohairs and the glass surface keep the nanohairs stuck fast to the surface. When lifted off the surface in a direction parallel to the main body of the nanotubes (and perpendicular to the surface), only the tips remain in contact, minimizing the attraction forces, and allowing the MWNT nanohairs to be detached from the glass surface. See Figure 19.21b. Similar to Velcro, the gecko detaches its setae layer by layer.

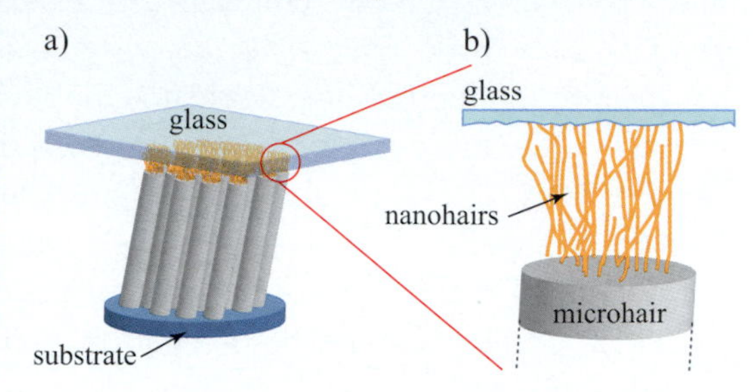

Figure 19.21: Possible design of artificial gecko seta.

In tests done on a variety of surfaces including glass, polymer sheets and rough surfaces, an adhesive force of up 100 N/cm^2 was measured if the nanohairs were pulled in a direction along the surface. The adhesive force in a direction perpendicular to the surface was 10 N/cm^2 — about the same as that for a real gecko. The resistance to shear increased with the length of the nanotubes, while the resistance to a normal force was independent of tube length.

Adhesives based on carbon nanotubes have many applications. In particular, because carbon nanotubes conduct heat and electrical current well, dry adhesive arrays could be used to connect electronic devices.

19.9 Conductors, insulators and semiconductors

We have briefly discussed conductors in Section 9.5, and the resistivity of a few solids is given in Table 9.1.

The electrical properties of solids are the result of quantum mechanical effects and we briefly discuss conductors, insulators and semiconductors. In quantum mechanics, electrons are represented by a probability amplitude that encodes the likelihood of observing them in a particular state, as discussed in Chapter 24.

Consider a conducting solid; the atoms in the solid have valence electrons in their outer shells that they can share with other atoms in the solid by a mechanism similar to the formation of delocalized electrons discussed in Section 19.7. As discussed in Section 5.5 on metallic bonds, the atoms of a conductor, such as a metal, have their valence electrons in a common pool of delocalized electrons that are free to move in the entire solid and in doing so create a sea of valence electrons. In contrast, the atoms of an insulator contribute very few electrons to the common sea. In an insulator, all the electrons are bound to atoms. Hence, compared to an insulator, in a conductor many more electrons are available to move and carry current in response to an applied voltage.

In terms of energy levels of the electrons in a solid, the electronic individual energy levels (of a single atom) combine with the energy level of other electrons. The resultant energy levels for the electrons in a solid form continuous energy bands of allowed and forbidden states that the electrons can occupy, as shown in Figure 19.23a.

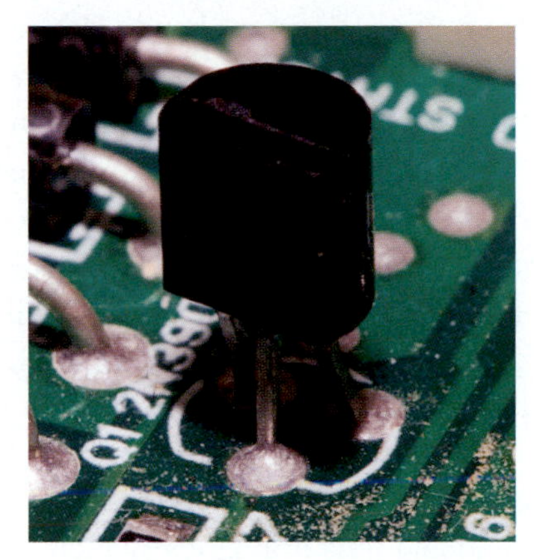

Figure 19.22: The transistor, a typical semiconductor device.

The bands of allowed electronic energy states can overlap or be separated by forbidden bands; in the forbidden band, there are no energy states available for the electrons. For a conductor, the energy band occupied by the electrons overlaps with the empty allowed energy bands, whereas for semiconductors and insulators the occupied energy band is separated by a forbidden energy band from the empty allowed energy states.

The Pauli exclusion principle states that no two electrons can be in the same (identical) state; hence, electrons in a solid can occupy only the available states, starting from the lowest energy state. The highest energy electrons occupy a state that is said to have the Fermi energy for the said solid; typically, the Fermi energy is of the order of a few electronvolts; the Fermi level is shown in Figure 19.23a. The electrons in the solid are said to occupy the *valence band*.

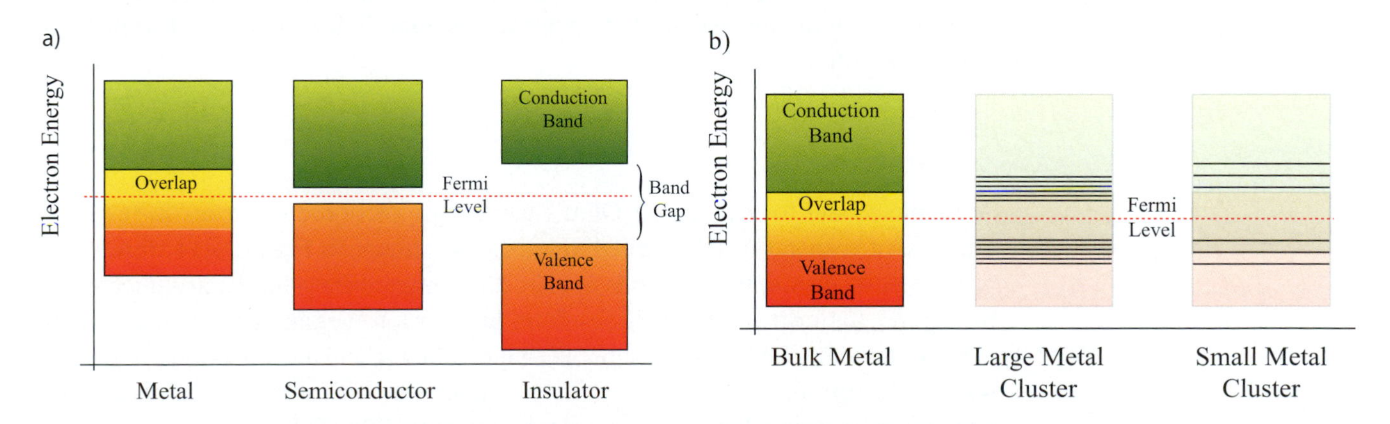

Figure 19.23: a) Energy bands. b) Energy levels versus number of atoms.

Conductors, of which metals are the prime example, have electrons that are available to flow when subjected to an external voltage. The conductor's electrons form a sea and the electrons with the highest energy occupy an energy band that overlaps with energy states of the *conduction band* that are not occupied, as shown in Figure 19.23a. Hence, when an external voltage is applied, the valence electrons of the conductor can respond by making transitions to higher energy states and thus move in response to the voltage. In contrast, as shown in Figure 19.23a, for an insulator, such as Teflon® or glass, the valence band is separated by a large gap from the conduction band. Hence, there are no available (empty) quantum states that the electrons can occupy in response to an external voltage. For an insulator, there is consequently no current in response to an applied voltage.

A semiconductor lies in-between a conductor and an insulator. For a semiconductor, the energy band gap between the valence band and the conduction band is small, comparable to the thermal energy $k_B T$ of the electrons (at temperature T); see Figure 19.23a. Hence, a small number of electrons are lifted across the band gap by their energy of thermal motion, giving it a conductivity better than that of an insulator, as given in Table 9.1. The semiconductor's electrical resistivity is sensitive to external means of excitations; for example, shining light on a semiconductor or applying a voltage bias can excite valence electrons into the conduction band and thus dramatically increase the semiconductor's conductivity.

19.10 Nanowires: modification of Ohm's law

An ordinary copper wire is usually a conductor that is narrow but long; in other words, a wire can be thought of as a long cylinder, with the radius being small, usually of the order of a few millimeters, as shown in Figure 9.11. Nanowires are one-dimensional conductors that can be arbitrarily long and have a radius of the order of 1 nm. Silicone and carbon nanotubes and other new systems are widely studied nanowires.

For ordinary circuits, Ohm's law determines the current in a conductor due to an applied voltage. In particular, for a conductor with resistance R, Ohm's law states that the current I is proportional to the applied voltage V, the proportionality constant being the inverse of R. In other words, Ohm's law states, as discussed in Section 9.5, that

$$I = \frac{V}{R}.$$

A simple model for explaining Ohm's law is that there are electrons in the conductor that respond to the applied voltage. As shown in Figure 9.11, through a process of diffusion, the electrons acquire an average drift velocity, as indicated in the figure; the drift of the electrons from high to low voltage creates the current in the conductor. The difference between a conductor and a semiconductor can be seen in the difference in the resistivity of a material, with some typical values for conductors, semiconductors and insulators given in Table 9.1.

Bulk graphite is a good electrical conductor, and in its bulk state there are plenty of electrons that can move through the material due to the vast electron delocalization within a carbon layer, as discussed in Section 19.7. These electrons are free to move in a given plane, so are able to conduct electricity when an external voltage is applied. Electricity is conducted by the electrons that move in a given plane of graphite.

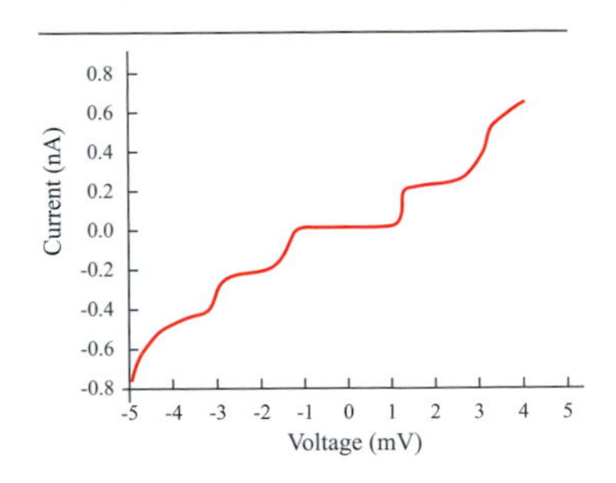

Fig. 19.24: Conduction in a carbon nanotube.

Carbon nanotubes can be made into electrical conductors and are a prime example of a nanowire. Carbon nanotubes do not obey Ohm's law. As shown in Figure 19.24 — instead of a linear dependence of the current on the applied voltage — the current in the nanowire increases in *discrete* steps when the voltage is increased, showing a behavior that cannot be explained by classical physics. In fact, the electrical properties of a carbon nanowire are a direct reflection of the quantum properties of the electrons in the carbon nanotube. If an external voltage is applied to the carbon nanotube, the rolling vector (m, n) determines how many electrons are able to carry the current. The electrical properties of an SWNT are determined as follows.

- If $m - n = 3 \times$ integer, then the SWNT is a conductor.
- Otherwise, the SWNT is a semiconductor.

When a single sheet of graphite is rolled into a carbon nanotube, electrons are completely confined to a single plane. The reason that not all SWNTs are conductors is because on being rolled into a tube, the delocalized electrons in a graphite layer now have a new *boundary condition* in the rolled direction that is fixed by the rolling vector. Only for the case of $m - n = 3 \times$ integer does the carbon nanotube continue to have delocalized electrons and hence is a conductor.

An electric current in an SWNT conductor consists of the following two components:

- *Diffusive electrons* behave like classical charges responding to an applied voltage. Diffusive electrons travel like classical particles and have a typical mean free path of about 10^{-7}–10^{-8} m, which is the distance the electrons travel before colliding with the atoms of the conductor. In ordinary conductors, flowing electrons continually collide with the atoms making up the material, slowing down the electrons and causing the material to heat, effectively creating resistance. Only those SWNTs which are conductors have charged carriers that are diffusive electrons.

- *Ballistic electrons* are electrons with a mean free path that is (much) bigger than the length of the wire in which they are moving; such electrons alter their motion only by hitting against the walls. Ballistic electrons arise entirely due to quantum mechanical reasons and propagate like a wave in the conductor. The conductor must be free of impurities that electrons could collide with. In the case of an SWNT conductor, ballistic electrons propagate like a wave with a definite momentum in a quantum-mechanically coherent manner and travel through the conductor without losing any energy due to scattering off the atoms of the conductor.

The discrete increase in the current as voltage is increased is because for a conducting SWNT more ballistic electrons become available for contributing to the current, in discrete steps, as the voltage is increased. Furthermore, the current can increase without any significant increase of voltage, as shown in Figure 19.24, since ballistic electrons contribute very little to resistance; this is the reason that Ohm's law does not apply to ballistic electrons.

19.11 Energy, entropy and nanophenomena

As discussed in Section 18.6, for a large system in contact with the environment at temperature T, the probability to be in its ith microstate, its average energy, and entropy are given by Equations 18.6, 18.8 and 18.9

$$p_i = \frac{1}{Z} e^{-E_i/k_B T}$$

$$E = \sum_{i=1}^{\Omega} p_i E_i$$

$$S(E, V) = -k_B \sum_{i=1}^{\Omega} p_i \ln p_i$$

where Ω is the number of microstates of the large macroscopic-size system.

For a nanoscale system, a modification of entropy and of thermodynamics in general seems to be necessary. The reason is the following: Macroscopic systems have a large number of particles and hence many microstates; for this reason, the system's energy and entropy is almost always near its average value. In contrast, nanosystems are made out of small numbers of particles and the energy and entropy of the system have large deviations away from the average values. This leads to a number of new features for nanosystems.

In particular, nanosystems do not scale with the number of particles. For example, if a macroscopic system is made out of two (large) subsystems, say A and B, then the energy and entropy of the large system is the sum of the values for the two subsystems; this property is called additivity. For nanosystems, additivity generally breaks down since the number of particles is very small, a far cry from the Avogadro's number of particles that is typical for a macroscopic system. In particular, it is thought that interactions taking place across the boundary between nanosystems A and B — unlike the case for macroscopic systems — yield a significant contribution to the total energy and entropy of the combined system and lead to the breakdown of scaling.

It is not known what should be the correct definition for the thermodynamics of a nanosystem. We discuss a possible generalization of thermodynamics and statistical mechanics due to Tsallis; this proposal provides a very specific model of how a nanosystem can have a lack of additivity of the physical quantities. Tsallis introduced a new natural constant $q > 0$ that needs to be experimentally determined for a nanosystem. All the results for a large system are recovered in the limit of

$q \to 1$. The following are the generalizations for a small nanosystem:

$$\tilde{p}_i = \frac{1}{Z_q}\left[1 - (1-q)\frac{E_i}{k_B T}\right]^{1/(q-1)} \quad ; \quad \sum_{i=1}^{\Omega} \tilde{p}_i = 1$$

$$E_q = \sum_{i=1}^{\Omega} \tilde{p}_i^q E_i$$

$$S_q = k_B \frac{1 - \sum_{i=1}^{\Omega} \tilde{p}_i^q}{1-q}.$$

Note as $q \to 1$, $\tilde{p}_i, E_q, S_q \to p_i, E, S$ respectively.

To get a feel for the new features that arise for $q \neq 1$, consider two systems A and B that are components of the total system. The probability $\tilde{p}_{ij}^{A+B}$ that the total system is simultaneously in microstate i of system A and in microstate j of system B is given by

$$\tilde{p}_{ij}^{A+B} = \tilde{p}_i^A \cdot \tilde{p}_j^B.$$

This yields the entropy of the total system $S_q(A+B)$ as follows:

$$\begin{aligned} S_q(A+B) &= k_B \frac{1 - \sum_{i,j} (\tilde{p}_{ij}^{A+B})^q}{1-q} \\ &= S_q(A) + S_q(B) - \frac{1-q}{k_B} S_q(A) S_q(B). \end{aligned}$$

Since $S_q(A+B) \neq S_q(A) + S_q(B)$, Tsallis entropy is not additive and hence does not behave like a macroscopic system. The entropic index q is a measure of the nonadditivity of the system and has three distinct cases, namely for $q < 1$, called superadditive, $q = 1$ that is additive, corresponding to the regular Boltzmann distribution discussed in Section 18.6, and $q > 1$, called subadditive.

Studies are underway to better understand the theoretical and experimental implications of the Tsallis generalization of statistical mechanics and thermodynamics.

19.12 Nanoparticles

The carbon nanotube is an extremely thin cylinder and is approximately a one-dimensional object. One can ask: Are there any nanostructures that are like a single particle, having only nanometer extension in all three space directions and hence occupying nm^3 volume in space? Such structures do indeed exist and are called nanoparticles, to emphasize their similarity to the behavior of a single quantum particle.

A nanoparticle is a collection of atoms or molecules that are bonded together and have a radius from 1–100 nm. Figure 19.25 shows in a schematic manner the classification of matter into atoms, nanoparticles and molecules.

What makes nanoparticles very interesting is that their size is smaller than many physical characteristics of a bulk material. A nanoparticle behaves in many ways like a single atom. The energy of the simplest atom, namely, the hydrogen atom, is in discrete levels; the energy levels of boron nanoparticles made out of 6, 8 and 12 atoms are shown in Figure 19.26. One can see that the nanoparticle has energy levels that are similar to a single atom. The shapes of boron nanoparticles are shown in Figure 19.27; since there are only a few atoms in a nanoparticle, all the atoms in effect are on the "surface". A cluster of atoms of 1-nm radius has about 94 atoms of which about 51% are on the surface, a cluster of 6-nm radius has about 9,000 atoms with about 14% on the surface, whereas a cluster of a size of 60 nm has about 10 million atoms of which only 1% are surface atoms. Nanoparticle atoms are mostly on the surface, showing the importance of surfaces in nanophenomena.

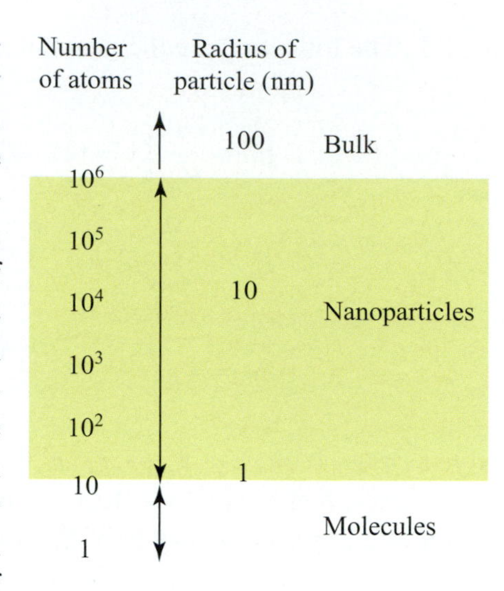

Figure 19.25: Size of the nanoworld.

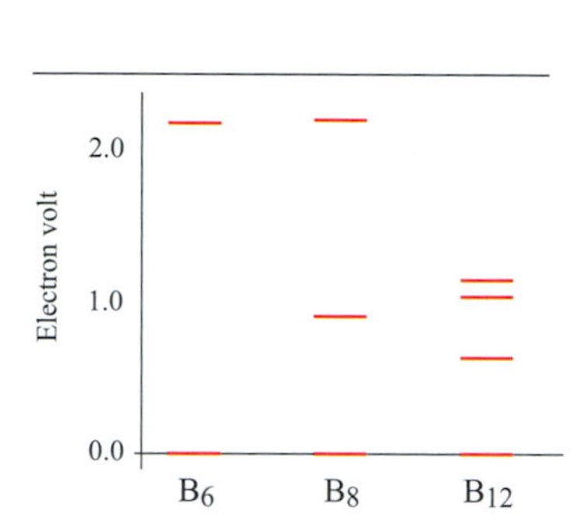

Fig. 19.26: Energy levels of boron nanoparticles.

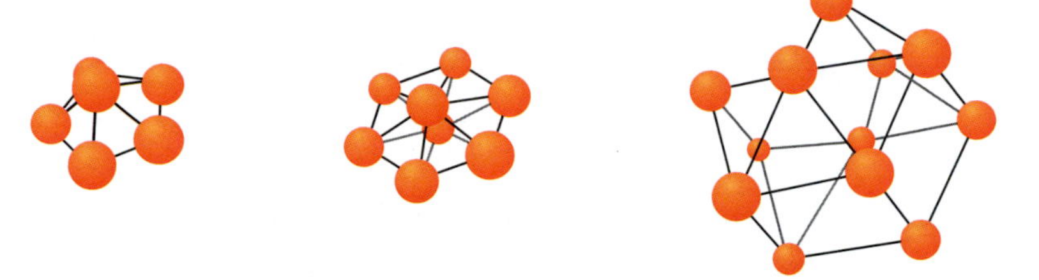

Figure 19.27: Structures of boron nanoparticles.

A nanoparticle can be thought of as being the result of subdividing a bulk material until one reaches the nanoscale.

Figure 19.28 shows how rhenium's magnetic moment per atom changes as one makes the sample smaller and smaller; at about 10 atoms there is a dramatic increase in sample's magnetic moment, showing the onset of nanoeffects. Similarly, Figure 19.29 shows the effect of size on the melting temperature of gold nanoparticles; there is a sharp drop for clusters smaller than 100 gold atoms. From these diagrams, we can conclude that when the size of a cluster of atoms reaches 10–100 atoms, the transition from bulk to nanobehavior takes place. It has been experimentally observed that certain "magic" numbers of atoms form stable nanoparticles. For example, stable gold nanoparticles are made from 5, 8, 13, 23 or 31 atoms.

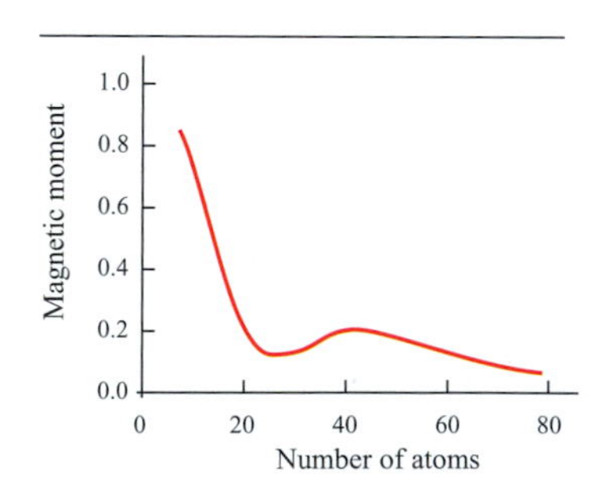

Fig. 19.28: Magnetic moment per atom of rhenium versus the number of atoms in the particle.

There are many nanoparticles which are *not* the result of subdividing a bulk material, the most famous being carbon buckyballs, namely C_{60}. As shown in Figure 19.30, C_{60} is a typical nanoparticle with a diameter of 0.5 nm. It is made out of 60 carbon atoms, all of which are on the surface, forming a domelike surface.

There is no hard and fast distinction between a nanoparticle and a large molecule. A single large biomolecule can be composed of 100 atoms, whereas a nanoparticle having a radius of 1 nm can have only 25 atoms. One of the distinctions of a large molecule is that, unlike many nanoparticles, it cannot be obtained by subdividing a bulk material. Of course, as mentioned above, not all nanoparticles can be obtained by subdividing a bulk material.

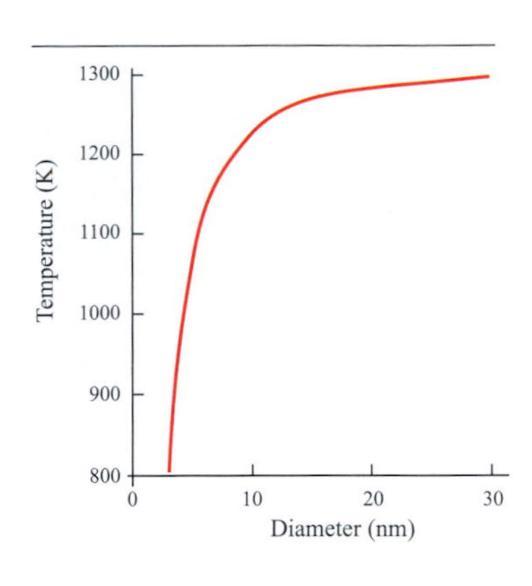

Fig. 19.29: Melting temperatures of 100 gold atoms' nanoparticles.

A simple criterion for a nanoparticle is that its energy levels should have a spacing that is larger than the thermal energy $k_B T$, which at room temperature of $T = 300$ K is equal to 0.026 eV. The spacing between the energy levels of a nanoparticle should go as $1/L^2$, where L is the linear size of the nanoparticle.

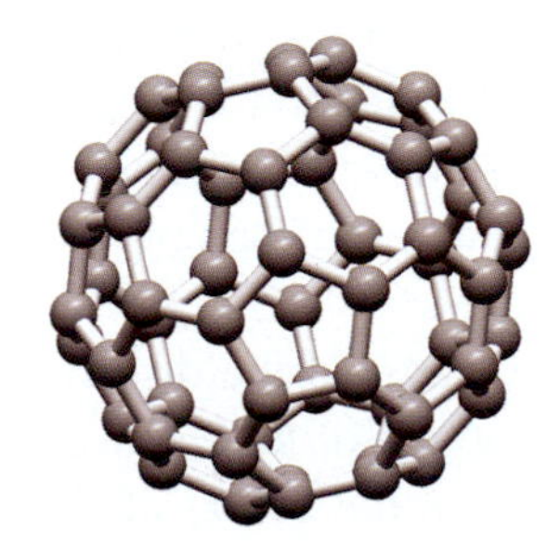

Figure 19.30: Carbon 60 "buckyball".

Nanoparticles have many novel properties leading to myriad applications. Nanosilver particles can kill bacteria and are being used to disinfect clothes and houses. Nanoiron can neutralize many toxins and tons of it have been pumped into contaminated groundwater in the United States. Many nanoparticles, such as those made by aggregating ZnO molecules, are being used in cosmetics. Nanoparticles can move through cell membranes and are ideal for drug delivery into the nucleus of cells; in particular, nanoparticles could carry medicine for killing cancerous cells and for repairing damaged cells and malfunctioning genes of the DNA.

19.13 Quantum dots: nano semiconductors

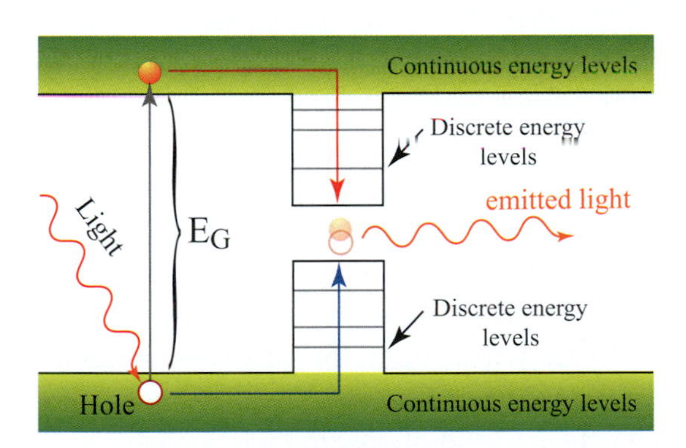

Figure 19.31: Quantum dot energy levels.

The behavior of electrical resistivity was seen, in Section 19.10, to change dramatically for a carbon nanotube. *Quantum dots* are a special class of nanoparticles that, in addition to having properties that are typical of a nanoparticle, are also *semiconductors*; we study how the concept of electrical conduction has special features for the quantum dot.

Figure 19.23b shows how the energy bands of a bulk conductor, such as a metal, are changed as one reduces its size down to a small cluster of atoms, namely a quantum dot: its energy band breaks up into discrete energy levels. The discrete states for a quantum dot arise due to its nanometer size and the fact that it behaves like a nanoparticle.

The energy spectrum of a quantum dot is shown in Figure 19.31. A typical quantum dot is a semiconductor with an *energy gap* E_G and other characteristics that are very close to its bulk material; the lowest energy states of the quantum dot are in *discrete* energy levels; as one goes to higher energy states of the quantum dot, the energy levels become continuous and form an energy band, as shown in the figure.

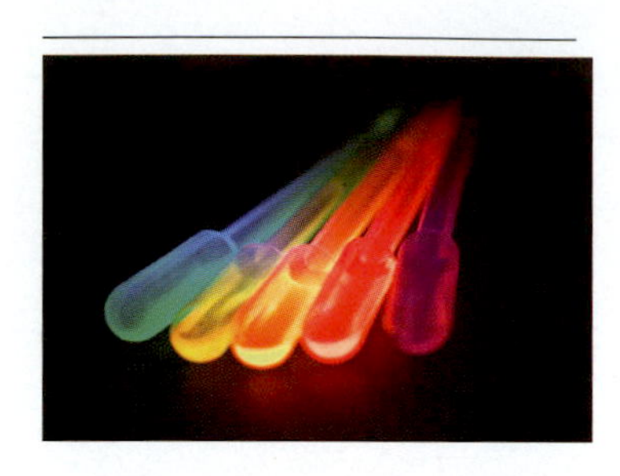

Fig. 19.32: Quantum dot colors. Source: Argonne National Laboratory.

On shining visible light on a quantum dot, an electron in the valence band absorbs energy and makes a transition to the conduction band, leaving a "hole" (absence of an electron) in the valence band, as shown in Figure 19.31. The electron loses energy due to collisions and migrates to the lowest (discrete) energy state available, which is one of the discrete energy levels of the quantum dot; the hole also, similarly, migrates to the highest available energy in the valence band, which is also a discrete energy level of the quantum dot. The electron-hole pair then combine — and in effect the excited electron returns to the valence band — with the energy of the electron-hole pair being emitted as radiation; the color of the radiation is usually in the visible range and near the infrared.

Quantum dots can be designed, by adjusting the energy gap and the discrete energy levels, to emit a range of distinct colors. The energy gap of an electron-hole pair for a quantum dot determines the energy (and hence the color) of the fluoresced light; for a quantum dot of linear dimension L, it can be shown that the energy gap goes as $1/L^2$, that is, inversely proportional to the square of the size of the quantum dot; L is usually 4 or 5 nm. A 1.67-nm quantum dot made from 123 silicon atoms emits green light under ultraviolet illumination; a quantum dot containing 29 silicon atoms glows with a bright fluorescent blue. Figure 19.32 shows the different colors emitted by different-size quantum dots when ultraviolet light shines on them.

Larger quantum dots have more closely spaced energy levels in which the electron-hole pair can be trapped. Therefore, electron-hole pairs in larger dots take a longer time to recombine. A large quantum dot is cadmium selenide (CdSe) coated with zinc sulphate and which, in turn, is coated with a polymer. The core size varies from 2–10 nm. Transmission electron microscope images show the quantum particles form a regular lattice giving rise to perfectly ordered (nano-sized) crystal — containing almost 50,000 atoms and having almost all the properties of the bulk crystal. The CdSe quantum dot also behaves as a nanoparticle, having discrete energy levels.

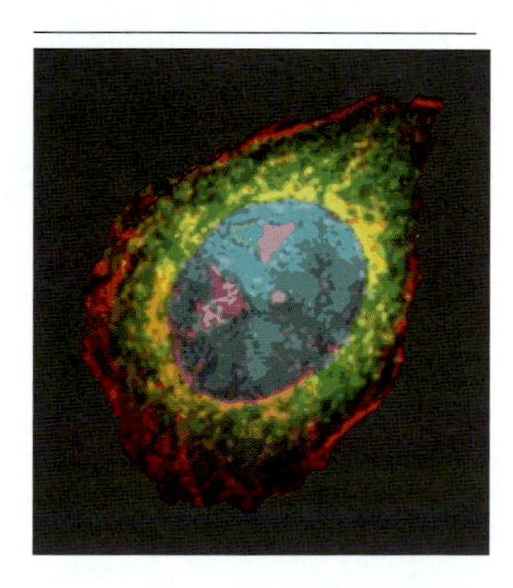

Fig. 19.33: Illustration of how the structure of a cell can be illuminated with quantum dots.

Since quantum dots can emit different characteristic colors, they provide ideal markers for studying biological processes at the molecular scale. Latex beads filled with several colors of quantum dots can potentially serve as unique labels for any number of different probes. In response to ultraviolet light, the beads would identify themselves (and, thus, their linked probes) by emitting light that separates into a distinctive spectrum of colors and intensities, a kind of spectral bar code.

CdSe has important applications in biology. Researchers have used CdSe quantum dots with different signature colors to study the detailed structure of the cell; for example, the outer cell wall can be illuminated by green quantum dots and with different components of the inner cell nucleus being illuminated in blue and pink quantum dots, as shown in Figure 19.33.

19.14 Nanoscience and nanotechnology

The gecko's ability for climbing walls shows the workings of "nanotechnology" in nature. The example of miniaturizing the Encyclopedia Britannica illustrates the possibilities of man-made technologies working at the nanoscale.

There are two noteworthy features of nanoscience and nanotechnogy. First, the fundamental principles required for nanoscience are already well established; the challenge is to understand how these principles are to be applied in a multifaceted situation and in a novel manner. Second, nanotechnology involves the fabrication and shaping of objects and devices at the atomic and molecular scale — in some cases atom by atom. By manipulating the very atoms and molecules that compose matter, a whole new world of phenomena as well as cutting-edge technologies becomes accessible.

The nanoworld may seem familiar, but there are many new effects that are unique to this scale. For example, gold nanoparticles can be oxidized, whereas a large piece of gold cannot. Many well-known concepts such as energy, resistance, current and so on are realized in the nanoworld quite differently from their behavior for macroscopic objects.

Nanoscience encompasses a wide range of phenomena and is an exciting subject that brings together the traditional disciplines of mathematics, information science, physics, chemistry, biology and engineering. Scientists are discovering mesoscale laws by fashioning new and original nanosystems and measuring their rather unexpected behavior.

A large object usually obeys the laws of classical physics. We know that classical physics fails for the microscopic world of the atom, which is, instead, described by the laws of quantum mechanics, as discussed in Chapters 23 and 24. One may ask, at what approximate scale do the laws of classical physics fail? It turns out that it is for nanoscale objects — ranging in size from 1–100 nm — that a purely classical description begins to fail.

An exact and entirely quantum description of the nanoworld, from first principles, should indeed give the correct results. However, due to the complexity of a nanoscale object — composed as it is of a large collection of atoms or molecules — a completely quantum calculation of its properties and behavior is, unfortunately, intractable. Furthermore, a description of the nanosystem by simply extrapolating from the well-understood quantum behavior of systems with few atoms or molecules also fails.

The nanoworld is best described by a *combination* of both classical and quantum principles: due to its size, there are some aspects of nanophenomena for which a classical description is adequate and for others a purely quantum approach is necessary. It is the interplay of classical and quantum behavior that is unique to the nanoworld and results in many new and interesting properties.

19.15 Conclusions

The gecko's feats are macroscopic expressions of nanoscale processes and structures. The advent of nanotools, such as the scanning electron microscope and the atomic force microscope, has opened the way for a quantitative and empirical study of nanophenomena. Given the small number of atoms that constitute nanostructures, the role of surfaces is very important. All the four kinds of nanotools we discussed are most effective when one is examining the surface of a sample and hence are ideally suited for studying the nanoworld.

The study of the gecko's ability leads to the study of the van der Waals and Casimir forces, which become strong only at the nanoscale. The possibility of emulating the gecko's feats using synthetic materials has lead us to the study of the amazing properties of carbon nanotubes. The study of nanowires made from carbon nanotubes shows how our understanding of macroscopic laws needs to be expanded and enhanced. Nanowires lead us naturally to the study of small clusters of atoms that behave as a single entity, namely the nanoparticles. The quantum dot is an example of a nanoparticle and was analyzed to exemplify the behavior of such nanostructures.

The nanoworld holds the possibility of miniaturization that could change the entire technical basis of human civilization. It is difficult to overestimate the importance of nanotechnology. It is the opinion of most of the experts that the total social impact of nanotechnology is expected to be much greater than that of the silicon-based integrated circuit because nanotechnology is applicable to many more fields than semiconductor-based hardware.

One can hope that we are on the threshold of a technological revolution far greater than anything that mankind has so far imagined or experienced. The fruits of nanotechnology will, hopefully, improve the life of the great majority of mankind as well as protect the earth and our environment.

19.16 The answer

The principles underlying the nanoworld are the reason that a gecko can walk on the ceiling. The gecko toe pad has nanosize hair that are attracted to almost any surface by nanoscale van der Waals and Casimir forces. The gecko's behavior is a demonstration of how man-made nanotechnology can arise from studying biological systems.

19.17 Exercises

1. Which type of fundamental force is mainly used when a gecko hangs on the ceiling?
2. What is the basic mechanism of an atomic force microscope?
3. Name one way in which a quantum dot can be modified to change its color.
4. Would it be possible to use quantum dots to generate infrared radiation?
5. What happens to the energy gap of a quantum dot when its size is increased?
6. Would it be possible to have a quantum dot with a diameter of 0.01 nm?
7. Is the theoretical adhesion strength of a single gecko foot enough to hold up a little child?
8. Can carbon nanotubes be conductors?

9. Give a sketch with the rolling vector R(7,7). Is this a zigzag formation?
10. Why is the Casimir force a short-range force?

20: Complexity

Q - Why Do We Need Only a Small Number of Genes?

Chapter Map

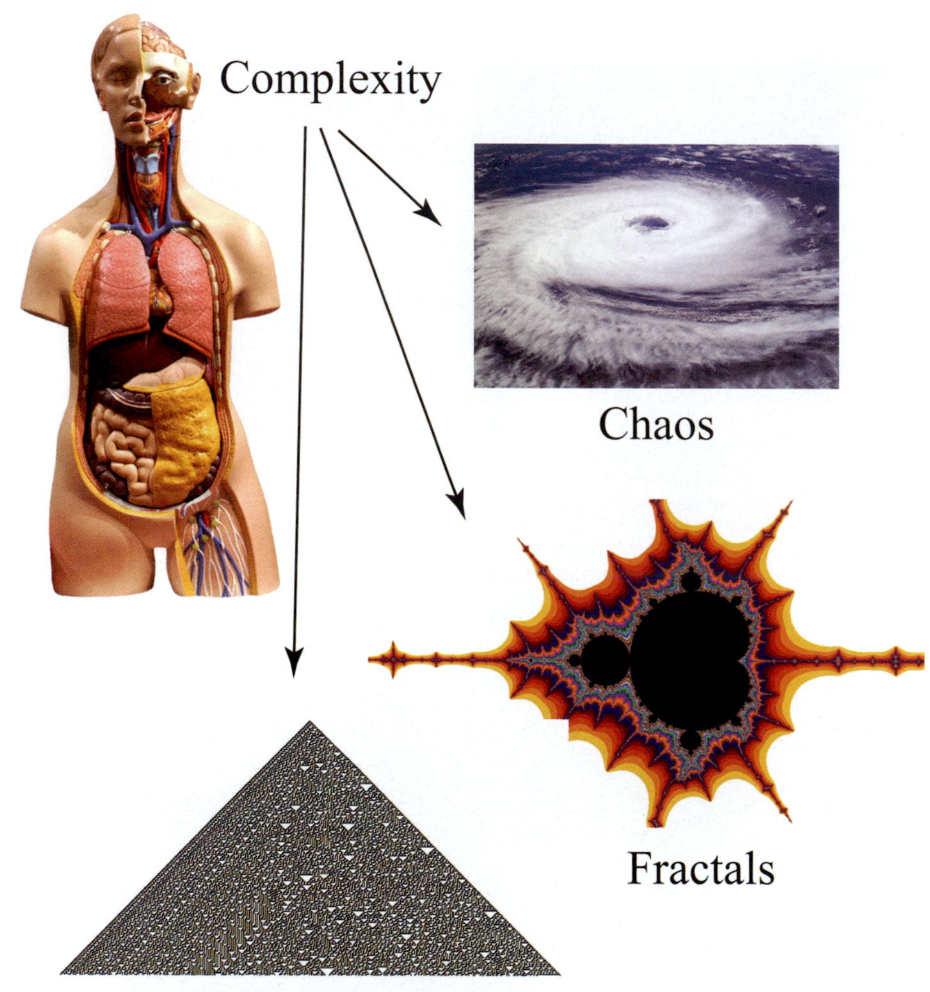

20.1 The question

Figure 20.1: Intricate patterns characterize many leaves.

Fig. 20.2: Is there a blueprint for a leaf?

In the natural world, both living and non-living, we often encounter intricate patterns of stunning complexity and beauty. A leaf vein pattern as in Figure 20.1 shows both a high level of organization and regularity as well as a significant level of irregularity. For example, larger veins branch into smaller veins but the branching is neither perfectly symmetrical nor very evenly spaced. How could such a leaf be constructed? Is there a basic "engineering" blueprint which specifies where veins (roughly) should branch? Isn't it so that all hereditary information is passed on from generation to generation in genes, and since shapes and vein patterns can be radically different for different plant species must these not be part of this hereditary information? Surely then, there must be some rules for this encoded *in* the DNA of the plant in the form of genes. Although that is basically the case, the question is: What *kind* of rules would that be? Indeed one may wonder how many and what kind of rules are necessary to code for the patterns in a leaf, or indeed to code for an entire leaf.

Before we pursue this matter further, let us think beyond the leaf and consider the entire plant. It is immediately clear that a blueprint must be enormously complicated if there is indeed a blueprint composed of genes for everything. Now if it's complicated for a plant, it must be incredibly complicated for a human being. After all, a human being not only has all kinds of organs but also self-awareness. That can't be simple! The stunning discovery of the late twentieth century was that plants can have more, and some worms nearly as many, genes as human beings. Human beings turned out to have only about 25,000 genes, an almost unimaginably small number considering how complex a human being is. Thus, we arrive at the question for this chapter: **Why do we only need a small number of genes?**

To answer that question, we first have to realize that every complex organism (be that a bacterium or an elephant) can only be what it is due to the interactions between its constituent parts. For example, just putting together the "parts" that make up an elephant (a brain, a kidney, etc.) will not bring it to life.

Hence, we need to investigate how things that interact behave collectively. For this there are two key aspects: the interacting units, and the interaction rules. As so often in science, it is a good idea to start with as simple a system as possible. For the interacting units, that would mean that they are all the same. As we know from a solid built from identical atoms, just having identical interacting units does not easily give us complex behavior. After all, in terms of complexity, a solid like a block of iron is nothing when compared to a bacterium or a human being. Well if simple units do not give complex behavior, how about simple interaction rules? Would it be possible to have simple interaction rules and complex behavior? To gain some insight into this question, let us go back to readily observable patterns in nature, such as the vein pattern in Figure 20.1.

20.2 Fractals

Figure 20.3: Broccoli romanesco. Photo: Jon Sullivan

We have already discussed the vein patterns in leaves. Another very nice example of a pattern in nature is shown in Figure 20.3 that gives a beautiful closeup of a broccoli romanesco. What is especially striking is that the smaller buds appear to look exactly like the bigger buds they are part of. In other words, parts of the structure look like the whole but smaller.

Would it be possible to create this kind of pattern with the help of simple rules? This was a question investigated by the Hungarian biologist **Aristid Lindenmayer** who, in 1968, first developed what are now called **L-systems**. Basically, L-systems are a set of (often simple) rules that are applied over and over again.

Let us have a brief look at one of the simplest L-systems. We have two symbols, A and B, and use the following replacement rules:

$$\begin{aligned} A &\rightarrow ABA \\ B &\rightarrow BBB. \end{aligned} \tag{20.1}$$

Start with A and apply the rules over and over again as shown in Figure 20.4 where the colors were added for easy visibility of the emerging pattern.

Step	Pattern
1	A
2	ABA
3	ABABBBABA
4	ABABBBABABBBBBBBBBABABBBABA

Figure 20.4: Time evolution of an L-system.

This may appear a children's game, devoid of a deeper scientific meaning, perhaps nothing more than an illustration of what an L-system does. Surprisingly though, this extraordinarily simple L-system has a direct connection to some very fundamental late nineteenth century mathematics!

So before continuing our discussion of L-systems, let us digress a bit into a mathematical object called a **fractal**. In the late nineteenth century, the German mathematician **Georg Cantor** was concerned with sets of numbers and, for example, proved that there are more real numbers than natural numbers (a far-reaching notion as there are infinitely many natural numbers!). In one of his studies he introduced what is now known as the **Cantor set**. The construction of the standard Cantor set is quite simple and illustrated in Figure 20.6. Start with a line and remove the middle third. This leaves us with two shorter line segments. Next, remove the middle third of those two line segments to obtain four line segments. Continue this process ad infinitum to obtain a Cantor set.

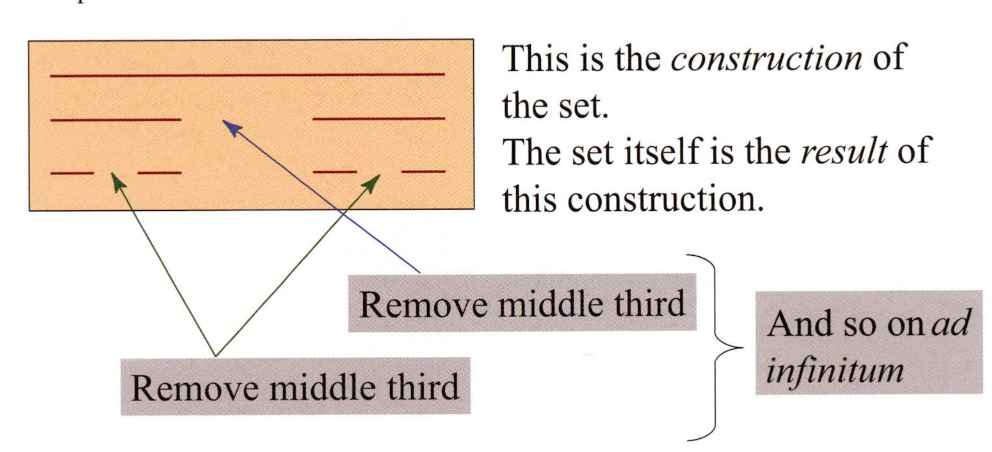
This is the *construction* of the set.
The set itself is the *result* of this construction.
Remove middle third
Remove middle third
And so on *ad infinitum*

Figure 20.6: Construction of the Cantor set.

Fig. 20.5: Georg Cantor, the creator of set theory.

Some rather counterintuitive things can be proved regarding the Cantor set. For example, it can be shown that infinitely many points remain despite having removed an infinite number of line segments. However, what is more important here is that the Cantor set has a dimension of about 0.63 and is an example of a fractal. This is in stark contrast to a point that has the integer dimension 0, a line that has the integer dimension 1 or a plane that has the integer dimension 2. It is rather surprising, and indeed it was shocking to the mathematical world at the time of the discovery, that geometric objects exist whose dimension lies between a point and a line. Although the actual definition of a fractal is quite hard, a rough definition is that a fractal is a geometric object with a noninteger dimension. Of course, this assumes that it is clear how to define a noninteger dimension, which is not entirely the case. Indeed, there are several possible definitions, a common one being the so-called **Hausdorff dimension** that was used above in the calculation of the dimension for the Cantor set giving 0.63. Although many fractals do have a noninteger Hausdorff dimension, there are some notable exceptions, with among them one of the most famous fractals, the Mandelbrot set, whose boundary has a Hausdorff dimension of 2 (see also p. 453).

Let us reflect on that one more time: A dimension of 0.63, being between 0 and 1, means that the Cantor set is neither a point nor a line but something in between. This is a rather amazing property. Another key characteristic of the Cantor set, which is very common among many fractals, is that parts can look the same or similar to the whole (does this remind us of the broccoli romanesco picture?).

Now, if we return to our L-system above in Equation 20.1, we see that it is exactly the same as a Cantor set if we take B to stand for "no line" and A for "line" and furthermore at each step shrink the string of letters to a given fixed length as illustrated in Figure 20.7. Thus we already see from this example that simple rules can give rise to unexpected properties. This is an encouraging start to our quest of finding out a possible reason for why one can do with only a small number of genes.

Figure 20.7: The L-system described is equivalent to a Cantor set.

As is the case for L-systems, the notion of fractals is an excellent tool for modeling biological patterns. However, while the Cantor set may be elegant in its simplicity, can more naturelike patterns be created with L-systems? Figure 20.8 shows the result of a still fairly simple L-system where the appearance of some weeds is modeled. The resemblance is stunning, and from a distance one can very well imagine these weeds to be pictures or illustrations of real weeds.

Figure 20.8: Fractal weeds. Image: Wikipedia user SolKoll.

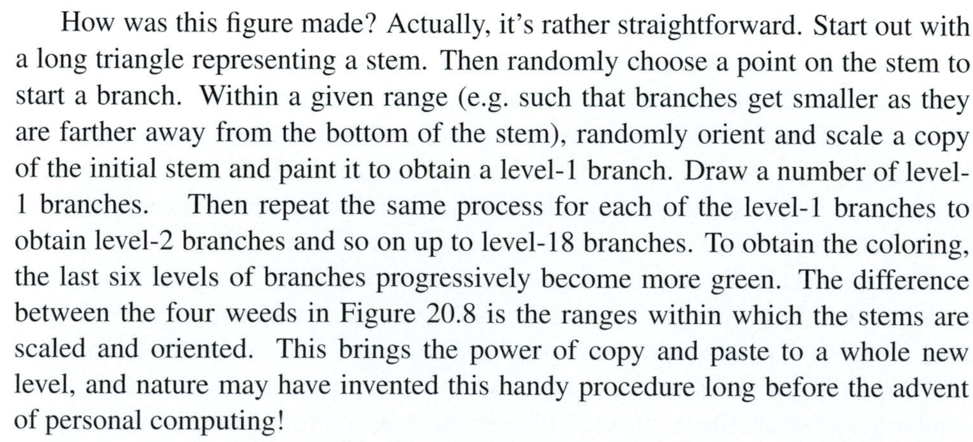

Fig. 20.9: Real weeds at sunset.

How was this figure made? Actually, it's rather straightforward. Start out with a long triangle representing a stem. Then randomly choose a point on the stem to start a branch. Within a given range (e.g. such that branches get smaller as they are farther away from the bottom of the stem), randomly orient and scale a copy of the initial stem and paint it to obtain a level-1 branch. Draw a number of level-1 branches. Then repeat the same process for each of the level-1 branches to obtain level-2 branches and so on up to level-18 branches. To obtain the coloring, the last six levels of branches progressively become more green. The difference between the four weeds in Figure 20.8 is the ranges within which the stems are scaled and oriented. This brings the power of copy and paste to a whole new level, and nature may have invented this handy procedure long before the advent of personal computing!

Thus we see that a simple rule can lead to what appears to be a biologically relevant outcome. Would there be other types of rules that lead to unexpected behavior? The answer is yes! In so-called chaotic systems, certain types of very simple equations give astonishingly complex outcomes. Let us therefore investigate chaos a bit. But before doing so, let us digress briefly into the history of chess.

20.3 Chaos

Chess is one of the most popular board games played all around the world. It has a square board with a total of 64 fields and a number of different figures that can move on the board. Chess is a strategy game that was probably invented in India or China around two millennia ago, and there is a nice fable surrounding its invention that illustrates how small differences can become large differences very fast.

Somewhat paraphrased, the fable roughly proceeds as follows: When the king was very bored, he called upon his subjects to invent a new game and he promised a huge reward for the person who would bring him a truly entertaining game. Eager to earn the reward, many subjects showed their invention to the king but nothing pleased him. Then one day, the game of chess was shown to him and he was most delighted. When the king asked the inventor what he'd like for his reward, the inventor apparently rather modestly replied "Just one grain of rice" and then after a short pause "on the first field, two on the second field, four on the third field, and so on until the board is full". The king immediately accepted and ordered rice to be brought over from the city store. However, it soon turned out that the king had been fooled and the inventor was executed for his impudence.

What went wrong? Why was the king so angry? Well, let's calculate how many grains the inventor would end up with. If we have 64 fields and double the number of grains from field to field, we end up with

$$
\begin{aligned}
N_{\text{grains}} &= \sum_{i=0}^{63} 2^i = 2^{64} - 1 \\
&= 18{,}446{,}744{,}073{,}709{,}551{,}615 \approx 18 \times 10^{18}
\end{aligned}
$$

i.e. about 18 exa (quintillion) grains. Note that "exa" is the SI unit for 10^{18}. Well, that's obviously a lot! But how can we picture that? The grain of a typical medium grain rice is about 5 mm long and about 2 mm wide with (very roughly) a volume of 10 mm^3 = 10^{-8} m^3. Therefore, 10^{18} grains require about 10^{10} m^3 (of course, in reality, there would be some space between the grains and hence the volume occupied would be larger). This volume is more than that of a cube that is 2,000 m wide, 2,000 m deep and 2,000 m high — dwarfing anything humankind had ever built. Or put differently, this quantity of rice could fill up around 4,000 great pyramids of Giza. With the area of modern India being about 3×10^{12} m^2, this would mean that there would be enough grain to cover all of India with a grain layer about 3 mm thick! Nobody has that much grain and it is little wonder that the king was angry.

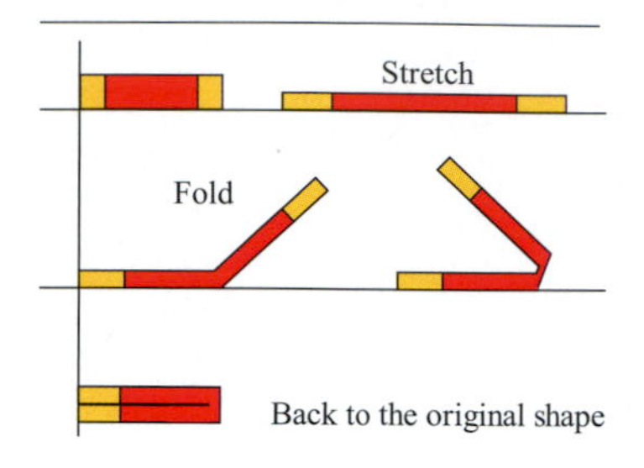

Fig. 20.10: The process of stretching and folding lies at the heart of chaos.

Now what does this have to do with chaos (the title of this section)?. The thing is that it took only 63 doublings to obtain the 9 exa grains on the last field. Just imagine how long it will take to say "double" 63 times. Less than a minute! So, if we have a system in which some kind of doubling or something similar occurs, even a minuscule value can become very large in a very short time. In other words, if, for example, we have two points that are 10^{-18} m apart, that is around 1/1000th the size of a nucleus of a single atom, the distance will grow to 9 m in those 63 doublings. That is amazing indeed, but the question remains: What does this have to do with chaos? And what is chaos?

We'll get to that shortly, but first let us reflect on this rapid growth a bit more. If we have a real system in which something like a doubling occurs, then clearly if a tiny distance becomes large, after repeating the process with this large distance a number of times it will become enormous. Or continuing the previous example, if we further double the 9 m 63 times, we obtain a distance of nearly 1,000 light-years or about the thickness of our Milky Way outside of its galactic core (i.e. by doubling a mere 126 times we go from the subnuclear to the galactic). Obviously this wouldn't work for any system on earth. Or would it?

The thing is that in real systems, growth needs not to continue in one single *spatial* direction. It is perfectly legitimate for folding to occur. Let us imagine our two points above to be on a near infinitely thin sheet and separated by a distance of 10^{-18}m. If we stretch out the sheet to double its size, the distance between the points will also double. Next we then fold over the sheet as shown in Figure 20.10. In two dimensions, the distance between the points may very well decrease as a result of the folding process, but *within* the sheet the distance does not change due to the folding process, as illustrated in Figure 20.11. Therefore, if we stretch and fold repeatedly, we can have a small system and still have two points separating from each other very fast.

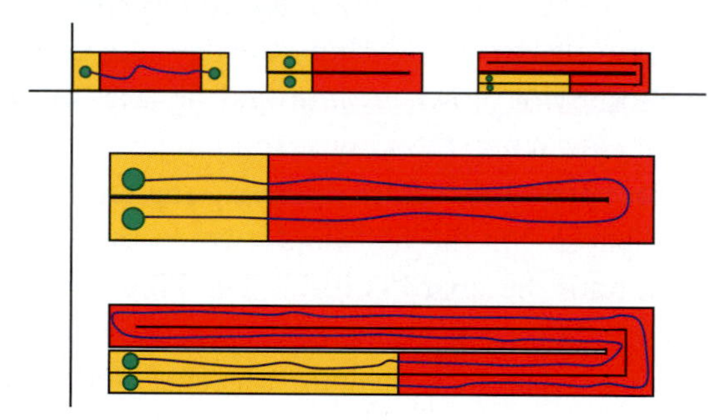

Figure 20.11: The distance *within* a layer grows exponentially fast.

Indeed, this is exactly what happens in **chaotic systems**. It is the principle of **stretch** and **fold**. Any tiny error, however small, will be amplified to the size of the system in a comparatively short time as is illustrated in Figure 20.12, where it can be seen that two nearby points could be spread out anywhere on the interval after some stretches and folds. The consequence is that the location of a point cannot be predicted unless its initial location (i.e. its starting position) is *exactly* known. Of course in the real world, nothing is exactly known!

There is a commonly used technical phrase that expresses the notion of the preceding paragraph: **"Sensitive dependence on initial conditions"**. The outcome "sensitively" depends (i.e. depends on the tiniest of tiniest differences) on the initial (starting) condition. This phrase is probably the best short way to characterize a chaotic system.

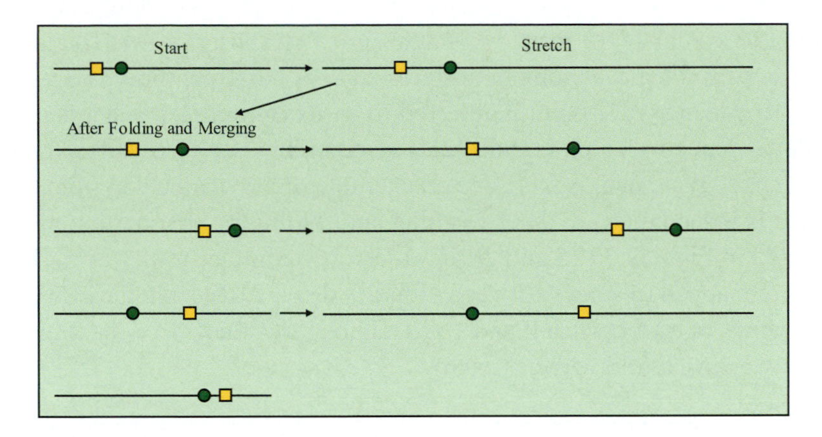

Figure 20.12: Stretching and folding on a line. Note the circle and dot have swapped positions from initial to final state.

Before we continue to look at how stretching and doubling can be achieved mathematically, let us illustrate the process with a "kitchen" example: the making of croissants. To make a good croissant, one needs pastry with many layers. To obtain that pastry, one starts with a simple piece of dough. Next, the dough is rolled out to, say, two or four times its size and then it is folded over once or twice to reduce the size of the sheet. This process is repeated a few times and the pastry is ready.

Fig. 20.13: How many layers could one make in three minutes?

How many layers would the pastry have? It is easy to roll out and fold the dough about 10 times. If each time we roll it out to four times its size and then fold it back twice so that the number of layers is multiplied by four, we will have a staggering $4^{10} = 1{,}048{,}576$ layers in the end. Yes, that's more than 1 million layers! If one observes the folding process carefully, it can also be seen that bits of flour from one layer rarely cross over to another layer and hence that the distance between them within the layer becomes very large.

A chaotic system can display sensitive dependence on initial conditions.

In science, one always strives to describe phenomena, notions or ideas mathematically. Hence we would like to do so here, too, and at first one would be tempted to think that the mathematics of stretch and fold must be pretty complicated. In fact, while this is certainly so for a more in-depth analysis, the basics are pretty simple. Perhaps the most generic example of a chaotic equation is the so-called **logistic map** given by

$$x_{n+1} = 1 - \alpha x_n^2 \qquad \textbf{Logistic map.} \tag{20.2}$$

A map is a mathematical equation that is usually **iterated** (i.e. the outcome of a step is the input for the next step as will be further illustrated below) and hence has discrete time steps. Often maps have something called a **parameter** like α in Equation 20.2. The idea of a parameter is usually to choose it once before the calculation starts and keep its value fixed in subsequent calculations.

In a map (or a function) we have an outcome based on some input. In a function plot, also known as a return map, the possible outcomes are graphed versus some or all of the possible inputs. The return map for the logistic map Equation 20.2 is given in Figure 20.14 where the value of x_{n+1} is plotted versus all possible values of x_n. At step one, we calculate x_1 by Equation 20.2 from a given starting value

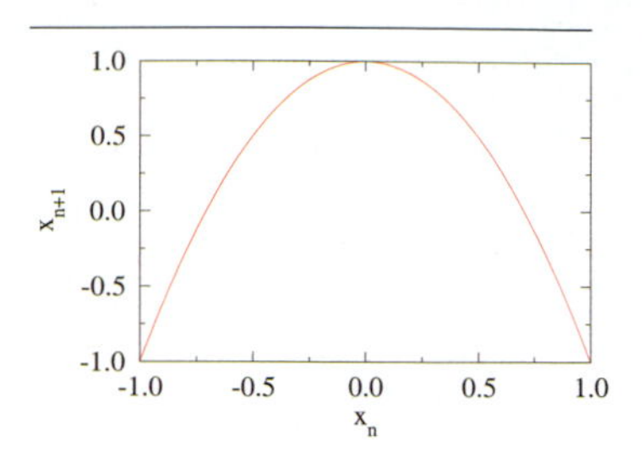

Fig. 20.14: Function plot of Equation 20.2 for $\alpha = 2.0$.

x_0. Next, we calculate x_2 from x_1 and so on. The subscript n is like time that moves in steps and hence is sometimes referred to as "discrete time". As in general the time dependence of a point is referred to as its **dynamics**; systems that have a time dependence based on certain rules or equations are also called dynamical systems. The logistic map is therefore an example of a dynamical system.

That it is indeed the case that stretching and folding occurs in the logistic map is illustrated in Figures 20.15 and 20.17. It can be seen that values of x_n between -1 and 0 are mapped to values between -1 and 1 for x_{n+1} (i.e. we have stretching), and that values of x_n between 0 and 1 are mapped to values between 1 and -1 for x_{n+1} (i.e. we have a fold with a stretch).

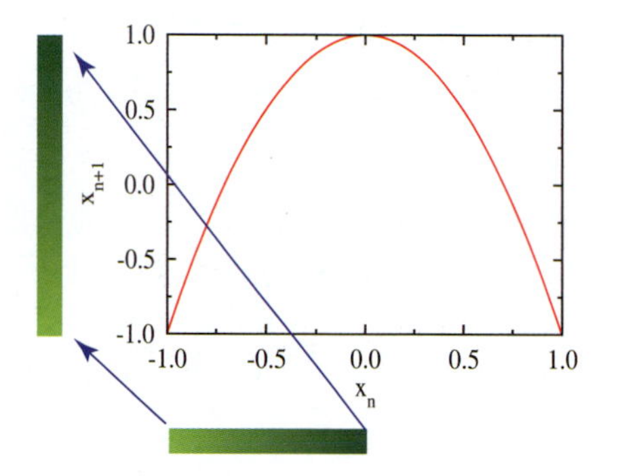

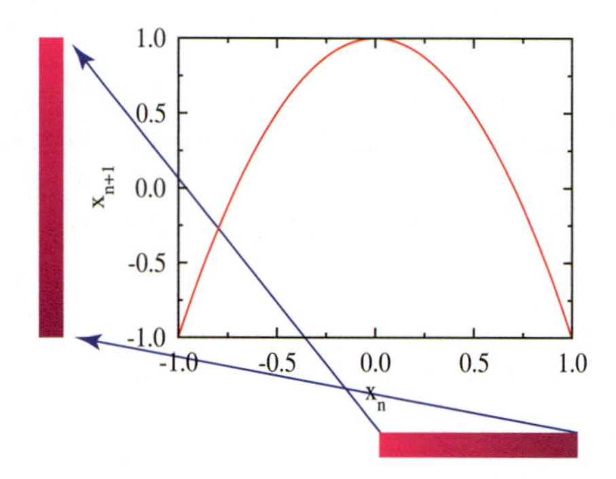

Figure 20.15: The stretching and folding as it occurs in a logistic map.

In more detail, Equation 20.2 means the following: The value of our variable x at time $n + 1$ is given by subtracting from 1 α times x^2 at time n. Well, that's pretty simple. To start off, we need to provide an initial value at time 0, for example $x_0 = 0.3$, and then we can calculate x_1 from that with the help of Equation 20.2 as $x_1 = 1 - \alpha x_0^2 = 1 - 2 \times 0.3^2 = 0.82$, where the parameter α was chosen to be $\alpha = 2.0$. Next, we take $x_1 = 0.82$ as our new "initial" value and calculate x_2 as $x_2 = 1 - \alpha x_1^2 = 1 - 2 \times 0.82^2 \approx -0.34$. Thus we continue as long as we wish.

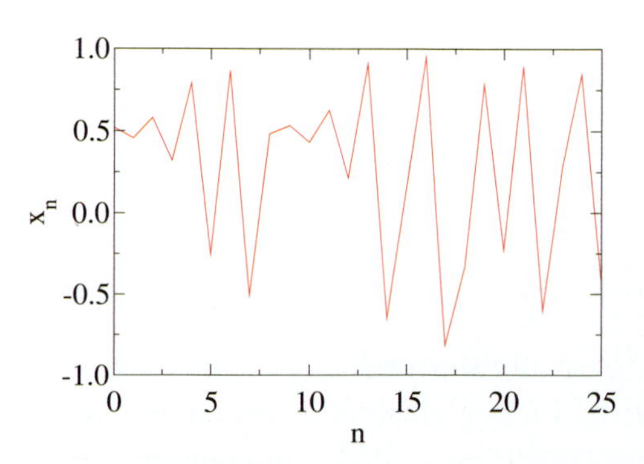

Figure 20.16: Time series of the logistic map for $\alpha = 2$.

Of course, it would be nice to have a graphical representation of the successive values $x_0, x_1, x_2, \ldots$. This is done in Figure 20.16 (a so-called time series). As clearly can be seen, the values jump all over the place. The way the values jump around looks pretty "chaotic" and indeed that is why chaos has its name.

Recall we had chosen $\alpha = 2.0$. Does this choice matter? Indeed it matters a lot as can immediately be seen by considering $\alpha = 0$. Doing so, we have the equation $x_{n+1} = 1 - 0 \times x_n^2 = 1$. Ergo all values are 1 and in a time series as in Figure 20.16 we obtain a straight line. This is clearly a completely different behavior from the chaotic jumping around we see for $\alpha = 2$, and we should therefore wonder what would happen for different parameter values α between 0 and 2.

One thing we could do is to graph the time series for different values of α and see what happens. Certainly, that would be interesting, but if we like to have an overview of the effect of a changing parameter α then looking at a large number of time series is not quite ideal.

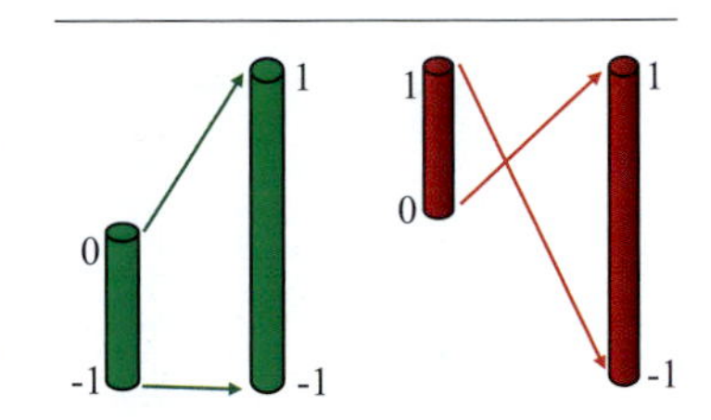

Fig. 20.17: Stretching and folding in the logistic map.

We'd like to have all the key information in one graph. Since we are looking at the effects of a changing α, this sort of automatically implies that the x-axis of that graph needs to be α. But then, what should be plotted on the y-axis? As we are trying to represent information from the time series, it must be a value of x_n. The question of course is: Which one? A little bit of reflection shows that a single value would not be enough because that would not tell us whether the value jumps around or not. Fortunately, there is nothing to stop us from plotting a number of successive values of x_n for each value of α; and that is exactly what is depicted in Figure 20.18.

Figure 20.18 is called a **bifurcation diagram** for it has many lines that split into two (in mathematics this splitting in called a bifurcation). It was obtained in the following way: Start with $\alpha = 0$, calculate x_{500} to x_{700} and plot the 200 points (i.e. the pairs $(0, x_{500})$, $(0, x_{501})$, ...). The starting point x_0 of the calculation can in principle be any value between -1 and 1 but an odd number works best and hence we've chosen $x_0 = 0.3$.

Since the map may need some time steps to "settle" into the final lasting pattern, we don't plot x_1 to x_{499}; instead, we plot the value of x_n for large n, which we have taken to be equal to 500 onwards. Note that the exact number of time steps to discard and the number of steps to plot is not important. After we're done plotting the 200 points for $\alpha = 0.0$, we continue in the same fashion but for a slightly increased value of α (here we chose to increase α by 0.01). When done, we again increase α and keep on doing so until we reach $\alpha = 2.0$.

The structure of the bifurcation diagram is more than just a little bit intriguing! First, we see that for small α, Equation 20.2 has a so-called **fixed point**. That is to say, for α having values in the range $[0, 0.75)$, for large n the successive values of

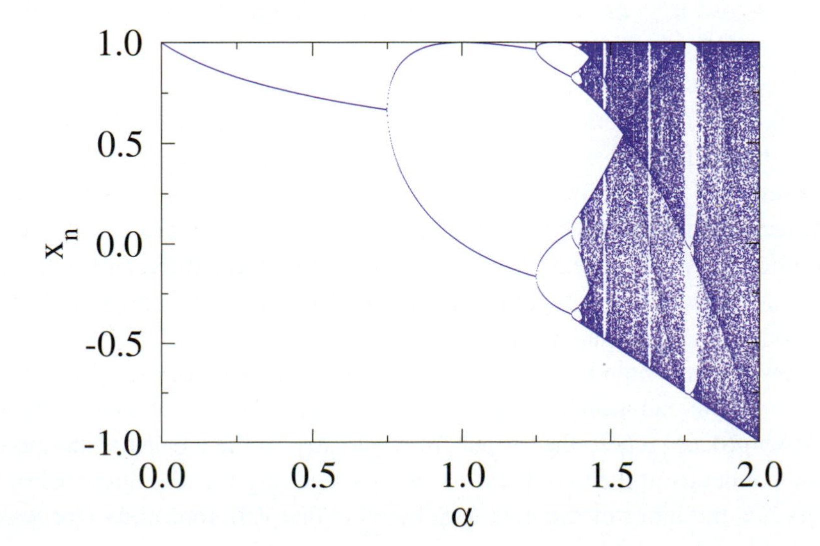

Figure 20.18: Bifurcation diagram of the logistic map.

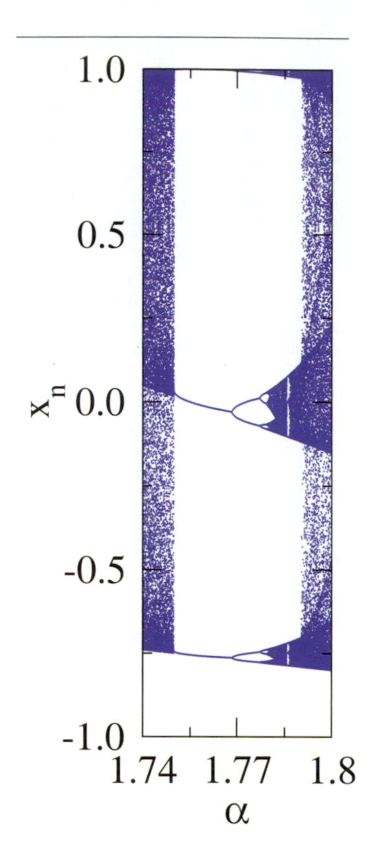

Fig. 20.19: The window opening at $\alpha = 1.75$. Note the three mini bifurcation diagrams.

x_n are all the same, that is, ($x_{n+2} = x_{n+1} = x_n \ldots$). Then at $\alpha = 0.75$, the fixed point splits (bifurcates) and successive values of x_n obey a period two pattern. That is to say, for large n, we have $x_{n+2} = x_n$ and $x_{n+3} = x_{n+1}$ and so on. In other words we see that the period of the logistic map *doubles* at $\alpha = 0.75$.

As can be seen by closely inspecting Figure 20.18, the doubling of periodicity for increasing α continues by becoming 4 at $\alpha = 1.25$, 8 at $\alpha \simeq 1.37$ and so on until infinity at $\alpha \simeq 1.40$. There are two things to note here. First, an infinite number of period doublings occurs within a *finite* range of α. Second, when the periodicity reaches infinity, the time series is chaotic.

What is particularly interesting about the period-doubling cascade is that it turns out to be a **universal** property of maps and functions with a shape similar to the one of the logistic map given in Figure 20.14 (note that the word "universal" is used here to indicate a shared property among similar maps). Not only is the doubling as such universal, it can be proved that if one considers certain ratios of the parameter values at which successive bifurcations occur, these tend to a universal constant. This constant is roughly equal to 4.7 and named **Feigenbaum constant** after its discoverer **Mitchell J. Feigenbaum**.

The bifurcation diagram shown in Figure 20.18 has many remarkable properties. For example, at $\alpha = 1.75$, a so-called **window** opens where the chaotic time series suddenly becomes periodic again (see Figure 20.19). It turns out that there is an infinite number of such windows though most are too small to be seen in the figure. Another fascinating feature is that there are an infinite number of copies of the bifurcation diagram embedded in itself and hence it is a type of fractal.

For parameter values of α where the fixed points are not periodic, like, for example, for $\alpha = 2.0$, we encounter the above mentioned sensitive dependence on initial conditions. In other words, we start with, say, $x_0 = 0.3$ and look at the points obtained after some steps, e.g. $x_{100}, x_{101}, \ldots$. Then start from a very nearby point like $x_0 = 0.3000000001$ and look again at the points obtained after the same number of steps, i.e. $x_{100}, x_{101}, \ldots$. What we find is that the points in the first and second case are completely different! In a chaotic system, extremely small differences become very large very quickly.

Fig. 20.20: How different are chimpanzees from human beings?

Although one may debate how different chimpanzees and human beings are, it is perhaps not unreasonable to state that we see something similar. Genetically speaking, the difference between a chimpanzee and a human being is only about 1.2% in terms of base pairs, yet clearly human beings and chimpanzees are radically different in many aspects (society, technological prowess etc.). Thus we have an example of a small genetic difference leading to a large difference in outcome. From this, one could be tempted to conclude that the gene expressions for both human beings and chimpanzees are chaotic!

So now we have seen that a simple L-system can reproduce the shape of a weed and that a simple equation, such as the logistic map, can give chaotic behavior in an iterative process where the output from one step is the input for the next step. If we look at nature and genes, however, we see not only that the output of one step may serve as the input of the next step but also that different units (for example, different genes) interact.[1]

[1]Chaotic behavior is a special case of nonlinear equations in science. Nonlinearity means that

One of the simplest ways to model such interactions is with the help of systems called **cellular automata**.

20.4 Cellular automata

Usually when we have organisms in proximity, the behavior of one organism is influenced by other organisms in the neighborhood. Similarly, in multicellular organisms, cells are influenced by neighboring cells. Indeed, the same is also true for matter in general. For example, an atom by itself cannot be in a gaseous, solid or liquid phase — it can only do that by interacting with other atoms.

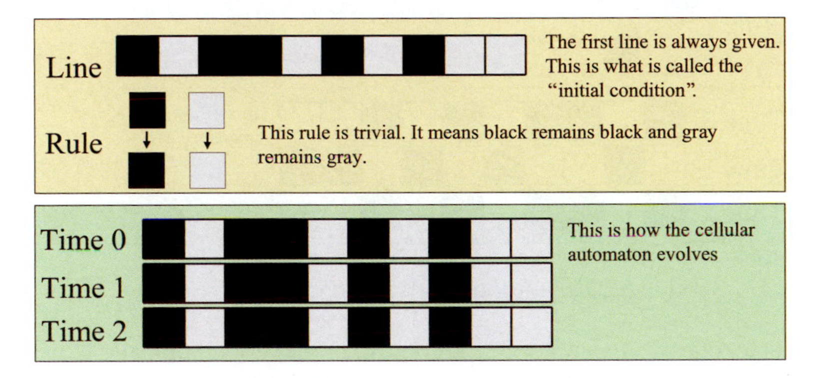

Figure 20.21: A trivial cellular automaton.

If we try to abstract those facts, we see that to model such systems, the very least we need are some discrete elements (called "cells" in this case) that can have some values (such a value is called the element's "state") and these elements somehow need to be linked. The simplest possible element is one with two states (for example, black and gray) and one of the simplest possible ways to link them is by arranging the elements on a line.

Of course, arranging a number of cells on a line isn't quite enough. We also need to have instructions on how the states of these cells evolve over time. One trivial instruction would be to simply remain the same as is shown in Figure 20.21. The idea of an automaton is that, once started, it evolves without further intervention by applying the set of instructions over and over again. Thus we see why a model as that of Figure 20.21 is called a **cellular automaton**.

As is the case with all models, one has to start somewhere. Therefore, the states at time 0 are assigned. In Figure 20.21, this is done randomly, but very often in the study of cellular automata, one starts with a single black cell in the center and keeps all the other cells gray. In order to obtain the states of the cells at time 1, the instructions are applied to the state of every cell at time 0, one by one until the entire line is done. Then, we can continue in the same fashion to obtain the states of the cells at time 2 by applying the instructions, but now to the states of

the equations do not depend linearly on the previous state, and in general nonlinear systems can produce results that are full of surprises since, unlike linear systems, they can generate results that one cannot guess by looking at the initial starting condition.

the cells at time 1. Thus, we see that we always obtain a new line by applying the instructions to the line directly before it.

Another trivial example, shown in Figure 20.22, is to swap states (for example, black becomes gray and gray becomes black). By the way, it should be noted that in this kind of one-dimensional (also called 1-D) cellular automaton, it is common to draw the starting line on top. The y-axis in a graph is then the time (in steps) going backwards (i.e. step 0 is on top and the most recent step on the bottom). That is just a convention and some researchers do it the other way around.

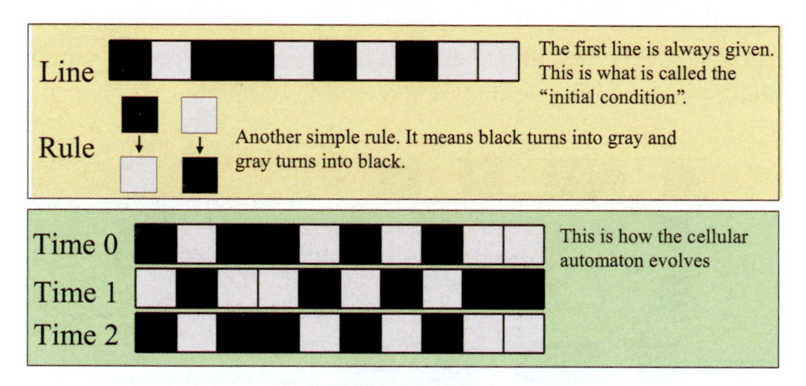

Figure 20.22: Another trivial cellular automaton.

Of course, our two examples in Figures 20.21 and 20.22 are called "trivial" because they don't really do what we set out to do. The cells do not interact with their neighbors. In these examples, the neighbors might just as well be non-existent. But that is not what we wanted and so the question should be: What are the simplest nontrivial cellular automata? The simplest would be to just take the direct neighbors into account. Two-state cellular automata (i.e. cellular automata whose cells have two states) with nearest neighbor interaction are often referred to as **elementary cellular automata** and investigated further in the next subsection.

Elementary cellular automata

In an elementary cellular automaton the new state of a cell is determined by a rule that takes into account the state of the cell itself and its direct (nearest) neighbors. An example of two possible instructions is shown in Figure 20.23. Therefore, including itself, the state of a cell is determined by a combination of the states of three cells. Three cells with each two possible states can give a total of $2^3 = 8$ possible combinations. All the possible combinations are shown in Figure 20.24 and hence if we want to define our cellular automaton, we need an instruction as to what happens to the state of the central cell for each one of these eight combinations.

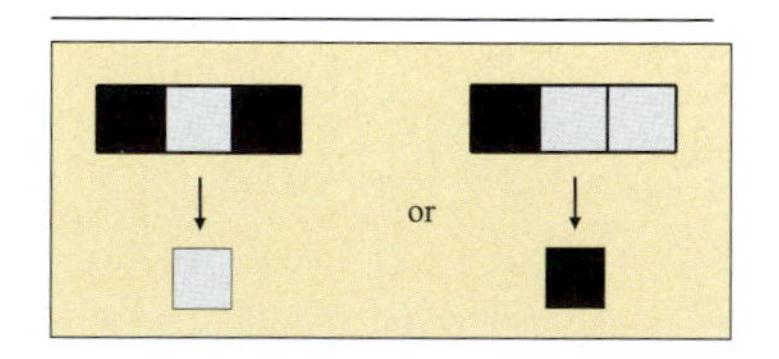

Fig. 20.23: Two possible instructions for an elementary cellular automaton.

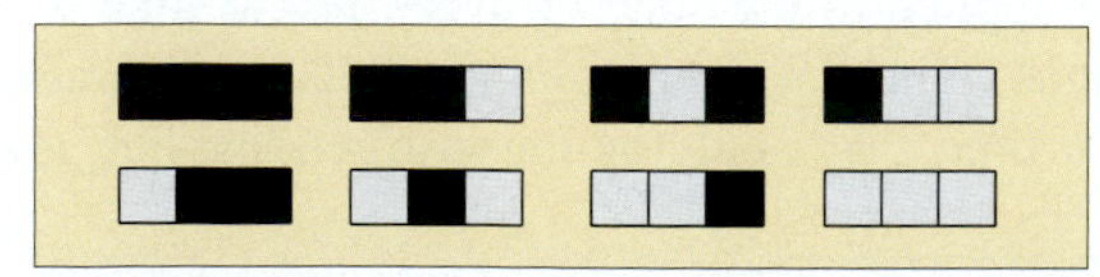

Figure 20.24: The eight possible input configurations of an elementary cellular automaton.

The set of instructions for all the combinations is usually termed a **rule**.

An example of a rule is shown in Figure 20.25. This rule has the name "254" but we will defer explaining what this name means until we have a better idea of what kind of elementary cellular automata there are.

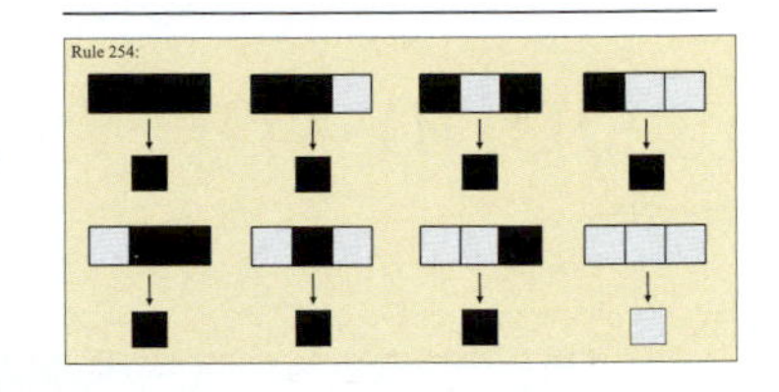

Fig. 20.25: Rule 254.

As can be seen, rule 254 is rather simple. All combinations of cells will keep or turn the central cell into a black cell, and only if all the cells are gray will the central cell remains gray. The evolution of rule 254 is shown in Figure 20.26.

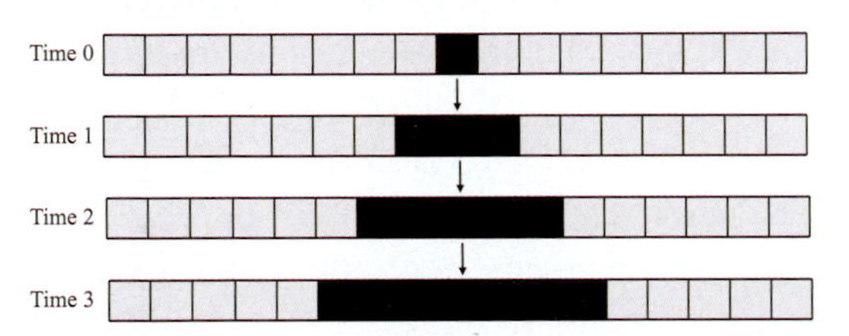

Figure 20.26: Rule 254 — evolution.

As mentioned above, it is often the convention to start with a single black cell in the center. Doing so, we see that we obtain a black central region that grows by one cell both to the right and left at each time step. Because there is really no need to explicitly number the steps and show the successive lines individually, it is furthermore common to omit the step numbers and blank spaces between the lines. The evolution of rule 254 shown in Figure 20.26 is then drawn as depicted in Figure 20.27.

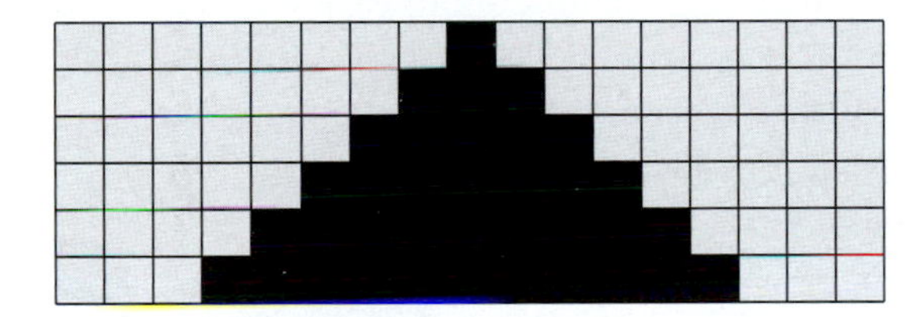

Figure 20.27: Rule 254 — result.

It is important to note that this makes it look as if it were a two-dimensional pattern but one has to keep in mind that it is a one-dimensional pattern showing its evolution in time.

Thus far, we have seen that elementary cellular automata can generate a *simple* pattern. If that were all that could be done, probably nobody would be particularly interested in them! As can be imagined, there must be quite a few rules possible. In fact we will show below that the number of different rules for elementary cellular automata is exactly 256, for those who are used to the numbering in computer languages not surprisingly numbered from 0 to 255 (why this is so will become clear later on).

Let us now look at the rather fascinating rule 90 that is defined as depicted in Figure 20.32. Clearly, this time, the rule is a bit more complicated than was the case for rule 254, but then again it does not look all too strange either. Certainly, it does not involve more cells, interactions beyond the nearest neighbor or more

John von Neumann

Figure 20.28: Portrait of John von Neumann.

John von Neumann was born in 1903 in the Hungarian city of Budapest, the first child of a wealthy family, and was in many ways a child prodigy with remarkable mental and mathematical abilities. He obtained his PhD in mathematics with minors in physics and chemistry from the University of Budapest in 1926 after also studying chemical engineering at the ETH in Zürich, Switzerland. From 1926 until 1930 he worked at the University of Berlin, and in 1930 he emigrated to the United States where he joined Princeton University. In 1933, he was selected as one of the first four faculty members of Princeton's Institute for Advanced Study, together with Albert Einstein and Kurt Gödel. John von Neumann died at a young age in 1957 from cancer possibly due to radiation exposure from nuclear tests (he contributed to the Manhattan Project and witnessed the first nuclear detonation in New Mexico, USA).

All in all, John von Neumann was one of the most influential scientists of the twentieth century, with major contributions to set theory, game theory, quantum mechanics, and computer science, among others.

In 1928, he published the minimax theorem that shows how two players in certain zero sum games can minimize their maximum losses. Besides its relevance for game theory, the minimax theorem has important implications in economics, and he coauthored the book "Theory of Games and Economic Behavior" that was published in 1944.

In the 1940s, while working at the Los Alamos National Laboratories on self-replicating systems, von Neumann designed the first cellular automaton (with pencil and paper!) and showed that in principle a certain pattern should be able to make copies of itself.

Figure 20.29: Modern computers use the so-called von Neumann architecture.

In 1945 von Neumann wrote a paper describing a computer architecture where instructions and data are stored in the same memory device. Although this type of computer architecture was not his invention, the paper's enormous impact resulted in this computer architecture generally being referred to as **von Neumann architecture**. All modern computers use this architecture and this design is a specific implementation of a universal Turing machine (see also Chapter 18). In the von Neumann architecture, both data and instructions (the program) are stored in a single storage structure like RAM as opposed to having separate structures for the instructions and the data.

Mandelbrot set

A good example of a fractal is the Mandelbrot set that can be obtained from the logistic map if complex instead of real numbers are used. Complex numbers are an extension of the real numbers such that the root of -1 is defined and given by $\sqrt{-1} = \mathrm{i}$. Complex numbers can be visualized as points in a plane where the x-axis represents the usual real numbers and where the y-axis represents multiples of i. Since their introduction in the sixteenth century, complex numbers have had an enormous impact on mathematics and science in general. For example, quantum mechanics is impossible without them.

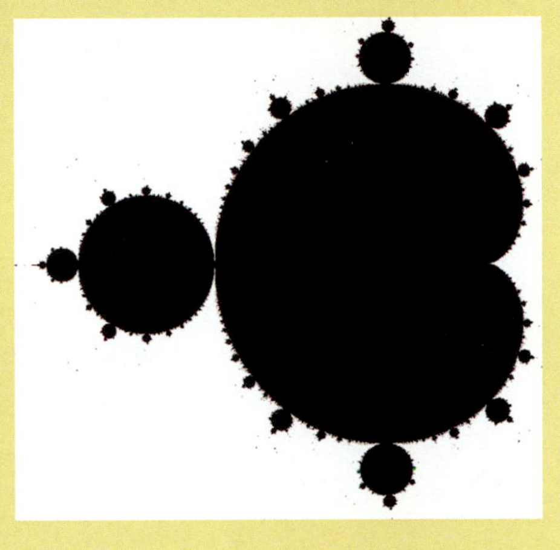

Figure 20.30: The black area is the Mandelbrot set.

In mathematics, the logistic map as we used it in Equation 20.2 is often rewritten in the equivalent but somewhat more convenient form $x_{n+1} = x_n^2 + c$. In this form, it is usually called the quadratic map but it should be stressed that as such it is exactly the same — all one needs is a coordinate transform. If complex numbers are used, it is called the complex quadratic map. The Mandelbrot set is then commonly obtained in the following way: Take a point in the complex plane and use it as the constant c of the complex quadratic map. Starting from $z_0 = 0$, on applying the map repeatedly, it turns out that there are two types of points: points that escape to infinity and points that stay near the origin. The set of all the points that remain near the origin is called the **Mandelbrot set** that roughly looks like a heart with some spheres of various sizes attached to it.

Figure 20.31: Deep zoom into the boundary with color coded escape time. Note the "baby" Mandelbrot set in the center.

The Mandelbrot set is shown in Figure 20.30. What is fascinating is that the Mandelbrot set has a fractal boundary, and the length of this boundary can be shown to be infinite. This boundary can spectacularly be illustrated by color coding the escape time to infinity, leading to stunningly beautiful pictures, as for example Fig 20.31. This figure was obtained by zooming far into the boundary of Figure 20.30 thus magnifying a tiny region. As before, points that are bounded near zero are painted black. Points that escape to infinity are colored according to the number of steps it takes to cross a circle of radius 2; it can be proved that all points outside of this circle escape to infinity. Which color is assigned to what number of steps has no direct mathematical meaning and is based on aesthetics. Interestingly enough, the dimension of the boundary of the Mandelbrot set is 2 and not a fractional number as is the case for the Cantor set.

The Mandelbrot set is closely related to the Julia set where instead of varying the constant c, one keeps the constant c fixed and varies the starting point z_0.

states. Yet is turns out that this rule has a deep connection with the mathematics of Cantor!

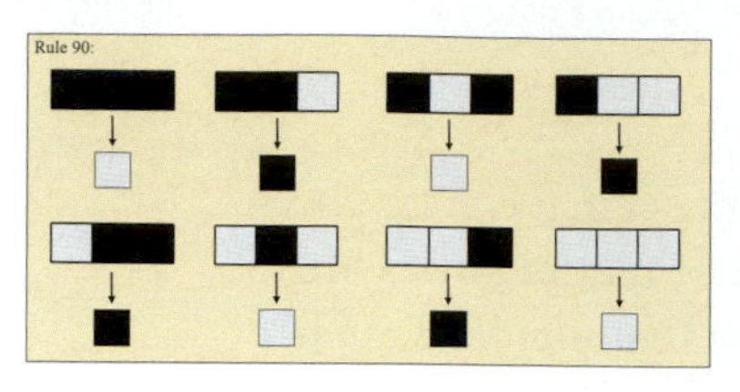

Figure 20.32: Rule 90.

The time evolution of rule 90 is shown in Figure 20.33 and as can be seen a rather non-trivial patterns appears to emerge.

If we continue the evolution of rule 90, we obtain the **Sierpinski gasket** shown in Figure 20.34.

The usual construction of the Sierpinski gasket is similar to that of the Cantor set described in Section 20.2. Start with a triangle, then remove the center triangle formed by connecting the middles of the three sides. This leaves us with three smaller triangles. For those smaller triangles, do the same thing to obtain nine triangles.

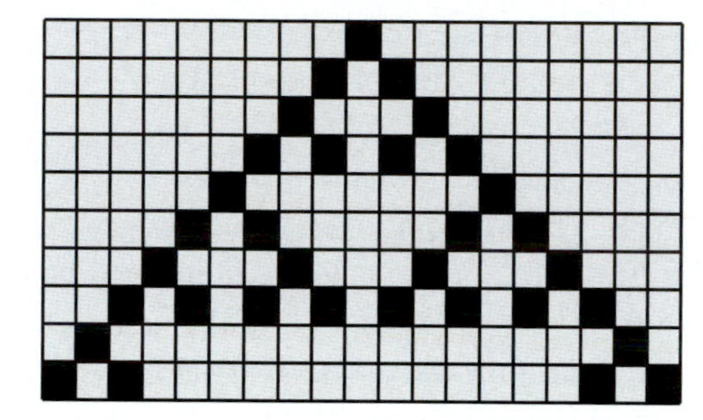

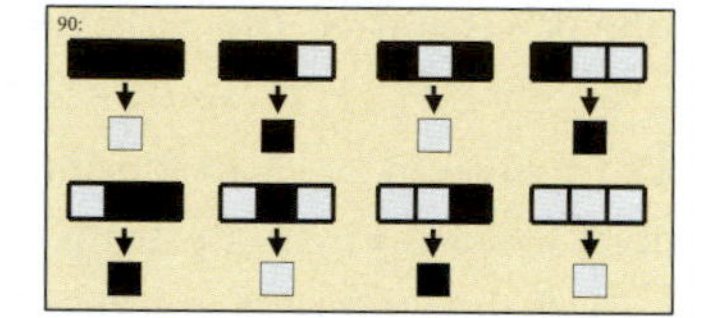

Figure 20.33: The elementary cellular automaton with rule 90.

Fig. 20.34: Sierpinski gasket generated by rule 90 after 256 steps.

Continue the process ad infinitum. Just like for the Cantor set, many interesting properties can be proved for the Sierpinski gasket. Not surprisingly it also has a fractal dimension, and this fractal dimension lies between the dimension of a line and of a square being 1.585. Thus, we find that a simple cellular automaton can generate a fractal!

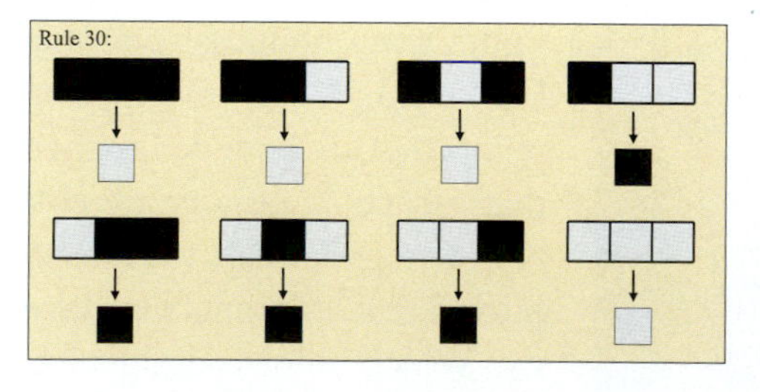

Figure 20.35: Rule 30.

So now we have seen that elementary cellular automata can produce both simple and fractal patterns. Are there any other types of patterns that can be generated? The answer to that is a resounding "yes". To see that let us consider rule 30 that is shown in Figure 20.35.

If we start again with a single black cell in the center, we obtain the rather curious pattern shown in Figure 20.36. While there appear to be regular patterns on the left hand side of the figure, the right hand side looks rather irregular. Indeed, careful analysis shows that the pattern generated at the right hand side is chaotic.

All in all, we therefore observe that despite their enormous simplicity, elementary cellular automata are able to display three fundamentally different types of behavior (simple, fractal and chaotic). But that's not all; as discussed in a bit more detail in Chapter 18, some elementary cellular automata can even be used for computing arbitrary complex algorithms. Not that that would be practical but, conceptually, the important thing is that, as such, it is possible. Specifically, it turns out that rule 110 is a universal Turing machine (see Chapter 18 for details).

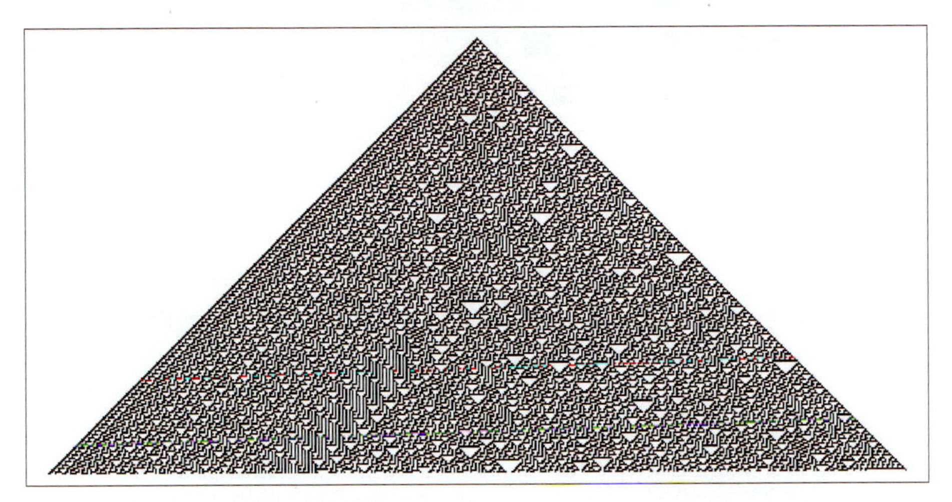

Figure 20.36: A rule 30 CA after 256 steps.

Now let us briefly explain the numbering scheme used for elementary cellular automata. First, the sequence of the eight possible input patterns shown in Figure 20.24 needs to be defined. A convenient ordering is to interpret a black cell as a 1 and a gray cell as a 0. Then the order as depicted in Figure 20.24 is just like counting in binary from right to left. Two examples are shown in Figure 20.37.

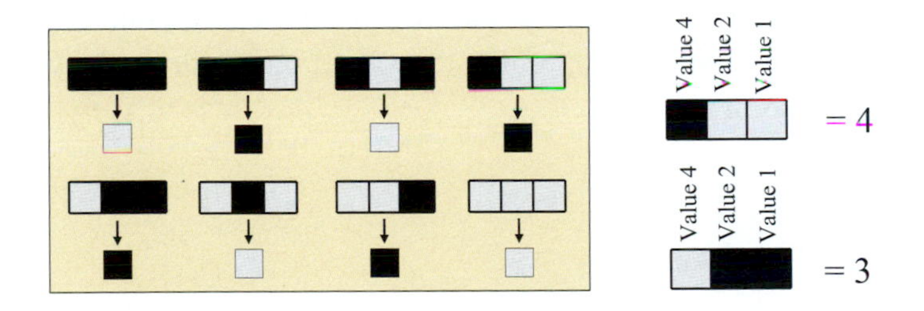

Figure 20.37: The sequence of the input pattern follows binary counting from 0 to 7. Illustrated here is rule 90.

Now that we have a given order for the input combinations we can look at the new states of the central cells. In the same spirit as before, we can again read the new states from right to left as a binary number associating a black cell with a 1 and a gray cell with a 0. An example is shown in Figure 20.38. We can now also understand why there are in total exactly 256 elementary cellular automata. In term of combinatorics one can say that we have 2^8 possible combinations (number of different states to the power of number of input configurations), or equivalently, we can say that with eight bits we can count from 0 to 255.

While it is of course nice to get a grip on the somewhat abstract notions that help us understand patterns in nature, it is even nicer to see an actual mechanism at work. An excellent example is that of the cone snail (*Conus textile*) shown in Fig-

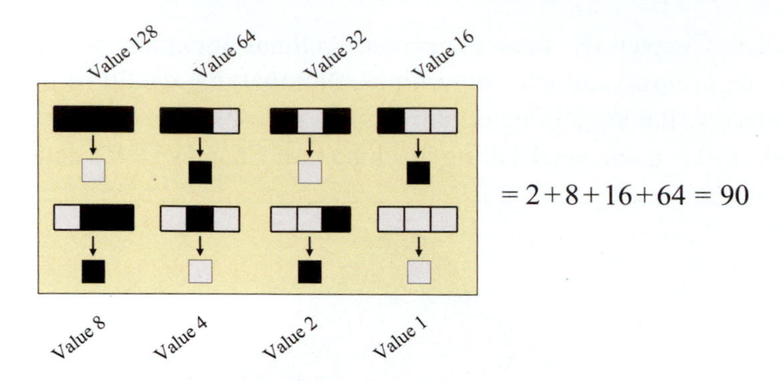

Figure 20.38: The sequence of the output pattern follows binary counting from 0 to 255.

ure 20.39, whose shell is the result of a natural cellular automaton that *resembles* the rule 30 elementary cellular automaton shown in Figure 20.35.

Figure 20.39: Seashell pattern created by a natural process. Source: NSF.

The pigment cells of the cone shell secrete pigments based on activating and inhibiting states of neighboring pigment cells leading to the formation of the intricate pattern. We thus see how an abstract and more-or-less "striped-to-the-bones" model like an elementary cellular automaton can generate patterns that are similar to an apparently complex biological feature such as a cone shell pattern.

What is even more important, however, is that this helps us understand how it can be possible to have only a small number of interacting units and yet have a complex outcome. It is interesting to illustrate this point with the help of a brain-inspired neural network model.

20.5 Neural networks

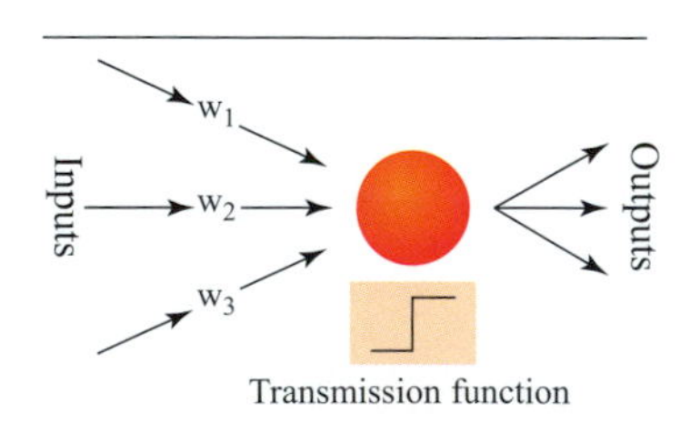

Fig. 20.40: A node in a neural network.

In Section 18.11, we briefly discuss the brain and how it is composed of a large number of relatively simple neurons. Since we know that brains are capable of extremely complex behavior, we have a situation not quite unlike that of cellular automata: complex collective behavior as a result of the interactions between (relatively) simple elements.

Would a simple brain-inspired model also display nontrivial behavior? We will show now that this indeed turns out to be the case by having a brief look at so-called **neural networks**. Neural networks consist of nodes (neurons) that receive a number of input signals, which are processed, possibly leading to the node sending a signal to a number of output channels. A basic node is illustrated in Figure 20.40. One of the simplest ways to process the inputs is to simply sum up all the inputs and generate an output if the sum is greater than some value. More generally, a transmission function can be used to determine whether or not a signal should be sent to the output channels.

Nodes can then be connected and one can study what happens. One interesting way of connecting nodes is by arranging them in layers and then linking every node in one layer with all the nodes in the next layer as shown in Figure 20.41 where the three nodes on the left hand side receive an input.

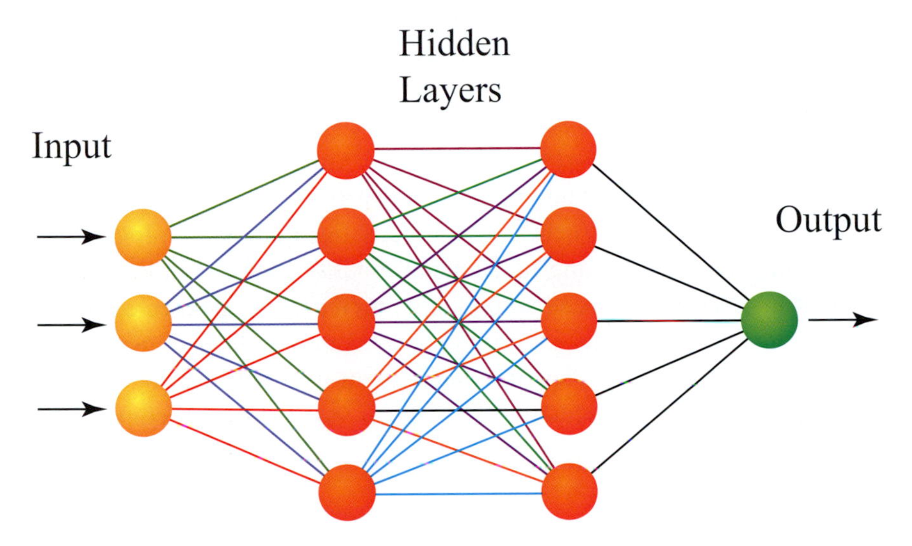

Figure 20.41: Example of a neural network with two hidden layers. The colors are for illustrative purposes only and have no specific meaning.

Each of the three nodes processes its input and, depending on the outcome of its transmission function, sends a signal to all the nodes in the first hidden layer. The transmission function can be a simple step function that gives no output when the input is below a certain threshold and a 1 otherwise. For simplicity, the transmission function can be the same for all the nodes, but that does not necessarily need to be the case. Hence, each of the nodes in the first hidden layer receives an input from three nodes. The nodes in the first hidden layer then process their inputs and, depending on the outcomes of their threshold functions, send outputs to each of the nodes in the second hidden layer. The same happens for the second hidden layer, except that now there is only a single output node. Of course there could also be several output nodes.

Now that we have a simple neural network, we can apply some inputs and investigate the outputs. Although that can be quite interesting, a fairly straightforward modification has some far reaching consequences. There is one feature in Figure 20.40 that we have not mentioned yet. Each input can have a weight, labeled as w_1 etc. The weight is the number with which an input is multiplied before it is processed by a node. By having weights, neural networks gain the ability to "learn" in the following way: At first, the weights are assigned some random value. Then a pattern to be learned is shown to the neural net and while proceeding from the input to the output layer. The output is then compared to a desired output — in the case of training a neural net to recognize a face, this could be a matching name. If the output does not match the desired output, the weights are adjusted and a new output based on the new weights is obtained. This is done until the desired output is obtained.

What makes neural nets so interesting is that a neural network can be trained to recognize not just a single input like, for example, a single face — but a number of inputs. For each different input the training procedure is repeated and the weights are adjusted. However, no new weights are added! Hence with a single network and a single set of weights, it is possible to obtain well-defined outputs from a set of given inputs. In other words, a single neural net can, for example, be trained to recognize a (large) number of faces and match them to the corresponding names. Remarkably, neural nets can often perform this task well even when the input is not perfect; for example, two pictures of a face taken under different conditions may still correctly be identified as being of the same person.

Thus we see that a model (neural net) directly inspired by a biological system (brain) exhibits exactly the same fundamental quality as the cellular automaton. Namely that interactions between simple units give rise to unexpected and fascinating properties. This helps us understand how a small number of genes can give rise to a complex organism. Not only that, it can also help us understand why similar numbers of genes can lead to organisms of vastly different complexities.

20.6 The answer

If, for a moment, we forget about nature and look at machines and objects created by humankind, we could easily come to think that simple systems lead to simple behavior and that complicated or complex systems lead to complicated or complex behavior. Indeed, if we consider, for example, a simple pendulum, we not only know that its oscillations are very regular, we also know that small changes in the system (e.g. by making the pendulum a bit longer) will lead to relatively small changes in behavior. This true for many of the linear scientific models currently in common use, for which small changes have small effects.

From an engineering point of view, that makes life a whole lot easier since it allows one to rather accurately predict what happens in a given range of possible settings. And this may very well be the reason why we see so many systems that have this property: It works well!

If, however, we look at the natural world we see intricate patterns and complex behavior, both in living and nonliving contexts. In trying to describe such contexts, we find that fractals are a powerful notion for describing many of the *geometric* patterns encountered in nature, that chaos can show us how simple equations can yield complicated *dynamics*, and that cellular automata demonstrate how interacting units give rise to completely unexpected *overall behavior*. The whole is truly more than the sum of its parts! In all three instances, small differences in the initial conditions can lead to huge differences in the final outcomes. Indeed, the size of a trivial rule and the size of a chaotic rule in elementary cellular automata are exactly the same. So, in general, one cannot state that complexity is a function of the number of interacting units per se but of the type of interaction and type of unit.

Above, we considered only simple rules and simple interacting units. It is rather conceivable that genes and the interactions between genes are vastly more complicated than those discussed. If, however, we already obtain astonishingly complex behavior from simple systems, then we can reasonably expect much more complex outcomes from the more complicated natural systems.

We can now answer the chapter question as to why we only need a small number of genes: **Rather than the number of genes of an organism, it is the nonlinear manner in which the genes interact that gives rise to how complex the organism is.**

20.7 Exercises

1. When starting from a single black cell, give two steps in the evolution of rule 27 for elementary cellular automata.
2. If instead of only nearest neighbor we also consider second nearest neighbor interactions in the elementary cellular automata discussed in this chapter, how many different rules will there be?
3. Based on the notion that swapping white for black or vice versa doesn't change anything, how many different rules for elementary cellular automata would that leave us with?
4. What does rule 0 do?
5. There are four different kinds of nucleotides in DNA. If we consider each nucleotide to represent the state of a cell in a cellular automaton and take only nearest neighbor interaction into account, how many rules can there be?
6. How many windows does the bifurcation diagram of the logistic map have?
7. Starting with $x_0 = 0.6$, what is x_3 in the logistic map if $\alpha = 2$?
8. How many points does a Cantor set contain?
9. Is nearest neighbor interaction an accurate reflection of how genes interact?
10. Symmetries are a common and important notion in physics. Are there symmetries in the 256 rules of the elementary cellular automata?

21: Evolution

Q - Why Are There Many Species?

Chapter Map

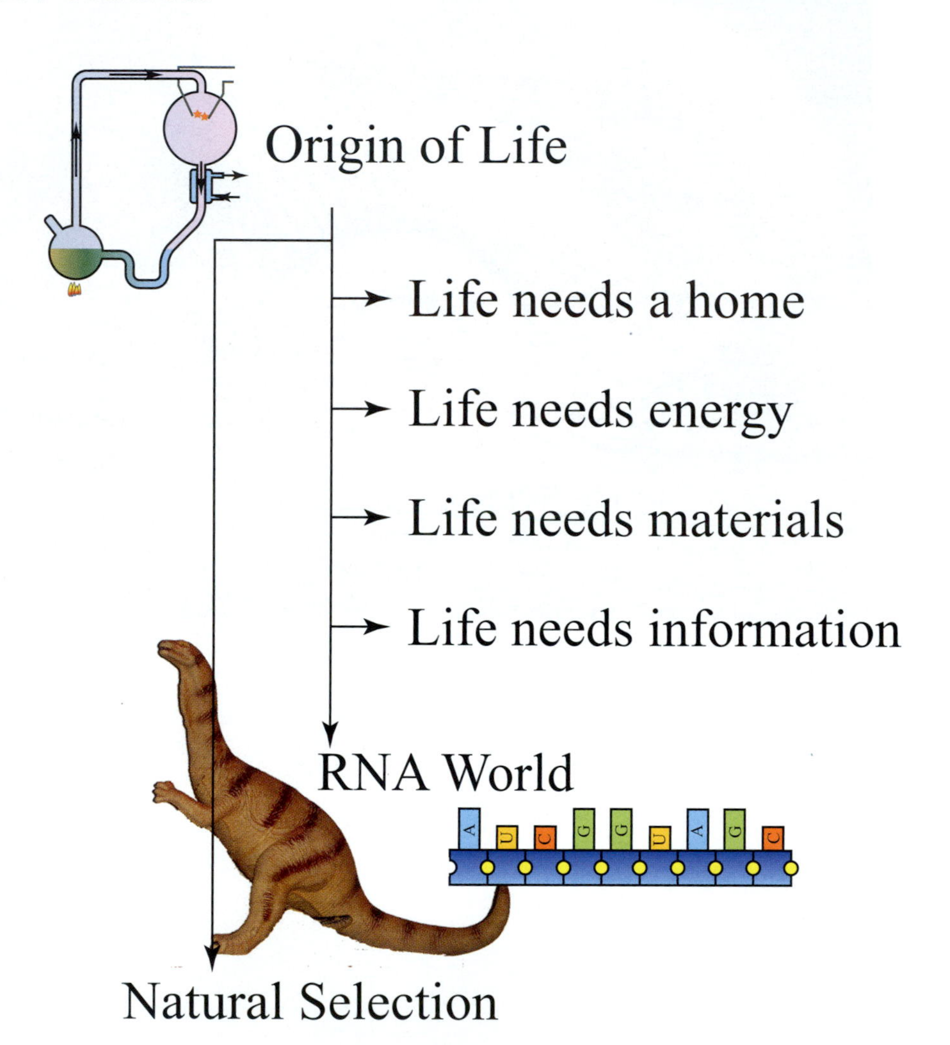

Evolution
Origin of Life
Life needs a home
Life needs energy
Life needs materials
Life needs information
RNA World
A
U
C
G
G
U
A
G
C
Natural Selection

21.1 The question

Figure 21.1: Life can be found even in apparently inhospitable places. Sahara dabb lizard.

Life on earth is ubiquitous — in the sea, on land, in the soil of the earth and even in the air. In possession of a microscope, we find that there is even more life invisible to the naked eye and that living organisms range in size from the very small to the rather large. Many of these organisms compete for limited food resources not only with their own kind but also with other species, who furthermore might prey on them. Clearly, different life-forms have different capabilities and strengths, making some more adapted to a particular environment than others. One would expect well-adapted species to overcome less well-adapted ones leading to a small number of specialized species, each with their own specific niche. However, this turns out not to be the case; while ill-adapted species do indeed die out, many competing species remain and one can hence wonder: **Why are there so many species?**

In order to answer this question, we need to first ask why there are species in the first place. We cannot do that without investigating the origin of life itself.

21.2 Origin of life I

Clearly we are alive. But when and where did life start? From fossil records, it is clear that life is surprisingly old. Indeed, the currently oldest known fossils are found in a fine-grained quartz called "chert". Quartz consists mainly of silicon dioxide, the same molecule which is also the major constituent of glass. In the formation of chert, tiny quartz grains accumulated while trapping microbes. However, solidification only proceeded very slowly over thousands of years, such that rather than crushing the trapped microbes, the quartz crystals penetrated the cell walls of the microbes, preserving their three-dimensional structure.

The microbes found in chert are about 3.5 billion years old and already display key features of modern microbes. This is rather astounding since our planet is only about 4.6 billion years old and was uninhabitable during the first few hundred million years. Since it is, furthermore, reasonable to assume that life did not come into existence in the rather complete form as found in the chert fossils, one may estimate that the origins of life date back to about 4 billion years ago. This means that in effect, almost as soon as our planet was inhabitable, life began.

Was it just a matter of luck or perhaps extraterrestrial seeding? Either could be possible but rather than engaging in wild speculation, it is scientifically more fruitful to investigate the circumstances under which life arose and try to find the natural laws which govern its emergence.

Although the natural laws underlying the origin of life and indeed the very possibility of the existence of such laws are still an area of active research that includes notions and methodologies from complex systems as discussed in Chapter 20, much more is known about the environment in which the first life-forms thrived. Furthermore, from modern organisms we can infer some general requirements that must have played key roles even for the earliest life-forms.

We therefore first look at how the environment of the early earth looked in Section 21.3, then consider that life needs energy in Section 21.4, that it requires materials for constructing a body in Section 21.5, and finally that it needs information for preserving adaptations from one generation to the next in Section 21.6.

These considerations then form a basis for discussing natural selection in Section 21.8 and the diversity of life in Section 21.9. Before answering the chapter question we discuss a theory that tries to explain the sudden rapid evolution of some species in Section 21.10 and the many pieces of evidence that support the notion of evolution in Section 21.11.

21.3 Life needs a home

Fig. 21.2: Life needs a home.

Although our discussion is limited to life on earth, as such it is rather reasonable to assume that in general, all life needs a home. The question is what kind of home that would be. To pursue this matter further, we need to distinguish two aspects. First, the processes involved in sustaining life need to be confined to a spatial region by, for example, a cell wall and second, an organism needs to live in a more-or-less hospitable environment. Since basically everything we consider to be alive has some form of a cell wall, let us focus on the environment and, for the origin of life, that means the geology of the early earth.

Composition of the first atmosphere

The first atmosphere had virtually no oxygen!

When the earth was formed around 4.6 billion years ago, gravity moved the heavier elements inwards relative to the lighter gaseous elements that moved outwards. Since the composition and relative abundances of the materials that formed the earth are quite well known, the composition of the early atmosphere can directly be inferred.

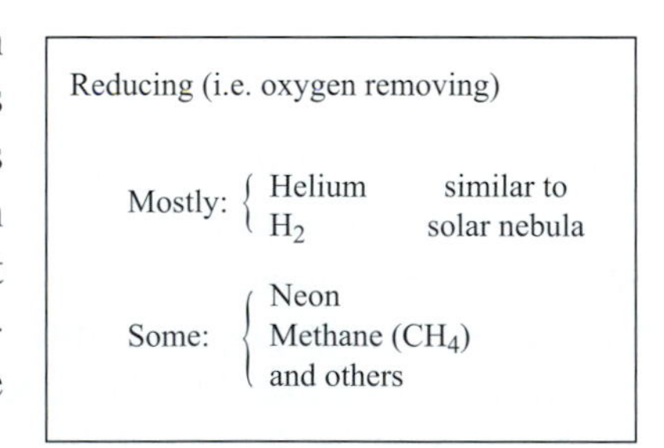

Reducing (i.e. oxygen removing)
Mostly:
Helium
H_2
similar to solar nebula
Some:
Neon
Methane (CH_4)
and others

Figure 21.3: Composition of the early atmosphere.

The main components of the early atmosphere are shown in Figure 21.3, and it is immediately clear that it's rather different from our current atmosphere. There is no oxygen! Stronger still, the atmosphere contained molecular hydrogen (H_2) and methane (CH_4), compounds that remove oxygen from their environment and are hence termed *reducing*. However, due to earth's relatively small size and consequent small gravitational attraction, and the immense heat resulting from the compactification of the heavier materials, this first atmosphere was lost to space, leaving our planet as a more-or-less barren rock.

Since the solar system as a whole was furthermore still being formed, there was also relentless bombardment of asteroids that probably continued for several hundred million years to largely end only about 4 billion years ago with a less severe bombardment until about 3.8 billion years ago. There weren't even any oceans yet. All in all, earth was not a very hospitable place!

But, one may wonder, is it really so that the first atmosphere was lost? There are two solid pieces of evidence for this. First, old rocks clearly show that the earth had a *neutral* (non-reduction) atmosphere between about 3.8 and 2 billion years ago (our current atmosphere is oxydizing due to its large percentage of about 20% oxygen). Second, there is a significant lack of inert neon gas in the current atmosphere even though it was rather abundant among the original gases that formed our solar system. Consequently, the second atmosphere was likely generated by volcanic activity. Let us now look at its composition.

Composition of the second atmosphere

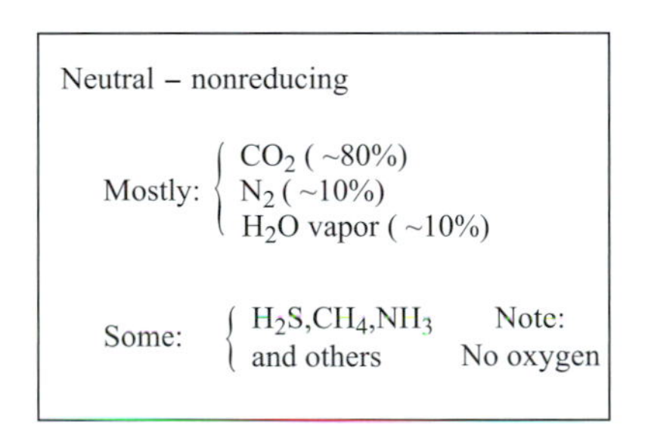

Neutral – nonreducing
Mostly:
CO_2 (~80%)
N_2 (~10%)
H_2O vapor (~10%)
Some:
H_2S,CH_4,NH_3
and others
Note:
No oxygen

Figure 21.4: Composition of the second atmosphere.

While the early earth circled the sun as a barren desolate rock, gravity continued to do its work compacting the planet and leading to a heating up. Assisted by radioactive decay of very heavy elements, save for a thin outer crust, most of the interior melted giving rise to a period of intense volcanic activity. Concurrently, immense amounts of gases including water were driven out to form the second atmosphere that, in turn, mostly due to biological activity, eventually turned into our current atmosphere. The formation of the second atmosphere began about 4.4 billion years ago and its composition was likely very similar to that of the gases still escaping volcanoes today, shown in Figure 21.4. At the same time, the surface of the earth had cooled enough to allow for the emergence of a crust, the existence of which is a prerequisite of having an ocean.

As the surface of the earth was very hot, most of its water was in vaporous form while intense rains slowly assisted in cooling the crust and eventually leading to the formation of oceans. These oceans, however, were quite a bit different from today's. So let us see how they looked.

Composition of the first oceans

From a human being's point of view, probably the most important difference between the early and current oceans was that they were completely devoid of life. On a macroscopic scale, it is already clear that there is a lot of life in the oceans, but on a microscopic scale it is absolutely teeming with living organisms. Indeed, even the teeniest drop of ocean water contains nearly uncountably many individual cells.

It is estimated that the earliest large oceans were in existence about 4.2 billion years ago. As is the case nowadays, they contained some neutral gases like CO_2

Fig. 21.5: Approximate color of the early ocean

and N_2, but importantly for the origin of life, virtually no oxygen. They also contained vast amounts of ferrous ions like Fe^{2+} making the ocean look greenish yellow as approximated in Figure 21.5. The first oceans also contained the salt ions Na^+ and Cl^- but only very little phosphate PO_4^{3-} that is important for the biochemistry of life.

Near hydrothermal vents, the seawater contained fairly large amounts of H_2S that can serve as an energy source (for more details, see Section 21.4).

A snapshot of the early earth

Now that we have an idea of the early atmosphere and ocean, let us have a brief look at how our planet looked about 4.2 billion years ago. Some key features are summarized in Table 21.1.

- Weak sun, no ozone layer, no UV shielding
- Shorter day, about 15 hours per revolution
- Extreme weather, strong winds, big waves
- Moon about one third closer to earth, tides up to 30 m high
- Very little land

Table 21.1: 4.2 billion years ago, our planet looked quite differently.

21.4 Life needs energy

Fig. 21.6: Life needs energy.

Thus far, we have found out how our planet roughly looked about 4.2 billion years ago. From this description, it is not clear that it was a suitable home for life. Since we know that life did take a hold, we do know that in fact at some stage the earth did become suitable for life.

A key property of life is that it creates a certain order from randomness and hence that it decreases entropy. It is important to elaborate a bit on this. In all known cells and indeed in all life-forms realistically imaginable, at least part of the atoms and molecules are arranged in very specific ways. For example, a cell wall consists of specifically arranged phospholipids (even though the individual phospholipid molecules can generally move within the membrane and the membrane as a whole can change shape, the overall structure of the membrane remains intact), DNA consists of long sequences of nucleotides etc. While there is always some randomness left, such arrangements are a far cry from the randomly dispersed atomic constituents dissolved in water. Also, cells maintain a fairly narrow range of atomic concentrations inside that can differ significantly from the concentrations outside.

From the basic laws of statistical physics, we know that arrangements as described in the previous paragraph need energy to be maintained in a dynamical environment. Experimentally, this can easily be verified by removing all the energy supply of an organism which will surely lead to its eventual death and the breakup of the intricately arranged cellular organization. Of course, one could argue that a deep frozen cell in a well-protected environment can survive for a very long time. However, although this is certainly the case, one would first need to consider that energy was necessary for the formation of the cell, and second that without energy supplied, the deep frozen organism would never come to life again.

Therefore, we need to wonder what kind of energy sources were available on the early earth. Clearly, there was sunlight but harvesting it is not trivial and hence may not have played a key role in the first stages of the origin of life. Another, relatively abundant source was chemical energy.

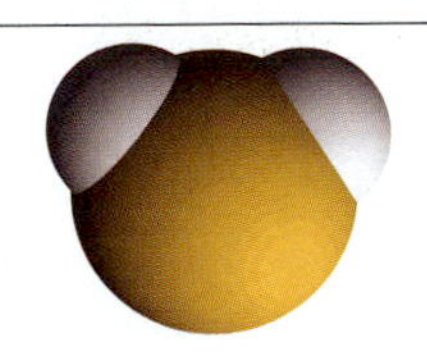

Fig. 21.7: Hydrogen sulfide may have been an important energy source for early life.

In the early days when life just emerged the organisms couldn't count on other organisms to serve as or produce food for them, the first living cells must have been so-called **autotrophs**, a term derived from Greek words "auto" meaning self and "tropos" meaning nutrition (in other words, an autotroph is an organism that does not need any organic food and produces biomolecules from simple inorganic molecules itself). If these organisms further used inorganic compounds as an energy source (rather than light), they are more specifically referred to as **chemoautotrophs** .

Some possible inorganic compounds that can serve as an energy source are hydrogen sulfide (H_2S) shown in Figure 21.7, ammonia, nitrites, hydrogen gas or iron.

21.5 Life needs materials

Besides inorganic compounds that include a carbon source like carbon dioxide (CO_2) and an energy source like hydrogen sulfide (H_2S), life could probably not have emerged without the presence of some basic building blocks in the form of simple amino acids. The question that immediately arises then is whether such building blocks can be found or formed in the environment of the early earth.

Fig. 21.8: Life needs materials.

In this context, some famous experiments seem to support that notion. Stanley Miller showed in 1953 that amino acids can quite easily be synthesized in a reducing atmosphere while Joan Oró did so for basic nucleotides in 1963.[1] Stanley Miller's experiment is depicted in Figure 21.9. Unfortunately, geological records show that the atmosphere was not reducing when life first emerged but was neutral instead.

Nevertheless, not all hope is lost. The key building blocks for RNA strands are nucleotides, so let us consider the structure of a nucleotide as shown in Figure 21.10a. Phosphate ions occur naturally and hence are readily available as an ingredient. Sugar can naturally be synthesized from formaldehyde (CH_2O) given sufficiently high concentration. A nitrogenous base is a bit more difficult to obtain but can be the result of an Oró-type reaction as long as the environment is devoid of oxygen. Figure 21.10b shows the possible synthesis of adenine from hydrogen cyanide (HCN), a compound amply available on the early earth.

However, especially considering the fact that there was much less land 4 billion years ago as is illustrated in Figure 21.11, perhaps focusing on the environment where crust and atmosphere meet when searching for the origin of life is a classical case of anthropocentrism. It is indeed tempting to reason that since we're alive and most of the visible easily recognizable life-forms are somewhat similar in needs,

[1]The amino acids produced were of both chirality; see also Section 21.11.

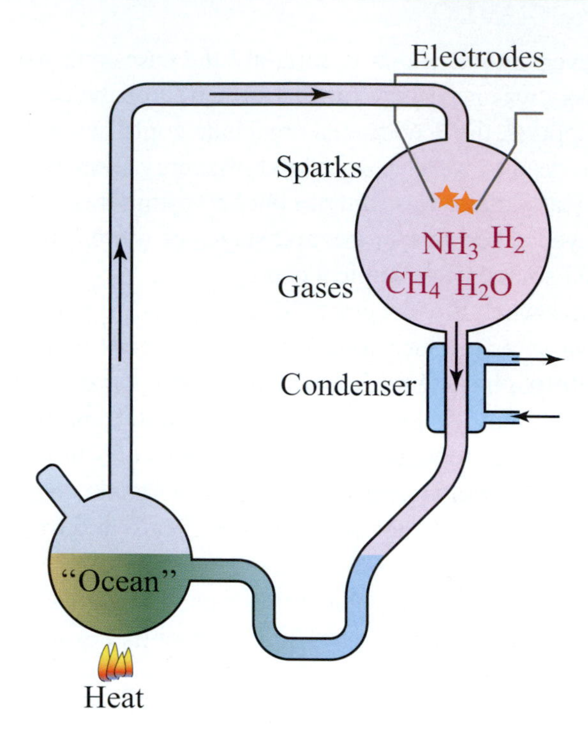

Figure 21.9: Stanley Miller's experiment for synthesizing amino acids in a reducing atmosphere.

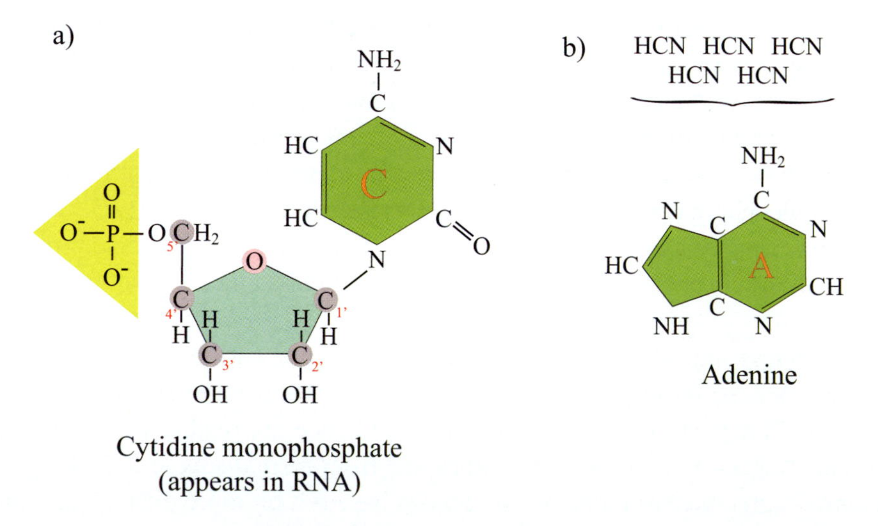

Figure 21.10: a) Nucleotides have three parts, two of which are relatively easy to synthesize in nature. b) Oró showed that the third part, a nitrogenous base, can in principle be synthesized from molecules available on the young earth.

what's good for us must be good for life in general. But this is of course not necessarily the case. Early life could have looked rather different from modern life, and what was a good environment for the emergence of life might not be the best environment for modern day organisms that not only have adapted to how earth looks now but also partially created it.

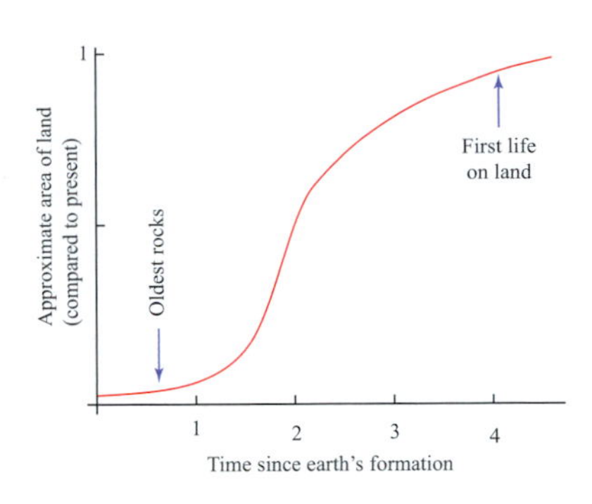

Figure 21.11: The amount of land mass was vastly smaller when life emerged about 4 billion years ago.

It is therefore very well conceivable that life formed in the crust or in the oceans. One particularly fascinating environment is that surrounding **hydrothermal vents**, an example of which is shown in Figure 21.12.

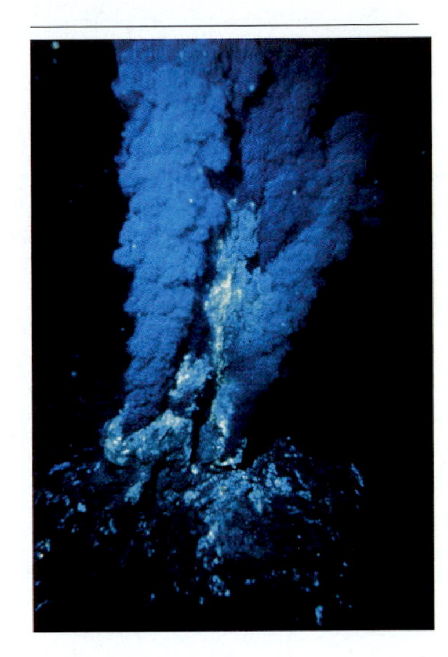

Fig. 21.12: Hydrothermal vent. Source: NOAA.

Hydrothermal vents are underwater structures that are formed as a result of superheated water escaping from below the crust into the ocean as illustrated in Figure 21.13. They are found near volcanically active sites like the ridge between Europe and North America that is part of a midocean ridge system more than 70,000 km in length (thus forming the longest mountain chain on earth). The water is rich in minerals and can be up to 400°C hot. Since the vents usually occur rather deep (the average is about 2 km) and the water surrounding them is very hot, and also very acidic (in some cases close to a pH of 2.8, or similar to vinegar), one would not expect there to be much life if any at all. However, nothing could be further from the truth!

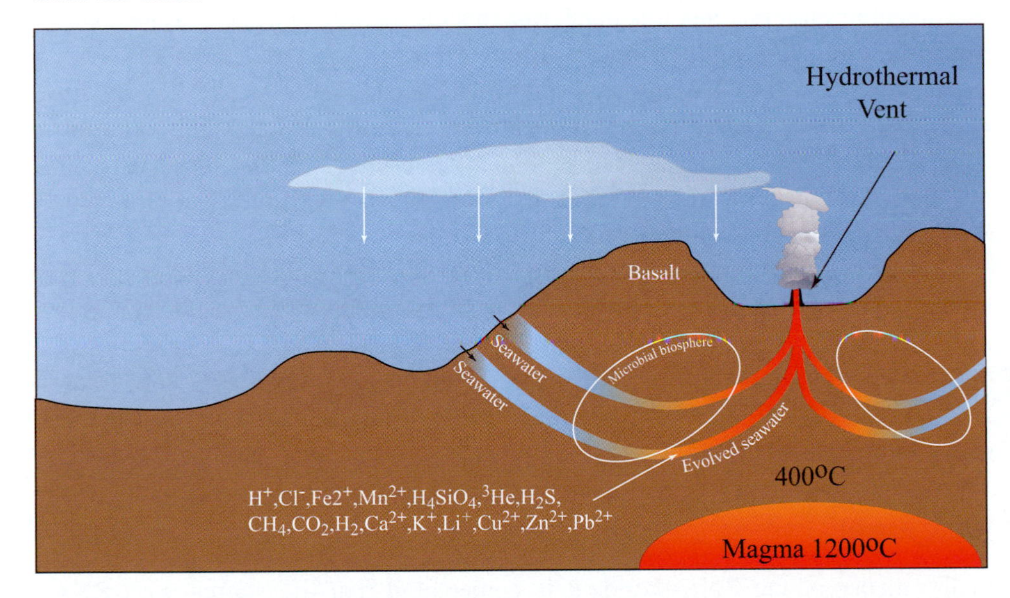

Figure 21.13: Schematic illustration of the environment near a hydrothermal vent.

The surroundings of hydrothermal vents are teeming with life that form complete ecosystems composed of a large number of creatures ranging from the microscopic to the macroscopic. The food chain begins with chemoautotrophic bacteria that feed on the minerals in the vent water and ends in snails, fish and other large predators. Interesting organisms are the tube worms shown in Figure 21.14 that have no mouth and no digestive tract. Instead they live in symbiosis with enormous quantities of bacteria (about 10 billion per gram of tube worm tissue).

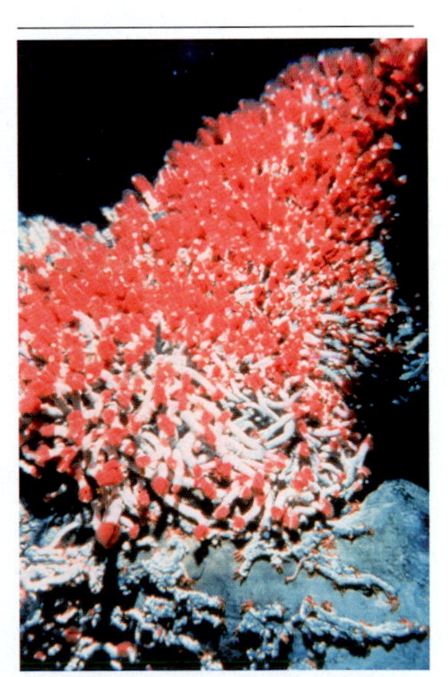

Fig. 21.14: Tube worms found near hydrothermal vents. Source: NOAA.

Now that we have seen where life can have a home, energy supply and materials, let us consider one last aspect before returning to the origin of life: information.

21.6 Life needs information

Generally, we know that life needs information. We need to know where we can find food and mates and, in general, be able to identify dangers and opportunities. However, this type of information is at quite a high level as it requires, at the very least, some sensors — like eyes, olfaction or touch — and minimally some rudimentary information processing (though perhaps not necessarily with a brain). Clearly, no such things could have been available to early life.

Perhaps, information is not necessary then. However, without information, how can a trait be transferred from one generation to the next? Although it is very well conceivable that early cells were not highly structured, somehow their composition would need to be preserved when replicating, as otherwise there would be no relationship between parent and offspring.

Let us now briefly consider some theories regarding the emergence of early cells. In order to discuss this, it is useful to note an observation first made by **John von Neumann** that metabolism and replication are logically separable. He introduced an analogy between a kind of simple computer programs called automata and living organisms while stressing the distinction between software and hardware. As argued in Chapter 17, this distinction is essential for understanding how life functions currently and it is likely that this distinction, at least to some degree, was also relevant during the origin of life.

If one considers what an early cell would need, there are two key aspects. First, there must be some kind of an enclosure, a cell wall. Second, some form of replicative mechanism is needed so that an offspring resembles its parents. The question then will be, which one came first, or perhaps they came at the same time? And what kind of cell could that have been?

Let us begin by looking at a possibility for the replicative mechanism. Currently, we know that our genetic makeup is stored in DNA but would this be the most suitable molecule to start out with? As such, DNA is not very versatile and, due to its double-stranded nature, structurally not overly simple. The Nobel laureate **Manfred Eigen** therefore proposed that early life started out with RNA in what is commonly called the **RNA world**.

RNA world

In the theory of the RNA world, it is conjectured that life started with RNA-based genes, then acquired the ability to generate enzymes and then finally formed cells. This theory has several strong points. First, Eigen showed that nucleotide monomers can give rise to nucleic acid polymers that replicate and mutate *without a template*. The fact that this can happen without a template is essential in the context of the origin of life since current life replicates by copying genetic material. But if there were no life, then there's nothing to copy, so the RNA polymer strands need to emerge by a separate mechanism.

In Eigen's experiments, an enzyme was necessary, and since RNA can act as an enzyme, one could imagine that if there are several types of RNA strands, these could catalyze each other's replication. Since it is clear that an early replicative system could not have been very accurate (accuracy requires refinement and hence significant evolutionary progress), Eigen proposed a combination of so-called **quasi species** with **hypercycles**.

A quasi species is a population of similar but not identical RNA strands that can replicate but only in an imperfect way. A hypercycle is formed by the interaction of several quasi species where one species assists (that is catalyzes) the replication of another species. A basic hypercycle is shown in Figure 21.15.

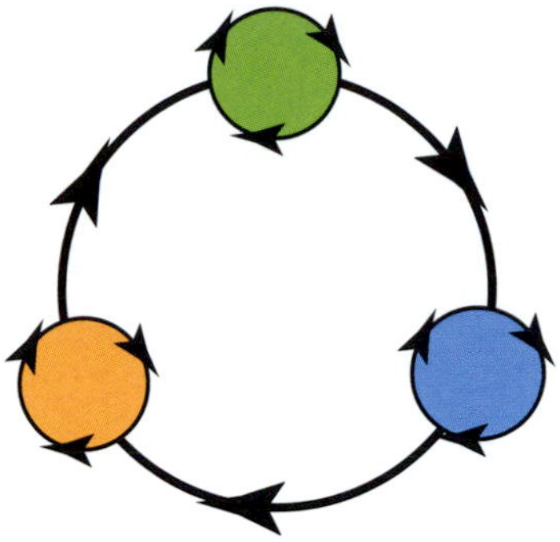

Fig. 21.15: Hypercycle.

Unfortunately, extensive numerical simulations showed that hypercycles are rather unstable and prone to error collapses. Although this may not be an insurmountable obstacle, it is nevertheless likely that Eigen's theory will need some modifications.

Since the need for enzymes plays a key role, it is important to note that **Leslie Orgel** showed the reverse from Eigen, namely, that RNA can replicate without the help of enzymes if given a template. In summary, Eigen demonstrated RNA strands can form without a template if there is an enzyme and Orgel showed that RNA strands can replicate without an enzyme if there is a template.

Garbage bag world

Another theory, proposed by **Alexander Oparin** in 1924 long before the genetic apparatus was discovered, starts with cellular enclosures rather than replicating units — as is the case in the RNA world. It is based on the observation that some organic molecules can spontaneously form little spheres under the right circumstances. A typical example is given by phospholipids, the same type of molecules that are the major constituent of cell membranes.

Doron Lancet modernized Oparin's theory and surmised that the little spheres at first contained random collections of molecules and could reproduce by growing past a certain size that led to a split. At some stage RNA strands entered the sphere, and slowly a replicative apparatus emerged.

Although no theory can give a clear and definite answer as to the origin of life, it does appear that information plays a key role. Indeed, it was perhaps the very ability to transfer information that was the crux for the emergence of living cells versus simple replicating units.

21.7 Origin of life II

Exactly how similar the offspring of early cells were to their parents is still unclear but at some stage, perhaps simply as the consequence of a random fluctuation, cells acquired the ability to transfer some of their composition and hence their traits when replicating. Clearly, without an intricate replication apparatus, offspring would very unlikely be nearly identical to their parents, giving rise to populations of perhaps similar but nevertheless also rather differing cells.

In an environment of limited resources, this immediately implies that some cells will be better at utilizing scarce nutrition than others. Thus, cells with certain traits will thrive better than others. Still, unless they can pass those traits on to future generations, having such a trait is not particularly useful since no cell has eternal life.

From this, we can see that it is indeed the very trait of information transfer that confers a key advantage to an organism. And once rudimentary information transfer is established, natural selection inevitably follows.

21.8 Natural selection

In its simplest form, **natural selection** refers to the observed fact that traits favorable to survival spread through a population and hence are selected, while unfavorable traits die out and hence are not selected. The term was introduced by Charles Darwin in his seminal masterpiece "On The Origin of Species" published in 1859.

For a somewhat more careful look, we need to distinguish between the concepts **phenotype** and **genotype**. The phenotype of an individual organism is its overall appearance and constitution but sometimes also refers to a specific trait or behavior. Examples of phenotypes are the patterns on a butterfly's wings or the size of a deer's antlers. Genotype, on the other hand, refers to the composition of an organism's genome. The phenotype is determined mostly by the genotype but is also influenced by the environment and random fluctuations.

Since genes are buried inside cells, it is clear that natural selection as such acts on phenotypes. The effect on genes is therefore indirect. A genome that yields a favorable phenotype is more likely to be transferred to a next generation. Also, selection acts on individuals, but the survival of a genome is dependent on the average success of a trait as with regards to the entire population. In this context, it is important to consider the notion of **fitness**.

Although commonly fitness and natural selection are summarized in the phrase "survival of the fittest", in reality, rather than survival rates, reproduction rates (of offspring that itself can reproduce) are more essential. After all, a very fit individual that doesn't reproduce will eventually die and with it its "fit" genome.[2]

The (generally) slow change of a population's traits over generations is usually referred to as **evolution**. Since natural selection acts on the individual and not the population as a whole, individuals of the same population with (initially possibly slightly) different beneficial phenotypes can pass these on to their offspring who in turn pass it on to their offspring with some more differences until eventually **speciation** occurs. Speciation refers to the point in time when organisms with a common ancestor have diverged sufficiently to be both phenotypically and genotypically distinct. In the case of multicellular organisms that reproduce sexually, that means the cessation of the possibility of interbreeding, but for other organisms the distinction is somewhat harder to make.

Although one effect of natural selection is (better) adaptation to new or niche environments, another less directly visible effect is optimization of key processes at the cellular level. Any organism that can carry out a certain cellular function more efficiently than another, even when its phenotype is otherwise the same, will likely have a higher chance of reproducing successfully.

Thus, a rather astonishing thing happened about 3.5 billion years ago. One cell was the end of an evolutionary process that made its genetic apparatus so efficient

[2]Of course one can argue that in this case the genome is not fit at all but only appears fit due to its phenotype's ability to survive.

that it would never need to be significantly changed again! This cell then became the starting point of a new evolutionary process giving rise to *all* life known today.

It is undoubtedly one of the greatest scientific discoveries of all time that all living cells function in basically the same way, having a universal genetic apparatus. Whether it is a bacterium or a human being, it does not matter; the way genes operate is the same. Technologically speaking, this is rather a blessing since it makes it possible to develop processes with bacteria that will still work for human cells.

Since all life is descendant from a common ancestor, it makes sense to try to trace all life that lives and ever lived back to that ancestor. This leads to the so-called tree of life whose main branches are the ranks of modern taxonomy. A **rank** is a level in the hierarchy of evolutionary relationships or in the picture of the tree of life a type of branch. The hierarchy of ranks is shown in Figure 21.16 while a basic tree of life is shown in Figure 21.17.

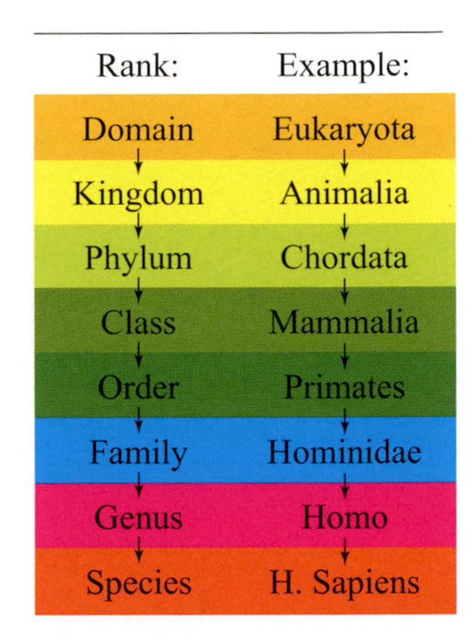

Rank:
Example:
Domain
Eukaryota
Kingdom
Animalia
Phylum
Chordata
Class
Mammalia
Order
Primates
Family
Hominidae
Genus
Homo
Species
H. Sapiens

Fig. 21.16: Ranks in the classification of life.

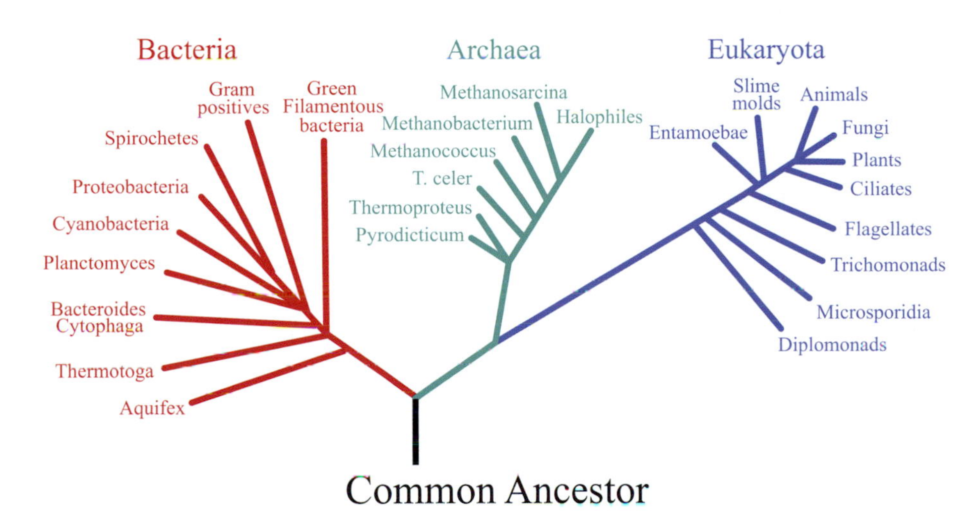

Figure 21.17: Tree of life.

21.9 Diversity

After the modern cellular apparatus emerged about 3.5 billion years ago, speciation continued but it would roughly be another 2 billion years until the first multicellular life-forms evolved. It is worthwhile to reflect on the timing a bit as illustrated in Figure 21.18. Life more or less *immediately* appeared as soon as our planet became sufficiently hospitable. It then took a relatively short time of probably less than 500 million years to evolve the modern genetic apparatus with its intricate interplay of DNA, RNA and proteins. Then it took 2 billion years for the first multicellular life to emerge and close to another 1 billion years for the first animals to appear. Human beings and their cousins, like the bonobo depicted in Figure 21.19, only arrived an evolutionary instant ago.

This context is important not only because it illustrates what processes life apparently finds easy and difficult, it also shows that despite our "natural" prefer-

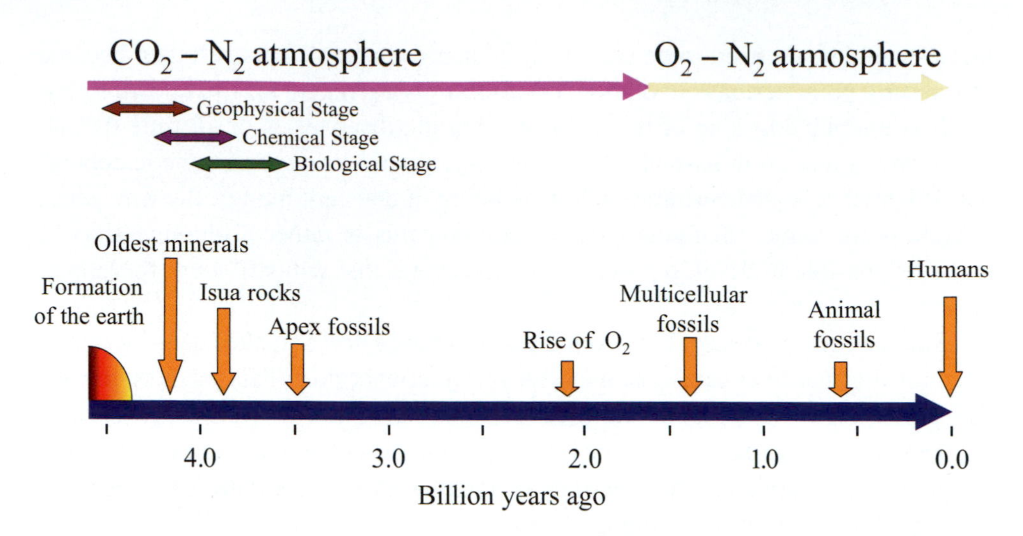

Figure 21.18: Some of the main events for the evolution of life on earth.

ences life as a whole is about neither whales nor human beings. Table 21.2 shows the number of species among various major groups.

Group	Rank	Number of Species
Bacteria	Domain	10,000,000
Fungi	Kingdom	1,500,000
Plants	Kingdom	300,000
Animals	Kingdom	1,300,000
Insects	Class	1,000,000
Mammals	Class	6,000

Table 21.2: Approximate number of species for various major groups of life.

What we can see is that there are more than three times as many species of bacteria than there are of fungi, plants and animals combined. We can also see that apparently more complex and larger creatures are very small in number. This is somewhat suggestive of a random process where the mean lies somewhere near moderately complex single-cell organisms. By random variations, organisms can then become more complex or less complex in a stochastic process. The probability to have many successive (and due to natural selection necessarily also evolutionary successful) variations in the direction of a more complex organism is then very small and hence it is perhaps understandable why there are so many insects but only a few mammals.

In what one can perhaps call a naïve view of the evolutionary process, one may consider it as processing in a very gradual way. At every instance in time, individuals in a population will be somewhat different due to random mutations in their genes and random influences from their environment.

In sexually reproducing species, beneficial traits will spread through the population and the species as a whole can adapt ever better to its environment. If due to geographical isolation or personal preferences (some female birds like males with a red beak better, others with a yellow etc.) groups of individuals are isolated, speciation occurs and little by little the diversity increases.

21.10 Punctuated equilibrium

It had been long observed in the fossil record, however, that many species apparently remain stable over long periods of time — and then suddenly change rapidly. This has led to the theory called **punctuated equilibrium**, a term coined by the eminent evolutionary biologist **Stephen Jay Gould**.

Figure 21.19: Bonobos are close cousins of our species.

The theory of punctuated equilibrium can be seen as a refinement of certain aspects of Darwin's theory in that it takes into account the stabilizing effect of large interbreeding populations. In such populations, gene variations spread fairly slowly and, quite likely, much slower than the speed with which the environment changes. A variation is not a real advantage as long as the population as a whole has not acquired it. Hence, there is no reason for the potential advantage to make progress and interbreeding will likely dilute its effects over time.

When parts of a population are isolated for some time, however, advantageous traits can spread rapidly giving rise to new species. When subsequently, the environment changes to the detriment of the original population, the new species can then take over the vacated ecological space, and it appears as if there is a jump in evolution at geological timescales. What is important to realize is that geological timescales are rather long compared to the life times of individual organisms. Indeed, 10,000 years is a geological instant but can easily comprise thousands of generations. Hence, even when the fossil record shows a jump, the change from generation to generation need not have been very big.

21.11 Evidence for evolution

In science, a theory's standing is entirely based on empirical evidence since the sole criterion of scientific truth is experiment. In short, empirical evidence needs to allow for reproducibility when experiments can be carried out and consistency both with factual observations as well as with other pieces of scientific evidence.

Some scientific theories like the special theory of relativity (see Chapter 22) are extremely well supported by many different pieces of evidence. Consequently, such theories are considered to be correct within their scope. Einstein's theory was an improvement of Newton's equations of motion. Newton's theory was shown to be an approximation to Einstein's better theory. In turn, it is very well possible that someone will discover a refinement to Einstein's theories. There are other scientific theories that do not stand on as firm a ground, although they are well supported by piecemeal evidence and logical consistency. One such case is, for example, a quantum theory of gravity; it is not clear whether such a theory will stand the test of experiment.

The theory of evolution is similar to theories on the formation and development of the solar system — with both theories spanning billions of years. Evolution is a scientific theory that is supported by a large number of interlocking pieces of evidence from various scientific fields including paleontology, genetics and biology. Let us now summarize a few of these.

Paleontological evidence

Fig. 21.20: Fossilized trilobite.

Most living organisms leave no trace when they die, either decomposing rapidly or being consumed by scavengers. Under some special circumstances, however, e.g. when an animal is buried in mud or sediments, mineral salts can gradually fill up the pores of its bones thus preserving its structure through petrification. Fossils can therefore often be found in sedimentary rock.

One of the most famous fossil sites is the **Burgess Shale** in the Canadian Rockies where a large number of organisms from around 505 million years ago can be found. When it was formed, the Burgess Shale was in shallow tropical water rich in nutrients that supported a diverse and extensive ecosystem. During the middle Cambrian (lasting from about 513 to 501 million years ago) when the organisms found in the shale lived, there were no plants on land and all multicellular life lived in the ocean. The barren land was therefore often subject to rapid erosion and mudslides that could bury the creatures rapidly, creating ideal circumstances for fossilization.

One key feature about sedimentary rocks is that they are formed in layers with the younger layers on top of the older layers. Thus, by pure physical arrangement, the succession of species can be found. Also, many rocks can be dated reasonably accurately by means of radiometric dating where the decay rates of naturally occurring radioactive elements are used to obtain an estimate of their age. Thus, even when similar fossils are found at different locations it can be determined which one came first. To date, more than 250,000 fossil species and a nearly uncountable number of fossils have been found.

All in all, despite being incomplete, by itself paleontological evidence provides strong support for the theory of evolution as it shows how one phenotype is succeeded by another in the tree of life. However, there is more evidence. Let us now consider anatomical evidence.

Anatomical evidence

In the theory of evolution, organisms higher up in the evolutionary tree share a common ancestor somewhere down the tree. Consequently, groups of organisms are expected to have common features more so when closer together and less so when farther apart. Taxonomically speaking, all mammals are relatively closely related. However, body shape, habitat and behavior can differ radically. Rather than looking at obvious similarities like the production of milk in females (the defining trait of mammals), let us consider the limbs.

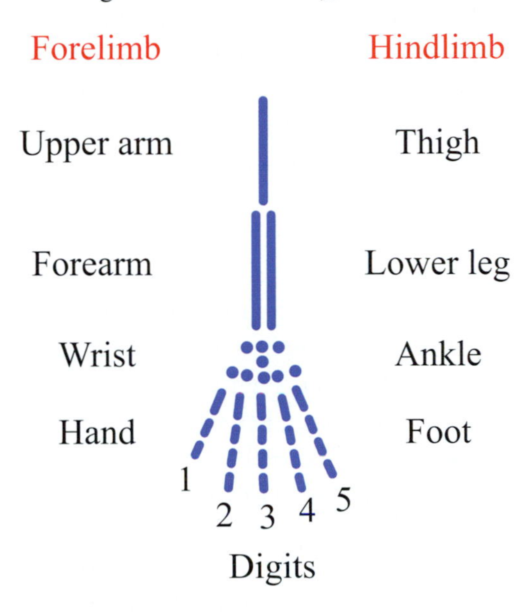

Figure 21.21: The limbs of all mammals share this basic structure.

Mammals include animals as diverse as bats, dolphins and monkeys. Now, if it is so that in evolution phenotypes change as adaptations to the environment, it is likely that these adaptations are modifications to existing structures rather than completely new features since that might require a rather radical genetic change.

Indeed, it turns out that the basic structure of the limbs of all mammals is identical: A single bone (e.g. upper arm) attached to the main body is followed by two bones (e.g. lower arm) then followed by the wrist bones to which the digits are attached. In bats, the forelimbs have turned into wings while in dolphins they became flippers, but the structure is the same. This basic structure is shown in Figure 21.21

Biochemical evidence

Finally, let us consider the biochemical evidence for evolution. Although, as noted above, the genetic apparatus is universal, as such this is not a requirement for the theory of evolution. It could very well have been that life emerged multiple times with a different basic cellular organization depending on the environment. After all, *a priori*, there is no need for a bacterium living 5,000 m below sea level under immense pressure in the perennial dark to replicate in the same way as a bacterium swimming in a sunlit tropical pool.

As it turns out, though, the genetic apparatus is universal, and that is a very strong argument for a common origin of all life. The genetic apparatus consisting of DNA, RNA, and a whole host of proteins and enzymes is a finely tuned machinery of enormous complexity. It is therefore unlikely that nature would have invented it numerous times. This is especially so since there are many possible similar arrangements with somewhat different amino acids that are conceivable.

Another interesting piece of evidence is the chirality (handedness) of DNA and RNA. Many organic molecules can as such be either right handed or left handed, an example of which is shown in Figure 21.22. Functionally, these molecules are the same but all living organisms use only one type, namely left-handed molecules. If life did not have a common origin, this would not be very likely. The very least one would expect is some organisms to use one chirality and some the other, but this is not the case.

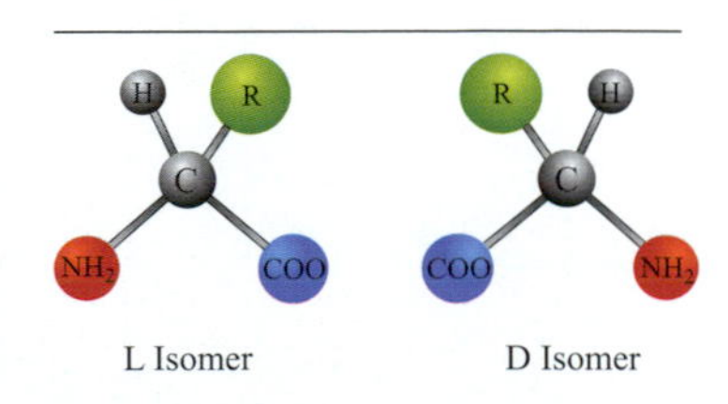

Fig. 21.22: Eucaryotes only use L isomeric amino acids.

However, for the theory of evolution to hold, we need more than just a common origin for life. We also need a path from that origin to the species alive today. Since we know that the genetic apparatus is universal and since it is improbable that all the genetic apparatus of all organisms developed in parallel (if not it would not be universal), one must conclude that it has remained unchanged for the last 3.5 billion years. Since the construction of the apparatus itself is encoded in the genome, we therefore know that some genes have remained the same over that period of time as well. Furthermore, since obviously there are big differences among the species in existence today, we also know that some genes must have changed a lot.

Life only uses left-handed amino acids.

Thus, we obtain a test for the theory of evolution and the process of natural selection in that if organisms adapt, then the change in their genome should not be too radical. Exactly what is a radical change is somewhat of a subtle point since genes can duplicate or exchange sequences of base pairs and so on. Nevertheless, closely related organisms would be expected to show fewer differences between their genomes than distantly related organisms. Taken together with the fact that the number of mutations in certain genes that are very old can also be used to judge the genetic distance between organisms, it becomes possible to construct a tree of life based on genetic data.

If evolution holds, this tree must be similar to the one obtained from paleontology and anatomy. The result is nothing less than spectacular: the trees are for all intent and purposes identical!

21.12 The answer

We started this chapter by asking why there are many species. This naturally leads us to wonder where species came from in the first place, and consequently where and how life originated. We found that life started on earth about 4 billion years ago and that the key notion for its continuity is the process of natural selection, which necessarily leads to speciation. Thus, we can answer our chapter question as: **There are many species because the evolutionary process intrinsically leads to diversification and specialization.**

21.13 Exercises

1. Which substance essential to human life was produced by cyanobacteria?

2. What was the approximate color of the early ocean (around 4.2 billion years ago)? Why did it have that color?

3. If the sun rises at around 7 am every day, as is the case for some places near the equator, approximately at what time would the sun have risen at the same location during the early earth (around 4.2 billion years ago)?

4. During the early earth (around 4.2 billion years ago), in many places building a house close to the ocean would have been problematic. Why?

5. Name a few possible energy sources for early life.

6. Although one can never exclude anything, why would an atmosphere like ours be an unlikely place for the origin of life (i.e. within the atmosphere such as within clouds — not at the interface between the surface and the atmosphere)?

7. If we had a time machine to transport us back 3.5 billion years, minimally what would we need to bring on a 1-hour expedition?

8. If there would have been humanlike intelligent life 3.5 billion years ago, what would have been the maximum possible size of a country?

9. To what phylum do humans belong?

10. What is the chirality of the amino acids found in *E. coli* bacteria? How does that compare to the amino acids found in humans?

22: Relativity

Q - Why Does $E = mc^2$?

Chapter Map

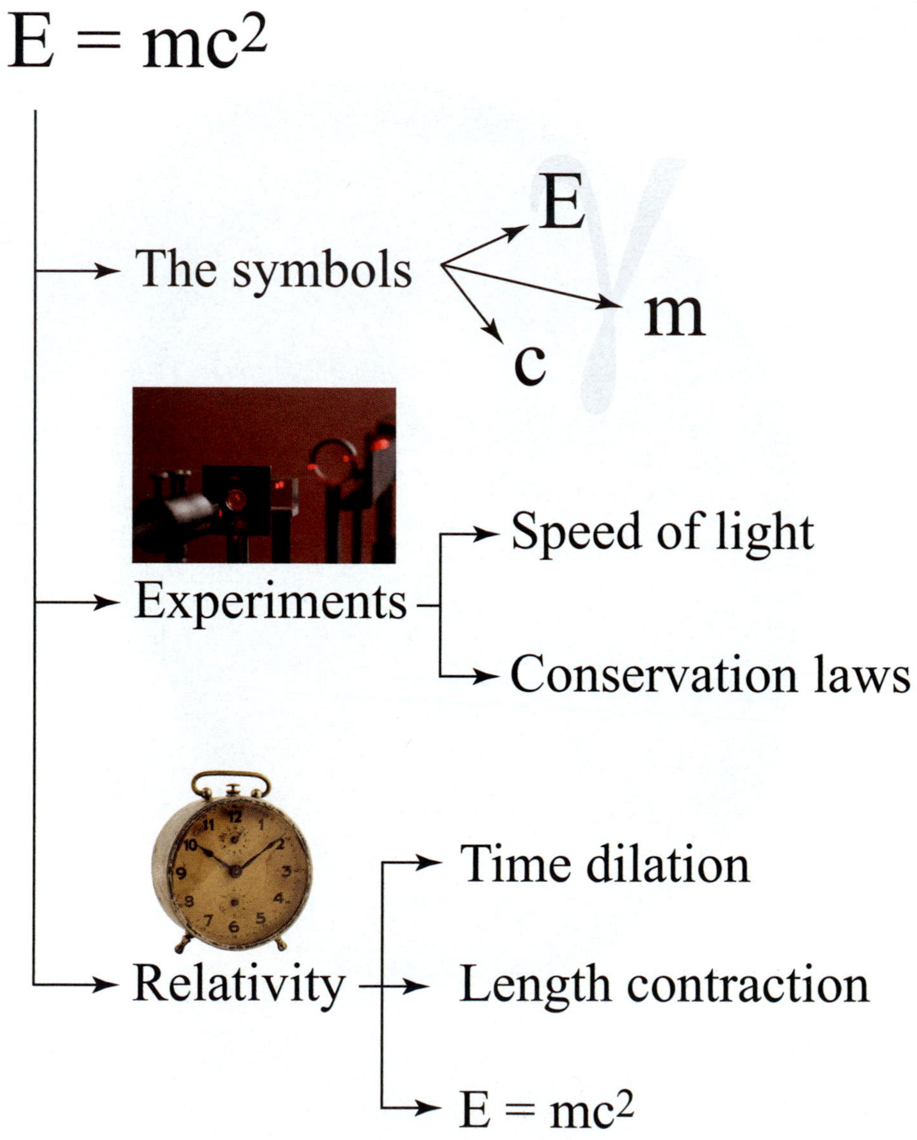

$E = mc^2$

- The symbols → E, m, c
- Experiments
  - Speed of light
  - Conservation laws
- Relativity
  - Time dilation
  - Length contraction
  - $E = mc^2$

22.1 The question

Figure 22.1: Einstein memorial in Washington D.C.

Between the images of rock or movie stars, the entertaining or perhaps not so entertaining slogans, and amusing patterns commonly found on popular T-shirts, there is an odd occurrence: a formula! Not just any formula but one of science's most important formulae. It is the expression discovered by **Einstein** in 1905 relating energy and mass. It is the famous $E = mc^2$. But what does this formula actually mean and how did Einstein find it? Indeed, why are energy and mass related in the first place? Is there a fundamental reason for that? Hence we should ask: **Why does $E = mc^2$?**

Well before we can dig into this question a bit further, we first need to know what the three letters actually stand for.

E "E" stands for ***energy***. We have encountered energy many times and discussed, for example, the kinetic and potential energies. However, there are many types of energy, and in this chapter we will find a rather surprising one: mass.

m "m" stands for ***mass***. Mass represents the amount of matter an object contains. In classical (i.e. Newtonian) physics, mass is equivalent to either inertial or gravitational mass, depending on the context. Inertial mass is the resistance to acceleration while gravitational mass is due to the effect of the gravitational field.

Mass and weight are not the same. Weight = mass × acceleration and hence at the surface of the earth we have: weight = 9.8 × mass. For convenience, in daily life, one often uses a nonscientific definition of the kilogram such that mass = weight by setting $kg_{\text{dailylife}}$ = kg/9.8.

c "c" stands for the ***speed of light***. As we have seen earlier, light is a phenomenon that has a speed. While this speed is very large, it is also known (from the various indices of refraction) that the speed of light is not constant, i.e. it is different in different materials (although for a given material and given wavelength, the speed of light is constant). It is slower in objects like glass than it is in a vacuum.

22.2 The speed of light

The speed of light has been known to be around 3×10^8 m/s since 1849 due to experiments by Arman H.L. Fizeau. His technique can, in principle, also be used to measure the speed of light in transparent objects, and indeed it was soon discovered that the speed of light depends on the medium it is traveling through. A discovery of enormous importance is the relationship between the index of refraction and the speed of light. Consequently, in the late 1800s, it was common knowledge that light

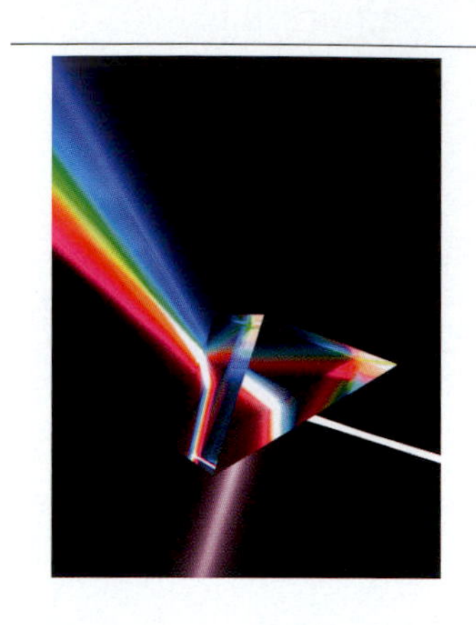

Fig. 22.2: Fact: The speed of light is medium and wavelength dependent.

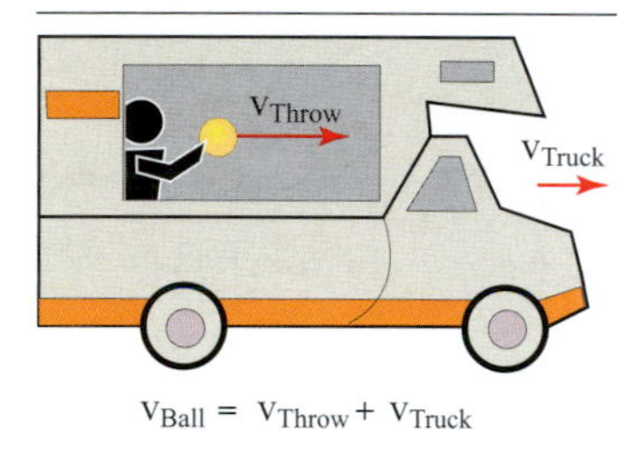

Fig. 22.3: In daily life, velocities can be added.

has a nonconstant speed. It was also known that light can travel through a vacuum (by shining it through a glass bottle whose air had been evacuated) and it was hence thought that light travels in a special stationary medium called **ether** that permeates all of space. The earth was, therefore, thought to move through stationary ether and that, consequently, the so-called ether-wind would have a certain direction at any given time — although, relative to the earth's surface, this direction would change slowly over time, due to the rotation of the earth, the rotations around the sun, the center of the Milky Way and so on.

Before continuing, we need to make a small detour into some notions regarding the speed of daily objects. If we have a tennis ball and shoot it off with a velocity of 1 m/s, what does that actually mean? In the most basic case, we assume that we are standing still and determine the speed of the tennis ball by measuring how far away from us it flies in 1 second. Indeed, most of the time when we talk about speed in daily life, we take the surface of the earth as standing still (usually a perfectly sensible thing to do). However, we also know that if we shoot off the tennis ball when standing on the back of a moving truck, with exactly the same force as before, the tennis ball will move much faster in the eyes of an observer on the ground. Generally, in daily life, speeds are additive. If we shoot out a tennis ball with a speed of 1m/s in the direction of the truck's motion, while standing on a truck moving at 0.2 m/s, then the speed of the tennis ball compared to the ground is 1.2 m/s. If the tennis ball were shot in the opposite direction to the truck's motion, the speed of the tennis ball compared to the ground would be 0.8 m/s.

So combining the above two paragraphs, we know that a) the speed of light is not fixed between various media and b) the speeds of objects are additive. As, in the nineteenth century it had long been accepted that the earth encircles the sun (with a reasonably well-known speed), it was natural to expect that an accurate measurement of the speed of light (relative to the still-standing ether) would depend on the direction in which the measurement was performed. If the measurement were to be made in the direction of the planet's motion, the speed of light should be somewhat bigger than when measuring its speed against the direction of the planet's motion.

In practice, measuring the difference in the speed of light in different directions accurately is not that straightforward. In 1881, however, Albert A. Michelson (in what is nowadays referred to as the Michelson-Morley experiment) first did so by cleverly employing the fact that light creates interference patterns when two light beams of the same wavelength meet. The idea is the following: If light is sent through a beam splitter such that two perpendicular beams are created as shown in Figure 22.4, then when those two beams are reflected by two mirrors, passing again through the beam splitters, the two beams will create a specific interference pattern according to the path difference of the two light beams. The exact value of the path difference depends not only on the distance between the mirrors and the beam splitters, but also on the speed of light between the splitter and the mirror. Consequently, if the speed of light changes for one or both of the beams, unless the changes exactly compensate each other — which is unlikely in the given setup — then the interference pattern will change by an amount that can be observed.

So under what circumstances can the speed of light change? By rotating the table on which the experimental setup is mounted! Since the earth is moving, the

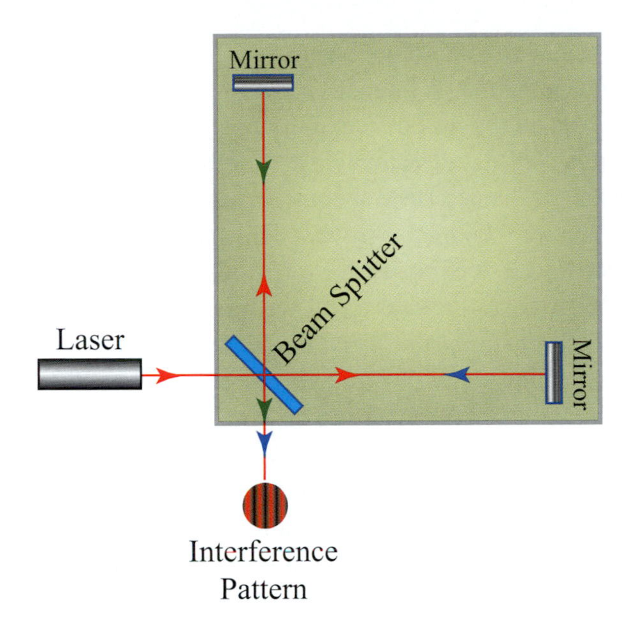

Figure 22.4: The Michelson interferometer.

speed along a path should be different from the speed perpendicular to the path. Even without establishing in which direction the earth is moving, or what the exact distance between mirrors and beam splitters is, rotating the table should lead to *shifting* interference patterns (note that the distance between the mirrors and the beam splitters is fixed).

Even though the Michelson interferometer had an accuracy of 0.01 fringe and the maximum shift was expected to be 0.4 fringe, absolutely nothing meaningful was found (here, a fringe is a dark or light band in the interference pattern). No measurable shift whatsoever! These findings were nothing but shocking to the world community of physicists. The speed of light is the same in *all* directions. In other words, the speed of light is constant regardless of the direction of the emitting source with respect to the direction of the light beam.

22.3 Frames

At this point, we should return to the additivity of velocities as observed in daily life since we have thus far not elaborated further on what is implied by considering the velocity of shooting off a tennis ball from a moving truck.

Really, when we talk about shooting a tennis ball on a truck, we should imagine that we're inside a camper van moving in a constant direction with a constant speed as illustrated in Figure 22.5. The back area of the van is large enough for a physicist to conduct experiments and do all kinds of things. Now imagine further that there is a two-way mirror so that an outside observer can see what we do but we can't see what happens on the outside of the van.

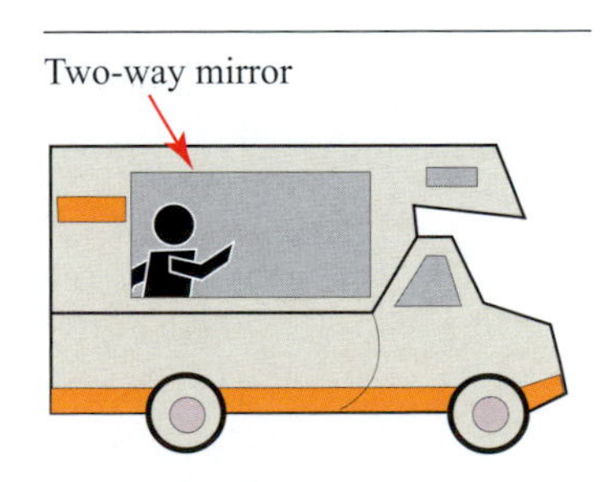

Fig. 22.5: We are inside a lab in a camper van. The window is a two-way mirror so the outside can see us but not vice versa.

Since we can only see what's inside the van, we measure everything with respect to the van. The outside observer, standing perhaps on a sidewalk, will measure everything with respect to the ground. Hence for us the van is the reference frame, and for the observer the ground is the reference frame. For general discus-

The principle of Galilean relativity is that the laws of mechanics are the same in all inertial frames of reference.

sions it is more convenient to drop notions like "van" and "ground" and simply talk about "frames" (dropping the "reference" as well).

Now since we cannot see the outside of the van, as long as the truck is moving at a constant speed in a constant direction we have in fact no clue that we are moving. All experiments of mechanics (like shooting a tennis ball) will give exactly the same results as when we'd be on the ground. This fact is referred to as the principle of **Galilean relativity**. Indeed, if we have two identical machines that shoot out tennis balls with a given force, one being located inside the van and one on the ground, then a measurement of the speed of the tennis ball 1 m from the machine by an experimenter in the van will give exactly the same result as a measurement of the speed of the tennis ball 1 m from the machine by an experimenter on the ground.

But is it really impossible to know whether one is inside a van or not? The answer to this question depends on the motion of the truck. If the truck is accelerating in one way or another (e.g. by increasing its speed or by turning), then we can measure this acceleration (e.g. by observing a pendulum or a gyroscope) and thus know that we are in the van (indeed, this is exactly the same principle with which Foucault demonstrated the rotation of the earth — see p. 5). However, if the truck is moving steadily so that there is no acceleration, then we cannot use the laws of mechanics to determine whether we are in the van. In order to be specific and accurate in discussions, it is of benefit to have a distinct name for a frame that does not accelerate: **inertial frame**.

So according to Galilean relativity all the laws of mechanics are the same within an inertial frame. Now let us consider the outside observer who sees us performing the experiments inside the van through the two-way mirror. In the coordinate system of the observer (i.e. the ground frame), the motion of the tennis ball is simply described by adding the motion of the tennis ball with respect to the van (i.e. the van frame) to the motion of the van with respect to the observer (i.e. the ground frame). In other words: If an object moves with velocity $\vec{v}'$ with respect to a frame S' which in turn is moving with velocity $\vec{v}_{S'}$ with respect to a reference frame S, then the velocity $\vec{v}$ of the object with respect to the reference frame S is given by $\vec{v} = \vec{v}' + \vec{v}_{S'}$.

More generally, motions and positions in the two frames can be related to each other by the so-called **Galilean transformation equations** given by

$$\begin{aligned} \vec{x} &= \vec{x}' + \vec{v}_{S'}t \\ \vec{v} &= \vec{v}' + \vec{v}_{S'} \end{aligned} \qquad (22.1)$$

where $\vec{x}'$ and $\vec{v}'$ are vectors measured with respect to frame S', and $\vec{x}$ and $\vec{v}$ are vectors measured with respect to frame S. For convenience, it is assumed that at time $t = 0$, the origins of the two frames were at the same location. To simplify the equations, consider a particle moving in one dimension. It has a velocity v in frame S and a velocity v' in frame S'; suppose frame S' is moving with velocity u with respect to frame S; then Equation 22.1 yields the special case that

$$v = v' + u. \qquad (22.2)$$

22.4 Einstein's postulates

We now translate the results of the Michelson-Morley experiment to the situation described in the section on frames where we have a moving van with a two-way mirror. Let us assume that we have a laser with us in the van and that the observer on the ground also has a laser. Think of the laser as a photon shooting machine. By direct inference from our tennis ball experiments, we would certainly expect that when we measure the speed of light inside the van we obtain the same value as the observer obtains when he measures the speed of light with his laser on the ground. Careful experiments prove this expectation to indeed be true, and there's no surprise in this at all.

What, then, will the observer find if he measures the speed of our laser light that we have inside the van by looking through the two-way mirror? If we had the tennis ball situation, the speed he would measure would be the previously determined speed of light plus the speed of the van. However, that is not what happens! Surprisingly, his measurement through the two-way mirror of the speed of light in the moving van yields exactly the same result as his measurement of the speed of light coming from his own laser. Whether the van moves at 1 m/s, 10 m/s or 100 m/s, he'll always get the same results. **It is an empirical fact that the speed of light is constant**, regardless of the speed of the source that emits the light.

What is going on here? Why are the tennis ball and light so different in this respect? The conundrum was resolved by **Einstein** with his **special theory of relativity**. The theory is based on two postulates that seem reasonable based on the empirical facts:

Einstein postulates

1. **The principle of relativity: The laws of physics must be the same in all inertial frames.**

2. **The constancy of the speed of light: The speed of light is an absolute quantity that is the same in all inertial frames, regardless of the velocity of the observer or the velocity of the light source. See Figure 22.6.**

The second postulate is simply the elevation of the experimental results to the status of a physical law. The first postulate states that *all* physical laws are always the same irrespective of which inertial frame they are measured in. This is a vastly wider ranging statement than the principle of Galilean relativity that only applies to mechanics. With Einstein's postulate, not only mechanics but electromagnetism, optics, thermodynamics etc. all need to be independent of which inertial frame they are measured in. Consequently, there is no preferred absolute frame with respect to that everything else needs to be measured.

While the postulates may follow more or less directly from observations, the questions is, what are their implications? Let us now look at some remarkable phenomena.

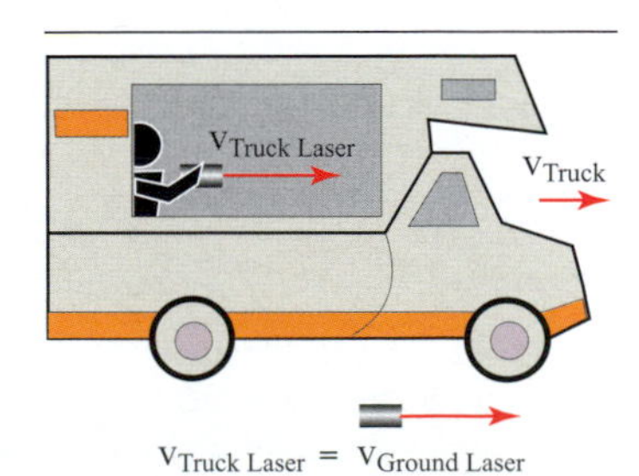

Fig. 22.6: Fact: The speed of light is constant and does not depend on the speed of the source.

22.5 Simultaneity, dilations and contractions

In classical (Newtonian) mechanics the analysis of time is universal and identical for all. However, as will be shown below, postulate 2 contradicts this. Hence, either postulate 2 is wrong or time is not absolute (that is, time depends on the observer). Since postulate 2 is well supported by facts, however strange it may seem, time must be relative.

Simultaneity

Consider a moving camper van and an observer standing on the ground. At some time in the past, we took two clocks, synchronized them precisely and then put one clock in the front of the van and one clock at the end of the van. At time $t = 0$, the middle of the van is exactly in front of the observer and the clocks in the van trigger the emission of a light flash as illustrated in Figure 22.7.

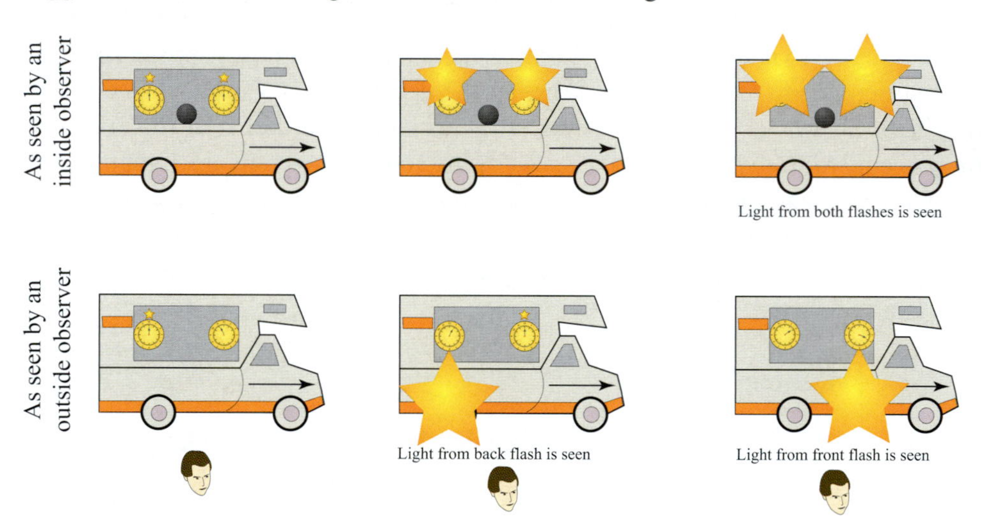

Figure 22.7: The flashes arrive at different times for the observer on the ground while they arrive at the same time for the observer in the van.

Since the observer in the van is exactly in the middle of the van we expect that he will see the flashes occur at exactly the same time. This would be expected because the two clocks were synchronized and set such that they emit a flash at the same time. What, however, will the observer on the ground actually see? Since the van is exactly in front of the observer, the clock in the back is approaching him and the clock in the front is moving away from him. If we look at two ticks of an approaching clock, then between the first and the second tick the van will have moved a bit, and consequently the light from the second tick will need to travel a shorter distance. We therefore will see the ticks closer together and the clock appears to be running faster. Similarly, the clock in the front appears to be running slower. Therefore, the observer on the ground will see the flash from the rear of the van first and then the flash from the front of the van.

Hence, we have to conclude that simultaneity is not an absolute concept but one that depends on the relative motion of inertial frames.

Time dilations

Let us now investigate the relativity of time a bit further. Suppose we want to measure the height h of our van with the help of a laser. The procedure is easy: We place a laser source on the floor of the van, a mirror on the ceiling of the van and measure the time Δt_p it takes for the laser light to travel back and forth. The subscript in Δt_p stands for "proper" and indicates that the quantity is measured with respect to the rest frame of the object (that is, the object and the person making the measurement are in the same frame as in Figure 22.8). The height of the van is then given by

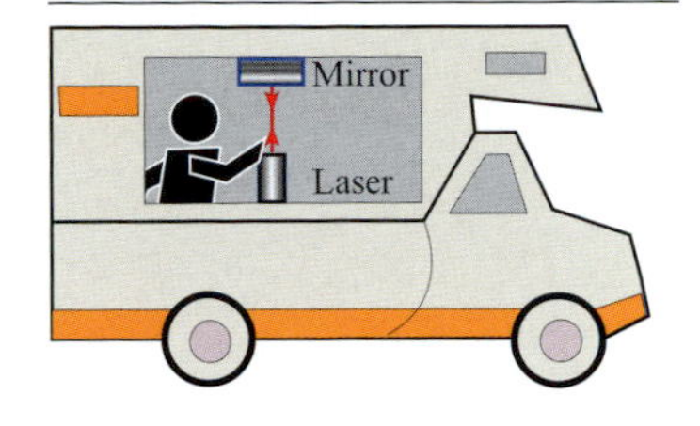

Fig. 22.8: The path of light in the rest frame.

$$h = \frac{\Delta t_p}{2} c. \tag{22.3}$$

Well that's what *we* measure. How about the observer who sees us doing the experiment through the two-way mirror? From his perspective, the light is not going straight up but diagonally up and then diagonally down. Let us assume that the observer measures Δt for the light to move back and forth. How can he calculate h from that? The thing to realize is that the van has moved a distance $\Delta t v$ in the mean time as illustrated in Figure 22.9.

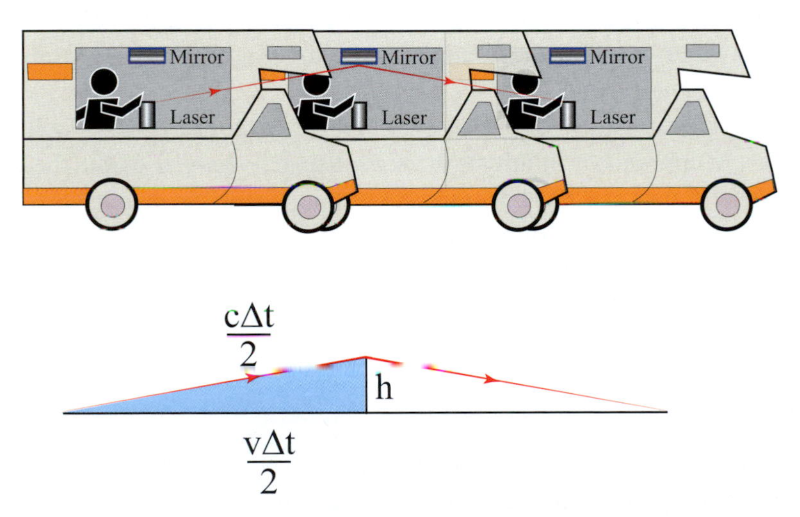

Figure 22.9: The path of light when viewed from a frame in which the camper van is moving.

Since the speed of light is the same for the observers on the ground and in the van, Figure 22.9 yields, from the Pythagorean theorem, the following:

$$\left(\frac{c\Delta t}{2}\right)^2 = \left(\frac{v\Delta t}{2}\right)^2 + h^2. \tag{22.4}$$

Solving this for h we obtain

$$h = \frac{\Delta t}{2}\sqrt{c^2 - v^2}. \tag{22.5}$$

Since the height is perpendicular to the direction of motion of the van, it must be the same whether measured from the inside or outside and hence we can combine this with Equation 22.3 to obtain

$$\frac{\Delta t_p}{2} c = \frac{\Delta t}{2} \sqrt{c^2 - v^2}. \tag{22.6}$$

Or, by reshuffling the terms a bit we have

$$\Delta t = \Delta t_p \frac{c}{\sqrt{c^2 - v^2}} \tag{22.7}$$

that is usually written as

$$\Delta t = \Delta t_p \gamma \quad \textbf{Time dilation} \tag{22.8}$$

where we use the common notation

$$\gamma = \frac{c}{\sqrt{c^2 - v^2}} = \frac{1}{\sqrt{1 - \frac{v^2}{c^2}}}. \tag{22.9}$$

Thus, we obtain the relationship between the proper time Δt_p and the externally observed time Δt. Since γ is larger than 1, we find that a clock in motion measuring time interval Δt appears to tick slower than the time interval Δt_p measured in the rest frame of the van.

As we can see for $v = 0$, γ will be 1. However, for large v, γ will become large (in fact, it approaches infinity as $v \to c$). Consequently, the time the observer measures (see Equation 22.8) becomes larger and larger for increasing speeds of the van. This effect is called **time dilation**.

Note the following points:

- The direction of motion of the van is perpendicular to the height of the van. Consequently, there cannot be any effect of the speed of the van on the height of the van.

- As the time dilation goes to infinity for $v \to c$ we can already strongly suspect that c must be the maximum speed possible. (Note that Einstein's postulates only require that c is constant and not that it is the maximum possible speed.)

Length contractions

Just like time, length also changes when observing moving frames. The proper length L_p of an object is the length of the object when measured by someone who is at rest relative to that object (i.e. the experimenter is in the rest frame of the object).

Let us consider the moving van again. This time we place a source in the back and a mirror in the front of the van and measure the time Δt_p for the light beam to travel back and forth. Clearly, being inside the van, we obtain for the length

$$L_p = \frac{\Delta t_p}{2} c. \tag{22.10}$$

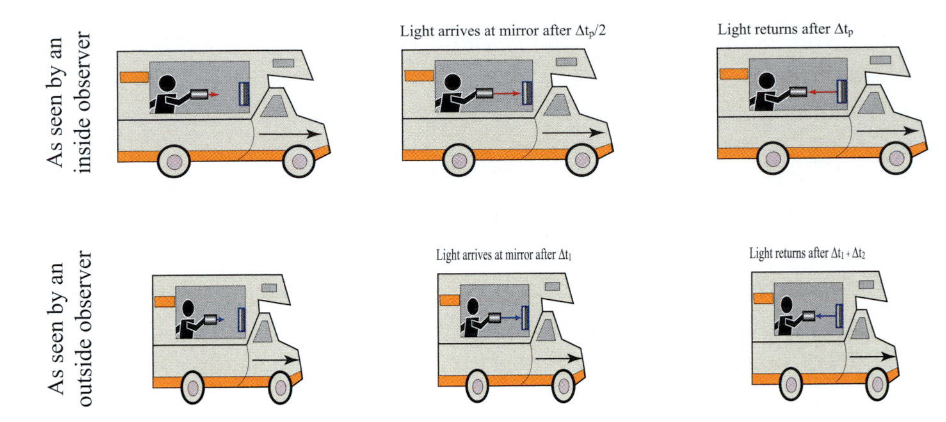

Figure 22.10: An observer on the ground measures contracted lengths for a moving object.

The outside observer will measure a different time Δt between the light leaving the source and returning via the mirror back to its starting point. If the source is pointing in the direction of motion, then after the light has been emitted the mirror will move away by a distance $v\Delta t_1$ such that the total distance traveled by the light equals $(c + v)\Delta t_1$, where Δt_1 is the time it takes for the light to reach the mirror. From the outside observer's perspective, however, the light simply moved from the source to the mirror and therefore covered a distance L and we thus have $L = (c + v)\Delta t_1$. On the return path, we similarly have $L = (c - v)\Delta t_2$. Hence for the total time it takes for the light to travel to the mirror and back we obtain

$$\Delta t = \Delta t_1 + \Delta t_2 = \frac{L}{c+v} + \frac{L}{c-v} = \frac{2L}{c}\gamma^2 \tag{22.11}$$

that after moving L to the left becomes

$$L = \frac{\Delta t}{2\gamma^2}c. \tag{22.12}$$

Combining Equations 22.10 and 22.12 on the basis that c is constant we have

$$L = \frac{\Delta t}{2\gamma^2}L_P\frac{2}{\Delta t_p} = L_P\frac{\Delta t}{\Delta t_p\gamma^2} \tag{22.13}$$

that, with the help of Equation 22.8, can be rewritten as

$$L = \frac{L_P}{\gamma} \quad \textbf{Length contraction}. \tag{22.14}$$

Consequently, the measured length of an object in motion will always be shorter than the measured length of the same object at rest. This effect is called **length contraction**.

Addition of velocities

We return to our ealier discussion on the addition of velocities and, in particular, we re-examine Equation 22.2. Consider again a particle that is moving in one dimension with a velocity v in frame S and a velocity v' in frame S'; frame S' moves with velocity u with respect to frame S; then the theory of special relativity tells us that the velocities are related in the following manner:

$$v = \frac{v' + u}{1 + \frac{uv'}{c^2}}. \tag{22.15}$$

For uv' much less than c^2 we recover the earlier result given in Equation 22.2. Now note the following remarkable feature of Equation 22.15: If light is emitted in frame S', then $v' = c$; hence, from Equation 22.15, the velocity of light observed in the S frame is

$$v = \frac{c + u}{1 + \frac{uc}{c^2}} = c.$$

In other words, in frame S, the velocity of light is also $v = c$ regardless of the value of u, which recall is the velocity of frame S' as measured by frame S!

The speed of light is independent of the velocity of the frame in which it is emitted; since only the relative velocity of two frames is in question, we have that — regardless of the velocity of the frame that emits light or observes it — the speed of light is always the same. In other words, the speed of light is the same in all inertial frames. It is also an empirical fact that the speed of light is the maximum possible speed that any physical object can have.

22.6 Newton revisited

Now, let us get back to the first symbol in the equation $E = mc^2$, E for Energy.

Relativistic momentum

In an isolated system, linear momentum must be conserved. Since this is a law of physics, according to postulate 1, it must hold for all inertial frames. That is to say, if we put a billiard table inside our moving van, then the linear momentum must be conserved whether it is measured from inside the van by us or whether it is measured through the two-way mirror by the observer on the ground. If one works this out, one finds that mass depends on velocity as

$$m = m_0\gamma \quad \textbf{Relativistic mass} \tag{22.16}$$

where m_0 is the rest mass (i.e. the mass of an object at rest with respect to the observer who measures that mass). Momentum is generally defined as $p = mv$ and this is still the case. However, now we need take the mass to be the **relativistic mass** so that

$$p = mv = m_0 v\gamma \qquad \textbf{Relativistic momentum.} \tag{22.17}$$

With the help of the relativistic momentum, we can now calculate the relativistic kinetic energy.

Relativistic kinetic energy

The requirement for the conservation of linear momentum forced us above to modify the definition of an object's momentum as $p = m_0 v\gamma$ (Equation 22.17). Since the nonrelativistic kinetic energy contains the same two quantities as the nonrelativistic momentum (i.e. the quantities mass and velocity such that $p = m_0 v$ and $K = \frac{1}{2}m_0 v^2$), it is natural to expect that the definition of the kinetic energy also needs to be modified.

While it is common to simply define the kinetic energy as $K = \frac{1}{2}m_0 v^2$, in order to see how it needs to be adapted to the special theory of relativity it is convenient to start with an expression for which the relativistic form is already known. A suitable choice is momentum since it can be used to obtain the kinetic energy. Indeed, in classical mechanics, the kinetic energy is equal to the work done by a force F on an object when accelerating it from rest to some velocity v through a distance x so that a momentum of p is attained. Therefore, we have from Equation 3.3

$$K = W = \int_0^x F dx = \int_0^x \frac{dp}{dt} dx = \int_0^p \frac{dx}{dt} dp = \int_0^p v dp. \qquad (22.18)$$

To proceed we need to evaluate dp. From Equation 22.17 we have that $p = mv = m_0 v\gamma$ and hence[1]

$$\begin{aligned} dp &= \frac{dp}{dv} dv \\ &= m_o \gamma^3 dv. \end{aligned}$$

Substituting this into Equation 22.18 we then have

$$K = \int_0^v m_0 \left(1 - \frac{v^2}{c^2}\right)^{-\frac{3}{2}} v dv = m_0 c^2 \left(\left(1 - \frac{v^2}{c^2}\right)^{-\frac{1}{2}} - 1 \right). \qquad (22.19)$$

In other words, using the definition of γ again we have

$$K = \gamma m_0 c^2 - m_0 c^2 \qquad \textbf{: Relativistic kinetic energy,} \qquad (22.20)$$

where we see that for $v = 0$ we indeed have $K = 0$ as required. Therefore, we find that the relativistic kinetic energy Equation 22.20 has two terms; one which is

---

1

$$\begin{aligned} \frac{dp}{dv} &= \frac{d}{dv}(m_0 v\gamma) = \frac{d}{dv}\left(m_0 v \left(1 - \frac{v^2}{c^2}\right)^{-\frac{1}{2}} \right) \\ &= m_0 \left(1 - \frac{v^2}{c^2}\right)^{-\frac{1}{2}} + m_0 v \left((\frac{-1}{2}) \left(1 - \frac{v^2}{c^2}\right)^{-\frac{3}{2}} (\frac{-2v}{c^2}) \right) \\ &= m_0 \gamma \left(1 + \frac{\frac{v^2}{c^2}}{1 - \frac{v^2}{c^2}} \right) = m_0 \gamma^3 \end{aligned}$$

dependent on the velocity v of the object and one which is not. The part independent of the speed is called the **rest energy** E_0 where the subscript indicates that the velocity is 0. Hence, we have

$$E_0 = m_0 c^2 \quad \text{Rest energy.} \tag{22.21}$$

The total relativistic energy must then be the sum of this rest energy and the kinetic energy so that we have

$$E_{\text{Total}} = E = K + E_0 = \gamma m_0 c^2 \tag{22.22}$$

that in terms of the relativistic mass is

$$E = mc^2 \quad \textbf{: Einstein relationship for energy}. \tag{22.23}$$

Although this result may seem strange, it has been confirmed by countless experiments and is universally accepted by the scientific community. Indeed, besides the notion of conservation laws, special relativity and its far-reaching implication of the equivalence of mass and energy is one of the solid cornerstones of modern physics.

22.7 The answer

We can now answer the question why mass and energy are equivalent or why does $E = mc^2$. In nature, there is no absolute reference frame and the speed of light is constant regardless of the speed of the experimenter who measures it or the speed of the light-emitting source. Einstein translated this observation into two powerful postulates and worked out what their consequences are when obeying the necessary conservation laws in his special theory of relativity. Hence we can say that: $E = mc^2$ **because nature considers all inertial frames to be equal leading to light being an absolute quantity**.

22.8 Exercises

1. What is the rest mass of a bowling ball that weighs 100 N on earth in space?

2. How long does it take light to travel 1000 km? During this time how many cycles does a 2.7 GHz microprocessor go through?

3. If the lifetime of an unstable particle is 1 s in its rest frame, then how long is the observed lifetime if the particle is moving at 0.95 c with respect to the observer?

4. How much slower is a clock traveling at 0.9 c?

5. How much mass needs to be converted into pure energy to run a hypothetical 9,000 MW "Einstein" reactor for 1 s?

6. By what percentage does a 1 m stick contract if it travels at 0.8 c?

7. We observe a standard clock to be running at half the speed from an identical clock in our lab. At what speed is the clock moving?

8. Elementary particles called muons have a very short lifetime τ of about 2.2×10^{-6} s. They are produced in the upper atmosphere when it is hit by cosmic rays. Despite this short lifetime, these muons can be detected in labs at the surface of the earth. Why is this so?

9. What is the relativistic mass of an electron moving at 0.9999 c? Compare this to the rest mass of a proton.

10. What is the relativistic kinetic energy of a tennis ball with a rest mass of 100 g moving at 0.1 c? Compare this to the regular (nonrelativistic) kinetic energy.

23: Quantum Mechanics I

Q - Why Are There Black Lines in the Spectrum of the Sun?

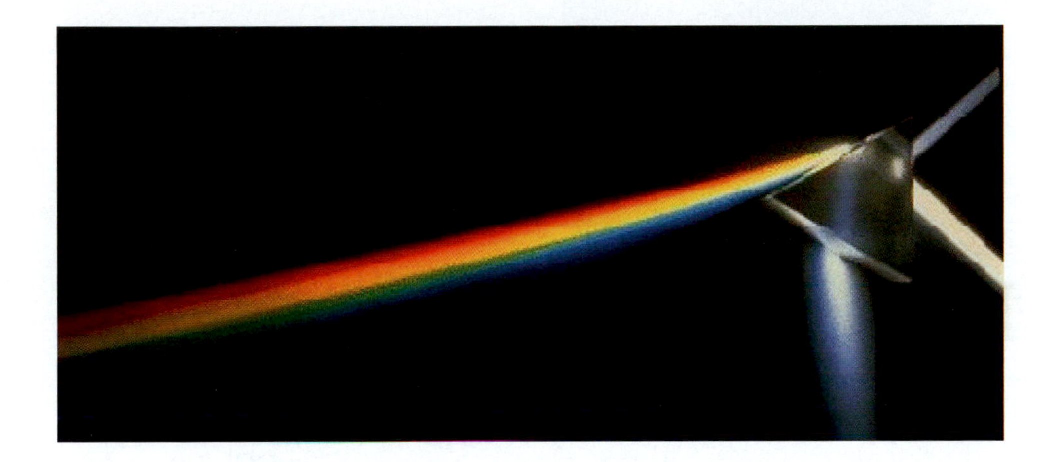

Chapter Map

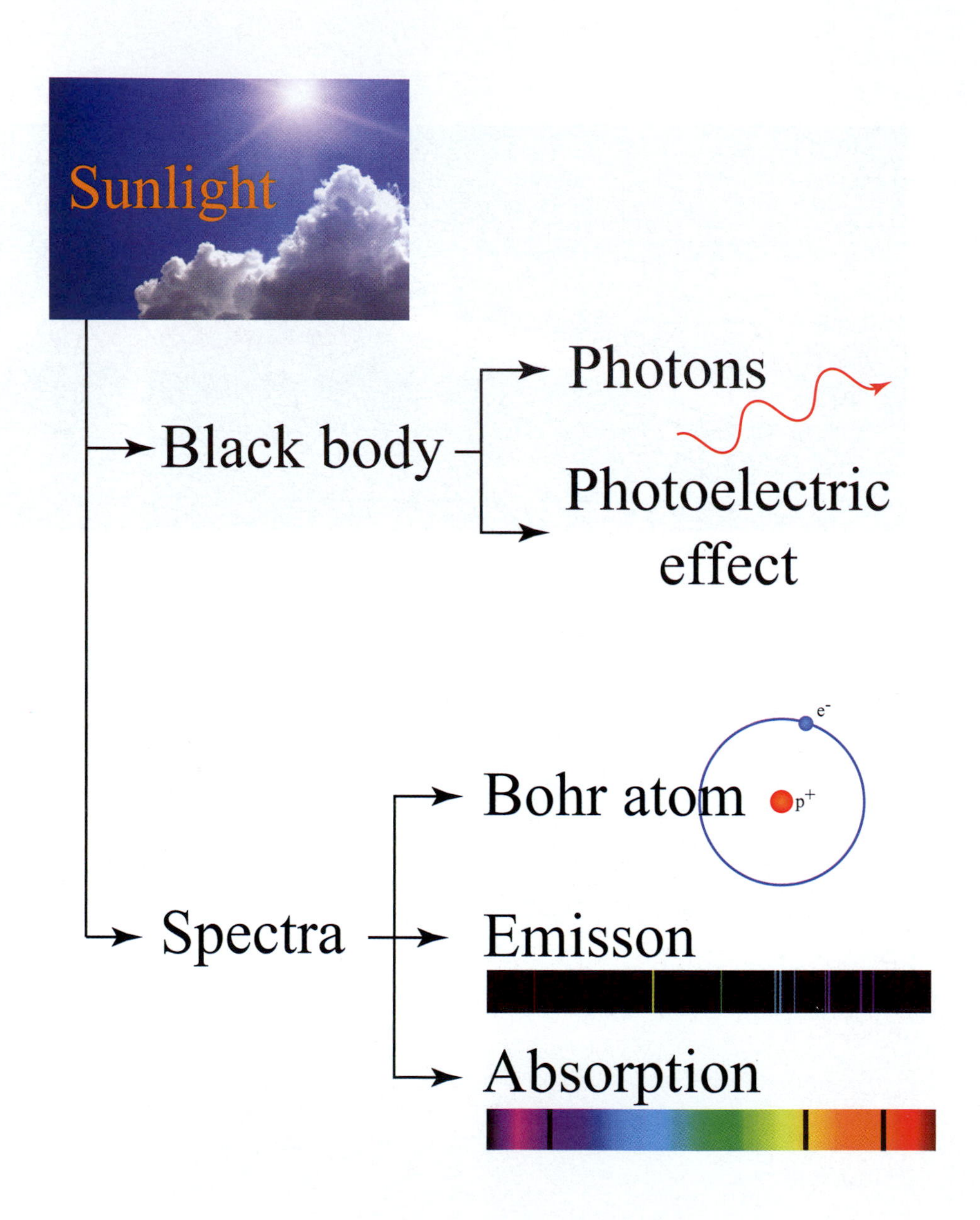

Sunlight
Black body
Photons
Photoelectric effect
Spectra
Bohr atom
e-
p+
Emisson
Absorption

23.1 The question

Figure 23.1: The spectrum of the sun has many unexpected black lines.

The light from the sun consists of waves with different wavelengths which we perceive as colors with 400 nm roughly corresponding to purple and 700 nm to dark red. Initially, one would expect that all wavelengths, at least in some allowed range, are present in the sun's spectrum since an analysis with a simple prism yields all the colors one usually encounters in daily life. However, a careful analysis using a more accurate prism shows that there are a large number of black lines in the spectrum as can be seen in Figure 23.1. What is the origin of these black lines? **Why are there black lines in the spectrum of the sun?**

The first step in understanding the spectrum of the sun is understanding optical spectra in general and the kinds of objects that emit waves that can be analyzed by a spectrum. Again from analysis with the help of prisms, we know that standard light sources like burning candles or the sun have a spectrum. Are such kinds of objects the only objects that have a spectrum? At the end of the nineteenth century, it had been observed that all bodies radiate, and that the relative intensities of the colors depend on the body's temperature.

How can one measure the radiation of an object? In the ideal case, one would want to consider only the radiation truly emanating from a body without any reflections from other sources; for example, one would want to measure the radiation of a so-called **black body**. If one heats a black body to a high enough temperature it will glow red, emitting radiation mostly with infrared wavelengths. In practice, an excellent approximation to a black body is a small opening of a cavity (hollow object), as illustrated in Figure 23.2, that is maintained at temperature T. The radiation that enters the hole will be scattered around on the inside and is very unlikely to exit via the hole unaltered, whereas the internal radiation generated by heating the material composing the cavity can escape freely from the cavity via the small opening. Indeed, a very hot cavity (body) radiates with a spectrum just like that of the sun (without the black lines) and one can hence conclude that the sun is very hot.

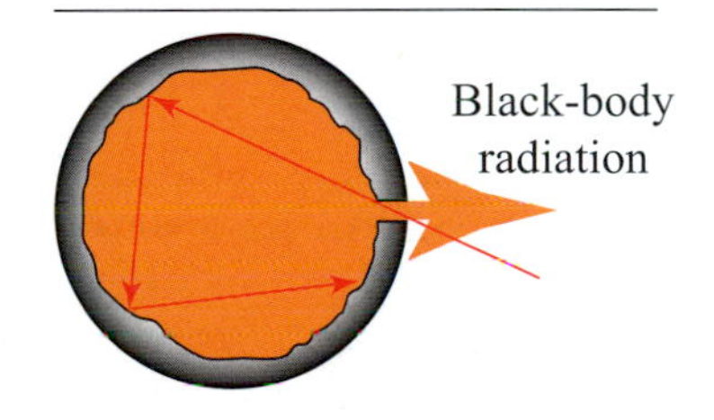

Fig. 23.2: Black-body radiation emanates from a hollow object.

Since James Clerk Maxwell had developed a very successful theory of electromagnetism and light was known to be an electromagnetic phenomenon, it was only natural to attempt to explain the black-body (intensity) spectrum with the help of Maxwell's theory. Although partially successful, in the nineteenth century a major discrepancy between experiments on black-body radiation and the predictions based on Maxwell's theory of radiation was found. With the help of classical electromagnetic theory, i.e. Maxwell's theory, the so-called Rayleigh-Jeans law for the intensity of black-body radiation was derived as

$$I(\lambda, T) = \frac{2\pi c k_B T}{\lambda^4} \tag{23.1}$$

where $I(\lambda, T)$ is the intensity of the emitted radiation having a wavelength of λ, T the temperature of the black body, c the speed of light, and k_B the Boltzmann constant.

While this formula matches experimental data reasonably well for longer wavelengths, it diverges for shorter wavelengths. That is to say, for shorter wavelengths the intensity becomes very large so that short wavelength radiation should dominate black-body radiation. This is, however, not the case in nature. Rather, the experimental data clearly show that the radiation emitted does not go to infinity for short wavelengths, but goes to zero as can be seen in Figure 23.3.

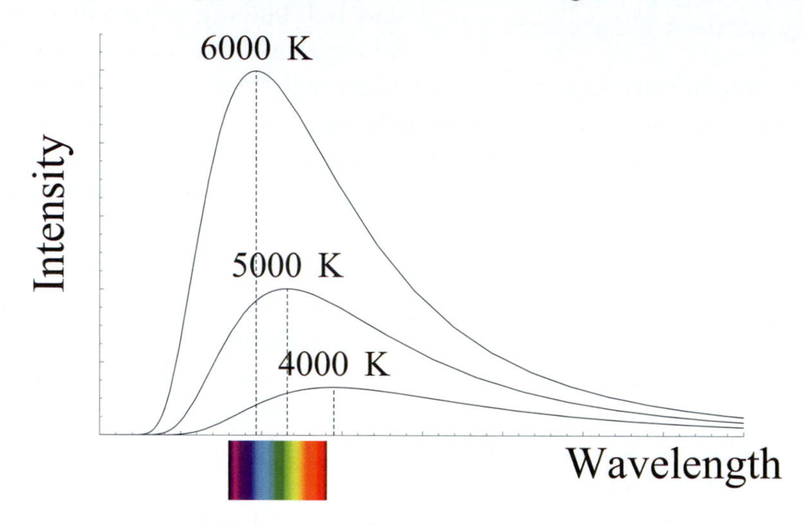

Figure 23.3: Black-body radiation. Note how much of the spectrum falls outside of the visible range.

When trying to find a solution to the discrepancy between theory and experiment, the German physicist **Max Planck** discovered that a formula exactly matching the experimental results for the spectrum of radiation can be derived by considering the energy of the emitted radiation as being a discrete quantity. In contrast, classical electromagnetic theory implies that the energy radiated is continuous and, consequently, all attempts to find an expression for black-body radiation to that point used a continuous distribution for the emitted radiation. The publication by Planck of his formula for black-body radiation was an epoch-making event and marks the beginning of quantum theory. For this, Planck was rewarded with the 1918 Nobel Prize in Physics.

23.2 Planck postulate

In his calculation of an expression for the intensity spectrum of black-body radiation, Planck set the energy of the emitted radiation to an integer multiple of a frequency-dependent chunk as

$$E = nhf \tag{23.2}$$

where f is the frequency of the radiation, n an integer and h a proportionality constant. The constant h is an empirical constant of nature and nowadays called **Planck constant**. Since energy has the unit joule J, the consistency of units requires that the constant h has the unit J s. Its numerical value is given by

$$h = 6.62618 \times 10^{-34} \text{J s} \quad \textbf{Planck constant} \tag{23.3}$$
$$= 6.58211 \times 10^{-16} \text{eV s}. \tag{23.4}$$

For the sake of convenience, it is customary to work with

$$\hbar \equiv \frac{h}{2\pi} = 1.054589 \times 10^{-34} \text{Js} \tag{23.5}$$

since in the majority of formulas in quantum physics it is $\hbar$ that appears.

The spectrum of radiation emitted by the cavity was derived by Planck to be

$$I(\lambda, T) = \frac{2hc^2}{\lambda^5 (e^{hc/\lambda k_B T} - 1)}. \tag{23.6}$$

Planck's postulate of quantized radiation packets solved the problem of the black-body spectrum and ushered in one of the greatest revolutions in science — the development of quantum mechanics.

Radiation with increasingly high (ultraviolet) frequency was incorrectly predicted by classical physics to make an increasingly large contribution to the energy radiated by the atoms of the cavity. In Planck's formalism, however, to emit even a single chunk of ultraviolet radiation would require an amount of energy much larger than the typical thermal energy that is available for emission, and hence would not be present in the radiated spectrum of a black body.

Photons

When he derived his formula for black-body radiation, Planck considered the energy chunks to be a purely mathematical construct in order to obtain a result that matches experiment. The idea that the physical radiation itself actually consists of packets of energy comes from Einstein. In one of his seminal papers of 1905 **Einstein** proposed that radiation propagated in the form of discrete energy packets nowadays called **photons**, with the energy E and momentum p of each photon being related to its frequency f and wavelength λ by the **Planck relation**:

$$E = hf = \frac{hc}{\lambda} \quad \textbf{Planck relation} \tag{23.7}$$
$$p = \frac{h}{\lambda} \; ; \; f\lambda = c. \tag{23.8}$$

Photons are rather exceptional if one thinks of them as particles, since they need to travel at the speed of light. The only way of avoiding conflict with the special theory of relativity, that forbids ordinary matter to travel at the speed of light, is for the photons to be massless (quantum) particles.

The source of the energy radiated by the atoms of the cavity is the thermal energy of the atoms. The average energy of an atom at temperature T is given by $3k_B T/2$, and determines the most likely frequency emitted; for example, one expects the peak of the intensity spectrum shown in Figure 23.3 to occur for photons with this energy. Since the energy of a single quantum of radiation at frequency $f = c/\lambda$ is $E = hf = hc/\lambda$, the most likely wavelength λ_M emitted by the cavity at temperature T has energy given by

$$\frac{hc}{\lambda_M} \simeq \frac{3}{2}k_B T$$
$$\Rightarrow \lambda_M T = \frac{2}{3}\frac{hc}{k_B}. \tag{23.9}$$

Clearly our derivation oversimplifies the problem since the atom can sometimes emit radiation with energy that is more or less than the atom's average thermal energy. A more careful derivation finds the wavelength λ_M, for which the intensity of light (given by the Planck distribution in Equation 23.6) is at a maximum, is given by[1]

$$\lambda_M T = 0.20141\frac{hc}{k_B} = 2.89776 \times 10^{-3}\,\mathrm{m\,K} \quad \text{Wien's law.} \tag{23.10}$$

This equation is generally referred to as Wien's law of displacement or simply Wien's law and matches experimental observations; one can see from Figure 23.3 that the maximum of the intensity moves toward smaller wavelengths, precisely as predicted by Wien's law. As the temperature of the black body increases, the wavelength for which the intensity of the radiation is maximal decreases. Note the constant $0.20141hc/k_B$ is a remarkable expression since it combines three of the pillars of contemporary physics, namely the quantum principle (through the constant h), relativity (through the constant c) and thermodynamics (through the constant k_B); the numerical factor 0.20141 is the result of the Planck distribution.

23.3 The photoelectric effect

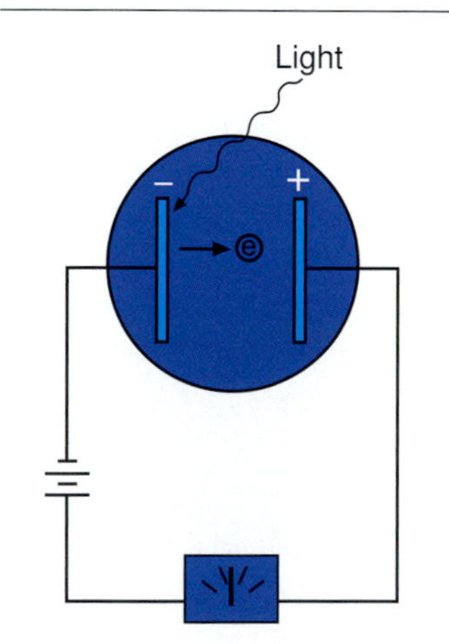

Fig. 23.4: Electrons are released when light shines on the metal.

It turned out that the notion of the photon as the fundamental quantum of light explained not only black-body radiation but also the then puzzling **photoelectric effect**. In the photoelectric effect, when light shines on some materials, the energy of light is transferred to the electrons that then have enough kinetic energy to leave the material, as depicted in Figure 23.4.

What was puzzling was that the energy of the ejected electrons depended only on the frequency of the light as illustrated in Figure 23.5. In fact, low-frequency light ejected no electrons, regardless of its intensity. On the other hand, when higher frequency light did eject photons, the kinetic energy of the electrons increased with the frequency of the light while the number of ejected electrons was proportional to the intensity. If light behaved as a wave then its intensity would depend not only on its frequency but also on its amplitude, and so even a low-frequency wave of sufficient intensity should have caused electrons to be emitted,

[1] The value of λ_M is obtained from Equation 23.6, which yields the following:

$$\left.\frac{\partial I(\lambda,T)}{\partial \lambda}\right|_{\lambda=\lambda_M} = 0.$$

contrary to experimental findings. Furthermore, the dependence of the kinetic energy of the emitted electrons only on frequency was difficult to understand in the light-is-a-wave picture.

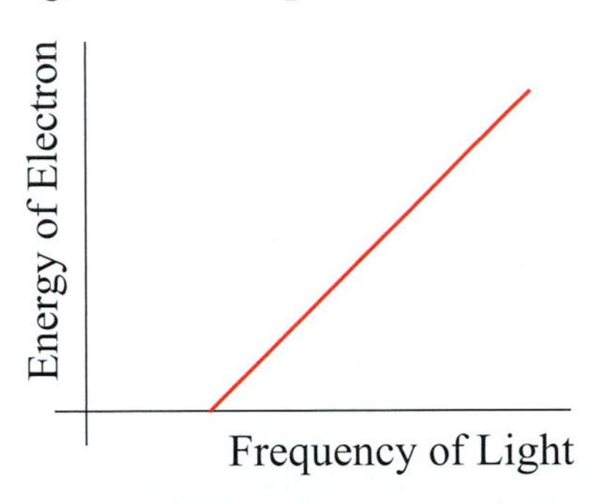

Figure 23.5: The photoelectric effect.

The above difficulties disappear in the photon interpretation. According to the photon theory, the energy carried by each photon is determined by its frequency as expressed by the Planck relation Equation 23.7. When a photon hits an electron, it can transfer only that quantum of energy, and so whether an electron is ejected and what its consequent kinetic energy is depends only on the frequency of the photon. On the other hand, the intensity or brightness of light is determined by the number of photons. A high-intensity beam has many photons while a low-intensity beam has few photons, and so the number of ejected electrons depends on the intensity of the beam. These facts agree not only qualitatively with experiment but also quantitatively. A practical application of the photoelectric effect is the transformation, in a solar cell of light energy, into an electric current. It is for his explanation of the photoelectric effect that Einstein was awarded the Nobel Prize in 1921.

All phenomena involving electromagnetic radiation can be fully explained by the quantum theory of photons. The phenomenon of electromagnetic radiation is a classical approximation to the quantum theory of photons and is in general a valid approximation when the number of photons is large enough.

23.4 Atomic spectra

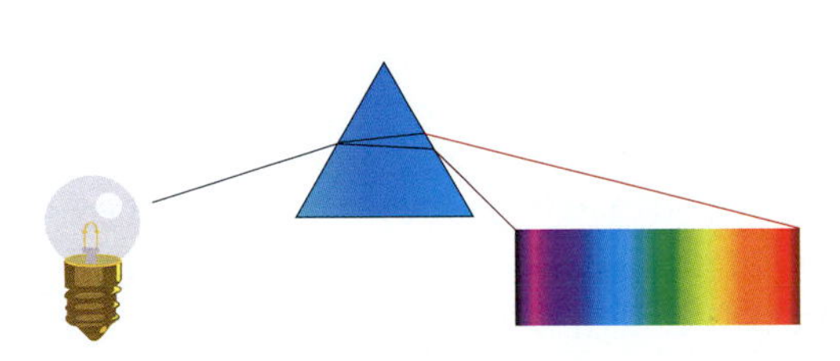

Figure 23.6: The spectrum of a hot solid like the bulb's filament is continuous.

Let us recall from our discussion of sunlight that a prism can split light into the various colors and hence wavelengths as illustrated in Figure 23.6. This is in accordance with our expectations as we have seen above that blackbody radiation consists of light of all wavelengths (though with vastly varying intensities).

Of course, one will wonder what the spectra look like when observing the spectra of many materials. What is particularly interesting is the observation that if one looks at the spectrum of a hot and low-density gas like hydrogen, the spectrum is no longer continuous but breaks up into a few well-defined lines as is shown in Figure 23.7.

For comparison, the spectra of hydrogen (H), helium (He) and carbon (C) are shown together in Figure 23.8. But one may object, carbon is not a dilute gas. In general that is true of course, but one can carefully vaporize carbon for the measurements.

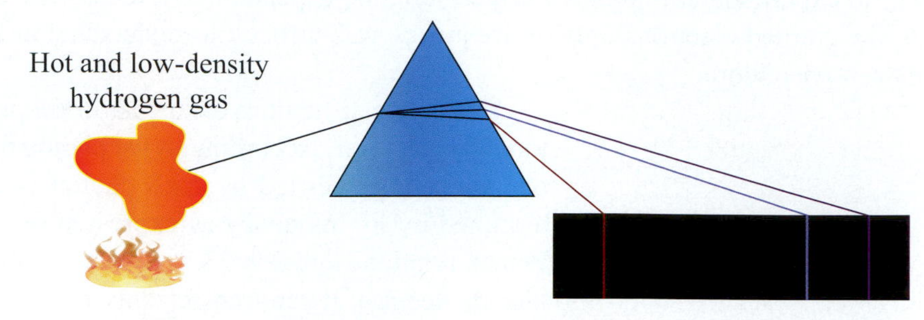

Figure 23.7: The emission spectrum of hydrogen in the visible range.

Figure 23.8: Atomic spectra for H, He and C side by side.

This type of a spectrum is called an **emission spectrum**.

As can be seen, the spectral lines for each of the elements are at different locations and therefore are like an "elementary" signature. Spectral analysis is a great way to analyze unknown materials, and indeed this is the method used to identify elements in space. After all, we can't get any samples of materials that are light-years away!

23.5 Stability of atoms

Various experiments toward the end of the nineteenth century indicated that atoms are made out of a positively charged nucleus with negatively charged particles moving around it. It was also clear that atoms interact with light. A simple manifestation of this interaction is the phenomenon that light is "reflected" off material bodies composed of atoms.

From the spectrum of hydrogen and other elements, one reaches the remarkable conclusion that the energy that the atom can emit only takes certain discrete values. This in turn implies that the atom's energy is quantized, and that the atom can have only a set of discrete energies. Figure 23.9 shows the discrete energy levels of a hydrogen atom.

How can this be explained? It was initially thought that the atom was a microscopic version of the solar system. The idea is mistaken for the following reason: An electron, bound to the nucleus of an atom, moves in a closed orbit and hence has to classically keep on accelerating since the direction of its velocity is constantly changing. The problem with this idea is that according to Maxwell's electromagnetic theory, an accelerating charge always radiates, and that consequently, an electron circling the nucleus then continuously loses energy, soon spiraling into the nucleus (effectively, the atom would collapse). Hence, according to classical electromagnetic theory, atoms are inherently unstable and cannot exist. This classical "prediction" contradicted the entire body of knowledge that chemistry had developed based on the idea that there is a different kind of an atom for each of the different elements.

The question that confronted physicists at the turn of the twentieth century was the following: Are atoms real; are they made out of material particles? If so, why don't they obey the laws of classical mechanics and electromagnetism?

A complete explanation of the atom emerged in 1926. In the absence of an alternative to Newton's law, the pioneers of quantum theory had to reason intuitively and metaphorically. By the early 1900s scientists like Niels Bohr, Max Born, Arnold Sommerfeld and Louis de Broglie had developed ad hoc rules to explain the existence of the atom, inspired by the works of Planck and Einstein.

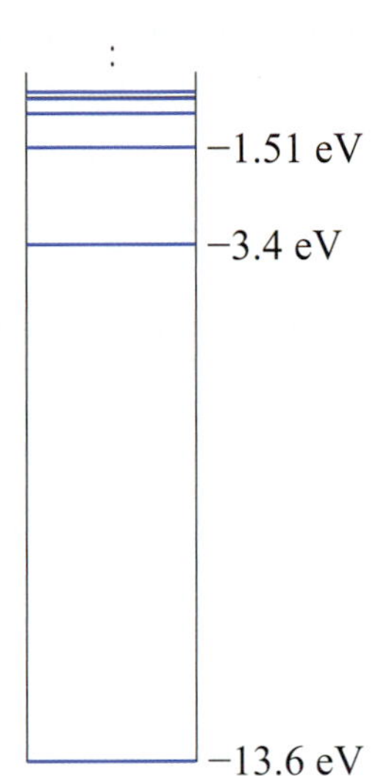

Figure 23.9: Discrete energy levels of hydrogen.

To drive physics forward, Bohr, de Broglie and others made a number of conjectures regarding the atom, but it is important to realize that at that moment there was no underlying theory from which these could be deduced. Nevertheless, these conjectures would turn out to completely revolutionize our understanding of nature.

Let us now step through the initial reasoning that led to some early key results of quantum theory; curiously enough, even though we now know that much of this reasoning was conceptually wrong, many of the outcomes of the calculations were nevertheless correct.

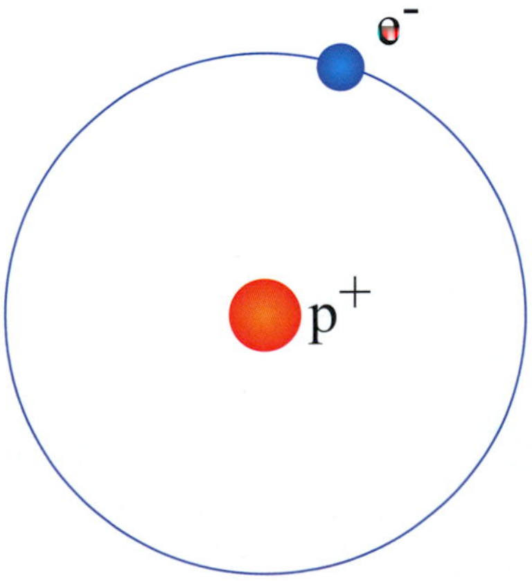

Figure 23.10: Pictorial representation of the hydrogen atom.

An electron inside an atom moves under the attractive influence of the Coulomb potential of the positively charged nucleus. Figure 23.10 schematically shows the hydrogen atom, that consists of a nucleus made out of a proton, and an electron "circling" around it. More precisely, a hydrogen atom consists of an electron of mass m_e and charge $-e$ that is bound to a nucleus (proton) having charge $+e$. The classical energy of an electron in a hydrogen atom, moving with velocity v, is given by

$$E = K + U = \frac{1}{2} m_e v^2 - \frac{k_e e^2}{r} \qquad (23.11)$$

where k_e is the Coulomb constant.

Bohr conjectured that the size of an atom should be determined by the fundamental constants involved in binding an electron to the nucleus, namely the electron mass m_e and charge e respectively, and since the existence of the atom is a quantum phenomenon, the Planck constant h should also appear.

Fig. 23.11: Niels Bohr.

By merely analyzing units, it can easily be seen that the combination $h^2/(m_e k_e e^2)$ has the unit of length and has a value of near 10^{-11} m. A more careful analysis shows that the radius of the hydrogen atom is approximately given by $\hbar^2/(m_e k_e e^2) = 0.529 \times 10^{-10}$ m.

What are the special allowed states of the electron? To get the correct energy levels, Bohr was led to the (correct) conjecture that the **angular momentum** L of the electron in the atom is quantized such that

$$L_{Bohr} = n\hbar. \tag{23.12}$$

Bohr further made the (incorrect) conjecture that the electron inside the hydrogen atom moves in an exactly circular orbit. It then follows from the expression for the angular momentum in classical mechanics, namely $L = rmv$, that

$$L_{Bohr} = n\hbar = m_e v r_n \tag{23.13}$$

which in effect means that the radii r_n for the electrons' motion are quantized.

Furthermore, for a particle moving in an exact circle, we have that the attractive force of the Coulomb potential is exactly balanced by the centrifugal force so that

$$\begin{aligned} F_{\text{Coulomb}} &= F_{\text{Centrifugal}} \\ k_e \frac{e^2}{r_n^2} &= \frac{m_e v^2}{r_n} \end{aligned} \tag{23.14}$$

that implies that the kinetic energy of the electron is

$$K = \frac{1}{2} m_e v^2 = \frac{k_e e^2}{2r_n}. \tag{23.15}$$

If we combine this with Equation 23.11, we obtain that the total energy of the electron is given by

$$E_n = -\frac{k_e e^2}{2r_n}. \tag{23.16}$$

In order to actually calculate this energy, we need to find an expression for the radius r_n. To do so, we first find v^2 from Equation 23.13 and then equate this to the v^2 we can find from Equation 23.15 so that we have

$$v^2 = \left(\frac{n\hbar}{m_e r_n}\right)^2 = \frac{k_e e^2}{m_e r_n}. \tag{23.17}$$

Rearranging the terms we thus obtain

$$r_n = \frac{n^2 \hbar^2}{m_e k_e e^2} \tag{23.18}$$

enabling us to calculate the energies in terms of n and natural constants as

$$E_n = -\frac{m_e k_e^2 e^4}{2n^2 \hbar^2}. \tag{23.19}$$

The result yields the energy levels of the hydrogen atom as given by

$$E_n = -\frac{1}{n^2} E_{\text{Rydberg}} \tag{23.20}$$

with

$$E_{\text{Rydberg}} = \frac{m_e k_e^2 e^4}{2\hbar^2} = 13.6 \text{ eV} \tag{23.21}$$

that were empirically discovered by the Swedish physicist Janne Rydberg. The energy of the hydrogen atom comes out to be negative, as expected since the electrons are in a **bound state** with the nucleus. The energy levels of the hydrogen atom given by Bohr are correct, although Bohr made a number of correct *and* incorrect assumptions and conjectures to come up with this result.

Now that we have a basic picture of the atom, can we explain the experimentally observed emission spectrum of hydrogen? The perhaps surprising answer is yes! When a dilute gas is hot or receives energy in some other way (e.g. through an electrical potential), some electrons go from the ground state to an excited state. In general, electrons do not remain in an excited state for long. When returning to the ground state they will emit a photon with an energy exactly equal to the energy difference between the initial and final electron state as illustrated in Figure 23.12. In other words, the Bohr atom explains the emission of radiation by the electron making "quantum transitions" from one of its allowed states to another one.

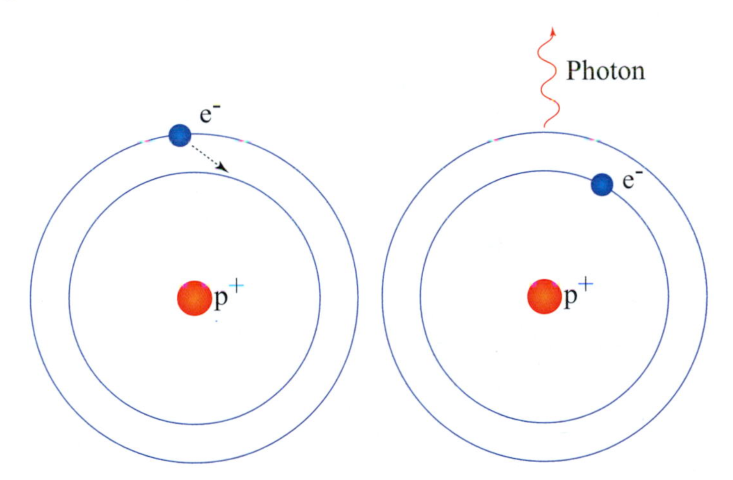

Figure 23.12: When the excited electron returns to the ground state, a quantum of light is emitted.

It should be noted that there is a maximum amount of energy that can be transferred to an electron while it remains part of the atom. If more energy is transferred, the electron will escape and no longer be a part of the atom. If that happens, the atom is called "ionized" — a situation that is illustrated in Figure 23.13.

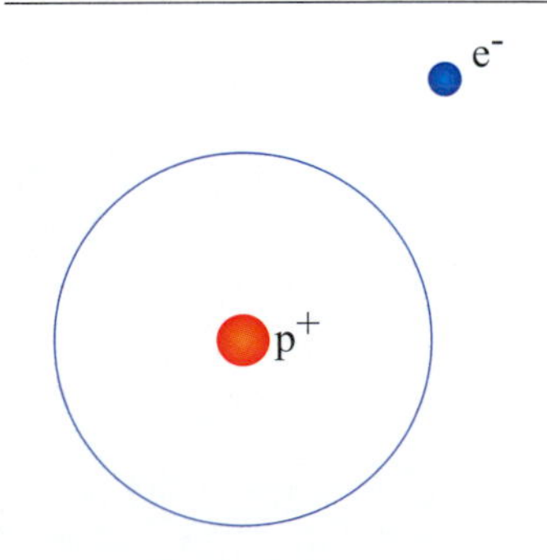

Fig. 23.13: If an electron receives too much energy, it will escape and the atom will be ionized.

23.6 Absorption spectra

Light from a lightbulb contains all the colors and can be used to obtain a continuous spectrum as was shown in Figure 23.6. If we insert a container with a somewhat dense gas in-between the bulb and the prism, one will see that the spectrum is characterized by some black lines as shown in Figure 23.14. Such a spectrum is called an **absorption spectrum**.

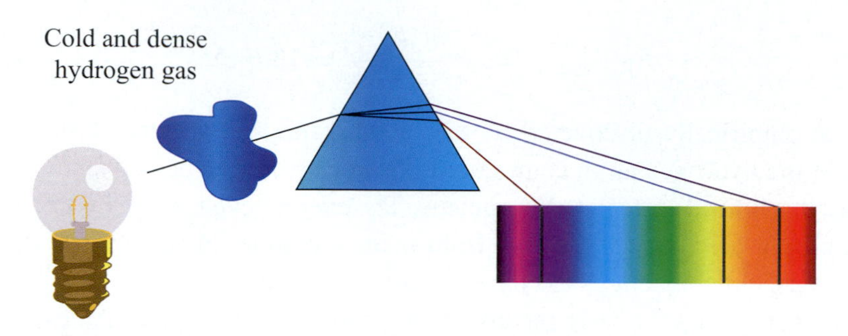

Figure 23.14: The absorption spectrum of hydrogen in the visible range.

Fig. 23.15: Emission (top) and absorption (bottom) spectra of hydrogen.

Indeed, if one carefully observes the absorption spectrum of hydrogen, one finds that the black lines are exactly at the same locations as the (nonblack) spectral lines in the emission spectrum of hydrogen, as can be seen in Figure 23.15.

It would be rather surprising if this were a mere coincidence and of course it isn't. Absorption is the opposite of emission. When light of many wavelengths enters the gas, or in other words when photons with many different energies enter the gas, those photons with exactly the right energy to boost an electron from one orbit to another will be absorbed as illustrated in Figure 23.16 and hence be removed from the spectrum, thus leaving a black line.

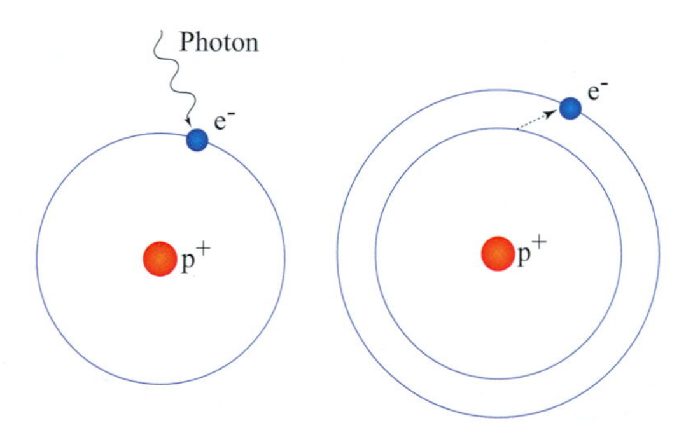

Figure 23.16: A photon (quantum of light) is absorbed by the electron, and in doing so the electron makes a quantum transition to an excited state.

But one may object, won't the electrons that were excited by absorbing a photon return to the ground state by emitting a photon as in Figure 23.12? Yes, they will, but the emission is in a *random* direction. Consequently, the number of photons emitted in the original direction of the light beam will be very small (though not zero) such that the spectrum appears black for those energies.

If we look at the discussion above, the photon is treated very much as if it were a particle. Since light consists of photons, doesn't this mean that light is therefore a (very large) collection of particles? If so, wouldn't that be rather odd? As discussed in Chapter 13, light is generally very accurately described as a wave. Clearly, as far as basic concepts go, particles and waves are very different. A water wave is nothing like a billiard ball and vice versa. How can this be? In order to obtain some idea of where the answer lies, let us now briefly elaborate on the particle nature of light.

23.7 Light is a (quantum) particle

In Chapter 13, we found that light is a classical electromagnetic wave. Let us summarize the empirical route that led scientists to first believe in the wave nature of light.

In the seventeenth century, Newton actually proposed a particle (corpuscular) theory of light to explain in the simplest way the observation that light seemed to travel in straight lines. However, this theory was not universally accepted, and some preferred Huygens' idea that light was a sort of wave similar to sound.

Fig. 23.17: Huygens: "Light is a wave!"

By the nineteenth century, empirical evidence had accumulated showing that light in fact behaved as a wave. It had been shown that light displays the phenomena of **diffraction** and **interference** which could not be explained in a particle picture but were well known as simple consequences of interacting waves from studies of water and sound waves. However, it was also recognized by Young that unlike sound waves, light waves must be transverse in order to explain the observed polarization behavior.

The transverse waves of Young were nonetheless considered to be some sort of mechanical waves propagating in an invisible medium called the "ether". In 1862, Maxwell predicted theoretically, and Hertz confirmed experimentally later, that light waves were in fact electromagnetic waves. But it was not until the work of Einstein and his theory of relativity that the need for an ether could be dispensed with.

But, having wave properties or not, light does travel in a straight line, a fact not so easily explained by considering it a wave; furthermore, the reflection and refraction of light through transparent material also seems to indicate that light behaves like a particle. Hence we should not dismiss Newton's corpuscular theory of light all too easily.

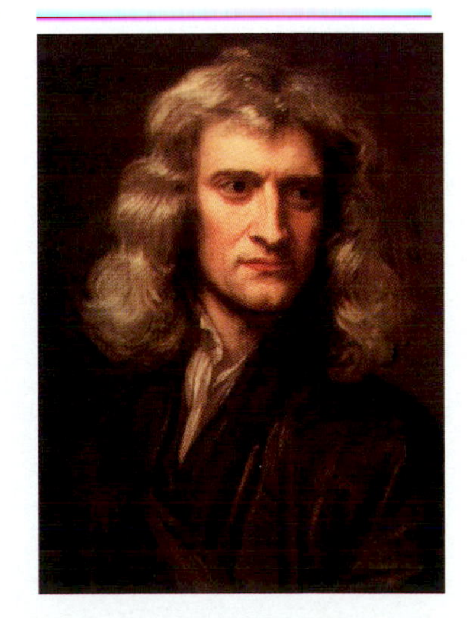

Fig. 23.18: Newton: "Light is a particle!"

The resolution provided by quantum theory is that at the most fundamental level light consists of packets of energy and momentum which are called photons. Since we associate the concept of a discrete packet of energy and momentum with a particle, we therefore think of the photon as a particle. But the photon is a rather special particle in that it has zero mass, travels at the speed of light and can (collectively) behave like a wave as well. Hence, although the photon is particle-like, it is not a classical (Newtonian) particle but, rather, is a completely relativistic quantum particle.

The wavelike behavior of light is the result of having a large collection of photons. For very low-intensity light that consists of only a few photons as well as for very short-wavelength radiation, the wave description fails and the correct description of light requires the photon. Geometrical optics — in which light behaves like rays — is valid for the case where the wavelength of light is very small when compared to the sizes of the devices involved.

Generally we can say that there are three different descriptions of light corresponding to different domains of phenomena:

- **Light is a ray:** If the wavelength of light is much smaller than the optical devices and the energy of the photon is small compared to the sensitivity of the equipment, then the description of geometrical optics, as discussed in Chapter 13, is adequate.

- **Light is a wave:** If the wavelength of light is comparable to the optical devices but the energy of the photon is still small compared to the sensitivity of the equipment, then light is described quite well by a wave.

- **Light is a photon:** If the intensity of the light is very low or the energy of the photons very high, then the full quantum theory of light, based on the idea of the photon, is necessary.

All three descriptions can be derived from the quantum theory of light in which light is made out of photons.

23.8 The answer

The black lines in the spectrum of the sun were first systematically investigated by the German physicist Joseph von Fraunhofer from 1814 onwards and are nowadays called **Fraunhofer lines**. The number of such lines is very large and Fraunhofer himself discovered more than 500 of them. These lines are mostly absorption lines of chemical elements in the upper layers of the sun while there is also some effect of our atmosphere. Hence, the answer to our question is: **The black lines are due to absorption** and the fact that nature in essence is quantum mechanical.

23.9 Exercises

1. Why does UV light more readily lead to skin damage when compared to visible light?

2. The neutral hydrogen atom has only one electron but many spectral lines. How is this possible?

3. What is the difference between a planet moving around the sun and the motion of an electron around a nucleus?

4. The Planck relation relates which physical quantities?

5. Does negative energy exist? Then why is there a minus sign for the Rydberg energies of the hydrogen atom?

6. At what wavelength is the maximum intensity of a black body with a temperature of 37°C?

7. What is the energy of an electron in the third shell of hydrogen following Bohr's line of reasoning in this chapter?

8. What is the energy of a photon with a wavelength of 550 nm?

9. What is the intensity of the green light emitted by a black body at a temperature 3000 K? (Take a frequency of 580 THz for the green light.)

10. Why would the visible emission spectrum of helium have more lines than hydrogen?

11. Why does the theory of relativity predict that the photon cannot have any finite mass?

24: Quantum Mechanics II

Q - Is Nature Counterintuitive?

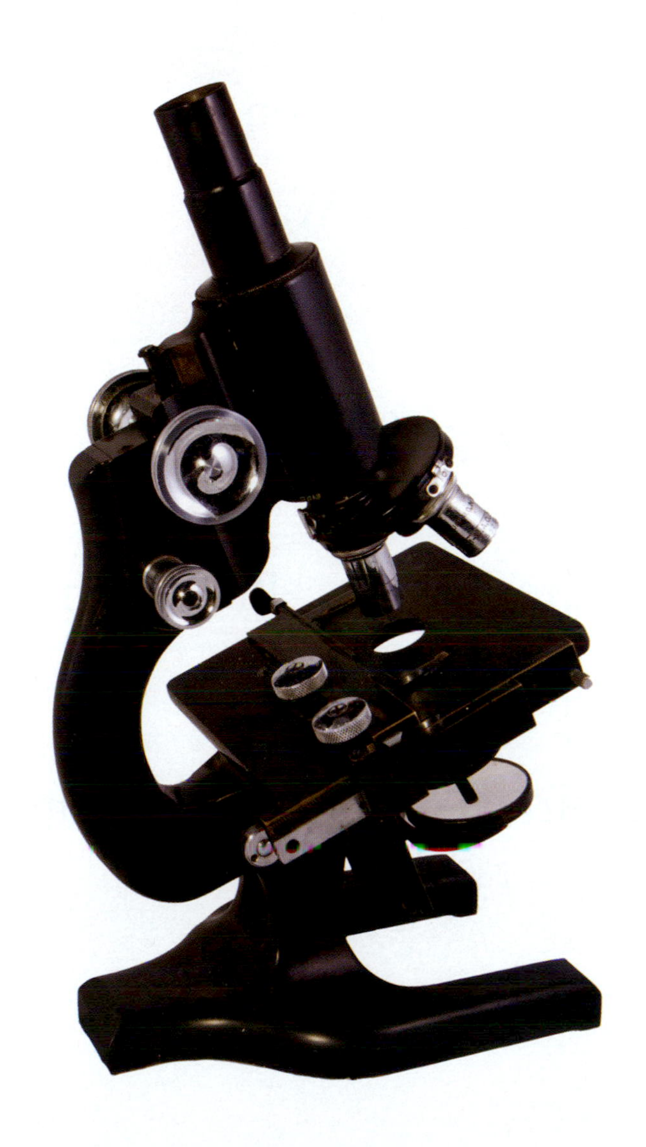

Chapter Map

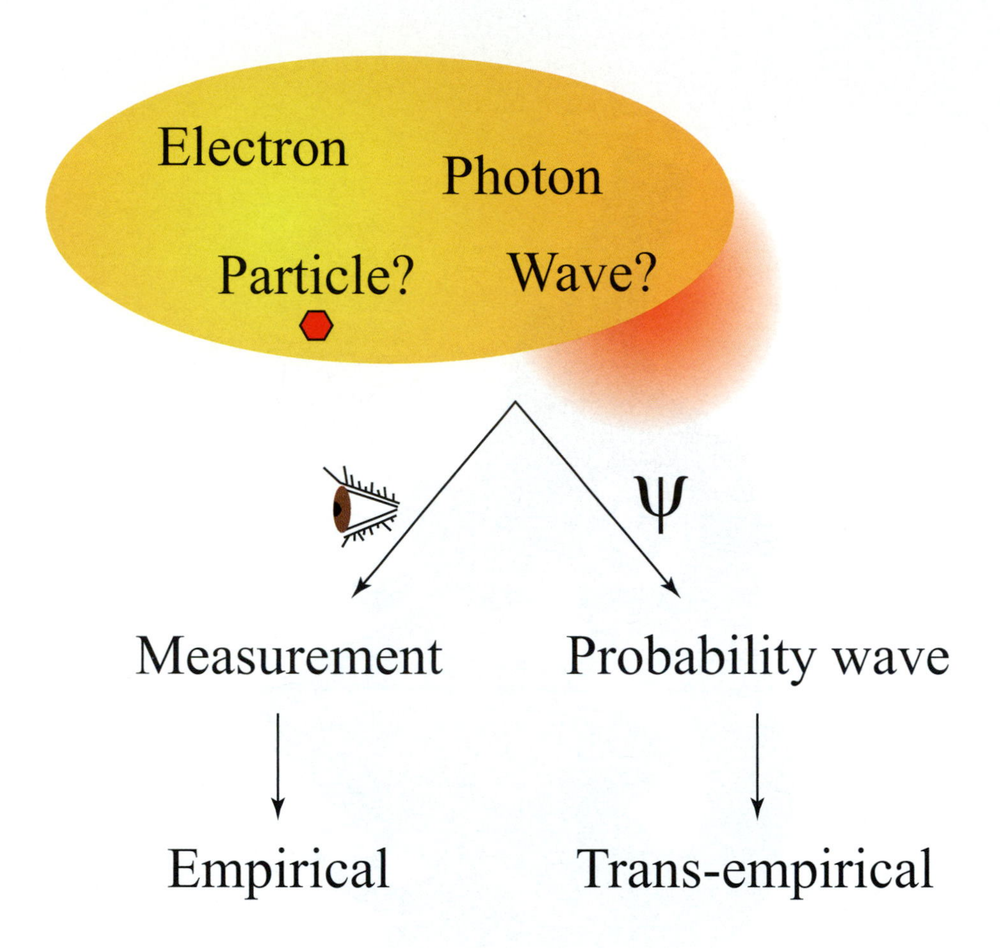

Electron
Photon
Particle?
Wave?
ψ
Measurement
Probability wave
Empirical
Trans-empirical

24.1 The question

Figure 24.1: Inspiration by an apple may have come to an end.

Is nature counterintuitive? In all the preceding chapters, we have appealed to one's direct experience of nature to introduce various scientific concepts. Making a few reasonable assumptions, we have shown how a questioning mind can explore natural phenomena and intuitively discover for oneself many of the fundamental concepts of science. The Oxford Dictionary defines intuition as "*the ability to understand or know something immediately, without conscious reasoning*". The question that arises is whether all natural phenomena are within the ambit of human intuition. Can the deep and underlying principles of nature be grasped using human intuition?

Quantum mechanics is undoubtedly the most important and ubiquitous scientific theory that has so far successfully stood the test of almost innumerable experiments. It has, however, until now defied all attempts to reach any intuitive understanding of its inner workings. The founders and leading proponents of quantum mechanics all agree on its paradoxical and opaque workings that do not conform to our intuition. The reason for this is because our physical intuition is derived from the experiences of everyday life that cannot be extrapolated to phenomena at the very small as well as the very large scale.

Niels Bohr had the following to say about quantum mechanics: *Those who are not shocked when they first come across quantum mechanics cannot possibly have understood it.* Richard Feynman was of the opinion that no one understands quantum mechanics: *I think it is safe to say that no one understands quantum mechanics. Do not keep saying to yourself, if you can possibly avoid it, "But how can it be like that?" because you will get "down the drain" into a blind alley from which nobody has yet escaped. Nobody knows how it can be like that.*

In this chapter, we examine the foundations of quantum mechanics and show that the best way to understand quantum mechanics is not to refer to one's experience of nature, but rather to study the mathematical representation of nature that quantum mechanics provides. We show that it is only by using a conscious reasoning process that one can build up an understanding of nature that is consistent with the principles of quantum mechanics.

24.2 The electron is a wave

In Chapter 23 we discussed that the electron in the hydrogen atom can only have certain discrete states. Why would that be so? Bohr had no explanation for this experimental fact, and it was only in 1923, 10 years after Bohr proposed his model for the hydrogen atom, that French physicist Louis de Broglie offered the following explanation: Just like an electromagnetic wave has particle-like characteristics, as realized by the photon, one may conversely conjecture that a *particle*, say, an electron, also has a *wavelike* properties. Recall from Equation 23.8 that a photon with wavelength λ_{photon} has momentum p given by

$$p = \frac{h}{\lambda_{\text{photon}}}.$$

A particle of mass m moving at velocity v has momentum $p = mv$. De Broglie used metaphoric reasoning by comparing a photon with a particle. He conjectured that, in analogy to the case for light, a particle is also a *wave*, with a wavelength λ given by the following:

$$p = \frac{h}{\lambda} \quad \Rightarrow \lambda = \frac{h}{p} = \frac{h}{mv}. \tag{24.1}$$

The wave, or more precisely the wave packet, as shown in Figure 24.2, indicates that the electron is "spread out" in space, with the extent of the spreading being given by the wavelength λ. As discussed in Section 19.3, since the mass of the electron is 0.511 MeV/c^2, for an electron with kinetic energy = 1 eV, the associated de Broglie wavelength is 1.23 nm.

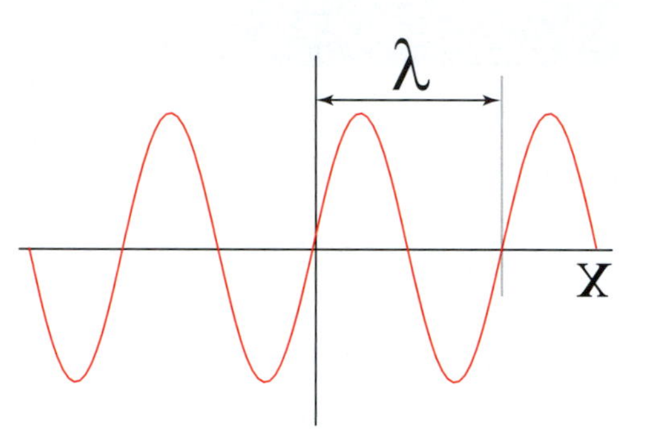

Figure 24.2: Wave of an electron.

How does the idea of de Broglie explain the behavior of the electron inside the hydrogen atom? At the scale of the atom, which is of the order of 10^{-10}m, de Broglie conjectured that the idea of an electron being a classical particle having a definite position and velocity is no longer valid. Furthermore, reasoning by analogy with the concept of resonant waves (as discussed in Chapter 11), de Broglie conjectured that the electron in the hydrogen atom is a resonant wave for which only certain frequencies are allowed.

To see how Bohr's conjectures follow from the idea of an electron wave, for a circular orbit of radius r_n, a state of the electron with n complete wavelengths yields as in Figure 24.3 — similar to Equation 11.32 for the case of resonance for a circular object — the following:

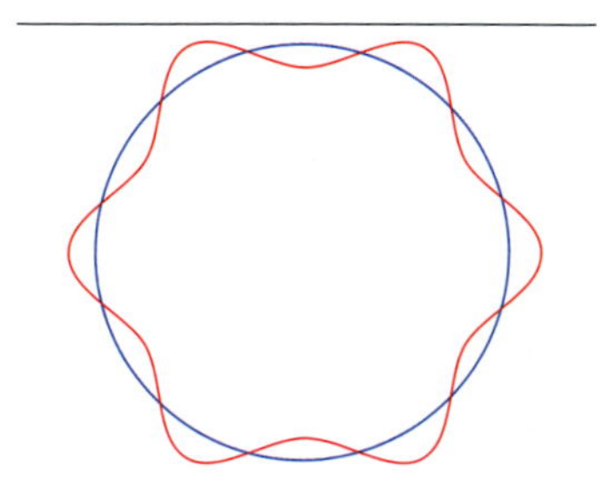
Fig. 24.3: A wave on a circle.

$$\begin{aligned} 2\pi r_n &= n\lambda = n\frac{h}{mv} \\ \Rightarrow L_{Bohr} &= mvr_n = n\hbar \ ; \ \hbar = \frac{h}{2\pi}. \end{aligned} \tag{24.2}$$

The last equation reproduces Bohr's conjecture, given in Equation 24.1, that the angular momentum of the electron is quantized.

The restriction of the states of the electron to be only resonant waves in turn provides an explanation of the discrete energy levels observed for atoms — which are interpreted as corresponding to the electron's allowed resonant frequencies.

The electron can make "quantum transitions" from a higher energy state to a lower energy state by radiating photons, and vice versa by absorbing photons. If an electron inside the atom is in an excited state, it will emit a photon and make a transition to the lowest available energy state. For example, Figure 24.4 shows two discrete energy levels of the electron in a typical atom, with the wavy line showing an electron emitting a photon in making a transition from the higher to the lower energy state.

Furthermore, the fact that there is a lowest frequency for a resonant wave explains why an electron can be in a bound state with the nucleus without radiating, since there is no lower state into which the electron can make a transition.

In summary, an electron can have a definite energy in an atom and be in a stable stationary state, but the price that we must pay is that we do not know its exact position. Instead, the electron has a wavelike behavior in an atom. This is the only way we can avoid the classical result that an accelerating charge must radiate. Note the inability to know the position and velocity of an electron in an atom is not like our ignorance in *statistical mechanics* (where an averaging method is used to study the properties of a large number of atoms that make up a gas, for example); rather, *the ignorance in quantum phenomena is an inherent limitation that is placed by nature on what can in principle be known, in this case about the electron's position and velocity*.

This wave concept of de Broglie yields a stable atom. But there is the following paradox inherent in de Broglie's postulate of an electron wave: Each and every time an electron is observed in an experiment, it is seen to be a point-like particle; on the other hand, a wave is spread over space. So this is the paradox: How can the electron be a point-particle and at the same time be a "wave"? This is the famous *"particle-wave" duality* that permeates quantum physics. What is the nature of the wave that de Broglie postulated? Is the electron wave a physical wave, like a sound wave or an electromagnetic wave?

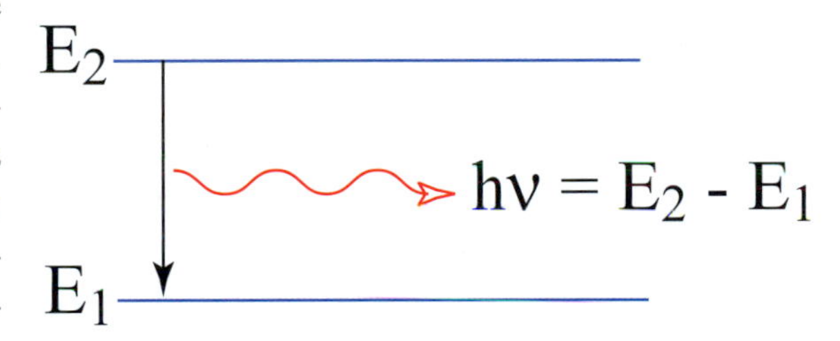

Figure 24.4: Allowed energy levels of an atom.

For 3 years, the electron wave that de Broglie conjectured was simply an interesting metaphor, without any sound theoretical or mathematical foundation. In 1926, it was finally understood that the electron wave of de Broglie is not a physical wave but instead a **probability wave**. What do we mean by a probability wave, and how does this relate to the behavior of an electron? The answer to these questions leads one directly to the formulation of quantum mechanics that is discussed in the sections below.

Direct empirical evidence was obtained for de Broglie's wave. Davisson and Germer did an experiment in 1927 that confirmed the wave nature of electrons. They scattered electrons off a crystal, and by looking at the resultant intensity they confirmed de Broglie's theoretical predictions.

24.3 Basic postulates of quantum mechanics

The modern formulation of quantum mechanics rests primarily on the ideas of Erwin Schrödinger, Werner Heisenberg and Paul A.M. Dirac. In the period 1926–1929 they laid the theoretical and mathematical foundations for a self-consistent quantum mechanics, and this theory has successfully stood the test of innumerable experiments over the last century. Unlike Einstein's theory of relativity, which reinterpret's the meaning of classical concepts such as time, position and mass, quantum theory introduces brand new and radical ideas such as the wave function, uncertainty and the quantum theory of measurement that are absent in classical physics.

Wave function

In Section 24.2 above, we discuss de Broglie's metaphoric conjecture that an electron described by a wave leads to the following paradox: Every time a measurement is made of the electron, it is always observed to be a point-like particle, but where is the wave? The resolution to this paradox is the following: The electron wave that de Broglie conjectured is not a classical wave, but rather a probability wave. This probability wave is a fundamental quantity in quantum theory and usually referred to as the Schrödinger *wave function* that, in general, is a complex valued function (of the state of the system) and at time t is denoted by ψ_t. Complex numbers are an extension of the real numbers and are briefly discussed on p. 557.

The wave function ψ_t contains all the possible information that can be extracted from the object by a process of repeated measurements.

A classical particle is fully described by Newton's laws. In particular, if we specify the position and velocity of a particle at some instant, its future trajectory is fully determined by Newton's second law. The wave function ψ replaces the role played by the position and velocity of a particle in Newtonian mechanics. In quantum mechanics one gives up any attempt to know both the particle's position and velocity; rather, one can only know of the probabilities that the particle has a certain position or a certain momentum.

In the quantum mechanical framework, nature is intrinsically and inherently indeterminate and uncertain: a quantum description of nature, say, a quantum particle, requires that we first enumerate *all the possible states* of the particle. All physical reality, at time t, is described by the wave function ψ_t and the *probability* of finding a quantum system in a particular and unique state at time t is given by $|\psi_t|^2$. The wave function ψ_t is also called the **probability amplitude** to emphasize that it is something more fundamental and deeper than the observed probabilities. More precisely, quantum probabilities, denoted by $P_t \geq 0$, are related to the wave function ψ_t by the following relation:

$$P_t(\text{some specific outcome}) \equiv |\psi_t(\text{some specific outcome})|^2. \qquad (24.3)$$

A complete description of a quantum system requires specifying the probability P_t for all the possible states of the quantum system. For example, for a quantum particle in space, its possible quantum states are to be at different positions x in space at time t. The positions of the quantum particle are indeterminate

Schrödinger equation for a free particle

With the benefit of hindsight, the Schrödinger equation for a free particle can readily be "derived" with the help of the de Broglie wavelength ($\lambda = h/p$), the Planck relation ($E = hf$) and the expressions for energy of photons ($E = pc$) and the Newtonian energy for atoms ($E = p^2/2m$).

From the Planck relation and the energy of photons we find

$$E = hf = \hbar\omega = \frac{hc}{\lambda} = pc \quad \rightarrow p = \frac{h}{\lambda} \quad \text{for photons} \tag{24.4}$$

where the angular frequency ω is defined as $\omega = 2\pi f$.

In 1923, de Broglie made the radical assumption that just like photons, particles with mass such as electrons or whole atoms have a wavelength, given by exactly the same expression as for photons:

$$\lambda = \frac{h}{p} \quad \text{for particles with mass (same as for photons!).} \tag{24.5}$$

Now if particles have a wavelength, then there must be a function describing this wave. A standard (classical) monochromatic plane wave is given by the real part of

$$\psi = e^{-i(\omega t - \mathbf{k}\cdot\mathbf{r})} \quad \text{real part = classical monochromatic wave.} \tag{24.6}$$

With the definition $k = 2\pi/\lambda$ and some reshuffling of the terms in Equations 24.4 and 24.5 we find that

$$\omega = E/\hbar \qquad \mathbf{k} = \mathbf{p}/\hbar$$

which we can insert into Equation 24.6 to obtain

$$\psi = e^{-\frac{i}{\hbar}(Et - \mathbf{p}\cdot\mathbf{r})} \quad : \quad \text{quantum wave.} \tag{24.7}$$

If we take the time derivative and the Laplacian of Equation 24.7, we obtain, for $\mathbf{r} = (x, y, z)$

$$\frac{\partial}{\partial t}\psi = -\frac{iE}{\hbar}\psi \qquad \nabla^2\psi = \frac{-p^2}{\hbar^2}\psi \text{ where } \nabla^2 = (\frac{\partial^2}{\partial x^2}, \frac{\partial^2}{\partial y^2}, \frac{\partial^2}{\partial z^2}).$$

And here it comes! The classical energy is given by $E = p^2/2m$, where p is the momentum of a particle; we hence find

$$i\hbar\frac{\partial}{\partial t}\psi = E\psi = \frac{p^2}{2m}\psi = -\frac{\hbar^2}{2m}\frac{-p^2}{\hbar^2}\psi = -\frac{\hbar^2}{2m}\nabla^2\psi$$

or in other words we have

$$i\hbar\frac{\partial}{\partial t}\psi \quad = \quad -\frac{\hbar^2}{2m}\nabla^2\psi \;: \quad \textbf{free particle Schrödinger equation.} \tag{24.8}$$

Indeed, the only new key ingredient in the derivation is de Broglie's formula for the wavelength of a particle. Nevertheless, it took nearly 3 years until 1926 that Schrödinger published his famous equation (albeit with the inclusion of the potential which is an additional important step). Furthermore, the wave function ψ is not a classical wave and its probabilistic interpretation goes beyond de Broglie's formula.

and $P_t(x) = |\psi(t,x)|^2$ is the probability of finding it at position x at time t. As expected, the sum of $P_t(x)$ over all possible positions x is equal to 1; more precisely

$$P_t(x) \geq 0 \ ; \quad \int_{-\infty}^{+\infty} dx P_t(x) = 1.$$

Equation 24.3 above is the great discovery of quantum theory, that *behind what we directly observe*, namely, the outcome of experiments from which we can compute the probabilities $P_t(x) = |\psi(t,x)|^2$, lies an *unobservable world of the probability amplitude* that is fully described by the wave function $\psi(t,x)$.

Measurement

The quantum particle's position is inherently indeterminate (random) when it is not being observed. In the sense of probability, it exists everywhere in space and hence is wavelike. When a measurement is performed to ascertain the position of the quantum particle, it is always found to be pointlike and hence the particle-like behavior of the quantum particle. The indeterminate virtual state of the electron *can never*, in principle, be directly observed by any experiment.

To clarify and understand the underpinnings of quantum mechanics, and in particular, to understand the concept of a quantum measurement, we introduce a new term, namely *trans-empirical*, into the lexicon of quantum physics. The term trans-empirical denotes the virtual state of the quantum system; we avoid "virtual" that has become associated with software-based simulations such as "virtual reality", "virtual machines" that are not related in any manner with quantum mechanics. The physical state that results from a measurement is termed as the *empirical* state of the quantum system.

The connection of the empirical state of the quantum system with its trans-empirical state is made by the process of measurement, which causes a discontinuous transition from the trans-empirical to the empirical state. It should be noted that the effect of measurement on a quantum system in *not* contained in the Schrödinger equation and is an *independent* postulate of quantum mechanics.

In quantum mechanics, the behavior of a quantum particle is radically different from a classical particle. The essence of the difference lies in the concept of measurement, which results in an **observation** of the state of the system. A classical particle, whether it is observed or unobserved, is in the same state. By contrast, a quantum particle has two completely different modes of existence. When a quantum particle is *observed* it appears to be a classical particle having, say, a definite position, and is said to be in a *physical* or *empirical* state. However, when it is *not observed*, it exists in a counterintuitive state, that is, the *trans-empirical* state.

Measurement plays an indispensable role in quantum mechanics since it connects the quantum state, as described by the wave function $\psi(t,x)$, with the probability of actually finding the particle in a particular state, which is given by $|\psi(t,x)|^2$. It should be emphasized that the probability amplitude $\psi(t,x)$ of a quantum particle is *not* an ordinary classical wave; rather, the only thing it has in common with a classical wave is that it is sometimes spread over space. There are quantum states that are not spread over space — for example, the spin of a particle at a given point; the description of such states is described by a wave function although it is *not* spread over space. If one is more precise, the term *state function* is a more appropriate name for $\psi(t, x)$ than the term *wave function*.

The nonclassical nature of $\psi(t, x)$ can seen from the following fact: The moment the quantum particle is observed (measured) to be at a definite position x, the wave function of the quantum particle instantaneously becomes zero (**"collapses"**) *everywhere in space*, since once we find the particle at position x there is zero likelihood of finding it at any other point! The process of the collapse of the wave-function causes an irreversible change in the system and is called **decoherence**. The nonlocal collapse of $\psi(t, x)$ has puzzled physicists since the dawn of quantum mechanics. For it to collapse instantaneously apparently requires that "information" about the particle being found at x is communicated at infinite speed to the rest of space — and would seem to violate the special theory of relativity. However, a detailed analysis has shown that quantum measurement theory is consistent with the special theory of relativity; and more importantly, the nonlocal nature of the wave function $\psi(t, x)$ is consistent with all the experiments that have been devised to test this aspect of quantum mechanics.

"*A careful analysis of the process of observation in atomic physics has shown that the subatomic particles have no meaning as isolated entities, but can only be understood as interconnections between the preparation of an experiment and the subsequent measurement.*" — Erwin Schrödinger.

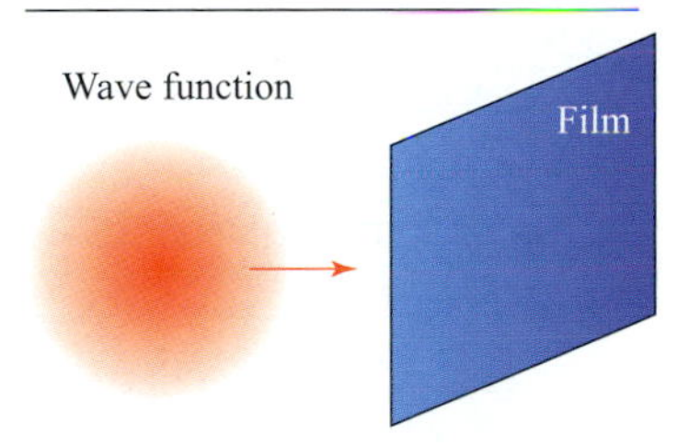

Fig. 24.5: A quantum particle approaches a detector, which is a photographic plate. The wave function shows that the particle has the likelihood of being at many points.

The process of measurement is illustrated in Figures 24.5 and 24.6. An incoming particle represented by a wave function is shown to be spread out to represent the fact that the particle has a finite probability of being found at many different points of space, with the photographic plate representing a typical detector carrying out a measurement. On hitting the photographic plate, the particle deposits all of its energy at a single point (or more precisely, in a volume having a dimension much, much smaller than the size of the spreading of its wave function). On carrying out the process of measurement, we conclude that we have observed the particle at a specific point on the photographic plate. If we now repeat the same experiment many, many times, we will find that the frequency with which the particle ends up at different points on the photographic plate is given by $|\psi(t, x)|^2$.

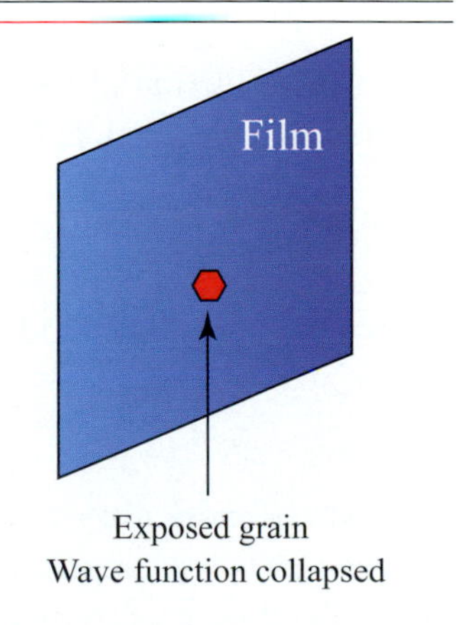

Fig. 24.6: The wave packet collapses and the electron is located at the position where the grain is exposed.

In summary, we see that the classical particle, which is point-like and fully described by its trajectory specified by $x(t)$, has been replaced by a quantum particle that is described by the wave function $\psi(t, x)$. Quantum theory has enhanced our understanding of the physical world by its discovery of the trans-physical domain of nature. The dichotomy of the empirical and trans-empirical realms of nature is shown in Figure 24.7. The term *particle-wave duality*, used by the earlier founders of quantum mechanics, is an expression of the duality of the empirical and trans-empirical domains. One loosely states that the quantum system "sometimes behaves like a particle and sometimes behaves like a wave". It should be kept in mind that the particle and wave in particle-wave duality do not *both* refer

The Schrödinger equation

The Schrödinger equation, discovered by Erwin Schrödinger, is the equation that determines how $\psi(t, x)$ changes with time; given the initial wave function of the particle at $t = 0$, i.e. $\psi(x, 0)$, the future behavior is then fixed by the Schrödinger equation. Or in the case of a time-independent system, solving the Schrödinger equation gives the stationary result. The Schrödinger equation is given by

$$i\hbar\frac{\partial\psi(t,x)}{\partial t} = H\psi(t,x)$$

where H is the so-called quantum Hamiltonian that represents the energy of the object being described. For example, in the case of a particle moving in a potential given bt $V(x)$, the quantum Hamiltonian is given by

$$H = -\frac{\hbar^2}{2m}\nabla^2 + V(x)$$

where ∇^2 is the Laplacian operator. The Schrödinger equation is given by

$$i\hbar\frac{\partial\psi(t,x)}{\partial t} = \left(-\frac{\hbar^2}{2m}\nabla^2 + V(x)\right)\psi(t,x). \tag{24.9}$$

Recall from Equation 23.11 that the classical energy of an electron in the hydrogen atom is given by

$$E = \frac{1}{2}m_e v^2 - \frac{k_e e^2}{r}$$

where k_e is the Coulomb constant. There is a close correspondence between the classical energy and the quantum Hamiltonian. In particular, the quantum Hamiltonian acts on the wave function ψ, and it can be shown that the classical expression $(m_e v)^2$ corresponds to the operator $-\nabla^2$; more precisely, the Hamiltonian for the hydrogen atom is given by

$$H = -\frac{\hbar^2}{2m_e}\nabla^2 - \frac{k_e e^2}{r}.$$

The wave function of the hydrogen atom can be factorized into a time-dependent term and a function independent of time; hence, for the hydrogen atom having energy E, its wave function is written as

$$\psi(t,x) = e^{-\frac{i}{\hbar}Et}\psi_E(x) \tag{24.10}$$

where $\psi_E(x)$ is independent of time. The Schrödinger equation then yields

$$H\psi_E(x) = E\psi_E(x).$$

The wave function $\psi_E(x)$ is a special state since, under the action of H — up to an overall factor denoted by E — it is unchanged and is called an eigenfunction. The allowed bound states of the hydrogen atom are determined by $\psi_E(x)$ that have only discrete energies E_n — as in Equation 23.21 — and are given by

$$H\psi_n(x) = E_n\psi_n(x) \ ; \ E_n = -\frac{1}{n^2}E_{\text{Rydberg}}$$

where $\psi_n(x)$ are the "resonant states" of the electron that de Broglie had conjectured.

to a classical particle and a classical wave; in our terminology, the particle usually refers to the empirically observed state and the wave refers to the trans-empirical wave function.

The radical and new concepts that have been introduced by quantum mechanics are studied in some detail in the following sections:

- The wave function embodies the exhaustive and complete description of nature. The wave function describes the trans-empirical state of the quantum system. To understand how the wave function's quantum description differs from the classical description, the example of a particle in a box is studied in Section 24.4

- The role of measurement is central in the discontinuous transition of the quantum system from the trans-empirical to the empirical domain and vice versa. Once a quantum system is prepared and no longer observed, it makes a smooth and continuous transition from the empirical to the trans-empirical state. We discuss in Section 24.5 how the Planck quantum postulate completely changes the classical concept of measurement and leads to Heisenberg's uncertainty principle.

- A classical wave, like the waves on an ocean, remains in its state whether one observes it or not. In contrast, the wave function is *not* a classical wave; the moment a particle is observed at position x, its wave function *instantaneously* becomes equal to zero everywhere, which in physics is called the "collapse" of the wave function. The collapse of the wave function in studied for the quantum Zeno effect and is discussed in Section 24.6.

24.4 Wave function: particle in a box

We examine the behavior of a classical and quantum particle confined inside a box since it is one of the simplest quantum systems. Consider a particle of mass m, confined to a one-dimensional box of length L with infinitely high and perfectly reflecting walls.

Classical description

Suppose the particle has a velocity v, and hence momentum $p = mv$. The classical (Newtonian) description of the particle is that it travels along a well-defined path, with a velocity v. Since the box has perfectly reflecting boundaries, every time the particle hits the wall, its velocity is reversed from v to $-v$, and it continues to travel until it hits the wall and bounces back and so on. The momentum and energy of the particle are given by

$$p_{\text{classical}} = mv \tag{24.11}$$

$$E_{\text{classical}} = \frac{1}{2}mv^2 = \frac{p^2}{2m}. \tag{24.12}$$

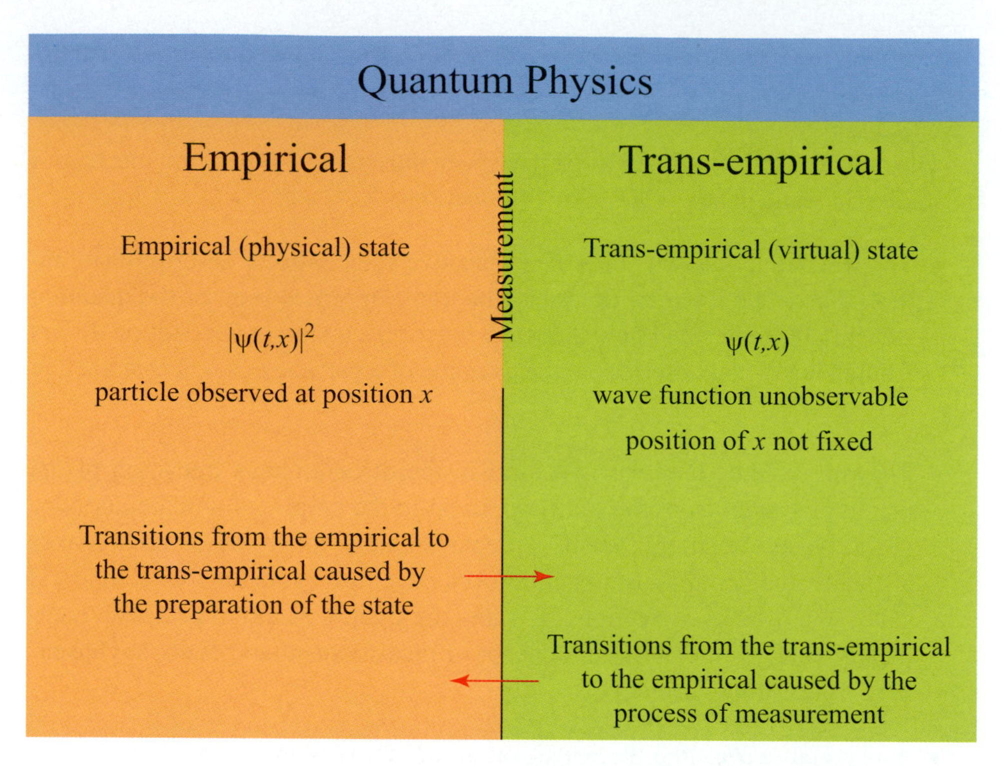

Figure 24.7: The dichotomy and duality of quantum physics.

The position of the particle can be predicted exactly. Suppose the particle starts its motion at time $t = 0$; then at time t its position is given by

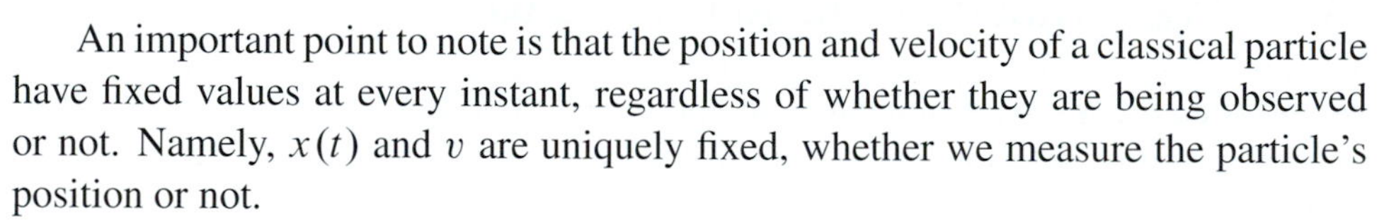

$$x(t) = \begin{cases} vt & 2nL/v < t < (2n+1)L/v \\ L - vt, & (2n+1)L/v < t < 2(n+1)L/v \end{cases}$$
$$n = 0, 1, \ldots$$

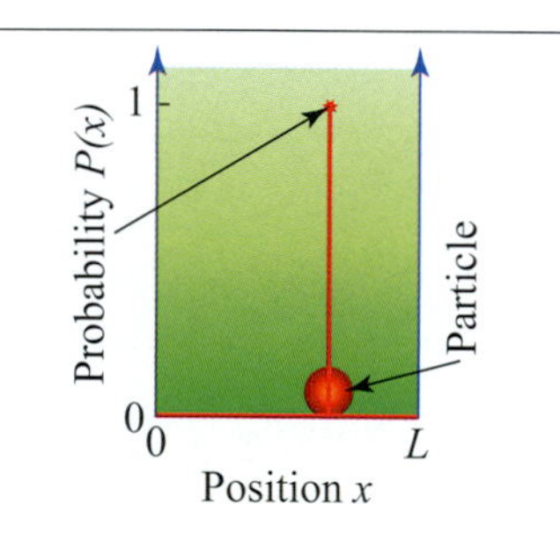

Fig. 24.8: Classical particle: stationary.

An important point to note is that the position and velocity of a classical particle have fixed values at every instant, regardless of whether they are being observed or not. Namely, $x(t)$ and v are uniquely fixed, whether we measure the particle's position or not.

If we plot the likelihood of finding, at any given instant t, the particle at different points inside the box, we are certain to find it at only *one* position, as shown in Figure 24.8.

Classical statistical description

In our discussion of the ideal gas in Section 10.4 we introduced the kinetic theory of gases in order to analyze a system that was far too complicated to describe by determining the position and velocity of each and every atom or molecule in the gas. We had instead introduced a statistical and probabilistic description of the ideal gas in which the position and velocity of each atom were taken to be random. We can take a similar approach to the motion of classical particle in a box. Suppose the particle has been bouncing around in the box for a sufficiently long time and we have no idea of when the motion started. Where do we expect to find the particle?

Since we don't know anything about the particle we postulate that all positions are equally likely and that there is an equal likelihood of finding the particle at any point inside the box, as shown in Figure 24.9. Hence

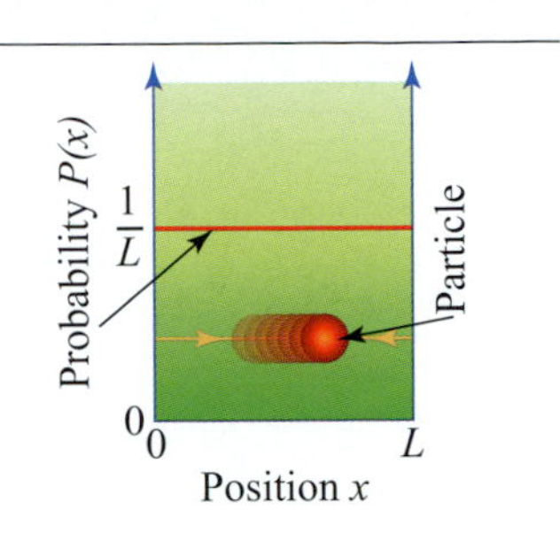

Fig. 24.9: Classical particle: moving.

$$P_C(x) = \text{Probability that particle is observed at point x} = \frac{1}{L}.$$

The velocity of the particle is a fixed quantity in the classical statistical description and does not add any further information about the position of the particle.

Quantum description

A particle inside a potential well is described by the wave function $\psi(t,x)$. The particle's position is indeterminate and we only have probabilistic and statistical information about its position. Consider the case when the particle in the box has a definite energy E; then, from Equation 24.10, one can write the wave function as follows:

$$\psi(t,x) = e^{-\frac{i}{\hbar}Et}\psi_E(x)$$

where, for the particle with momentum p, we have $E = p^2/(2m)$. The wave function $\psi_E(x)$ must be 0 at the boundaries, namely $\psi_E(0) = 0 = \psi_E(L)$ and yields, up to a normalization, that $\psi_E(x) = \sin(n\pi x/L)$. Hence

$$\psi_E(x) = \sqrt{\frac{2}{L}}\sin\left(\frac{px}{\hbar}\right)$$

$$\begin{aligned} P_{QM}(x) &= \text{Probability that particle is observed at point x} \\ &= |\psi_E(x)|^2 = \frac{2}{L}\sin^2\left(\frac{px}{\hbar}\right) \ ; \ x \text{ inside the box} \end{aligned} \quad (24.13)$$

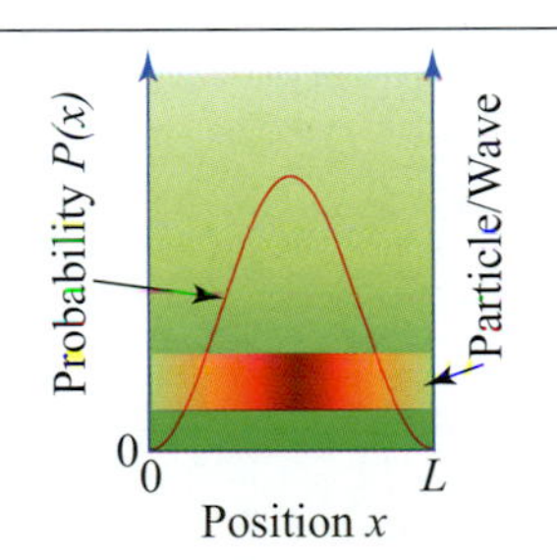

Fig. 24.10: Quantum particle: ground state.

where the momentum $p = p_n$ is discretized due to the boundary condition and yields

$$p_n = \frac{n\pi}{L}\hbar$$

$$E_n = \frac{p_n^2}{2m} = \frac{\hbar^2}{2m}\left(\frac{n\pi}{L}\right)^2 \propto \frac{1}{L^2}.$$

Figures 24.10 and 24.11 are plots of $P_{QM}(x)$, as a function of the particle's possible positions inside the box x for two different momenta. Note a number of remarkable features of the quantum description of the particle in a box.

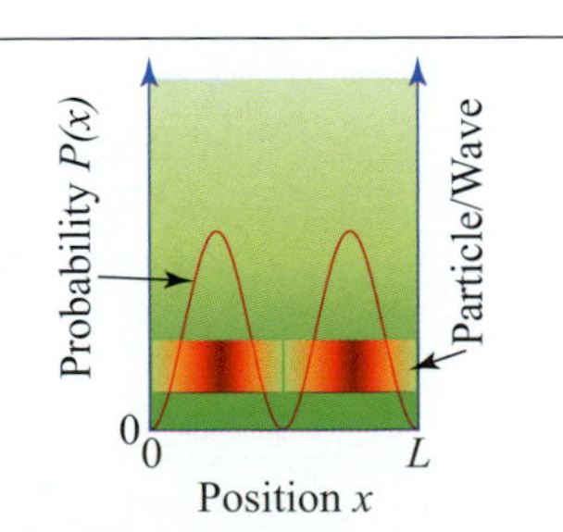

Fig. 24.11: Quantum particle: excited state.

- The position x inside the box at a given time t is *not* fixed; the particle is in a trans-empirical (virtual) state, being simultaneously everywhere inside the box.
- The likelihood of finding the particle at different points inside the box is not a constant like that predicted by the classical statistical description given by $P_C(x)$; instead, a well-defined and nontrivial prediction is made as to the likelihood of finding the particle at different points x.

- There is a null result contained in the quantum behavior of the particle. The expression for $P_{QM}(x)$ states that one will *never* find the particle at points where $P_{QM}(x_M) = 0$ — in other words at points where the argument of the sine function is a multiple of π. Hence the particle will never be found at points x_M given by

$$\sin\left(\frac{p_n x_M}{\hbar}\right) = 0 \ \Rightarrow \ x_M = \left(\frac{M}{n}\right)L \ ; \ M = 0, 1, ..., n.$$

- The simple example of a particle in a box has been experimentally tested, and all the predictions of quantum mechanics are born from experiment. In particular the concept of a nanoparticle, discussed in Section 19.12 and which plays an important role in nanotechnology, is described quite well by a quantum particle in a three-dimensional box.

We conclude that the particle inside a box has a definite momentum (and energy), but its position inside the box is indeterminate and is a *random variable*. When it is *not* observed, the quantum particle exists in a (random) trans-physical state or virtual state.

Quantum physics leads us to the conclusion that the position of a quantum particle is indeterminate and is a random variable. Although we do not know what the particle's precise position is, we *do* know what the probability or, equivalently, the likelihood is of the particle being at different possible positions within the interval L. The indeterminate trans-physical state of a particle is *fully* described by specifying the wave function of the particle.

What happens if we perform a measurement to actually "see" what is the position occupied by the particle? The measurement will find the electron to always be at some *definite* position; the act of measurement causes the electron to make a quantum transition from its trans-empirical state to an empirical state. If the measurement is repeated a great many times for the same system, that is, for a particle in a box always prepared to have momentum p, then the probability of finding the particle at position x will be given by $P_{QM}(x) = |\psi(x)|^2$.

In summary, the quantum particle has *two* forms of existence: a trans-empirical state, denoted by $\psi(t, x)$, when it is not being observed, and an empirical state – given by $|\psi(t, x)|^2$ – which is observed when a measurement is performed on the particle. This is the duality that normally goes under the name of particle-wave duality.

24.5 Measurement and Heisenberg's uncertainty principle

As can be inferred from our discussion on the wave function, measurement plays a central role in connecting the empirical to the trans-empirical state of a quantum system. The process of measurement in quantum mechanics is fundamentally different from classical measurements, and we illustrate this difference by discussing the famous **Heisenberg uncertainty principle**.

Planck's quantum postulate was made in 1900. It took over 25 years and the insight of German physicist Werner Heisenberg to understand that the Planck postulate completely overturned the classical theory of measurement — which is implicitly assumed in Newtonian mechanics. The inherent limits to the precision of measurement that follow from Planck's quantum postulate were encoded, in 1926, by Heisenberg's principle of uncertainty.

Heisenberg's uncertainty principle follows from the quantum view that nature is inherently indeterminate. The principle can also be seen to follow from the fact that all measurements must follow the quantum postulate. We give a heuristic derivation of uncertainty principle from the perspective of a particle's motion. It should, however, be kept in mind that Heisenberg's uncertainty principle is a universal principle that is valid for all circumstances.

Consider the motion of a classical particle, say, a piece of stone that one has thrown. At each instant, Newton's law determines the position and velocity of the particle. Knowing the position and velocity of the particle at each instant, one can construct the unique path (trajectory) that the particle follows. Suppose for simplicity that the particle is pointlike and we try to actually *measure* its position and velocity at each instant. We fundamentally need to make very precise measurements of the position of the particle at each instant, from which we can construct the position of the particle at each instant and hence its velocity as well. Hence, we focus on how to make a precise measurement of the position of the particle.

One can shine light on the particle and locate its position by observing the light that is reflected by the particle. Suppose one wants to know the position of the particle to a very high degree of precision, say, Δx; then, since light of a given wavelength λ can only resolve distances greater than roughly $\lambda/2$, we will have to shine light on the particle with wavelength given by $\lambda \simeq \Delta x$. To determine the particle's position more and more precisely, we will need to shine light with smaller and smaller wavelength. And this is where we run into quantized energies: The minimum amount of light that we can shine on the particle has to have **at least** one quantum, which has energy hc/λ and momentum h/λ; as we make λ smaller and smaller, the energy and momentum of the single quantum become larger and larger; we will discuss the consequence of this quantum conundrum.

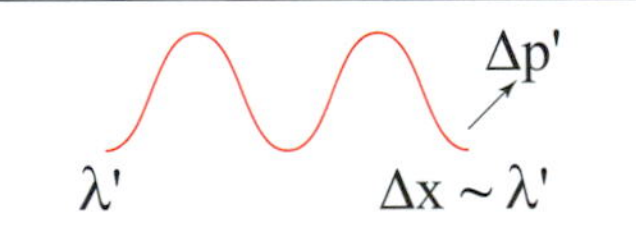

Fig. 24.12: Shining light with *large* wavelength.

From our discussion above, to make a very precise measurement of the position of the particle, we are forced to use light having a very small wavelength λ; by the quantum postulate, the quantum of light will have to impart a high amount of momentum and energy to the measured particle, causing a large change in the particle's velocity and hence kinetic energy. Thus, the velocity of the particle changes in an uncontrollable and irreversible manner, and we end up with a final velocity of the particle that is different from the value it had before we made the measurement.

What is the uncertainty in the velocity that results from the fact that the minimum quantity of light is one quanta? For the case shown in Figures 24.12 and 24.13, the precision of the particle's position is approximately equal to the wavelength λ of light that we are shining on it. Hence we have

$$\Delta x \sim \lambda. \quad (24.14)$$

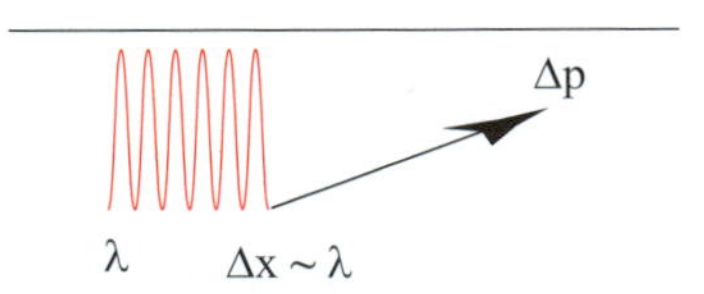

Fig. 24.13: Shining light with *small* wavelength.

In the process of measurement, we have to scatter off the particle at least one photon with wavelength λ. The photon, from Equation 23.8, is carrying momentum given by $p = h/\lambda$, which it will impart to the particle; hence, the change in the momentum of the particle due to the photon bouncing off it is given by

$$\Delta p = \frac{h}{\lambda}. \tag{24.15}$$

From Equations 24.14 and 24.15, we obtain

$$\Delta x \Delta p \sim h.$$

In other words, the accuracy of the measurement that we make of the particle's position and momentum is intrinsically limited; the *least* disturbance to the particle that we can cause by a measurement is to probe the particle with *one photon*, and it leads to the uncertainty derived above. A more precise statement is the following:

$$\Delta x \Delta p \geq \frac{\hbar}{2} \quad : \textbf{Heisenberg's Uncertainty Principle}. \tag{24.16}$$

Note that the Heisenberg uncertainty principle states that if one knows the position of a particle exactly, that is, $\Delta x = 0$, then $\Delta p = \infty$ meaning that we have no information regarding the momentum of the particle and vice versa. This is in sharp contrast to the classical case, where both the position and momentum of the classical particle are exactly known at every instant.

One might object that even in a classical measurement a certain amount of energy has to be imparted to the particle being observed. However, in principle, in classical physics the energy imparted in a process of classical measurement can be made *arbitrarily* precise; in contrast, in quantum physics, to carry out the measurement with a precision of, say, Δx will necessarily involve at least *one photon*, having a *minimum* momentum of h/λ and *minimum* energy hc/λ. In other words, in quantum mechanics, the momentum and energy required to make a measurement of a given precision *cannot* be made vanishingly small. This, in essence, is the dividing line between classical and quantum measurement theory: namely, the nonvanishing value of Planck's constant.

To recapitulate, we started by trying to precisely measure the position of the particle with no desire to disturb its velocity. But we discovered that, due to the quantum principle, the more precisely we measured the position of the particle the more we uncontrollably disturbed the velocity of the particle. Hence we ended up with a precise measurement of the position of the particle and, due to this very measurement, we lost information on the precise value of the particle's velocity.

Note that one is not assuming that the particle had an intrinsic position and momentum p and x before the measurement was made and that we could not measure these precisely because the probe, namely the photon, obeyed the quantum principle. Rather, the purpose of this discussion is to emphasize that the quantum system is inherently indeterminate and possible outcomes of any measurement will always reflect this inherent indeterminacy.

What happens when we *do not* make any measurement, for example, to determine the position or the momentum of the particle? Does it have a definite position or a definite momentum? The answer is no; the particle is in a trans-empirical probabilistic state in which both its position and momentum are indeterminate, having a likelihood of having a whole range of values, as has been discussed in Section 24.3.

The trans-empirical state of a quantum particle implies the *counterintuitive* result that the outcome of an observation depends on what we decide to measure! For example, if we decide to make a very precise measurement of the position of the particle we will end up with a large uncertainty in its momentum, whereas if we decide to make a very precise measurement of the momentum of that same particle we will end up with a large uncertainty in its position!

24.6 Wave function collapse: the quantum Zeno effect

One of Zeno's paradoxes is discussed in Section 2.6 and is resolved by using concepts from calculus. Zeno has another paradox that states that motion is impossible by discussing the case of an arrow in flight. He reasons that at a given instant of time, for the arrow to be moving it must either move to where it is, or it must move to where it is not. It cannot move to where it is not, because this is a single instant, and it cannot move to where it is because it is already there. Hence, the arrow cannot move and in general there is no motion possible in nature.

Consider an atom that has only two possible states, called the two-state system, with ground state ψ_0 and excited state ψ_1, as shown in Figure 24.14a.

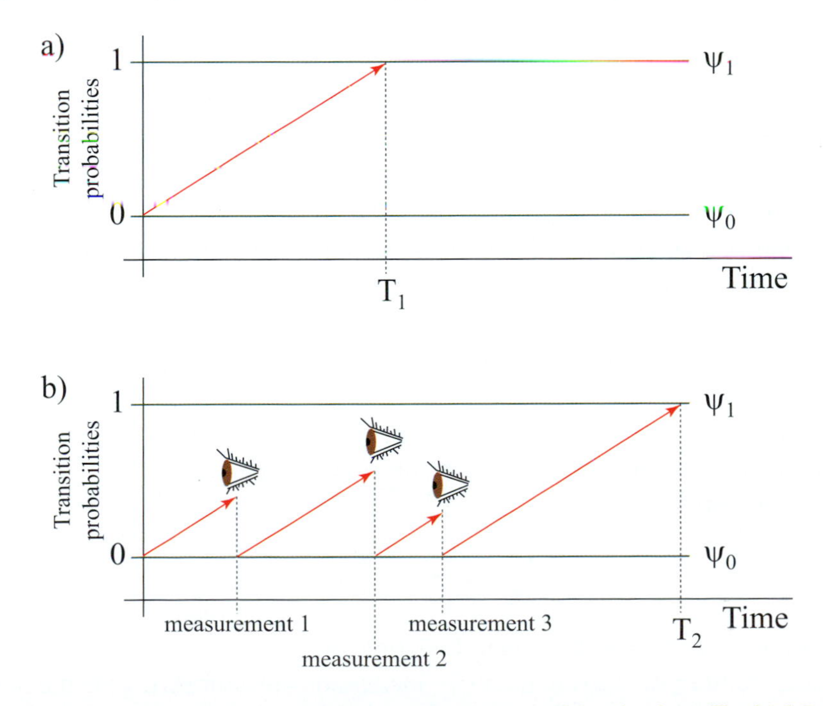

Figure 24.14: a) The atom is certain to make the transition in time T_1. b) Measurements of the state of a system reset its state and hence delay its transition to time T_2.

Let the system be in the lowest energy (ground) state, that is

$$\psi = \psi_0.$$

If one shines electromagnetic radiation on the atom within a given time interval and with the "right" frequency, then it can be shown from Schrödinger's equation that, in time T_1, the atom is *certain* to make a transition from the ground state to the excited state. Define P to be the probability for a transition from

$$\psi_0 \rightarrow \psi_1.$$

Hence, if *one* measurement is made after time T_1, the probability that the atom has made the transition is certain. In other words

$$P = 1 \text{ for one measurement after time } T_1. \tag{24.17}$$

The act of measurement collapses a quantum system to one of its possible states. What will happen if we carry out a measurement of the atom's state at a time $t < T_1$? The atom is most likely still in state ψ_0. The measurement has two possible outcomes: that state that the atom is in, namely, ψ_0 may collapse ψ_1 thus hastening the transition or else ψ_0 may be collapse to ψ_0. But note collapsing to the ground state ψ_0 means that the atom's state, at time t, has been reset to the ground state.

Hence, after the measurement is performed, the time the atom will take to make a transition to the excited state ψ_1 is now $t + T_1$. In other words, it will take a longer time to make the transition ψ_g to ψ_1. Now suppose we make many measurements of the atom's state at time intervals much shorter than T_1; most of the time, we will find the atom in its ground state and hence each act of measurement will reset the state of the atom to be its ground state, as shown in Figure 24.14b, thus lengthening the time of the transition to $T_2 >> T_1$.

Suppose measurements of the atom's state are made at regular time intervals given by $\Delta t = T_1/n$; after n measurements, the time that has elapsed is $n\Delta t = T_1$. It can be shown that the probability of the atom making a transition from the ground state ψ_0 to the excited state ψ_1 is now given by

$$P = \frac{1}{2}\Big[1 - \cos^n\Big(\frac{\pi}{n}\Big)\Big] \text{ for } n \text{ measurements in time interval } T_1. \tag{24.18}$$

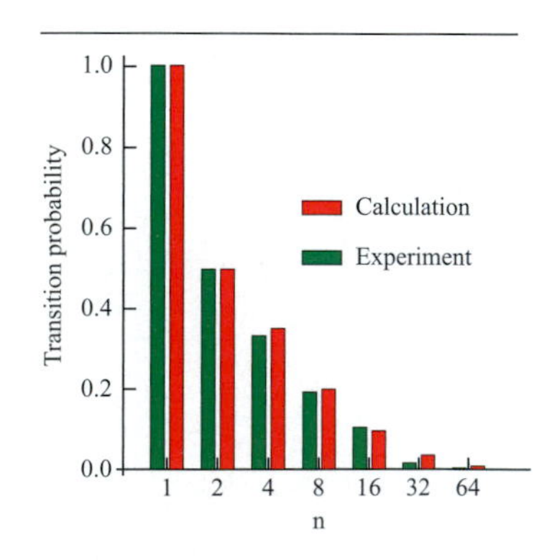

Fig. 24.15: Quantum Zeno effect. Theory matches experiment very well.

Recall if one measurement is performed after time T_1, the system is certain to have made a transition from ψ_0 to ψ_1; this result is contained in Equation 24.18 since, for $n = 1$, we have

$$P = \frac{1}{2}\Big[1 - \cos\big(\pi\big)\Big] = 1.$$

Hence, we recover the result given in Equation 24.17.

As the number of measurements is increased, the probability of the system making the transition in time T_1 becomes smaller and smaller, as shown in Figure 24.15 (see also Figure 24.16). We also have the remarkable result that continuously performing measurements on the state of the atom entails taking the limit of

$n \to \infty$ and yields the following:

$$P \to \frac{1}{2}\Big[1 - 1\Big] = 0.$$

Namely, the atom will *never* make the transition from the ground to the excited state if the system is continuously observed. Hence, Zeno's paradox of there being no motion is realized in quantum mechanics by acts of repeated measurements that reset the state of the system to its initial state!

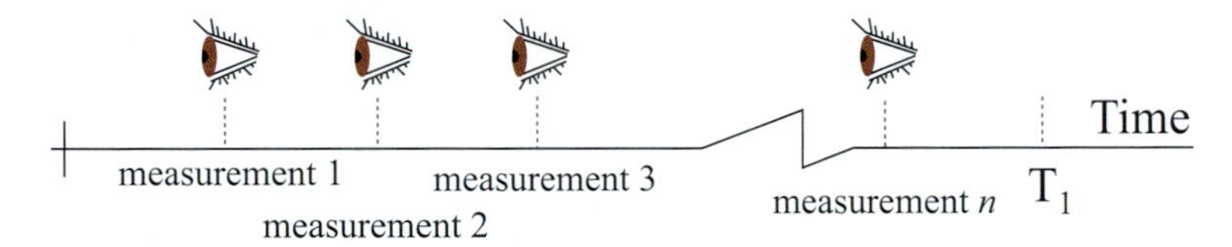

Figure 24.16: Making n measurements, separated by $\Delta t = T_1/n$, in the time interval T_1.

The theoretical result given in Equation 24.18 is compared with experiment and is validated to a high degree of accuracy.

24.7 Trans-empirical state: tunneling

Quantum mechanics predicts that it is possible for particles to move into regions of space which are classically forbidden. We consider the example of a free quantum particle crossing a region of space that requires a potential energy U_0 that is higher than its kinetic energy T. The reason that the particle cannot be in a region of space where $T - U_0 < 0$ is because the classical particle would have negative kinetic energy that requires imaginary velocity, something that is not allowed by classical physics.

The quantum particle can be in a classically forbidden region by being in a trans-empirical state; the quantum particle cannot be directly observed in the trans-empirical state and hence can never be observed to be violating energy conservation.

Consider the following experiment: There is potential barrier of height U_0 with width W; this could be a wall. A particle of mass m is incident on this barrier with velocity v and kinetic energy $T = mv^2/2$. As stated, suppose that the kinetic energy of the particle is lower than the energy of the barrier, that is $T < U_0$.

In classical physics, if the energy of the particle is lower than the barrier, that is $T < U_0$, the particle cannot break through the wall, and hence it will bounce back.

What happens in quantum physics? A particle incident on the barrier is represented by a propagating wave function; this trans-empirical wave function encodes the *quantum likelihood* that, on hitting the potential barrier, the particle goes through or bounces back. What this means in practice is the following: If N identical particles — all with the same energy T — are incident, one by one, on the barrier, a certain fraction, say, N_R, will be reflected, and the remaining $N - N_R$ will pass through the barrier. This passing through the barrier is called **tunneling**.

Hence, in *quantum mechanics energy is propagated in a probabilistic manner*, sometimes being reflected and sometimes being transmitted across the barrier.

The probability wave is indicated as a wavy incoming line in Figure 24.17.

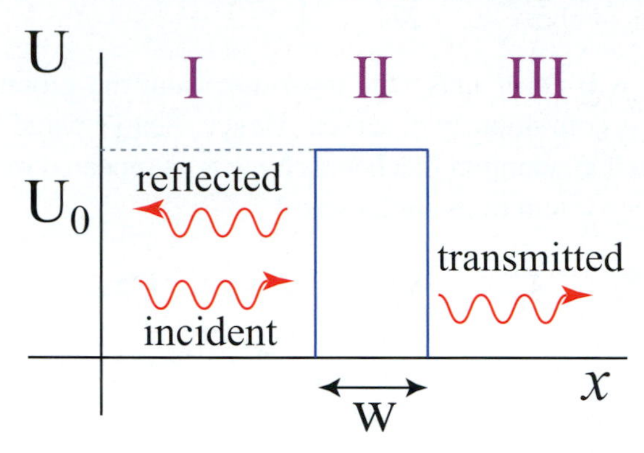

Figure 24.17: Quantum tunneling through an energy barrier.

On hitting the barrier, one fraction of the wave is reflected and the other is transmitted across the barrier. Note the reflection and transmission is a trans-physical process that can never be directly observed — the particle never splits into two — and the wave function encodes the *probability amplitude* that the particle undergoes either reflection or transmission. This process is impossible in classical physics where the particle always occupies one position at one instant.

Energy is exactly conserved in quantum mechanics. Every time a particle's energy is measured, it is *always* exactly equal to its initial energy T. In region I and III the energy of the wave is consistent with a classical particle traveling with some velocity. But what happens to the particle in region II — inside the barrier? Since being inside the barrier requires the particle to have energy of at least $E > T$, this is a region of space forbidden for the classical particle. Inside the barrier, the quantum particle is in a nonclassical trans-empirical state and *cannot* be directly observed. One may, of course, insist on observing the particle while it is in region II. To do so would entail using a probe to detect the particle transiting region II by, say, shining light on it. And this is where we come up against the quantum conundrum discussed in Section 24.5.

To detect the particle in region II, we will need to impart to the particle at least one photon worth of energy, denoted by E_{photon}. The energy of the particle is then $T + E_{\text{photon}}$ and it can be shown that to detect the particle in region II the photon must have an energy such that we have, as a result of the experiment, $T + E_{\text{photon}} > U_0$! In other words, if we do indeed manage to "see" the particle in the forbidden region II, then the very act of measurement imparts just enough energy to the particle so that it has energy greater than the barrier. The empirically observed state of the quantum particle has the requisite energy needed to be in region II.

The phenomenon of tunneling is not just a curiosity but is actually required to explain physical processes such as alpha decay and many other processes. Tunneling is also used in the design of such modern devices as the tunneling and scanning electron microscopes discussed in Section 19.3.

24.8 Quantum superposition

We have encountered the superposition of classical waves in Chapter 11. We now study the nonclassical operation of the superposition principle in quantum mechanics.

In quantum theory, a particle is described by specifying all the possible states it can have and the probability amplitudes for these states given by the wave function ψ. To simplify our discussion, consider a particle that can have only *two* possible states, which is the simplest quantum system possible and is also called, for obvious reasons, a two-state system. An example of a two-state system is the spin of an electron. In addition to moving around in space, following O. Stern and W. Gerlach's experimental discovery in 1922 that the state of the electron is quantized, G.E. Uhlenbeck and S. Goudsmit proposed in 1925 that the electron has an *intrinsic* angular momentum called *spin*. The spin of the electron can point either up or down, and hence forms a two-state system. A quantum two-state system is described by determining its probability amplitude to be in one of the two states.

The history of the discovery of the electron spin is quite interesting and illustrates how anybody, even the greatest minds, can be completely wrong. Figure 24.18 nicely illustrates Pauli's opposition to the idea. Wolfgang Pauli was a Nobel laureate and one of the founding fathers of quantum mechanics.

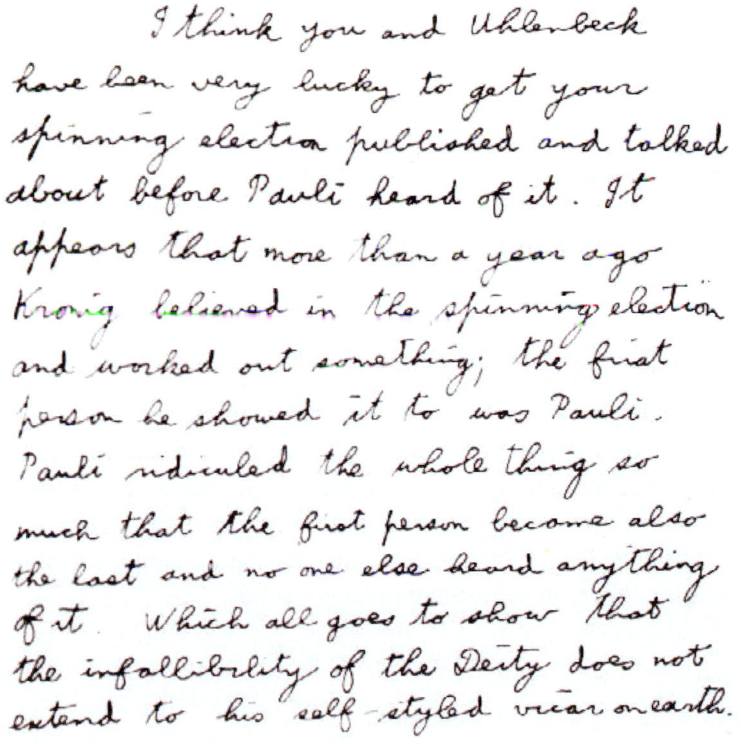

I think you and Uhlenbeck
have been very lucky to get your
spinning electron published and talked
about before Pauli heard of it. It
appears that more than a year ago
Kronig believed in the spinning electron
and worked out something; the first
person he showed it to was Pauli.
Pauli ridiculed the whole thing so
much that the first person became also
the last and no one else heard anything
of it. Which all goes to show that
the infallibility of the Deity does not
extend to his self-styled vicar on earth.

Figure 24.18: Part of a letter by L.H. Thomas to Goudsmit (March 25, 1926).

How should we mathematically describe our two-state system? The two different states should be "orthogonal" to each other, in the sense that being in one state is completely different from being in the other state; in basic geometry, the two vectors are at right angles. The simplest way to realize this expected orthogonality of the two states is to represent them by two-dimensional *vectors*, and the idea of orthogonality translates exactly into the concept of vectors being perpendicular. Hence, we can represent the wave function of a two-state system by *two-dimensional vectors*. One should note that the two-dimensional vector space has got nothing to do with a physical two-dimensional space, but rather should be viewed as a mathematical construction for describing the spin of an electron.

In some textbooks, a vector is represented by boldface, and in others an arrow is placed on the letter denoting a vector. At the expense of introducing new notation, we follow Dirac's notation that is common in quantum physics and denote a vector by the symbol $|\psi\rangle$. Consider a two-dimensional space with unit-vectors pointing along the x-axis or y-axis. The wave functions are denoted, respectively, by

$$|u\rangle \equiv \text{unit-vector pointing along x-axis} \tag{24.19}$$

$$|d\rangle \equiv \text{unit-vector pointing along y-axis.} \tag{24.20}$$

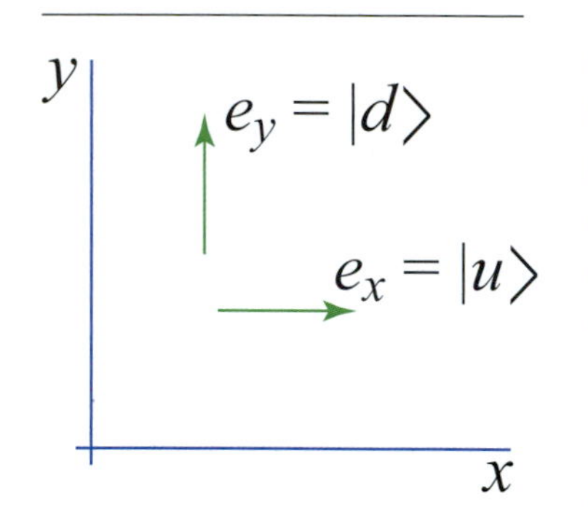

Fig. 24.19: The basis is expressed in terms of generic vectors.

The reason that the symbols u (for up) and d (for down) are used is to free one from thinking of these vectors as referring to physical space, which are usually labeled by x- and y-axes. The space to which the wave function belongs is referred to as state space, and there are quantum systems for which u and d do not refer to space. In the case of spin, the the basis states reduce to the basis of the usual two-dimensional vector space, see Figure 24.19. An arbitrary vector in two-dimensional state space, denoted by $|\psi\rangle$, is given by

$$|\psi\rangle = \alpha|u\rangle + \beta|d\rangle \tag{24.21}$$

where, in general, α and β are complex numbers.

The spin can point either "up" or "down". The wave functions for these two special cases are the following:

$$|\psi_{\text{up}}\rangle = |u\rangle \tag{24.22}$$

$$|\psi_{\text{down}}\rangle = |d\rangle \tag{24.23}$$

So far we could have been discussing classical physics, since a spin pointing up or down with 100% certainty is a classical concept. A quantum mechanical spin is more subtle, since we can *superpose* two states and obtain a trans-empirical state that simulataneously points up or down with only a certain *likelihood*. For such a quantum mechanical state, obtained by superposing a quantum spin pointing up with one pointing down, its wavefunction is given by

$$|\psi\rangle = \alpha|u\rangle + \beta|d\rangle, \tag{24.24}$$

with the following physical interpretation:

$$|\alpha|^2 = \text{Probability that the spin is pointing up} \tag{24.25}$$

$$|\beta|^2 = \text{Probability that the spin is pointing down.} \tag{24.26}$$

The state $|\psi\rangle$ in Equation 24.24 is obtained by superposing two different possible states (outcomes) of the quantum system. Note this is radically different from superposition for classical waves, where the amplitudes of the oscillation's two physical waves are added to yield the amplitude of oscillation for the resultant wave.

Thus we see that in quantum mechanics the particle is described by the sum of possible states (that is, by superposing them), in contrast to classical physics where a particle is always in one or the other state. The superposition principle has rather interesting consequences, one of them being colorfully illustrated by the story of Schrödinger's cat.

Quantum superposition results in a trans-empirical that is made by superposing other possible states of the quantum state. Quantum superposition can result only in trans-empirical states described by the wave function ψ; the empirical (observed) states of the quantum system, described by $|\psi|^2$, do not obey the quantum superposition principle. The mathematical basis for the quantum superposition is that the Schrödinger equation is a *linear equation* for the wave function ψ; if ψ_1 and ψ_2 are solutions of the Schrödinger equation, then so is the linear combination $\psi = \alpha\psi_1 + \beta\psi_2$, where α, β are complex numbers.

The paradox implicit in quantum superposition is exemplified by Schrödinger's cat.

Schrödinger's cat

The fact that a quantum particle can *simultaneously* be in two states that are *mutually exclusive* is highly counterintuitive and paradoxical. Since there is nothing special about spin, one can replace spin-up and spin-down with any two independent states. To illustrate the paradox, Schrödinger proposed the following experiment: Suppose a cat is put inside a sealed and opaque box, with a radioactive substance inside the box as well. The radioactive material randomly emits alpha particles, and if it emits a strong burst of alpha particles it will trigger the release of a poisonous gas, causing the cat to die.

The problem Schrödinger posed is the following: As long as we do not open the box (technically speaking: perform a measurement), we do not know what has transpired, and there is some likelihood that the cat is either dead or alive. Hence the cat's wave function will be

$$|\psi(\text{cat})\rangle = \alpha|\text{cat dead}\rangle + \beta|\text{cat alive}\rangle. \tag{24.27}$$

This famous cat, called Schrödinger's cat, illustrates the counterintuitive and bizarre world of quantum mechanics, namely that the cat can be alive and dead at the same time! Schrödinger felt this was an absurd situation, since — regardless of whether a measurement is performed — the cat should either be dead or alive; after all, how can the cat be dead and alive at the same time? The paradox that Schrödinger's cat brings out is the need to understand what is the *physical* meaning of the trans-empirical state that we obtain by superposing two or, for that matter, many states.

We offer no solution to the paradox of Schrödinger's cat since it involves a macroscopic object. Instead, we explore the consequences of the trans-empirical state obtained by superposing two distinct quantum states.

24.9 Empirical and trans-empirical: the double-slit experiment

We study the paradoxes of quantum mechanics that result from the duality of the empirical and trans-empirical states of the quantum system. The central experimental setup to explore this duality is the double-slit experiment, where an object impinges on a barrier (metal plate) with two small slits in it. After passing through the slits, the object then hits a wall, where its position is measured by a detector. The result of repeating the experiment a great many times will show certain characteristic patterns that will reveal the empirical or trans-empirical properties of the object.

The double-slit experiment is one of the deepest and most important experiments in quantum mechanics as well as for the rest of physics. As explained by Richard Feynman, the double-slit experiment "*has been designed to contain all of the mystery of quantum mechanics, to put you up against the paradoxes and mysteries and peculiarities of nature one hundred percent*". Feynman further explains other paradoxical situations in quantum mechanics can always be explained by this experiment, which reveals "*nature in her most elegant and difficult form*".

Interference and traveling along well-defined trajectories are the key properties that differentiate a wave from a particle. We investigate what the notions of waves and particles mean in quantum mechanics by considering several scenarios involving classical and quantum particles and waves. The quantum superposition principle, which leads to Schrödinger's cat paradox, can be understood in light of this experiment.

We analyze the double-slit experiment first for a classical particle and a classical wave and then for the case of electrons.

Classical particles

In the macroscopic world, that is, the world that we can directly perceive with our senses, particles can be taken to be bullets or billiard balls. In the double-slit experiment illustrated in Figure 24.20, a bullet from the firing gun can go through either slit 1 or slit 2 and it is detected by the detector at the screen. The experiment is first done by covering slit 2 so that the bullet can go through only slit 1. After firing a suitable number of bullets, the distribution of the bullets detected is plotted along the x-direction and the distribution curve $P_1(x)$ is obtained. The experiment is repeated, but with slit 1 being covered instead of slit 2. The result is the distribution curve $P_2(x)$.

When *both* slits 1 and 2 are open, we find that the combined distribution curve for particles arriving at the screen is in fact the *sum* of $P_1(x)$ and $P_2(x)$, namely $P_1(x) + P_2(x)$. The joint probability $P_{12}(x)$ of a bullet arriving at the screen through either slit 1 or slit 2 is given by

$$P_{12} = P_1 + P_2. \tag{24.28}$$

In this way we identify particle-like behavior for a physical system in the microworld — by the distribution curves $P_1(x)$, $P_2(x)$ and $P_{12}(x)$. There is no

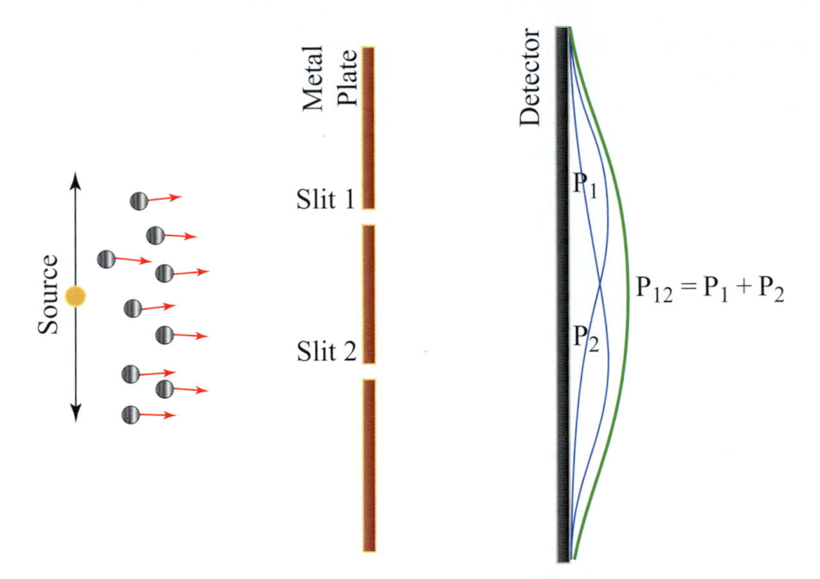

Figure 24.20: Double-slit experiment with bullets.

interference in the sense that the probability that a bullet will reach a point x is by taking a path *either* through slit 1 *or* through slit 2, and this explains the final result $P_{12}(x)$ is the sum of $P_1(x)$ and $P_2(x)$.

Classical waves

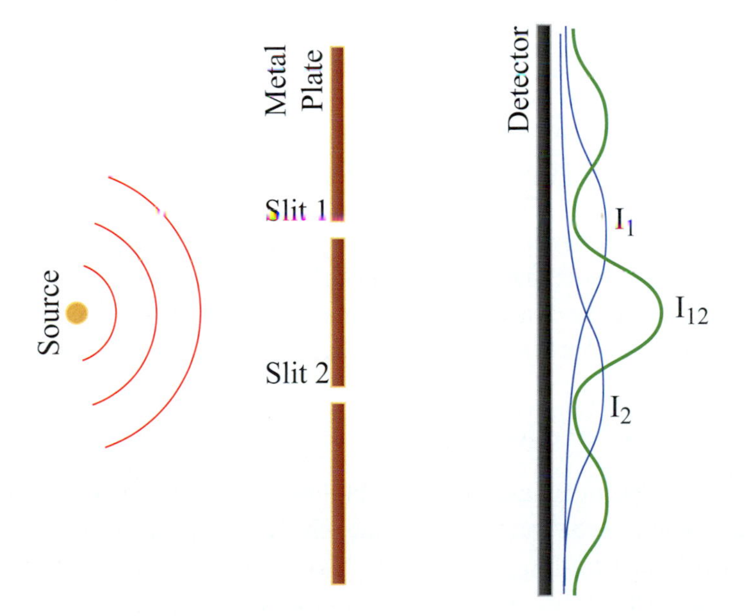

Figure 24.21: Double-slit experiment with waves.

A similar experiment is performed with water waves, as shown in Figure 24.21. The detector can only measure the intensity I of the wave, which is proportional to the square of the height h of the wave. Let h_1 and h_2, respectively, be the amplitudes (heights) of the waves arriving at the detector when slit 2 (in the case of h_1) or slit 1 (in the case of h_2) is closed; we then have the intensity distributions $I_1 = h_1^2(x)$ and $I_2 = h_2^2(x)$ of the water wave when either one of the two slits is closed. Now, when both slits 1 *and* slit 2 are open, the resultant intensity is given by the superposition of the individual waves:

$$I_{12} = (h_1 + h_2)^2 = h_1^2 + h_2^2 + 2h_1h_2 \quad (24.29)$$

$$\neq I_1 + I_2. \quad (24.30)$$

The reason that the intensity of the interference pattern, namely $(h_1 + h_2)^2$, is *not* the sum of the individual amplitude is due to (constructive and destructive) interference, and it is this that gives the characteristic minima and maxima of interference. Interference, as originally used by Young for light, is the best indication for a phenomenon being wavelike.

Interference is the best indication for a phenomenon being wavelike.

24.10 Quantum double-slit experiment

Now that we have seen how particles and waves behave in a classical double-slit experiment, let us see what happens when we have a quantum object. We investigate the empirical and trans-empirical behaviors of a typical "quantum particle", namely, the electron.

The experimental arrangement consists of an electron gun that sends identical electrons through a barrier with two slits to a screen where an apparatus keeps track of the point at which the electrons stop. The electron gun produces the electrons *one by one*, so that at any given time there is only *one* electron traveling from the electron gun to the wall.

First, we consider the case where one of the two slits is closed. Not surprisingly we obtain exactly the same result as in Figure 24.20, with the distribution curves $P_1(x)$ and $P_2(x)$, respectively. The next step is to find out the distribution curve when *both* slits are open. We have two cases for both slits open, namely a) *with* detectors to determine which slit the electron passed through and b) with *no* detectors next to the slits. Note that cases a) and b) have no influence on the outcome in the classical setup above, since it does not matter whether we know or don't know which slit the bullet went through; the classical results remain the same.

Experiment with detection

We perform the experiment as depicted in Figure 24.22 with *both* slits 1 and 2 open and with the *requirement* that we determine which slit the electron actually passes through. This can be arranged by fixing two detectors at the back of the slits, as shown in Figure 24.22. Since we know which slit the electron goes through we can plot three distribution curves. P_1 and P_2 are the distribution curves for electrons going through slit 1 and slit 2, respectively. P_{12} is the distribution curve for electrons that pass through either slit 1 or 2. Similar to the result obtained for bullets, the probability of the electron arriving at a point on the wall when both slits are open, denoted by P_{12}, is given by an expression similar to the case of classical particles given in Equation 24.28.

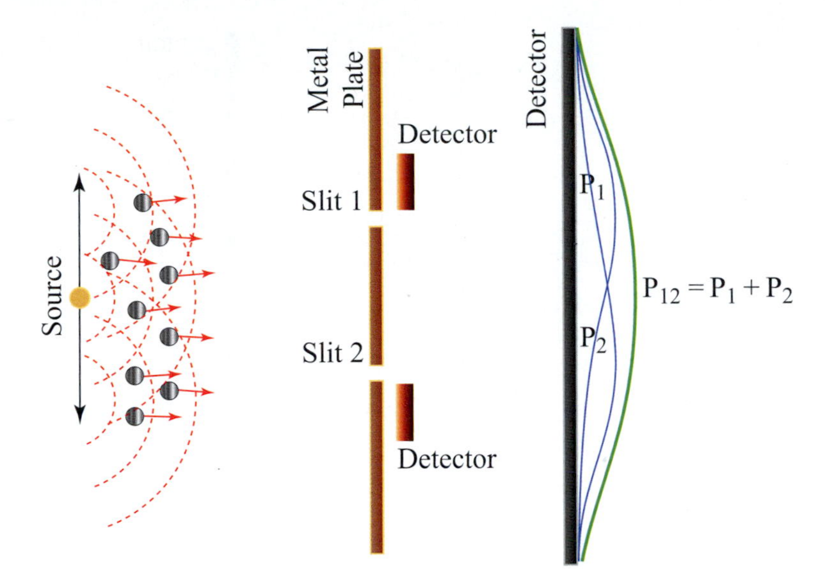

Figure 24.22: Double-slit experiment with detectors for electrons.

Introduce a wave amplitude Ψ_1 for electrons when slit 2 is closed and an amplitude Ψ_2 for electrons when slit 1 is closed. When the detectors are switched on, we know which slit the electron went through. The joint probability of finding an electron at the detector, as given in Equation 24.31, is

$$P_{12} = P_1 + P_2 \tag{24.31}$$

$$P_1 = |\Psi_1|^2 \quad ; \quad P_2 = |\Psi_2|^2 . \tag{24.32}$$

Thus we see that the behavior of the electron in this scenario is exactly the same as that for bullets discussed above and we consequently can state with some confidence that **an electron displays particle-like behavior**.

Experiment *without* detection

Consider now exactly the same experiment, but with the detectors removed. In other words, everything else remaining exactly the same, we send electrons one by one and *do not* make any measurement to determine which slit the electron goes through. The result of this experiment is illustrated in Figure 24.23 and shows that a *single* electron gives rise to interference. The interference pattern P_{12} is exactly like I_{12} as obtained for water waves. Hence we can say with some confidence that **an electron displays wavelike behavior**.

A single electron can give rise to interference.

An actual interference experiment for atoms instead of electrons is given in Figure 24.24 and leads to the conclusion that the atom's probability distribution is wavelike since it is in a trans-empirical state since no measurement was made regarding which slit the atom passed through.

Quantum explanation

What is going on? How can an electron be a particle and also a wave? Could there be a mistake in the experiments? The answer to this question is a resounding no.

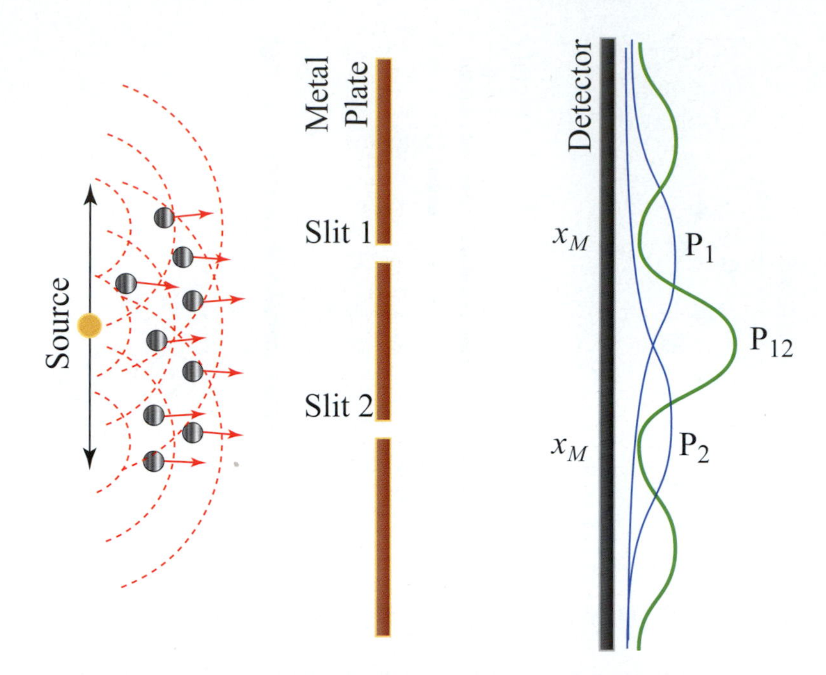

Figure 24.23: Double-slit experiment without detectors for electrons.

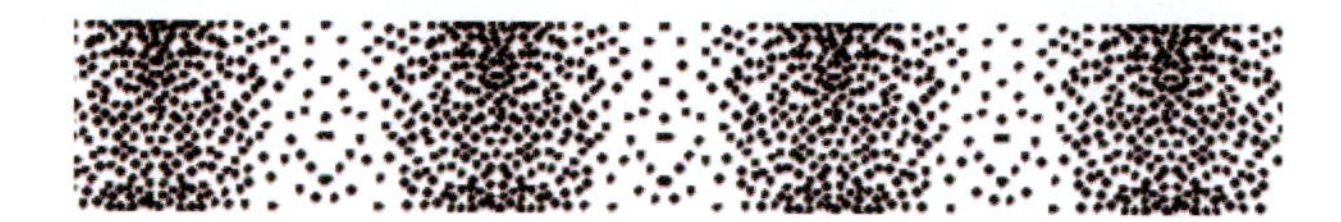

Figure 24.24: The double-slit experiment for atoms. The observed intensity of atoms received at the detector shows interference, with the heavy shades indicating the maxima in the intensity observed at the detectors and the lighter shades indicating the minima in the intensity.

The experiments have been conducted many times by many groups with many variations and with always the same results. The explanation lies in the Schrödinger wave function description of the electrons.

When the electrons are *not* observed, the probability amplitude obeys the *superposition principle*; the wave function for this trans-empirical state is represented by *superposing* (adding) the wave function for the two possibilities, namely the wave function is now given by $\Psi_1 + \Psi_2$. Note, as required by the quantum superposition principle, we are *adding* different *possible outcomes* and not physical waves as in the case of classical waves. Hence

$$\Psi_{12} = \Psi_1 + \Psi_2. \tag{24.33}$$

The superposition principle has a rather unique realization of quantum mechanics. Unlike classical waves, for which superposition entails adding the physical amplitude of oscillations, in quantum mechanics a trans-empirical state Ψ_{12} is constructed by superposing two possible states, namely Ψ_1 and Ψ_2 of the particle. Ψ_{12} graphically shows that, when the particle's state (path taken) is not detected, the particle behaves as a probability wave.

When no slit detection is performed, the probability that determines the outcome of detecting an electron P_{12} is the superposed amplitude Ψ_{12}. Hence, in analogy with classical waves, as given in Equation 24.29, the resultant distribution P_{12} is the square of modulus of the sum of $\Psi_1 + \Psi_2$, and we have

$$\begin{aligned} P_{12} &= |\Psi_{12}|^2 = |\Psi_1 + \Psi_2|^2 \\ \Rightarrow P_{12} &= |\Psi_1|^2 + |\Psi_2|^2 + \Psi_1^* \Psi_2 + \Psi_1 \Psi_2^* \\ &\neq |\Psi_1|^2 + |\Psi_2|^2 = P_1 + P_2 \end{aligned} \tag{24.34}$$

where P_1, P_2 are defined as in Equation 24.32.

Noteworthy 24.1: Trans-empirical state

P_{12} is the *probability* to find a particle at a certain point on the screen when a measurement is made. Consider the term $\Psi_1^* \Psi_2 + \Psi_1 \Psi_2^*$ in Equation 24.34 that contributes to P_{12}; how are we to interpret this term?

Note that $\Psi_1^* \Psi_2 + \Psi_1 \Psi_2^*$ contains *cross terms* of the electron going through slit 1, represented by Ψ_1 with the term Ψ_2 that represents the amplitude for going through slit 2. If one is to *interpret* this term, then we conclude that when the electron is *not* observed, it interferes with *itself*! One interpretation of this is to conclude that the electron is in a trans-empirical state in which a single electron *simultaneously* goes through *both slits*.

The trans-empirical state of the electron *in principle* can never be observed and is clearly counterintuitive since one can never directly observe a trans-empirical state using one's five senses. The existence of the trans-empirical state has to be inferred from the mathematical symbols of quantum mechanics.

Note from Figure 24.23 that the points of minima, say, x_M, of the interference pattern indicate that *no* electrons will *ever* be detected at those points. This is remarkable, since if, say, only one slit were open there is a finite likelihood of an electron arriving at x_M, but with both slits open, unlike the case for bullets, no electron can arrive there. This result is counterintuitive since one would expect, as in the case of bullets, that for both slits open the electron would have *two* ways of arriving at point x_M and hence be more likely to be found at x_M.

In short, when we *do not* observe which path the electron takes, it is in a trans-empirical state that it behaves like a probability wave. Hence, in the quantum context, fundamental entities like an electron have a dual nature. Depending on whether it is observed or not, the electron is either in an empirical particle-like state described by $|\psi|^2$, or in a trans-empirical wavelike state described by ψ.

24.11 The Mach-Zehnder interferometer

The results of the double-slit experiment are reproduced by the Mach-Zehnder interferometer using photons instead of electrons. We repeat the double-slit experiment using photons since we can then go on to examine more closely the role of measurement in the duality between the empirical and trans-empirical states of a quantum system.

In the Mach-Zehnder interferometer, as shown in Figure 24.25, a single photon is directed toward a half-silvered mirror, called a beam splitter, b_1, where the photon has an equal probability of passing through the mirror or being reflected downwards. The photon travels along a fiber optic cable and is reflected by mirrors. Mirror 1 reflects the photon along the x-axis, and after going through a phase shifter the photon ends up at detector d_1; similarly, the photon reflecting off mirror 2 ends up at detector d_2. The phase shifter is simply a means of slightly changing the path length so that the photon reflecting off mirror 1 travels a distance different from the photon reflecting off mirror 2. A photon traveling along the two paths has a slightly different relative phase when it arrives at the detectors. Let ϕ denote the relative phase of the photon arriving via the two different paths.

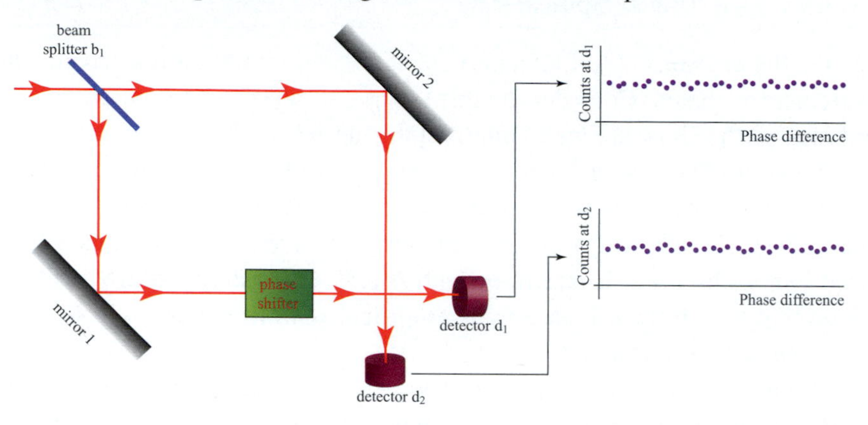

Figure 24.25: A Mach-Zehnder interferometer arranged to show no interference.

In this section, all the experimental arrangements that we will consider consist of only a single photon that is either in a (superposed) trans-empirical state or in a definite empirical state.

Empirical state: no interference

In the arrangement shown in Figure 24.25, which path the photon takes can be found by which detector clicks, since detector d_1 will only click if the photon was reflected off mirror 1 and similarly for detector 2. The setup in Figure 24.25 is identical to the case of an electron traveling through the two slits with detection, and the role of the phase shifter in changing the intensity is analogous to moving the detector up and down the screen as in Figure 24.22.

Now, as we vary the phase difference, we see that the state of the photons in the interferometer is an empirical state since the path taken by the detected photon is unique and known. Hence, there is *no interference* as the relative phase is varied since the information of which path was taken by the photon is known. The lack of interference is shown in the graphs showing the counts in detectors d_1 and d_2. Hence, if P_1, P_2 is the probability that the electron will be detected by detector d_1 and d_2, respectively, then we have for all phase differences ϕ

$$\begin{aligned} & P_1 = \frac{1}{2} \; ; \; P_2 = \frac{1}{2} \quad : \quad \text{No interference} \\ & P_1 + P_2 = 1 \end{aligned} \tag{24.35}$$

Trans-empirical state: interference

Suppose we put a second beam splitter b_2 just before the photon reaches the detectors, as shown in Figure 24.26. By putting in the second beam splitter b_2, the information about which path the photon took has been lost and the photon is now in a superposed trans-empirical state. The photon arriving at the detectors could have come from either path and hence what is received at the detector is a *superposed* state of the photon, with equal amplitude to have taken from either path. The detectors hence now show interference, with the probability for the photon — having relative phase difference of ϕ — being detected at detectors d_1 and d_2 given by

$$P_1 = \sin^2\left(\frac{\phi}{2}\right) \;;\; P_2 = \cos^2\left(\frac{\phi}{2}\right) \quad : \quad \text{Interference} \tag{24.36}$$
$$P_1 + P_2 = 1.$$

The graphs in Figure 24.26 show the probability of receiving a photon; since there is only a single photon in the system, only one of the detectors receives a photon.

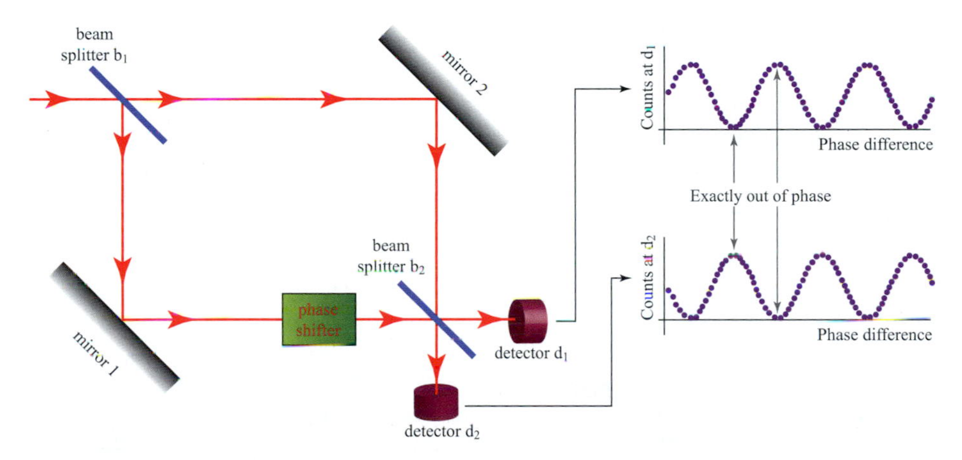

Figure 24.26: A Mach-Zehnder Interferometer arranged to show interference.

The difference between the classical and quantum states of the photon can now be seen clearly and is summarized in Figure 24.27. For the case of two beam splitters b_1 and b_2, the classical explanation is that the photon is in an empirical state taking *either* the path reflected off mirror 1 *or* the path that reflects off mirror 2; this gives the result of the resulting intensity having no interference, which is shown by experiments to be wrong. In contrast, the quantum mechanical explanation is that the single photon is in a trans-empirical state in which it *simultaneously* takes *both paths*, with the probability amplitude given in obvious notation by $\psi_{12} = \psi_1 + \psi_2$; since the two paths have a phase difference of ϕ, similar to the difference positions on the screen for the electrons, the photon from the two paths interferes either constructively or destructively, giving rise to the interference pattern.

Needless to say, one can never *actually* observe the photon taking both paths simultaneously but can only detect the results of the trans-empirical state by making a transition to an empirical state of the photon by actually detecting it.

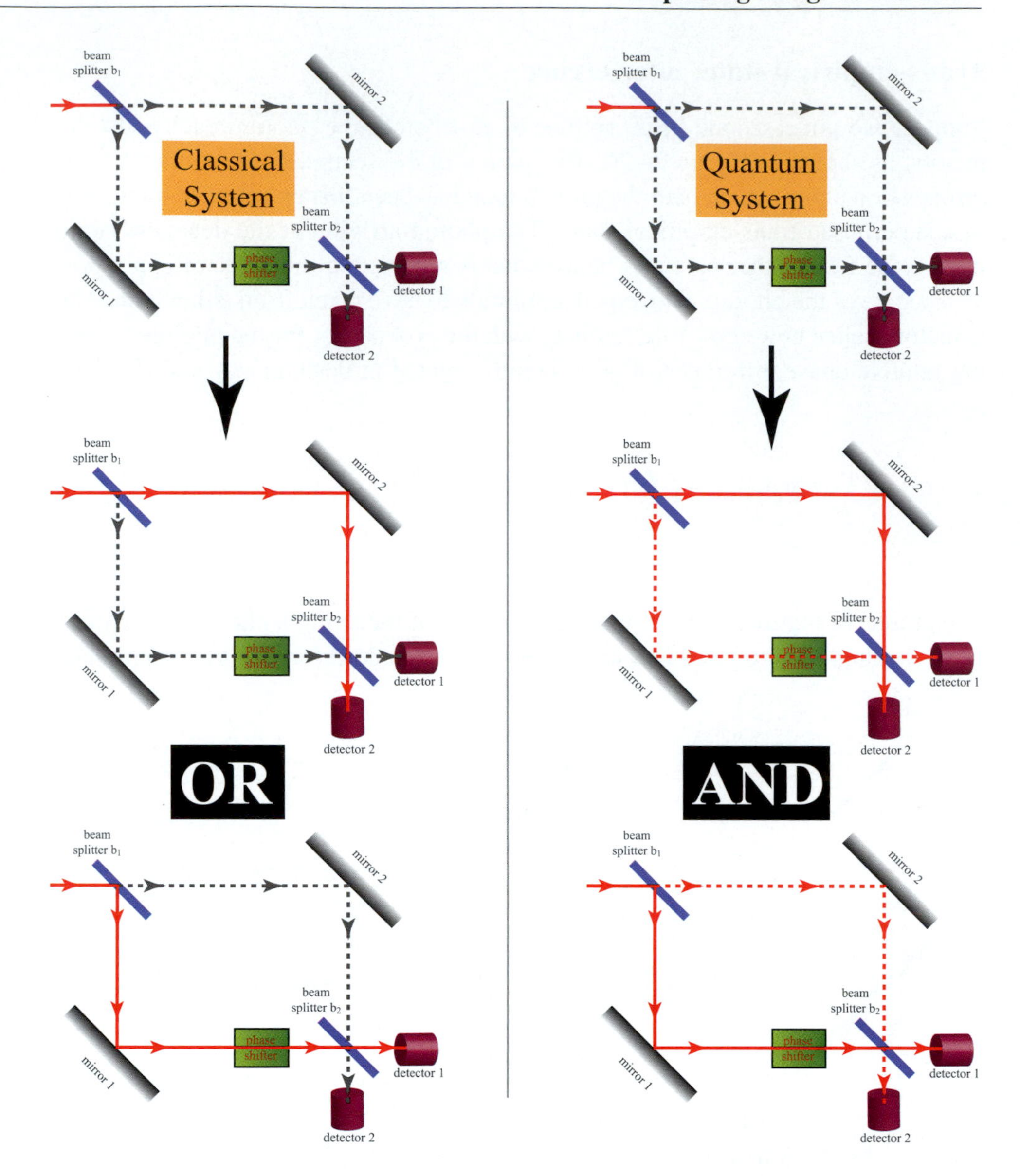

Figure 24.27: Possible paths in a classical and a quantum system. In the classical case, the photon will choose either one *or* the other but never both. Paradoxically, in the quantum case, the photon is in a trans-empirical state and takes *both* paths simultaneously.

The interference that results from the superposed trans-empirical state has been experimentally confirmed time and again for a great variety of cases.

Delayed choice experiment

There are further refinements that one can do for the double-slit experiment. In the Mach-Zehnder experiment shown in Figure 24.26 with both beam splitters in place, one may naïvely think that the photon has "decided" which path it will take once it has crossed beam splitter b_1. To see that this is not the case, one can insert beam splitter b_2 *after* the photon has crossed beam splitter b_1 but *before* the photon has reached the detectors. This is the *delayed choice experiment* and has become experimentally possible only recently. One should not be surprised to find that interference is restored the moment the second beam splitter is inserted, showing that the interference pattern does not depend on time but, rather, on whether one has information on which path the photon took. In other words, there is no *temporal* element in the interference or lack of it; the photon does not take any definite path *until* the detection is performed.

The following is an illustration of the apparent paradox that is implied by the delayed choice experiment: Consider a photon emitted by a distant galaxy billions of light-years away. The momentum of a photon emitted by an atom of one of the galaxy's stars can point in any direction and hence the atom emits a photon in a superposed state; experiments in astrophysics have confirmed that photons emitted by stars are correlated as a consequence of being emitted in a superposed state.

Suppose the photon passes a strong gravitational field of another galaxy that is between the emitting galaxy and earth. Due to bending of light caused by gravitational attraction, the photon has two possible paths in arriving at earth, as shown in Figure 24.28a. If we detect the photon with a detector on earth, we can determine which path it has taken by the angle it makes at the detector. As shown in Figure 24.28b and c, detectors placed on earth can detect which path the photon has taken. However, if we put a beam splitter *before* the photon reaches the detector, as shown in Figure 24.28d, the detectors can no longer tell which path the photon took. The observed photon with a beam spliter in place shows interference as the beam splitter's position is varied. One is led to conclude that the observed interference pattern implies that the photon has *simultaneously* taken *both the paths* around the galaxy — as shown in Figure 24.28d. In other words, similar to the case of the Mach-Zehnder interferometer, having no beam splitter, as in Figure 24.28b and c, is equivalent to knowing which path the photon took. On the other hand, putting in the beam splitter before detecting the photon, as shown in Figure 24.28d, implies that it cannot be determined which path the photon has taken and hence, in the trans-empirical sense, the photon takes both paths simultaneously — similar to the Mach-Zehnder case as shown in Figure 24.27.

In summary, when we detect the photon, we have two choices: We can either put in the beam splitter or not put in the beam splitter. In case we don't put in the beam splitter, the photon takes only one path and there is no interference. However, when we place a beam splitter before measuring the photon, there is interference showing that the photon has taken both paths simultaneously. Until we make the measurement, the photon does not "know" in what manner the measurement is going to be carried out, namely, whether we will put in the beam splitter or not. How does the photon decide on which path it takes? Does it need to "alter" the path it has taken if the beam splitter is put in and which entails going back billions of years in time? The answer is no. The photon is not in an empirical state before the measurement is performed and does not have any unique trajectory. Depending on the state of the detector on earth, we are measuring either an empirical photon that has a well-defined path or a trans-empirical photon that simultaneously took both paths. The measurement itself decides what is the state of the photon that we measure; before the photon is measured, it cannot be said to have any *a priori* state.

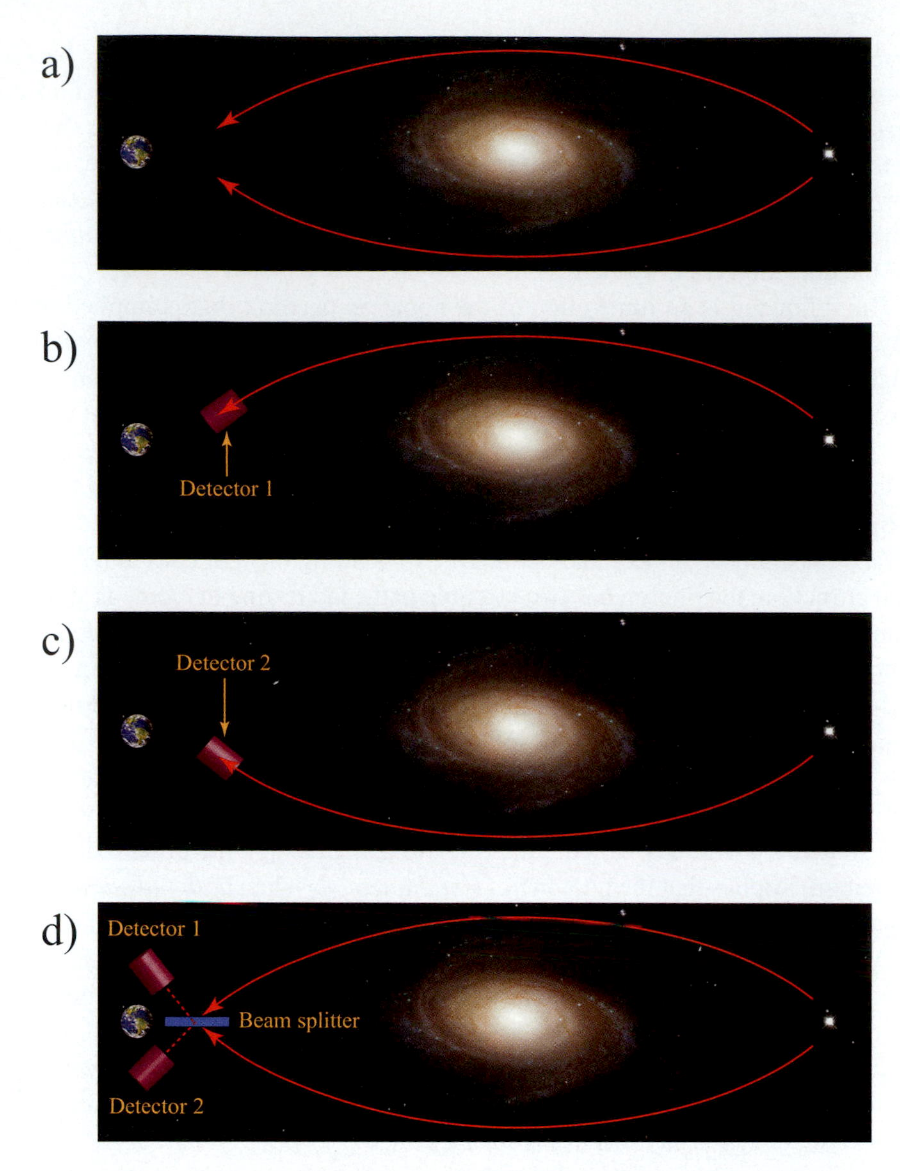

Figure 24.28: A photon simultaneously taking two possible paths that span a distance of billions of light-years.

When the path of a photon is *not* observed, it is in a trans-empirical state simultaneously taking all possible paths from the emitting galaxy to earth; when a measurement is performed that can fix which path the photon takes, it is in an empirical state and takes a unique path. This explanation is the best that we can come up with to understand the paradoxes of quantum mechanics.

24.12 Quantum eraser

We take a closer look at the process of measurement to decide if we can erase the "which path" information even *after* we have an experimental arrangement that has measured which path the photon has taken. We proceed by first starting with the earlier experiments and then showing how modifications are made to introduce the idea of erasure of information.

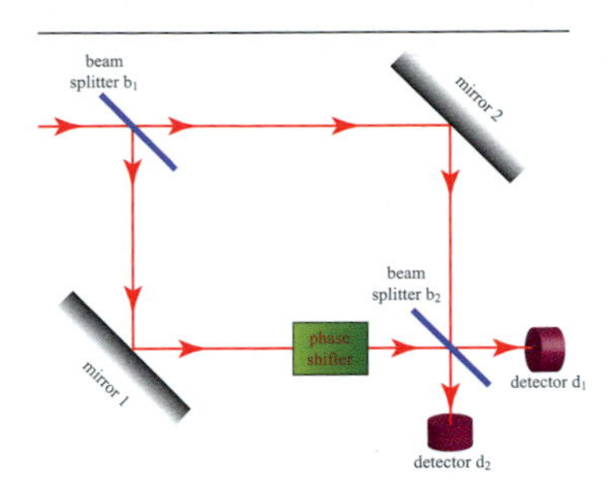

Fig. 24.29: Interference in a Mach-Zehnder setup.

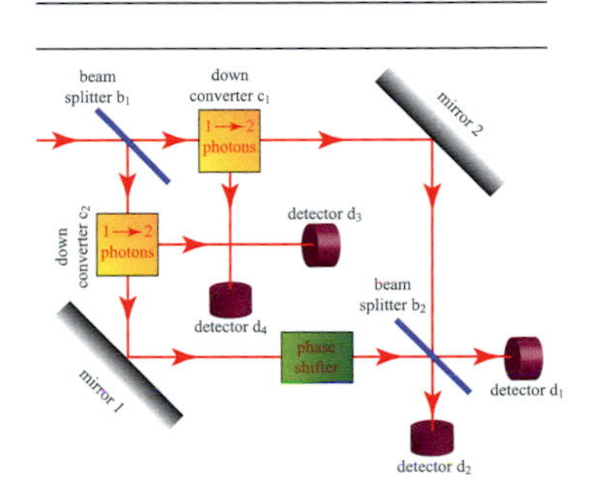

Fig. 24.30: Creation of additional photons with a down converter.

- Consider the experimental setup in Figure 24.29 where two beam splitters b_1 and b_2 have been placed, as discussed in Section 24.11, resulting in the interference pattern shown in Figure 24.26. The photon arriving at detectors d_1 and d_2 is in a trans-empirical superposed state and hence exhibits interference.

- We now have an experimental arrangement, as shown in Figure 24.30, where a photon that is reflecting off either mirror 1 or mirror 2 goes through a device, called a down converter, where a nonlinear process *splits* the *single* photon — of course conserving energy and momentum — into *two* photons. One of the resulting photons continues along the original path and the other photon heads in an orthogonal direction, as shown in Figure 24.30, with detectors d_3 and d_4 detecting the down-converted photon.

  After down conversion, the Mach-Zehnder interferometer has *two photons* in the apparatus, unlike the case discussed in Section 24.11.

  Now, even though we still have beam splitters b_1 and b_2 in place at the source of light and at detectors d_1 and d_2, respectively, when a photon is detected in, say, detector d_4, we know that the photon detected in either d_1 or d_2 has taken the path of being reflected off mirror 2. Hence, in spite of having beam splitter b_2 in place, detectors d_1 and d_2 will *not* show any interference! And this is precisely what experiments confirm to be the case.

We can go further and place a third beam splitter b_3 between detectors d_3 and d_4, as shown in Figure 24.31. The down-converted photon that is received could have come from *either* of the two paths taken by the photon. Hence, detecting the down-converted photon no longer tells us the which-path information that led to the absence of interference for a photon detected by detectors d_1 and d_2. Hence, *interference* is *restored* since the which path knowledge has been erased by beam splitter b_3.

The interference does not exist for the single photon detected by detectors d_1 and d_2; rather, if one does a *coincidence measurement* by counting only those cases when *two* photons are simultaneously detected by *both* detectors d_1 and d_3 (or equivalently, by *both* detectors d_2 and d_4), then, as one varies the relative phase ϕ, one obtains an interference pattern similar to the one given in Equation 24.36. One may object that in the case of the quantum eraser, one has shifted one's attention from single-photon interference to the case of interference involving two photons. This shift is unavoidable since we are probing the path of the photon with another photon and both photons have to be observed to obtain the which-path information.

The experimental arrangement of the quantum eraser is not simply that the initial photon exists in either an empirical or a trans-empirical state; to the contrary, after down conversion there are now two photons in the apparatus and they are in an empirical or trans-empirical state depending on how one arranges the measurements being carried. In particular, the interference that exists for the coincidence

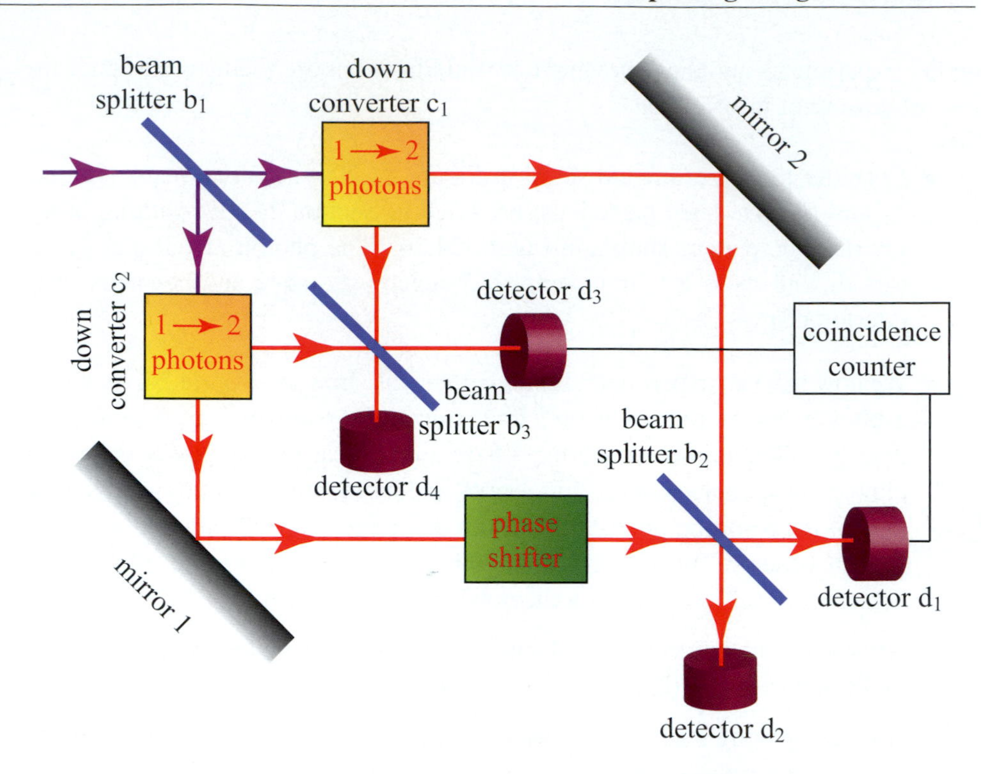

Figure 24.31: Quantum eraser; the down converters allow for knowing which path the photon takes.

counts shows that it is the *two-photon* system that is now in the trans-empirical state and gives rise to the interference; if one were to measure only a single photon one would completely miss the interference effect.

24.13 Partial quantum erasure

In the case of the quantum eraser, discussed in Section 24.12, we had either no interference or a complete erasure of information leading to full interference. Can we partially erase the information and smoothly go between no interference and full interference? The answer is yes, and we examine an experiment that does so.

As before, the Mach-Zehnder interferometer has two beam splitters b_1 and b_2 and two detectors d_1 and d_2, with a relative phase shifter ϕ. The general idea is to couple a measuring device, which consists of another quantum system, say, a spin variable $|s\rangle$, to the photon's path. Measuring what state the spin $|s\rangle$ is in will let us know if the photon has taken a particular path. We can turn on the coupling (interaction) slowly and examine the influence of the interaction on the interference pattern. The experimental arrangement is shown in Figure 24.32. The spin $|s\rangle$ has an interaction U with the photon. If the photon travels on the path that reflects off mirror 2, then it interacts with $|s\rangle$ and makes it undergo a transition to another state $|s'\rangle = U|s\rangle$. The probability amplitude for the transition is given by the complex number $\alpha = \langle s|U|s\rangle$; let $\alpha = |\alpha|\exp(i\chi)$.

Consider a coincidence measurement in which the state $|s'\rangle = U|s\rangle$ is measured as well detecting a photon in d_1 (or d_2). It can be shown, for the coincidence

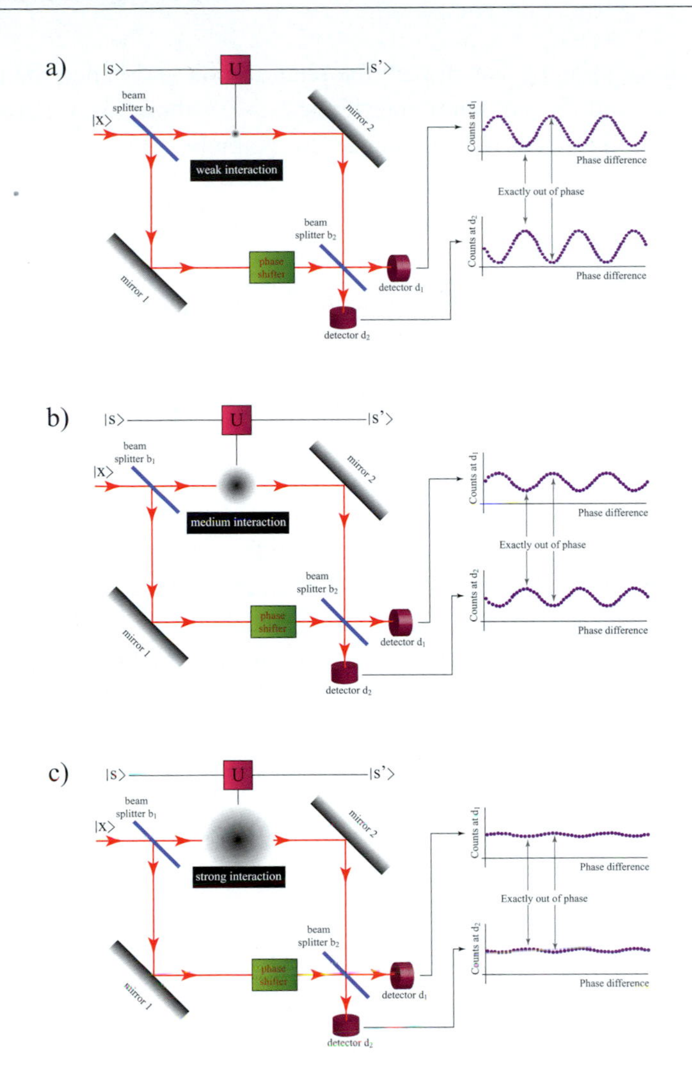

Figure 24.32: Partial quantum erasure of which-path information.

measurement, that the probability of detecting a photon in d_1 and d_2 is given by

$$\langle s|U|s\rangle = |\alpha| \exp(i\chi) \tag{24.37}$$

$$P_1 = \frac{1}{2}[1 - |\alpha| \cos(\phi + \chi)] \tag{24.38}$$

$$P_2 = \frac{1}{2}[1 + |\alpha| \cos(\phi + \chi)] \tag{24.39}$$

$$P_1 + P_2 = 1.$$

For the general case $|\alpha|$, χ given in Equation 24.37, the quantum system is in a state that is partly empirical and partly trans-empirical; we explain this in the discussion given below.

Consider the case of $U = 1$ (no interaction), which yields $\alpha = \langle s|U|s\rangle = \langle s|s\rangle = 1$ and hence $|\alpha| = 1$; $\chi = 0$. Since there is no interaction, we have not gained any information on which path the photon took and hence the readings of the detectors d_1 and d_2 will show interference — as shown in Figure 24.32a — and Equation 24.38 reduces to Equation 24.36. Namely,

$$\langle s|U|s\rangle = 1 \;\Rightarrow\; |\alpha| = 1;\; \chi = 0$$

$$P_1 \to \frac{1}{2}[1 - \cos(\phi)] = \sin^2\left(\frac{\phi}{2}\right)$$

$$P_2 \to \frac{1}{2}[1 + \cos(\phi)] = \cos^2\left(\frac{\phi}{2}\right)$$

$$P_1 + P_2 = 1.$$

If the final state $|s'\rangle$ can be distinguished with *certainty* from the initial state $|s'\rangle$, then we are certain that the photon took the path that reflects off mirror 2, and which should in turn lead to the destruction of interference in the detectors d_1 and d_2. In quantum mechanics, two states are distinguishable with 100% confidence if the states are orthogonal, that is, if $\langle s|s'\rangle = \langle s|U|s\rangle = 0$, which in turn implies $|\alpha| = 0 = \chi$. The strength of the interaction U is clearly strong since it can cause spin $|s\rangle$ to flip to an orthogonal state $|s'\rangle$ such that $\langle s|s'\rangle = 0$. We have from Equation 24.38 the following:

$$\langle s|U|s\rangle = 0 \;\Rightarrow\; |\alpha| = 0;\; \chi = 0$$

$$P_1 \to \frac{1}{2}[1 - 0] = \frac{1}{2}$$

$$P_2 \to \frac{1}{2}[1 + 0] = \frac{1}{2}$$

$$P_1 + P_2 = 1$$

which is the result for the case of no interference given in Equation 24.35 and shown in Figure 24.32c.

As one varies the strength of the interaction in Equation 24.38 by varying the value of the complex number α, we smoothly interpolate between the case of interference and no interference.

- For the case when $\langle s|U|s\rangle = 1$ the photon is in a trans-empirical state since path information has been completely lost. Figure 24.32a shows full interference when there is no coupling with the spin variable.

- For the case when $0 < |\langle s|U|s\rangle| < 1$ the photon exists in a state that is a *mixture* of the trans-empirical and empirical states corresponding to a partial erasure of information. Figure 24.32b shows a partial erasure of interference. We conclude that a quantum system can exist in a state that is partly in the empirical state and partly in the trans-empirical state, as expressed in Equation 24.37.

- For the case of $\langle s|U|s\rangle = 0$ the photon is in an empirical state since the path taken by the photon is known with 100% certainty. Figure 24.32c shows that interference is completely removed.

24.14 Theories of quantum measurement

There is at present no consensus among quantum theorists as to what constitutes a measurement. We have used the concept of the "collapse of the wave function" to explain the result of quantum measurements, but all the results can be obtained without invoking the collapse concept. There are presently five main schools of thought on what constitutes a quantum measurement.

- The projection of a quantum state to one of its possible specific states is a more technical statement of the wave function collapsing to one of its state. Measurement is equivalent to the projection from the trans-empirical state onto a unique state and is an assumption that needs to be made in addition to the Schrödinger equation.

- All measurements of a quantum state invariably need a device that greatly amplifies, by some irreversible process, the signal from the quantum state. Consider measuring, using a Geiger counter, the emission of an alpha particle by a radioactive material. The Geiger counter has an internal high voltage such that the presence of an alpha particle causes a cascade of electrons from the detector's material, releasing up to $10^7 - 10^8$ electrons. The recording of the cascade of electrons is the signal indicating the presence of an alpha particle. A large detector creating decoherence of the quantum system and leading to an observed classical state is another view on how measurements take place.

- There is a "many-world" interpretation of quantum mechanics in which every measurement is thought to bifurcate the wave function into new branches, depending on the outcome of the measurement. In other words, one measures a superposed state, such as $\psi_{12} = \psi_u + \psi_d$, then sometimes one observes ψ_u and at other times one observes ψ_d; one then obtains two universes, the first in which the wave function is ψ_u and the other with wave function ψ_d. Each branch is taken to be equally real and has a subsequent temporal evolution determined by the Schrödinger equation until another measurement is made.

- Another approach is to introduce a random (fluctuating) classical force that acts on macroscopic objects and that causes superposed states to continually fluctuate. This random force causes the trans-empirical superposed states to evolve into empirical states that are governed by classical probability.

- Finally, there is view that the process of measurement does not take place in any apparatus, which are in any case governed by quantum principles, but measurement occurs only when a consciousness-like entity becomes aware of the measurement. In this view, the human *brain* is thought to be described by quantum mechanics but the human *mind* — taken to be an exemplar of consciousness — is thought to be outside the workings of physical laws.

Measurement lies at the heart of quantum mechanics, but paradoxically, it is not yet fully understood. Although the *procedures* for carrying out measurements on a quantum system are garden variety, well understood and carried out everyday, the *theoretical* understanding of measurement is far from clear. There are some "pragmatic" scientists who think this question is unimportant, focusing on the operational side of measurement. But the fact is that measurement remains outside the Schrödinger equation and points to a need for either a deeper understanding of quantum mechanics or a new theory that goes beyond quantum mechanics.

24.15 Quantum mechanics and other disciplines

A total and complete break with the paradigms of classical physics was required for engaging in quantum reasoning. Quantum theory required a completely new framework for thinking about nature before the questions of quantum measurement and uncertainty could become clear. This *shift of paradigm* took the physicists many years to achieve.

As we discussed earlier, it is our view that the reason that quantum mechanics is so radical is because in the quantum framework nature has *two forms* of existence, a trans-empirical state when it is not observed and the actual empirical state when it is observed. The trans-empirical and empirical states have completely different behaviors — the trans-empirical being counterintuitive and leading to all the paradoxes of quantum mechanics; the empirical state is the actually observed state and appears to be classical-like. Since in classical physics all reality is empirical, it is only with the introduction of the trans-empirical realm that the concept of a quantum measurement and uncertainty could be addressed.

Although quantum mechanics has revolutionized our view of nature, our understanding is still far from complete and one can be sure there are a lot of surprises awaiting us.

The unusual properties of semiconductors and electronic chips in general are due to quantum phenomena. Electronic devices, from computers, television, to mobile phones, are all based on semiconductors, and airplanes, ships and cars all use semiconductors in an essential manner. More complex technologies such as scanning electron microscope, MRI (magnetic resonance imaging), lasers, fabrication of new drugs and modern materials science all draw on the principles of the quantum world. Indeed, most of what goes under the name of high technology is a direct result of the workings of quantum mechanics, and most modern conveniences that we take for granted today would be impossible without it.

About 150 years ago, chemistry had almost no connection with physics, and concepts of chemistry such as valency, activity, solubility and volatility had more of a qualitative character. The first application of physics to chemistry started in the nineteenth century with the theory of heat, and was led by the hope of understanding the laws of chemistry in terms of the mechanics of atoms. One of the most successful applications of quantum mechanics is the explanation of all the atoms that form the periodic table, and which is the starting point of all of chemistry.

The present relation of biology to physics and chemistry is similar to that of chemistry to physics 100 years ago. Biological concepts such as life, organ, cell function and adaptation presently have no explanation in terms of physical and chemical laws. However, with the rapid advances in molecular biology this will hopefully change soon.

Cosmology studies the large-scale structure of the universe, and one would have thought that quantum mechanics, which apparently is concerned with the microworld, would not have any relevance to cosmology. In fact, the current hot big-bang theory of cosmology relies solely on quantum mechanics to explain the events which occurred within the first 100 seconds and that are the determinate events shaping all the later evolution of the universe. If one probes even closer to within a few trillionths of a second after the big bang, then one needs to understand even more deeply the quantum behavior of the universe, and this leads one to the study of the microscopic constituents of nature that are also described by the quantum principle.

The human being's five physical senses are based on natural processes that can perceive only a finite range of physical phenomena. In the case of electromagnetic radiation, only a tiny and limited range of the wavelength of electromagnetic radiation is visible to the human eye; radiation of much longer and much shorter wavelengths such as x-rays are invisible to the human eye. Since the smallest allowed quantum of energy for light and atoms is truly minuscule when compared to the energies we encounter in daily life , there are only a few physical process, most of them being man made, where one can directly observe quantum phenomena using one's five senses. When we extend our five senses with experimental devices and instruments, we can probe more deeply into nature's secrets and the quantum aspect of nature becomes more apparent.

Noteworthy 24.2: "No one understands quantum mechanics": Feynman

According to Niels Bohr, if one does not "see" the paradox of quantum mechanics, namely the clash of quantum mechanics with our everyday intuition of how nature should behave, then one can be sure that one has not understood quantum mechanics.

Bohr and Feynman, quoted in the beginning of this chapter, use the word "understand" at two different levels. Bohr's statement refers to the operational aspect of quantum mechanics, the fact that before one performs an experiment the system does not have a unique state, the fact that the quantum state is counterintuitive and nonclassical. Feynman on the other hand is referring to something deeper. Understanding usually means that once we know some physical principle, the phenomenon that this principle explains then becomes intuitively obvious. For example, once we know of Fermat's principle of least time, we "understand" why light bends in going from one medium to another.

In the case of quantum mechanics, even when we know all its principles the behavior of a quantum system is still unfathomable. For example, the result of an experiment can never, in principle, be predicted beforehand — only the likelihood of a result occurring can be computed from quantum mechanics; we have no clue as to how, when the quantum system is observed, it "chooses" one of the possible final states. Feynman's advice is that a complete "understanding" of quantum mechanics, similar to what we have for classical systems, can never be achieved, and one should stay away from trying to have such an "understanding". We are a bit more optimistic than Feynman and feel that future generations may develop new forms of understanding such that quantum mechanics may seem most natural and reasonable!

24.16 Summary

Classical mechanics works very well for the kinds of objects one encounters in daily life that are moving much slower than the velocity of light. Once objects start to move very fast, we need to modify Newton's equations to Einstein's relativistic equations. On the other hand, for objects that are very small, such as electrons and atoms, quantum theory becomes necessary. If one attempts to extend Newton's laws to domains that are far from daily experience, they start to fail and give incorrect results.

The following aspects of quantum theory provide a stark contrast to our intuitive expectations regarding nature:

- Nature has a *counterintuitive trans-empirical domain* that can never be directly perceived by our five senses in addition to the intuitively obvious and directly observable empirical domain.

- *The trans-empirical state of a quantum system is described by the wave function.* The electron behaves like a wave, not as an ordinary wave but rather as a probability wave. The wave function of quantum mechanics is a probability wave that describes the trans-empirical state of the system; in particular, the wave function provides the likelihood of the quantum system being in one of its possible (determinate) classical states.

- *Quantum particles have two forms of existence, one trans-empirical and the other empirical.* When a particle is *observed* it is in an empirical state, and when it is *not observed* it is in a trans-empirical state. In contrast, a classical particle has no trans-empirical level of reality and exists only in the empirical state. In the discussion on the quantum superposition principle in Section 24.8, the difference and the interplay between the trans-empirical and empirical states were elaborated upon.

- *The transition from the trans-empirical state to the empirical state is accomplished by the act of measurement.* In contrast to classical physics, measurement plays a central role in rendering the theory internally consistent. The fact that energy is quantized directly leads in the particle picture to the necessity of the Heisenberg uncertainty principle.

- *Cause and effect* are radically transformed in quantum mechanics. The *same* cause, namely the act of measurement, leads to a *multiplicity* of effects. In particular, consider a quantum system being prepared in an identical manner; for example, sending in an electron in the double-slit experiment. Every time we do the same experiment, namely have the same cause, we get a different effect, with the electron being found at different points on the screen. Unlike classical physics, we *cannot* make a prediction as to where the electron will be detected in a particular experiment; instead, we can only make a statistical prediction on the likelihood of the electron being found at different positions.

- *Quantum energy propagates in a probabilistic and trans-empirical manner*, becoming a physical and empirical entity only when a measurement is performed. The trans-empirical propagation of energy allows for phenomena that are classically forbidden.

- The *quantum eraser* and its variants exemplify the concept of quantum measurement. In particular, the quantum eraser displays the counterintuitive concept of erasing or restoring information about a quantum system and in doing so driving the system between the trans-empirical and empirical states.

24.17 The answer

The mathematical and symbolical representation of nature, as exemplified in quantum mechanics, provides the means for an understanding of nature that direct intuition and the five senses can never provide. The process of reasoning, reflection and symbolical thinking comes to the fore in engaging with physical phenomena that are far removed from our everyday life. The fact that nature is counterintuitive is borne out in other scientific disciplines at the frontiers of human knowledge — such as the study of relativity, curved space-time and black holes as well as the ultramicroscopic realm of elementary particles and fundamental interactions. Our study of quantum mechanics leads to the conclusion that nature at the deepest and most fundamental level is indeed counterintuitive.

24.18 Exercises

1. What is the de Broglie wavelength of an electron with an energy of 5 eV?

2. Is the electron wave a physical wave or a probability amplitude? Explain the difference.

3. Is the probability wave part of the trans-empirical domain?

4. If we know that an electron has an energy of 10 eV, what is its momentum?

5. Consider textbook Figure 24.10. If the width of the box is $L = 1\ \mu$m and the particle is an electron, what is the probability to find the electron at $x = 0.5\ \mu$m if the electron is in the ground state?

6. If I know that the energy of a photon is exactly 2 eV, how accurately can I determine its position?

7. What is the key step in the quantum Zeno effect that prevents an atom to transition to a different state?

8. If I know that a photon is somewhere in a cubic box with a side of 10^{-9} m, how precisely can we determine its momentum?

9. If I have a barrier with a certain energy and the knowledge that quantum particles can tunnel through it with ease, is it possible to make any conclusive statements about whether a classical particle can cross this barrier?

10. If a quantum particle were described by the sum of the two waves $\sin(x)$ and $\sin(x + \pi)$, what would be the probability to observe that particle at $x = \pi/8$?

11. How can the state of a quantum mechanical particle be made empirical?

12. I make a special and very accurate Mach-Zehnder-type interferometer for electrons. If I point an electron gun at it that shoots out one electron per second, is it possible to observe interference?

Complex numbers

A real number x is such that its square is always positive, that is, $x^2 > 0$. One can extend the real numbers to a larger set by introducing the unit imaginary number i, defined such that

$$i^2 = -1 \Rightarrow i = \sqrt{-1}.$$

A complex number z is defined by

$$z = x + iy$$

where x and y are real numbers. The complex conjugate z^* of z is defined by

$$z^* = x - iy.$$

Hence, for two complex numbers $z_1 = x_1 + iy_1$ and $z_2 = x_2 + iy_2$, we have

$$z_1^* z_2 = x_1 x_2 + y_1 y_2 + i(x_1 y_2 - x_2 y_1).$$

In particular

$$|z|^2 \equiv z^* z = x^2 + y^2 > 0.$$

One can verify that adding, subtracting, multiplying and dividing two complex numbers results in yet another complex number. Just like powers of real numbers, a complex number can be raised to a complex power. We have, for real numbers a, x and y

$$a^{x+iy} = a^x a^{iy} \tag{24.40}$$
$$= a^x e^{iy \ln a}. \tag{24.41}$$

So we need to understand what does $e^{i\theta}, \theta \equiv y \ln a$ stand for? Since $e^{i\theta}$ is a complex number, there are values of a and b such that we have

$$e^{i\theta} = a + ib \tag{24.42}$$
$$e^{i\theta} \times e^{-i\theta} = (a + ib)(a - ib) = a^2 + b^2 \tag{24.43}$$
$$\Rightarrow 1 = a^2 + b^2. \tag{24.44}$$

Motivated by the well-known trigonometric identity $1 = \sin^2\theta + \cos^2\theta$ we have the fundamental result

$$e^{i\theta} = \cos\theta + i \sin\theta. \tag{24.45}$$

Complex numbers form a system of arithmetic similar to real numbers. For the more mathematically minded, it can be shown that to solve for the roots of an arbitrary nth-order polynomial equation, it is necessary and sufficient to extend the real numbers to complex numbers.

Laws

Name	Page	Equation
Coulomb's law	188	$F = k_e \frac{q_1 q_2}{r^2}$
Hooke's law	59	$F = -kx$
Hubble's law	14	$v = H_0 d$
Ideal gas law	215	$PV = nRT$
Newton's laws		
First law	18	Inertia
Second law	19	$F = ma$
Third law	19	Action = - Reaction
Law of gravitation	22	$F = G \frac{m_1 m_2}{r^2}$
Ohm's law	195	$V = IR$
Snell's law	289	$n_1 \sin \Theta_1 = n_2 \sin \Theta_2$

Equations

Name	Page	Equation
Doppler effect	12	$f' = \left(\frac{v}{v + v_r}\right) f$
Momentum	18	$p = mv$
Work	52	$W = F \times d$
Work-energy theorem	53	$\Delta W = \Delta E$
Kinetic energy	54	$T = \frac{1}{2} m v^2$
Power	55	$P = \frac{\Delta W}{\Delta t}$
Gravitational potential energy	57	$U = mgh$
Helmholtz free energy	65	$F = E - TS$
Michaelis–Menten equation	320	$V = \frac{k_{\mathrm{cat}}[\mathrm{E_T}][\mathrm{S}]}{K_m + [\mathrm{S}]}$
Mass-energy equivalence	494	$E = mc^2$
Planck relation	501	$E = h\nu$
Schrödinger equation	522	$i\hbar \frac{\partial \psi(t, x)}{\partial t} = \left(-\frac{\hbar^2}{2m} \nabla^2 + V(x)\right) \psi(t, x)$
Navier-Stokes equation	150	$\rho\left(\frac{\partial \mathbf{v}(t, \mathbf{x})}{\partial t} + \mathbf{v} \cdot \nabla \mathbf{v}(t, \mathbf{x})\right) = -\nabla p(t, \mathbf{x}) - \nabla U(\mathbf{x}) + \eta \nabla^2 \mathbf{v}(t, \mathbf{x})$

Units

Système International (SI) base units

Symbol	Name	Quantity	Original inspiration
m	meter	length	1/10,000,000 the distance from the equator to the north pole
kg	kilogram	mass	Mass of 1 L of water
s	second	time	1/60 of a minute
A	ampere	electric current	Current required to deposit 1.118 mg of silver from a silver nitrite solution
K	kelvin	temperature	Absolute zero is 0, 100 units from frozen to boiling water
mol	mole	amount of substance	Same number of "elemental entities" as there are in 12 g of carbon 12
cd	candela	luminous intensity	Light from a candle

Selected Système International (SI) derived units

Symbol	Name	Base units	Quantity	Example
J	joule	$kg\ m^2\ s^{-2} = N\ m$	energy, work	1 J is roughly the energy required to lift 1 kg up by 10 cm.
N	newton	$kg\ m\ s^{-2}$	force	At earth's surface, 1 kg exerts a force of about 9.8 N.
Pa	pascal	$kg\ m^{-1}\ s^{-2} = N\ m^{-2}$	pressure	At earth's surface, the pressure is about 10^5 Pa.
T	tesla	$kg\ s^{-2}\ A^{-1}$	magnetic field	The magnetic field of the earth is about 10^{-4} T.
C	coulomb	s A	electric charge	1 C is the charge of about 6.2×10^18 protons.
V	volt	$m^2\ kg\ s^{-3}\ A^{-1}$ = J/C	electrical potential difference	A standard alkaline AA battery has about 1.5 V.
Ω	ohm	$m^2\ kg\ s^{-3}\ A^{-2}$ = V/A	electrical resistance	The resistance of 50 m of 1-mm diameter copper wire is about 1 Ω.
F	farad	$s^4\ A^2\ m^{-2}\ kg^{-1} = C\ V^{-1}$	capacitance	1 F produces a 1-V potential difference for an electric charge of 1 C.
W	watt	$m^2\ kg\ s^{-3}$ = J/s = V A	power	A typical incandescent lightbulb uses between 25 W and 100 W.

Prefixes

10^n	Prefix	Symbol	Name	Number
10^{24}	yotta	Y	Septillion	1,000,000,000,000,000,000,000,000
10^{21}	zetta	Z	Sextillion	1,000,000,000,000,000,000,000
10^{18}	exa	E	Quintillion	1,000,000,000,000,000,000
10^{15}	peta	P	Quadrillion	1,000,000,000,000,000
10^{12}	tera	T	Trillion	1,000,000,000,000
10^{9}	giga	G	Billion	1,000,000,000
10^{6}	mega	M	Million	1,000,000
10^{3}	kilo	k	Thousand	1,000
10^{2}	hecto	h	Hundred	100
10^{1}	deca	da	Ten	10
10^{0}	(none)	(none)	One	1
10^{-1}	deci	d	Tenth	0.1
10^{-2}	centi	c	Hundredth	0.01
10^{-3}	milli	m	Thousandth	0.001
10^{-6}	micro	μ (u)	Millionth	0.000 001
10^{-9}	nano	n	Billionth	0.000 000 001
10^{-12}	pico	p	Trillionth	0.000 000 000 001
10^{-15}	femto	f	Quadrillionth	0.000 000 000 000 001
10^{-18}	atto	a	Quintillionth	0.000 000 000 000 000 001
10^{-21}	zepto	z	Sextillionth	0.000 000 000 000 000 000 001
10^{-21}	yocto	y	Septillionth	0.000 000 000 000 000 000 000 001

Constants

Symbol	Name	Value	Units
c	speed of light in vacuum	2.998×10^{8}	m s^{-1}
G	gravitational constant	6.6742×10^{-11}	N m^{2} kg^{-2}
g	standard gravity	9.80665	m s^{-2}
h	Planck constant	6.626×10^{-34} 4.136×10^{-15}	J s eV s
k	Boltzmann constant	1.381×10^{-23}	J K^{-1}
N_A	Avogadro's number	6.022×10^{23}	mol^{-1}
R	molar gas constant	8.314	J mol^{-1} K^{-1}
e	electric charge	1.602×10^{-19}	C
m_e	electron rest mass	9.109×10^{-31} 0.5110	kg MeV/c^{2}
m_p	proton rest mass	1.673×10^{-27} 938.3	kg MeV/c^{2}
m_n	neutron rest mass	1.675×10^{-27} 939.6	kg MeV/c^{2}
μ_0	magnetic constant (vacuum permeability)	$\pi \times 10^{-7}$	N A^{-2}
ϵ_0	electric constant (vacuum permittivity)	8.854×10^{-12} $= \frac{1}{\mu_0 c^2}$	F m^{-1}
k_e	electrostatic constant	8.988×10^{9} $= \frac{1}{4\pi\epsilon_0}$	m^{2} N C^{-2}
T_0	absolute zero	-273.15 0	°C K

Sizes

Item	Approximate size
Size of hydrogen atom	0.1 nm
Size of one amino acid molecule	0.8 nm
Diameter of the DNA helix	2 nm
Thickness of cell membranes	10 nm
Size of large molecules	100 nm
Size of small procaryote	1 μm
Size of *Escherichia coli* (*E. coli*)	2 μm
Most animal eukaryotic cells	10 μm – 30 μm
Most plant eukaryotic cells	10 μm – 100 μm
Diameter of relatively thick human head hair	100 μm
Size of some ants	1 mm
Medium-sized coin	1 cm
Size of large human	2 m
Size of small building	10 m
Size of large building	100 m
Height of small mountain	1 km
Diameter of earth	12.8 Mm
Diameter of the solar system	8 Tm
Diameter of the sun	1.4 Gm
Diameter of the Milky Way	1 Zm
Diameter of the visible universe	120 Ym

Credits

Chapter 1

Page 1: Source: NASA. Hubble Heritage Team. Image of the spiral galaxy NGC 4414.

Page 1: Source: NSF. The image was taken at the Cerro Tololo Inter-American Observatory in Chile and shows the view toward the center of the galaxy.

Fig. 1.1: Source: NASA. Jupiter and its four planet-size moons.

Fig. 1.2: Source: NASA. The galaxy cluster Abell 1689.

Fig. 1.3: Source: US National Park Service. Panoramic view of our Milky Way from Death Valley's Racetrack Playa.

Fig. 1.11: Source: Spitzer Space Telescope. Reconstruction of the top view of our Milky Way.

Fig. 1.12: Source: NASA. Spiral galaxy NGC 1232, elliptical galaxy M87, barred spiral galaxy NGC1300, irregular galaxy sextans A.

Fig. 1.19: Source: NASA. Planets in our solar system.

Fig. 1.20: Source: NASA. Moon behind the earth.

Fig. 1.21: Source: NASA and the Hubble Deep Field Team. Distant galaxies: the Hubble deep field.

Chapter 6

Fig. 6.3: Source: USGS. Ganges River delta.

Chapter 15

Fig. 15.13: Tubulin Dimer. Source: Lawrence Berkeley National Laboratory.

Fig. 15.17: Source NIH. Cells: bovine pulmonary arthery endothelial cells. Blue: nucleus stained with DAPI. Green: Tubulin stained with Bodipy FL goat anti-mouse IgG. Red: F-Actin stained with Texas Red X-Phalloidin.

Fig. 15.23: Adapted from P. Janmey et al., *J. Cell Biol.* 113:155.

Chapter 17

Fig. 17.22: From Phillips et al., *Physical Biology of the Cell*, Garland Science/Taylor & Francis LLC, 2009. With permission.

Periodic Table

Atomic Number / Element Symbol / Element Name / Atomic Mass (u) — key cell: 1 H Rutherfordium 1.00794; shells K 2, L 8, M 18, N 32, O 32, P 18, Q 8

- s-Block
- p-Block
- d-Block
- f-Block

State at room temperature (298 K): X Liquid, X Solid, X Gaseous

Each cell below: atomic number, symbol, element name, atomic mass, (electrons per shell as printed).

Period \ Group	1	2	3	4	5	6	7	8	9	10	11	12	13	14	15	16	17	18
1	1 H Hydrogen 1.00794 (1)																	2 He Helium 4.00260 (2)
2	3 Li Lithium 6.941 (2-1)	4 Be Beryllium 9.01218 (2-2)											5 B Boron 10.811 (2-3)	6 C Carbon 12.0107 (2-4)	7 N Nitrogen 14.0067 (2-5)	8 O Oxygen 15.9994 (2-6)	9 F Fluorine 18.9984 (2-7)	10 Ne Neon 20.1797 (2-8)
3	11 Na Sodium 22.98976 (2-8-1)	12 Mg Magnesium 24.3050 (2-8-2)											13 Al Aluminium 26.9815 (2-8-3)	14 Si Silicon 28.0855 (2-8-4)	15 P Phosphorus 30.9738 (2-8-5)	16 S Sulfur 32.065 (2-8-6)	17 Cl Chlorine 35.453 (2-8-7)	18 Ar Argon 39.948 (2-8-8)
4	19 K Potassium 39.0983 (2-8-8-1)	20 Ca Calcium 40.078 (2-8-8-2)	21 Sc Scandium 44.9559 (2-8-9-2)	22 Ti Titanium 47.867 (2-8-10-2)	23 V Vanadium 50.9415 (2-8-11-2)	24 Cr Chromium 51.9961 (2-8-12-2)	25 Mn Manganese 54.9380 (2-8-13-2)	26 Fe Iron 55.845 (2-8-14-2)	27 Co Cobalt 58.9332 (2-8-15-2)	28 Ni Nickel 58.6934 (2-8-16-2)	29 Cu Copper 63.546 (2-8-17-2)	30 Zn Zinc 65.38 (2-8-18-2)	31 Ga Gallium 69.723 (2-8-18-3)	32 Ge Germanium 72.64 (2-8-18-4)	33 As Arsenic 74.9216 (2-8-18-5)	34 Se Selenium 78.96 (2-8-18-6)	35 Br Bromine 79.904 (2-8-18-7)	36 Kr Krypton 83.798 (2-8-18-8)
5	37 Rb Rubidium 85.486 (2-8-18-8-1)	38 Sr Strontium 87.62 (2-8-18-8-2)	39 Y Yttrium 88.9059 (2-8-18-9-2)	40 Zr Zirconium 91.224 (2-8-18-10-2)	41 Nb Niobium 92.9063 (2-8-18-12-1)	42 Mo Molybdenum 95.96 (2-8-18-13-1)	43 Tc Technetium 97.9072 (2-8-18-14-1)	44 Ru Ruthenium 101.07 (2-8-18-15-1)	45 Rh Rhodium 102.906 (2-8-18-16-1)	46 Pd Palladium 106.42 (2-8-18-18-0)	47 Ag Silver 107.868 (2-8-18-18-1)	48 Cd Cadmium 112.411 (2-8-18-18-2)	49 In Indium 114.818 (2-8-18-18-3)	50 Sn Tin 118.710 (2-8-18-18-4)	51 Sb Antimony 121.760 (2-8-18-18-5)	52 Te Tellurium 127.60 (2-8-18-18-6)	53 I Iodine 126.904 (2-8-18-18-7)	54 Xe Xenon 131.293 (2-8-18-18-8)
6	55 Cs Caesium 132.905 (2-8-18-18-8-1)	56 Ba Barium 137.327 (2-8-18-18-8-2)	57 La Lanthanum 138.905 (2-8-18-18-9-2)	72 Hf Hafnium 178.49 (2-8-18-32-10-2)	73 Ta Tantalum 180.948 (2-8-18-32-11-2)	74 W Tungsten 1.00794 (2-8-18-32-12-2)	75 Re Rhenium 186.207 (2-8-18-32-13-2)	76 Os Osmium 190.23 (2-8-18-32-14-2)	77 Ir Iridium 192.217 (2-8-18-32-15-2)	78 Pt Platinum 195.084 (2-8-18-32-17-1)	79 Au Gold 196.967 (2-8-18-32-18-1)	80 Hg Mercury 200.59 (2-8-18-32-18-2)	81 Tl Thallium 204.3833 (2-8-18-32-18-3)	82 Pb Lead 207.2 (2-8-18-32-18-4)	83 Bi Bismuth 208.980 (2-8-18-32-18-5)	84 Po Polonium 208.982 (2-8-18-32-18-6)	85 At Astatine 209.987 (2-8-18-32-18-7)	86 Rn Radon 222.018 (2-8-18-32-18-8)
7	87 Fr Francium 223 (2-8-18-32-18-8-1)	88 Ra Radium 226 (2-8-18-32-18-8-2)	89 Ac Actinium 227 (2-8-18-32-18-9-2)	104 Rf Rutherfordium 261 (2-8-18-32-32-10-2)	105 Db Dubnium, 262 (2-8-18-32-32-11-2)	106 Sg Seaborgium 266 (2-8-18-32-32-12-2)	107 Bh Bohrium 264 (2-8-18-32-32-13-2)	108 Hs Hassium 277 (2-8-18-32-32-14-2)	109 Mt Meinerium 268 (2-8-18-32-32-15-2)	110 Ds Darmstadtium 271 (2-8-18-32-32-17-1)	111 Rg Roentgenium 272 (2-8-18-32-32-18-1)	112 Uub Ununbium 285 (2-8-18-32-32-18-2)	113 Uut Ununtrium 284 (2-8-18-32-32-18-3)	114 Uuq Ununquadium 289 (2-8-18-32-32-18-4)	115 Uup Ununpentium 288 (2-8-18-32-32-18-5)	116 Uuh Ununhexium 292 (2-8-18-32-32-18-1)	117 Uus Ununseptium 290? (2-8-18-32-32-18-7)	118 Uuo Ununoctium 294 (2-8-18-32-32-18-8)

58 Ce Cerium 140.116 (2-8-18-19-9-2)	59 Pr Praseodymium 140.908 (2-8-18-21-8-2)	60 Nd Neodymium 144.242 (2-8-18-22-8-2)	61 Pm Promethium 145 (2-8-18-23-8-2)	62 Sm Samarium 150.36 (2-8-18-24-8-2)	63 Eu Europium 151.964 (2-8-18-25-8-2)	64 Gd Gadolinium 157.25 (2-8-18-25-9-2)	65 Tb Terbium 158.925 (2-8-18-27-8-2)	66 Dy Dysprosium 162.500 (2-8-18-28-8-2)	67 Ho Holmium 164.930 (2-8-18-29-8-2)	68 Er Erbium 167.259 (2-8-18-30-8-2)	69 Tm Thulium 168.934 (2-8-18-31-8-2)	70 Yb Ytterbium 173.04 (2-8-18-32-8-2)	71 Lu Lutetium 174.967 (2-8-18-32-9-2)
90 Th Thorium 232.038 (2-8-18-32-18-10-2)	91 Pa Protactinium 231.036 (2-8-18-32-20-9-2)	92 U Uranium 238.029 (2-8-18-32-21-9-2)	93 Np Neptunium 237 (2-8-18-32-22-9-2)	94 Pu Plutonium 244 (2-8-18-32-24-8-2)	95 Am Americum 243 (2-8-18-32-25-8-2)	96 Cm Curium 247 (2-8-18-32-25-9-2)	97 Bk Berkelium 247 (2-8-18-32-27-8-2)	98 Cf Californium 251 (2-8-18-32-28-8-2)	99 Es Einsteinium 252 (2-8-18-32-29-8-2)	100 Fm Fermium 257 (2-8-18-32-30-8-2)	101 Md Mendelevium 258 (2-8-18-32-31-8-2)	102 No Nobelium 259 (2-8-18-32-32-8-2)	103 Lr Lawrencium 262 (2-8-18-32-32-9-2)

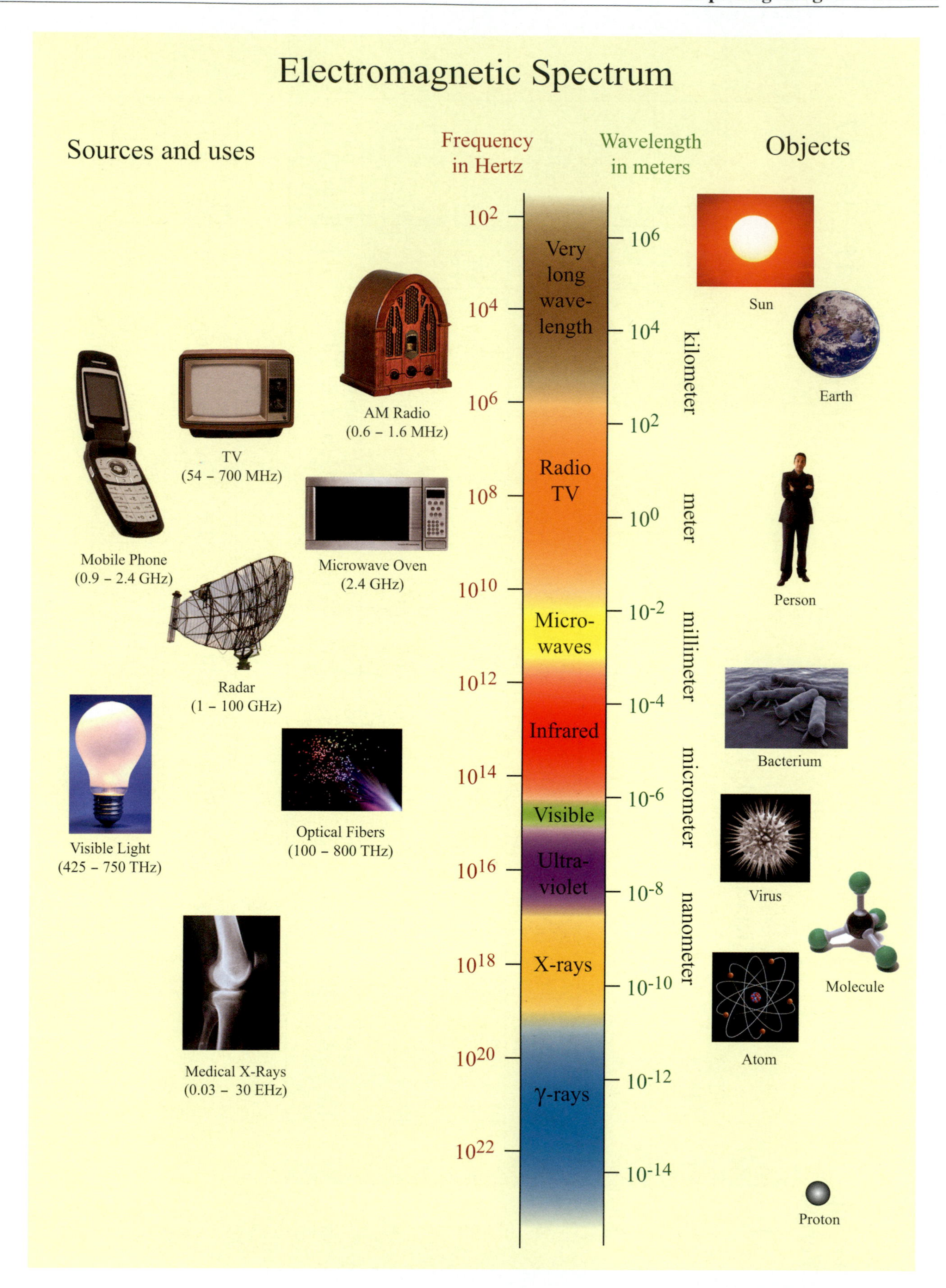

Electromagnetic Spectrum
Sources and uses
Frequency in Hertz
Wavelength in meters
Objects
10^2
10^4
10^6
10^8
10^10
10^12
10^14
10^16
10^18
10^20
10^22
Very long wave-length
Radio TV
Micro-waves
Infrared
Visible
Ultra-violet
X-rays
γ-rays
10^6
10^4
10^2
10^0
10^-2
10^-4
10^-6
10^-8
10^-10
10^-12
10^-14
kilometer
meter
millimeter
micrometer
nanometer
AM Radio (0.6 – 1.6 MHz)
TV (54 – 700 MHz)
Mobile Phone (0.9 – 2.4 GHz)
Microwave Oven (2.4 GHz)
Radar (1 – 100 GHz)
Visible Light (425 – 750 THz)
Optical Fibers (100 – 800 THz)
Medical X-Rays (0.03 – 30 EHz)
Sun
Earth
Person
Bacterium
Virus
Molecule
Atom
Proton

Index